T0134492

Communications
in Computer and Information Science 1801

Rationale

The CCIS series is devoted to the publication of proceedings of computer science conferences. Its aim is to efficiently disseminate original research results in informatics in printed and electronic form. While the focus is on publication of peer-reviewed full papers presenting mature work, inclusion of reviewed short papers reporting on work in progress is welcome, too. Besides globally relevant meetings with internationally representative program committees guaranteeing a strict peer-reviewing and paper selection process, conferences run by societies or of high regional or national relevance are also considered for publication.

Topics

The topical scope of CCIS spans the entire spectrum of informatics ranging from foundational topics in the theory of computing to information and communications science and technology and a broad variety of interdisciplinary application fields.

Information for Volume Editors and Authors

Publication in CCIS is free of charge. No royalties are paid, however, we offer registered conference participants temporary free access to the online version of the conference proceedings on SpringerLink (http://link.springer.com) by means of an http referrer from the conference website and/or a number of complimentary printed copies, as specified in the official acceptance email of the event.

CCIS proceedings can be published in time for distribution at conferences or as post-proceedings, and delivered in the form of printed books and/or electronically as USBs and/or e-content licenses for accessing proceedings at SpringerLink. Furthermore, CCIS proceedings are included in the CCIS electronic book series hosted in the SpringerLink digital library at http://link.springer.com/bookseries/7899. Conferences publishing in CCIS are allowed to use Online Conference Service (OCS) for managing the whole proceedings lifecycle (from submission and reviewing to preparing for publication) free of charge.

Publication process

The language of publication is exclusively English. Authors publishing in CCIS have to sign the Springer CCIS copyright transfer form, however, they are free to use their material published in CCIS for substantially changed, more elaborate subsequent publications elsewhere. For the preparation of the camera-ready papers/files, authors have to strictly adhere to the Springer CCIS Authors' Instructions and are strongly encouraged to use the CCIS LaTeX style files or templates.

Abstracting/Indexing

CCIS is abstracted/indexed in DBLP, Google Scholar, EI-Compendex, Mathematical Reviews, SCImago, Scopus. CCIS volumes are also submitted for the inclusion in ISI Proceedings.

How to start

To start the evaluation of your proposal for inclusion in the CCIS series, please send an e-mail to ccis@springer.com.

Linqiang Pan · Dongming Zhao · Lianghao Li ·
Jianqing Lin
Editors

Bio-Inspired Computing: Theories and Applications

17th International Conference, BIC-TA 2022
Wuhan, China, December 16–18, 2022
Revised Selected Papers

 Springer

Editors

Linqiang Pan
Huazhong University of Science
and Technology
Wuhan, China

Lianghao Li ⓘ
Huazhong University of Science
and Technology
Wuhan, China

Dongming Zhao
Wuhan University of Technology
Wuhan, China

Jianqing Lin
Huazhong University of Science
and Technology
Wuhan, China

ISSN 1865-0929   ISSN 1865-0937 (electronic)
Communications in Computer and Information Science
ISBN 978-981-99-1548-4   ISBN 978-981-99-1549-1 (eBook)
https://doi.org/10.1007/978-981-99-1549-1

This Springer imprint is published by the registered company Springer Nature Singapore Pte Ltd.
The registered company address is: 152 Beach Road, #21-01/04 Gateway East, Singapore 189721, Singapore

Preface

Bio-inspired computing is a field of study that abstracts computing ideas (data structures, operations with data, ways to control operations, computing models, artificial intelligence, multisource data analysis, etc.) from living phenomena or biological systems such as cells, tissue, the brain, neural networks, the immune system, ant colonies, evolution, etc. The areas of bio-inspired computing include Neural Networks, Brain-inspired Computing, Neuromorphic Computing and Architectures, Cellular Automata and Cellular Neural Networks, Evolutionary Computing, Swarm Intelligence, Fuzzy Logic and Systems, DNA and Molecular Computing, Membrane Computing, Artificial Intelligence and its Application in other disciplines such as machine learning, deep learning, image processing, computer science, and cybernetics, etc. Bio-Inspired Computing: Theories and Applications (BIC-TA) is a series of conferences that aims to bring together researchers working in the main areas of bio-inspired computing, to present their recent results, exchange ideas and cooperate in a friendly framework.

Since 2006, the conference has taken place at Wuhan (2006), Zhengzhou (2007), Adelaide (2008), Beijing (2009), Liverpool and Changsha (2010), Penang (2011), Gwalior (2012), Anhui (2013), Wuhan (2014), Anhui (2015), Xi'an (2016), Harbin (2017), Beijing (2018), Zhengzhou (2019), Qingdao (2020), and Taiyuan (2021). Following the success of previous editions, the 17th International Conference on Bio-Inspired Computing: Theories and Applications (BIC-TA 2022) was held in Wuhan, China, during December 16-18, 2022, organized by Wuhan University of Technology with the support of the Operations Research Society of Hubei.

We would like to thank the keynote speakers for their excellent presentations: Junbiao Dai (Institute of Synthetic Biology, Shenzhen Institute of Advanced Sciences, Chinese Academy of Sciences, China), Di Liu (Wuhan Institute of Virology, Chinese Academy of Sciences, China), Sergey Verlan (Université Paris-Est Créteil, France), Yi-Lun Ying (Nanjing University, China), and Mengjie Zhang (Victoria University of Wellington, New Zealand).

A special thank you is given to the honorable chair, professor Chaozhong Wu, for his guidance and support of the conference.

We thank Jianqing Lin and Lianghao Li for their help in collecting the final files of the papers and editing the volume and maintaining the website of BIC-TA 2022 (http://2022.bicta.org/). We gratefully thank Guotong Chen, Xuan Guo, Shichen Wang, Haiyu Zhang, and Hao Zhou for their contribution in organizing the conference. We would like to thank Cheng He, Jie Song, and Fei Xu, and Gexiang Zhang for hosting the meetings. We also thank all the other volunteers, whose efforts ensured the smooth running of the conference.

Although BIC-TA 2022 was affected by COVID-19, we still received 148 submissions on various aspects of bio-inspired computing, of which 56 papers were selected in a single-blind review process with each submission receiving on average three reviews for the volume of *Communications in Computer and Information Science*. We are grateful to all the authors for submitting their interesting research work. The warmest thanks should be given to the external referees for their careful and efficient work in the reviewing process.

Special thanks are due to Springer-Nature for their skilled cooperation in the timely production of these volumes.

January 2023

Linqiang Pan
Dongming Zhao
Lianghao Li
Jianqing Lin

Organization

Steering Committee

Xiaochun Cheng	Middlesex University London, England
Guangzhao Cui	Zhengzhou University of Light Industry, China
Kalyanmoy Deb	Michigan State University, USA
Miki Hirabayashi	National Institute of Information and Communications Technology, Japan
Joshua Knowles	University of Manchester, UK
Thom La Bean	North Carolina State University, USA
Jiuyong Li	University of South Australia, Australia
Kenli Li	University of Hunan, China
Giancarlo Mauri	Università di Milano-Bicocca, Italy
Yongli Mi	Hong Kong University of Science and Technology, Hong Kong
Atulya K. Nagar	Liverpool Hope University, UK
Linqiang Pan (Chair)	Huazhong University of Science and Technology, China
Gheorghe Paun	Romanian Academy, Romania
Mario J. Perez-Jimenez	University of Seville, Spain
K. G. Subramanian	Liverpool Hope University, UK
Robinson Thamburaj	Madras Christian College, India
Jin Xu	Peking University, China
Hao Yan	Arizona State University, USA

Honorable Chair

Chaozhong Wu	Wuhan University of Technology, China

General Chair

Zhiyong Pei	Wuhan University of Technology, China

Program Committee Chairs

Dongming Zhao   Wuhan University of Technology, China
Linqiang Pan   Huazhong University of Science and Technology,
                             China

Local Chair

Xuan Guo   Wuhan University of Technology, China

Registration Chair

Haiyu Zhang   Wuhan University of Technology, China

Program Committee

Muhammad Abulaish   South Asian University, India
Andy Adamatzky   University of the West of England, UK
Chang Wook Ahn   Gwangju Institute of Science and Technology,
                             Republic of Korea
Adel Al-Jumaily   University of Technology Sydney, Australia
Bin Cao   Hebei University of Technology, China
Junfeng Chen   Hoahi University, China
Wei-Neng Chen   Sun Yat-Sen University, China
Shi Cheng   Shaanxi Normal University, China
Xiaochun Cheng   Middlesex University, UK
Tsung-Che Chiang   National Taiwan Normal University, China
Sung-Bae Cho   Yonsei University, Korea
Zhihua Cui   Taiyuan University of Science and Technology,
                             China
Kejie Dai   Pingdingshan University, China
Ciprian Dobre   University Politehnica of Bucharest, Romania
Bei Dong   Shanxi Normal University, China
Xin Du   Fujian Normal University, China
Carlos Fernandez-Llatas   Universitat Politecnica de Valencia, Spain
Shangce Gao   University of Toyama, Japan
Marian Gheorghe   University of Bradford, UK
Wenyin Gong   China University of Geosciences, China

Shivaprasad Gundibail	Manipal Academy of Higher Education (MAHE), India
Ping Guo	Beijing Normal University, China
Yinan Guo	China University of Mining and Technology, China
Guosheng Hao	Jiangsu Normal University, China
Cheng He	Southern University of Science and Technology, China
Shan He	University of Birmingham, UK
Tzung-Pei Hong	National Univesity of Kaohsiung, China
Florentin Ipate	University of Bucharest, Romania
Sunil Kumar Jha	Banaras Hindu University, India
He Jiang	Dalian University of Technology, China
Qiaoyong Jiang	Xi'an University of Technology, China
Licheng Jiao	Xidian University, China
Liangjun Ke	Xian Jiaotong University, China
Ashwani Kush	Kurukshetra University, India
Hui Li	Xi'an Jiaotong University, China
Kenli Li	Hunan University, China
Lianghao Li	Huazhong University of Science and Technology, China
Yangyang Li	Xidian University, China
Zhihui Li	Zhengzhou University, China
Jing Liang	Zhengzhou University, China
Jerry Chun-Wei Lin	Western Norway University of Applied Sciences, Norway
Qunfeng Liu	Dongguan University of Technology, China
Xiaobo Liu	China University of Geosciences, China
Wenjian Luo	University of Science and Technology of China, China
Lianbo Ma	Northeastern University, China
Wanli Ma	University of Canberra, Australia
Xiaoliang Ma	Shenzhen University, China
Francesco Marcelloni	University of Pisa, Italy
Efrén Mezura-Montes	University of Veracruz, Mexico
Hongwei Mo	Harbin Engineering University, China
Chilukuri Mohan	Syracuse University, USA
Abdulqader Mohsen	University of Science and Technology Yemen, Yemen
Holger Morgenstern	Albstadt-Sigmaringen University, Germany
Andres Muñoz	Universidad Catòlica San Antonio de Murcia, Spain

G. R. S. Murthy	Lendi Institute of Engineering and Technology, India
Akila Muthuramalingam	KPR Institute of Engineering and Technology, India
Yusuke Nojima	Osaka Prefecture University, Japan
Linqiang Pan	Huazhong University of Science and Technology, China
Andrei Paun	University of Bucharest, Romania
Gheorghe Paun	Romanian Academy, Romania
Xingguang Peng	Northwestern Polytechnical University, China
Chao Qian	University of Science and Technology of China, China
Balwinder Raj	NITTTR, India
Rawya Rizk	Port Said University, Egypt
Rajesh Sanghvi	G. H. Patel College of Engineering and Technology, India
Ronghua Shang	Xidian University, China
Zhigang Shang	Zhengzhou University, China
Ravi Shankar	Florida Atlantic University, USA
V. Ravi Sankar	GITAM University, India
Bosheng Song	Hunan University, China
Tao Song	China University of Petroleum, China
Jianyong Sun	University of Nottingham, UK
Yifei Sun	Shaanxi Normal University, China
Handing Wang	Xidian University, China
Yong Wang	Central South University, China
Hui Wang	Nanchang Institute of Technology, China
Hui Wang	South China Agricultural University, China
Gaige Wang	Ocean University of China, China
Sudhir Warier	IIT Bombay, China
Slawomir T. Wierzchon	Polish Academy of Sciences, Poland
Zhou Wu	Chongqing University, China
Xiuli Wu	University of Science and Technology Beijing, China
Bin Xin	Beijing Institute of Technology, China
Gang Xu	Nanchang University, China
Yingjie Yang	De Montfort University, UK
Zhile Yang	Shenzhen Institute of Advanced Technology, Chinese Academy of Sciences, China
Kunjie Yu	Zhengzhou University, China
Xiaowei Zhang	University of Science and Technology of China, China
Jie Zhang	Newcastle University, UK

Contents

Machine Learning and Deep Learning

Intelligent Control and Simulation

Evolutionary Computation and Swarm Intelligence

Surrogate Model-Assisted Evolutionary Algorithms for Parameter Identification of Electrochemical Model of Lithium-Ion Battery: A Comparison Study

Yan-Bo He[1] , Bing-Chuan Wang[1](✉) , and Zhi-Zhong Liu[2]

[1] School of Automation, Central South University, Changsha 410083, China
{21461111,bingcwang}@csu.edu.cn
[2] College of Information Science and Electronic Engineering, Hunan University, Changsha 410083, China
liuzz@hnu.edu.cn

Abstract. Lithium-ion batteries are widely used in various fields due to their high energy density and long cycling life. However, over-charge, over-discharge, and over-heating will cause the battery's performance to drop rapidly and even cause safety crises. The parameters of an electrochemical model are critical to making lithium-ion batteries operate safely because they can help indicate the internal states of the battery. Thus, it is significant to identify the parameters of the electrochemical model of lithium-ion batteries. The parameter identification of the electrochemical model of lithium-ion batteries is a complex expensive optimization problem intrinsically. Surrogate model-assisted evolutionary algorithms have been designed to solve this problem, but the choice of surrogate models is still an open question. A suitable surrogate model can reduce the number of time-consuming simulations of the electrochemical model; thus, it can improve the identification accuracy significantly. However, how to select a proper surrogate model has not been studied adequately. In view of this, this paper compares seven different surrogate models' performance for parameter identification and aims to provide some insights for future researchers when choosing surrogate models. Extensive simulations show that the support vector regression (SVR) model would be a good choice to aid the parameter identification.

Keywords: Lithium-ion battery · Parameter identification · Electrochemical model · Evolutionary algorithm · Surrogate model

1 Introduction

Due to the deplete of fossil fuels and the deterioration of the environment, it is urgent to develop clean energy. Among various clean energy resources, lithium-ion batteries are widely used as power sources of electric vehicles (EVs), hybrid EVs, and plug-in hybrid EVs [12,16]. Because they own the following advantages:

L. Pan et al. (Eds.): BIC-TA 2022, CCIS 1801, pp. 3–16, 2023.
https://doi.org/10.1007/978-981-99-1549-1_1

high energy density, high power density, no memory effect, low self-discharge rate, long cycling life, etc. However, in practical utilization, lithium-ion batteries would suffer from over-charge, over-discharge, and over-heating which will lead to the rapid degradation of battery's performance and even safety issues [14]. Battery management systems (BMSs) are of great significance to guarantee that lithium-ion batteries are operated reliably and safely [28].

In order to design an efficient BMS, batteries' behaviors and internal characteristics should be simulated and analyzed through a well-performed model. Traditionally, the models of lithium-ion batteries can be divided into two classes: the empirical models [22] and the electrochemical models [21]. Due to their simplicity and high speed, the empirical models such as equivalent circuit models and neural network models are widely used in BMSs [15,19]. However, empirical models cannot reflect the internal physical phenomena in batteries. Electrochemical models are derived based on the concentrated solution theory and the porous electrode theory. Thus, they are more sophisticated and accurate than empirical models [11]. Additionally, the parameters of electrochemical models can be matched with that of lithium-ion batteries exactly. The performance of electrochemical models is greatly affected by their model parameters. If the parameters are not set properly, electrochemical models would fail to simulate batteries' internal changes. Thus, the key parameters of electrochemical models should be accurately identified through an effective parameter identification method and then an effective BMS can be obtained by using the electrochemical models [29].

During last decades, many parameter identification methods have been proposed to seek suitable parameters for electrochemical models, e.g., least square methods [5], recursive least square methods [25], Gaussian-Newton methods [2], evolutionary algorithms (EAs) [8,9,20,29], etc. For highly complex and strongly nonlinear electrochemical models, gradient-free EAs that can be used without a closed-form expression of the objective function become more and more favored by researchers [9,29]. Unfortunately, numerical simulations of electrochemical models are computationally expensive, and EAs always require a lot of real simulations to seek a satisfying solution, which would consume much time. In this case, a surrogate model is usually used to assist EAs to optimize the parameters of electrochemical models where a key point is to construct an accurate surrogate model to approximate the objective function of parameter identification [29]. An accurate surrogate model can improve the identification accuracy as well as the identification efficiency. However, how to select a proper surrogate model has not attracted researchers' much attention yet. Although there are many surrogate models such as Kriging [18], radial basis function neural network (RBFNN) [23], support vector regression (SVR) [1], and artificial neural network (ANN) [24], it is non-trivial to choose a proper surrogate model for identifying the parameters of electrochemical models of lithium-ion batteries. Because electrochemical models involve strong nonlinearities and time/space coupled dynamics. Thus, it is of significance to compare the performance of several representative surrogate models in assisting parameter identification of electrochemical models of lithium-ion batteries.

Based on the above observations, we seek to compare the performance of different surrogate models and focus on four most commonly used models including SVR [1], RBFNN [23], ANN [24], and Kriging [18]. Note that we use them as our primary models. Meanwhile, in order to achieve a balance between exploration and exploitation, we borrow the idea of importing uncertainty estimation from Kriging to other models [27]. As a result, we propose other three models: SVR with uncertainty estimation (SVRue), RBFNN with uncertainty estimation (RBFNNue), and ANN with uncertainty estimation (ANNue). To implement a fair comparison of the performance of surrogate models, all methods adopt differential evolution (DE) as the identification algorithm and its parameters are set to default values [4]. Based on the prediction accuracy and running time of the above seven surrogate model-assisted identification algorithms, we summarize some concluding remarks which will provide some insights of selecting a proper surrogate model for parameter identification of electrochemical models of lithium-ion batteries.

The rest of this paper is organized as follows. Section 2 gives the problem description. The details of surrogate model-assisted DE are presented in Sect. 3. Extensive simulations and discussions are given in Sect. 4. Section 5 summarizes the conclusion.

2   Problem Description

2.1   Pseudo-Two-Dimensional Model

The pseudo-two-dimensional (P2D) model has become one of the most popular electrochemical models of lithium-ion batteries because it can accurately describe the internal physical phenomena of lithium-ion batteries. To be specific, the P2D model contains a set of coupled nonlinear partial differential equations (PDEs) and algebraic equations, the details of which can be found in [6]. Due to the complex structure and strong nonlinearities, the calculation of the P2D model requires a lot of time and storage space. Given the input current $I(t)$, the output voltage $V(t)$ can be expressed as follows:

$$V(t) = \varphi_s|_{x=0} - \varphi_s|_{x=L_p+L_{sep}+L_n} \tag{1}$$

where t denotes the time, φ_s denotes the potential in the solid phase, and L_p, L_{sep}, and L_n are the thicknesses of the positive electrode, the separator, and the negative electrode, respectively.

2.2   Objective Function

The parameter identification of the P2D model of lithium-ion batteries can be formulated as an optimization problem intrinsically. Its task is to seek the optimal parameters that minimize the objective function. Traditionally, the battery parameters can be roughly divided into two categories, i.e., dynamic parameters and static parameters. Compared with static parameters, dynamic parameters will vary significantly with the change of lithium-ion concentration or reaction temperature [29]; thus, they have greater impact on the output voltage. In this paper, we will focus on identifying the dynamic parameters of the P2D model of lithium-ion batteries. In general, the objective function can be defined as follows:

$$f(\boldsymbol{\theta}) = \left\| \boldsymbol{V}^{ref} - \boldsymbol{V} \right\|_2 = \sqrt{\frac{1}{N} \sum_{i=1}^{N} \left(V_i^{ref} - V_i \right)^2} \qquad (2)$$

where $\| \cdot \|_2$ denotes the Euclidean norm, $\boldsymbol{\theta} = \{\theta_1, \cdots, \theta_d\}$ contains the parameters to be identified, $\boldsymbol{V}^{ref}$ is the vector of measured voltages, $\boldsymbol{V}$ represents the output voltages generated by the P2D model, V_i^{ref} and V_i are the ith samples in $\boldsymbol{V}^{ref}$ and $\boldsymbol{V}$, respectively, and N is the size of sampling time window. Besides, each parameter $\theta_j \in \boldsymbol{\theta}$ is limited to a feasible region:

$$\underline{\theta}_j \leq \theta_j \leq \bar{\theta}_j, j = 1, \cdots, d \qquad (3)$$

where $\underline{\theta}_j$ and $\bar{\theta}_j$ are the lower and upper bounds, respectively.

Note that each simulation of the P2D model is time-consuming, so $f(\boldsymbol{\theta})$ is an expensive objective function. In view of this, we use surrogate model-assisted DE to solve this problem.

3   Surrogate Model-Assisted DE

3.1   General Framework

The framework of the surrogate model-assisted DE is described in Algorithm 1. For the sake of simplicity, the meanings of the symbols in the algorithm are given as follows:

- t: number of iterations.
- $\boldsymbol{\Theta}$: training data set containing the solutions that have been evaluated by the P2D model, $\boldsymbol{\Theta} = (\boldsymbol{\theta}_1, \cdots, \boldsymbol{\theta}_n)^T$.
- $\boldsymbol{Y}$: evaluation values of $\boldsymbol{\Theta}$ by simulating the P2D model.
- $\hat{f}(\boldsymbol{\theta})$: predicted value of the surrogate model.
- $\hat{s}(\boldsymbol{\theta})$: uncertainty estimation of the surrogate model.
- $\boldsymbol{X}$: initial population.
- $\boldsymbol{\theta}_{best}^t$: the best individual at the tth iteration based on the surrogate model.
- f_{loss}^t: evaluation value of $\boldsymbol{\theta}_{best}^t$ by simulating the P2D model.
- $\boldsymbol{\theta}_{target}$: the best solution found during last iterations.
- f_{loss}: evaluation value of $\boldsymbol{\theta}_{target}$ by simulating the P2D model.
- f_{loss}^{min}: value of the halting criterion.

As shown in Algorithm 1, a database is first constructed by using Latin hypercube sampling (LHS) method (Line 2). Note that the solutions in the database are evaluated by simulating the P2D model. At each iteration, a set of solutions is first selected from the database to construct a surrogate model (Lines 4–5).

Algorithm 1. SURROGATE MODEL-ASSISTED DE

1: $t \leftarrow 1$;
2: Sample solutions from decision space to construct the database;
3: **while** $t \leq t_{max}$ **do**
4:     Select a set of solutions (Θ, Y) from the database as the training set;
5:     $[\hat{f}(\theta), \hat{s}(\theta)] \leftarrow$ surrogate_construction (Θ, Y);
6:     $X \leftarrow$ LHS_initialization;
7:     $\theta_{best}^t \leftarrow$ DE(X);
8:     $f_{loss}^t \leftarrow$ P2D_simulation(θ_{best}^t);
9:     Add θ_{best}^t, f_{loss}^t to the database;
10:    **if** $f_{loss}^t \leq f_{loss}$ **then**
11:        $\theta_{target} = \theta_{best}$;
12:        $f_{loss} = f_{loss}^t$;
13:    **end if**
14:    **if** $f_{loss}^t \leq f_{loss}^{min}$ **then**
15:        **break**
16:    **end if**
17:    $t \leftarrow t + 1$;
18: **end while**
19: Output the best solution θ_{target};

Thereafter, a population of solutions X is initialized by using the LSH method (Line 6). Subsequently, DE is applied to search for a satisfying solution θ_{best}^t (Line 7). Note that the surrogate model is used to evaluate solutions in DE. Then the P2D model is used to evaluate θ_{best}^t (Line 8). Next, add θ_{best}^t and the corresponding evaluation value f_{loss}^t to the database (Line 9). If θ_{best}^t is better than the best solution found during last iterations θ_{target}, then replace θ_{target} by θ_{best}^t (Lines 10–13). If θ_{target} satisfies the halting criterion, then break the loop (Lines 14–16); otherwise, repeat the above steps (Lines 3–18) until the maximum number of iterations is achieved. Finally, output the best solution θ_{target}. For convenience, the flowchart of the algorithm is described in Fig. 1.

Particularly, Kriging, SVRue, RBFNNue, and ANNue will provide the uncertainty estimation $\hat{s}(\theta)$. In this case, to fully search the solution space with a balance between exploration and exploitation, the lower confidence bound (LCB) criterion [3] is used as the prescreening objective function:

$$\hat{f}_{lcb}(\theta) = \hat{f}(\theta) - \omega\hat{s}(\theta) \tag{4}$$

where ω is generally set to 1 based on the suggestions in [13]. In addition, SVR, RBFNN, and ANN don't provide the uncertainty estimation, so the value of w is set to 0.

Subsequently, a brief introduction to the surrogate models and DE is given.

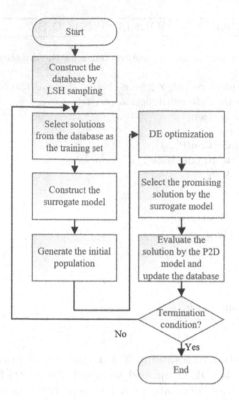

Fig. 1. The flowchart of the framework.

3.2   Surrogate Models

Kriging. Kriging (i.e., Gaussian process regression) [18] is a statistical interpolation model in which the statistical correlations among the measured points are used for prediction. Kriging assumes that the distance between each two samples can reflect a spatial correlation used to explain variation in the response surface. To be specific, the correlation between two samples (i.e., $\boldsymbol{\theta}_i$ and $\boldsymbol{\theta}_j$) is defined as follows:

$$\text{Corr}\,(\boldsymbol{\theta}_i, \boldsymbol{\theta}_j) = \exp\left(-\sum_{k=1}^{d} p_k\,|\theta_{i,k} - \theta_{j,k}|^2\right) \tag{5}$$

where $\theta_{i,k}$ and $\theta_{j,k}$ are the kth dimensions of $\boldsymbol{\theta}_i$ and $\boldsymbol{\theta}_j$, respectively, $p_k\,(0 \le p_k < \infty)$ is the weight coefficient, and d is the dimension of $\boldsymbol{\theta}$. Based on this assumption, Kriging can produce a prediction surface and measure the predictions' uncertainty. The prediction and uncertainty estimation provided by Kriging are formulated as follows:

$$\hat{f}\,(\boldsymbol{\theta}) = \hat{\mu} + \mathbf{r}'\mathbf{R}^{-1}(\mathbf{f} - \mathbf{1}\hat{\mu}) \tag{6}$$

$$\hat{s}^2(\boldsymbol{\theta}) = \hat{\sigma}^2 \left[1 - \mathbf{r}'\mathbf{R}^{-1}\mathbf{r} + \frac{(1 - \mathbf{1}'\mathbf{R}^{-1}\mathbf{r})^2}{\mathbf{1}'\mathbf{R}^{-1}\mathbf{1}} \right] \tag{7}$$

where $\hat{\mu}$ denotes the mean of the stochastic process, $\hat{\sigma}^2$ is the variance, $\mathbf{R}$ is the correlation matrix including the correlations between each two samples, $\mathbf{r}$ denotes the correlation vector including the correlations between $\boldsymbol{\theta}$ and each sample, $\mathbf{f}$ includes the outputs of samples, and $\mathbf{1}$ is a vector of ones. We can obtain the values of $\hat{\mu}$ and $\hat{\sigma}^2$ that maximize the likelihood function in a closed form:

$$\hat{\mu} = \frac{\mathbf{1}\mathbf{R}^{-1}\mathbf{f}}{\mathbf{1}'\mathbf{R}^{-1}\mathbf{1}} \tag{8}$$

$$\hat{\sigma}^2 = \frac{(\mathbf{f} - \mathbf{1}\hat{\mu})'\mathbf{R}^{-1}(\mathbf{f} - \mathbf{1}\hat{\mu})}{n} \tag{9}$$

where n denotes the number of samples.

SVR. As its name refers, support vector regression (SVR) [1] generalizes support vector machine (SVM) to solve regression problems. SVR is characterized by the use of kernels, sparse solution, Vapnik-Chervonenkis (VC) bound control of the margin, and the number of support vectors. The general form of SVR is given as follows:

$$\hat{f}(\boldsymbol{\theta}) = b + \boldsymbol{w}^\top \boldsymbol{\theta}. \tag{10}$$

where b denotes the intercept and w is the weight vector. Both b and w can be calculated through a mathematical optimization problem. The details can be found in [1]. The core technique for extending SVR to high dimensional problems is the kernel function. By using the kernel technique, SVR can be rewritten as follows:

$$\hat{f}(\boldsymbol{\theta}) = \sum_{i=1}^{n} w_i \psi(\boldsymbol{\theta}, \boldsymbol{\theta}_i) + b \tag{11}$$

where $\psi(\cdot, \cdot)$ is the kernel function, n is the number of samples, $\boldsymbol{\theta}_i$ is the ith sample, and w_i is the ith element in $\boldsymbol{w}$. In view of its ability of universal approximation [7], the radial basis function (RBF) is adopted as the keneral function in our study.

ANN. Artificial neural network (ANN) is an efficient data-driven modeling tool. It is widely used for dynamic modeling and optimization of nonlinear systems due to its universal approximation ability and flexible structure [24]. Traditionally, ANN comprises an input layer, an output layer, and multiple hidden layers. The mathematical formula of an ANN containing one hidden layer can be expressed as follows:

$$\hat{f}(\boldsymbol{\theta}) = \sum_{i=1}^{n} v_i g(\sum_{j=1}^{d} w_{i,j}\theta_j + d_i) \tag{12}$$

where n is the number of neurons in the hidden layer, v_i is the weight connecting the ith hidden neuron with the output neuron, $w_{i,j}$ is the weight connecting the ith hidden neuron and the jth input neuron, d_i is the intercept of the ith hidden neuron, and $g(\cdot)$ is the activation function. ANN is trained to approximate a particular function by adjusting its weights and intercepts between neurons until the network can capture the distribution of data well [17].

RBFNN. Radial basis function neural network (RBFNN) [23] is a kind of feedforward neural network. The basic idea of RBFNN is to map low-dimensional linearly inseparable data to high-dimensional space, making it linearly separable in high-dimensional area. RBFNN typically consists of three layers: an input layer, a hidden layer where each neuron has a RBF activation function, and a linear output layer. The input of the network is a vector of real numbers $\boldsymbol{\theta} \in \mathbb{R}^d$. The output of the network is a scalar function of the input vector $\varphi : \mathbb{R}^d \rightarrow \mathbb{R}$:

$$\varphi(\boldsymbol{\theta}) = \sum_{i=1}^{N} w_i \rho\left(\|\boldsymbol{\theta} - \mathbf{c}_i\|_2\right) \tag{13}$$

where N is the number of neurons in the hidden layer, $\mathbf{c}_i$ is the center vector of the ith neuron, and w_i is the weight of the ith neuron in the output layer. The commonly used RBF is described as follows:

$$\rho\left(\|\boldsymbol{\theta} - \mathbf{c}_i\|_2\right) = \exp\left[-\beta\|\boldsymbol{\theta} - \mathbf{c}_i\|_2^2\right] \tag{14}$$

where $\|\cdot\|_2$ denotes the Euclidean norm and β is a parameter that should be set properly.

Based on the above descriptions, ANNue, RBFNNue, and SVRue consist of the original models of ANN, RBFNN, and SVR, respectively. Besides, they also include the uncertainty estimation imported from the Kriging model [27]. The prediction of the original model and the uncertainty estimation will be used to calculate the LCB criterion in our study.

3.3   DE

DE is a simple yet effective evolutionary algorithm. It consists of three steps: mutation, crossover and selection. In the framework described in Algorithm 1, we use the traditional DE settings without any modifications. First, an initial population of NP solutions $\boldsymbol{X} = \{\boldsymbol{x}_1, \cdots, \boldsymbol{x}_{NP}\}$ is generated by using LSH method. Then, the mutation operator DE/rand/1 [26] is adopted to generate the mutant vector for each solution $\boldsymbol{x}_i$:

$$\boldsymbol{v}_i = \boldsymbol{x}_{r_1^i} + F \cdot \left(\boldsymbol{x}_{r_2^i} - \boldsymbol{x}_{r_3^i}\right) \tag{15}$$

where $\boldsymbol{v}_i$ denotes the mutant vector, the indices r_1^i, r_2^i, r_3^i are mutually exclusive integers randomly generated within the range $[1, NP]$, and F is a positive control parameter to scale the difference vector.

After the step of mutation, the crossover operator is applied to generate a trial vector as follows:

$$u_{i,j} = \begin{cases} v_{i,j} & \text{if } (\text{rand}_j[0,1] \leq CR) \text{ or } (j = j_{\text{rand}}) \\ x_{i,j} & \text{otherwise} \end{cases} \quad (16)$$

where $\boldsymbol{u}_i$ denotes the trial vector, $x_{i,j}$, $v_{i,j}$, and $u_{i,j}$, are the jth dimensions of $\boldsymbol{x}_i$, $\boldsymbol{v}_i$, and $\boldsymbol{u}_i$, respectively, $rand_j[0,1]$ is a random value generated between 0 and 1 uniformly, $j = j_{\text{rand}}$ is a random integer selected from $\{1, \cdots, d\}$, and CR is the crossover control parameter.

Finally, at the step of selection, the LCB criterion defined in Eq. (4) is used as the prescreening objective function. If the LCB value of $\boldsymbol{u}_i$ is better than that of $\boldsymbol{x}_i$, we will replace $\boldsymbol{x}_i$ by $\boldsymbol{u}_i$.

4   Results and Discussion

4.1   Simulation Settings

The main task of this paper is to seek a proper surrogate model for parameter identification of the P2D model of a lithium-ion batteries. Thus, we have established a simulation system of a 20 Ah LiMn2O4/Graphite pouch lithium-ion battery in the COMSOL multiphysics to compare the performance of different surrogate models. Since we focus on identifying dynamic parameters, the discharge rate of 3C was used to obtain the reference voltage curve [10]. Note that the static parameters were set as the same in [10] and described in Table 1. The identification algorithm was run in MATLAB based on a workstation with Intel (R) Xeon (R) E5-1620 CPU and 8 GB RAM. The specific settings of the identification algorithm were summarized as follows:

- Size of the training set: $N_s = 100$.
- Maximum number of iterations: $t_{max} = 40$.
- Size of the population: $NP = 60$.
- Number of generations of DE: $G = 60$.
- Halting criterion: $f_{loss}^{min} = 0.001$
- Scaling factor: $F = 0.9$.
- Crossover control parameter: $CR = 0.5$.

4.2   Result Analysis

The performance of different surrogate models were compared based on the root mean square error (RMSE) on the testing data and the running time. As shown in Table 2, SVR obtains the smallest RMSE and it runs faster than the other competitors. As shown in Fig. 2, the discharge voltage curve produced by the P2D model using the parameters estimated by SVR can agree with the reference voltage curve well. Among different surrogate models, SVR obtains the smallest relative error for most of the time. ANN and RBFNN perform worse

Table 1. Settings of Static Parameters

Parameters	Unit	Value
L_p	m	1.81E-04
L_{sep}	m	5.36E-05
L_n	m	1.04E-04
R_p	m	6.97E-06
R_n	m	1.37E-05
$\varepsilon_{s,p}$	–	2.51E-01
$\varepsilon_{s,n}$	–	4.24E-01
$\varepsilon_{e,p}$	–	2.57E-01
$\varepsilon_{e,n}$	–	2.33E-01
σ_p	$mol\ m^{-3}$	3.36E+01
σ_n	$mol\ m^{-3}$	1.47E+02
cl_0	$mol\ m^{-3}$	1.95E+03

Table 2. Simulation Results

Surrogate Model	RMSE	Running Time(s)
ANN	0.007836	1648
ANNue	0.036359	2588
Kriging	0.004161	458
RBFNN	0.013782	1507
RBFNNue	0.040038	1614
SVR	**0.002593**	**453**
SVRue	0.010561	564

than SVR and Kriging in terms of RMSE. Besides, they cost far more time than SVR and Kriging. ANN and RBFNN may suffer from the underfitting problem because few samples were used for training. ANNue, RBFNNue, and SVRue cannot outperform their original models (i.e., ANN, RBFNN, and SVR) in terms of both RMSE and running time. It implies that the uncertainty estimation imported from Kriging cannot improve the performance of ANN, RBFNN, and SVR. The reason may be that the uncertainty estimation provided by Kriging is mismatched with the original models.

As shown in Table 3, most of the parameters estimated by the SVR-assisted DE are close to the reference parameters. Compared with other parameters, the identified values of k_p, k_n and $D_{s,p}$ are less accurate. The reason may be that they are not very sensitive to the output voltage [29]. Note that the discharge current used for parameter identification is constant. However, the applied current in practice usually varies randomly according to actual working conditions. Therefore, we further evaluated the performance of the parameters estimated

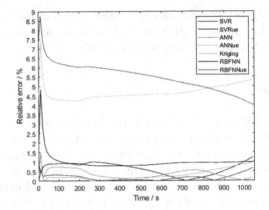

Fig. 2. Relative errors of the discharge voltages of the seven different surrogate models.

Table 3. Identified Results of the SVR-Assisted DE

Notation	Unit	Effective Ranges	Reference Values	Identified Results
D_e	m^2s^{-1}	[1e-11,2e-12]	7.5e-11	8.310e-11
$D_{s,p}$	m^2s^{-1}	[8.5e,2e-13]	1.025e-13	9.488e-14
$D_{s,n}$	m^2s^{-1}	[3e-13,7-13]	5.323e-13	3.391e-13
k_p	$m^{2.5}mol^{-0.5}s^{-1}$	[8e-12,4e-11]	2.303e-11	9.809e-12
k_n	$m^{2.5}mol^{-0.5}s^{-1}$	[8e-12,4e-11]	8e-12	1.798e-11
t_1	-	[0.3,0.5]	0.3	0.3

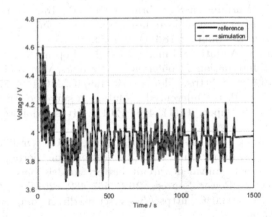

Fig. 3. Simulation voltage of the P2D model using the parameters identified by the SVR-assisted DE under UDDS operating mode.

by the SVR-assisted DE under the complex urban dynamometer driving schedule (UDDS) operating mode. As shown in Fig. 3, the identified parameters have achieved outstanding performance.

In summary, SVR can provide satisfying results in parameter identification of the P2D model of lithium-ion batteries. Additionally, the uncertainty estimation provided by Kriging would not be effective in other surrogate models.

5   Conclusion

In the parameter identification of the electrochemical model of lithium-ion batteries, it is of great significance to select a suitable surrogate model. In this paper, we have compared the performance of seven representative surrogate models for parameter identification. Four of them (i.e., ANN, RBFNN, SVR, and Kriging) are frequently used in surrogate model-assisted EAs. In addition, we borrowed the idea of importing uncertainty estimation from Kriging to other models and constructed three new surrogate models (i.e., ANNue, RBFNNue, and SVRue). The seven surrogate models were embedded into the framework of a surrogate model-assisted DE for parameter identification. The simulation results show that SVR can provide the best results in terms of both RMSE and running time. Additionally, the uncertainty estimation provided by Kriging would not be effective in other surrogate models. In future work, we will focus on designing SVR-assisted EA to identify the parameters of the P2D model at multiple temperatures.

References

1. Awad, M., Khanna, R.: Support vector regression. In: Efficient Learning Machines, pp. 67–80. Springer, Cham (2015). https://doi.org/10.1007/978-1-4302-5990-9_4
2. Boovaragavan, V., Harinipriya, S., Subramanian, V.R.: Towards real-time (milliseconds) parameter estimation of lithium-ion batteries using reformulated physics-based models. J. Power Sources **183**(1), 361–365 (2008)
3. Cox, D.D., John, S.: A statistical method for global optimization. In: Proceedings of the 1992 IEEE International Conference on Systems, Man, and Cybernetics, vol. 2, pp. 1241–1246 (1992). https://doi.org/10.1109/ICSMC.1992.271617
4. Das, S., Suganthan, P.N.: Differential evolution: a survey of the state-of-the-art. IEEE Trans. Evol. Comput. **15**(1), 4–31 (2011). https://doi.org/10.1109/TEVC.2010.2059031
5. Fleischer, C., Waag, W., Heyn, H.M., Sauer, D.U.: On-line adaptive battery impedance parameter and state estimation considering physical principles in reduced order equivalent circuit battery models part 2. Parameter and state estimation. J. Power Sources **262**, 457–482 (2014). https://doi.org/10.1016/j.jpowsour.2014.03.046. https://www.sciencedirect.com/science/article/pii/S0378775314003590
6. Guo, M., Kim, G.H., White, R.E.: A three-dimensional multi-physics model for a Li-ion battery. J. Power Sources **240**, 80–94 (2013)
7. Han, S., Qubo, C., Meng, H.: Parameter selection in SVM with RBF kernel function. In: World Automation Congress 2012, pp. 1–4 (2012)
8. Jokar, A., Rajabloo, B., Désilets, M., Lacroix, M.: An inverse method for estimating the electrochemical parameters of lithium-ion batteries. J. Electrochemical Soc. **163**(14), A2876–A2886 (2016)

9. Kim, M., et al.: Data-efficient parameter identification of electrochemical lithium-ion battery model using deep Bayesian harmony search. Appl. Energy **254**, 113644 (2019)

10. Li, J., Zou, L., Feng, T., Dong, X., Zou, Z., Yang, H.: Parameter identification of lithium-ion batteries model to predict discharge behaviors using heuristic algorithm. J. Electrochemical Soc. **163** (2016)

11. Li, W., et al.: Parameter sensitivity analysis of electrochemical model-based battery managemient systems for lithium-ion batteries. Appl. Energy **269**, 115104 (2020)

12. Lipu, M.H., et al.: A review of state of health and remaining useful life estimation methods for lithium-ion battery in electric vehicles: Challenges and recommendations. J. Clean. Prod. **205**, 115–133 (2018)

13. Liu, B., Yang, H., Lancaster, M.J.: Global optimization of microwave filters based on a surrogate model-assisted evolutionary algorithm. IEEE Trans. Microw. Theory Tech. **65**(6), 1976–1985 (2017). https://doi.org/10.1109/TMTT.2017.2661739

14. Liu, B., et al.: Safety issues caused by internal short circuits in lithium-ion batteries. J. Mater. Chem. A **6**(43), 21475–21484 (2018)

15. Liu, K., Li, K., Peng, Q., Zhang, C.: A brief review on key technologies in the battery management system of electric vehicles. Front. Mech. Eng. **14**(1), 47–64 (2019)

16. Lu, L., Han, X., Li, J., Hua, J., Ouyang, M.: A review on the key issues for lithium-ion battery management in electric vehicles. J. Power Sources **226**, 272–288 (2013)

17. Nasr, M.S., Moustafa, M.A., Seif, H.A., El Kobrosy, G.: Application of artificial neural network (ANN) for the prediction of EL-AGAMY wastewater treatment plant performance-Egypt. Alexandria Eng. J. **51**(1), 37–43 (2012)

18. Oliver, M.A., Webster, R.: Kriging: a method of interpolation for geographical information systems. Int. J. Geogr. Inf. Syst. **4**(3), 313–332 (1990). https://doi.org/10.1080/02693799008941549

19. Plett, G.L.: Battery Management Systems, Volume II: Equivalent-Circuit Methods. Artech House (2015)

20. Rahman, M.A., Anwar, S., Izadian, A.: Electrochemical model parameter identification of a lithium-ion battery using particle swarm optimization method. J. Power Sources **307**, 86–97 (2016)

21. Ramadesigan, V., Northrop, P.W.C., De, S., Santhanagopalan, S., Braatz, R.D., Subramanian, V.R.: Modeling and simulation of lithium-ion batteries from a systems engineering perspective. J. Electrochemical Soc. **159** (2010)

22. Seaman, A., Dao, T.S., McPhee, J.: A survey of mathematics-based equivalent-circuit and electrochemical battery models for hybrid and electric vehicle simulation. J. Power Sources **256**, 410–423 (2014)

23. She, C., Wang, Z., Sun, F., Liu, P., Zhang, L.: Battery aging assessment for real-world electric buses based on incremental capacity analysis and radial basis function neural network. IEEE Trans. Ind. Inf. **16**(5), 3345–3354 (2019)

24. Shokry, A., Espuña, A.: The ordinary kriging in multivariate dynamic modelling and multistep-ahead prediction. In: Friedl, A., Klemeš, J.J., Radl, S., Varbanov, P.S., Wallek, T. (eds.) 28th European Symposium on Computer Aided Process Engineering, Computer Aided Chemical Engineering, vol. 43, pp. 265–270. Elsevier (2018). https://doi.org/10.1016/B978-0-444-64235-6.50047-4. https://www.sciencedirect.com/science/article/pii/B9780444642356500474

25. Singh, A., Izadian, A., Anwar, S.: Model based condition monitoring in lithium-ion batteries. J. Power Sources **268**, 459–468 (2014)

26. Storn, R.: On the usage of differential evolution for function optimization. In: Proceedings of North American Fuzzy Information Processing, pp. 519–523 (1996). https://doi.org/10.1109/NAFIPS.1996.534789
27. Viana, F., Haftka, R.: Importing Uncertainty Estimates from One Surrogate to Another. https://doi.org/10.2514/6.2009-2237. https://arc.aiaa.org/doi/abs/10.2514/6.2009-2237
28. Xiong, R., Li, L., Tian, J.: Towards a smarter battery management system: a critical review on battery state of health monitoring methods. J. Power Sources **405**, 18–29 (2018)
29. Zhou, Y., Wang, B.C., Li, H.X., Yang, H.D., Liu, Z.: A surrogate-assisted teaching-learning-based optimization for parameter identification of the battery model. IEEE Trans. Ind. Inform. (2020)

Research on Multi-modal Multi-objective Path Planning by Improved Ant Colony Algorithm

Juan Jing[1], Ling Zhang[1], Chaonan Shen[1(✉)], and Kai Zhang[1,2]

[1] School of Computer Science and Technology, Wuhan University of Science and Technology, Wuhan 430065, China
449777215@qq.com
[2] Hubei Province Key Laboratory of Intelligent Information Processing and Real-Time Industrial System, Wuhan University of Science and Technology, Wuhan 430065, China

Abstract. A research about the path planning problem has been popular topic nowadays and some effective algorithms have been developed to solve this kind of problem. However, the existing algorithms to solve the path problem can only find a single optimal path, cannot satisfactorily find multiple groups of optimal solutions at the same time, and it is very necessary to propose as many solutions as possible. So this paper carries out a research on the Multi-modal Multi-Objective Path Planning (MMOPP), the objective is to find all sets of Pareto optimal path solutions from the start point to the end point in a grid map. This paper proposes a multi-modal multi-objective ant colony path planning optimization algorithm based on matrix preprocessing technology and Dijkstra algorithm (MD-ACO). Firstly, a new method of storing maps that reduces the size of the map and reduces the size of the decision space has been proposed in this paper. Secondly, using the characteristics of the Dijkstra algorithm that can quickly find the optimal path, generate an initial feasible solution about the problem, and improve the problem that the initial pheromone of ant colony algorithm is insufficient and searching for solutions is slow. Thirdly, a reasonable threshold is set for the pheromone to avoid algorithm getting stuck in local optimal solution. Finally, the algorithm is tested on the MMOPP test sets to evaluate the performance of the algorithm, and the results show that MD-ACO algorithm can solve MMOPP and get the optimal solution set.

Keywords: Path Planning · Multi-modal Multi-objective Optimization · Ant Colony Optimization (ACO) · Dijkstra

1 Introduction

Multi-Objective problem means that solving a problem needs to consider multiple objectives, and the objectives are conflicting and restricting each other. Generally speaking, optimizing one of the objectives will lead to the weakening of other objectives. For example, in the path planning problem, it may be necessary to consider multiple objectives such as the quantity of blocked paths, path length, and the quantity of intersections. There may be exist where the length of path is short but the quantity of blocked paths or the quantity of intersections is relatively large. Therefore, there may be more than one optimal solution for the Multi-objective optimization problem (MOP) [1–5], and there may be a set composed of multiple optimal solutions. The front composed of these optimal solutions in the objective space are called Pareto Front (PF). In MOP, there may be two or more global or local Pareto optimal sets, some of which may correspond to the same PF, which is the Multi-modal Multi-objective Optimization Problem (MMOP) [6].

MMOP can be described as

$$min_x^{f(x)} = min_x^{[f_1(x), f_2(x), \dots f_m(x)]}, x \in \Omega, f_i(x) \in R^m \tag{1}$$

$$PS = \{PS_1, \dots PS_k\}, k > 1 \tag{2}$$

where the decision vector $X = (x_1, x_2, \dots x_n)$ belongs to the non-empty decision space Ω, the objective function vector $f : \Omega \rightarrow \Lambda$ consists of $m(m \geq 2)$ objectives and Λ is the objective space. A few relevant definitions are briefly described below: a solution $x \in \Omega$ in a decision space that satisfies these constraints is called a feasible solution. Given two solutions $x, y \in \Omega$ and their corresponding objective vectors $f(x), f(y) \in R^m$, x dominates y (denoted as $x \prec y$) when $\forall i \in \{1, 2, \dots, m\}, f_i(x) \leq f_i(y)$ and $\exists j \in \{1, 2, \dots, m\}, f_j(x) < f_j(y)$. A solution that is not dominated by any other solution is defined as Pareto optimal solution. In Eq. (2), it is assumed that PS_k is the k-th Pareto optimal solution, $\{PS_1 \dots PS_k\}$ denotes multiple equivalent Pareto optimal solutions, corresponding to the same PF, PS is a set consisting of PS_k.

In this paper, a new path planning combinatorial algorithm is proposed to solve the MMOPP problem. Firstly, to improve the speed of the global path search of ACO, Dijkstra is used to improve the initial pheromone concentration allocation for different costs and increase the accuracy of the initial search. Secondly, in order to maintain the diversity of solutions in the decision space, non-dominated solutions are retained in the ant colony algorithm by judging dominant relationships. Finally, in order to maintain the diversity of the objective space and avoid getting stuck in the local optimal solution, the pheromone threshold is set.

2 Related Works

2.1 MMOPP Example

The path problem is a typical MOP problem. As shown in Fig. 1, from the start point to the end point, there are three optimal paths. These three paths correspond to the same path length and the same congestion point. The optimal solutions of these paths correspond to the same PF in the objective space. The purpose of this study is to obtain multiple paths with optimal and equal target values, such as the length of multiple paths, the number of blocked road sections, and the number of cross-road conditions.

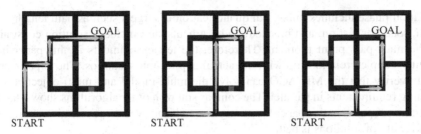

Fig. 1. Three paths corresponding to the same point in PF

2.2 Research on MMOPP Problems

From the perspective of practical application, the research of MMOPP problem is usually divided into three types. One is path planning that simulates real-life road conditions [6], decision makers need to consider planning requirements such as path length, congestion points, and must-pass points at the same time. Two is the vehicle path planning considering the logistics distribution requirements [7], which usually needs to consider the load constraints of the distribution vehicles and the time constraints of the distribution points. Another is to consider robot path planning [8], which needs to consider planning requirements such as path length and obstacles. This paper considers the first MMOPP problem, and studies the Multi-objective Path Planning problem under the requirement of multi-modal optimization.

From the point of view of selection algorithm, classical path planning algorithms such as Dijkstra algorithm and A* algorithm only consider the optimization of path length, which is hard to resolve the MMOPP problem. Genetic algorithm [9], particle swarm algorithm [10], evolutionary algorithm NSGA-II [11], etc. can also solve the MMOPP problem, but these algorithms focus more on approaching the PF, ignoring the diversity of solution distributions in the decision space. Therefore, most scholars choose to introduce swarm intelligence algorithms with self-organization, self-adaptation and self-learning characteristics to realize MMOPP. The ACO used in this paper is used by many scholars to solve the path problem. People are committed to improving ACO to improve its effectiveness and applicability. At present, it mainly focuses on two aspects: process improvement and parameter setting. In terms of process improvement, Mao [12] proposed an improved A* algorithm to quickly find a better path, and adjust the pheromone concentration on the path through coefficients. MOHAMED et al. [13] proposed a solution method combining local search and ACO to solve the vehicle routing problem with loading capacity constraints. In terms of parameter setting, Yuan et al. [14] studied the path problem of the tea picking robot, improved the ant colony algorithm, and improved the global search ability and computational efficiency by changing the adaptive adjustment pheromone concentration value and iterative termination conditions. Liu et al. [15] proposed adaptive pheromone concentration and dynamic pheromone volatilization factor when designing a path navigation system for indoor service robots. The improved ACO has higher global search ability.

The MD-ACO used in this paper combines the process improvement and parameter setting to solve the MMOPP problem. The MMOPP test set used in the algorithm validation phase includes numerical quantification of targets such as path length, road width, road congestion, and blocked road segments, and includes solution constraints for the must-pass point problem. Therefore, the test experiments in this paper have a tightly coupled relationship with the actual application scenarios. The experimental results verify that the MD-ACO can solve the multi-modal and multi-objective path planning requirements in parallel. The comparison with other algorithms shows that the algorithm proposed in this paper has stronger search ability and optimization ability, and the diversity of solutions is better.

3 Problem Description

In this chapter, we describe the MMOPP problem based on the documentation provided by the test set [16–18]. This test set models the road features in the actual road network map into a regular grid map, which is divided into three categories according to the types of optimization objectives, with a total of 12 test questions. The first type of test problem simulates the congested road sections in the actual road network map, and its optimization goal is to minimize the path length, the quantity of congested road sections, and the quantity of intersections. According to the size of the problem, this type of problem contains 5 test questions. In the second type of test problem, different F values are used to simulate information such as road congestion or road width, and the optimization objective is to minimize the path length and each F value. According to the number of optimization objectives, this type of problem contains 5 tests question. The third type of test problem simulates the path planning requirements that have must-pass points constraints, and allows the existence of a re-entrant path. The optimization goal is to minimize the path length and each F value. According to the different scale of the problem, this type of problem contains two test problems.

The test set provides some definitions. The map is a two-dimensional matrix consisting of 0 or 1. The matrix value of 0 means that this area is passable in the map, and the value of 1 means that this area is not passable. Walking in four directions is permitted, repeating the same path is not allowed, and it is forbidden to go beyond the boundary of the map. As shown in Fig. 1, the map also gives the starting point and ending point, red congestion points, yellow must-pass points. At the same time, passing a certain area in the map will generate different costs, this cost has multiple dimensions, including road width, road congestion, the number of intersections, and must-pass points.

The cost of each area is expressed as (3), the cost of the path is expressed as (4).

$$cost \rightarrow [f_1, f_2, \ldots f_m], f_m \in F \tag{3}$$

$$cost_{path}[i] = \sum_{j=1}^{n} cost[j][i], i \in M \tag{4}$$

In formula (3), the cost array consists of the costs of different optimization objectives. In formula (4), $cost_{path}[i]$ is defined as the i-th cost sum of this path, $cost[j][i]$ is defined as the i-th cost of the j-th region on this path (Fig. 2).

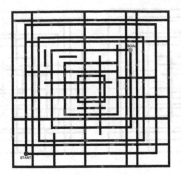

Fig. 2. Map example

4 Proposed Approach

4.1 Matrix Pre-processing

Since the test set provides us with a 0–1 matrix, we need to preprocess the matrix at first. The optimization of matrix preprocessing also helps to reduce the scale and complexity of the problem. Matrix preprocessing includes three steps. First, we need to get rid of useless roads, it means to turn some 1 in the matrix that are considered as useless areas into 0 in a 0–1 matrix. The result obtained at this time is still a 0–1 matrix. Second, we need to get the set of valid points, it means that we need to get a set of some key points in the matrix, then check the feasibility of the problem. Third, we need to turn the 0–1 matrix into a weighted directed graph.

Removal of Unless Areas. We can process the 0–1 matrix and delete unnecessary areas in the map to shrink the problem scale. Abridged regions can be divided into two types. First, disconnected regions belonging to two road segments are useless. Second, we get all the key points from the map. If the must pass point is not in this group of areas separated by key points, we can ignore this group. The following Fig. 3(a) is the original map generated by the two-dimensional matrix, and the following Fig. 3(b) is the reduced map after matrix preprocessing.

I will then describe the algorithms and algorithm-related auxiliary arrays used to implement this section. Depth[x][y] represents the depth of DFS search. Minor[x][y] represents depth of the nearest common ancestor in DFS search. Children[x][y] represents the subset of children of area (x, y) in DFS search. Must[x][y] represents whether the current point is a necessary point.

After creating the auxiliary matrix, we can build a DFS spanning tree by first traversing the map, and then traversing the DFS spanning tree to do the following operations. If the child node (x', y') of (x, y) belongs to children[x][y], and $minor[x'][y'] \geq depth[x][y]$ and (x', y') is not a necessary point, then the subtree rooted at (x', y') can be deleted.

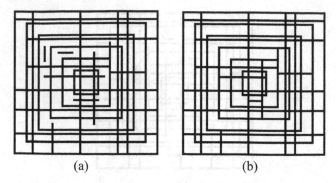

(a)   (b)

Fig. 3. Comparison of original and reduced maps

Algorithm 1   Removal of unless area
___
Input: Two-dimensional matrix map

Output: Reduced two-dimensional matrix map

1: procedure $INITTREE(x, y, h)$
2:   $depth[x][y] \leftarrow h$ $minor[x][y] \leftarrow h$
3:   for (x', y') : children list of (x, y)
4:     if (x', y') is not in matrix map then continue
5:     if $depth[x', y'] = 0$ then
6:       $children[x][y]$ append (x', y')
7:       $INITTREE(x', y', h + 1)$
8:       if $must[x'][y'] == true$ then
9:         $must[x][y] \leftarrow true$
10:      end if
11:      $minor[x][y] \leftarrow \min(minor[x][y], depth[x'][y'])$
12:    end if
13: end for
14: end procedure
15: procedure $REDUCETREE(x, y)$
16:   set (x, y) as retained point
17:   for (x', y') : $children[x][y]$ do
18:     if $minor[x'][y'] < depth[x][y]$ or $must[x][y] == true$
19:       $REDUCETREE(x', y')$
20:     end if
21:   end for
22: end procedure

Get Key Points and Check Problem Feasibility. We first traverse the retained points obtained in the previous step, and then select the start point, end point, must-pass point, and intersection point among the retained points to form a key point array. In addition, before starting to solve the MMOPP problem, we need to check whether the problem has a feasible solution. For some scenarios with necessary points, we need to check whether all feasible solutions have passed the necessary points. This verification is achieved by traversing all retained points and determining whether all necessary points are included.

Build a Weighted Directed Graph. We first traverse the key point set obtained in the previous step, calculate all the costs between the key points, and construct a weighted directed graph, where the weights refer to different types of costs. The data structure of the edge includes the starting point, the ending point, the cost array of the edge, the path composed of all points, and the unique identifier id of the edge. The data structure of the edge is as follows (Fig. 4).

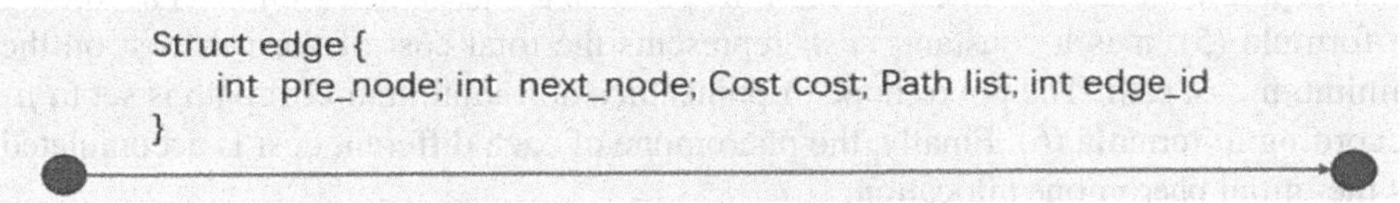

```
Struct edge {
        int  pre_node; int  next_node; Cost cost; Path list; int edge_id
}
```

Fig. 4. The data structure of the edge

Algorithm 2   Build a weighted directed graph
Input: key point set
Output: edges sets
1: create Path e, Edge edge
2: for current point : key point set do
3:    for next point near to current point do
4:      if next point is not in matrix map then continue
5:        e.add(current point)
6:      while next point is not in valid point sets do
7:        e.add(next point)
8:        calculate the cost according to formula (4)
9:        continue traversing the four directions to get next point
10:     end while
11:     edges.add({current point, next point, e, cost arrays, edges id})
12:   end for
13: end for

4.2 Initial Pheromone Assignment Based on Dijkstra Algorithm

ACO is designed to simulate the action mode of ant colony looking for food. Ant colonies secrete different pheromones along the way when searching for food, and the amount of pheromone is related to the distance of the path. Other ants will make corresponding decisions according to the concentration of pheromone on the path, and finally the ant colony will gradually gather on the shortest path. Since the initial pheromone concentration of the traditional ACO is evenly distributed, the initial search of the algorithm has strong blindness, which leads to a slow convergence speed and easily makes the algorithm get stuck in local optimum and affects the optimization ability of the algorithm. In order to avoid the algorithm getting stuck in local optimum due to pheromone and reduce the misleading of ants caused by wrong heuristic information, improve the path planning ability of the algorithm, this paper improves the allocation of initial pheromone based on the minimum cost planned by Dijkstra algorithm.

The Dijkstra algorithm is a classic algorithm for finding the shortest path. Since this paper studies multiple objectives, the objectives to be optimized are not only related to the path length, but also related to different types of cost. The algorithm steps are as follows. First, for each type of cost, we use Dijkstra algorithm to traverse the map, then we can obtain the minimum cost path and the minimum cost sum from the start point to the end point. Second, the pheromone of the point on the least-cost path is set according to formula (5). π is a constant, $cost_i$ represents the total cost of the i-th cost on the minimum cost path, The pheromone of points on a non-minimum-cost path is set to μ_i according to formula (6). Finally, the pheromone of each different cost is accumulated as the initial pheromone allocation.

$$pheromone[i] = \frac{\pi}{cost_i}, i \in costType \tag{5}$$

$$\mu_i = pheromone[i] - \theta \tag{6}$$

Algorithm 3 Dijkstra algorithm

Input: weighted directed graph

Output: Least-cost path and least-cost

```
 1: for i = 1 : cost type
 2:     create auxiliary matrix cost, P, path
 3:     for each point v in graph
 4:         cost[v] ← Infinity
 5:         add v to P
 6:     end for
 7:     while P is not empty
 8:         choose u with the least cost
 9:         remove u from P and add u to path
10:         for each v near to u
11:             if v is not visited and cost[v][i] > cost[u][i] + Cost(u → v)
12:                 cost[v][i] → cost[u][i] + Cost(u → v)
13:             end if
14:         end for
15:     end while
16:     for each point k in graph
17:         if k in path
18:             update pheromone according to formula (5)
19:         else update pheromone according to formula (6)
20:     end for
```

4.3 Select the Next Node According to the State Transition Rules

In the process of ants walking, the ants will select the next point to reach according to the probability rule, and record the next point that the ants walk through in the taboo table. The probability of choosing the next path is expressed as the formula (7).

$$P_{ij}^k(t) = \begin{cases} \dfrac{[\pi_{ij}(t)]^\alpha * [\varphi_{ij}(t)]^\beta}{\sum_{s \in allowed_k}[\pi_{is}(t)]^\alpha * [\varphi_{is}(t)]^\beta} & j \in allowed_k \\ 0 & other \end{cases} \tag{7}$$

$$\varphi_{ij} = \frac{1}{cost_{ij}} \tag{8}$$

where $P_{ij}^k(t)$ represents the state transfer probability of ants k from point i to point j at time t. $\pi_{ij}(t)$ denotes the pheromone concentration on the path from point i to point j.

In the process of ant walking, if it reaches the end point and has passed all the necessary points, and if the current solution is dominated by the solutions in the PS, it will not be added to the PS. If the current solution is not dominated by the solutions in the PS, it will join the PS.

4.4 Update Pheromone

After all ants have completed a traversal, the residual information on the paths is updated and the pheromones on each path are adjusted according to the following formula.

$$\pi_{ij}(t+1) = \pi_{ij}(t) * (1 - \rho) + \Delta\pi_{ij}(t), 0 < \rho < 1 \tag{9}$$

$$subject\ to\ a \le \pi_{ij}(t+1) \le b$$

$$\Delta\pi_{ii}(t) = \sum_{k=1}^{m} \Delta\pi_{ij}^k(t) \tag{10}$$

$$\Delta\pi_{ij}^k = \begin{cases} \dfrac{Q}{cost_k} \\ 0 \end{cases} \tag{11}$$

In formula (9), ρ denotes the pheromone volatility coefficient, $\Delta\pi_{ij}(t)$ denotes pheromone increment on path (i, j) after this cycle, if $\pi_{ij}(t+1)$ exceeded a threshold, we take the threshold as its value. In formula (10), $\Delta\pi_{ij}^k(t)$ denotes the pheromone content left on the path (i, j) by the kth ant in this cycle. In formula (11), if ant k passes through (i, j) in this cycle, then execute $\Delta\pi_{ij}^k = \frac{Q}{cost_k}$. $cost_k$ represents the total cost of all paths taken by ant k. Q is the total amount of pheromones released by ant k after completing a complete path search.

5 Experimental Results

All algorithms are implemented in C++ language and compiled on CLion. The test set of the experiment, the Pareto Set and the Pareto Front obtained by running the algorithm are as Table 1.

The evaluation criteria of the experimental results are as follows. The point in the PF is assigned 1, and N found paths corresponding to the point which actually existed M paths is regarded as an additional $(N - 1)/M$ score. As shown in the Fig. 5, the winner's score is $2 \times 1 + \frac{4-1}{6} + \frac{3-1}{6} = 2.8$. While the loser's score is $3 \times 1 + \frac{1-1}{6} + \frac{2-1}{6} + \frac{2-1}{6} = 3.3$. Compared with other algorithms, MD-ACO algorithm scores are as Table 2.

Table 1. Problem scale and results

Test problems	Map size	Objectives	Pareto Set	Pareto Front
Problem1	40 * 40	2	9	4
Problem2	40 * 40	3	24	7
Problem3	50 * 50	3	13	4
Problem4	50 * 50	3	9	7
Problem5	84 * 84	5	19	5
Problem6	40 * 40	2	5	3
Problem7	40 * 40	3	16	12
Problem8	50 * 50	4	47	35
Problem9	50 * 50	5	103	79
Problem10	84 * 84	7	1217	1010
Problem11	40 * 40	2	4	2
Problem12	40 * 40	3	11	6

Table 2. The score of the problem1–12

Test problems	MD-ACO	MSCL	ClusteringGA	NAN	INSGA-III
Problem1	5.3	5.3(=)	3.8(−)	5.3(=)	4.3(−)
Problem2	9.532	9.532(=)	7.5(−)	9.532(=)	8.207(−)
Problem3	5.875	5.875(=)	5.875(=)	5.875(=)	3.75(−)
Problem4	8.0	8.0(=)	7.0(−)	8.0(=)	7.0(−)
Problem5	7.583	6.333(−)	3.0(−)	6.333(−)	3.167(−)
Problem6	4.0	4.0(=)	3.5(−)	2.0(−)	3.5(−)
Problem7	14.0	14.0(=)	14.0(=)	12.5(−)	12.0(−)
Problem8	41.0	37.5(−)	42.0(+)	15.5(−)	14.5(−)
Problem9	90.0	77.5(−)	88.0(−)	12.5(−)	28.5(−)
Problem10	1102.5	199.0(−)	0.0(−)	9.0(−)	0.0(−)
Problem11	2.667	2.667(=)	2.0(−)	2.667(=)	1.0(−)
Problem12	8.5	10.5(+)	0.0(−)	7.5(−)	4.0(−)

"+" means the algorithm works better than MD-ACO, "−" means the algorithm works worse than MD-ACO, "=" indicates that the comparison algorithm behaves similarly to MD-ACO.

From the above experimental results, it can be concluded that the evaluation score of the MD-ACO algorithm is higher than other algorithms. Especially for problem 5, problem 9 and problem 10, MD-ACO algorithms can find more solutions than other algorithms.

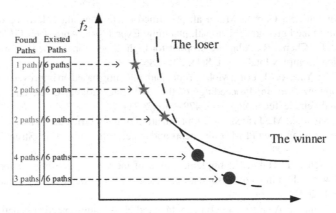

Fig. 5. An example of calculating fractions

6 Conclusions

The main research content of this paper is to solve the MMOPP problems through the MD-ACO. First, matrix preprocessing is performed to process the initial data. Second, Dijkstra is used to optimize the initial pheromone distribution of the ACO. Finally, the ACO is used to traverse the map. The experiment proves the feasibility and effectiveness of the MD-ACO algorithm to solve the MMOPP problems.

Acknowledgment This work was supported by the National Natural Science Foundation of China (Grant No. 62176191 and 62272355).

References

1. Ma, Y., Hu, M., Yan, X.: Multi-objective path planning for unmanned surface vehicle with currents effects. ISA Trans. **75**, 137–156 (2018)
2. Ajeil, F.H., Ibraheem, I.K., Sahib, M.A., et al.: Multi-objective path planning of an autonomous mobile robot using hybrid PSO-MFB optimization algorithm. Appl. Soft Comput. **89**, 106076 (2020)
3. Wang, J., Weng, T., Zhang, Q.: A two-stage multiobjective evolutionary algorithm for multi-objective multidepot vehicle routing problem with time windows. IEEE Trans. Cybern. **49**(7), 2467–2478 (2018)
4. Wang, J., Zhou, Y., Wang, Y., et al.: Multiobjective vehicle routing problems with simultaneous delivery and pickup and time windows: formulation, instances, and algorithms. IEEE Trans. Cybern. **46**(3), 582–594 (2015)
5. Xiang, D., Lin, H., Ouyang, J., et al.: Combined improved A* and greedy algorithm for path planning of multi-objective mobile robot. Sci. Rep. **12**(1), 1–12 (2022)
6. Zhang, K., Shen, C.N., Gary, G.Y., et al.: Two-stage double niched evolution strategy for multimodal multi-objective optimization. IEEE Trans. Evol. Comput. **25**(4), 754–768 (2021)
7. Liu, J., Ji, J., Ren, Y., et al.: Path planning for vehicle active collision avoidance based on virtual flow field. Int. J. Automot. Technol. **22**(6), 1557–1567 (2022)

8. Liu, J., Anavatti, S., Garratt, M., et al.: Modified continuous ant colony optimisation for multiple unmanned ground vehicle path planning. Expert Syst. Appl. **196**, 116605 (2022)
9. Lin, J., He, C., Cheng, R.: Adaptive dropout for high-dimensional expensive multiobjective optimization. Complex Intell. Syst. **8**(1), 271–285 (2021)
10. Jing, Y., Chen, Y., Jiao, M., et al.: Mobile robot path planning based on improved reinforcement learning optimization. In: Proceedings of the 2019 International Conference on Robotics Systems and Vehicle Technology, pp. 479–486 (2021)
11. Korus, K., Salamak, M., Jasinski, M.: Optimization of geometric parameters of arch bridges using visual programming FEM components and genetic algorithm. Eng. Struct. **241**, 112465 (2021)
12. Ozsari, S., Uguz, H., Hakli, H.: Implementation of meta-heuristic optimization algorithms for interview problem in land consolidation: a case study in Konya/Turkey. Land Use Policy **108**, 105511 (2021)
13. Deb, K., Pratap, A., Agarwal, S., et al.: A fast and elitist multiobjective genetic algorithm: NSGA-II. IEEE Trans. Evol. Comput. **6**(2), 182–197 (2002)
14. Mao, J.Q.: Research on robot path planning based on improved ant colony algorithm. Comput. Appl. Softw. **38**(5), 300–306 (2021)
15. Bullnheimer, B., Hartl, R.F., Strauss, C.: A new rank based version of the ant system-a computational study. Cent. Eur. J. Oper. Res. **7**(1), 25–38 (1999)
16. Li, B., Chiong, R., Gong, L.G.: Search-evasion path planning for submarines using the artificial bee colony algorithm. In: 2014 IEEE Congress on Evolutionary Computation (CEC). IEEE, pp. 528–535 (2014)
17. Ishibuchi, H., Imada, R., Setoguchi, Y., et al.: Performance comparison of NSGA-II and NSGA-III on various many-objective test problems. In: 2016 IEEE Congress on Evolutionary Computation (CEC). IEEE, pp. 3045–3052 (2016)
18. Liang, J., Yue C.T., Li, G.P., et al.: Problem definitions and evaluation criteria for the cec 2021 on multimodal multiobjective path planning optimization. Technical Report (2021)

Local Path Planning Algorithm Designed for Unmanned Surface Vessel Based on Improved Genetic Algorithm

Yi Liu[1]([✉]), Huizi Li[1], Xinlong Pan[2]([✉]), Haipeng Wang[2], Yong Chen[3], Heng Fang[4], and Hao Liu[3,5]

[1] China Ship Development and Design Center, Wuhan 430064, China
ly2021@hust.edu.cn
[2] Naval Aviation University, Yantai 264001, China
airadar@126.com
[3] Wuhan Institute of Digital Engineering, Wuhan 430205, China
[4] Wuhan University of Technology, Wuhan 430070, China
[5] Shanghai Jiao Tong University, Shanghai 200240, China

Abstract. At present, the algorithms used in local path planning of unmanned ships mainly include simulated annealing algorithm, artificial potential field method and genetic obstacle is above the globally planned algorithm. Among them, genetic algorithm has strong spatial search ability and strong adaptive ability. However, due to the low efficiency of the traditional genetic algorithm, it cannot meet the needs of the real-time path planning of unmanned ships. To solve this problem, this paper designs an improved genetic algorithm based on dynamic fitness function to make up for the shortcomings of the traditional genetic algorithm. This method can improve genetic manipulation by guiding the direction of population evolution optimization and meet the needs of unmanned ship obstacle avoidance. Aiming at the complex dynamic environment in the local path planning of unmanned ship, a simulation experiment combining static environment and dynamic environment is designed. Simulation results show that the improved algorithm has a better obstacle avoidance effect in dynamic environment.

Keywords: Unmanned surface vessel · Partial path planning · Genetic algorithm

1 Introduction

Surface unmanned ship is an important part of Marine intelligence system. The research directions of unmanned ship are various, but the path planning is the core function module of unmanned ship system. It is the basis of the track control of unmanned ships, and also the key for unmanned ships to safely reach the target destination from the target starting point under complex sea conditions and maintain the optimal path during the whole voyage. The research on the path planning of unmanned ship is of great significance to the development of unmanned ship technology [1, 2].

L. Pan et al. (Eds.): BIC-TA 2022, CCIS 1801, pp. 29–43, 2023.
https://doi.org/10.1007/978-981-99-1549-1_3

Path planning is divided into global path planning and local path planning. At present, the most widely used algorithms in the field of path planning are particle swarm optimization algorithm, ant colony algorithm, Cuckoo algorithm, etc., because these algorithms are often used in the actual path planning problems such as insufficient optimization accuracy, slow convergence speed, so they cannot meet the needs of the actual control objects [3, 4]. Global path planning is a path planning that is carried out on the premise that the global environment is known, while local path planning is a dynamic process of global path planning that perceives the environment in real time according to the sensor of the control object and dynamically changes the route according to the changes of the environment information when the environment is unknown. Dynamic collision avoidance is the main task of local path planning.

The common algorithms for local path planning include simulated annealing, artificial potential field and genetic algorithm. The main idea of genetic algorithm is to simulate the evolutionary process of natural organisms, and to build a random search algorithm based on the survival of the fittest. The main steps are population initialization, fitness function calculation, selection, crossover and variation. Genetic algorithm has the random search ability of variation, which makes full use of the characteristics of group optimization. Instead of a single point-to-point evolution, genetic algorithm can be evolved in parallel to speed up the search efficiency of the algorithm. In addition, genetic algorithm will rely on fitness function to carry out genetic operations such as replication, mutation and crossover in each evolution to carry out algorithm operation. This evolutionary mode has low requirements on constraints and objective functions, and can be searched in the whole population, so the optimal solution can be obtained with high probability after multiple iterations. However, genetic algorithm also has natural defects: the evolutionary calculation of the whole population requires a large amount of memory and storage space, and consumes longer time. When solving complex multi-objective optimization problems, the initial decoding and encoding are very tedious, and the algorithm convergence speed is slow.

Due to the low search efficiency of traditional genetic algorithm, it is unable to meet the requirements of real-time path planning of unmanned ships. In this paper, dynamic fitness function is designed to guide the population evolution to find the optimal direction, and genetic operation is improved to meet the needs of unmanned ships to avoid obstacles.

2 Improved Genetic Algorithm

The main idea of local path planning is to plan the optimal obstacle avoidance path according to the position and speed of real-time obstacles. While maintaining a safe distance from obstacles, the local path should be restored to the initial global path and ensure the shortest path. Due to the low search efficiency of traditional genetic algorithm, it is unable to meet the requirements of real-time path planning of unmanned ships. A dynamic fitness function is designed to guide the population evolution to find the optimal direction, and genetic manipulation is improved to meet the requirements of unmanned ships to avoid obstacles.

2.1 Standard Particle Swarm Optimization

The initial setup includes two parts: coding setup and population initialization [5–7].

(1) Code setting

To apply genetic algorithm to path planning, it is necessary to encode the actual passage path of the unmanned ship, convert the path into a chromosome that the genetic algorithm can operate, and design the fitness function based on the actual collision avoidance requirements of the unmanned ship. Considering that the encoding mode of chromosomes greatly affects the speed of the algorithm, this section adopts the grid method for modeling, and the grid is encoded, as shown in Fig. 1.

Since the starting point and ending point are determined, the global path planning is decomposed into multiple interconnecting path intervals, and local path planning is carried out on all interval paths. The default starting point and ending point are coordinates (0.5, 9.5) and (9.5, 0.5), and random search is carried out on each intermediate path. Therefore, chromosome representation of genetic algorithm is 09 → L → 90. The gene locations of chromosomes are grid numbers, and the encoding of chromosomes is decimal, and grid numbers are the path of the unmanned ship.

The search method of genetic algorithm is random search, and the path length generated by each evolution is different, so the length of chromosome is changed. The length of chromosome is closely related to the solving accuracy. The longer the chromosome is, the higher the solving accuracy of the genetic algorithm will be, but the greater the computational amount of the corresponding algorithm will be, which will affect the convergence speed of the algorithm. Therefore, the limit length of chromosomes, namely the longest path, was set to avoid excessive length of chromosomes generated during population initialization, which would affect the convergence speed of genetic algorithm.

Therefore, the local path planning problem based on genetic algorithm is transformed into the shortest path composed of the non-collision grid from the target starting point to the target ending point [8, 9].

The corresponding relation between grid and rectangular coordinate system is shown in formula (1):

$$P = x + 10y \tag{1}$$

$$\begin{cases} x = \text{rem}(P, 10) \\ y = fix(P, 10) \end{cases} \tag{2}$$

where *rem* is the remainder, *fix* is the integer, and P is the number of rasters.

(2) Initialization of population

Local path planning is a dynamic collision avoidance based on the global path, so the environment models are all grid environments, passable grid is 0 and impassable grid is 1. When initializing the population, only free search is carried out in the free

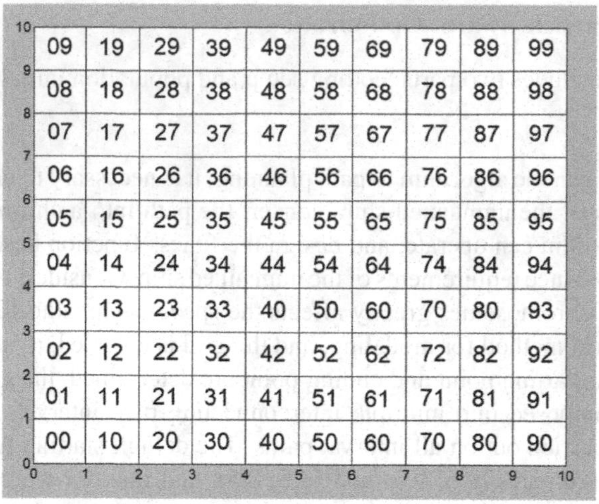

09	19	29	39	49	59	69	79	89	99
08	18	28	38	48	58	68	78	88	98
07	17	27	37	47	57	67	77	87	97
06	16	26	36	46	56	66	76	86	96
05	15	25	35	45	55	65	75	85	95
04	14	24	34	44	54	64	74	84	94
03	13	23	33	40	50	60	70	80	93
02	12	22	32	42	52	62	72	82	92
01	11	21	31	41	51	61	71	81	91
00	10	20	30	40	50	60	70	80	90

Fig. 1. Environment information number

grid, which greatly improves the initial operation speed of the algorithm and greatly increases the excellent and good rate of the initial population.

2.2 Standard Particle Swarm Optimization

In the process of iterative evolution of genetic algorithm, only the fitness function dominates the evolution direction of the whole genetic algorithm. Therefore, a reasonable setting of fitness function can greatly improve the convergence speed of genetic algorithm.

Considering the special sailing environment of the unmanned ship, the unmanned ship may encounter static obstacles missing from the global environment in the early stage and dynamic obstacle information obtained by its own sensor in real time during local path planning. The static fitness function and dynamic fitness function are set respectively.

(1) Static fitness function

The static fitness function mainly considers the static obstacles in the environment information and is only related to the path length. In the path planning process of genetic algorithm, the smaller the fitness function, the better the local path optimization. Because the longer path the unmanned ship moves, the more grids it passes through, and the lower the fitness value of the corresponding genetic population. In the process of genetic algorithm evolution, the higher the elimination rate of the offspring with higher fitness value in the survival of the fittest, the less inherited to the next generation. On the contrary, the shorter the path of the unmanned ship, the greater the fitness value of the population, and the greater the chance of being passed on to the next generation. Therefore, the larger the fitness function, the faster the evolutionary iterative process of genetic algorithm and the faster the algorithm convergence [10–13].

To sum up, the static fitness function is shown in Formula (3), where n represents the total number of grids passed by the current path, and r represents the length of the current path.

$$f_i = k \times r \tag{3}$$

$$k = 1 + \sqrt{n - 1} \tag{4}$$

$$r = \sum_{i=0}^{n-1} \sqrt{(x_{i+1} - x_i)^2 + (y_{i+1} - y_i)^2} \tag{5}$$

(2) Dynamic fitness function

The dynamic fitness function set in the algorithm mainly considers the moving obstacles in the environment information and is related to the moving speed and direction of obstacles. The main purpose of dynamic obstacle avoidance of unmanned ship is to change its speed and direction of movement when its own sensor detects moving obstacles, so as to safely avoid obstacles and ensure the shortest obstacle avoidance path. Therefore, the dynamic fitness function is mainly related to the three indexes of maintaining a safe distance, dynamic obstacle avoidance and shortest path, and the specific contents are as follows:

a. Keep a safe distance

Assume that the moving obstacle is a particle, the coordinate in the environmental model is (x_o, y_o), and the real-time coordinate of the unmanned ship is (x_r, y_r), then the distance between the unmanned ship and d_{or} can be calculated by formula (6), and the nearest safe distance between the moving obstacle and the unmanned ship is D. When the distance between the unmanned ship and the moving obstacle is greater than D, the unmanned ship is considered to be in a safe position. If it's less than D, it's in a dangerous position.

Therefore, whether the unmanned ship is in a safe position can be judged according to formula (7). According to formula (7), when the unmanned ship and the moving obstacle are in a safe position, the fitness value of the genetic algorithm population is 1. And vice versa. If at a dangerous distance, the fitness value is 0.

$$d_{or} = \sqrt{(x_r - x_o)^2 + (y_r - y_o)^2} \tag{6}$$

$$f_1 = \begin{cases} 1, & d_{or} \geq D \\ 0, & d_{or} \leq D \end{cases} \tag{7}$$

b. Dynamic collision avoidance

During local path planning, the unmanned ship can dynamically obtain the obstacle information of the surrounding environment in real time according to its own sensors, including the number, location and speed of moving obstacles. In the process of local path planning, the sensor will update the position and speed information of obstacles at a fixed frequency. In order to ensure the real-time performance of environmental information to meet the requirement that the unmanned ship can achieve dynamic obstacle avoidance under the condition of high-speed motion, the update frequency is generally set relatively high, and the moving distance is short. Therefore, it is assumed that in this control period, the unmanned ship and moving obstacles both move in a uniform and straight line.

The current speed of the unmanned ship is v_r, the position is p_o, the time to move to $p_{i-1}(x_{i-1}, y_{i-1})$ is t_{i-1}, and the time to move to $p_i(x_i, y_i)$ is t_i. Therefore, the time for the unmanned ship to move from local waypoint p_{i-1} to waypoint p_i is shown in Formula (8). Meanwhile, since the unmanned ship moves uniformly within a single frequency period, Therefore, formula (9) can also obtain the motion time of the unmanned ship moving from waypoint p_{i-1} to p_i. The movement time of the unmanned ship between the two path points is a control period.

$$T_{i-1} = t_i - t_{i-1} \tag{8}$$

$$T_{i-1} = \frac{\sqrt{(x_i - x_{i-1})^2 + (y_i - y_{i-1})^2}}{v_r} \tag{9}$$

Meanwhile, assuming that there are multiple obstacles in the current control period, whose position information is $O_n(x_n, y_n)$ and velocity is v_n, the current x-axis moving velocity component is v_{nx} and y-axis moving velocity component is v_{ny}. Therefore, the moving position of the obstacle in the current control period can be calculated according to formula (10) and formula (11):

$$x_n(t_i) = x_n(t_{i-1}) + v_{nx} \times T_{i-1} \tag{10}$$

$$y_n(t_i) = y_n(t_{i-1}) + v_{ny} \times T_{i-1} \tag{11}$$

In order to ensure that the unmanned ship can safely avoid obstacles in local path planning, the distance between the unmanned ship and moving obstacles can be decomposed into x axis and y axis direction distance component greater than the safety distance D. The position relationship between the unmanned ship and the moving obstacle is shown in Fig. 2. Therefore, the path component of the unmanned ship and the moving obstacle in the x-axis direction can be expressed as $v_oT \cos \alpha + v_rT \cos \beta$, and the distance component in the y-axis direction is $v_oT \sin \alpha + v_rT \sin \beta$.

Therefore, the fitness function considering dynamic obstacle avoidance is shown in Formula (12):

$$f_2 = \begin{cases} 0, \dfrac{v_oT \cos \alpha + v_rT \cos \beta}{d_{or}} \leq 1 \text{ and} \dfrac{v_oT \sin \alpha + v_rT \sin \beta}{D} \leq 1 \\ 2 - \dfrac{v_oT \cos \alpha + v_rT \cos \beta}{d_{or}} - \dfrac{v_oT \cos \alpha + v_rT \cos \beta}{d_{or}} \text{ other} \end{cases} \tag{12}$$

c. Shortest path

Even if the unmanned ship is conducting dynamic obstacle avoidance, the shortest path is still an important standard to measure the quality of path rules. Therefore,

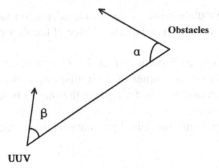

Fig. 2. Schematic diagram of unmanned ship and obstacles

the fitness function of the shortest path index is shown in Formula (13). The shorter the path, the higher the fitness.

$$f_3 = \frac{20}{\sum_{i=1}^{n-1} d(m, m_{i+1})} \tag{13}$$

Comprehensive safety distance, dynamic obstacle avoidance and shortest path three fitness functions, the set fitness function is shown in Formula (14), weighted evaluation of the three fitness functions can be obtained:

$$f_d = a \sum f_1 + b \sum f_2 + c \sum f_3 \tag{14}$$

According to the different weights of each part and combined with the actual situation of unmanned ship navigation, the values of the three weight coefficients are respectively $a = 0.1, b = 0.1, c = 0.8$.

2.3 Standard Particle Swarm Optimization

(1) Operator of selection

In order to accelerate the initial evolution speed of the algorithm, a fitness value was set during the initialization of the population, and individuals with high fitness values in the population were equally replaced with individuals with low fitness values. The procedure is as follows:

Step 1: After population initialization, the adaptation value F of each progeny is calculated respectively. The adaptation value of the shortest path is F_{max} and the longest path is F_{min}.

Step 2: Take the average adaptation value F', $F' = \frac{F_{max} + F_{min}}{2}$;

Step 3: Compare the fitness values of each offspring in the initial population with F', if $F > F'$, then retain; if $F < F'$, replace the current fitness values with the first retained fitness values;

Step 4: Repeat Step 3 until the initial population size is reached.

(2) Crossover operator

The selected single-point crossover operator is shown in Table 1. A gene is randomly selected from the paired chromosomes for cross-replacement to obtain two new chromosomes, and the crossover probability is set as 0.25.

Table 1. Examples of crossover algorithm

	Before exchange	After exchange
Single point crossing	Parent generation A:1 1 1 1 1 1	Parent generation A:1 1 1 1 0 0
	Parent generation A:0 0 0 0 0 0	Parent generation A:0 0 0 0 1 1
Multiple point crossing	Parent generation A:1 1 1 1 1 1	Parent generation A:1 1 0 1 1 0
	Parent generation A:0 0 1 0 0 0	Parent generation A:0 0 1 0 0 1

(3) Operator of variation

The mutation operator was set as shown in Table 2. A chromosome gene location was randomly selected, and the mutation probability was generally very low. It was set as 0.001 for mutation. The function of variation is to generate new genes in the population and increase the diversity of the population. Replication and crossover can only select the best among the existing populations and may fall into the local optimal. Therefore, mutation operators can improve the global optimization ability of genetic algorithm. When applied in the field of path planning, randomly mutated genes may lead to ineffective path planning of unmanned ships and pass-through obstacle areas. Therefore, the following two mutation methods are designed:

a. When the current fitness is high, that is, when the unmanned boat path is feasible, the variation range of gene locus is controlled in a small range, and the grid number corresponding to the mutated gene locus is guaranteed to be passable.

b. When the current fitness is low, that is, when the path of the unmanned ship is poor, the variation range of the gene locus is controlled within a large range and the current poor path can be jumped out.

Table 2. Examples of mutation operators

Before mutation	1 0 1 1 0 1 0
After mutation	1 1 1 1 0 0 0

(4) Operator of optimization

The main function of the optimization operator is to preserve each individual with high fitness, so that the excellent individuals in the population are preserved genetically. After each iteration evolution, the current optimal fitness value needs to be compared with the optimal fitness value of the population. If the optimal fitness value of the population is low, the evolution will be updated and replaced to ensure that the optimal individual can survive the evolution of the genetic algorithm.

(5) Smoothing operator

In the actual sailing process of unmanned ship, due to its motion characteristics, the steering Angle is small, and the forward direction cannot be changed greatly. Therefore, it is necessary to design a smoothing operator to deal with the progressive smoothing of sharp corners in path sailing.

Algorithm termination condition: the number of iterations reaches 100. If the algorithm has not converged, the individual with the highest fitness in the last generation population is selected as the global optimal solution.

2.4 Flow of Algorithm

The flow chart of local path planning for unmanned ships based on the improved genetic algorithm is shown in Fig. 3.

The main steps are as follows:

Step 1: Initialize the code.
Step 2: Population initialization.
Step 3: Conduct individual evaluation according to the improved fitness function.
Step 4: Population evolution according to improved genetic manipulation.

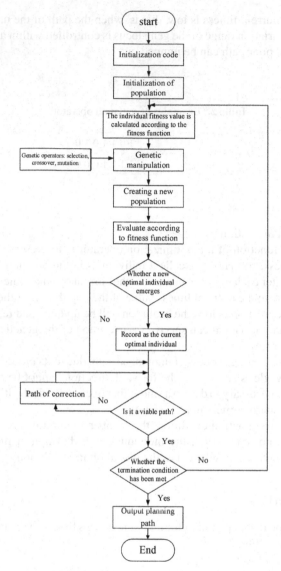

Fig. 3. Flowchart of local path planning based on genetic algorithm

Step 5: Judge whether there is a new optimal individual and retain the optimal individual if there is one.
Step 6: Determine whether the current optimal individual is a feasible path.
Step 7: If the termination condition is reached, output the current optimal path; otherwise, return to Step 3.

3 Simulation Test Results and Analysis

In order to verify the feasibility of the improved genetic algorithm in local path planning, the algorithm principal simulation experiment is needed. Local path planning is carried out on the basis of global path planning. According to the motion form of obstacles, static obstacle avoidance simulation and dynamic obstacle avoidance simulation are carried out respectively, and dynamic obstacle avoidance simulation experiments are carried out in a variety of different obstacle environments.

3.1 Initial Global Path Settings

In the initial grid environment of 10*10, ant colony algorithm is used for global path planning in the current environment [14–18]. The chromosome code in the genetic algorithm is shown on the right in Fig. 4 as 09-18-17-16-25-24-23-32-42-52-62-72-81-90. The gene locations of the chromosome are grid numbers, namely, the moving path of the unmanned ship.

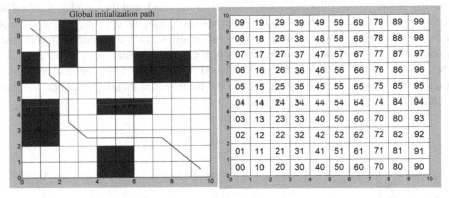

Fig. 4. Global initial path and path sequence diagram

3.2 Static Obstacle Avoidance Simulation

The parameters of the genetic algorithm are set as follows: the selection operator is set to select a total of 200 offspring to form the initial population; The number of iterative evolutions was 100 times, the probability of chromosome crossing was 0.3, and the probability of chromosome gene mutation was 0.001. Static obstacle avoidance simulation is divided into the following two situations:

(1) The obstacle is outside the globally planned path

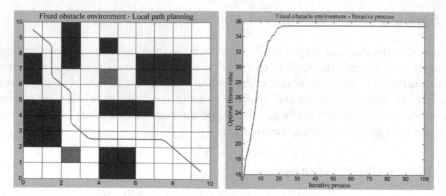

Fig. 5. Fixed obstacle environment (1) Local path planning

As can be seen from Fig. 5 of the iterative process, due to the improvement of the selection operator, the evolution process of the genetic algorithm is very fast. In about 20 generations, the global environment is known, at this time, the fitness value of the offspring in the genetic population has reached a high level, and the path has reached an optimal level. After 5 generations of continuous evolution, the optimal fitness value of the population reaches the maximum, the evolution stops, and the optimal path is output. By analyzing the local path planning diagram, it can be seen that when static obstacles appear on the non-global path, the path planned by the improved genetic algorithm is roughly the same as the global path. Due to the improved smoothing operator, the path is relatively smooth, and the design effect is achieved.

The obstacle is above the globally planned path

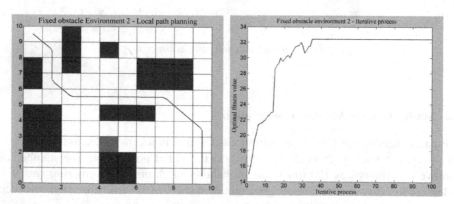

Fig. 6. Fixed obstacle environment (2) Local path planning

By analyzing the iterative process Fig. 6, it can be seen that in the interval from 0 to 30 generations, since the newly emerged obstacles are on the global planning path, the genetic algorithm keeps evolving and searching. As a result, the optimal fitness value in the population has many twists and turns, and the optimal fitness value keeps rising until the beginning of the 35th generation, there are more and more optimal solutions, and the path planning tends to be optimal. The optimal fitness value is reached. According to the analysis of the local path simulation diagram, when static obstacles appear on the global path, the improved genetic algorithm will re-plan the local route, maintain a safe distance from the obstacles, and finally return to the end of the global planned path, while making the path smooth.

3.3 Simulation of Dynamic Obstacle Avoidance

The setting of dynamic obstacle environmental genetic algorithm is the same as that of static obstacle simulation. The setting selection operator selects 200 children to form the initial population. The number of iterative evolutions was 100 times, the probability of chromosome crossing was 0.3, and the probability of chromosome gene mutation was 0.001.

Red represents moving obstacles, which move back and forth between grid number 16-26-36-46-56. Real-time path planning is shown in Fig. 7:

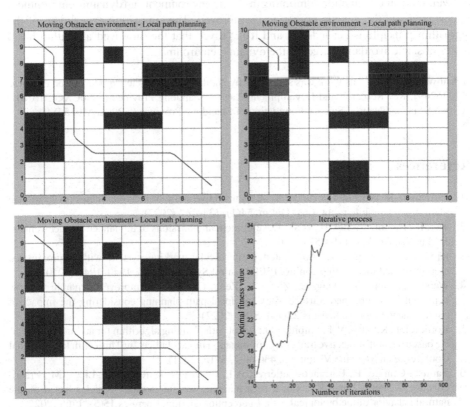

Fig. 7. Local path planning in moving obstacle environment

As can be seen from Fig. 7, when the obstacle moves, the genetic algorithm is divided into three steps for dynamic path planning. When the current row path is completely blocked, the algorithm falls into local optimal, and the path is stagnant. As the obstacle continues to move, the population keeps evolving to obtain the optimal path until the optimal adaptive value is reached. Therefore, in the early stage, the optimal fitness value fluctuated sharply, and the genetic algorithm evolved to find the optimal path. In the late stage, the global environment information was known, and the optimal fitness value of the population rose sharply. Finally, the optimal path reached a stable state, and the current optimal path was output.

4 Conclusion

In view of the low efficiency of the genetic algorithm commonly used in local path planning of unmanned ships, which cannot meet the requirements of real-time path planning of unmanned ships, this paper designed a dynamic fitness function to guide the population evolution to find the optimal direction and improved the genetic operation to meet the needs of unmanned ships to avoid obstacles. Aiming at the complex dynamic environment of the local path planning of unmanned ship based on the genetic algorithm, the weighted fitness function combining the static environment and dynamic environment is designed and the genetic operation is improved. The simulation experiment of the algorithm principle is carried out, and it is proved that the improved algorithm has a better obstacle avoidance effect in the dynamic environment.

Acknowledgement. This work was supported by National Nature Science Foundation of China (62076249), Key Research and Development Plan of Shandong Province (2020CXGC010701, 2020LYS11), and Natural Science Foundation of Shandong Province (ZR2020MF154).

References

1. Pattnaik, S.K., Mishra, D., Panda, S.: A comparative study of meta-heuristics for local path planning of a mobile robot. Eng. Optim. **54**(1), 134–152 (2022)
2. Yan, R., Pangm, S., Sun, H., et al.: Development and missions of unmanned surface vehicle. J. Mar. Sci. Appl. **9**, 451–457 (2010)
3. Kim, D.J., Chung, C.C.: Automated perpendicular parking system with approximated clothoid-based local path planning. IEEE Control Syst. Lett. **5**(6), 1940–1945 (2020)
4. Wang, Z., Liang, Y., Gong, C., Zhou, Y., Zeng, C., Zhu, S.: Improved dynamic window approach for Unmanned Surface Vehicles' local path planning considering the impact of environmental factors. Sensors **22**(14), 5181 (2022)
5. Kornev, I. I., Kibalov, V. I., Shipitko, O.: Local path planning algorithm for autonomous vehicle based on multi-objective trajectory optimization in state lattice. In: Thirteenth International Conference on Machine Vision, pp. 430–437. SPIE (2021)
6. Bautista-Camino, P., Barranco-Gutiérrez, A.I., Cervantes, I., Rodríguez-Licea, M., Prado-Olivarez, J., Pérez-Pinal, F.J.: Local path planning for autonomous vehicles based on the natural behavior of the biological action-perception motion. Energies **15**(5), 1769 (2022)

7. Lin, J., Pan, L.: Multiobjective trajectory optimization with a cutting and padding encoding strategy for single-UAV-assisted mobile edge computing system. Swarm Evol. Comput. **75**, 101163 (2022)
8. Kim, M., Yoo, S., Lee, D., Lee, G.H.: Local path-planning simulation and driving test of electric unmanned ground vehicles for cooperative mission with unmanned aerial vehicles. Appl. Sci. **12**(5), 2326 (2022)
9. Li, J., Sun, J., Liu, L., Xu, J.: Model predictive control for the tracking of autonomous mobile robot combined with a local path planning. Measur. Control **54**(9–10), 1319–1325 (2021)
10. He, C., et al.: Accelerating large-scale multiobjective optimization via problem reformulation. IEEE Trans. Evol. Comput. **23**(6), 949–961 (2019)
11. Yang, G., Yao, Y.: Vehicle local path planning and time consistency of unmanned driving system based on convolutional neural network. Neural Comput. Appl. **34**(15), 12385–12398 (2021)
12. Hu, Y., Yang, S.X.: A knowledge based genetic algorithm for path planning of a mobile robot. In: IEEE International Conference on Robotics and Automation, 2004. Proceedings. ICRA'04, pp. 4350–4355. IEEE (2004)
13. Tu, J., Yang, S. X.: Genetic algorithm based path planning for a mobile robot. In: 2003 IEEE International Conference on Robotics and Automation (Cat. No. 03CH37422), pp. 1221–1226. IEEE (2003)
14. Huang, P.Q., Wang, Y., Wang, K., Liu, Z.Z.: A bilevel optimization approach for joint offloading decision and resource allocation in cooperative mobile edge computing. IEEE Trans. Cybern. **50**(10), 4228–4241 (2019)
15. Huang, P.Q., Wang, Y.: A framework for scalable bilevel optimization: identifying and utilizing the interactions between upper-level and lower-level variables. IEEE Trans. Evol. Comput. **24**(6), 1150–1163 (2020)
16. Cai, Z., Peng, Z.: Cooperative coevolutionary adaptive genetic algorithm in path planning of cooperative multi-mobile robot systems. J. Intell. Rob. Syst. **33**, 61–71 (2002)
17. Wen, Z.-Q., Cai, Z.-X.: Global path planning approach based on ant colony optimization algorithm. J. Cent. South Univ. Technol. **13**(6), 707–712 (2006). https://doi.org/10.1007/s11 771-006-0018-4
18. Tsai, C.C., Huang, H.C.: Parallel elite genetic algorithm and its application to global path planning for autonomous robot navigation. IEEE Trans. Industr. Electron. **58**(10), 271–275 (2011)

A Multi-objective Particle Swarm Algorithm
Based on a Preference Strategy

Yi Wang[✉], KangShun Li, and Yong Fan

College of Computer Science, Guangdong University of Science and Technology,
Dongguan 523083, China
gust312@163.com

Abstract. To solve the problems of premature convergence and insufficient diversity in high-dimensional multi-objective optimization. A preferred area-based multi-objective particle swarm optimization (PAMOPSO) is proposed. The algorithm first determines the preferred regions through reference points, divides the target space into preferred regions and non-preferred regions, and uses different dominance rules in the different regions to improve the survival pressure of the algorithm in the high-dimensional space. Secondly, we design the selection mode of the global optimal position Gbest in two stages, comprehensively considering the convergence and distribution of the algorithm. Experimental results show that the proposed algorithm has good convergence and diversity compared with other improved multi-objective algorithms.

Keywords: High-dimensional multi-objective optimization · multi-objective particle swarm algorithm · preferred multi-objective algorithm

1 Introduction

In practical optimization problems, decision-makers often have to consider many conflicting goals, namely, multi-objective optimization problems (MOPs). Intelligence optimization algorithms prove to be effective for solving multi-objective optimization problems. Particle swarm optimization (PSO) is a well-known evolutionary algorithm to simulate the interaction during bird group predation. Due to its simple mechanism, few parameters, and fast convergence speed, PSO has been successfully applied to solve MOPs and expanded to a multi-objective particle swarm optimization algorithm. However, MOPSO can easily fall into local optimal. Besides, a single search mode of particle swarm quickly leads to poor convergence accuracy and diversity [1].

With the deepening of the research, the researchers explore the introduction of the particle swarm algorithm into the process of multi-object problem solving, combined with the rapid convergence characteristics of the use of the particle swarm algorithm and the excellent performance in the multi-object optimization problem. Domestic and foreign researchers mainly divided the multi-target particle swarm algorithms into the following groups according to the selection mechanism:

L. Pan et al. (Eds.): BIC-TA 2022, CCIS 1801, pp. 44–53, 2023.
https://doi.org/10.1007/978-981-99-1549-1_4

(1) **Multi-objective particle swarm algorithm dominated by Pareto.** Zhang et al. proposed the CMOPSO algorithm to change the particle speed through the update mechanism based on a learning strategy. The particles in the current population select the winning particle through pairwise competition. This particle guides another particle to perform speed updates, better balancing convergence and population diversity and reduces unnecessary memory overhead and computing complexity [2]. By introducing two-stage strategies, Hu et al. emphasized convergence and distribution in different stages. They maintained the variety of external archives by introducing parallel cell systems to support the selection of more diverse solutions [3]. Lin et al. facing the multi-target particle group algorithm under high target space selection pressure drop, put forward the NMPSO algorithm, and introduced a new fitness evaluation method to overcome the Pareto-dominated ranking difficulty or the limitation of decomposition method, combining the convergence distance and diversity distance to balance the algorithm convergence ability and population diversity ability, and adopted the new speed update way, not research provides another direction [4].

(2) **Multi-objective particle swarm algorithm based on decomposition.** This kind of method is a multi-objective optimization problem into a set of single-objective optimization problems, by solving each subproblem without using Pareto control. Coello et al. use the decomposition method using a global optimal position solution set to update example locations, simultaneously maintaining the diversity of the population [5]. Dai et al. proposed the MPSO/D algorithm to maintain diversity. The target space of the multi-objective optimization problem is divided through the direction vector and incorporates the idea of differential evolution [6].

(3) **Index-based multi-target particle swarm algorithm.** The algorithm is in the process of updating by evaluating the index to guide the search direction of the algorithm. Garcia et al., by introducing the HV index guide algorithm search and convergence, according to the optimal global location and individual historical optimal HV index, the external archive update according to the size of HV index, its low dimensional target space can be a good guide algorithm search, but in high target problem, HV solution is very difficult, the calculation complexity of the algorithm is further improved [7–9].

This study proposes a multi-objective particle swarm optimization algorithm based on a preference strategy to divide the target space into preferred and non-preferred regions and improves the archiving mode and the optimal global location selection mechanism, verifying the algorithm performance on different test sets.

2 Relevant Theoretical Basis

2.1 Particle Swarm Optimization

The particle swarm optimization algorithm is initialized into a group of random particles, in which the optimal solution is found through iteration. During each iteration, the particle updates its speed and position by tracking two extreme values (Pbest, Gbest). the updated formula is [8]:

$$v_t(t+1) = w \times v_t(t) + r_1 \times c_1 \times (pbest_i(t) - x_i(t)) + r_2 \times x_2 \times (gbest(t) - x_i(t)) \tag{1}$$

$$x_i(t+1) = x_i(t) + v_i(t+1) \tag{2}$$

where $x_i(t)$ = (xt1, xt2,..., xtn) is the position information of the i-th particle at the t iteration, n represents the dimension of the decision variable in the particle solution $v_i(t)$ = (vt1, vt2,..., vtn) is the velocity information of the i-th particle in the t-th iteration. Pbest represents the historical optimal position of the i-th particle, the local optimal position at the t generation, while Gbest represents the best location explored in the whole population.

c1 and c2 are the learning factors, c1 is the individual learning factor, c2 is the individual social learning factor, if c1 = 0 means the particle only group experience, its convergence is fast, but easy to fall into local optimal, if $c_2 = 0$, means the particle share no group information, a scale for M group run M are particles, the chance of solution is very small, so generally c_1 and c_2 is set to the same size. r_1 and r_2 are the random numbers between [0,1], w is the inertial weight factor when w is larger, the next generation speed is larger, so can expand the global search range, w small, the next generation speed will decrease, thus local search ability enhancement, so can choose larger w value, prevent the algorithm missing the possible optimal solution, into local optimal, and later choose smaller w value, improve the convergence speed of the algorithm.

2.2 Multi-objective Optimization Problem

Due to the complexity of practical problems, our solution is not limited to single-objective optimization but to two or more issues, calling these including two or more objective problems as multi-objective optimization problems. These goals to be optimized include the problems of maximization, minimizing, or both, and to facilitate the solution, we can transform the goal problems into the maximization or minimization solution. This paper discusses minimization. Then the mathematical representation of the minimized multi-objective optimization problem is as follows [10, 11]:

$$\min f(x) = (f_1(x), f_2(x), \ldots, f_m(x)) \tag{3}$$

$$subject \text{ to } \quad x \in S \subset R^n \tag{4}$$

where $x = (x_1, x_2, \ldots, x_n)$ is n dimension decision variables, S is the feasible domain of x, and m represents the number of objective functions. $f_i(x)$ the i-th objective function.

3   The Proposed Algorithm

3.1   Improve the Preference Areas

The MOPSO algorithm follows strict Pareto dominance rules when evaluating the merits of the solutions, but it works against the algorithm convergence in a high-dimensional space. Lopez et al. proposed a new preference dominance relationship, dividing the entire target space into two subspaces, one subspace of the individual is compared according to the Pareto dominance relationship, and the payoff scalar function reaches one subspace. In the proposed algorithm τ value is set in the algorithm initialization stage in advance for the value of the [0,1] interval. Still, for multi-target particle algorithm itself has "precocious" characteristics. If fixed the size of the preferred area is, it is easy to make the algorithm converge to the local optimal, and it may miss other better solutions and not be conducive to the diversity of the solution. Therefore, this paper proposes dynamic value adjustment to ensure that the algorithm can expand the search range in the first and mid stages, safeguard the diversity of solutions, and narrow the search range in the late stage so that the algorithm can converge effectively.

The value of τ decreases gradually with the number of iterations but then linear. If the decrease rate is fast before the algorithm has fully searched the feasible area, the τ value is defined by formula (5):

$$\tau = \tau_{max} - (\tau_{max} - \tau_{min}) * (current_{iter}/\max_{iter})^c \qquad (5)$$

τ_{max} and τ_{min} set the maximum and minimum values of the decision maker for the search stage, currentiter is the current number of iterations, maxiter is the maximum number of iterations, and c is the descent speed.

3.2 Gbest Selection Strategy

The Gbest selection strategy of the MOPSO algorithm depends on the density of the particles, but the algorithm only considers maintaining the uniformity of the solution distribution while ignoring strengthening the convergence ability of the algorithm. This paper proposes a two-stage Pbest selection strategy to consider both the algorithm's convergence ability and distribution ability when selecting the Pbest.

In the first stage, according to the position of the particle in the target space and the target space to calculate the corresponding similar distance (Similarity Distance, SD), according to the similar distance set selected from the particles far from the noninferior solution, keep the noninferior solution as the selection set of the second stage, SD solution method as shown in Eq. (6) and Eq. (7):

$$d(x_i, y_j) = \sqrt{\sum_{k=1}^{M} (f_k(x_i) - f_k(y_i))^2} \qquad (6)$$

$$SD_i\{d(x_i, y_1), d(x_i, y_2), \ldots, d(x_i, y_t)\} \qquad (7)$$

where x_i represents the i-th particle in the population, y_j represents the j-th non-inferior solution (j = 1, 2, ..., t), t is the number of non-inferior solutions in the external archive Archive, $f_k(x_i)$ represents the function value of the i-th particle on the k-th target, M is the number of an objective function, SD_i is the collection of stored i-th particle and each particle in the external archive Archive in the target space.

Similar distances represent the distance of the particles to be updated from the particles in an external archive. In the first stage, the algorithm can mainly search for more solutions. The x_i particles should tend to select particles more prominent than the average similarity distance in the external archive, as shown in Eqs. (8) and (9).

$$AVG_{SD_i} = \frac{\sum_{j=1}^{t} SD_{i,j}}{t} \tag{8}$$

$$\sum_{j=1}^{t} SD_{i,j} = d(x_i, y_1) + d(x_i, y_2) + \ldots + d(x_i, y_t) \tag{9}$$

Through the selection of the first stage, the set of non-inferior solutions to be processed in the second stage, such as the formula (10).

$$S = \{y | y \in Archive \ \& \ d(x_i, y) > AVG_{SD_i}\} \tag{10}$$

In the first stage, it considers the diversity of the algorithm solution. To strengthen the convergence ability of the algorithm, the second stage focuses on realizing the uniform distribution of the solution set and selects the individuals with excellent distribution performance in the S set. The entire calculation step of the improved algorithm is as follows: The whole calculation step of the improved algorithm is shown as Algorithm 1:

Algorithm 1 Improves the multi-objective particle swarm algorithm

Step 1: Set the relevant parameters, and initialize the population speed, position, Pbest and Gbest;

Step 2: Calculate the value of the particles on each target function in the population and select the preferred region according to formula (5).

Step 3: evaluate the advantages and disadvantages of particles according to the control rules, and save the non-inferior solution set to the external archive.

Step 4: Select Gbest for each particle according to formula (6-10).

Step 5: Update the position and speed of each particle.

Step 6: execute the variation, evaluate the quality of the particles, update the particle Pbest.

Step 7: Update the external archive and remove particles if the external archive exceeds the storage capacity.

Step 8: Determine whether the maximum number of iterations or the stop condition of the algorithm is reached. If not reached, return to step 4, if reached, the cycle ends, output the solution set in the external archive, and the end of the algorithm.

---

4 Experimental Results and Analysis

4.1 Evaluation Indicators

Different from single objective optimization, which can compare the size of the target function value to evaluate the algorithm performance, multi-objective optimization needs some performance indicators to evaluate the algorithm effect, mainly from two aspects to evaluate the algorithm, on the one hand, evaluate the algorithm solution set and the real Pareto edge, on the other hand, evaluate the distribution of the algorithm solution set.

Therefore, the performance indicators of multi-objective optimization can be divided into three categories:

(1) The first type of indicator considers the distance between the solution set obtained by the algorithm and the real Pareto solution set, such as the GD index.
(2) The second type of indicator considers the distribution of the solution set obtained by the algorithm in the target space, such as the Spacing index.
(3) The third category is the comprehensive index, which considers the algorithm solution's convergence while considering the algorithm's distribution, such as IGD and HV index.

This paper uses GD, IGD, HV, and distance performance indicators from the reference point to evaluate the algorithm [12–14].

(1) The GD calculation formula is shown in formula (11):

$$GD(P, P^*) = \frac{\sum_{\mu \in P} d(\mu, P^*)}{|P|} \tag{11}$$

where P^* is the set of sampling points uniformly distributed on the optimal edge of real Pareto, P is the Pareto non-dominant solution set solved by the algorithm, $d(u, P)$ represents the minimum value of the distance between all solutions from u to P. from formula (10) can be found that the smaller the GD value, the more the algorithm converges.

(2) The IGD calculation formula is shown in formula (12):

$$IGD(P^*, P) = \frac{\sum_{v \in P} d(v, P^*)}{|P^*|} \tag{12}$$

(3) The HV calculation formula is shown in formula (13):

$$HV = \bigcup_i vol_i| \in PF \tag{13}$$

4.2 Algorithm Parameter Setting

To verify the performance of the proposed algorithm PAMOPSO, several commonly used benchmark functions (DTLZ2, DTLZ4, DTLZ6–7) will be used to test the proposed algorithm, and the test results will be compared with the more popular multi-objective optimization algorithms, including MMOPSO, CMOPSO, g-NSGA. In the experiment, the number of decision variables of all test functions is M+9, and M is the number of objective functions.

Comparing the population size of the algorithm, the external archive capacity, and the maximum number of iterations are shown in Table 1. The other parameters of each algorithm are set from the relevant references. The algorithm was run 50 times independently in each case test function, and the performance index was averaged 50 times.

Table 1. Initial conditions

Algorithms	Population size	Archive	Iterations
PAMOPSO	4 Objectives:150	Reference population size	800
MMOPSO	6 Objectives:150	——	800
CMOPSO	8 Objectives:200	——	800
g-NSGAII	10 Objectives:20	——	800

4.3 Experimental Results

Table 2, Table 3, and Table 4, respectively, show the measured values of the three test functions on 4, 6, 8, and 10 targets, which include the average values of GD, IGD, and HV. The best results are marked using the bold font method. Also, the number of PAMOPSO better, worse, and equal to other algorithm test examples is given in the last row of the table.

Table 2. The mean GD values obtained on the DTLZ problem

Test functions	M	PAMOPSO	MMOPSO	CMOPSO	g-NSGAII
DTLZ2	4	3.7738E−02	5.3122E−02+	1.5387E−01+	3.7821E−02=
	6	5.3536E−02	1.3321E−01+	1.5442E−01+	5.3235E−02=
	8	5.9284E−02	1.4647E−01+	1.5598E−01+	5.9343E−02=
	10	6.6561E−02	1.6158E−01+	1.5996E−01+	6.6670E−02=
DTLZ4	4	2.6924E−02	5.1425E−02+	1.0257E−01+	1.4394E−01+
	6	5.0021E−02	1.4825E−01+	1.3869E−01+	1.2707E−01+
	8	5.5324E−02	1.3526E−01+	1.4015E−01+	1.8321E−01+
	10	7.6291E−02	1.4061E−01+	1.2667E−01+	1.2787E−01+
DTLZ6	4	**6.1534E−02**	5.6114E−01+	6.9726E−01+	6.3809E−01+
	6	**6.5981E−02**	7.0722E−01+	6.7222E−01+	6.2658E−01+
	8	**5.4311E−02**	6.2307E−01+	6.8057E−01+	6.2985E−01+
	10	**5.2320E−02**	6.5274E−01+	6.5195E−01+	6.3943E−01+
±/=			12/0/0	12/0/0	8/0/4

Tables 2, 3 and 4 show that the proposed PAMOPSO algorithm performs well on both the DTLZ4 and DTLZ6 test functions. The best convergence is achieved with the other three algorithms, and the performance is comparable to the g-NSGA algorithm on DTLZ2. Overall, this also proves that the PAMOPSO algorithm can be guaranteed

Table 3. The mean IGD values obtained on the DTLZ function

Test functions	M	PAMOPSO	MMOPSO	CMOPSO	g-NSGAII
DTLZ2	4	1.5221E−01	1.3221E−01=	1.0954E−01−	5.7264E−01+
	6	5.3263E−01	3.83485E−01−	1.5987E+00+	1.5961E+00+
	8	7.3661E−01	7.6232E−01=	2.3621E+00+	2.0513E+00+
	10	8.2562E−01	1.0299E+00+	2.4361E+00+	1.1236E+00+
DTLZ4	4	2.2331E−01	1.2103E−01−	1.3954E−01−	1.8324E−01=
	6	4.7395E−01	3.4351E−01−	6.2101E−01+	9.3651E−01+
	8	7.1324E−01	8.2571E−01+	1.3215E+00+	1.9876E+00+
	10	8.0281E−01	1.5631E+00+	1.3947E+00+	1.4764E+00+
DTLZ6	4	1.8540E−02	4.2164E−02+	2.5468E−01+	2.8674E+00+
	6	3.2440E−02	2.6345E−01+	6.2422E+00+	8.5981E+00+
	8	3.1230E−02	4.8451E−01+	9.4966E+00+	8.6568E+00+
	10	3.2931E−02	4.6954E−01+	9.5032E+00+	8.9614E+00+
±/=			7/3/2	10/2/0	11/0/1

Table 4. Average HV values obtained on the DTLZ problem

Test functions	M	PAMOPSO	MMOPSO	CMOPSO	g−NSGAII
DTLZ2	4	5.7643E−01	6.3689E−01−	6.5911E−01−	1.4214E−01+
	6	3.1364E−01	5.3652E−01−	4.5623E−01−	3.2462E−04+
	8	2.9981E−01	2.8511E−01+	0.0000E+00+	3.2141E−05+
	10	1.4695E−01	1.1364E−01+	0.0000E+00+	7.3124E−03+
DTLZ4	4	6.5684E−01	6.6542E−01+	6.2965E−01+	4.6912E−01+
	6	4.6381E−01	6.0369E−01−	1.7361E−01+	3.5243E−02+
	8	4.8954E−02	2.2456E−01−	2.5124E−03+	8.9631E−03+
	10	2.9961E−02	4.1245E−02−	1.3512E−04+	9.7512E−04+
DTLZ6	4	1.5631E−01	1.5212E−01=	1.2365E−01+	0.0000E+00+
	6	1.1023E−01	9.7864E−02+	0.0000E+00+	0.0000E+00+
	8	9.8964E−02	8.8624E−02+	0.0000E+00+	0.0000E+00+
	10	9.7521E−02	8.9657E−02+	0.0000E+00+	0.0000E+00+
±/=			6/5/1	10/2/0	12/0/0

faster in the high-dimensional target space by introducing the reference points and the preferred regions, adopting the new dominance rules according to the preferred regions, and adopting the new Gbest strategy.

5 Conclusion

The particles of traditional multi-target particle clusters all use a single search pattern, resulting in premature population convergence and poor diversity while causing poor particle generalization ability.

In this study, we set up reference points and divided the target space into two subspaces according to the preferred area. In different spaces, the solution evaluation method improved the survival pressure of the population and searched the algorithm for the region of interest of the decision maker. Secondly, the two-stage selection Gbest method is proposed. In the first stage, the algorithm can search for a wider range of solutions, and the second stage ensures that the understanding can be distributed more evenly. Experimental results show that the improved algorithm has excellent performance.

Acknowledgment. This work is supported by the Guangdong Youth Characteristic Innovation Project (2021KQNCX120), the Natural Science Foundation of Guangdong Province of China (2020A1515010784), the Natural Science Project of Guangdong University of Science and Technology (GKY-2021KYYBK-20), the General Project of Science and Technology of Dongguan Social Development (20231800910352), the Natural Science Project of Guangdong University of Science and Technology (XJ2022003501), the University Distinguishing Innovation Project of Guangdong Provincial Department of Education (2021KTSCX149), and the Key Project of Science and Technology of Dongguan Social Development (20211800905512).

References

1. Habib, M., Aljarah, I., Faris, H., Mirjalili, S.: Multi-objective particle swarm optimization: theory, literature review, and application in feature selection for medical diagnosis. In: Mirjalili, S., Faris, H., Aljarah, I. (eds.) Evolutionary Machine Learning Techniques. AIS, pp. 175–201. Springer, Singapore (2020). https://doi.org/10.1007/978-981-32-9990-0_9
2. Zhang, X., Zheng, X., Cheng, R., Qiu, J., Jin, Y.: A competitive mechanism based multi-objective particle swarm optimizer with fast convergence. Inf. Sci. **427**, 63–76 (2018)
3. Hu, W., Yen, G.G., Luo, G.: Many-objective particle swarm optimization using two-stage strategy and parallel cell coordinate system. IEEE Trans. Cybern. **47**(6), 1446–1459 (2016)
4. Lin, Q., et al.: Particle swarm optimization with a balanceable fitness estimation for many-objective optimization problems. IEEE Trans. Evol. Comput. **22**(1), 32–46 (2016)
5. Coello, C.A.C., Pulido, G.T., Lechuga, M.S.: Handling multiple objectives with particle swarm optimization. IEEE Trans. Evol. Comput. **8**(3), 256–279 (2004)
6. Dai, C., Wang, Y., Ye, M.: A new multi-objective particle swarm optimization algorithm based on decomposition. Inf. Sci. **325**, 541–557 (2015)
7. García, I.C., Coello, C.A.C., Arias-Montano, A.: Mopsohv: A new hypervolume-based multi-objective particle swarm optimizer. In: 2014 IEEE Congress on Evolutionary Computation (CEC), pp. 266–273. IEEE (2014)
8. Poli, R., Kennedy, J., Blackwell, T.: Particle swarm optimization: an overview. Swarm Intell. **1**(1), 33–57 (2007). https://doi.org/10.1007/s11721-007-0002-0
9. Deb, K.: Multi-Objective Optimization, pp. 403–449. Search Methodologies. Springer, Boston, MA (2014)
10. Tian, Y., et al.: Evolutionary large-scale multi-objective optimization: a survey. ACM Comput. Surv. (CSUR) **54**(8), 1–34 (2021)

11. Zhang, Y., Wang, S., Phillips, P.: Detection of Alzheimer's disease and mild cognitive impairment based on structural volumetric MR images using 3D-DWT and WTA-KSVM trained by PSOTVAC. Biomed. Signal Process. Control **21**(8), 58–73 (2015)
12. Chen, L., Wang, H., Ma, W.: Two-Stage multi-tasking transform framework for large-scale many-objective optimization problems. Complex Intell. Syst. **7**(3), 1499–1513 (2021)
13. Tian, Y., Zheng, X., Zhang, X.: Efficient large-scale multi-objective optimization based on a competitive swarm optimizer. IEEE Trans. Cybern. **50**, 3696–3708 (2019)
14. Das, S., Mullick, S., Suganthan, P.: Recent advances in differential evolution – an updated survey. Swarm Evol. Comput. **27**, 1–30 (2016)

An Improved Harris Hawk Optimization Algorithm Based on Spiral Search and Neighborhood Perturbation

Yanfeng Wang[1,2], Yuhang Xia[1,2], Dan Ling[1,2(✉)], and Junwei Sun[1,2]

[1] School of Electrical and Information Engineering,
Zhengzhou University of Light Industry, Zhengzhou 450002, China
lingdan@zzuli.edu.cn
[2] Henan Key Lab of Information-Based Electrical Appliances,
Zhengzhou 450002, China

Abstract. Aiming at the problems of the basic Harris Hawk optimization (HHO), such as insufficient global search ability, slow convergence speed, low convergence accuracy and easy to fall into local optimization, this paper proposes an improved HHO (IHHO) algorithm based on spiral search and neighborhood perturbation. First, Tent chaotic mapping is used to initialize the population, enhance the diversity of the population and improve the initial convergence speed of the proposed algorithm. Secondly, the collaborative updating strategy based on spiral search and neighborhood perturbation is used in the exploration stage to expand the search space and improve the global search ability of the proposed algorithm. In the exploitation stage, the adaptive inertia weight is introduced to improve the convergence speed. Then, the dimensional cross-variation is used to improve the ability of the algorithm to jump out of the local optimum. CEC2021 test function and 8 classic test functions are used to demonstrate the effectiveness of the proposed method. Compared with the basic HHO and the other improved swarm intelligence algorithms, the results show that the proposed algorithm has better global search ability, faster convergence speed and higher accuracy. Finally, the proposed IHHO algorithm is used to solve tension spring design problems.

Keywords: Harris Hawk optimization · Spiral search · Neighborhood perturbation · Dimensional cross-variation

1 Introduction

Swarm intelligence algorithms, as a class of metaheuristic algorithms that simulates the behavior of social insects or animal clusters, have received extensive

This work was supported in part by the National Key Research and Development Program of China for International S and T Cooperation Projects (2017YFE0103900), in part by the Open Fund of State Key Laboratory of Esophageal Cancer Prevention & Treatment (K2020-0010 & K2020-0011), and in part by the Promotion Special Project Science and Technology in Henan Province (222102210091).

attention in solving various scientific computing and application problems. Compared with traditional optimization algorithms, swarm intelligence algorithms were characterized by initial value-independent, low function requirement and good solution performance. Since Holland proposed genetic algorithms based on Darwin's theories of evolution in 1975 [1], researchers have proposed various intelligence optimization algorithms by analyzing different biological populations and physical phenomena, such as the Salp Swarm Algorithm (SSA), Ant Lion Optimizer (ALO), Dragonfly Algorithm (DA) [2–4].

Inspired by the cooperative behavior and chasing strategy of Harris hawks, Harris hawks optimization (HHO) algorithm was proposed by Heidari [5]. It performs well in color image segmentation, neural network training, motor control and other fields [6–8]. However, the basic HHO algorithm has some shortcomings, e.g., weak global search capability, poor optimization accuracy, slow convergence speed and insufficient search vitality. In order to solve those problems, a large number of improved HHO algorithms have been proposed. In [9], Cauchy function variation was introduced into the basic HHO algorithm to improve the population diversity and the random contraction index function was used to correct the prey energy. A Quasi-Reflected HHO (QRHHO) algorithm based on anti-learning mechanism was proposed to increase the population diversity and improve the optimization accuracy [10]. A modified variant with a Long-term Memory HHO (LMHHO) was proposed to improve the search efficiency of the algorithms [11]. In [12], an improved information exchange mechanism and a chaotic disturbance nonlinear escape factor are introduced in the basic HHO algorithm, enhancing the convergence accuracy and robustness of the algorithm. In [13], an improved Harris hawk optimization algorithm based on cauchy distribution inverse cumulative function and tangent flight Operator was proposed to improve the search efficiency and convergence of the algorithms.

Although the above algorithm improved the performance of HHO to some extent, there was still room for improvement, such as insufficient search space and weak global search ability in the exploration stage. In this paper, an improved Harris hawk optimization (IHHO) algorithm based on spiral search and neighborhood perturbation is proposed. A spiral search and neighborhood perturbation collaborative updating strategy is used in the exploration stage to improve the search space and enhanced the global exploration ability of the algorithm. The population is initialized by Tent chaotic mapping to increase the diversity of the population and enhance the initial convergence speed of the algorithm. In the exploitation stage, the adaptive inertia weight [14] is introduced to improve the convergence speed and accuracy of the proposed algorithm. To avoid HHO falling into the local optimal, the dimensional cross variation [15] is used to enhance the interpopulation communication. The superiority of the proposed IHHO algorithm is demonstrated by a comparative analysis on the CEC2021 test function and 8 classical test functions.

The rest of this paper is structured as follows: The basic HHO algorithm is introduced in Sect. 2; In Sect. 3, the improved HHO is proposed; In Sect. 4, 8 classical test functions and CEC2021 test functions are used to verify the improved HHO algorithm; In Sect. 5, IHHO algorithm is applied to engineering applications; conclusions are provided in Sect. 6.

2 Basic HHO Algorithm

HHO simulates the real situation of Harris hawk hunting its prey. HHO is mainly composed of three parts: exploration stage, exploitation stage and a transformation between exploration and exploitation.

2.1 Exploratory Stage

In the exploration stage, Harris hawk randomly perches at a location, tracks and detects prey through its keen eyes, and hunts with two equal opportunity strategies. The specific formula is as follows:

$$x(t+1) = \begin{cases} x_{rand}(t) - r_1 \left| x_{rand}(t) - 2r_2 \right| & q > 0.5 \\ (x_{rabbit}(t) - x_m(t)) - r_3 \left(r_4 \left(ub - lb \right) + lb \right) & q < 0.5 \end{cases} \tag{1}$$

where x_{rand} denotes the location of random individuals in the current population; x_{rabbit} is the position of the prey; $x_m(t) = \frac{1}{N} \sum_{i=1}^{N} x_i(t)$ is the average position of the Harris Hawk; r_1, r_2, r_3, r_4, q, are the random numbers in $(0, 1)$ and updated in each iteration; ub, lb represent the upper and lower bounds of the search space respectively.

2.2 Transformation Between Exploration and Exploitation

In nature, when Harris hawks chases its prey, the energy of the prey decreases during the escape process. The HHO algorithm realizes the conversion from exploration to exploitation by simulating this feature. The specific formula is as follows:

$$E = 2E_0(1 - t/T) \tag{2}$$

where E represents the escaped energy of prey; E_0 is the initial value of its energy and varies randomly within the interval $(-1, 1)$; T is the maximum iteration number, t is the current iterations. In HHO algorithm, escape energy E tends to decrease during iteration. When the escape energy is $|E| \geq 1$, Harris hawk searches for prey positions in different regions and executes the exploration stage; When $|E| < 1$, Harris hawk searches the adjacent solutions locally, corresponding to the exploitation stage.

2.3 Exploitation Stage

In the exploitation stage, the Harris hawk forms a circle around its prey, waiting for the opportunity to make a surprise attack. However, the actual predation process is complex, the prey under siege will also react in some way, and the Harris hawk must then make the necessary adjustments according to the prey's behavior. In order to better simulate the hunting behavior, HHO simulate four update strategies and decide which strategy to use by and a random number within $(0, 1)$.

Soft Besiege. When $R \geq 0.5$, $|E| \geq 0.5$, Harris hawk circles around the prey, consuming its physical strength to make it tired, and finally succeeds in capturing. The specific formula is as follows:

$$x(t+1) = \Delta x(t) - E\,|Jx_{rabbit}(t) - x(t)| \tag{3}$$

where $\Delta x(t) = x_{rabbit}(t) - x(t)$, represents the distance between prey and individual; $J = 2(1 - r_5)$, represents the random jump of prey during escape, r_5 is the random numbers in $(0, 1)$.

Hard Besiege. When $R \geq 0.5$, $|E| < 0.5$, the escape energy of the prey is low, and unable to protrude. Harris hawk takes a hard encirclement to attack the target prey. The specific formula is as follows:

$$x(t+1) = x_{rabbit}(t) - E\,|\Delta x(t)| \tag{4}$$

Soft Besiege with Progressive Rapid Dives. When $R < 0.5, |E| \geq 0.5$, the prey can break through the encirclement of Harris hawk and enough escape energy. In view of this situation, Harris hawk needs to form a soft besiege with Progressive rapid dives before attacking. The specific formula is as follows:

$$x(t+1) = \begin{cases} Y, F(Y) < F(x(t)) \\ Z F(Z) < F(x(t)) \end{cases} \tag{5}$$

where $Y = x_{rabbit}(t) - E\,|Jx_{rabbit}(t) - x(t)|$, the fitness value of Y, and the current position $x(t)$ are compared to detect whether the Harris hawk is a good raid; If it is not a successful raid, the prey will be besieged by the dive mode based on Levy flight, $Z = Y + S \times LF(D)$, LF represents the Levy flight, D represents dimension.

Hard Besiege with Progressive Rapid Dives. When $R < 0.5, |E| < 0.5$, he prey can escape from the enclosure but has insufficient energy . Before raiding and capturing the prey, Harris hawk adopts a hard besiege with progressive rapid dives to capture the prey. The specific formula is as follows:

$$x(t+1) = \begin{cases} Y, F(Y) < F(x(t)) \\ Z F(Z) < F(x(t)) \end{cases} \tag{6}$$

where $Y = x_{rabbit}(t) - E\,|Jx_{rabbit}(t) - x_m(t)|$, $Z = Y + S \times LF(D)$. The way of siege is similar to that of soft besiege with progressive rapid dives.

3   Improved HHO Algorithm (IHHO)

An excellent swarm intelligence algorithm should not only have strong global search ability in the exploration stage, which can quickly locate the scope of the global optimal solution in the search space, but also meet the mining ability in the exploitation stage, so as to improve the optimization accuracy of the algorithm. Based on this idea, this paper proposes an improved Harris hawk optimization. Tent chaotic mapping is used to initialize the population, increase the diversity of the population, and improve the initial convergence speed of the algorithm; In the exploration stage, the spiral search and neighborhood perturbation collaborative updating strategy is adopted to expand the search scope of the algorithm, so as to improve the global search ability of the algorithm; In the exploitation stage, the adaptive inertia weight is introduced to enhance the mining capacity of the algorithm, so as to improve the optimization accuracy of the algorithm; Finally, dimensional cross-variation and greedy strategy are used to strengthen the communication between populations, so as to avoid the algorithm falling into local optimization. The algorithm flow chart is shown in Fig. 1.

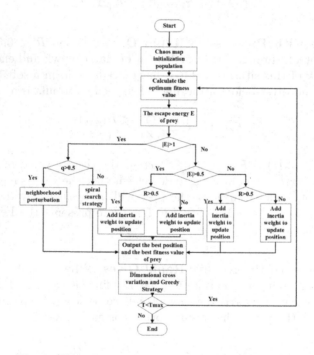

Fig. 1. Flow chart of the proposed IHHO algorithm

3.1   Tent Chaotic Mapping

The initial position of individual population is crucial to the optimization performance of swarm intelligence algorithm, and the quality of initial population will determine the convergence speed and accuracy of the algorithm. The individual generated randomly with the basic HHO reduces the search efficiency of the population and slows the convergence rate at the initial stage of computation. Due to the randomness, regularity and ergodicity of chaotic motion, it can maintain the diversity of population to a great extent and improve the global search ability. Common chaotic maps include Logistic map and Tent chaotic map. Shan Liang et al. proves that the ergodic uniformity of Tent chaotic mapping is better than that of Logistic mapping [16]. The formula of Tent mapping is as follows:

$$\lambda_{t+1} = \begin{cases} \lambda_t/\alpha, & 0\lambda_t < \alpha \\ (1 - \lambda_t)/(1 - \alpha), & \alpha\lambda_t 1 \end{cases} \quad (7)$$

where λ_t is the chaos number generated by iteration, α is a constant of $(0,1)$, which is taken as 0.7 in this paper.

3.2 Collaborative Updating Strategy of Spiral Search and Neighborhood Disturbance

In the exploration stage of the HHO, when $R < 0.5$, no Harris hawk individual finds the location of prey, and Harris hawk will search randomly. Because the whale optimization algorithm [17] has strong search ability, this paper introduces the spiral search strategy of the whale optimization algorithm to improve the global search ability of the algorithm. The specific formula is as follows:

$$x(t + 1) = x_{rand}(t) - r_1 |x_{rand}(t) - 2r_2| e^l \times \cos(2\pi l) \quad (8)$$

where l is a constant of $(0, 1)$.

In the basic HHO algorithm, when $R > 0.5$, the Harris hawk finds the target, circling the prey according to the position of other individuals and updates the position. However, in the exploration stage of the algorithm, the optimal positions searched by Harris hawk are often lack of diversity, which may lead to other undiscovered optimal solutions around the searched positions. In this paper, the neighborhood perturbation strategy is introduced, when Harris hawk finds prey, it will search again in its neighborhood, and select the optimal location by comparing the fitness values of two individual positions. The specific formula is as follows:

$$X_{new} = x_{rabbit}(t) + 0.5 \times r_6 \times x_{rabbit}(t) \quad (9)$$

$$x' = (x_{rabbit}(t) - x_m(t)) - r_3 (r_4 (ub - lb) + lb) \quad (10)$$

$$x(t + 1) = \begin{cases} X_{new} & fobj(X_{new}) < fobj(x') \\ x' & else \end{cases} \quad (11)$$

where r_6 is a constant of $(0, 1)$. X_{new} is new individuals generated by neighborhood disturbance strategy.

3.3 Adaptive Inertia Weight

In particle swarm optimization (PSO) [18], particle velocity directly determines particle position update. Inertia weight is an important parameter in the updating formula of PSO, which reflects the influence of previous generation of particles on the current particles. In the exploitation stage of swarm intelligence algorithm, the algorithm needs more mining capacity. Inspired by PSO, this paper introduces adaptive inertia weight. An inertia weight that varies with the number of iterations is added to the position update of Harris hawk. As the number of iterations increases, the influence of the optimal Harris hawk position is gradually increased, thus improving the convergence speed and accuracy of the whole algorithm. The adaptive inertia weight formula is as follows:

$$w = 0.2 \times \cos(\pi \times (1 - t/T)/2) \tag{12}$$

In this paper, the adaptive inertia weight is added to the four updating methods to improve the exploitation ability of the proposed algorithm. When $R \geq 0.5$, $|E| \geq 0.5$, the location update is shown in (13); When $R \geq 0.5$, $|E| < 0.5$, the location update is shown in (14); When $R < 0.5$, $|E| \geq 0.5$, the location update is shown in (15) and (16); When $R < 0.5, |E| < 0.5$, the location update is shown in (17) and (18); The specific formula is as follows:

$$x(t+1)' = \Delta x(t) \times w - E |Jx_{rabbit}(t) - x(t)| \tag{13}$$

$$x(t+1)' = x_{rabbit}(t) \times w - E |\Delta x(t)| \tag{14}$$

$$Y_1 = x_{rabbit}(t) \times w - E |Jx_{rabbit}(t) - x(t)| \tag{15}$$

$$x(t+1) = \begin{cases} Y_1, F(Y_1) < F(x(t)) \\ ZF(Z) < F(x(t)) \end{cases} \tag{16}$$

$$Y_2 = x_{rabbit}(t) \times w - E |Jx_{rabbit}(t) - x_m(t)| \tag{17}$$

$$x(t+1) = \begin{cases} Y_2, F(Y_2) < F(x(t)) \\ ZF(Z) < F(x(t)) \end{cases} \tag{18}$$

3.4   Dimensional Cross Variation

Similar with other intelligent optimization algorithms, HHO is easy to fall into local optimization. Comparing the value of the objective function within one dimension, the individual with the best fitness is selected, while ignoring the differences between individuals of different dimensions. At the same time, there may be connections between individuals of different dimensions, thus influencing each other. The effect of crossover is to enhance the diversity of the population. The more diverse the population is, the more abundant the search space will be, the stronger the ability to find the optimal will be, while it can avoid falling into the local optimal. Therefore, in this paper, individuals with two random dimensions are introduced to carry out dimensional cross-variation in the optimal position, and the variation mode is as follows:

$$x(t+1) = x_{best} + r_7 \times (x_{rand1} - x_{rand2}) \tag{19}$$

where r_7 is a constant of $(0, 1)$, x_{best} is the current optimal individual, and x_{rand1}, x_{rand2} are random individuals in $(1, d)$ dimension interval. Greed strategy is used to judge whether new individuals are accepted or not for the dimensional cross-variation. If the results are better, replace them; otherwise, retain the original optimal position.

4   Experimental Simulation and Analysis

4.1   Experimental Design

In order to verify the performance of the IHHO proposed in this paper, CEC2021 function and 8 classic functions are selected for testing. The test environment adopts Microsoft's 64 bit operating system, and the programming and implementation of all algorithms are completed by MATLAB software. The proposed IHHO is compared with the basic HHO algorithm, elite opposition learning-based dimension by dimension improved dragonfly algorithm (EDDA) [19], improved SSA [20], improved ALO [21,22] and Harris Hawk optimization algorithm Based on Cauchy Distribution Inverse Cumulative Function and Tangent Flight Operator (CTHHO). The parameters of each algorithm are set according to the original literature. The population size of all algorithms is set to 30, and the maximum number of iterations is set to 500. At the same time, in order to avoid the contingency of the experiment, each algorithm runs 30 times independently.

4.2   Results of Classical Test Functions

In this paper, eight classical test functions are selected for comparative analysis. The test function is shown in Table 1, which gives information about the test function, including the function name, search range and the optimal value. In Table 1, F1, F2, F3 and F4 are the unimodal test functions and F5, F6, F7 and F8 are the multimodal test functions.

Table 1. Classical test function

NO.	Function	Range	Optimal value
F1	Sphere	$[-100,100]$	0
F2	Schwefel 2.22	$[-10,10]$	0
F3	Schwefel 1.2	$[-100,100]$	0
F4	Schwefel 2.21	$[-100,100]$	0
F5	Quartic	$[-1.28,1.28]$	0
F6	Rastrigin	$[-5.12,5.12]$	0
F7	Ackley	$[-32,32]$	0
F8	PGriewank	$[-600,600]$	0

The optimization accuracy of the algorithm is reflected by the mean value, and the robustness and stability of the algorithm are reflected by the standard deviation. Table 2 shows the comparison of test results of functions in 30 dimensions, including the comparison of solution quality based on optimal value, mean value and standard deviation. The convergence curve can show the number of times the algorithm falls into the local optimum and the convergence speed. In this paper, the iterative optimization effects of some functions are selected and the convergence curve is drawn, as shown in Fig. 2, where the horizontal axis is the number of iterations and the vertical axis is the optimal fitness value.

It can be seen from the test results in Table 2 that in the unimodal test function, the basic HHO has better optimization effect than other improved swarm intelligence algorithms, but no theoretical optimal value has been found, there is some room for improvement. Compared with HHO, the improvement effect of CTHHO is dozens of orders of magnitude, but it still does not converge to the optimal. The IHHO proposed in this paper finds the theoretical optimal value in all unimodal test functions, and the standard deviation is 0, which proves that the IHHO has good optimization accuracy and strong stability. In multimodal test functions, IHHO searches the optimal solution on F5 test function better than other swarm intelligence algorithms. In F6-F8 test function, IHHO is superior to EDDA and IALO, and has found the same solution as CTHHO, HHO and ISSA. According to Fig. 2, although IHHO and CTHHO, HHO, ISSA have found the same solution, IHHO can find the optimal solution within 100 iterations on the convergence curve, with good convergence speed.

Table 2. Comparative data of function optimization under different algorithms.

Function		IHHO	HHO	EDDA	ISSA	IALO	CTHHO
F1	Best	**0.00E+00**	3.88E−101	2.18E−55	6.07E−122	4.03E−09	1.93E−169
	Mean	**0.00E+00**	6.04E−98	2.41E−49	6.33E−121	8.39E−09	6.79E−140
	Std	**0.00E+00**	2.47E−97	9.76E−49	1.49E−120	9.01E−09	3.66E−139
F2	Best	**0.00E+00**	2.36E−55	2.07E−29	5.38E−62	5.94E−05	5.55E−83
	Mean	**0.00E+00**	9.02E−50	3.02E−28	2.05E−61	6.07E−05	3.37E−82
	Std	**0.00E+00**	4.52E−49	6.57E−28	3.10E−61	4.09E−06	1.73E−81
F3	Best	**0.00E+00**	2.32E-80	1.90E+04	3.42E−173	3.99E−08	2.26E−141
	Mean	**0.00E+00**	9.83E−74	1.75E+04	2.27E−163	3.78E−08	7.95E−109
	Std	**0.00E+00**	5.29E−73	3.40E+03	0.00E+00	8.94E−09	4.28E−108
F4	Best	**0.00E+00**	3.14E−57	2.46E−01	6.45E−62	5.98E−05	1.52E−76
	Mean	**0.00E+00**	6.76E−50	1.38E−01	1.88E−61	3.37E−05	1.14E−69
	Std	**0.00E+00**	3.35E−49	9.47E−02	3.58E−61	1.61E−05	4.29E−69
F5	Best	**1.03E−07**	2.47E−04	4.94E−02	1.22E−05	2.17E−04	2.35E−05
	Mean	**5.23E−06**	2.35E−04	4.53E−02	6.20E−05	2.40E−04	1.88E−04
	Std	**4.14E−06**	2.34E−04	1.51E−02	4.21E−05	1.59E−04	2.46E−04
F6	Best	**0.00E+00**	**0.00E+00**	**0.00E+00**	**0.00E+00**	5.72E−09	**0.00E+00**
	Mean	**0.00E+00**	**0.00E+00**	3.79E−15	**0.00E+00**	7.66E−09	**0.00E+00**
	Std	**0.00E+00**	**0.00E+00**	1.42E−14	**0.00E+00**	5.08E−09	**0.00E+00**
F7	Best	**8.88E−16**	**8.88E−16**	3.64E−14	**8.88E−16**	1.20E−05	**8.88E−16**
	Mean	**8.88E−16**	**8.88E−16**	3.18E−14	**8.88E−16**	2.84E−05	**8.88E−16**
	Std	**0.00E+00**	**0.00E+00**	3.20E−15	**0.00E+00**	1.05E−05	**0.00E+00**
F8	Best	**0.00E+00**	**0.00E+00**	0.00E+00	**0.00E+00**	3.06E−08	**0.00E+00**
	Mean	**0.00E+00**	**0.00E+00**	3.30E−04	**0.00E+00**	1.84E−08	**0.00E+00**
	Std	**0.00E+00**	**0.00E+00**	1.77E−03	**0.00E+00**	1.43E−08	**0.00E+00**

4.3   Analysis of the Results of CEC2021 Test Functions

CEC2021 test functions are 10 complex test functions that can be expanded in 10 and 20 dimensions. There are three different transformation operations can

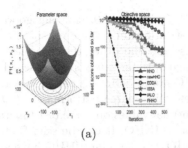 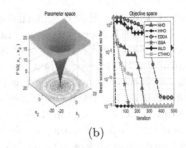

(a)   (b)

Fig. 2. Convergence curves for function optimization under different algorithms ((a) is F1 Function, (b) is F7 Function)

be applied, including bias, shift, and rotation. For each operation, there are two options: add and not add. There are eight different combinations. In this paper, the basic CEC2021 without adding operation and the CEC2021 with rotation operation are selected to test the 20 dimensional function. The CEC2021 test function is shown in Table 3, which provides information about the test function, including function name, dimension and optimal value. In this paper, the difference between the optimization results of each algorithm and the theoretical optimal value is calculated, and the difference 0 is taken as the theoretical optimal value of each function.

Table 3. CEC2021 test function

NO.	Function	dimension	Optimal value
CEC01	Shifted and Rotated Bent Cigar Function	10, 20	100
CEC02	Shifted and Rotated Schwefel's Function	10,20	1100
CEC03	Shifted and Rotated Lunacek bi-Rastrigin Function	10, 20	700
CEC04	Expanded Rosenbrock's plus Griewang's Function	10, 20	1900
CEC05	Hybrid Function1 (N = 3)	10, 20	1700
CEC06	Hybrid Function2 (N = 4)	10, 20	1600
CEC07	Hybrid Function3 (N = 5)	10, 20	2100
CEC08	Composition Function1 (N = 3)	10, 20	2200
CEC09	Composition Function2 (N = 4)	10, 20	2400
CEC10	Composition Function3 (N = 5)	10, 20	2500

In order to further verify the optimization performance of IHHO, the same comparison algorithm is selected for verification in CEC2021 test function. Tables 4 and 5 shows the test results, and Fig. 3 shows the comparison of some function convergence curves. It can be seen that the HHO can only converge to the theoretical optimal value in functions CEC02, CEC03, CEC04, and CEC08, and the convergence effect is poor in other test functions. The IHHO can converge to the theoretical optimal value in the eight test functions CEC01-CEC08. Although it does not converge to the theoretical optimal value in the test functions CEC09 and CEC10, it is improved by hundreds of orders of magnitude compared with the HHO, and the standard deviation is 0, showing good stability. The CTHHO converges to the theoretical optimal value on the CEC02, CEC03, CEC04, and CEC08 test functions, and the other test functions are also improved by dozens. When the theoretical optimal value is not reached, the optimization effect is not as good as the IHHO proposed in this paper, which further proves that the IHHO has more comprehensive search performance. EDDA, ISSA, and IALO can only find theoretical optimal values on some test functions, and their search performance is also inferior to IHHO. It can be seen from the convergence curve in Fig. 3 that although HHO, CTHHO and IHHO have found the same solution on CEC02, IHHO can find the theoretical optimal value within 50 iterations, with good convergence speed. On CEC09 function, HHO curve and CTHHO curve are relatively gentle in the late stage and fall into local optimum, but IHHO curve continues to explore, proving that it has certain ability to jump out of local optimum and strong search ability.

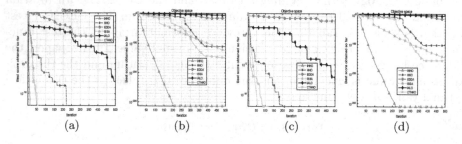

Fig. 3. Convergence curves of function optimization under different algorithms ((a) is CEC02 without operation, (b) is CEC09 without operation. (c) is CEC02 with rotation, (d) is CEC09 with rotation)

5   Optimization of Engineering Problems

In order to verify the performance of IHHO algorithms, compared with EDDA, ISSA, CTHHO and IALO in optimizing the tension spring problem. The purpose of the tension spring design problem is to minimize the weight of the tension spring under the four inequality constraints of minimum deflection, shear stress, oscillation frequency and outside diameter limit. The problem consists of three continuous decision variables, namely the diameter of the spring coil, the diameter of the spring coil and the number of winding coils. Its mathematical model is shown below. Objective function: $\min f(x) = (x_3 + 2)x_2 x_1^2$. Constraint condition: $g_1(x) = 1 - \frac{x_2^3 x_3}{71785 x_1^4} \le 0$ $g_2(x) = \frac{4x_2^2 - x_1 x_2}{12566(x_2 x_1^3 - x_1^4)} + \frac{1}{5108 x_1^2} - 1 \le 0$ $g_3(x) = 1 - \frac{140.45 x_1}{x_2^2 x_3} \le 0$ $g_4(x) = \frac{x_1 + x_2}{1.5} - 1 \le 0$ where, $0.05 \le x_1 \le 2$, $0.25 \le x_2 \le 1.3$, $2 \le x_3 \le 15$.

In order to make the experimental results more real and fair, all experiments were independently run for 30 times. Table 6 shows the comparison between IHHO algorithm and CTHHO, EDDA, ISSA and IALO algorithm in the design

Table 4. Comparative data of function optimization under different algorithms. (CEC2021, Basic)

Function		IHHO	HHO	EDDA	ISSA	IALO	CTHHO
CEC01	Best	**0.00E+00**	2.42E−91	4.93E−52	2.67E−117	7.42E−03	2.24E−161
(Basic)	Mean	**0.00E+00**	1.01E−86	1.25E−47	4.27E−115	3.69E−03	2.58E−114
	Std	**0.00E+00**	5.42E−86	6.72E−47	1.38E−114	3.26E−03	1.48E−112
CEC02	Best	**0.00E+00**	0.00E+00	1.62E−01	0.00E+00	1.36E−07	0.00E+00
(Basic)	Mean	**0.00E+00**	0.00E+00	6.25E−01	0.00E+00	9.35E−08	0.00E+00
	Std	**0.00E+00**	0.00E+00	6.22E−01	0.00E+00	6.65E−08	0.00E+00
CEC03	Best	**0.00E+00**	0.00E+00	0.00E+00	0.00E+00	5.11E−09	0.00E+00
(Basic)	Mean	**0.00E+00**	0.00E+00	6.08E+00	0.00E+00	3.44E−08	0.00E+00
	Std	**0.00E+00**	0.00E+00	8.61E+00	0.00E+00	2.85E−08	0.00E+00
CEC04	Best	**0.00E+00**	0.00E+00	9.58E−01	0.00E+00	0.00E+00	0.00E+00
(Basic)	Mean	**0.00E+00**	0.00E+00	7.76E−01	0.00E+00	0.00E+00	0.00E+00
	Std	**0.00E+00**	0.00E+00	3.17E−01	0.00E+00	0.00E+00	0.00E+00
CEC05	Best	**0.00E+00**	2.32E−94	1.04E−01	2.31E−109	5.34E−05	5.92E−101
(Basic)	Mean	**0.00E+00**	8.33E−87	7.27E−02	4.53E−89	2.12E−05	1.38E−93
	Std	**0.00E+00**	3.69E−86	7.13E−02	2.26E−88	1.20E−05	1.57E−92
CEC06	Best	**0.00E+00**	7.31E−05	3.23E−01	0.00E+00	3.85E−04	4.55E−13
(Basic)	Mean	**0.00E+00**	6.90E−05	4.93E−01	1.78E−04	4.35E−04	3.19E−04
	Std	**0.00E+00**	2.41E−04	1.71E−01	3.70E−04	2.22E−04	7.42E−04
CEC07	Best	**0.00E+00**	4.72E−07	4.16E−01	2.95E−116	3.20E−04	2.42E−05
(Basic)	Mean	**0.00E+00**	1.34E−06	3.65E−01	1.23E−05	7.54E−04	5.19E−06
	Std	**0.00E+00**	4.31E−06	1.56E−01	3.13E−05	2.98E−04	4.58E−05
CEC08	Best	**0.00E+00**	0.00E+00	0.00E+00	0.00E+00	3.29E−07	0.00E+00
(Basic)	Mean	**0.00E+00**	0.00E+00	0.00E+00	0.00E+00	1.32E−07	0.00E+00
	Std	**0.00E+00**	0.00E+00	0.00E+00	0.00E+00	7.92E−08	0.00E+00
CEC09	Best	**4.03E−314**	5.62E−116	3.55E−14	3.67E−123	8.73E−05	2.64E−160
(Basic)	Mean	**3.28E−313**	1.16E−98	3.61E−14	1.34E−122	1.91E−04	9.40E−112
	Std	**0.00E+00**	6.25E−98	3.93E−15	3.45E−122	7.11E−05	9.67E−109
CEC10	Best	**1.12E−315**	4.89E−10	5.58E+01	2.66E−55	5.14E−03	7.37E−11
(Basic)	Mean	**1.16E−315**	2.01E−04	5.34E+01	1.28E−41	5.14E−03	3.59E−06
	Std	**0.00E+00**	5.19E−04	8.37E+00	4.39E−41	1.18E−03	4.29E−06

of tension spring problem. The IHHO algorithm gets the minimum value in the mean value, the optimal value and the standard deviation, which indicates that the optimization ability of IHHO algorithm is better than other algorithms, and can solve the tension spring design problem well.

Table 5. Comparative data of function optimization under different algorithms. (CEC2021, Rotation)

Function		IHHO	HHO	EDDA	ISSA	IALO	CTHHO
CEC01	Best	**0.00E+00**	7.32E−100	4.02E+03	4.24E−117	1.42E−02	4.70E−142
(Rotation)	Mean	**0.00E+00**	2.97E−92	5.69E+03	6.58E−11	4.89E−03	4.83E−112
	Std	**0.00E+00**	7.47E−92	4.52E+03	2.30E−114	4.06E−03	3.80E−103
CEC02	Best	**0.00E+00**	**0.00E+00**	2.09E+02	**0.00E+00**	3.68E−08	**0.00E+00**
(Rotation)	Mean	**0.00E+00**	**0.00E+00**	3.32E+02	**0.00E+00**	8.05E−08	**0.00E+00**
	Std	**0.00E+00**	**0.00E+00**	1.11E+02	**0.00E+00**	5.10E−08	**0.00E+00**
CEC03	Best	**0.00E+00**	**0.00E+00**	1.36E+01	**0.00E+00**	3.65E−08	**0.00E+00**
(Rotation)	Mean	**0.00E+00**	**0.00E+00**	2.71E+01	**0.00E+00**	5.51E−08	**0.00E+00**
	Std	**0.00E+00**	**0.00E+00**	1.08E+01	**0.00E+00**	0.00E+00	**0.00E+00**
CEC04	Best	**0.00E+00**	**0.00E+00**	2.25E+00	**0.00E+00**	0.00E+00	**0.00E+00**
(Rotation)	Mean	**0.00E+00**	**0.00E+00**	2.00E+00	**0.00E+00**	0.00E+00	**0.00E+00**
	Std	**0.00E+00**	**0.00E+00**	4.23E−01	**0.00E+00**	0.00E+00	**0.00E+00**
CEC05	Best	**0.00E+00**	7.56E−82	2.52E+06	1.16E−114	2.92E−05	1.43E−103
(Rotation)	Mean	**0.00E+00**	1.17E−65	1.32E+06	2.13E−111	1.44E−05	2.41E−85
	Std	**0.00E+00**	6.29E−65	9.54E+05	8.42E−111	1.05E−05	1.86E−84
CEC06	Best	**0.00E+00**	1.02E−07	2.68E+00	**0.00E+00**	4.11E−04	**0.00E+00**
(Rotation)	Mean	**0.00E+00**	2.65E−05	5.10E+00	9.85E−04	5.60E−04	1.08E−04
	Std	**0.00E+00**	7.94E−05	2.76E+00	2.35E−03	3.49E−04	2.97E−04
CEC07	Best	**0.00E+00**	5.99E−06	1.97E+05	3.22E−116	7.06E−04	4.67E−92
(Rotation)	Mean	**0.00E+00**	8.32E−07	5.53E+05	2.87E−112	7.52E−04	3.50E−63
	Std	**0.00E+00**	2.80E−06	4.55E+05	4.32E−111	2.72E−04	2.34E−61
CEC08	Best	**0.00E+00**	**0.00E+00**	2.23E+03	**0.00E+00**	2.22E−07	**0.00E+00**
(Rotation)	Mean	**0.00E+00**	**0.00E+00**	2.11E+03	**0.00E+00**	1.56E−07	**0.00E+00**
	Std	**0.00E+00**	**0.00E+00**	2.39E+02	**0.00E+00**	8.44E−08	**0.00E+00**
CEC09	Best	**4.88E−312**	1.31E−120	1.78E−14	2.06E−126	8.59E−05	1.06E−146
(Rotation)	Mean	**3.49E−311**	3.11E−93	7.12E−08	1.12E−122	1.79E−04	1.97E−128
	Std	**0.00E+00**	1.67E−92	3.83E−07	2.39E−122	6.80E−05	1.67E−119
CEC10	Best	**1.12E−315**	1.46E−09	6.44E+01	5.93E−04	2.41E−03	1.52E−94
(Rotation)	Mean	**1.66E−315**	2.06E−05	6.96E+01	1.16E−03	3.36E−03	1.68E−76
	Std	**0.00E+00**	6.49E−05	5.22E+00	1.50E−03	9.03E−04	5.13E−77

Table 6. Convergence optimization data of tension/compression springs problem

Algorithms	optimal	mean	std
IHHO	0.013482	0.013631	0.0028
CTHHO	0.014335	0.014667	0.0039
EDDA	0.015517	0.015996	0.0043
ISSA	0.015439	0.015879	0.0046
IALO	0.016003	0.016737	0.0058

6   Conclusion

Aiming at the shortage of HHO, this paper proposes an IHHO algorithm. A spiral search and neighborhood perturbation collaborative updating strategy is to enhance the global search ability of the algorithm. Tent chaotic mapping is used to initialize the population. In the exploitation stage, the adaptive inertial weight is introduced to update the position to improve the exploitation ability of the algorithm. Dimensional cross variation improves the ability of the algorithm to jump out of the local optimum. The IHHO proposed in this paper has more stable global search ability and stronger convergence speed than EDDA, ISSA, IALO, and CTHHO on eight classical test functions and CEC2021 test function. In engineering application, the IHHO algorithm is applied to the design of tension spring design problem, which also shows better optimization performance.

References

1. Holland, J.H.: Genetic algorithms. Sci. Am. **267**(1), 66–73 (1992)
2. Mirjalili, S., Gandomi, A.H., Mirjalili, S.Z., et al.: Salp swarm algorithm: abio-inspired optimizer for engineering design problems. Adv. Eng. Softw. **114**(1), 163–191 (2017)
3. Mirjalili, S.: The ant lion optimizer. Adv. Eng. Softw. **83**, 80–98 (2015)
4. Mirjalili, S.: Dragonfly algorithm: a new meta-heuristic optimization technique for solving single-objective, discrete, and multi-objective problems. Neural Comput. Appl. **27**(4), 1053–1073 (2015). https://doi.org/10.1007/s00521-015-1920-1
5. Heidari, A.A., Mirjalili, S., Faris, H., et al.: Harris hawks optimization: algorithm and applications. Future Gener. Comput. Syst. **97**(8), 849–872 (2019)
6. Jia, H., Peng, X., Kang, L., Li, Y., Jiang, Z., Sun, K.: Pulse coupled neural network based on Harris hawks optimization algorithm for image segmentation. Multimedia Tools Appl. **79**(37), 28369–28392 (2020). https://doi.org/10.1007/s11042-020-09228-3
7. Fan, C., Zhou, Y., Tang, Z.: Neighborhood centroid opposite-based learning Harris hawks optimization for training neural networks. Evol. Intell. **14**(4), 1847–1867 (2021). https://doi.org/10.1007/s12065-020-00465-x
8. Saravanan, G., Ibrahim, A.M., Kumar, D.S., et al.: IoT based speed control of BLDC motor with Harris hawks optimization controller. Int. J. Grid Distrib. Comput. **13**(1), 1902–1915 (2020)
9. Yuxin, G., Sheng, L., Wenxin, G., et al.: Improved Harris hawks optimization algorithm with multiple strategies. Microelectron. Comput. **38**(7), 18–24 (2021)
10. Fan, Q., Chen, Z., Xia, Z.: A novel quasi-reflected Harris hawks optimization algorithm for global optimization problems. Soft Comput. **24**(19), 14825–14843 (2020). https://doi.org/10.1007/s00500-020-04834-7
11. Hussain, K., Zhu, W., Salleh, M.N.M.: Long-term memory Harris' hawk optimization for high dimensional and optimal power flow problems. IEEE Access **7**, 147596–147616 (2019)
12. Qu, C., He, W., Peng, X., et al.: Harris hawks optimization with information exchange. Appl. Math. Model. **84**, 52–75 (2020)

13. Wang, M., Wang, J.-S., Li, X.-D., Zhang, M., Hao, W.-K.: Harris hawk optimization algorithm based on Cauchy distribution inverse cumulative function and tangent flight operator. Appl. Intell. **52**, 10999–11026 (2021). https://doi.org/10.1007/s10489-021-03080-0
14. Mirjalili, S., Lewis, A.: The whale optimization algorithm. Adv. Eng. Softw. **95**, 51–67 (2016)
15. Liu, L., Keqiang, B., Song, Z., et al.: Whale optimization algorithm based on global search strategy. J. Chin. Comput. Syst. **41**(09), 1820–1825 (2020)
16. Mohamed, A.W., Hadi, A.A., Agrawal, P., et al.: Gaining-sharing knowledge based algorithm with adaptive parameters hybrid with IMODE algorithm for solving CEC 2021 benchmark problems. In: 2021 IEEE Congress on Evolutionary Computation (CEC), pp. 841–848. IEEE (2021)
17. Liang, S., Hao, Q., Jun, L., et al.: Chaotic optimization algorithm based on Tent map. Control Decis. **2**, 179–182 (2005)
18. Marini, F., Walczak, B.: Particle swarm optimization (PSO). A tutorial. Chemometr. Intell. Lab. Syst. **149**, 153–165 (2015)
19. Qing, H., Minming, H., Xu, W.: Elite opposition learning-based dimension by dimension improved dragonfly algorithm. J. Nangjing Normal Univ. (Nat. Sci. Ed.) **42**(3), 65–72 (2019)
20. Zhiqiang, Z., Xiaofeng, L., Liansheng, S., et al.: A Salem swarm algorithm integrating random inertia weights and differential mutation operation. Comput. Sci. **47**(08), 297–301 (2020)
21. Hegazy, A.E., Makhlouf, M.A., El-Tawel, G.S.: Improved Salp swarm algorithm for feature selection. J. King Saud Univ.-Comput. Inf. Sci. **32**(3), 335–344 (2020)
22. Huang, P.-Q., Zhang Q., Wang Y.: Bilevel optimization via collaborations among lower-level optimization tasks. IEEE Trans. Evol. Comput. (2023)

S-Plane Controller Parameter Tuning Based on IAFSA for UUV

Zheng Wang, Yang Yang, Shuai Zhou, and Houpu Li[✉]

School of Electrical Engineering, Naval University of Engineering, Wuhan 430033, China
lihoupu1985@126.com

Abstract. Path following control based on S-plane controller is one of the key technologies of UUV motion control. Aiming at the problem of high UUV path following error caused by manually setting S-plane control parameters, the artificial fish swarm algorithm is improved by adopting methods such as predatory behavior, adaptive step size, and field of view with attenuation factor to improve the optimization performance of the artificial fish swarm. The improved fish swarm algorithm (IAFSA) is used to tune the control parameters of the S-plane forward speed controller and the yaw angular speed controller. Through simulation and experimental analysis, the IAFSA has a faster convergence speed, and the ability to jump out of the local optimal value is significantly enhanced. The index of the S-plane controller using the tuned parameters is reduced compared with that before tuning.

Keywords: UUV · Path following Control · S-Plane Controller · Fish Swarm Algorithm · Parameter Tuning

1 Introduction

Unmanned underwater vehicle (UUV) is a kind of underwater mobile carrier with small volume, strong endurance, good maneuverability, and can carry various sensors. The characteristics of unmanned ride and low cost enable it to serve different fields such as marine science and technology, marine military and marine economy.

Due to the strong nonlinear and coupling dynamic characteristics of UUV and the uncertainty of underwater environment, the research on its path following control technology is difficult to refer to the relatively mature unmanned systems such as unmanned vehicles and unmanned aerial vehicles, which has become the basic bottleneck of other important functions.

Literature [1] designed an S-plane controller for underwater robot based on the concept of fuzzy controller and PD controller, which has the characteristics of less input parameters, high fitting degree for underwater system and strong practicability, and has proved its effectiveness in many subsequent field tests. Li Y M, et al. borrowed from the characteristics of sliding mode variable controller, designed an S-plane speed controller with adaptive adjustment of controller parameters as the motion state changes, and verified the control effect of this method in marine experiments [2]. However, the

L. Pan et al. (Eds.): BIC-TA 2022, CCIS 1801, pp. 69–80, 2023.
https://doi.org/10.1007/978-981-99-1549-1_6

control parameters in the traditional S-plane controller are often fixed, do not have adaptive ability, and need to be set and adjusted manually. This kind of approximation method makes the UUV still have large following error in the underwater movement process, which is not in line with the application practice. Therefore, applying intelligent optimization algorithm to the tuning of control parameters has become an effective solution [3–7].

In recent years, some scholars have studied the application of intelligent algorithms in the parameter tuning of S-plane controller, including particle swarm optimization [8–13], neural network [14, 15], simulated annealing algorithm [16], immune genetic algorithm [17], etc. Among many optimization algorithms, the artificial fish swarm algorithm was proposed by Li Xiaolei in 2002 [18]. It has the ability of global optimization, has the characteristics of fast convergence and good fitting to uncertain models, and is suitable for such a complex nonlinear mathematical model as UUV path following control. However, the artificial fish swarm algorithm also has some problems, such as blindness in the late iteration period and slow convergence speed. In the later stage, it is difficult to jump out once falling into the local optimal value.

Aiming at the problems of the artificial fish swarm algorithm, an improved artificial fish swarm algorithm (IAFSA) is proposed. The improvement measures include adding the predatory behavior into the four basic behaviors of the artificial fish, using the adaptive step size when the fish swarm moves, and using the field of view with attenuation factor when searching for food. These measures can improve the optimization performance of the artificial fish swarm. In order to avoid the problem of high path following error caused by manually setting the control parameters of the UUV, IAFSA is used to tune the control parameters of the forward speed controller and the yaw angular speed controller for the UUV path following task. Through simulation analysis and underwater experiment, the feasibility and effectiveness of the IAFSA in the parameter tuning of the UUV path following controller are verified.

2   Path Following Control Based on S-plane Controller

The biggest difference between UUV motion control and surface ship's is that the motion space of UUV is three-dimensional. However, considering most of the task requirements in the actual research, UUV often needs to keep constant depth navigation. This paper studies the UUV path following control techniques in the horizontal plane.

In the horizontal plane, the motion controller of UUV is mainly divided into the forward speed controller for thruster's thrust, and the yaw angular speed controller for controlling the vertical rudder angle. Therefore, the path following control problem is decomposed into the two controllers mentioned above to be designed and solved separately.

2.1   S-Plane Control Theory

When designing a fuzzy controller for UUV path following, the method of loose on both sides and dense in the middle is often used, that is, when the deviation e is large, the control is rough, and when e is small, the control is gradually fine. When the resolution of the fuzzy control table approaches infinity, countless small folds can be simplified into smooth surfaces, and the change trend of the sigmoid plane is similar to this design idea.

Therefore, sigmoid plane can be used to replace the rule table of this kind of fuzzy control to form an S-plane control law. In the process of using the traditional S-plane controller, there is often a problem that the steady-state error caused by forward damping cannot be eliminated. Therefore, an integral control is introduced into the exponential term of the sigmoid function to eliminate such steady-state error, as shown in Formula (1):

$$u = \frac{2.0}{\left(1.0 + e^{-k_1 e - k_2 \int edt - k_3 \dot{e}}\right)} - 1.0 \tag{1}$$

where u represents the control output, in UUV path following control, u represents the vertical rudder angle of the yaw controller and the propulsion motor voltage of the forward controller, e and $\dot{e}$ respectively represent the deviation and its derivative, $k_1 - k_3$ represents the control parameter, k_1 represents the coefficient of the proportional term, which affects the rise time and overshoot of the control; k_2 is the integral coefficient to eliminate the steady-state error caused by forward damping; k_3 is the coefficient of the differential term, which affects the overshoot amplitude and stability of the control. In engineering practice, it is often required to be given manually based on experience.

In complex underwater environment, unknown interference factors such as ocean current can be simplified as a fixed interference force applied to the UUV within a certain period of time. On the basis of formula (1), an item is added to offset such steady state error by incremental output. The expression is as follows.

$$u = \frac{2.0}{\left(1.0 + e^{-k_1 e - k_2 \int edt - k_3 \dot{e}}\right)} - 1.0 + \Delta u \tag{2}$$

The thrust controller and heading controller required by the UUV to complete the path following task can be constructed through the above methods, but the selection of control parameters $k_1 - k_3$ is often done manually by experienced personnel, which brings great difficulties in engineering practice. Therefore, intelligent optimization algorithms such as artificial fish swarm algorithm can be used to achieve the self-tuning of control parameters.

2.2 Design of Forward Speed Controller and Yaw Angular Speed Controller

Since the forward error caused by damping exists in the forward speed control process, it is necessary to add an integral term to the exponential term according to the idea of formula (2) to eliminate the error. The S-plane front speed controller is designed as follows:

$$u_1 = \frac{2.0}{\left(1.0 + e^{-k_1 e_1 - k_2 \int edt - k_3 \dot{e}_1}\right)} - 1.0 + \Delta u_1 \tag{3}$$

where u_1 is the forward speed output of the thrust controller, $k_1 - k_3$ are the control parameters, and e_1 represents the difference between the actual forward speed and the expected value, and $\dot{e}_1$ represents its derivative. Δu_1 represents the incremental output of the forward speed controller.

$$e_1 = v_b - v \tag{4}$$

where v_b represents the expected forward velocity and v represents the current UUV velocity.

Since there is no steady-state error caused by forward damping in heading control, the integral coefficient is set to zero, and the controller is designed as follows:

$$u_2 = \frac{2.0}{\left(1.0 + e^{-k_4 e_2 - k_5 \dot{e}_2}\right)} - 1.0 + \Delta u_2 \tag{5}$$

where u_2 is the output of heading angular velocity, k_4 and k_5 are the control parameters, and e_2 represents the difference between the actual value and the expected value of heading angular velocity, and $\dot{e}_2$ represents its derivative. Δu_2 represents the incremental output of the yaw rate controller.

$$e_2 = \dot{\psi}_b - \dot{\psi} \tag{6}$$

where $\dot{\psi}_b$ represents the expected heading angular velocity of UUV in geodetic coordinate system, and $\dot{\psi}$ represents the actual heading angular velocity of UUV.

From the design process of the two controllers, it can be seen that there are control parameters $k_1 - k_5$ that need to be given. However, the selection of control parameters is often done manually by experienced personnel, which brings great difficulties in engineering practice. Therefore, intelligent optimization algorithms such as artificial fish swarm algorithm can be used to achieve the tuning of control parameters.

3   Improved Artificial Fish Swarm Algorithm

3.1   Theory of Artificial Fish Swarm Algorithm

In a water area, the fish in a natural fish school can often find a place with higher food concentration by themselves or following other fish, so the place with the largest number of individuals in the fish school is generally the place with the highest food concentration in the water area. According to this characteristic, the classical artificial fish swarm algorithm simulates the various survival behaviors of natural fish by constructing artificial fish to achieve optimization.

In the classical artificial fish swarm algorithm, the artificial fish's perception of the external environment is realized by the simulated biological vision. The artificial fish model uses the method shown in Fig. 1 to realize the virtual vision of the artificial fish.

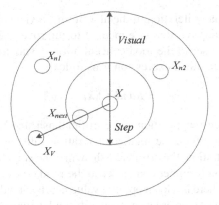

Fig. 1. Artificial fish visual concept map

In the figure, X represents the current position of the fish. If the food concentration at X_V at the next view point is higher than that at X at the current position, the artificial fish will move one step in the direction of X_V to the X_{next} position. The relationship between X_V, X_{next} and X can be expressed by Eqs. (7) and (8) respectively:

$$X_V = X + Visual * r \tag{7}$$

$$X_{next} = X + \frac{X_V - X}{X_V - X} * Step * r \tag{8}$$

where *Step* represents the moving step, *Visual* represents the visual range, and r is the random number of $[-1, 1]$ interval.

Similar to fish in nature, artificial fish have typical behaviors including foraging behavior, clustering behavior, tail chasing behavior, random behavior. The explanation of the four basic behaviors can refer to the references, and will not be repeated here.

Based on the above four basic behaviors, the artificial fish will gradually find the place with the highest food concentration within a certain range in the iteration, record the position and the highest food concentration, and then complete the function optimization.

However, the artificial fish swarm algorithm also has some problems, such as blindness in the late iteration period and slow convergence speed; In the later stage, it is difficult to jump out once falling into the local optimal value. Therefore, it is necessary to use appropriate measures to improve these problems.

3.2 Algorithm Improvement Measures

In view of the main problems of the classical fish swarm algorithm, the following measures can be used to improve:

① When the artificial fish swims to a place with high food concentration, it selects a random position in this direction. If the current position is close to the place with the highest food concentration, it will oscillate near the optimal value, which is easy to cause the algorithm to take a long time, which is also the main reason for the slow convergence rate in the later iteration of the algorithm. In view of this problem, we can consider adopting the adaptive step size method to improve, that is, when the current position is higher and closer to the food concentration, a smaller step value should be used. The specific mathematical expression is as follows:

$$Step = Rand * \|X_i - X_V\| \qquad (9)$$

where X_i represents the current position of the artificial fish, and X_V represents a position within its field of vision. When the food concentration value at X_V is greater than the current position concentration, the artificial fish swims the step distance in its direction. Since the current position is higher and closer to the food concentration, a smaller $Step$ value is used, and when there is a local optimal value nearby, it will be trapped in its local optimal value due to too fast convergence and it is difficult to jump out. Therefore, $Rand()$ function is added to provide randomness and facilitate jumping out. If the artificial fish does not move to a higher food concentration after performing the tail chasing behavior, the step will still use a fixed value in order to increase the randomness and jump out of the local optimal value.

② The foraging behavior of fish schools plays an important role in solving discrete optimization problems. When the fish are looking for food, their vision remains unchanged. When the artificial fish is gradually approaching the optimal solution, its position state X_i is only 1–2 dimensions different from the optimal solution. At this time, using the original fixed Visual value will lead to the artificial fish blindly searching for optimization, which greatly increases the complexity of the later iteration of the algorithm. At the same time, if a fixed Visual value is used, too large field of view will lead to too slow convergence speed, and too small field of view will increase the possibility of falling into the local optimal value. To avoid this shortcoming, the following strategies can be adopted to improve the foraging behavior:

When the algorithm starts iteration, use a larger $Visual$ value to expand the optimization range. As the iteration process progresses, gradually reduce the $Visual$ value appropriately to speed up the convergence. The mathematical expression is:

$$Visual_k = \alpha Visual_{k-1} \qquad (10)$$

where $\alpha \in (0, 1)$ represents the attenuation rate of the field of view and k represents the number of iterations.

③ In real nature, when the concentration of surrounding food reaches a certain level, fish will become aggressive due to hunger. If there are other fish occupying more food in the field of vision, the hungry fish will try to seize the position of the fish with more food, which is called predatory behavior. Therefore, we can consider adding predatory behavior to the basic behavior of artificial fish. The specific mathematical expression is as follows:

$$\frac{Y_i}{Y_V} < \beta n_j \delta \qquad (11)$$

where Y_i represents the food concentration at the current location X_i. When the ratio of Y_i to the food concentration Y_V at a location X_V in the field of vision is less than a certain value, the artificial fish will perform the predatory behavior. This threshold is used for product of execution possibility $\beta = rand(0,1)$ and congestion factor.

The purpose of adding predatory behavior is to strengthen the trend of artificial fish flocks gathering towards the better value of the objective function, while introducing randomness, which is more conducive to jumping out of the local optimal value.

After improving the main problems, the IAFSA algorithm process can be summarized as follows:

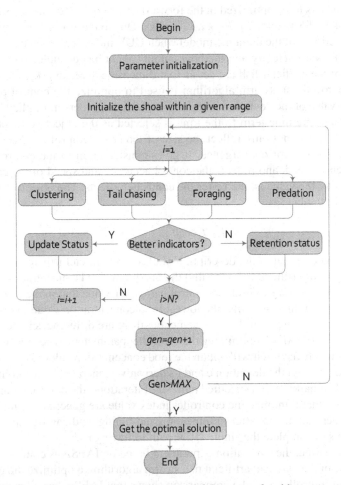

Fig. 2. Flow chart of improved fish swarm algorithm

In Fig. 2, the artificial fish will use the current coordinates to calculate the food concentration value to judge whether it is better or not every time it selects the current behavior through judgment. The specific calculation process is to complete a path following task to obtain the index value. This link is the main link of IAFSA.

4 Simulations

4.1 Simulation Environment and Basic Parameters

Based on the above theoretical analysis, in order to avoid the problem of high horizontal plane path following error caused by manually setting control parameters, the code simulation of single UUV path following control technology based on IAFSA-S-plane controller is carried out in this chapter.

The parameters to be optimized in the forward speed controller and the yaw angular speed controller of S include k_1, k_2, k_3, k_4 and k_5. Due to the strong coupling and non-linear characteristics of the dynamic model when UUV moves underwater, the coupling between the forward velocity and the bow output cannot be decoupled, so the position coordinates of the artificial fish are set as five-dimensional vectors $(k_1, k_2, k_3, k_4, k_5)$ when the improved fish swarm algorithm is used to optimize the control parameters. The absolute value of the deviation in the two S-plane controllers multiplied by the sum of the integral of the time term to the time is selected as the objective function of the algorithm, which can not only reflect the size of the error (control accuracy), but also reflect the speed of error convergence. It gives consideration to the control accuracy and convergence speed, and reflects the control accuracy and speed of the control system. The smaller the value is, the better the controller effects. The objective function is expressed in Formula (12).

$$Y = \int_0^\infty t|e_1|dt + \int_0^\infty t|e_2|dt \tag{12}$$

It is worth noting that in the description of classical artificial fish swarm algorithm, the ultimate goal of artificial fish is to find the maximum food concentration, but in this simulation, it is necessary to find the control parameter value that makes the controller index the minimum, that is, to find the lowest food concentration value of artificial fish $(k_1, k_2, k_3, k_4, k_5)$. The principle is the same, but there are differences in description.

In the process of IAFSA optimizing the controller parameters, when the artificial fish selects the current behavior, it will obtain the food concentration value (i.e. the controller index Y value) through the deviation e and its derivative generated by the complete path following task. Finally, after a specified number of iterations, the control parameters K_1, K_2, K_3, K_4, K_5 that minimize the controller index value are generated, which are used as the controller parameters after optimization and tuning, and can be replaced by the two controllers to complete the single UUV path following task.

At the same time, the simulation experiment based on IAFSA is compared with the same example of the classical artificial fish swarm algorithm to optimize the parameters of the S-plane controller, and the comparison shows that IAFSA has advantages in the application of optimizing the parameters of the UUV path following control.

This simulation uses MATLAB R2017a software platform. The PC system is Windows 7 with 12 GB running memory and Intel i7-4710MQ CPU. The parameters used in the simulation are shown in Table 1.

Table 1. IAFSA parameter selection

Parameter name	Alphabetic representation	Numerical value
Number of artificial fish	N	300
Initial field of vision	*Visual*	1
Number of attempts	*Try_number*	100
Crowding degree	δ	0.618
Maximum Iterations	*MAXGEN*	100
Attenuation factor	α	0.95

4.2 Simulation Results and Analysis

In the same simulation environment, IAFSA and classical artificial fish swarm algorithm are respectively used to tune the control parameters. The change of controller index Y value during iteration is shown in Fig. 3:

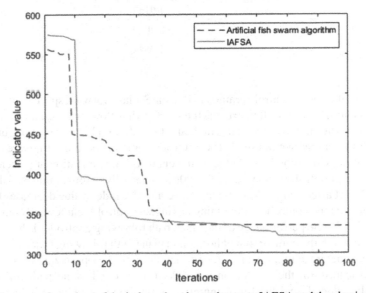

Fig. 3. The comparison chart of the index value change between IAFSA and the classic fish swarm algorithm

It can be seen from Fig. 3 that the convergence speed of the classical fish school algorithm is significantly slower than that of the improved fish school algorithm in the range of 10–40 iterations. After 40 iterations, the index value stagnates around 334, making it difficult to jump out of the local optimal value, and the optimal value of the objective function cannot be reached in the later stage. After the above three measures, the convergence speed of IAFSA in the middle of iteration has been improved, and after 60 iterations, the local optimal value has been jumped out, and the index value has been optimized to 321.4, which verifies the superiority of IAFSA.

After a complete iteration of *MAXGEN* times, the IAFSA output result is the control parameter value with the smallest controller index Y (four decimal places reserved) (Table 2):

Table 2. Control parameter values after optimization and tuning

Control parameters	Final value
K_1	1.4068
K_2	0.2463
K_3	1.9167
K_4	2.4029
K_5	1.0544

Substitute the above control parameters into the S-plane forward speed controller and the yaw angular speed controller, execute the UUV path following task, and compare the obtained track with the planned path graph and the path graph before setting under the same conditions, as shown in Fig. 4. The control parameters before setting are obtained by manual trial, which conforms to the actual engineering application at present.

In Fig. 4, the controller index value $Y = 8062.7$ manually before tuning, and the index value $Y = 321.4$ after tuning. After tuning, the controller index value decreases by 96%, which means that the controller after tuning has higher control accuracy and convergence speed. Obviously, the performance of the tuned path following controller is better. At the same time, through the comparison of horizontal plane path following trajectories before and after parameter tuning, it can be found that, through the optimization of improved fish swarm algorithm, the oscillation of UUV in horizontal plane path following is significantly reduced, and the following error is significantly reduced, which verifies the correctness and effectiveness of IAFSA in the application of UUV S-plane path following control parameter tuning.

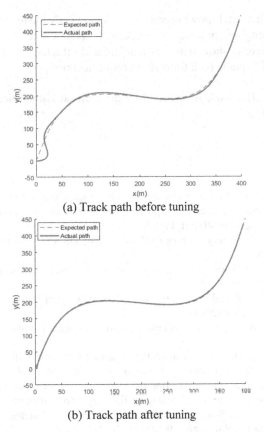

(a) Track path before tuning

(b) Track path after tuning

Fig. 4. Comparison of path following track before and after control parameters tuning

5 Conclusions

Aiming at the problem of path following control of UUV under the condition of fixed depth, this paper analyzes the principle of UUV path following control, and studies the path following control technology based on S-plane controller, including the design of S-plane forward speed controller and S-plane yaw angular speed controller. The classical artificial fish swarm algorithm is analyzed, and combined with the UUV dynamic model, the classical algorithm is optimized and improved from three measures: adaptive step size, field of view with attenuation factor, and predatory behavior, and IAFSA is proposed. At the same time, the simulation experiment based on IAFSA-S-plane path following control is carried out. Using IAFSA, appropriate controller indicators are designed, and the control parameters that need to be manually adjusted in the two controllers are optimized. In the same environment, the algorithm is compared with the classical artificial fish swarm algorithm. After analysis, IAFSA's ability to jump out of the local optimal value is significantly enhanced, and it has a faster iteration speed in the later stage. The optimized control parameters enable the forward speed controller and the yaw angular speed controller to jointly complete the path following task of the virtual navigator UUV, with small following error. In order to verify the feasibility of IAFSA-S-plane controller in engineering practice, underwater physical tests are also carried out. The experimental results show that it is feasible and effective to apply IAFSA to the parameter tuning of S-plane path following control technology.

Acknowledgements. This work is supported by the National Natural Science Foundation of China (No. 41974005).

References

1. Liu, X., Xu, Y.: S control of automatic underwater vehicles. Ocean Eng. **19**(3), 81–84 (2001)
2. Li, Y., Pang, Y., Wan, L.: Adaptive s-plane control for autonomous underwater vehicle. J. Shang Hai Jiao Tong Univ. **46**(02), 195–200+206 (2012)
3. Zhou, Z.: AUV 3D trajectory tracking method based on DRNN-S control. Ship Sci. Technol. **43**(21), 96–99 (2021)
4. Yang, Q.: Research on Trajectory Tracking Control of Underactuated Ships Based on S-plane. Dalian Maritime University (2021)
5. Northwestern Polytechnical University: A quadrotor UAV cooperative control method based on expert S-plane control (2021)
6. Shandong University: AUV path tracking method and system based on S-plane control and TD3 (2021)
7. Pan, W.: Research on Three-Dimensional Path Tracking Control of Fully-Driven Autonomous Underwater Robot Recovery. Jiangsu University of Science and Technology (2021)
8. Lu, C., Pang, Y., Wang, B., et al.: Improved S-surface control and hardware-in-the-loop simulation of underwater robots. J. Shanghai Jiaotong Univ. **44**(7), 957–961, 967 (2010)
9. Guo, B., Xu, Y., Li, Y.: S-surface controller for underwater vehicles using particle swarm optimization. J. Harbin Eng. Univ. **29**(12), 1277–1282 (2008)
10. Li, H., Wang, Y., Lu, Z., et al.: Attitude Coordination Control of Underwater Robot Based on PSO-GA Algorithm and Neural Network (2022)
11. Liu, S., Ren, D., Li, B.: Submarine S-plane control based on SA-PSO algorithm. Control Eng. **18**(05), 710–714 (2011)
12. He, C., Li, L., Tian, Y., Zhang, X., Cheng, R., Jin, Y., Yao, X.: Accelerating large-scale multiobjective optimization via problem reformulation. IEEE Trans. Evol. Computat. **23**(6), 949–961 (2019)
13. Huang, P.Q., Wang, Y., Wang, K., Liu, Z.Z.: A bilevel optimization approach for joint offloading decision and resource allocation in cooperative mobile edge computing. IEEE Trans. Cybern. **50**(10), 4228–4241 (2019)
14. Tang, X., Pang, Y., Wang, J.: Adaptive motion control of S-surface of underwater robot based on single neuron. Comput. Appl. **27**(12), 2899–2901 (2007)
15. Wan, L., Tang, W., Li, Y.: BP neural network S-plane control for autonomous underwater vehicle. Ind. Instrum. Autom. Devices **2019**(2), 13–17 (2019)
16. Sun, Y., Li, Y., Zhang, Y., et al.: Application of improved simulated annealing algorithm in motion control parameter optimization of s-plane of underwater robot. Acta Armamentarii **34**(11), 1418–1423 (2013)
17. Li, Y., Pang, Y., Wan, L., et al.: Immune genetic optimization of underwater vehicle S-surface control. J. Harbin Eng. Univ. **27**(7), 324–330 (2006)
18. Li, X.: A Novel Intelligent Optimization Method-Artificial Fish Swarm Algorithm. Zhejiang University (2003)

A Reinforcement-Learning-Driven Bees Algorithm for Large-Scale Earth Observation Satellite Scheduling

Yan-jie Song[1]([envelope])[iD], Jun-wei Ou[1], D. T. Pham[2], Ji-ting Li[3], Jing-bo Huang[1], and Li-ning Xing[4]

[1] College of Systems Engineering, National University of Defense Technology, Changsha, Hunan, China
songyj_2017@163.com
[2] College of Engineering and Physical Sciences, The University of Birmingham, Birmingham, UK
[3] Chinese Academy of Military Science, Beijing, China
[4] School of Electronic Engineering, Xidian University, Xi'an, China

Abstract. The Earth Observation Satellite Scheduling Problem (EOSSP) is difficult to solve due to its scale and constraints. Through analysing the problem, we build a mathematical programming model of the EOSSP. After that, we propose a reinforcement-learning-driven bees algorithm (RLBA) to solve a large-scale EOSSP (LSEOSSP). The RLBA adopts a Q-learning method to select search operations from global search and neighbourhood search. We define a new state action combination in the Q-learning method to cooperate with the population search and obtain high-quality solutions. Through experimental verification, the performance of the proposed algorithm is obviously better than that of several comparison algorithms, and the RLBA can solve LSEOSSP well.

Keywords: reinforcement learning · bees algorithm · earth observation satellite · scheduling · intelligent system

1 Introduction

As an important space platform, satellites have been well applied in many scenarios, such as weather forecasts, high-speed Internet, and car navigation. Among them, an Earth Observation Satellite (EOS) is a kind of satellite that uses various loads, such as visible light, synthetic aperture radar, and infrared rays [14]. It can obtain image data from areas such as the ground, the air, and the sea [11]. Users put forward several observation requirements for these areas to a group of EOSs and hope that the satellites can successfully execute them within the

This work was supported by the Special Projects in Key Fields of Universities in Guangdong (2021ZDZX1019) and the Hunan Provincial Innovation Foundation For Postgraduate (CX20200585).

expected time ranges [8]. The scheduling process of EOSs is to arrange satellite resources for a series of observation tasks and to complete the tasks for satellite resources when their capabilities allow the satellite's maximum efficiency to be achieved. Different from many other classical planning and scheduling problems, such as the Vehicle Routing Problem (VRP) and Flexible Flow Shop Scheduling Problem (FJSP), the satellite must be within the angle range where the load can observe the task area if it wants to perform the observation task [6]. To construct the model and solve it conveniently, we describe this range from the time dimension and call it the visible time window (VTW) [10]. Each task must be completed within a VTW, which means a strict time limit to complete any tasks. When the number of tasks is small or the distribution is not dense, all tasks can be completed through simple scheduling. However, in recent years, the number of tasks has increased rapidly, and the distribution density is high. Due to the limited number of satellites, oversubscription makes it challenging to solve large-scale EOSSPs (LSEOSSPs).

The EOSSP problem, like other types of satellite scheduling problems, is NP-hard. This means that the exact solution algorithm can only be obtained for small-scale problems, and the solution time will increase exponentially as the problem scale increases. This situation is unacceptable for the LSEOSSP. It is more reasonable to use an evolutionary algorithm to solve the LSEOSSP, considering both the solution quality and the solution time. At present, there have been a series of studies on solving the EOSSP by evolutionary algorithms. Niu *et al.* considered the EOSSP problem after an earthquake and used an NSGA-II algorithm to quickly cover a large area after a disaster [7]. Zhu *et al.* adopted the idea of combining a genetic algorithm with a simulated annealing algorithm to find a suitable EOS task execution scheme through a two-stage algorithm search [12]. Wei *et al.* built a dual-objective EOS scheduling model with the optimisation objectives of minimising the number of failed tasks and minimising the resource load balance [15]. They designed a memetic algorithm framework to obtain a high-quality task execution scheme. Jiang *et al.* designed an evolutionary algorithm to solve the EOSSP problem of multiple satellites by using a multi-population idea [5]. The search efficiency of the algorithm is improved by information interaction among populations. Other studies on solving the EOSSP with evolutionary algorithms refer to Wang *et al.* [13]. Through relevant research, the evolutionary algorithm has good scheduling performance in solving the EOSSP, and the use of improved mechanisms is significantly helpful for the algorithm to deal with complex problem scenarios [2].

We use the bees algorithm (BA), which has a simple structure and good search performance, to solve the LSEOSSP [9]. The BA algorithm has the ability of global search and local search, which is helpful for solving complex scheduling problems. However, because each problem scenario is different, it will take considerable energy to find the best algorithm parameter configuration and search method suitable for the specific problem scenario. Therefore, it is meaningful to make the BA algorithm learn in the process of solving and to choose the search method adaptively. According to this algorithm design idea, we propose a reinforcement-learning-driven bees algorithm (RLBA). The main contributions of this paper are as follows:

1. We build a mathematical scheduling model of the EOSSP. In this model, an optimisation function with the maximum total task priority is proposed, and various constraints, such as satellite capability and user requirements, are considered.
2. We propose a reinforcement-learning-driven bees algorithm. The proposed algorithm effectively balances the relationship between exploration and exploitation. Reinforcement learning (RL) is used to decide whether bees should adopt global or local search methods. In RL, we define a new combination of <state, action>, so that the algorithm can decide the search strategy for the next round of population search independently according to the search performance. In addition, we also designed several search operations combined with reinforcement learning.

The structure of the rest of this paper is as follows. The next section introduces the mathematical programming model of the EOSSP. The third section presents the main flow of the reinforcement-learning-driven bees algorithm and the RL method used in the RLBA. The fourth section evaluates the performance of the proposed algorithm in solving the EOSSP through experiments. The final section summarises the paper and suggests further research directions.

2   Model

In a certain scheduling time range, there are a series of observation tasks T and satellite resources S, each satellite s_i has its orbit, and a visible window oct TW can be calculated according to the position relationship between satellites and tasks. To successfully execute the task $task_j$, it needs to start and finish within the time range $[evt_{ij}, lvt_{ij}]$ of the time window tw_k, and the required detection time of the task d_j. The angle θ_{ij}^t between the mission and the satellite cannot exceed the maximum allowable detection angle θ_i^{max} of the satellite. The successful execution of the task should be within the desired time range $[rst_{ij}, ret_{ij}]$ proposed by users. When a satellite completes a task, it takes conversion time η before it can perform the next task. In addition, any one task can be executed at most once.

According to the above problem description, we can construct the EOSSP model in the form of mixed integer linear programming (MILP). Our proposed EOSSP scheduling model is based on the following assumptions:

1. All tasks are in the projection area directly below EOS, regardless of the satellite slew;
2. The storage and power of the satellite are in the most ideal state when the satellite flies in each orbit;
3. The tasks have all been determined before the scheduling starts and will not change during the planning process and task execution;
4. Satellite Telemetry, Track & Command (TT&C) and data download links are smooth, task instructions can be successfully uploaded, and acquired image data can be successfully downloaded;
5. All satellites have been in normal working conditions within the time range.

There are two decision variables in the EOSSP scheduling model. x_{ijk} indicates whether the task j is executed in the time window k by satellite i. If the task is performed, $x_{ijk} = 1$, otherwise, $x_{ijk} = 0$. Another decision variable, st_{ij} represents the start time of the task j by satellite i, and st_{ij} is a non-negative integer.

Task priority p_j reflects the importance of the observation task, and the task with higher priority should be executed first. The optimisation goal of the EOSSP is to find an observation task execution scheme that can maximize the sum of task priorities.

Objective function:

$$\max \sum_{i \in S} \sum_{j \in T} \sum_{k \in TW} p_j \cdot x_{ijk} \tag{1}$$

Subject to:

$$rst_j \cdot x_{ijk} \leq st_{ij}, i \in S, j \in T \tag{2}$$

$$(st_{ij} + d_j) \cdot x_{ijk} \leq ret_j, i \in S, j \in T \tag{3}$$

$$evt_{ijk} \cdot x_{ijk} \leq st_{ij}, i \in S, j \in T, k \in TW \tag{4}$$

$$(st_{ij} + d_j) \cdot x_{ijk} \leq evt_{ijk}, i \in S, j \in T, k \in TW \tag{5}$$

$$\theta_{ij}^t \cdot x_{ijk} \leq \theta_i^{\max}, i \in S, j \in T, t \in [st_{ij}, st_{ij} + d_j] \tag{6}$$

$$(st_{ij} + d_j + \eta) \cdot x_{ijk} \leq st_{ij'} \cdot x_{ij'k}, i \in S, j, j' \in T \wedge st_{ij} < st_{ij'} \tag{7}$$

$$\sum_{j \in T} \sum_{k \in TW} x_{ijk} \leq 1, i \in S \tag{8}$$

Equation 2 and Eq. 3 indicate that the task should start and end within the time range expected by the user. Equation 4 and Eq. 5 indicate that the task should be executed within VTW. Equation 6 indicates that the angle between the satellite and the mission cannot exceed the maximum allowable detection angle. Equation 7 indicates that the same satellite needs to meet the conversion time requirement when performing two tasks continuously. Equation 8 indicates that a task can be executed at most once.

In the next part, we will introduce the reinforcement-learning-driven bees algorithm for solving the EOSSP.

3   Reinforcement Learning Driven Bee Algorithm

To obtain a reasonable implementation scheme, the scheduling algorithm is responsible for selecting the appropriate matching scheme from numerous tasks and VTWs [4]. We decided to use the bees algorithm (BA), a nature-inspired optimisation tool with strong global search and local search capabilities to solve the LSEOSSP problem. The BA simulates the process of bees searching for food sources. It is a population-based algorithm with a simple structure and good search performance. However, the traditional BA has the disadvantage of slow convergence speed, and the combination of global search and local search has a great influence on the scheduling results. Targeting the shortcomings of the traditional BA, we improve the algorithm structure and search operation and use an RL method to guide the algorithm search.

3.1 Overall Flow of Algorithm

This part will introduce the whole process of RLBA. Based on the simplest BA, we improve a series of algorithm structures and search strategies according to the characteristics of LSEOSSP. The RLBA mainly includes the following steps: initialising the population, strengthening the learning method to decide on search mode, global search, local search, etc. After the population is initialised, the algorithm will start the population search process. In the search process, the reinforcement learning method is adopted to dynamically decide the search strategy adopted by the next generation bee colony according to the information obtained from the search. The search strategy includes global search and local search. Global search is the main search method of the RLBA, while local search can further develop the local solution space and improve performance. The pseudocode of the RLBA is shown in Algorithm Table 1.

Algorithm 1: Reinforcement Learning Driven Bee Algorithm(RLBA)

Input: scout bee population size N_p, number of site N_s , learning rate α , discount coefficient γ, random search ratio ρ .

Output: task execution scheme *Solution*

1 Initialize parameters of the algorithm, Q-table;
2 Generate initial population;
3 Generate initial state S_0;
4 Set $t \leftarrow 0$;
5 **while** *termination criterion is not met* **do**
6   Select the action with the largest Q value according to the state;
7   **if** A_t *is the gobal search* **then**
8   According to two search methods, the scout bee population with ratio ρ and $1 - \rho$ is generated respectively;
9   Select N_s sites from the population;
10   **else**
11   Perform local search;
12   Update Q-table;
13      $t \leftarrow t + 1$;

As shown in algorithm Table 1, global search or local search can be used in the algorithm, and the reinforcement learning method is used to decide the specific search to be conducted before each generation of search starts. We use the Q-learning method in the reinforcement learning method to update the Q value by evaluating the performance of the population search and this search strategy and perform a new search according to the updated Q value.

3.2   Reinforcement Learning Method Based Search Mode Selection

The reinforcement learning method uses agent decision-making, evaluates decision-making performance, and adjusts subsequent decision-making by interacting with the environment. The basis of the RL method is that the decision-making problem needs to conform to Markov decision processes (MDPs). The reward generated by the RL method decision-making search mode is only related to the current state, and this decision-making is independent of the previous decision-making. This process meets the requirement of no after-effect, and an MDP can be constructed.

The Q-learning method we use consists of four parts: $< S, A, R, V >$. S represents the state of the Agent, A represents the action, R represents the reward, and V represents the value function. At time, the agent makes the corresponding decision, adopts the action and evaluates the performance of the action, calculates the reward, and updates the Q value through the value function. We define a new combination form of <state, action>, in which the state is divided into two cases, that is, the objective function value is improved or not, and the action is global search and local search, respectively. Therefore, there are four values for the corresponding state-action combination in the Q table.

In the Q-learning method we use, the formula for calculating reward is as follows:

$$R_t = F_t\left(S_t, A_t\right) - F_{t-1}\left(S_{t-1}, A_{t-1}\right) \tag{9}$$

Among them, F_t is the best objective function value obtained by the population after the bee colony search, and F_{t-1} represents the best objective function value obtained by the previous generation bee colony. Accordingly, we can update the Q value according to R_t, and the calculation formula is as follows:

$$Q\left(S_{t+1}, A_{t+1}\right) \leftarrow Q\left(S_t, A_t\right) + \alpha\left(R_t + \gamma \max_a Q\left(S_{t+1}, a\right) - Q\left(S_t, A_t\right)\right) \tag{10}$$

where γ represents the discount coefficient and α represents the learning rate.

The search process based on the reinforcement learning method is the core part of the whole RLBA. We respectively involve the corresponding global search and local search operations.

3.3   Population Evolution Operation

Population Initialization. Generating an initial population in the RLBA is the basis of searching. The initial population should have as high randomness as possible and a certain direction of search. First, we randomly generate a scout bee population with a size of N_p and then select N_s sites with high objective function values for subsequent search.

Global Search. In the RLBA, global search is the main form of population search. This search method should have strong randomness, which makes it easier for scout bees to search a wide area. The population size of scout bees is N_p. In the algorithm, we design two global search operations: one is to search for a brand-new site, and the other is to search for a new site that has a large difference from the existing site. For the first search operation, RLBA generates a new task sequence completely randomly. On the other hand, searching for a new location based on the existing location is completed by exchanging locations of some segments in the task sequence, as shown in Fig. 1. The proportion of each generation of bees searching for new places randomly is ρ ($\rho \in [0,1]$), while the proportion of those searching for new places based on existing sites is $1 - \rho$.

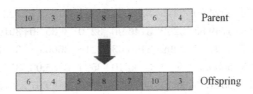

Fig. 1. Task Segment Location Exchange

Population Update. Because the size N_p of the scout bee population used in the RLBA for global search is larger than the number N_s of sites, the next generation of sites needs to be selected from within the population. We will keep the best individual in the population as a place. In addition, to ensure the diversity of locations, we randomly select N_s sites from the scout bee population as the basis of the next-generation search.

Local Search. In the RLBA, local search is also an important local development method to improve the search performance of the algorithm. Local search adopts the simplest 2-opt method; that is, two tasks are randomly selected from the task sequence to exchange their positions in the task sequence to obtain a new task sequence. Local search is only performed at the place with the best objective function value, and it stops when a better site cannot be found.

Termination. After searching for a certain time, the RLBA should stop running and output the best result found by searching as the final execution scheme of the task. Because the search methods used by the algorithm are different in each generation of bee colony search, it is more reasonable to use the evaluation algorithm search times to judge whether to terminate the algorithm. We set a maximum number of fitness evaluations MFE before each algorithm starts running, and when the number of fitness evaluations in the algorithm is equal to MFE, we end the algorithm and output the optimal scheme.

4 Experiment

To verify the performance of the RLBA, we designed a series of simulation experiments. Our experimental platform is a desktop computer with Core I7-7700 3.6 GHz CPU, 16 GB memory, Windows 11 operating system, and the coding environment is MATLAB 2021a.

Table 1. Results of Algorithms

Instance	RLBA		IGA		ALNS-I		NS	
	Best	Ave	Best	Ave	Best	Ave	Best	Ave
1000-1	4956	4821.97	3371	3308.33	3384	3220.73	3213	3052.73
1000-2	5072	4924.60	3375	3282.97	3331	3159.17	3224	3016.33
1000-3	4884	4749.83	3292	3146.90	3210	3035.30	3010	2911.63
1000-4	5027	4891.47	3299	3166.63	3216	3060.63	3101	2947.43
1100-1	5405	5030.67	3694	3506.10	3524	3385.03	3551	3281.67
1100-2	5479	5361.33	3845	3653.20	3719	3559.97	3612	3429.23
1100-3	5252	5101.17	3537	3405.10	3463	3278.13	3347	3144.80
1100-4	5253	5111.47	3592	3490.53	3552	3376.87	3398	3232.20
1200-1	5735	5546.73	3845	3745.60	3755	3627.20	3736	3472.60
1200-2	5767	5608.40	3845	3707.17	3759	3597.77	3705	3476.73
1200-3	5777	5515.13	3783	3695.30	3773	3584.60	3629	3440.90
1200-4	5625	5471.97	3921	3740.03	3753	3599.13	3662	3494.73
1300-1	6013	5889.13	4107	3988.80	4036	3885.23	4006	3721.20
1300-2	6311	6068.90	4441	4207.63	4262	4063.23	4208	3912.30
1300-3	5948	5725.60	4047	3927.10	3945	3771.97	3910	3583.77
1300-4	6134	5946.03	4004	3901.17	3969	3765.53	3846	3630.87
1400-1	6326	6097.13	4314	4237.40	4358	4120.10	4143	3890.13
1400-2	6258	6004.57	4288	4125.03	4303	4000.70	4006	3790.93
1400-3	6138	6014.90	4323	4114.23	4176	4011.83	4129	3823.47
1400-4	6214	6023.70	4317	4197.10	4278	4107.67	4157	3905.17

Because the LSEOSSP problem did not disclose the benchmark test set, we randomly generated a series of scenarios for the experiment. The task scale of the experiment is set to 1000, 1100, 1200, 1300, and 1400, and four groups of scenes are generated under each task scale.

We use the improved genetic algorithm [1], the improved adaptive large neighbourhood search algorithm (ALNS-I) [3], and the neighbourhood search algorithm (NS) as the experimental comparison algorithms. Based on the traditional genetic algorithm, the improved genetic algorithm improves the search strategy for the EOSSP. The improved adaptive large neighbourhood search algorithm uses a tabu search strategy in the adaptive large neighbourhood search framework to improve the search efficiency of the algorithm. The neighbourhood search algorithm obtains a good observation task execution scheme through continuous improvement of the neighbourhood structure.

As the proposed algorithm and comparison algorithm are random search algorithms, the results of running each algorithm 30 times are recorded, and the optimal value (denoted as Best) and average value (denoted as Ave) are used as the indexes to evaluate the search performance of the algorithm.

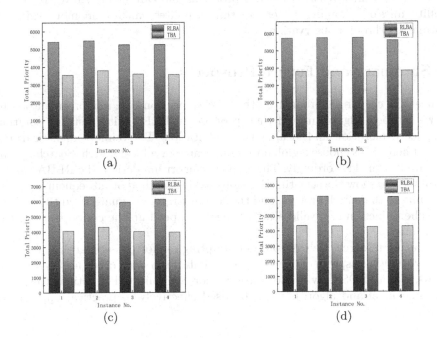

Fig. 2. Comparison result between traditional bee algorithm and RLBA

The experimental results are shown in Table 1. As seen from Table 1, our proposed RLBA is superior to the comparison algorithm in both the optimal value and the average value of the scheduling results. The algorithm has a good search performance, and the bees algorithm framework has obvious advantages in global search. Combined with the reinforcement learning method, the ability to mine new search performance can further improve the possibility of finding a better solution. With the increase in the task scale, the advantages of the RLBA are more obvious, which reflects that the RLBA may deal with more complicated situations more effectively.

After that, we also use the traditional bees algorithm and RLBA to compare and evaluate the role of reinforcement learning in the algorithm. The planning results of 1100 tasks to 1400 task scenarios are shown in Fig. 2. It can be seen from the figure that the reinforcement learning method can find a search strategy more suitable for the characteristics of the problem, which plays a more obvious role when the problem is large. The use of the learning method can make the algorithm search smart and make the algorithm find more ideal results through fewer searches.

It can be seen from the above experiments that the RLBA proposed by us is superior to the comparison algorithm in terms of planning performance, and the use of learning methods can help improve the performance of the algorithm. The performance of RLBA in solving the LSEOSSP reflects that the algorithm has a strong ability to solve practical problems and can be applied to the actual satellite mission planning system so that satellites can obtain more valuable information through observation.

5   Summary and Future Prospects

To meet the challenges brought by the EOSSP, we combine reinforcement learning with a bees algorithm and then propose an RLBA with a strong solution for space exploration and local space exploitation. The reinforcement learning method helps bees choose a global random search or a local search. Search operations are designed accordingly. Through this algorithm design, the RLBA can be guaranteed to show the scientific decision-making method of subsequent searches according to the search process, and the search strategy is guaranteed to be effective. The reduction of invalid search times can speed up the convergence of the algorithm.

In future research, we will consider adopting reinforcement learning methods to improve other aspects of the BA and will also introduce other learning mechanisms. The LSEOSSP will also consider more complicated situations to ensure that the model and algorithm can be used effectively in practical application scenarios.

Acknowledgement. Junwei Ou and Yanjie Song contribute equally to this article. Thanks to Prof. Cham for his valuable comments.

References

1. Barkaoui, M., Berger, J.: A new hybrid genetic algorithm for the collection scheduling problem for a satellite constellation. J. Oper. Res. Soc. **71**(9), 1390–1410 (2020)
2. Chang, Z., Zhou, Z., Xing, L., Yao, F.: Integrated scheduling problem for earth observation satellites based on three modeling frameworks: an adaptive bi-objective memetic algorithm. Memet. Comput. **13**(2), 203–226 (2021)
3. He, L., de Weerdt, M., Yorke-Smith, N.: Tabu-based large neighbourhood search for time/sequence-dependent scheduling problems with time windows. In: Proceedings of the International Conference on Automated Planning and Scheduling, vol. 29, pp. 186–194 (2019)
4. Huang, Y., Luo, A., Zhang, M., Zhang, X., Lin, M., Song, Y.: A novel mission planning model and method for combat system-of-systems architecture design. In: 2022 8th International Conference on Big Data and Information Analytics (BigDIA), pp. 239–245. IEEE (2022)

5. Jiang, X., Song, Y., Xing, L.: Dual-population artificial bee colony algorithm for joint observation satellite mission planning problem. IEEE Access **10**, 28911–28921 (2022)
6. Li, G.: Online scheduling of distributed earth observation satellite system under rigid communication constraints. Adv. Space Res. **65**(11), 2475–2496 (2020)
7. Niu, X., Tang, H., Wu, L.: Satellite scheduling of large areal tasks for rapid response to natural disaster using a multi-objective genetic algorithm. Int. J. Disaster Risk Reduct. **28**, 813–825 (2018)
8. Ou, J., et al.: Deep reinforcement learning method for satellite range scheduling problem. Swarm and Evolutionary Computation, p. 101233 (2023)
9. Pham, D.T., Ghanbarzadeh, A., Koç, E., Otri, S., Rahim, S., Zaidi, M.: The bees algorithm-a novel tool for complex optimisation problems. In: Intelligent Production Machines and Systems, pp. 454–459. Elsevier (2006)
10. Song, Y., Wei, L., Yang, Q., Wu, J., Xing, L., Chen, Y.: Rl-ga: a reinforcement learning-based genetic algorithm for electromagnetic detection satellite scheduling problem. Swarm and Evolutionary Computation, p. 101236 (2023)
11. Vasquez, M., Hao, J.K.: A logic-constrained knapsack formulation and a tabu algorithm for the daily photograph scheduling of an earth observation satellite. Comput. Optim. Appl. **20**(2), 137–157 (2001)
12. Waiming, Z., Xiaoxuan, H., Wei, X., Peng, J.: A two-phase genetic annealing method for integrated earth observation satellite scheduling problems. Soft. Comput. **23**(1), 181–196 (2019)
13. Wang, X., Wu, G., Xing, L., Pedrycz, W.: Agile earth observation satellite scheduling over 20 years: formulations, methods, and future directions. IEEE Syst. J. **15**(3), 3881–3892 (2020)
14. Wei, L., Song, Y., Xing, L., Chen, M., Chen, Y.: A hybrid multi-objective coevolutionary approach for the multi-user agile earth observation satellite scheduling problem. In: Pan, L., Cui, Z., Cai, J., Li, L. (eds.) BIC-TA 2021. CCIS, vol. 1565, pp. 247–261. Springer, Singapore (2022). https://doi.org/10.1007/978-981-19-1256-6_18
15. Wei, L., Xing, L., Wan, Q., Song, Y., Chen, Y.: A multi-objective memetic approach for time-dependent agile earth observation satellite scheduling problem. Comput. Ind. Eng. **159**, 107530 (2021)

Optimal Formation of UUV Groups Based on Shape Theory and Improved Ant Colony Algorithm Under Communication Delay

Fan Ye[1], Ziwei Zhao[2], and Xuan Guo[2](✉)

[1] Chang Jiang Communication Administration, Wuhan 430014, China
[2] Wuhan University of Technology, Wuhan 430070, China
guoxuanwhut@163.com

Abstract. Aiming at the path optimization of formation of Unmanned Underwater Vehicle (UUV) in the process of formation reorganization. Firstly, this paper describes the UUV formation on the basis of shape theory. Based on the UUV kinematic model. The master-slave UUV consensus controller, under time-delay condition, is designed to realize the formation and maintenance of multiple UUVs. Secondly, the local update method is introduced into the pheromone update of the standard ant colony algorithm. The improved ant colony algorithm is used to quickly generate the shortest formation path. Finally, the simulation experiment was designed based on python. After the UUVs complete autonomous obstacle avoidance, the improved ant colony algorithm was introduced to realize the selection of the recombinant formation route, according to the initial coordinates of the UUV and the relative target point. The simulation results show that the designed tracking controller and the improved ant colony algorithm are effective in the formation maintenance and reorganization of multi-UUV formation.

Keywords: Unmanned underwater vehicle · Communication delay · Formation control · Bio-inspired algorithm · Formation reconfiguration

1 Introduction

Unmanned Underwater Vehicle (UUV) plays a huge role in maritime combat. With the improvement of the complexity of the task, single UUV can no longer complete the multi-task work independently, so the formation of multiple UUV can better coordinate to complete all kinds of tasks [1]. And for tasks with a large regional range, multiple UUV formations can be combined into a large formation to improve the efficiency of task execution formation control can generally be divided into two stages: 1) determine a desired formation, and 2) design the corresponding control algorithm for reaching and keeping this formation. At present, there are relatively few research on formation and recombination, but more research on formation control, which mainly includes leader-following method [2, 3] Artificial potential field method [4], virtual structure method [5], consistency control [6] and so on.

© The Author(s), under exclusive license to Springer Nature Singapore Pte Ltd. 2023
L. Pan et al. (Eds.): BIC-TA 2022, CCIS 1801, pp. 92–105, 2023.
https://doi.org/10.1007/978-981-99-1549-1_8

The existing formation description methods mainly include three types of algorithms based on displacement, distance and angle. These three types of algorithms respectively use relative displacement, relative distance and relative Angle between UAVs as constraints to define the formation of multi-UUV formation [7–9].

In recent years, scholars have proposed some new methods to define formation [10–12]. Examples include centroid coordinate method [13], complex Laplacian matrix method [14], and shape theory method [15]. The use of these theories as constraints to define the formation improves the invariance of the formation constraints. Among them, shape theory is a simple method for defining formation proposed by scholars [16]. Because shape theory only cares about the geometric definition of the formation, it can be more flexible to describe the generation and transformation of the UUV formation.

When the multi-UUV formation system encounters an emergency situation, such as an obstacle or communication interruption during the process of moving, it will autonomtively avoid obstacles or interrupt communication, and the initial formation will be disrupted. In the process of solving the emergency situation and reorganizing the formation, reasonable coordination planning should be carried out when selecting the position of each UUV in the formation, so as to reduce the time-consuming of UUV formation recovery process. Therefore, in the optimization selection of position, the formation problem can be transformed into the problem of finding the maximum value of the nearest position between the UUV and the formation. In order to solve this problem, this paper proposes to use the biological- inspired algorithm to optimize and select the optimal route.

2 Preliminaries and Modelling

2.1 Shape Theory

Definition 1: Shape of formation refers to the geometric information retained after removing rotation, contraction and displacement of formation.

Supposing UUVs (the quantity is n) in three-dimensional space needs to form a fixed formation. Defining $p_i = \left(p_i^x, p_i^y, p_i^z\right)^T$ as the position of UUV in a fixed coordinate system, then the expected position of each UUV in the target formation can be expressed by the following matrix:

$$S = (s_1, s_2, \ldots, s_n) \in \mathbb{R}^{3 \times n} \tag{1}$$

In cases where relative relationships between UUVs are concerned rather than absolute positions, shape theory becomes a more appropriate way to describe UUV formations.

2.2 UUV Kinematic and Kinetic Models

The rectangular coordinate system is used to study the motion of underactuated UUV, and the coordinate system mainly includes the fixed coordinate system and the moving coordinate system, as shown in Fig. 1.

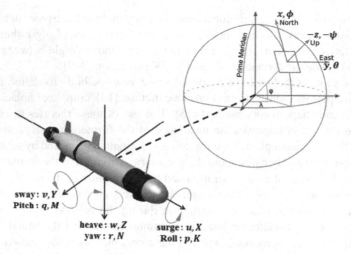

Fig. 1. UUV motion model

The UUV formations studied in this paper are located at the same depth. Considering that the UUVs keep a fixed formation moving in the horizontal plane, the kinematic and dynamic models of the UUVs without interference are established [17, 18].

$$\begin{cases} \dot{x} = u\cos\psi - v\sin\psi \\ \dot{y} = u\sin\psi - v\cos\psi \\ \dot{\psi} = r \\ \dot{u} = \frac{m_2 vr - X_u u + \tau_u}{m_1} \\ \dot{v} = \frac{-m_1 ur - Y_v v}{m_2} \\ \dot{r} = \frac{(m_1 - m_2)uv - N_r r + \tau_r}{m_3} \end{cases} \quad (2)$$

Among them, x, y respectively represent the coordinates of UUV in the inertial coordinate system, ψ represents the heading Angle, u, v, r respectively represent the vertical velocity, horizontal velocity and yaw angular velocity of UUV, X_u, Y_u, N_r represent the hydrodynamic coefficient, m_1, m_2, m_3 represent the mass inertia coefficient of UUV, τ_u, τ_r respectively represent the thrust and steering torque.

3   Design of UUV Formation Controller with Time Delay Based on Consistency

The premise of formation transformation is the control of formation keeping. In a multi-UUV system, one UUV is selected as the primary UUV, and the other UUV is the secondary UUV. The master/slave path tracking controller is designed so that these UUV tracks the primary UUV to keep a fixed formation forward. Since the propagation speed of underwater sonar signal is about 1480 m/s, the underwater communication capability of UUV is very limited, and underwater acoustic communication has delay. Therefore, this paper considers the delay of multiple UUV groups to design the controller.

It is assumed that UUV groups keep a fixed depth motion of z_i, and when there is a delay in communication between UUVs, the controller is designed as follows:

$$
\begin{aligned}
u_{x_i} = &\dot{f}^x(t) - \gamma(v_{x_i}(t - \tau_{ij}(t)) - f^x(t - \tau_{ij}(t))) \\
&- \sum_{j=1}^{m} a_{ij}\left[(x_{f_i}(t) - x_{f_j}(t - \tau_{ij}(t))) - \left(\delta_i^x - \delta_j^x\right) + \gamma(v_{x_i}(t - \tau_{ij}(t)) - v_{x_j}(t - \tau_{ij}(t)))\right]
\end{aligned}
\tag{3}
$$

$$
\begin{aligned}
u_{y_i} = &\dot{f}^y(t) - \gamma(v_{y_i}(t - \tau_{ij}(t)) - f^y(t - \tau_{ij}(t))) \\
&- \sum_{j=1}^{m} a_{ij}\left[(y_{f_i}(t) - y_{f_j}(t - \tau_{ij}(t))) - \left(\delta_i^y - \delta_j^y\right) + \gamma(v_{y_i}(t - \tau_{ij}(t)) - v_{y_j}(t - \tau_{ij}(t)))\right]
\end{aligned}
\tag{4}
$$

Among them, the control gain $\gamma > 0$, $\tau_{ij}(t)$ represents the UUV communication delay between the ith and jth UUV. $f^x(t)$ and $f^y(t)$ are continuously differentiable functions, representing the motion velocity characteristics of UUV. δ_i^x, δ_i^y indicates the desired location. If the upper limit of time delay is τ_0, there is $0 < \tau_{ij}(t) < \tau_0$, and it must meet:

$$
0 \leq \tau_{ij}(t) \leq \frac{1}{\omega_0} \arctan \frac{(1 + \lambda_i)\omega_0 \gamma}{\lambda_i}
\tag{5}
$$

In the above equation,

$$
\omega_0 = \sqrt{\frac{(1 + \lambda_i)^2 \gamma^2 \pm \sqrt{(1 + \lambda_i)^4 \gamma^4 + 4\lambda_i^2}}{2}}
\tag{6}
$$

where λ_i is the eigen root of the Laplacian matrix L of figure G. When the above conditions are met, the UUV group maintains a stable formation movement.

When Eq. (5) established, the second order UUV group system with time delay can achieve consistent stability and maintain stable formation navigation.

4 UUV Formation Recombination Based on Bio-inspired Algorithm

4.1 Description of the Problem

As shown in Fig. 2, the UUV formation system composed of two sub-formations maintains a longitudinal prismatic formation, UUV1 and UUV4 are the leading aircraft, and the rest are tracking slaves. After the obstacle avoidance process is completed, the UUV formation system needs to re-formation according to the formation controller, and the path will be significantly longer. Therefore, it is necessary to recalculate the optimized path, so that the multi-UUV system can re-plan the path according to the optimized route, and finally complete the formation with the shortest path.

Supposing there are UUV formation teams (the quantity is m) and UUVs (the quantity is n), then the desired formation (and target formation) of the jth formation can be represented by a vector $(s_{j.1}, s_{j.2}, \ldots, s_{j.n_j})$. Among them, n_j represents the number of

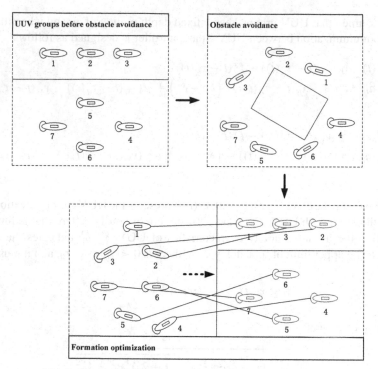

Fig. 2. Recombination process of multiple UUV formations

UUVs in the jth formation, representing the two-dimensional coordinates of the desired formation, then the desired formation is:

$$S_j = \left(s_{j.1}^T, s_{j.2}^T, \ldots, s_{j.n_j}^T\right)^T \in \mathbb{R}^{3 \times n_j} \quad (j = 1, \cdots m) \tag{7}$$

The desired formation of multiple formations can be expressed as:

$$S = \left(S_1^T, S_2^T, \ldots, S_m^T\right)^T \in \mathbb{R}^{3 \times (n_1 + n_2 + \ldots + n_m)} \tag{8}$$

The initial position of the multi-formation can be expressed as:

$$P = \left(P_1^T, P_2^T, \ldots, P_m^T\right) \in \mathbb{R}^{3 \times (n_1 + n_2 + \ldots + n_m)} \tag{9}$$

The target position of the multi- formation can be expressed as

$$Q = \left(Q_1^T, Q_2^T, \ldots, Q_m^T\right) \in \mathbb{R}^{3 \times (n_1 + n_2 + \ldots + n_m)} \tag{10}$$

For a UUV formation, it must be possible to find a target formation Q, which minimizes the total distance between the starting position and the target position $P - Q$ of all UUVs. Therefore, the multi-UUV formation control problem is transformed into an optimization problem with constraints through these transformations:

$$\min_{Q \in \mathbb{R}^{3 \times n}} \|P_j - Q_j\|, s.t. Q_j \in [S_j] \quad (or \quad Q_j \, S_j, j = 1, 2, \ldots, m). \tag{11}$$

Among them, $[S_j]$ represents a set of feasible formations with the same shape and different sizes and directions; m represents all the number of sub-formations.

According to the definition of shape theory, if Q_j S_j, then there are corresponding rotation R, translation d and scaling operation a, under the corresponding operation,

$$Q_j = \alpha_j R_j S_j + d_j \tag{12}$$

Since the formation needs to ensure the safe distance between UUVs, and the spacing distance should not be too far, the zoom ratio value a_j should be valued within a fixed range $a_{min} \leq a_j \leq a_{max}$. In order to avoid the overlap of the initial formation between formations, it is necessary to increase the allowable range $\Omega_j \in \mathbb{R}^3$ for each expected formation, and there is a safe distance between each formation's Ω_j to avoid collision. Therefore, considering the rotation, scaling and scope constraints of the target formation, the formation generation problem of the formation can be written in the following form:

$$\min_{Q \in \mathbb{R}^{3 \times n}} \|P_j - Q_j\|$$
$$s.t. \quad A_j Q_j = a_j (I_{n_j} \otimes R_j) S_j \quad (j = 1, 2, \ldots, m). \tag{13}$$
$$a_{min} \leq a_j \leq a_{max}$$
$$Q_j \in \Omega_j$$

The formation of the desired one, rotation angle of the target formation, scaling ratio and the restricted area are taken as constraint conditions, and the problem of the form of the formation is transformed into an optimization issue.

In the case that the initial position of UUV and the target position of formation are known, the target position of UUV can be assigned to each UUV by local adjustment, and the total path can be the shortest, then it will be converted to another optimization problem:

$$\min \sum_{i=1}^n \sum_j^n c_{ij} x_{ij}$$
$$s.t. \sum_{j=1}^n x_{ij} = 1$$
$$\sum_{i=1}^n x_{ij} = 1 \tag{14}$$
$$x_{ij} = \{0, 1\} \quad i, j = 1, 2, \ldots, N$$

Among them, n represents the sum of UUV; c_{ij} represents the expected distance from the ith UUV to jth; $x_{ij} = 1$ represents ith UUV's targeted coordinate is the coordinate of the j.

4.2 Standard Ant Colony Algorithm

The idea of the standard ant colony algorithm is to use the characteristics of the pheromone released by the ant colony in the process of reaching the target point, with the iteration of time, all ants concentrate on the shortest path to obtain the optimal solution.

Supposing the number of ants in the ant colony is m, the number of ants is n, the distance between waypoints i and waypoints j is $d_{ij}(i, j = 1, 2, \ldots, n)$, at the time of t, the pheromone concentration on the connecting path between ant i and ant j is $\tau_{ij}(t)$. At the initial moment, the pheromone concentration among $k(k = 1, 2, \ldots, m)$ all ants are the same, setting as $\tau_{ij}(0) = \tau_0$. Setting $P_{ij}^k(t)$ as a demonstration of probability for the ant k to travel from waypoint i to waypoint j at the time of t, then:

$$P_{ij}^k(t) = \begin{cases} \dfrac{[\tau_{ij}(t)]^\alpha \times [\gamma_{ij}(t)]^\beta}{\sum_{s \in allow_k} [\tau_{is}(t)]^\alpha \times [\gamma_{is}(t)]^\beta}, & s \in allow_k \\ 0, & s \notin allow_k \end{cases} \tag{15}$$

In Formula (15), $\gamma_{ij}(t) = 1/d_{ij}$ represents the heuristic function, that is, the expectation degree of the ant from the waypoint i to the waypoint j; α represents the factor of pheromone importance; β represents the important degree factor of heuristic function; $allow_k(k = 1, 2, \ldots, m)$ represents the set of waypoints to be accessed by the ant k, with the increase of time, the elements in $allow_k$ decrease continuously until the access is completed and it becomes an empty set. Ant k releases the pheromone, the pheromone on the connecting path of each waypoint gradually decreases. Supposing $\rho(0 < \rho < 1)$ represent the degree of pheromone evaporation. When the ant completes an iteration, the pheromone concentration between each waypoint needs to be updated according to the following formula:

$$\begin{cases} \tau_{ij}(t+1) = (1-\rho)\tau_{ij}(t) + \rho\Delta\tau_{ij} \\ \Delta\tau_{ij} = \sum_{k=1}^n \Delta\tau_{ij}^k \end{cases} \tag{16}$$

In the above formula, $\Delta\tau_{ij}^k$ denotes the pheromone concentration released by the kth ant at the conjunction waypoint of path i and path j, $\Delta\tau_{ij}$ denotes the sum of the pheromone concentration released by all ants at the conjunction waypoint of path i and path j. Generally speaking, the ant cycle system model is used to express $\Delta\tau_{ij}$:

$$\Delta\tau_{ij}^k = \begin{cases} Q/L_k, & \text{Ant k goes from i to j} \\ 0 & \text{else} \end{cases} \tag{17}$$

Among them, Q is a constant parameter, representing the total amount of pheromone released by the ant in one iteration; L_k represents the length of the path taken by the kth ant.

4.3 Improved Ant Colony Algorithm

In this paper, the standard ant colony algorithm is being improved by the introduction of local pheromone update, to accelerate the convergence speed and avoid falling into local optimal. The following formula (18) is used for local update:

$$\tau_{ij}(t+1) = (1-\theta)\tau_{ij}(t) + \theta\tau_0 \tag{18}$$

In Formula (18), τ_0 represents the initial pheromone value, θ represents the local pheromone volatile factor. In the present position i of UUV, selecting the next position j. By reducing the pheromone value on the path (i, j) through the volatilization factor, the probability of the rest of the UUV selecting the path (i, j) will be reduced, hence, avoiding the repetition of the search path, the probability of searching the rest of the path can be increased, and the search speed of the algorithm can be improved. The global update still adopts the standard ant colony algorithm

$$\tau_{ij}(t+1) = (1 - \rho)\tau_{ij}(t) + \Delta\tau_{ij}. \tag{19}$$

The steps of the improved ant colony algorithm to solve the UUV recombination formation are described as follows:

Step 1: Initializing parameters.

Initializing related parameters, including the number n of UUV in UUV formation, pheromone importance factor α important factor of inspiration function β, local pheromone volatile factor θ, global pheromone volatile factor ρ, total pheromone release Q, maximum number of iterations iter_max, initial value of iteration number iter, Obtaining the initial position coordinates of each UUV and the position coordinates of the target point.

Step 2: Calculating the distance between UUVs.

According to the initial position coordinates of each UUV, the mutual distance is calculated.

Step 3: Constructing the solution space.

Each UUV is randomly placed at different starting points, and the next target position is calculated according to the following formula until all UUV reach the end of the formation.

Step 4: Updating the path.

Calculating the path length $L_k (k = 1, 2, \ldots, n)$ of each UUV and recording the shortest length L_{min} of the formation path in the current number of iterations.

Step 5: Updating pheromones locally.

When each UUV selects the target position once, pheromone update is carried out on the last path information (i, j) will be updated according to the formula

$$\tau_{ij}(t+1) = (1 - \theta)\tau_{ij}(t) + \varepsilon\tau_0 \tag{20}$$

until the path construction is completed. Otherwise, returning to the Step 4 to continue updating the path.

Step 6: Updating pheromones globally.

Updating the global pheromone concentration according to the formula, until the number of iterations is reached.

Step 7: The algorithm terminates.

If the algorithm does not reach the maximum number of iterations iter_max, setting iter = iter + 1 and returning to step 3; Otherwise, the algorithm terminates, and the results are outputted.

The block diagram of the algorithm is shown as follows (Fig. 3):

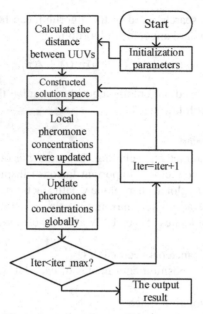

Fig. 3. Block diagram of improved ant colony algorithm

5 Simulation Experiment

In order to verify the effectiveness of the algorithm proposed above, this paper conducts simulation experiments based on python. The experiment was conducted on a PC with a 2.7 GHz Intel Core i7-5700HQ CPU (4 CPU cores) on a Windows 10 64-bit operating system. Parameters of the improved ant colony algorithm are shown in Table 1, formation configuration is shown in Fig. 4.

Table 1. Parameters of the improved ant colony algorithm.

Initialization parameters	Parameter value
n	10
α	1
β	5
θ	0.1
ρ	0.5
Q	1
iter_max	200
iter	1

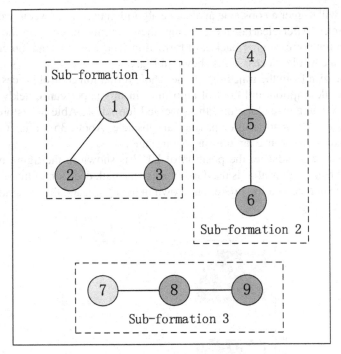

Fig. 4. Formation diagram.

9 UUVs are placed in the starting position, and the UUVs are divided into two sub-formations, among which, 1, 4 and 9 are the UUVs, which are the leaders. And the rest are the followers. After the simulation starts, the formation is formed at first, and the formation is converted into triangle and longitudinal shape respectively, then the formation is kept while sailing. If the unmanned ships encounter an obstacle while

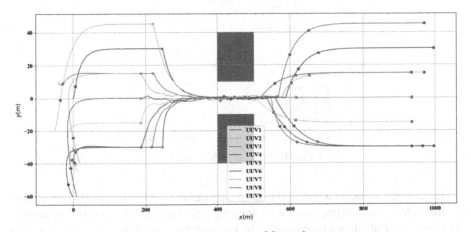

Fig. 5. Sailing chart of formation

sailing, they will trigger an obstacle avoidance algorithm to pass between two relatively distant obstacles. After clearing the obstacle, they continue to perform the formation algorithm, in time of changing back into formation. The whole formation controls the trajectory of the whole process, as is shown in Fig. 5.

As can be seen from the trajectory of the whole formation control process in Fig. 6, the formation description and control algorithm in this paper can quickly transform multiple UUV unmanned ships into the specified formation. Able to restore a broken formation to a set formation after passing an obstacle. About 35 s after the obstacle avoidance is over, the formation returns to a triangle.

After obstacle avoidance, the position of UUV is shown in the figure below. The improved ant colony algorithm is used to optimize the path decision of the reformation. The new formation generation process is shown in the figure, and the generated optimal path is shown in Fig. 7.

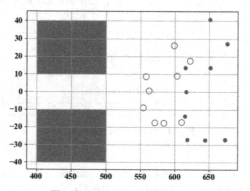

Fig. 6. Initial and target points

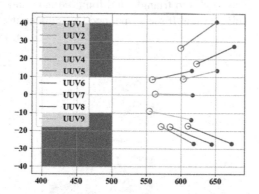

Fig. 7. Optimal path diagram

The standard particle swarm optimization algorithm (PSO), the standard ant colony algorithm (ACO) and the improved ant colony algorithm (IACO) in this paper are respectively used for simulation experiments. The convergence rate of the algorithm is shown in Fig. 8. It can be seen from the convergence curve that IACO has faster convergence speed and higher search accuracy. When the algorithm is carried out for about 9 times, the optimal solution can be searched. Similarly, PSO and ACO require about 17 and 11 iterations respectively. The UUV formation is in a scattered position after autonomous obstacle avoidance. After IACO calculation, the shortest average distance of the optimal path generated after the 9th iteration is 72.4 m.

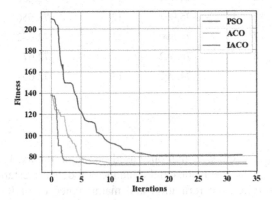

Fig. 8. Iterative process

The algorithm was tested for 10 times, and the distance of the optimization path and the number of iterations to reach the final result were recorded. The results are shown in Table 2. It can be seen from the analysis of the data in the table that the path optimization based on IACO is significantly better than that of ACO and PSO in terms of path length and number of iterations. It can find the path with good quality in a short time, with the highest search accuracy and the least number of iterations. The path searched by PSO fluctuates greatly, with the average path length remaining at about 81 m, that by ACO is about 74 m, and that by IACO is about 72 m. The search accuracy is the highest and the number of iterations is the least.

Table 2. Comparison of the effects of three algorithms

No	PSO		ACO		IACO	
	Length of path	Number of iterations	Length of path	Number of iterations	Length of path	Number of iterations
1	83.161	16	74.786	13	72.651	9
2	80.887	15	74.995	11	72.602	11
3	82.412	17	74.664	14	72.644	12
4	79.665	17	74.471	13	72.660	9
5	83.665	15	74.968	14	72.655	8

(continued)

Table 2. (*continued*)

No	PSO		ACO		IACO	
	Length of path	Number of iterations	Length of path	Number of iterations	Length of path	Number of iterations
6	83.189	16	74.432	13	72.686	8
7	81.332	16	75.001	12	72.656	8
8	82.501	17	74.890	12	72.602	10
9	83.881	18	75.012	10	72.679	10
10	79.627	20	74.697	13	72.604	10

6 Conclusion

In this paper, firstly, the UUV formation, based on shape theory, is described. Then the UUV horizontal plane kinematics and dynamics models are established. Based on the consistency controller, the formation and maintenance of multi-UUV formation control under the delay condition are realized. Secondly, the ant colony algorithm in the bio-inspired algorithm was improved, and the local pheromone updating method was proposed to accelerate the convergence speed of the ant colony algorithm and avoid falling into the local optimal. The improved ant colony algorithm was used to find the optimal path in the process of multi-UUV formation recombination. Finally, the simulation platform is used to verify the effectiveness of the proposed method. The proposed method can be applied to a variety of task scenarios. In the face of external interference and obstacles, the time of reformation is effectively reduced, the endurance of the multi-UUV formation system is improved, and the redundancy is enhanced.

References

1. Fang, X., et al.: Formation control for unmanned surface vessels: a game-theoretic approach. Asian J. Control **24**(2), 498–509 (2022)
2. Xia, G., Zhang, Y., Zhang, W., et al.: Dual closed-loop robust adaptive fast integral terminal sliding mode formation finite-time control for multi-underactuated AUV system in three-dimensional space. Ocean Eng. **233**, 108903 (2021)
3. Rui, G., Chitre, M.: Cooperative positioning using range-only measurements between two AUVs. In: OCEANS'10 IEEE SYDNEY, pp. 1–6. IEEE (2010)
4. Khatib, O.: Real-time obstacle avoidance for manipulators and mobile robots. In: Proceedings. 1985 IEEE International Conference on Robotics and Automation, pp. 500–505. IEEE (1985)
5. Tan, K.H., Lewis, M.A.: Virtual structures for high-precision cooperative mobile robotic control. In: Proceedings of IEEE/RSJ International Conference on Intelligent Robots and Systems. IROS'96, pp. 132–139. IEEE (1996)
6. Leonard, N.E., Paley, D.A., Lekien, F., et al.: Collective motion, sensor networks, and ocean sampling. In: Proceedings of the IEEE, vol. 95, no. (1), pp. 48–74. IEEE (2007)
7. Zhao, S., Zelazo, D.: Bearing rigidity and almost global bearing-only formation stabilization. IEEE Trans. Autom. Control **61**(5), 1255–1268 (2016)

8. Lin, Z., Wang, L., Chen, Z., Fu, M., Han, Z.: Necessary and sufficient graphical conditions for affine formation control. IEEE Trans. Autom. Control **61**(10), 2877–2891 (2016)

9. Lin, Z., Wang, L., Han, Z., Fu, M.: Distributed formation control of multi-agent systems using complex laplacian. IEEE Trans. Autom. Control **59**(7), 1765–1777 (2014)

10. Huang, P.Q., Wang, Y., Wang, K., Liu, Z.Z.: A bilevel optimization approach for joint offloading decision and resource allocation in cooperative mobile edge computing. IEEE Trans. Cybern. **50**(10), 4228–4241 (2019)

11. Lin, J., He, C., Cheng, R.: Adaptive dropout for high-dimensional expensive multiobjective optimization. Complex Intell. Syst. **8**(1), 271–285 (2021). https://doi.org/10.1007/s40747-021-00362-5

12. Ren, W.: On consensus algorithms for double-integrator dynamics. IEEE Trans. Automat. Contr. **53**(6), 1503–1509 (2008)

13. Sun, Z., Park, M.C., Anderson, B.D.O., Ahn, H.-S.: Distributed stabilization control of rigid formations with prescribed orientation. Automatica **78**, 250–257 (2017)

14. Zhao, S., Zelazo, D.: Translational and scaling formation maneuver control via a bearing-based approach. IEEE Trans. Control Netw. Syst. **4**(3), 429–438 (2017)

15. Han, T., Lin, Z., Zheng, R., Fu, M.: A barycentric coordinate-based approach to formation control under directed and switching sensing graphs. IEEE Trans. Cybern. **48**(4), 1202–1215 (2017)

16. Han, Z., Wang, L., Lin, Z., Zheng, R.: Formation control with size scaling via a complex Laplacian-based approach. IEEE Trans. Cybern. **46**(10), 2348–2359 (2016)

17. Derenick, J.C., Spletzer, J.R.: Convex optimization strategies for coordinating large-scale robot formations. IEEE Trans. Rob. **23**(6), 1252–1259 (2007)

18. Okamoto, A., Imasato, M., Hirao, S.C., et al.: Development of testbed AUV for formation control and its fundamental experiment in actual sea model basin. J. Robot. Mechatron. **33**(1), 151–157 (2021)

Global Path Planning for Unmanned Ships Based on Improved Particle Swarm Algorithm

Chang Liu$^{(\boxtimes)}$ and Kui Liu

Wuhan Digital Engineering Institute, Wuhan 430074, China
liuchang923@163.com

Abstract. In this paper, the global path planning research is carried out using particle swarm algorithm in combination with the characteristics of unmanned boat navigation environment. To address the problem that the particle swarm algorithm is easy to fall into local optimum at the later stage, we first integrate chaos theory into the basic particle swarm algorithm, and generate chaotic population and replace some particles that fall into local optimum by chaotic iteration of contemporary global optimum, which improves the problem of insufficient particle diversity at the later stage of population search; meanwhile, to strengthen the local search ability of the algorithm, we combine the particle swarm algorithm with the following bee. To enhance the local search capability of the algorithm, we combine the particle swarm algorithm with the following bee strategy in the swarm search algorithm and propose the chaotic particle swarm-bee swarm algorithm. The improved algorithm is applied to the global path planning, and the simulation verifies the advantages of the search algorithm in terms of convergence speed and search accuracy.

Keywords: Particle swarm · USV · swarm search algorithm

1 Introduction

Global path planning is to plan an optimal path that meets effective constraints given the starting point and target point of USVs with known environmental information [1]. Generally speaking, the complete global path planning process includes obstacle processing, map model construction, path optimization, path smoothing and other steps. The global path planning flow chart is shown in Fig. 1.

Most of the existing unmanned vehicle path planning methods are derived from the path planning of unmanned robots, and the global path planning capability is an important embodiment of its intelligence level [2]. Global path planning is to use the existing map information, combined with the task mission, and adopt the appropriate search algorithm to find the appropriate path in the global sense.

Rao Sen proposed a global path planning for unmanned craft by using hierarchical thought and genetic algorithm [3]. Fan Yunsheng et al. proposed a global path planning method combining electronic chart rasterization with genetic algorithm optimization, and adopted a comprehensive evaluation fitness function to improve the efficiency

© The Author(s), under exclusive license to Springer Nature Singapore Pte Ltd. 2023
L. Pan et al. (Eds.): BIC-TA 2022, CCIS 1801, pp. 106–116, 2023.
https://doi.org/10.1007/978-981-99-1549-1_9

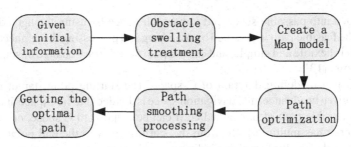

Fig. 1. Global path planning flowchart

of path optimization [4]. Liu Jian designed the grid dynamic refinement method to build the obstacle model, and improved the artificial potential field method to improve the traditional potential field easy stagnation shortcomings [5]. Chen Chao, Tang Jian et al. designed A global path planning method combining viewable and A* algorithm to improve the inadaptability of the traditional viewable method to the environment [6]. Zhuang Jiayuan et al. designed a global path planning algorithm for UVs with distance optimization, which was combined with electronic chart to improve the accuracy of path planning [7]. Shu Zongyu adopted multi-objective particle swarm optimization algorithm to optimize the initial path searched by Dijkstra algorithm, and introduced simulated annealing operator into the algorithm to obtain the global shortest path [8]. Liu Kun proposed an artificial potential field-ant colony path planning method to reduce the length and time of the planned path, aiming at the blindness of the artificial potential field method which is easy to fall into local optimal and ant colony algorithm search [9]

Path planning of unmanned craft is different from that of unmanned robot and unmanned vehicle [10]. First of all, the navigation environment of the unmanned boat is generally relatively wide, not restricted by the fixed traffic network, the number of obstacles is small, the volume of obstacles is large; In addition, the motion control of unmanned craft is not as real-time and simple as that of robot. The steering process also needs to consider the rotation radius and hull inertia and other factors. A* algorithm and Dijkstra algorithm need to build complex raster map, and need to search the raster part traversal, the efficiency is low. Ant colony algorithm (ACO) and genetic algorithm (GA) have many shortcomings such as difficulty in adjusting parameters [11]. Particle swarm optimization (PSO) is widely used in parameter optimization problems because of its advantages such as low adjustment parameters, high search efficiency, easy implementation, and easy combination with other algorithms [12]. Therefore, PSO is used in this paper to study the global path planning of unmanned craft, and the particle swarm optimization algorithm is improved to meet the requirements of path planning for search accuracy and search speed.

2 Path Planning Modeling and Problem Description

In the global path planning of unmanned vehicle, the volume of unmanned vehicle is very small compared with the whole navigation environment, so it can be regarded as a particle without volume and mass for research. The navigation environment of unmanned boat is divided into passable space and impassable space by using the agreed polygon shape, and the feasible path is obtained by searching and processing the connected space. This method is flexible and simple, and can adapt to the arbitrary change of starting point and target point [13].

Figure 2 is a modeling diagram of C space. The starting point, target point and obstacle are given in advance, and navigation space is divided into free space and obstacle space. Each feasible path of unmanned craft can be mapped to C space, or vice versa. Unmanned craft has multiple paths from the starting point to the end point, and each path is composed of multiple nodes. Finding the optimal path without crossing obstacles through certain methods is equivalent to the process of finding a group of appropriate path nodes.

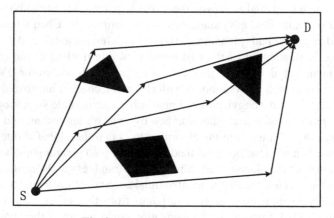

Fig. 2. C-space environment model

When using the particle swarm optimization algorithm to solve the path planning problem of unmanned craft, each particle represents a feasible path, and the dimension of the particle corresponds to the number of control nodes in the feasible path. The feasible path can be expressed as:

$$P[p_1(x_1, y_1), p_2(x_2, y_2), \cdots p_d(x_d, y_d)] \tag{1}$$

where $p_1, p_2 \cdots p_d$ is the path node. For a simple map, an optimal path can be found after 3–5 nodes.

3   Path Planning Based on Improved Particle Swarm Optimization Algorithm

3.1   Standard Particle Swarm Optimization

In 1995, American scholars Kennedy and Ebehart proposed the particle swarm optimization algorithm inspired by the bird foraging model [14]. PSO is a bionic intelligent algorithm based on the research of foraging behavior of birds [15–18]. It finds the optimal solution through the random search and information exchange of individuals in the population. In the early stage, the birds were randomly distributed in the search area. The birds only knew that there was food in the area but did not know the location of the food. The individuals assisted each other by judging the distance between the food and their position, and guided the birds to fly to the optimal individual and eventually gathered near the food.

The particle swarm optimization algorithm maps the solution of the problem to be solved to the position of each particle, guides the particle movement through the information transmission inside the particle swarm, and finally obtains the optimal solution of the problem to be solved. Particle swarm optimization can be used to describe the D-dimension problem to be solved as:

$$\begin{cases} \min f(x) = f(x_1, x_2, \cdots x_d) \\ s.t. x_{d\,min} \leq x_i \leq x_{d\,max}, i=1, 2, \cdots D \end{cases} \tag{2}$$

where $f(x)$ stands for particle fitness function, $X_i = [x_{i1}, x_{i2}, \ldots, x_{id}]$ is the position of the i particle, $x_{d\,max}$, $x_{d\,min}$ represents the upper and lower limits of the d-dimensional search space respectively. The velocity of the particle is expressed as $V_i = [v_{i1}, v_{i2}, \ldots, v_{id}]$. In the iteration process, the particle updates its speed and position by tracking the particle individual extreme value P_{best} and the global optimal value G_{best} of the population, the iterative formula is shown in Eq. (3) and Eq. (4) as follows:

$$v_i^{k+1} = \omega v_i^k + c_1 \times r_1 \times (P_{best}^k - x_i^k) + c_2 \times r_2 \times (G_{best}^k - x_i^k) \tag{3}$$

$$x_i^{k+1} = x_i^k + v_i^{k+1} \tag{4}$$

where v_i^{k+1} represents the velocity of particle i at the $(k + 1)$ th iteration, ω is the inertia weight factor, the inertia weight factor shows the correlation between the current generation particle velocity and the previous generation particle velocity and is an important parameter of the extended particle search area. If the inertia weight is too large, the particles will maintain a large motion speed, which improves the global search ability, but it is easy to oscillate at the optimal solution. If the value is too small, the particles will slow down or even fail to reach the optimal position. c_1, c_2 is the learning factor, it reflects the learning degree of particles to individual information and population information. If c_1 is too large and c_2 is too small, the particle tends to use its own experience to search, which is not conducive to global optimization. If c_1 is too small and c_2 is too large, the particle tends to accept the population experience and move towards the global extreme value, but it may fall into the danger of local optimization. r_1, r_2 is the random number between [0,1], which is used in combination with the learning factor to enable the particle to randomly use the individual extreme value and the global optimal extreme value, increasing the randomness of search.

3.2 Chaotic Disturbance Optimization

When PSO falls into the precocious convergence state, chaos optimization is used to match the search process to the traversal process of chaotic orbit, so as to improve the ability of particles to jump out of the local optimal. The algorithm in this paper determines the degree of population aggregation according to the fitness variance. When the calculated fitness deviation is less than the given deviation, the particle with the best fitness value is selected to perform chaotic search. The formula for calculating population fitness variance is:

$$\sigma^2 = \sum_{i=1}^{n} (\frac{f_i - f_{av}}{f})^2 \tag{5}$$

where f_i is the current fitness value of particle i, f_{av} is the average current fitness value of the population, and f is the normalization factor.

$$f = \begin{cases} \max|f_i - f_{av}| & \max|f_i - f_{av}| > 1 \\ 1 & \text{other} \end{cases} \tag{6}$$

The introduction of chaos theory is bound to increase the complexity and operation time of the algorithm. In this paper, only a certain number of chaotic searches are carried out for the global optimal position. The specific operation steps are as follows:

1) During each iteration, the global optimal position $x_g = (x_{i1}, x_{i2}, ..., x_{iD})$ is selected as the initial chaotic quantity for chaos optimization.
2) Each dimension component of the decision variable x_g is mapped into a chaotic variable within the range of [0,1] according to Eq. (5), and the corresponding chaotic variable $cx_d = (cx_{i1}, cx_{i2}, ..., cx_{id})$ is obtained.

$$cx_{id}^k = \frac{x_{id}^k - x_{\min,d}}{x_{\max,d} - x_{\min,d}}, d = 1, 2, \ldots, D \tag{7}$$

where $x_{\max,d}$ and $x_{\max,d}$ respectively represent the upper and lower bounds of the d-dimensional search space.

3) Using the Tent chaotic mapping formula, each dimension variable is iterated for K times in the chaotic space ($K \leq 0.5n$, n is the number of particle swarm particles), and a set of chaotic sequence $CX = (cx^1, cx^2, \ldots, cx^K)$ is obtained.

$$cx_{id}^{k+1} = \begin{cases} cx_{id}^k/0.4 & (0 \leq cx_{id}^k \leq 0.4) \\ (1 - cx_{id}^k)/(1 - 0.4) & (0.4 < cx_{id}^k < 1) \end{cases} \tag{8}$$

4) Map the chaos variables back to the decision space to generate the feasible solution sequence $X = (x^1, x^2, \ldots, x^K)$.

$$x_{id}^{k+1} = x_{\min,d} + cx_{id}^{k+1}(x_{\max,d} - x_{\min,d}) \tag{9}$$

5) The generated feasible solution is used to replace the K particles with the worst fitness value in the particle swarm, evaluate the fitness value of each particle in the population, and update the global optimal position.

3.3 Search Strategy Optimization for Follower Bees

Particle swarm optimization has a simple motion mode and can converge quickly. However, as the number of iterations goes on, the population will become more and more dependent on the information of the global optimal particle and the individual optimal particle, resulting in insufficient local search ability. Swarm algorithm is a kind of guided random search, which can generate a certain number of following bees to search near the honey source according to the quality of the solution, and has a good local search ability. The combination of PSO and bee colony algorithm can give consideration to the fast convergence and local search ability of the algorithm, and effectively improve the diversity of the population.

In the hybrid algorithm, the particles of the particle swarm optimization algorithm are regarded as the leading bees. After completing a particle swarm optimization, the better solution in the global scope is selected as the honey source for the following bees to search. The aggregation degree of the follower bees is determined by means of roulette. The better the honey source is, the more attractive it will be to the follower bees. The following probability is shown in Formula (10):

$$P_i = \frac{1/Fit_i}{\sum_{i=1}^{m} (1/Fit_i)} \tag{10}$$

where $1/Fit_i$ represents the inverse of the fitness of particle, the shorter the path, the smaller the fitness, the more attractive it is to the following bees, m represents the number of following bees.

The following bees randomly search the current nectar source, and the search radius is shown in Eq. (11):

$$R_d = \frac{X_{\max d} - X_{\min d}}{10}, \quad d = 1, 2, \cdots, D \tag{11}$$

where $X_{\max d}$ and $X_{\min d}$ respectively represent the upper and lower bounds of the d-dimensional search space. This formula indicates that each follower bee only searches for optimization within one cell of its selected nectar source. If a better nectar source is found within this range, this nectar source will be used to replace the previously selected nectar source. If no better nectar source is found, the original nectar source will be retained. The introduction of this search operator enhances the search of the local nectar source. Even if the better solution is not found, the accuracy of the original particle swarm optimization algorithm will not be damaged.

4 Simulation Experiment

Using Matlab to write algorithm program for path planning simulation research, compare the planning performance of different algorithms. The particle swarm size was set to 50, all algorithms were iterated 200 times, and 5 control nodes were selected. The starting point coordinates are (0,0), the target point coordinates are (50,50), and the map is a square map with 50 unit lengths. The coordinates and radii of obstacles are shown in the following table (Table 1):

Table 1. Obstacle information table

	Obstacle 1	Obstacle 2	Obstacle 3	Obstacle 4	Obstacle 5	Obstacle 6
Horizontal coordinate	5	12	17	20	37	41
Vertical coordinate	18	8	20	40	22	42
Radius	3	4	5	7	8	4

The standard PSO, chaotic Particle Swarm optimization (CPSO) and chaotic particle Swarm optimization (CPSO-ABC) algorithms were respectively used for path planning simulation experiments. The convergence rate of the algorithms is shown in Fig. 3.

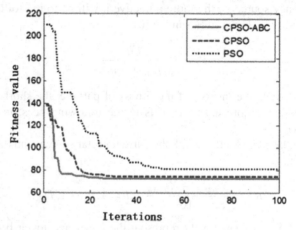

Fig. 3. Convergence curve of the algorithm

It can be seen from the convergence curve that the hybrid particle swarm optimization algorithm with chaotic strategy and following bee search operator has faster convergence speed and higher search accuracy, and the optimal solution can be found when the algorithm is carried out for about 26 times. In the same case, the standard particle swarm optimization algorithm and chaotic particle swarm optimization algorithm need about 48 and 30 iterations respectively.

The algorithm was tested 10 times to record the distance of the optimization path and the number of iterations to reach the final result. The results are shown in Table 2.

Table 2. Comparison of the effects of three algorithms

Number	Standard particle swarm		Improved particle swarm		Improved Particle swarm - Swarm	
	Length of path	Number of iterations	Length of path	Number of iterations	Length of path	Number of iterations
1	83.161	48	74.786	38	72.651	28
2	80.887	45	74.995	34	72.602	29
3	82.412	50	74.664	37	72.644	29
4	79.665	51	74.471	36	72.660	27
5	83.665	46	74.968	38	72.655	26
6	83.189	48	74.432	39	72.686	27
7	81.332	48	75.001	37	72.656	27
8	82.501	49	74.890	37	72.602	31
9	83.881	47	75.012	30	72.679	31
10	79.627	48	74.697	39	72.604	30

Based on the analysis of the data in the table, it can be seen that the path planning based on chaotic particle swarm optimization algorithm is significantly better than the standard particle swarm optimization algorithm and chaotic particle swarm optimization algorithm in terms of path length and iteration times. It can find the path with good quality in a short time, and its search arrival path length is stable. The path searched by the standard particle swarm optimization algorithm fluctuates greatly, with the average path length maintaining at 81 unit lengths, the path length searched by the improved particle swarm optimization algorithm is about 74 unit lengths, and the path length searched by the improved particle swarm optimization algorithm is about 72 unit lengths, with the highest search accuracy and the least number of iterations.

The typical path planning diagram obtained by selecting three algorithms is shown as follows (Figs. 4, 5 and 6):

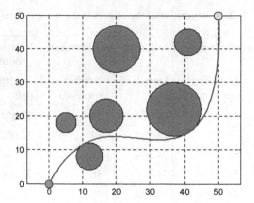

Fig. 4. Standard particle swarm path planning

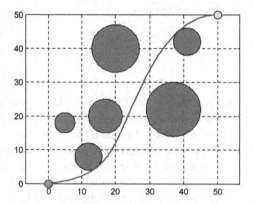

Fig. 5. Chaotic particle swarm path planning

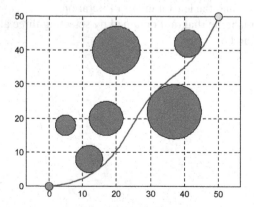

Fig. 6. Path planning with chaotic particle swarm - swarm hybrid algorithm

By comparing the path planning diagram of the three algorithms, it can be seen that the standard particle swarm convergence is slow, and it takes 68 iterations on average to find the optimal path, and the searched path has large randomness and low search accuracy. The improved PSO hybrid algorithm can quickly find the optimal solution, and its convergence speed is fast and the search results are stable. The path planned by the improved PSO algorithm is flat with fewer turns, which can take into account the safety and economy of unmanned boat navigation. Simulation results show the effectiveness and practicability of the proposed algorithm.

5 Conclusion

This paper mainly designs a global path planning method for unmanned craft. Firstly, the existing PSO algorithm is improved. Firstly, chaotic perturbation strategy is introduced into the algorithm. When the particles gather in the late stage of the algorithm, chaotic particles are used to replace some of the particles with poor fitness. Secondly, the search strategy of following bees in artificial bee colonies is integrated into the algorithm to increase the local search intensity. Simulation results show the effectiveness of the algorithm.

References

1. Ren, G., Liu, P., He, Z.: A global path planning algorithm based on the feature map. IET Cyber-Syst. Robot. **4**(1), 15–24 (2022)
2. Ganesan, S., Natarajan, S.K., Srinivasan, J.: A global path planning algorithm for mobile robot in cluttered environments with an improved initial cost solution and convergence rate. Arab. J. Sci. Eng. **47**(3), 3633–3647 (2022). https://doi.org/10.1007/s13369-021-06452-3
3. Tao, Y., et al.: A mobile service robot global path planning method based on ant colony optimization and fuzzy control. Appl. Sci. **11**(8), 3605 (2021)
4. Xiaofei, Y., Yilun, S., Wei, L., Hui, Y., Weibo, Z., Zhengrong, X.: Global path planning algorithm based on double DQN for multi-tasks amphibious unmanned surface vehicle. Ocean Eng. **266**, 112809 (2022)
5. Lin, J., Pan, L.: Multiobjective trajectory optimization with a cutting and padding encoding strategy for single-UAV-assisted mobile edge computing system. Swarm Evol. Comput. **75**, 101163 (2022)
6. Bai, J.H., Oh, Y.-J.: Global path planning of lunar rover under static and dynamic constraints. Int. J. Aeronaut. Space Sci. **21**(4), 1105–1113 (2020). https://doi.org/10.1007/s42405-020-00262-x
7. Ou, J., Hong, S.H., Ziehl, P., Wang, Y.: GPU-based global path planning using genetic algorithm with near corner initialization. J. Intell. Rob. Syst. **104**(2), 34 (2022)
8. Wang, D., Zhang, J., Jin, J., Liu, D., Mao, X.: Rapid global path planning algorithm for unmanned surface vehicles in large-scale and multi-island marine environments. PeerJ. Comput. Sci. **7**, e612 (2021)
9. Lin, X.M., et al.: research on global path planning method of mobile robot based on BAS. J. Phys: Conf. Ser. **1284**, 012014 (2019)
10. Zhao, H., Zhou, H., Yang, G.: Research on global path planning of artificial intelligence robot based on improved ant colony algorithm. J. Phys: Conf. Ser. **1744**(2), 022032 (2021)

11. Hu, Z., Wang, Z., Yin, Y.: Research on three dimensional global path planning of unmanned underwater vehicle. J. Phys: Conf. Ser. **1894**(1), 012014 (2021)
12. Tian, S., Li, Y., Li, J., Liu, G.: Robot global path planning using PSO algorithm based on the interaction mechanism between leaders and individuals. J. Intell. Fuzzy Syst. **39**(4), 4925–4933 (2020)
13. Al-Gabalawy, M.: Path planning of robotic arm based on deep reinforcement learning algorithm. Adv. Control Appl.: Eng. Ind. Syst. **4**(1), e79 (2022)
14. He, C., et al.: Accelerating large-scale multiobjective optimization via problem reformulation. IEEE Trans. Evol. Comput. **23**(6), 949–961 (2019)
15. Wang, X., Moriyama, K., Brooks, L., Kameyama, S., Matsuno, F.: Real-time global path planning for mobile robots with a complex 3-D shape in large-scale 3-D environment. Artif. Life Robot. **26**(4), 494–502 (2021). https://doi.org/10.1007/s10015-021-00706-x
16. Huang, P.Q., Wang, Y.: A framework for scalable bilevel optimization: identifying and utilizing the interactions between upper-level and lower-level variables. IEEE Trans. Evol. Comput. **24**(6), 1150–1163 (2020)
17. Peng, C., Qiu, S.: A decomposition-based constrained multi-objective evolutionary algorithm with a local infeasibility utilization mechanism for UAV path planning. Appl. Soft Comput. J. **118**, 108495 (2022)
18. Kyriakakis, N.A., Marinaki, M., Matsatsinis, N., Marinakis, Y.: A cumulative unmanned aerial vehicle routing problem approach for humanitarian coverage path planning. Eur. J. Oper. Res. **300**(3), 992–1004 (2022)

A Self-adaptive Single-Objective Multitasking Optimization Algorithm

Xiaoyu Li[1,3], Lei Wang[1,2(✉)], Qiaoyong Jiang[1], Wei Li[1], and Bin Wang[1]

[1] The Key Laboratory of Network Computing and Security Technology of Shaanxi Province, Xi'an University of Technology, Xi'an 710048, China
leiwang_lw@126.com
[2] The Key Laboratory of Industrial Automation of Shaanxi Province, Shaanxi University of Technology, Hanzhong 723001, China
[3] School of Electronic and Information Engineering, Ankang University, Ankang 725000, China

Abstract. Evolutionary multitasking optimization algorithms have been presented for dealing with multiple tasks simultaneously. Many studies have proved that EMTOs often perform better than conventional single-task evolutionary. Transferring knowledge plays a very important role in multitask optimization algorithms. Many existing methods transfer elite solutions between tasks to improve algorithm performance, however, these methods may or produce negative transfer if inter-task similarity is low or irrelevant. This paper presents a self-adaptive multitasking optimization algorithm, SAMTOA, to find more valuable transferred solutions between tasks. In SAMTOA, the solutions for the next generation transfer are adaptively determined based on the successful transfer solutions of the previous generation. The method can effectively reduce the probability of transferring useless solutions between tasks and effectively utilize valuable solutions in tasks to improve the efficiency of knowledge transfer between tasks. Experimental results on single-objective multitasking optimization benchmark problems indicate that SAMTOA outperforms the other the state-of-the-art EMTO algorithms.

Keywords: Multitasking Optimization · Knowledge transfer · Differential evolution

1 Introduction

Evolutionary multitasking optimization(EMTO) is a rising search method in the evolutionary community. The main feature of evolutionary multitasking is to obtain better optimization performance by using underlying complementary knowledge between different tasks. EMTO has attracted extensive attention from researchers due to its effectiveness and great potential to handle related tasks through knowledge transfer. The multifactorial evolutionary algorithm (MFEA) [9] was first proposed by Gupta et al. in 2016. Subsequently, to further improve the performance of the EMTO algorithm many EMTO algorithms

L. Pan et al. (Eds.): BIC-TA 2022, CCIS 1801, pp. 117–130, 2023.
https://doi.org/10.1007/978-981-99-1549-1_10

(EMTOAs) have been proposed by researchers [4,10,11,21]. In [8], the authors first introduced differential evolutionary algorithm and particle swarm algorithm into MTO. In order to achieve positive transfer across tasks, an EMTOA via explicit autoencoding was proposed in [7]. A multi-population multifactorial evolution framework was proposed in [12], which used an adaptive control mechanism based on the success improvement rate to adjust the mating probability between tasks. The correlation between tasks is another factor that affects the performance of EMTOAs. In [2], the authors used online transfer parameter estimation methods to improve the MFEA's performance. The approach used online learning and the correlation between different tasks to reduce negative transfer. In [23], a self-regulated evolutionary multitasking optimization algorithm was proposed, abbreviated as SREMTO, which adjusted the interaction intensity between tasks according to their correlation. In SREMTO, the capability vectors are used to evaluate the correlation between tasks. In [5], a many-task evolutionary algorithm (MaTEA) for solving dynamic many-tasking was proposed. MaTEA used an external archives to store task information and adaptively adjusted the probability of interaction between tasks using inter-task correlation and knowledge transfer success rates. In [15], the authors used a selective crossover to generate offspring solutions based on gene similarity between tasks and used mirror transformation to enlarge the search range of the algorithm and improve its diversity. Many EMTOAs have been presented [13,19] to solve real-world optimization problems, such as fuzzy system optimization, parameters extraction of photovoltaic models, and job shop scheduling.

EMTO mainly relies on the underlying complementarity between tasks to facilitate the co-evolution of multiple tasks. In EMTO, in the absence of a priori knowledge about complementarities between different tasks, it becomes crucial to capture the complementarities between tasks through the dynamics of evolving populations. Existing EMTO algorithms mainly focus on designing strategies to capture inter-task correlations and reduce negative transfer between tasks by developing effective transfer strategies [17]. However, as the search proceeds, how to find and transfer valuable knowledge across tasks has received little attention. To tackle this issue, this paper proposes an improved multitasking optimization algorithm (SAMTOA). In SAMTOA, we use a novel success history-based knowledge transfer strategy that adaptively selects knowledge for transfer between tasks. SAMTOA uses differential evolution algorithm as the task solver, and uses different mutation strategies to generate individuals to improve the performance of the algorithm.

The rest of this study is arranged as below. Section 2 describes the related work of SAMTOA. Section 3 introduces the proposed SAMTOA. Section 4 shows and analyzes experimental results. Finally, Sect. 5 draws the conclusions.

2   Related Work

2.1   Evolutionary Multitasking Optimization (EMTO)

EMTO algorithm handles multiple optimization tasks simultaneously with a single search [9]. The result obtained by the EMTO operation is the set of optimal solutions for each constituent task. Assume that there are K different minimization tasks to be processed simultaneously. Let x_k $(k = 1, 2, ..., K)$ be a feasible solution for the kth task. A single-objective multitasking problem can be expressed as

$$\{x_1^*, \cdots, x_K^*\} = \arg\min\{f_1(x_1), \cdots, f_K(x_K)\} \tag{1}$$

where $f_1(x_1) : \Omega_1 \rightarrow R$ is an objective function. Ω_1 is the decision space corresponding to task 1, x_1, $x_1^* \subseteq \Omega_1$.

2.2   Differential Evolution (DE)

Differential Evolution, firstly proposed by Storn and Price, is an efficient and robust EA for solving global numerical optimization [18]. DE produces new offspring individuals mainly through mutation and crossover.

(1) Mutation
The mutation strategy is a crucial factor affecting the performance of the difference algorithm. Different strategy choices and parameter settings will bring different results to the optimization of the problem [20]. Several widely used mutation strategies are shown below [16,18]:

DE/rand/1:
$$V_i^g = X_{r1}^g + F \cdot (X_{r2}^g - X_{r3}^g) \tag{2}$$

DE/rand/2:
$$V_i^g = X_{r1}^g + F \cdot (X_{r2}^g - X_{r3}^g) + F \cdot (X_{r4}^g - X_{r5}^g) \tag{3}$$

DE/best/1:
$$V_i^g = X_{best}^g + F \cdot (X_{r1}^g - X_{r2}^g) \tag{4}$$

DE/best/2:
$$V_i^g = X_{best}^g + F \cdot (X_{r1}^g - X_{r2}^g) + F \cdot (X_{r3}^g - X_{r4}^g) \tag{5}$$

DE/current-to-rand/1:
$$V_i^g = X_i^g + F \cdot (X_{r1}^g - X_i^g) + F \cdot (X_{r2}^g - X_{r3}^g) \tag{6}$$

DE/current-to-best/1:
$$V_i^g = X_i^g + F \cdot (X_{best}^g - X_i^g) + F \cdot (X_{r1}^g - X_{r2}^g) \tag{7}$$

where F is the scaling factor. X_{best} and X_i are the optimal and current individuals in the population, respectively. The indexes r1, r2, r3, r4, and r5 are randomly chosen from 1 to N, where $r1 \neq r2 \neq r3 \neq r4 \neq r5 \neq i$.

(2) Crossover

In DE, the binomial crossover is used to generate the components of the trial individual U. The binomial crossover operator can be expressed as below:

$$U_{i,j}^{g+1} = \begin{cases} V_{i,j}^g, & if \ rand < CR \ or \ j = j_{rand} \\ X_{i,j}^g, & otherwise \end{cases} \quad (8)$$

where, $U_{i,j}$ represents the jth component of the $i - th$ trial vector. CR denotes the crossover rate, j_{rand} is an integer randomly produced between 1 and D. rand is a random variable within $[0, 1]$.

The mutation operation will cause some components of U_i^{g+1} to violate the bounds constraints, so the mutation variable needs to reset [1].

$$U_{i,j}^{g+1} = \begin{cases} L_j + rand(0,1).(U_j - L_j), & if \ U_{i,j}^{g+1} \notin [L_j, U_j] \\ U_{i,j}^{g+1}, & otherwise \end{cases} \quad (9)$$

where U_j is upper bound of the jth dimension of search space. L_j is the lower bound of the jth dimension of search space.

Due to DE's simplicity and high efficiency, many DE-based MTO algorithms have been proposed by researchers over the past years [4,7,22]. These algorithms introduce the DE search mechanism into the EMTO field. In SAMTOA, we use the DE search mechanism to produce offspring.

3   The Proposed Method

To suppress negative transfer and effectively utilize the common knowledge between tasks, we introduce a historical success information-based transfer strategy into the MTO algorithm, called self-adaptive single-objective multitasking optimization Algorithm (SAMTOA). The SAMTOA algorithm uses DE as a solver combined with a multitasking and multi-population framework to implement. In this section, we firstly introduce our proposed transfer strategy, and then give the main framework of the SAMTOA algorithm.

3.1   Transfer Strategy

In this paper, a transferred solution is a positive transfer if its fitness value on the target task outperforms the average fitness value of the population where the target task is located. The positive transfer solutions have a high probability of improving the performance of the target task in the next generation. We take the positive transfer solution and its nearest neighbors generated by the gth generation in the source task as the transfer solutions of the $g + 1$ generation. Because these solutions are more likely to improve the performance of the target tasks. The specific operation process of knowledge transfer in SAMTOA is given below.

In the case of $g = 1$, the transferred solutions are elite individuals chosen from the source task. When $(g > 1)$, the acquisition of the transfer solutions of the gth generation are determined based on the positive transfer solutions obtained in the $(g-1)th$ generation. The transfer solutions are composed of the positive transfer solutions of the $(g-1)th$ generation and their neighborhoods. The neighborhood of a positive transfer solution is composed of several solutions near it in the search space of the source task where it is located. In the $(g-1)th$ generation, if the solution transferred from one task to another does not have a positive transfer solution, then the transfer solution of the gth generation is randomly selected from the current source task.

3.2   Main Framework of SAMTOA

Algorithm 1 demonstrates the framework of SAMTOA. First, we assign an independent population P_i to each task $task_i$, $i = 1, 2, \ldots, k$, and randomly initialize each population in the unified search space. Next, the individuals belonging to a specific population are assessed in an associated task, e.g., evaluate the individuals in P_i on $task_i$. Subsequently, perform the transfer operation from source task $task_j$ to current task $task_i$, using the strategy presented in section 3.1, and evaluate the transfer solution on the current task $task_i$. Use the method proposed in Sect. 3.1 to generate successful transfer solution, denoted as $STarc_{j \longrightarrow i}$. The unsuccessful transfer solution is denoted as $FTarc_{j \longrightarrow i}$. Update the neighborhood of the next generation transfer solutions using the successful transfer solutions. If no positive transfer solution satisfies the condition in the current generation, we randomly select the solution to be transferred in the new generation from the co-evolution tasks. Then, $Tarc_{j \longrightarrow i} = \{STarc_{j \longrightarrow i} \bigcup FTarc_{j \longrightarrow i}\}$ is incorporated into the target population P_i to form the joint population TP_i and performs mutation and crossover operations on TP_i to generate a new offspring population $Child_i$. Different mutation strategies of differential evolution algorithms have their unique advantages. Some mutation strategies focus on exploitation, some focus on exploration, and some can achieve a balance between exploration and exploitation. Different mutation strategies with different advantages can also complement each other, and the effective integration of different mutation strategies is an effective way to promote the performance of DE. Two mutation strategies are used in SAMTOA, "DE/pbest/1," which is good at exploitation, and "DE/rand/1," which is good at exploration. These mutation strategies for generating offspring individuals are adaptively adjusted according to the fitness value of the individuals. The specific operation is as follows.

$$
V_i = \begin{cases} x_{pbest} + F \cdot (x_{r1} - x_{r2}), & if \ fitness(i) < mean(fitness) \\ \\ x_{r1} + F \cdot (x_{r2} - x_{r3}), & otherwise \end{cases} \tag{10}
$$

where x_{r1}, x_{r2}, and x_{r3} are randomly picked up from TP_i. x_{pbest} is chosen from the elite solution set of TP_i. $p * N_i$ is the size of the elite solution set. F is the mutation factor.

Algorithm 1. Pseudo code of SAMTOA

Require: Maximum function evaluation FESmax; All task $task_1$, ..., $task_K$;
Population size N; Component population size N_i ($N_i = \frac{N}{K}$, i=1,2,...,K);
Number of transfer solutions $Tnum$;
Number of successfully migrated solutions $tempnum$;
Neighborhood size $Nnum$;
Ensure: The optimal solution for all problems;
 1: Initialize $Tnum$=4
 2: **for** each task $task_i$ **do**(i=1,2,...,K)
 3:    Initialize the population P_i corresponding to task $task_i$
 4:    Evaluate each individual in population P_i on $task_i$ (i=1,2,...,K)
 5: **end for**
 6: Set $Nnum$=3, g=1, FES=N;
 7: **while** FES $\leq$ FESmax **do**
 8:    **for** each task $task_i$ **do**
 9:       Set totalnumber=0
10:       **if** g $\leq$ 1 **then**
11:          Select $Tnum$ elite solutions from $task_j$ ($j \neq i, j \in 1 : K$) as transferred
    solutions denoted as $Tarc_{j \to i}$
12:       **else**
13:          **if** $tempnum$==0 **then**
14:             Randomly select $Tnum$ solutions from $task_j$ as transferred solutions
15:          **end if**
16:          Use the solution in $Tarc_{j \to i}$ as transferred solutions
17:       **end if**
18:       Evaluate transfer solutions on $task_i$ (i=1,2,...,K)
19:       Use the strategy presented in section 3.1 to generate successful migration
    solutions denoted as $STarc_{j \to i}$
20:       Use $P_i^g \bigcup Tarc_{j \to i}$ to generate offsprings $Child_i$
21:       Calculate combined population $UniPOP_i^g = P_i^g \bigcup Tarc_{j \to i} \bigcup Child_i$
22:       Evaluate all individuals in $UniPOP$
23:       Select N_i fittest solutions from $UniPOP_i^g$ into P_i^{g+1}
24:       Update the neighborhoods of successfully migrated solutions and $Tarc_{j \to i}$
25:       g=g+1;
26:       totalnumber=totalnumber+Tnum
27:    **end for**
28:    $FES = FES + N + totalnumber$
29: **end while**
30: return the best agent fitness ;

After then, the offspring population $Child_i$ is evaluated on the current task $task_i$ and incorporated into the population TP_i to produce a joint population $UniPOP_i$. Lastly, the $UniPOP_i$ is evaluated on the current task T_i, and the N_i fittest solutions are picked up from it to enter the next generation population. In case of the stopping criterion is satisfied, the final optimal solution is output. Otherwise, the iterative operation continues. $fitness(i)$ represents the fitness of the ith individual. $mean(fitness)$ denotes the average fitness of TP_i.

4   Experiments and Analysis

In this part, to comprehensively assess the superiority of SAMTOA, we compare *SAMTOA* with seven state-of-the-art *MTO* algorithms on nine single-object multitasking benchmark problems introduced in [6] and conduct an in-depth analysis of the results.

4.1   Test Benchmark and Experimental Settings

The single-objective multitasking test benchmarks is used to test SAMTOA performance in our experiments [6]. The test benchmark consists of 9 multitasking optimization problems, each consisting of two single-objective minimization tasks. The problems in the test benchmark according to the globally optimal degree of intersection and Spearman's correlation coefficient can be expressed as Completely Intersecting and Highly Similar (CIHS), Completely Intersecting and Medium Similarity (CIMS), Completely Intersecting and Low Similarity (CILS), Partially Intersecting and Highly Similar (PIHS), Partially Intersecting and Medium Similarity (PIMS), Partially Intersecting and Full Intersecting Low Similarity (PILS), No Intersecting and High Similarity (NIHS), No Intersecting and Medium Similarity (NIMS), No Intersecting and Low Similarity (NILS). In the following experiments, we compared SAMTOA to seven excellent *MTO* algorithms, i.e., MFEA [9], MFDE [8], SBO [14], LDA-MFEA [3], MTGA [19], GMFEA [11], and EMT-EGT [7] to show the comprehensive performance of SAMTOA. These algorithms employ different knowledge transfer strategies. MFEA is the first evolutionary multitasking optimization algorithm, which uses assortative mating to obtain knowledge transfer between tasks. MFDE first introduces the DE algorithm into multi-task optimization and uses the mutation strategy in the DE algorithm to perform the knowledge transfer between different tasks. SBO enables cross-task knowledge transfer through individual replacement between tasks. LDA-MFEA employs a linear space mapping strategy to relieve negative transfer between tasks. EMT-EGT utilizes autoencoding to realize explicit gene transfer across tasks. MTGA uses inter-task bias to facilitate knowledge transfer between different tasks. GMFEA facilitates inter-task knowledge transfer using decision variable transformation and recombination strategies. To take full advantage of the helpful information between tasks, MPEMTO employs two information transfer strategies for inter-task knowledge transfer.

The maximum function evaluation is set as 1×10^5 for all algorithms, and all algorithms are conducted 20 repetitions on each task. The experimental results are the mean value and standard deviation of the optimal objective value for each algorithm over 20 repetitions. The parameters setting fot MFEA, MFDE, SBO, LDA-MFEA, MTGA, GMFEA, MPEMTO, and EMT-EGT are based on their original publications. The parameters in SAMTOA are set as below.

NPi: Population size corresponding to the component task is set as 100.
$Tnum$: Number of initial transferred solution is set as 4.
F: Mutation factor is set as 0.5.
CR: Crossover rate is set as 0.6.

4.2   Experimental Results

The experimental results are the mean and standard deviation of each algorithm run 20 times independently on each task, as shown in Table 1. The smaller the values of mean and standard deviation, the more accurate and stable the algorithm is. The best results on each task are shown in bold. To achieve reliable statistical results, the statistical testing method is used to analyze the experimental results. Additionally, *Wilcoxon* signed-rank test with significance level $\alpha = 0.05$ is employed to test the statistical significance between SAMTOA and other comparison algorithms. In Table 1, the symbol "+" shows that SAMTOA performs better than the compared algorithm; the symbol "−" represents that SAMTOA is worse than the compared algorithm; the symbol "≈" indicates that the results obtained by SAMTOA are not statistically different from the comparison algorithm.

As shown in Table 1, compared with MFEA, MFDE, SBO, LDA-MFEA, MTGA, GMFEA, and EMT-EGT, SAMTOA obtains the optimal solutions on 15 out of 18 tasks. From this result, we can see that the proposed SAMTOA algorithm can effectively solve the single-objective MFO problem. SAMOTA achieves better performance compared to other algorithms on all CI problems. Compared with MFEA, SBO, LDA-MFEA, GMFEA and EMT-EGT, SAMTOA achieves better solutions on all optimization tasks, which means that the transfer mechanism in SAMTOA is more efficient than that in MFEA, SBO, and GMFEA. SAMTOA is slightly worse than MFDE, LDA-MFEA, and MTGA on the Rastrigin function in the partial intersection (PI) problem and no intersection(NI) problem, which indicates that the knowledge from other tasks is detrimental to SAMTOA. This situation is likely because the transfer solution and its neighborhoods are stuck in local optima, and SAMTOA evolves to a local optimum region under the guidance of these solutions. The Rastrigin function has multiple local optimums. As evolution progresses, the population of Rastrigin function gradually converges to multiple different search regions. DE uses a random transfer mechanism and randomly selects several individuals from the constituent tasks for transfer each time, which can better maintain the diversity of the population. Therefore, DE is not easy to fall into local optima, which makes DE has better performance than SAMTOA on PIHS(T1) and NILS(T1). For PI problems, the global best variables of the constitutive tasks are partially different. In the NI problems, the variables in the global optima of the two tasks are totally different. If the tasks in NI and PI interact directly, they will provide useless knowledge to each other, which is prone to negative transfer. To mitigate the negative transfer between unrelated tasks, LDA-MFEA introduces a linear transformation strategy to construct highly similar gene expression spaces between individuals belonging to different tasks. MTGA uses inter-task bias to reduce inter-task variance and suppress negative transfer. Hence LDA-MFEA and MTGA perform better on PIHS(T1) and NILS(T1). In addition, Fig. 1 depicts the average convergence graphs for 20 independent runs of the eight comparison algorithms on the nine benchmark SOMTO problems. In these convergence curves, the x-axis denotes the consumption of function evaluation, and the y-axis denotes the mean

Table 1. Comparison results of SAMTOA and seven comparison algorithms on SOMTO problems.

Problem	Task	MFEA Mean (std dev)	MFDE Mean (std dev)	SBO Mean (std dev)	LDA-MFEA Mean (std dev)	MTGA Mean (std dev)	GMFEA Mean (std dev)	EMT-EGT Mean (std dev)	SAMTOA Mean (std dev)
CIHS	Griewank(T1)	3.74E-01+ (6.64E-02)	8.94E-04+ (2.70E-03)	9.35E-01+ (5.64E-02)	3.38E-01+ (4.14E+03)	1.24E-02+ (8.83E-03)	1.05E+00+ (1.59E-02)	7.05E-01 + (8.22E-02)	**3.7017E-04** (1.6551E-03)
	Rastrigin(T2)	1.98E+02+ (5.16E+01)	1.89E+00+ (5.65E+00)	3.09E+02+ (2.94E+01)	1.53E-02+ (3.19E+01)	5.06E+01+ (1.64E+01)	3.68E+02+ (2.57E-01)	3.72E+02+ (6.23E+01)	**5.9731e-01** (2.6707e+00)
CIMS	Ackley(T1)	4.72E+00+ (5.49E-01)	4.45E-02+ (1.96E-01)	4.72E+00+ (2.90E-01)	3.15E-00+ (4.27E-01)	1.20E+00+ (8.22E-01)	7.15E+00+ (6.92E-01)	3.97E+00+ (2.64E-01)	**1.0631E-05** (6.2234E-06)
	Rastrigin(T2)	2.12E+02+ (6.29E+01)	1.50E-01+ (6.67E-01)	3.24E+02+ (3.90E+01)	1.69E-02+ (3.10E+01)	5.18E+01+ (1.84E+01)	4.69E+02+ (6.56E+01)	3.48E+02+ (3.78E+01)	**8.2382E-08** (9.4978E-08)
CI+LS	Ackley(T1)	2.02E+01+ (6.46E-02)	2.12E-01+ (3.10E-02)	2.11E+01+ (2.43E-01)	2.10E-01 + (2.07E-01)	2.02E+01+ (3.51E-01)	2.02E+01≈ (4.03E-02)	2.12E+01+ (3.79E-02)	**1.1051e+00** (4.7377e+00)
	Schwefel(T2)	3.71E+03+ (4.93E+02)	1.11E+04+ (1.62E+03)	4.56E+03+ (7.23E+02)	4.28E+03+ (5.11E+02)	3.41E+03+ (5.99E+02)	6.62E+03 + (6.30E+02)	4.49E+03+ (5.60E+02)	**3.5757e-01** (1.5529e+02)
PIHS	Rastrigin(T1)	5.81E+02+ (1.17E+02)	8.27E+01− (1.67E+01)	4.26E+02+ (6.73E+01)	3.17E+02− (4.88E+01)	**5.40E+01−** (1.91E+01)	9.09E+02+ (1.33E+02)	3.93E+02+ (5.11E+01)	3.9260e+02 (1.7710e+01)
	Sphere(T2)	8.82E+00+ (2.06E+00)	2.30E-05+ (2.14E-05)	1.03E+02+ (2.32E+01)	1.19E+01+ (2.63E+00)	1.17E-08+ (3.24E-08)	2.31E+02+ (5.15E+01)	5.63E+01+ (1.48E+01)	**1.3798e-08** (1.0944e-08)
PIMS	Ackley(T1)	3.53E+00+ (5.04E-01)	1.03E-01+ (4.50E-01)	5.04E+00+ (3.06E-01)	2.84E+00+ (4.63E-01)	1.51E-01+ (4.67E-01)	7.99E+00+ (8.12E-01)	3.96E+00+ (4.23E-01)	**5.7780e-05** (2.2208e-05)
	Rosenbrock(T2)	6.38E+02+ (1.96E+02)	7.22E+01+ (2.26E+01)	1.42E+04+ (4.53E+03)	5.31E+02+ (1.81E+02)	2.34E+02+ (4.00E+02)	1.15E+05+ (4.88E+04)	4.57E+03+ (2.64E+03)	**8.2951e-01** (1.3002e+01)
PILS	Ackley(T1)	2.00E+01+ (1.15E-01)	7.80E-01+ (7.39E-01)	5.43E+00+ (9.38E-01)	3.18E+00+ (4.19E-01)	1.73E+00+ (6.82E-01)	1.84E+01+ (4.78E+00)	3.94E+00+ (4.98E-01)	**1.7113e-04** (1.1306e-04)
	Weierstrass(T2)	2.11E+01+ (3.29E+00)	1.12E-01+ (2.73E-01)	5.75E+00+ (1.35E+00)	3.15E+00+ (8.29E-01)	2.64E+00+ (2.37E+00)	2.01E+01+ (5.58E+00)	7.39E+00+ (3.45E+00)	**6.9746e-04** (6.1102e-04)
NIHS	Rosenbrock(T1)	7.49E+02+ (2.68E+02)	9.53E+01+ (6.81E+01)	1.40E+04+ (5.39E+03)	8.04E+02 + (3.93E-02)	1.12E+02+ (1.09E+02)	5.15E+04+ (1.99E+04)	6.18E+03+ (3.61E+03)	**5.9825e-01** (2.2294e+01)
	Rastrigin(T2)	2.60E+02+ (4.39E+01)	3.03E+01− (1.34E+01)	3.66E+02+ (3.09E+01)	2.08E+02 + (6.78E-01)	5.96E-01− (1.94E+01)	5.00E+02+ (6.73E+01)	3.70E+02+ (6.80E+01)	5.9718e+01 (1.2234e+02)
NIMS	Griewank(T1)	4.09E-01+ (6.63E-02)	2.65E-03+ (4.18E-03)	9.24E-01+ (5.20E-02)	3.92E-C1+ (6.61E-02)	6.27E-03+ (9.74E-03)	1.05E+00+ (1.81E-02)	8.06E-01 + (8.76E-02)	**3.8957e-06** (2.2661e-06)
	Weierstrass(T2)	2.58E+01+ (3.05E+00)	3.21E+01+ (1.65E+00)	2.33E+01+ (5.19E+00)	1.41E+01 + (2.14E-00)	8.41E+00+ (6.83E+00)	2.88E+00+ (2.56E+00)	2.34E+01+ (5.02E+00)	**1.3484e+00** (1.0828e+00)
NILS	Rastrigin(T1)	6.06E+02+ (9.99E+01)	**1.01E+02+** (2.38E+01)	4.27E+02+ (5.47E+01)	3.20E+02 − (4.73E-01)	1.69E+02+ (6.90E+01)	8.71E+02+ (1.81E+02)	4.38E+02+ (8.68E+01)	3.8888e+02 (1.3976e+01)
	Schwefel(T2)	3.62E+03+ (4.60E+02)	4.07E+03+ (6.99E+02)	4.31E+03+ (5.90E+02)	4.14E-03 + (5.23E-02)	2.89E+03+ (6.43E+02)	6.65E+03+ (6.71E+02)	4.52E+03+ (5.55E+02)	**3.6111e+02** (1.9741e+02)

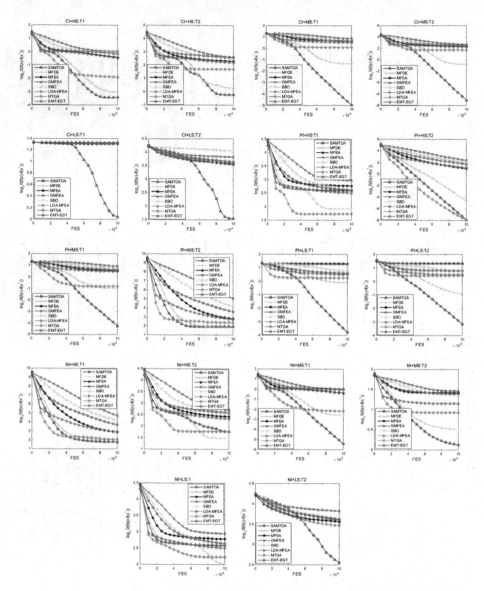

Fig. 1. The average convergence curves of SAMTOA, MFEA, MFDE, SBO, LDA-MFEA, MTGA, GMFEA, EMT-EGT on nine single-objective MFO problems

value on a log scale. It can learn from Fig. 1, SAMTOA converges faster than other compared algorithms on the majority of test problems. On the completely intersecting problems CIHS, CIMS, and CILS, SAMTOA outperforms all comparison algorithms. The CI problem consists of two tasks whose global optima have identical components in the unified search space. Therefore, transferring superior solutions between different tasks can facilitate the convergence of the algorithm. SAMTOA uses an adaptive transfer strategy to transfer the historical positive transfer solution and its neighborhoods between co-evolutionary tasks, which can effectively solve the CI problems. The PI problem consists of two tasks in which two global optimal solutions have a partial intersection. In other words, the components of the global optima solution in the co-evolution tasks are not identical. Therefore, transfer information between tasks for PI problems is not always useful. As seen from Fig. 1, SAMTOA does not achieve satisfactory convergence performance on PIHS(1), which indicates that the information transfer brings negative transfer. The reason is that the transfer solutions get stuck in a local optimum. SAMTOA outperforms other comparison algorithms on rest PI problems. The reason is that SAMTOA cannot completely avoid negative transfer. The NI problem consists of two tasks in which two global optima solutions do not have an intersection. Therefore, knowledge transfer directly between tasks may not improve task performance and is prone to negative transfer. In addition, different types of individuals in SAMTOA use different mutation strategies to generate mutant individuals, which can better play the role of each individual. As shown in Fig. 1, SAMTOA converges significantly faster than other EMT algorithms on most tasks for NI problems.

Overall, these experimental results manifest that SAMTOA has faster convergence and better solution accuracy than the compared algorithms.

4.3 Parameter Tuning

To analyze the effect of the initial number of transferred individuals $Tnum$ on SAMTOA, we set the $Tnum$ with different values (e.g., 3, 4, 5, 6, 7, 8, and 9). The Friedman test is used to verify the statistical significance among SAMTOA variants with different $Tnum$ values, and the results are illustrated in Table 2. According to the results in Table 2, it can be seen that the p-values on both

Table 2. Friedman-test results of SAMTOA with different parameter settings on SOMTO problems

	Task1		Task2	
	Tnum	Average Ranking	Tnum	Average Ranking
1	Tnum=3	3.56	Tnum=3	4.00
2	Tnum=4	2.56	Tnum=4	3.33
3	Tnum=5	3.22	Tnum=5	3.72
4	Tnum=6	4.33	Tnum=6	3.94
5	Tnum=7	4.22	Tnum=7	4.28
6	Tnum=8	5.17	Tnum=8	3.94
7	Tnum=9	4.94	Tnum=9	4.78
Statistic	11.958		2.347	
p value	0.063		0.885	

Task 1 and Task 2 are higher than 0.05, indicating that there is insignificant performance discrepancy among these SAMTOA variants. From this, we can deduce that the parameter $Tnum$ has no significant influence on the performance of SAMTOA. From the results, we can also see that $Tnum = 4$ obtains the best ranking on all tasks for all problems. Finally, we recommend $Tnum = 4$.

5   Conclusion

To effectively find common knowledge between tasks, we propose a self-adaptive multitasking optimization algorithm (SAMTOA) to address the single-objective MTO problems. SAMTOA uses the successfully transferred solutions of each generation and their neighborhoods as the transfer solutions for the next generation. Furthermore, SAMOT is compared with seven other remarkable EMTOAs on the SOMTO benchmark to assess the superiority of SAMTOA. Experimental results manifest that SAMTOA has faster convergence speed and higher solution accuracy on most test problems. The result proves that SAMTOA can effectively handle single-objective multitasking optimization problems. In our future work, we will seek an efficient neighborhood construction strategy to prevent elite individuals and their communities from simultaneously falling into local optima. Moreover, the adaptive interaction probability between tasks is designed to reduce the effect of negative transfer between unrelated problems and improve the effectiveness of multitasking optimization algorithms.

Acknowledgements. This work was supported by the National Natural Science Foundation of China under Grant (No· 62176146, No· 62272384), the National Social Science Foundation of China under Grant No· 21XTY012, the National Education Science Foundation of China under Grant No· BCA200083, and Key Project of Shaanxi Provincial Natural Science Basic Research Program under Grant 2023−JC−ZD-34.

References

1. Ahmad, M.F., Isa, N.A.M., Lim, W.H., Ang, K.M.: Differential evolution: a recent review based on state-of-the-art works. Alex. Eng. J. **16**, 22–33 (2021). https://doi.org/10.1016/j.aej.2021.09.013
2. Bali, K.K., Ong, Y., Gupta, A., Tan, P.S.: Multifactorial evolutionary algorithm with online transfer parameter estimation: MFEA-II. IEEE Trans. Evol. Comput. **24**(1), 69–83 (2020). https://doi.org/10.1109/TEVC.2019.2906927
3. Bali, K.K., Gupta, A., Feng, L., Ong, Y.S., Siew, T.P.: Linearized domain adaptation in evolutionary multitasking. In: 2017 IEEE Congress on Evolutionary Computation (CEC), pp. 1295–1302. IEEE (2017). https://doi.org/10.1109/CEC.2017.7969454
4. Cai, Y., Peng, D., Fu, S., Tian, H.: Multitasking differential evolution with difference vector sharing mechanism. In: 2019 IEEE Symposium Series on Computational Intelligence (SSCI), pp. 3039–3046 (2019). https://doi.org/10.1109/SSCI44817.2019.9002698

5. Chen, Y., Zhong, J., Feng, L., Zhang, J.: An adaptive archive-based evolutionary framework for many-task optimization. IEEE Trans. Emerg. Top. Comput. Intell. **4**(3), 369–384 (2020). https://doi.org/10.1109/TETCI.2019.2916051
6. Da, B., et al.: Evolutionary multitasking for single-objective continuous optimization: benchmark problems, performance metric, and baseline results. Technical report (2017). https://doi.org/10.48550/arXiv.1706.03470
7. Feng, L., et al.: Evolutionary multitasking via explicit autoencoding. IEEE Trans. Cybern. **49**, 1–14 (2018). https://doi.org/10.1109/TCYB.2018.2845361
8. Feng, L., et al.: An empirical study of multifactorial PSO and multifactorial DE. In: 2017 IEEE Congress on Evolutionary Computation (CEC), pp. 921–928 (2017). https://doi.org/10.1109/CEC.2017.7969407
9. Gupta, A., Ong, Y.S., Feng, L.: Multifactorial evolution: toward evolutionary multitasking. IEEE Trans. Evol. Comput. **20**(3), 343–357 (2016). https://doi.org/10.1109/TEVC.2015.2458037
10. Gupta, A., Mańdziuk, J., Ong, Y.-S.: Evolutionary multitasking in bi-level optimization. Complex Intell. Syst. 83–95 (2016). https://doi.org/10.1007/s40747-016-0011-y
11. Jing, T., Chen, Y., Deng, Z., Xiang, Y., Joy, C.P.: A group-based approach to improve multifactorial evolutionary algorithm. In: Twenty-Seventh International Joint Conference on Artificial Intelligence, IJCAI 2018, pp. 3870–3876 (2018)
12. Li, G., Lin, Q., Gao, W.: Multifactorial optimization via explicit multipopulation evolutionary framework. Inf. Sci. **512**, 1555–1570 (2020). https://doi.org/10.1016/j.ins.2019.10.066
13. Liang, J., et al.: Evolutionary multi-task optimization for parameters extraction of photovoltaic models. Energy Convers. Manag. **207**, 112509.1–112509.15 (2020). https://doi.org/10.1016/j.enconman.2020.112509
14. Liaw, R.T., Ting, C.K.: Evolutionary manytasking optimization based on symbiosis in biocoenosis. In: Proceedings of the AAAI Conference on Artificial Intelligence, vol. 33, pp. 4295–4303 (2019). https://doi.org/10.1609/aaai.v33i01.33014295
15. Ma, X., et al.: Improving evolutionary multitasking optimization by leveraging inter-task gene similarity and mirror transformation. IEEE Comput. Intell. Mag. **16**(4), 38–53 (2021). https://doi.org/10.1109/MCI.2021.3108311
16. Qin, A.K., Huang, V.L., Suganthan, P.N.: Differential evolution algorithm with strategy adaptation for global numerical optimization. IEEE Trans. Evol. Comput. **13**(2), 398–417 (2009). https://doi.org/10.1109/TEVC.2008.927706
17. Shang, Q., et al.: A preliminary study of adaptive task selection in explicit evolutionary many-tasking. In: 2019 IEEE Congress on Evolutionary Computation (CEC), pp. 2153–2159. IEEE (2019). https://doi.org/10.1109/CEC.2019.8789909
18. Storn, R., Price, K.: Differential evolution - a simple and efficient heuristic for global optimization over continuous spaces. J. Glob. Optim. **11**(4), 341–359 (1997). https://doi.org/10.1023/a:1008202821328
19. Wu, D., Tan, X.: Multitasking genetic algorithm (MTGA) for fuzzy system optimization. IEEE Trans. Fuzzy Syst. **28**(6), 1050–1061 (2020). https://doi.org/10.1109/TFUZZ.2020.2968863
20. Wu, G., Mallipeddi, R., Suganthan, P.N., Rui, W., Chen, H.: Differential evolution with multi-population based ensemble of mutation strategies. Inf. Sci. Int. J. **329**(C), 329–345 (2016). https://doi.org/10.1016/j.ins.2015.09.009
21. Xie, T., Gong, M., Tang, Z., Lei, Y., Liu, J., Wang, Z.: Enhancing evolutionary multifactorial optimization based on particle swarm optimization. In: 2016 IEEE Congress on Evolutionary Computation (CEC), pp. 1658–1665 (2016). https://doi.org/10.1109/CEC.2016.7743987

22. Yin, J., Zhu, A., Zhu, Z., Yu, Y., Ma, X.: Multifactorial evolutionary algorithm enhanced with cross-task search direction. In: 2019 IEEE Congress on Evolutionary Computation (CEC), pp. 2244–2251 (2019). https://doi.org/10.1109/CEC.2019.8789959
23. Zheng, X., Qin, A.K., Gong, M., Zhou, D.: Self-regulated evolutionary multitask optimization. IEEE Trans. Evol. Comput. **24**(1), 16–28 (2020). https://doi.org/10.1109/TEVC.2019.2904696

Improved Whale Optimization Algorithm by Multi-mechanism Fusion

Ronghang Liao[1,2], Yuanpeng Xu[1,2], Zicheng Wang[1,2], and Yanfeng Wang[1,2(✉)]

[1] School of Electrical and Information Engineering,
Zhengzhou University of Light Industry, Zhengzhou 450002, China
yanfengwang@yeah.net
[2] Henan Key Lab of Information-Based Electrical Appliances,
Zhengzhou 450002, China

Abstract. To solve the problems of insufficient global exploration ability, low convergence accuracy and slow speed of traditional whale optimization algorithm, an improved whale optimization algorithm by multi-mechanism fusion is proposed. Firstly, the algorithm uses the nonlinear parameter to coordinate the exploration and exploitation ability of the whale optimization algorithm. Secondly, combine with the Harris hawks optimization algorithm, it improves the global exploration and local optimization ability of the whale optimization algorithm. Finally, consider the important role of the fitness of the algorithm in the optimization, the Gaussian detection mechanism is proposed. The improved algorithm and other algorithms are simulated and tested on the eight variable dimension benchmark functions and design problems of tension spring. The results show that the improved whale optimization algorithm by multi-mechanism fusion has better robustness and stability, while ensure convergence accuracy and speed.

Keywords: Whale optimization algorithm · Nonlinear parameter · Harris hawks optimization algorithm · Gaussian detection mechanism

1 Introduction

The Whale Optimization Algorithm (WOA) is a population-based intelligent optimization algorithm proposed by Mirjalili et al., in 2016 [1]. The algorithm

This work was supported in part by the National Natural Science Foundation of China under Grant 62276239 and 62272424, in part by the Joint Funds of the National Natural Science Foundation of China under Grant U1804262, in part by Henan Province University Science and Technology Innovation Talent Support Plan under Grant 20HASTIT027, in part by Zhongyuan Thousand Talents Program under Grant 204200510003, in part by Zhongyuan Talents Program under Grant ZYY-CYU202012154, and in part by Henan Natural Science Foundation-Outstanding Youth Foundation under Grant 222300420095.

L. Pan et al. (Eds.): BIC-TA 2022, CCIS 1801, pp. 131–143, 2023.
https://doi.org/10.1007/978-981-99-1549-1_11

uses mathematical formulas to simulate the whale's predation behavior. Compared with the traditional meta-heuristic optimization algorithm, the whale optimization algorithm has the characteristics of simple principle, less parameter settings, and strong optimization ability. At the same time, there are also some defects such as falling into local optimum, slow convergence speed and low convergence accuracy [2,3]. Due to the existence of both advantages and disadvantages of the Whale Optimization algorithm, its applicability is limited. Therefore, this paper proposes an improved whale optimization algorithm by multi-mechanism fusion to address the problems existing on the traditional whale optimization algorithm.

For the defects of whale optimization algorithm, which is easy to fall into local optimal and has low convergence accuracy, scholars put forward many improved strategies. For example, Kaur et al. introduced chaos theory into WOA optimization process to adjust the parameters of whale optimization algorithm, enhance exploration and exploitation capacity [4]. Luo Jun et al. introduced a new position update strategy in the exploitation and exploration phases to avoid premature of the whale optimization algorithm [5]. Saha et al. adjusted the control parameters, used correction factors to reduce step size to improve the exploitation and exploration capability of the whale optimization algorithm [6].

In this paper, an improved whale optimization algorithm by multi-mechanism fusion is proposed. Firstly, introducing the nonlinear convergence parameter a, the improved whale algorithm can adapt to nonlinear problems by using the improved parameters. Secondly, referring to Harris hawks optimization algorithm [7], the shrinking encircling mechanism of whale optimization algorithm is improved to speed up individual whale's search for the optimal position, avoiding the waste of computational resources by one individual exploring at a useless position as far as possible. Finally, at the end of each iteration of the algorithm, the fitness of the whale position is updated using the Gaussian detection mechanism [8], so that the whale algorithm has a better exploration position and accelerates the convergence of the whale optimization algorithm.

2   Basic WOA Algorithm

The whale optimization algorithm simulates the whale's predation action. According to the characteristics of whale's predation, the whale's predation process is divided into three steps. The three types of position updating: contraction encircling, spiral position updating and random searching [9].

2.1   Shrinking Encircling Mechanism

The whale senses the area where the prey is located and surrounds it. Since the location of the optimal design in the hunting or search space is not consistent with the previous location, the WOA optimization algorithm assumes that the current best candidate solution is the target prey or close to the optimal solution. In this case, whales define the best search agent, the other search agents will try to change locations to the best search agent. The hunting behaviour of the shrinking encircling is described by the following formula:

$$X(t+1) = X^*(t) - A \cdot D_1 \tag{1}$$

$$D_1 = |C \cdot X^*(t) - X(t)| \tag{2}$$

where, t represents the number of current iterations, A and C are vector coefficients, $X(t)$ is the position at the current moment, $X(t+1)$ is the position at the next moment, D_1 is C times of the absolute value, that the difference between the prey position and the current whale position, and $X^*(t)$ is the position vector to obtain the optimal solution at present. If there is a better solution in the result of each iteration, the fitness value of the position at this point is less than the fitness value of $X^*(t)$, then the whale position vector should be set to the new X^* in the iteration.

2.2 Location Update

The position of exploring and updating methods of whales is divided into two types: spiral updating position and random searching. In order to simulate the position updating mode of whales at a certain time, the whales are guaranteed to choose spiral updating position or random searching mode with equal probability at the same time. Set a random number p with values in the range $[0, 1]$.

2.3 Spiral Updating Position

When $p \geq 0.5$, the spiral updating position method is selected, that is established to update the position of the whale next time by simulating the whale's spiral updating position surrounding the prey. The calculation formula is as follows:

$$X(t+1) = D_2 \cdot e^{bl} \cdot \cos(2\pi l) + X^*(t) \tag{3}$$

$$D_2 = |X^*(t) - X(t)| \tag{4}$$

where, D_2 represents the distance between prey and whale, b is the parameter controlling the shape of the spiral, set to 1 in this paper, and l takes values in the range of $[-2, 1]$.

2.4 Random Searching

When $p < 0.5$, select the position update formula of random searching. Random searching is divided into two ways, when $|A| < 1$, indicates that the whale is moving towards the prey position, then use the shrinkage encirclement formula to simulate the action behaviour of the whale. Use formula (1) to encircle the prey.

When the $|A| \geq 1$ indicates that the whale is moving beyond the location where the prey is present, right now the whales will give up before moving

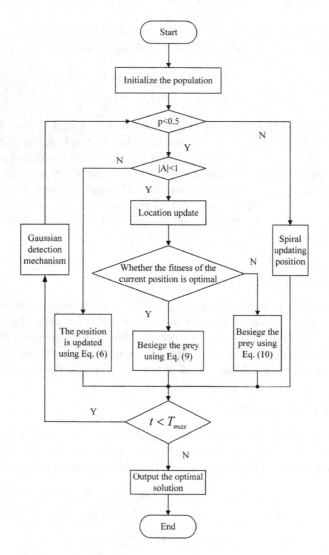

Fig. 1. The flow chart of improved whale optimization algorithm by multi-mechanism fusion

direction, random searching new update location to the other direction, avoid falling into local minima.

$$D_{\text{rand}} = |C \cdot X_{\text{rand}}(t) - X(t)| \qquad (5)$$
$$X(t+1) = X_{\text{rand}}(t) - A \cdot D_{\text{rand}} \qquad (6)$$

where, X_{rand} denotes the randomly chosen whale position vector, and D_{rand} denotes the absolute value of the difference between C times X_{rand} and $X(t)$.

3   Improved Whale Optimization Algorithm

In the basic whale optimization algorithm, the updating process of the whale position is by randomly selecting three updating mechanisms, so there is a problem that the most effective updating method cannot be selected in the whale position updating. Moreover, in the search process of the algorithm, there are many iterations but the leader $X^*(t)$ position is not changed, which leads to the early end of the convergence process [10–12], when solving the optimization problem, it may converge quickly to the local optimum, and the quality of the solution decreases. Aiming at the problems existing in the traditional whale optimization algorithm, this paper proposes an improved whale optimization algorithm by multi-mechanism fusion.

Firstly, a new nonlinear parameter a is proposed to make the whale optimization algorithm adapt to complex nonlinear problems and accelerate the convergence speed of the algorithm 0. The soft siege mechanism of the Harris hawks optimization algorithm is introduced to accelerate the hunting speed of whales. Finally, at the end of each whale hunting iteration, a position control mechanism using Gaussian detection is added to increase the optimization accuracy of the algorithm. The flow chart of the improved whale optimization algorithm by multi-mechanism fusion (IWOA) is shown in Fig. 1.

3.1   Nonlinear Parameter

For swarm intelligence optimization algorithms, exploration and exploitation capabilities are very important for their optimization performance. As for WOA, both shrinking encircling and random searching in position updating are related to the value of a. How to select an appropriate convergence factor a to coordinate the exploration and exploitation ability of WOA is a research problem worth further investigation. The exploration ability means that the population needs to detect a wider search area to avoid the algorithm falling into local optimum [13, 14]. The exploitation ability mainly uses the information already available to the population to conduct local search on certain neighbourhoods of the solution space, which has a decisive influence on the convergence speed of the algorithm. The convergence factor a with large variation has better global search ability and avoids the algorithm falling into local optimum. The smaller convergence factor a has a stronger local search ability, which can accelerate the convergence speed of the algorithm. However, the convergence factor a in the whale optimization algorithm decreases linearly from 2 to 0 with the number of iterations, which cannot fully reflect the exploration and exploitation process of WOA.

In this paper, a nonlinear decreasing convergence factor a with rapid change in the early stage and relatively slow change in the later stage is designed to balance the exploration and exploitation of WOA. The calculation formula is as follows:

$$a = 2 \cdot (1 - \sqrt{\frac{t}{T_{max}}}) \tag{7}$$

The parameter A is controlled by the coefficient a, and the change of the coefficient a leads to certain changes in both the random searching mechanism and the shrinking encircling mechanism. Where T_{max} is the maximum number of iterations and t is the current number of iterations.

3.2   Harris Hawks Optimization Algorithm

The Harris hawks Optimization algorithm simulates the predatory movements of the Harris hawks by using mathematical formulas to simulate its movements. The algorithm vividly simulates the siege predation mechanism of the Harris hawks, making the algorithm extremely powerful in global search.

In the traditional whale optimization algorithm, the process of finding the optimal position is a random exploration by individual whales. The lack of communication between individuals and the group makes some individuals carry out several useless explorations at a distance from the prey. Therefore, we will refer to the soft besiege strategy of the Harris hawks optimization algorithm to improve the location of the whale optimization algorithm as follows:

$$X(t+1) = \begin{cases} Y & f(Y) < f(X(t)) \\ Z & f(Z) < f(X(t)) \end{cases} \tag{8}$$

$$Y = X^*(t) - A \cdot D_1 \tag{9}$$

$$Z = Y + S * LF(D) \tag{10}$$

where, S is a D-dimensional random vector on the uniform distribution of $(1, D)$, $f(x)$ is the fitness function, which means that a certain position is substituted into the fitness function to calculate its fitness value. $LF(D)$ is a D-dimensional random vector generated by the Lévy flight. The Lévy flight formula is shown in formulas (11) and (12).

$$LF(D) = 0.01 \times \frac{u \times \sigma}{|v|^{\frac{1}{\beta}}} \tag{11}$$

$$\sigma = \left(\frac{\Gamma(1+\beta) \times \sin\left(\frac{\pi\beta}{2}\right)}{\Gamma\left(\frac{1+\beta}{2}\right) \times \beta \times 2^{\left(\frac{\beta-1}{2}\right)}} \right)^{\frac{1}{\beta}} \tag{12}$$

where, u and v are random values between $(0, 1)$, and β is set to 1.5.

3.3   Gaussian Detection Mechanism

This section uses the Gaussian variant for the current position and compares the position fitness of the variant with the position fitness before detection to select the optimal position. The main purpose is to improve the ability of the algorithm to jump out of the local optimum and enhance the optimization ability of the algorithm.

The formula of Gaussian detection mechanism is as follows:

$$X(N) = X(t) + X(t) * N(0,1) \tag{13}$$

$$X_{t+1} = \begin{cases} X(N) \ f(X(N)) < f(X(t)) \\ X(t) \ f(X(N)) > f(X(t)) \end{cases} \tag{14}$$

where, $N(0,1)$ generates a random number with Gaussian distribution between 0 and 1. $X(N)$ is the position vector generated after Gaussian mutation.

4 Experiment and Analysis

In this paper, the eight variable dimension benchmark functions are selected to test and evaluate the improved algorithm. All benchmark functions have a theoretical optimal value, which is the extreme value of this test function. The benchmark functions are shown in Table 1.

In this paper, the basic Whale Optimization Algorithm (WOA), Gray Wolf Optimization Algorithm (GWO) [15], Harris Hawks Optimization Algorithm (HHO), and other improved Whale Optimization Algorithms WOABAT [16], EWOA [17] are selected and compared with the improved whale optimization algorithm by multi-mechanism fusion in this paper, on F1-F8 variable dimensional benchmark functions.

Table 1. Benchmark function

Function number	Function name	Dimension	Interval	Theoretical optimal value
F1	Sphere	30	[−100,100]	0
F2	Schwefel 2.22	30	[−10,10]	0
F3	Rosenbrock	30	[−30,30]	0
F4	Step	30	[−100,100]	0
F5	Ackley	30	[−32,32]	0
F6	Griewank	30	[−600,600]	0
F7	Penalized1	30	[−50,50]	0
F8	Penalized2	30	[−50,50]	0

In order to ensure the fairness of the comparison experiment, the population and the number of iterations are set to the same value. The population is set to 30 and the number of iterations is set to 500. The convergence performance of different optimization algorithms in different dimensions is analyzed by testing multidimensional benchmark functions. To avoid randomness and ensure the accuracy of the experiments, all optimization algorithms are run 30 times independently. Since the mean value reflects the optimization accuracy of each algorithm, the standard deviation reflects the robustness and stability of each algorithm, the average convergence accuracy and stability of the optimization algorithms are analyzed by comparing the mean and standard deviation of the optimal fitness obtained by running each optimization algorithm separately for 30 times [18,19]. First, the mean results of the IWOA algorithms in different dimensions are analyzed by means of the Sign test to determine whether they are better, equal or inferior to the comparison algorithms. Second, the Friedman

Table 2. 30 dimension optimization results comparison

Functions	Statistical	WOA	GWO	HHO	WOABAT	EWOA	IWOA
F1	Mean	4.17E-75	8.85E-28	1.48E-94	1.82E-06	1.90E-147	0.00E+00
	Std	7.24E-75	1.13E-27	4.41E-94	7.71E-07	5.69E-147	0.00E+00
F2	Mean	6.86E-52	6.17E-17	2.58E-50	7.20E-03	2.43E-79	0.00E+00
	Std	2.83E-51	4.50E-17	1.04E-49	1.49E-03	1.31E-78	0.00E+00
F3	Mean	2.79E+01	2.71E+01	9.49E-03	8.46E+00	2.75E+01	1.52E-09
	Std	4.43E-01	6.44E-01	1.28E-02	1.29E+01	5.96E-01	4.23E-09
F4	Mean	4.05E-01	8.40E-01	1.86E-04	1.54E-06	9.81E-01	7.51E-13
	Std	2.11E-01	4.00E-01	3.64E-04	8.04E-07	5.59E-01	1.91E-12
F5	Mean	3.64E-15	9.70E-14	4.44E-16	9.05E-04	2.58E-15	4.44E-16
	Std	1.91E-15	1.67E-14	0.00E+00	2.34E-04	1.74E-15	0.00E+00
F6	Mean	3.93E-03	4.01E-03	0.00E+00	7.35E-08	2.41E-03	0.00E+00
	Std	2.12E-02	7.35E-03	0.00E+00	3.13E-08	1.30E-02	0.00E+00
F7	Mean	2.37E-02	5.14E-02	9.97E-06	1.54E-08	5.25E-02	1.04E-13
	Std	2.52E-02	3.18E-02	1.40E-05	7.40E-09	3.26E-02	2.22E-13
F8	Mean	5.18E-01	7.07E-01	5.86E-05	2.37E-07	1.12E+00	3.90E-13
	Std	2.86E-01	2.16E-01	7.79E-05	1.05E-07	3.51E-01	9.70E-13
Friedman		4.25	5	2.63	3.75	4.25	1.13
+/-/=		8/0/0	8/0/0	6/0/2	8/0/0	8/0/0	–

test is performed to compare the performance of the optimization algorithms by averaging them over eight benchmark functions.

By setting the same dimensionality, population and number of iterations, the algorithms were compared in 30 and 100 dimensions using the variable dimensional benchmark function F1–F8 to validate the WOA, GWO, HHO, WOABAT, EWOA and IWOA algorithms.

Figure 2 shows the convergence curves of the three basic optimization algorithms WOA, GWO, HHO and the improved whale optimization algorithm IWOA in 30 and 100 dimensions. Figure 3 shows the convergence curves of the three different improved whale optimization algorithms WOABAT, EWOA, IWOA and the whale optimization algorithm WOA in 30 and 100 dimensions, where the horizontal axis is the number of iterations and the vertical axis indicates the optimal adaptation values. It is clear from the figure that the convergence characteristics of the selected optimization algorithms do not change significantly in different dimensions, and the IWOA algorithm shows excellent convergence accuracy and convergence speed.

Table 2 and Table 3 show the mean and standard deviation of the optimal fitness values of the six algorithms run separately for 30 times in 30 and 100 dimensions. According to the Friedman test results in Table 2 and Table 3, under different dimensions, the mean test results of IWOA are all the optimal values. The standard deviation results are excellent, which proves that the IWOA algorithm has the overall optimal convergence accuracy and stability in 30 and 100 dimensions. The average result of the IWOA algorithm is the 8 test functions in different dimensions. Only in the 100-dimensional F3 function is lower than that

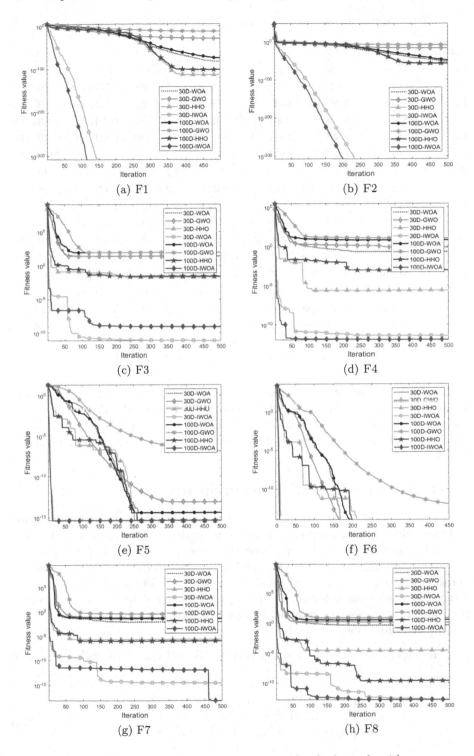

Fig. 2. Comparison of convergence curves for the base algorithm

Table 3. 100 dimension optimization results comparison

Functions	Statistical	WOA	GWO	HHO	WOABAT	EWOA	IWOA
F1	Mean	1.38E-71	1.64E-12	2.57E-94	2.56E-05	1.05E-138	0.00E+00
	Std	5.84E-71	9.72E-13	8.01E-94	5.54E-06	4.55E-138	0.00E+00
F2	Mean	2.96E-51	4.54E-08	1.56E-51	4.10E-02	3.22E-83	0.00E+00
	Std	6.05E-51	1.30E-08	2.86E-51	3.93E-03	6.09E-83	0.00E+00
F3	Mean	9.83E+01	9.82E+01	3.94E-02	2.94E+01	9.81E+01	9.72E+00
	Std	1.10E-01	3.02E-01	2.53E-02	4.48E+01	2.33E-01	2.91E+01
F4	Mean	4.22E+00	1.01E+01	2.83E-04	2.28E-05	6.03E+00	1.56E-12
	Std	1.11E+00	6.57E-01	3.73E-04	5.34E-06	1.17E+00	3.02E-12
F5	Mean	2.93E-15	1.27E-07	4.44E-16	1.99E-03	2.22E-15	4.44E-16
	Std	2.77E-15	4.80E-08	0.00E+00	1.96E-04	1.78E-15	0.00E+00
F6	Mean	0.00E+00	3.60E-03	0.00E+00	3.53E-07	0.00E+00	0.00E+00
	Std	0.00E+00	1.08E-02	0.00E+00	1.02E-07	0.00E+00	0.00E+00
F7	Mean	6.14E-02	2.69E-01	4.45E-06	5.10E-08	1.13E-01	6.30E-14
	Std	2.17E-02	1.04E-01	4.33E-06	1.18E-08	4.55E-02	1.38E-13
F8	Mean	3.10E+00	6.75E+00	1.39E-04	2.93E-06	5.01E+00	1.83E-13
	Std	1.36E+00	4.32E-01	1.08E-04	5.37E-07	9.78E-01	3.13E-13
Friedman		4.06	5.5	2.5	4	3.56	1.38
+/-/=		7/0/1	8/0/0	5/1/2	8/0/0	7/0/1	-

of the HHO, and all other conditions are better or equal to the other 5 algorithms. Summing up the results of the above three tests, the IWOA algorithm is superior and more stable than the WOA, HHO, IWOA, WOABAT, EWOA and IWOA algorithms in the 30-dimensional and 100-dimensional F1-F8 test functions.

5   Design Problems of Tension Spring

Design problems of tension spring [20] is a classic engineering design optimization problem, which is composed of four design variables: inner radius R, cylindrical length L, vessel thickness Ts and head thickness Th. The goal of this problem is to minimize the total cost while satisfying the production needs, which is equivalent to the problem of minimizing the objective function with constraints. If the four design variables are x_1, x_2, x_3 and x_4 respectively, the total cost is $f(x)$, and the mathematical model is shown in formulas (15) and (16).

Objective function:

$$\min f(x) = 0.6224x_1x_3x_4 + 1.7781x_2x_3^2 + 3.1661x_1^2x_4 + 19.84x_1^2x_3 \qquad (15)$$

Constraints:

$$\begin{cases} g_1(x) = -x_1 + 0.0193x_3 \leq 0 \\ g_2(x) = -x_2 + 0.00954x_3 \leq 0 \\ g_3(x) = -\pi x_3^2 x_4 - \frac{4}{3}\pi x_3^3 + 1296000 \leq 0 \\ g_4(x) = x_4 - 240 \leq 0 \end{cases} \qquad (16)$$

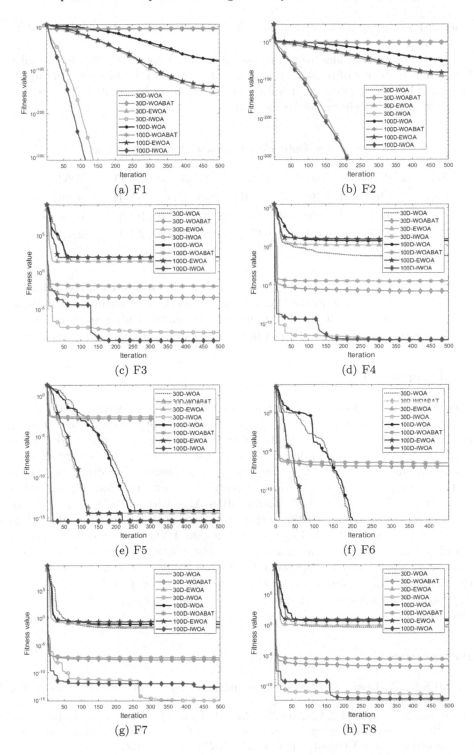

Fig. 3. Comparison of convergence curves for improved algorithms

where, the value range of x_1 and x_2 is $[0, 99]$, and the value range of x_3 and x_4 is $[10, 200]$.

The optimization algorithm mentioned in this paper was used to optimize the design problems of tension spring. All algorithms were run separately for 30 times, and the best value, mean value and standard deviation of the final results were compared, as shown in the Table 4.

Table 4. The Optimization Results of Tension Spring

Function	WOA	GWO	HHO	WOABAT	EWOA	IWOA
Best	5.97E+03	5.89E+03	6.56E+03	5.93E+03	6.27E+03	5.89E+03
Mean	9.72E+03	7.29E+03	7.00E+03	3.37E+04	9.14E+03	6.93E+03
Std	2.44E+03	6.98E+02	6.65E+02	6.10E+04	2.10E+03	2.69E+02

From Table 4, it shows that the improved whale optimization algorithm by multi-mechanism fusion proposed in this paper has the best optimal value, mean and standard deviation in the design problems of tension spring, so there is some effectiveness and stability of IWOA in the application of engineering problems.

6   Conclusion

Aiming at the performance deficiency of the traditional whale optimization algorithm, this paper proposes an improved whale optimization algorithm by multi-mechanism fusion, which introduces the Harris hawks optimization algorithm and Gaussian detection mechanism based on the improvement of non-linear parameters. The IWOA algorithm is analyzed by Friedman test, Sign test and design problems of tension spring. Comparing the Whale Optimization Algorithm (WOA), Gray Wolf Optimization Algorithm (GWO), Harris Hawk Optimization Algorithm (HHO), and other improved Whale Optimization Algorithms WOABAT and EWOA, it is proved that IWOA has excellent optimal stability and convergence accuracy under eight different benchmark functions. The experimental results show that IWOA has better optimization effect than the original Whale Optimization Algorithm, has better stability while ensuring convergence accuracy and speed, reflecting the effectiveness of the improved algorithm.

References

1. Mirjalili, S., Lewis, A.: The whale optimization algorithm. Adv. Eng. Softw. **95**, 51–67 (2016)
2. Rana, N., Latiff, M.S.A., Abdulhamid, S.I.M., et al.: Whale optimization algorithm: a systematic review of contemporary applications, modifications and developments. Neural Comput. Appl. **32**(20), 16245–16277 (2022)

3. Tubishat, M., Abushariah, M.A., Idris, N., et al.: Improved whale optimization algorithm for feature selection in Arabic sentiment analysis. Appl. Intell. **49**(5), 1688–1707 (2019)
4. Kaur, G., Arora, S.: Chaotic whale optimization algorithm. J. Comput. Des. Eng. **5**(3), 275–284 (2018)
5. Luo, J., Shi, B.: A hybrid whale optimization algorithm based on modified differential evolution for global optimization problems. Appl. Intell. **49**(5), 1982–2000 (2019)
6. Saha, N., Panda, S.: Cosine adapted modified whale optimization algorithm for control of switched reluctance motor. Comput. Intell.-Us. **38**(3), 978–1017 (2022)
7. Heidari, A.A., Mirjalili, S., Faris, H., et al.: Harris hawks optimization: algorithm and applications. Future Gener. Comp. Sy. **97**, 849–872 (2019)
8. Song, S., Wang, P., Heidari, A.A., et al.: Dimension decided Harris hawks optimization with Gaussian mutation: balance analysis and diversity patterns. Knowl.-Based Syst. **215**, 106425 (2021)
9. Rahnema, N., Gharehchopogh, F.S.: An improved artificial bee colony algorithm based on whale optimization algorithm for data clustering. Multimed. Tools Appl. (5), 32169–32194 (2020). https://doi.org/10.1007/s11042-020-09639-2
10. Lakshmi, A.V., Mohanaiah, P.: WOA-TLBO: whale optimization algorithm with teaching-learning-based optimization for global optimization and facial emotion recognition. Appl. Soft Comput. **110**, 107623 (2021)
11. Sayed, G.I., Darwish, A., Hassanien, A.E.: A new chaotic whale optimization algorithm for features selection. J. Classif. **35**(2), 300–344 (2018)
12. Hemasian-Etefagh, F., Safi-Esfahani, F.: Group-based whale optimization algorithm. Soft Comput. **24**(5), 3647–3673 (2020)
13. Chen, X.: Research on new adaptive whale algorithm. IEEE Access **8**, 90165–90201 (2020)
14. Gao, Z.M., Zhao, J.: An improved grey wolf optimization algorithm with variable weights. Comput. Intell. Neurosci **2019**, 1–13 (2019)
15. Mirjalili, S., Mirjalili, S.M., Lewis, A.: Grey wolf optimizer. Adv. Eng. Softw. **69**, 46–61 (2014)
16. Mohammed, H.M., Umar, S.U., Rashid, T.A.: A systematic and meta-analysis survey of whale optimization algorithm. Comput. Intell. Neurosci. **2019**, 1–25 (2019)
17. Feng, W., Song, K.: An enhanced whale optimization algorithm. Comput. Simul. **37**(11), 275–279+357 (2020). (in Chinese)
18. Mirjalili, S.: Moth-flame optimization algorithm: a novel nature-inspired heuristic paradigm. Knowl.-Based Syst. **89**, 228–249 (2015)
19. Mirjalili, S.: The ant lion optimizer. Adv. Eng. Softw. **83**, 80–98 (2015)
20. Bayzidi, H., Talatahari, S., Saraee, M., et al.: Social network search for solving engineering optimization problems. Comput. Intell. Neurosci. **2021**, 1–32 (2021)

Multi-stage Objective Function Optimized Hand-Eye Self-calibration of Robot in Autonomous Environment

Kaibo Liu[1,2], Wangli Zheng[3], Huasong Min[4], and Yunhan Lin[1,2,4]([✉])

[1] School of Computer Science and Technology,
Wuhan University of Science and Technology, Wuhan 430065, Hubei, China
yhlin@wust.edu.cn
[2] Hubei Province Key Laboratory of Intelligent Information Processing and
Real-Time Industrial System, Wuhan University of Science and Technology,
Wuhan 430065, Hubei, China
[3] State Grid Electric Power Research Institute, Nanjing 211106, Jiangsu, China
[4] Institute of Robotics and Intelligent Systems, Wuhan University of Science
and Technology, Wuhan 430081, Hubei, China

Abstract. In the application scenario of robot autonomous tasks, the robot needs to be able to complete calibration online and automatically to achieve self-maintenance, which differs from traditional robot hand-eye calibration in that the traditional robot hand-eye calibration requires a dedicated calibration board to assist offline completion. Aiming at the problem that the existing self-calibration methods cannot be optimized as a whole, which leads to low accuracy and instability of the solution, a multi-stage objective function optimization self-calibration algorithm is proposed, which describes the solution of hand-eye self-calibration as a minimization objective function problem involving multiple stages. An optimization method based on the minimization of re-projection error is designed to compensate for the results, which uses an efficient Oriented fast and rotated brief (ORB) feature extraction algorithm and introduces a scoring mechanism to retain more correct matching points in the feature matching stage. Two different types of experiments were designed to validate our method. One is a single camera dataset experiment, which shows that our method is more accurate and robust than the existing self-calibration method; the other is an application platform experiment, which verifies the feasibility and availability of our method.

Keywords: Autonomous operation · Hand-eye self-calibration ·
Multi-stage objective function · Re-projection error optimization

1 Introduction

The "composite robot" consists of mobile chassis, cooperative manipulators, and end effectors, which can not only flexibly operate the manipulators, but

This work is supported by National Natural Science Foundation of China (Grant No.: 62073249).

also expand its mobile range. And at the same time, it can combine human-robot collaboration, intelligent grasping, intelligent perception, and related AI technologies to achieve fine and complex autonomous work. The premise for autonomous operation of composite robots is the ability to perform accurate 3D environment perception. Before the environment perception, hand-eye calibration must be carried out to estimate the spatial position posture transformation relationship between the camera coordinate system and the end effector coordinate system. In addition, hand-eye calibration is required when the hand-eye connection of composite robot is deviated due to long time walking, or after the camera has been reinstalled. The ability of the robot to perform autonomous hand-eye self-calibration when necessary in an autonomous operating environment is an important aspect of its self-maintenance capability.

In recent years, researches on the improvement of hand-eye calibration accuracy, simplicity and intelligence mainly focuses on two aspects: camera pose acquisition and solving hand-eye calibration models [1]. In the solution of these two tasks, the hand-eye calibration problem can be divided into two types: traditional hand-eye calibration and extended hand-eye calibration, depending on whether the camera and the end effector are at the same scale, which means whether the camera poses need to be found by a known calibration tool.

Traditional hand-eye calibration uses known calibration objects, such as a checkerboard, dot calibration plates, and ARTag, etc. The solution methods can be divided into three categories: step-by-step solutions [2,3], synchronous solutions [4], and iterative solutions [5,6]. These traditional hand-eye calibration algorithms assume that the camera and the end effector are at the same scale, that is, they require a known object such as a checkerboard to solve for the camera poses, however, such objects often do not exist or are not easily produced in natural scenes, so the camera poses are generally determined by the Structure from motion(SFM), in this case, it is necessary to use an extended hand-eye calibration for the solution.

The calibration objects for extended hand-eye calibration are usually uncertain or feature points in special structures, thus enabling autonomous calibration for self-maintenance purposes in some specific applications. Andreff et al. [7] first proposed the definition of extended hand-eye calibration and derived a linearization formula based on the Kronecker product. Heller et al. [8] proposed the branch and bound search method, which can guarantee the global optimum under the L_∞-norm, but this method has more restrictions and is less practical. Recently, Xie et al. [9] proposed a calibration method based on binocular vision measurement, which uses three translational movements and one rotational movement of the end effector to obtain more accurate results in autonomous space. However, the method needs to be performed under the premise of a fixed viewpoint, and thus can only work on a specific scene. In International Conference on Intelligent Robots and Systems (IROS), Zhi et al. [10] proposed a hand-eye calibration method without a calibrated target. He used Scale-invariant feature transform (SIFT) [11] to extract the image feature points and perform the corresponding point matching, and solved with Singular value decomposition (SVD) to obtain the closed solution, but the feature extraction method used in this algorithm is time consuming and prone to feature point mismatching and under-matching. After that, Xu Guoshu et al. [12] proposed an automatic calibration method for industrial robots based on scene feature points, but this method is also highly demanding on scene features and cannot achieve stable calibration results. Xu et al. [13] used least squares to recover the rotation and translation transformations by detecting the straight edges of common objects, but this method is semi-autonomous and requires manual assistance to complete and cannot get better autonomy.

Among these calibration algorithms, the method of Zhi et al. [10] has better generalization and can be adapted to more application scenarios. Therefore, in order to obtain better generalization as well as to improve the coupling of each stage, taking the method of [10] as a reference, the self-calibration problem is transformed into a problem of minimizing the objective function from the structure of the projection camera system, to solve this problem, we propose a robot hand-eye self-calibration method based on multi-stage objective function optimization, and design an overall optimization method based on minimization of re-projection error. In the experiments we designed, we verify that our method has higher efficiency and solution accuracy. The main contributions of this paper are as follows:

(1) Aiming at the autonomous operation scenario of robots, a multi-stage objective function optimization algorithm is proposed to solve the problems of poor autonomy of traditional hand-eye calibration and low solution accuracy of existing self-calibration methods.
(2) In order to improve the robustness and calibration accuracy of our algorithm, an optimization method for minimizing re-projection error is designed to effectively compensate for the calibration results.
(3) To verify the feasibility and effectiveness of our algorithm in practical application scenarios, the performance of the algorithm is evaluated on the dataset and application platform, and compared with similar algorithms.

The rest of this paper is organized as follows: Sect. 2 introduces the proposed self-calibration algorithm; Sect. 3 presents experiments to analyze and compare the performance of our algorithm with others; Sect. 4 gives a conclusion of this work.

2   Self-calibration Algorithm

2.1   Self-calibration Problem Description

Figure 1 shows the schematic diagram of hand-eye self-calibration, from the geometric relationship in the figure, the mathematical model of hand-eye self-calibration is described by Eq. (1):

$$A_{ij}X = XB_{ij} \tag{1}$$

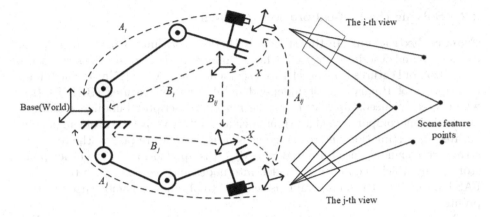

Fig. 1. The illustration of Hand-eye self-calibration

where A_i, B_i respectively represents the transformation from camera to world coordinate system and from end effector to robot base coordinate system, A_{ij} represents the relative positional transformation of camera coordinate system from position i to position j, and B_{ij} represents the relative positional transformation of end effector coordinate system from position i to position j, then X represents the desired hand-eye transformation.

To solve the problem of Eq. (1), the hand-eye self-calibration problem is represented by the following equation:

$$\underset{X}{\arg\min} , C\left(A_{ij}, B_{ij}, X\right) \tag{2}$$

where C is a minimization objective function, which including the hand-eye transformation matrix X. The solution of the objective function is mainly influenced by the relative position A_{ij} of the camera coordinate system and the relative position B_{ij} of the end effector coordinate system. Where the accuracy of B_{ij} is determined by the physical structure of the robot and is not affected by other factors. As can be seen from the self-calibration schematic in Fig. 1, A_{ij} is obtained by extracting and matching scene feature points, the scene feature points can be represented by a feature descriptor, and in order to better measure the accuracy of feature point matching, we introduce a matching point judging mechanism after coarse matching of feature points. Therefore, the objective function can be rewritten as:

$$\underset{X}{\arg\min} , C\left(f_i, f_j, S_{ij}, B_{ij}, X\right) \tag{3}$$

where f_i and f_j represent the description of feature points in the i-th and j-th views respectively, and S_{ij} represents the matching score of the corresponding matching point x_{ij} in the i-th and j-th views. Hence, we can minimize the error of the hand-eye transformation matrix by optimizing the parameters of the objective function.

2.2   Solving the Self-calibration Problem

Feature Extraction. In this stage, ORB [14] algorithm for fast feature point extraction and description is used, which combines and optimizes the fast feature extraction of Features from accelerated segment test (FAST) [15] with the binary descriptors of Binary robust independent elementary features (BRIEF) [16], which greatly speeds up feature extraction and descriptor building, making it more efficient compared to algorithms such as SIFT. To solve the problem of inaccurate pose estimation due to the concentration of feature points, the quadtree structure commonly used in ORB-SLAM [17] is applied for FAST corner point storage, by which the image is divided into nodes and the point with the largest FAST corner point response value is selected to obtain the homogenized feature points.

A more powerful feature descriptor is obtained by improving the BRIEF descriptor, whose description vector is defined as follows:

$$f_n(p) = \sum_{1 \leq i \leq n} 2^{i-1} \tau(p ; x_i, y_i) \tag{4}$$

$$\tau(p ; x_i, y_i) = \begin{cases} 1, & p(x) < p(y) \\ 0, & p(x) \geq p(y) \end{cases} \tag{5}$$

where $p(x)$ and $p(y)$ are the intensity of pixels in the image feature sampling region p, τ represents the Gaussian distribution at the center of the image block, and n represents the dimensionality of the pixel points. The original BRIEF descriptor is not rotation invariant and the matching performance is poor when the image is rotated. Therefore, to solve this problem, ORB uses the orientation of the previous keypoints to rotate the image blocks during the descriptor construction phase. Meanwhile, to obtain more discriminative descriptors, ORB performs a greedy search for all the needed values of τ to find out the values with high variance, superior mean, and low correlation to determine the final description vector, thus making the features rotationally invariant. By this method, the value of f in Eq. (3) is obtained.

Feature Matching. For methods such as SIFT and ORB, after obtaining feature points and feature descriptions, the next step is usually to use descriptors to perform violent matching on two images, that is, the quality of the match between two key points is judged by computing the relationship between the descriptors. However, this method makes it difficult to distinguish between correct and incorrect matches, which results in the need to eliminate a large number of correct matches to reduce the impact of incorrect matches. At present, many algorithms use the Random sampling consistent algorithm (RANSAC) to alleviate this problem, but this method requires most of the incorrect matches be eliminated in advance, and there are many iterations and instability shortcomings. Since the pixel points of the matched image have motion smoothness, this property makes the neighborhood of the correctly matched feature points have more matching points, while the neighborhood of the incorrectly matched points are mostly mis-matched. Therefore, in this paper, we introduce a scoring mechanism based on motion statistics [18] after ORB coarse matching, and scores the matching by counting the support in the domain of matching points, so as to quickly and robustly eliminate incorrect matches to get more correct matches.

Firstly, the matching image is divided into multiple grids, the number of matching points in each grid is counted, and the number of matches is scored by S_{ij}, which can be expressed as follows:

$$S_{ij} \sim \begin{cases} B\left(Kn, p_t\right), & x_{ij}=true \\ B\left(Kn, p_f\right), & x_{ij}=false \end{cases} \tag{6}$$

where x_{ij} is the matching point in the grid, K is the number of grids around matching point x_{ij}, n is the number of matching points in this region, p_t denotes the probability that matching point x_{ij} is a correct match, p_f denotes the probability that matching point x_{ij} is a incorrect match. Figure 2 shows the distribution of matching scores S_{ij}, it can be seen that S_{ij} obeys the binomial distribution, and in order to more conveniently represent the variability of correct and incorrect matches, the mean and variance of S_{ij} are introduced, after which the threshold between correct and incorrect matches is set to quantify the neighborhood scores S_{ij} as P. The mean and variance of correct and incorrect matches are expressed in Eq. (7) and P is expressed in Eq. (8):

$$\begin{cases} m_t = Knp_t, s_t = \sqrt{Knt\left(1-p_t\right)} \Big\}, x_{ij} = true \\ m_f = Knp_f, s_f = \sqrt{Knp_f\left(1-p_f\right)} \Big\}, x_{ij} = false \end{cases} \tag{7}$$

$$P = \frac{m_t - m_f}{s_t + s_f} = \frac{Knp_t - Knp_f}{\sqrt{Knp_t\left(1-p_t\right)} + \sqrt{Knp_f\left(1-p_f\right)}} \tag{8}$$

where m_t and s_t respectively represent the mean and variance of correct matches, m_f and s_f respectively represent the mean and variance of incorrect matches. Equation (8) shows that P is proportional to the number of surrounding grids and the number of regional feature point matches, so the discrimination between correct matches and incorrect matches can be increased by maximizing the value of P. Finally, Eq. (9) is used to judge whether a matching point is correct:

$$P\{i, j\} \in \begin{cases} true, & S_{ij} \geq \tau \\ false, & Others \end{cases} \tag{9}$$

where S_{ij} is the matching score of i, j matching pairs, and τ is the threshold value of correct matching and error matching, which is determined by the total number of characteristics of the grid area. This method is used to filter the matching points to obtain more correct matching point pairs.

Solving the Hand-Eye Transformation. By extracting the scene feature points and matching the feature points for both views, the relative pose A_{ij} of cameras at different positions can be obtained from the pose transformation between images. After that, the corresponding relative poses of the end effector

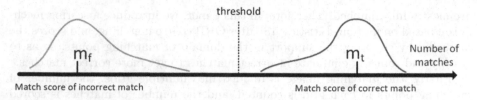

Fig. 2. Probability distribution of matching scores

are combined to form a sequence pair and the initial hand-eye transformation matrix is obtained using singular value decomposition [10]. The main steps are as follows:

Firstly, the translational part of the eye calibration model is separated from Eq. (1) and expressed by a skew-symmetric matrix as follows:

$$\hat{t}_{A_{ij}} R_{A_{ij}} t_x = \hat{t}_{A_{ij}} R_X t_{B_{ij}} + \hat{t}_{A_{ij}} t_X \tag{10}$$

where $\hat{t}_{A_{ij}}$ represents the skew-symmetric matrix that corresponds to the translation part of A_{ij}, $R_{A_{ij}}$ represents the rotation part of A_{ij}, t_x and R_x are the translation and rotation parts of the hand-eye transformation X to be solved respectively, and $t_{B_{ij}}$ represents the translation part of B_{ij}.

Then, in order to solve the spatial matrix more conveniently, Eq. (10) is written in the form of Kronecker product:

$$\underbrace{\begin{bmatrix} I_9 - R_{A_{ij}} \otimes R_{B_{ij}} & 0_{9 \times 3} \\ \hat{t}_{A_{ij}} \otimes t_{B_{ij}}^T & \hat{t}_{A_{ij}} \left(I_3 - R_{A_{ij}} \right) \end{bmatrix}}_{N} \begin{bmatrix} vec\left(R_x \right) \\ t_x \end{bmatrix} = \begin{bmatrix} 0_{9 \times 1} \\ 0_{3 \times 1} \end{bmatrix} \tag{11}$$

where $\otimes$ represents the Kronecker product of spatial matrix, $vec\left(R_x \right)$ represents an operation of vectorization of spatial matrix, where there are 12 unknowns in total, and the 12 × 12 matrix in the left equation is named N. For i motions, the following 12i × 12 matrix T can be constructed:

$$T = \left(N_1^T N_2^T \cdots N_i^T \right)^T \tag{12}$$

Next, the singular value decomposition of matrix T is performed. Since the rotation matrix R_X satisfies the constraint that the determinant is 1, we can use this property to find the unique solution X.

2.3   Optimizing Calibration Results

After obtaining the initial transformation matrix, in order to get higher calibration accuracy, the calibration results also need to be optimized, and since the camera and the end effector are not at the same scale in hand-eye self-calibration, the commonly used optimization methods are not well adapted to the self-calibration system. Therefore, for the hand-eye self-calibration problem, we design an optimization method based on the minimization of re-projection error to compensate the calibration results from the overall structure of the projection camera system. When considering the robot base coordinate system as coincident with the world coordinate system, the camera-to-base transformation A_j can be represented by the robot end effector-to-base transformation B_j and the hand-eye transformation X using Eq. (13), which transforms the minimization objective function C into Eq. (14):

$$A_j = B_j X^{-1} \tag{13}$$

$$\arg\min_X C = \arg\min_X \sum_{i=1}^{m} \sum_{i-1}^{n} v_{ij} \left\| P_{C_{ij}} - \left(X B_j^{-1} P_{W_{ij}} \right) \right\|_2^2 \tag{14}$$

where m and n are the number of feature points and views, $P_{C_{ij}}$ and $P_{W_{ij}}$ represent the coordinates of the i-th feature point in the j-th frame view under the camera coordinate system and the world coordinate system respectively, the coordinates of v_{ij} says whether the i-th feature point can be observed in the j-th frame view, $P_{W_{ij}}$ can be solved by triangulation, $P_{C_{ij}}$ is obtained from the camera. Finally, Levenberg-Marquardt (LM) optimization algorithm is used to solve the problem.

3   Experiments

In this section, we validate the performance of our method with a single camera dataset and application platform experiments. In the dataset validation experiments, the method proposed in [10] is named "ZHI", and our method is named "OURS" to compared with it. The algorithms involved in the experiments are run on Intel Core i7-7700 2.80 GHz processor and Ubuntu 16.04 system.

3.1   Single Camera Dataset Validation Experiment

In the hand-eye calibration, since the real transformation of the hand-eye cannot be measured directly, a suitable evaluation criterion needs to be determined. It is currently used to predict the camera motion by $\tilde{A}_i = X B_i X^{-1}$ and the difference between it and the actual camera motion A_i is used as the error of the hand-eye calibration, but since the camera motion itself is error-prone, there is error transfer in this validation method. Therefore, to solve the problem where there is no real value, the single camera experiment proposed in the literature [10] is referenced in this paper to evaluate the algorithm.

In the experiment, a scene containing a checkerboard is built, as shown in Fig. 3(a), assuming that "two" cameras are calibrated and images containing the complete checkerboard are acquired during the camera movement. After collecting the data, on the one hand, we identify the checkerboard to get the camera pose, and use this pose as the "hand"; on the other hand, ignoring the checkerboard in the scene, and extracting and match the scene feature points to get the camera pose, and using this pose as the "eye", the "hand" and "eye" are

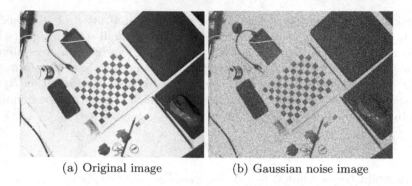

(a) Original image   (b) Gaussian noise image

Fig. 3. Single camera scene images

actually the same camera, so the true rotation value of the hand-eye calibration is a 3×3 unit matrix, and the true value of the translation is a three-dimensional column vector with value 0. By doing this, we solve the problem that there is no true value of the hand-eye calibration.

The experiments were performed using images captured by the Mech-Eye Nano 3D camera with an image resolution of 1280×1024 pixels, the size of each checkerboard calibration plate in the scene is 20×20 mm. In order to quantitatively evaluate the performance of the algorithm, the L2 norm of the difference between the calibration result obtained by the algorithm and the true value is used as the calibration error in this paper. The rotation error and translation error are expressed by Eqs. (15) and (16) respectively:

$$e_R = \left\| \tilde{R}_x - I_3 \right\|_2 \tag{15}$$

$$e_t = \left\| \tilde{t} \right\|_2 , \ \tilde{t} = (\tilde{x}, \tilde{y}, \tilde{z})^T \tag{16}$$

where $\tilde{R}_x$ is the rotation part of the hand-eye transformation matrix estimated by the algorithm, $\tilde{t} = (\tilde{x}, \tilde{y}, \tilde{z})^T$ is the translation part, and I_3 is 3×3 identity matrix. In order to better represent the generalization of the algorithm, we conduct comparison experiments from three different perspectives.

Image Noise. To test the effect of image noise on calibration error, Gaussian noise is added to the original images, where the mean value of noise is 0 and the standard deviation increased from 0 to 0.3 in steps of 0.05. For each different noise level, 100 sets of randomized experiments were performed, and each containing 18 hand-eye movements. The original image and the image corresponding to a Gaussian noise standard deviation of 0.3 are shown in Fig. 3. Where Fig. 4 shows the box line plots of the translation and rotation errors at different noise levels. It can be seen from the plots that our method is more densely distributed and has smaller errors, which indicates that the method is more resistant to noise and has better robustness.

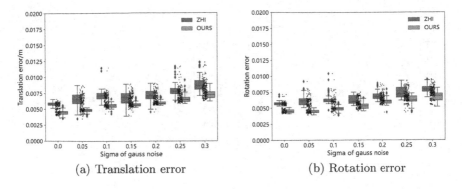

(a) Translation error   (b) Rotation error

Fig. 4. Comparison under different image noise

Number of Hand-Eye Motions. In order to test the effect of different number of hand-eye movements on the calibration error, 12 to 18 images and the corresponding "hand" posture were randomly selected from the original images, and 100 sets of randomized experiments were conducted for each number of movements. The box line plots of translation and rotation errors for different number of movements are shown in Fig. 5. As can be seen from the figure, the distribution of the "ZHI" method is discrete when the number of motions is too small, that is because the number of matching points obtained by the SIFT method is small, and the motions with fewer matching points than the set threshold will be eliminated in the subsequent steps. Thus, when the number of motions is too small, there are not enough matching points for stable optimization, which makes the obtained hand-eye transformation results fluctuate greatly. In contrast, the feature matching method used in this paper is robust and has a large number of matching feature points, and is able to obtain a sufficient number of correct matches even with a reduced number of movements. Meanwhile, from the trend of error transformation in the graph, the calibration error tends to be stable when the number of movements is above 16, therefore, in practice, more than 16 movements can be set to obtain better calibration results.

Data Type. To test the performance of the algorithm with different data types, the images captured by the Mech-Eye Nano camera is used as dataset 1 and the images used in the literature [10] is used as dataset 2. The dataset 2 is an image captured by the rear camera of a Samsung Galaxy S6 phone with a resolution of 3264 × 1836 pixels and a checkerboard calibration plate with each grid size of 28 × 28 mm. 100 repeated experiments were conducted on each dataset, and their average values were taken. The experimental results are shown in Table 1, from which it can be seen that for dataset 1, the average translation error accuracy of our method is improved by 25.76% and the average rotation error accuracy is improved by 17.34% compared with the "ZHI" method; while for dataset 2, the average translation error accuracy of our method is improved by 25.33% and the average rotation error accuracy is improved by 24.56% compared with the "ZHI"

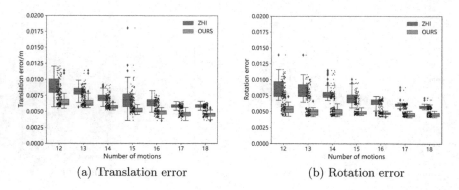

(a) Translation error   (b) Rotation error

Fig. 5. Comparison under the number of hand-eye motions

method. In terms of algorithm efficiency, our method takes less time than the "ZHI" method because the feature extraction and matching method proposed in this paper is more efficient than the SIFT method.

Table 1. Calibration results under different data types.

Evaluation metrics	Algorithms	Data type	
		Dataset 1	Dataset 2
Translation error/m	ZHI	0.00621	0.00075
	OURS	0.00461	0.00056
Rotation error	ZHI	0.00542	0.00057
	OURS	0.00448	0.00043
Time/s	ZHI	14.22	18.34
	OURS	11.73	13.45

3.2   Application Platform Experiment

In this section, our algorithm was verified on the experimental platform of power ground rod hanging and removing, which consists of three main stages: calibrating, hanging and removing. The calibrating stage is to obtain the positional transformation between the camera coordinate system and the end effector coordinate system; in the hanging stage, the task is to recognize the position of bare wire leakage and hang the ground rod to this position; in the removing stage, the task is to remove the grounding rod from the wire and put it back to the starting position. The main steps of these three stages are shown in Fig. 6.

To better validate our algorithm, we simulate the deviation of the composite robot after long time working or accidental collisions by artificially and randomly adjusting the deviation of the hand-eye connector. Two types of experiments were designed, the first one is the experiment of automatically hanging

(a) Self-calibration   (b) Grasping the grounding (c) Recognizing the target
                            rod

(d) Hanging up the ground- (e) Removing the ground-   (f) Putting the grounding
ing rod   ing rod   rod back

Fig. 6. The main process of the power ground rod hanging and removing experiment

and removing the power grounding rod without self-calibration, and the second one is using our self-calibration method, that is, the robot can perform autonomous calibration at idle time or the hand-eye connector is deviated. Each type of experiment is tested 50 times, among them, the autonomous calibration scenario is shown in Fig. 6(a), which is a random image acquisition in the working scene, and 18 images were acquired in the self-calibration process by changing the position and angle autonomously to realize the calculation of the hand-eye transformation matrix.

Table 2. Comparison experiment on before and after using the self-calibration method.

Methods	Number of experiments	Hanging success rate	Removing success rate
Without self-calibration	50	72%	56%
With self-calibration	50	88%	80%

Table 2 shows the experimental comparison before and after using self-calibration in the case of hand-eye connector deviation. It can be seen that the success rate of hanging the ground rod increased by 16% and the success rate of removing the ground rod increased by 24% after introducing the self-calibration method. It shows that by introducing a robot hand-eye self-calibration algorithm, the robot is able to achieve self-maintenance capability for stable autonomous operation over a long period of time.

4 Conclusions

In order to improve the accuracy and stability of the self-calibration method under the application scenario of robot autonomous operation, a robot hand-eye self-calibration algorithm based on multi-stage objective function optimization is proposed, which transforms the self-calibration problem into a problem of minimizing the objective function. Among them, the ORB method is used in the feature extraction stage to improve the efficiency of the algorithm, and the incorrect matching points are eliminated in the feature matching stage by calculating the matching scores of matching points, which in turn provides more correct matching points for the optimization of the objective function, after which the singular value decomposition and re-projection error optimization are used to obtain a higher calibration accuracy. In the single camera dataset experiment, it is verified that our algorithm has high accuracy and time efficiency by comparing with the same type of algorithm. After that, the method is applied to the power ground rod hanging and removing platform to verify the effectiveness and feasibility of our method. For future work, we will apply our method to more application scenarios and do more comprehensive verifications and analysis.

References

1. Jiang, J., Luo, X., Luo, Q., Qiao, L., Li, M.: An overview of hand-eye calibration. Int. J. Adv. Manuf. Technol. 1–21 (2021). https://doi.org/10.1007/s00170-021-08233-6
2. Shiu, Y.C., Ahmad, S.: Calibration of wrist-mounted robotic sensors by solving homogeneous transform equations of the form AX=XB. IEEE Trans. Robot. Autom. 5(1), 16–29 (1989). https://doi.org/10.1109/70.88014
3. Tsai, R.Y., Lenz, R.K.: A new technique for fully autonomous and efficient 3D robotics hand/eye calibration. IEEE Trans. Robot. Autom. 5(3), 345–358 (1989). https://doi.org/10.1109/70.34770
4. Zhang, Y., Qiu, Z., Zhang, X.: A simultaneous optimization method of calibration and measurement for a typical hand-eye positioning system. IEEE Trans. Instrum. Meas. 70, 1–11 (2021). https://doi.org/10.1109/tim.2020.3013308
5. Daniilidis, K.: Hand-eye calibration using dual quaternions. Int. J. Robot. Res. 18(3), 286–298 (1999). https://doi.org/10.1177/02783649922066213
6. Liu, Z., Liu, X., Duan, G., Tan, J.: Precise hand-eye calibration method based on spatial distance and epipolar constraints. Robot. Auton. Syst. 145, 103868 (2021). https://doi.org/10.1016/j.robot.2021.103868
7. Andreff, N., Horaud, R., Espiau, B.: On-line hand-eye calibration. In: Second IEEE International Conference on 3-D Digital Imaging and Modeling, pp. 430–436. IEEE, New York (1999). https://doi.org/10.1109/im.1999.805374
8. Heller, J., Havlena, M., Pajdla, T.: A branch-and-bound algorithm for globally optimal hand-eye calibration. In: IEEE Conference on Computer Vision and Pattern Recognition, pp. 1608–1615. IEEE, New York (2012). https://doi.org/10.1109/cvpr.2012.6247853
9. Xie, X., Peng, Z.: A hand-eye calibration method based on robot with stationary viewpoint. China Measur. Test (2018)

10. Zhi, X., Schwertfeger, S.: Simultaneous hand-eye calibration and reconstruction. In: IEEE/RSJ International Conference on Intelligent Robots and Systems (IROS), pp. 1470–1477. IEEE, New York (2017). https://doi.org/10.1109/iros.2017.8205949

11. Ng, P.C., Henikoff, S.: SIFT: predicting amino acid changes that affect protein function. Nucleic Acids Res. **31**(13), 3812–3814 (2003). https://doi.org/10.1093/nar/gkg509

12. Xu, G.S., Yan, Y.H.: A scene feature-based auto hand-eye calibration method for industrial robot. Mach. Des. Res. **4**, 179–191 (2021). https://doi.org/10.1007/978-3-030-43703-9_15

13. Xu, J., Hoo, J.L., Dritsas, S.: Hand-eye calibration for 2D laser profile scanners using straight edges of common objects. Robot. Comput.-Integr. Manuf. **73**, 102221 (2022). https://doi.org/10.1016/j.rcim.2021.102221

14. Rublee, E., Rabaud, V., Konolige, K., et al.: ORB: an efficient alternative to SIFT or SURF. In: 2011 International Conference on Computer Vision, pp. 2564–2571. IEEE, New York (2011). https://doi.org/10.1109/iccv.2011.6126544

15. Mair, E., Hager, G.D., Burschka, D., Suppa, M., Hirzinger, G.: Adaptive and generic corner detection based on the accelerated segment test. In: Daniilidis, K., Maragos, P., Paragios, N. (eds.) ECCV 2010. LNCS, vol. 6312, pp. 183–196. Springer, Heidelberg (2010). https://doi.org/10.1007/978-3-642-15552-9_14

16. Calonder, M., Lepetit, V., Strecha, C., Fua, P.: BRIEF: binary robust independent elementary features. In: Daniilidis, K., Maragos, P., Paragios, N. (eds.) ECCV 2010. LNCS, vol. 6314, pp. 778–792. Springer, Heidelberg (2010). https://doi.org/10.1007/978-3-642-15561-1_56

17. Campos, C., Elvira, R., Rodríguez, J.J.G., et al.: ORB-SLAM3: an accurate open-source library for visual, visual-inertial, and multimap SLAM. IEEE Trans. Rob. **37**(6), 1874–1890 (2021). https://doi.org/10.1109/tro.2021.3075644

18. Bian, J.-W., et al.: GMS: grid-based motion statistics for fast, ultra-robust feature correspondence. Int. J. Comput. Vis. **128**(6), 1580–1593 (2019). https://doi.org/10.1007/s11263-019-01280-3

An Improved Chaos-Based Particle Swarm Optimization Algorithm

Yi Wang[✉], Suping Liu, and Wenlong Su

College of Computer Science, Guangdong University of Science and Technology,
Dongguan 523083, Guangdong, China
gust312@163.com

Abstract. Particle swarm optimization (PSO) is a well-known swarm intelligence algorithm widely used to solve various numerical optimization problems. However, the PSO can easily fall into local optimum when it solves complex optimization problems. An improved chaotic particle swarm optimization algorithm (ICPSO) is proposed to address this problem. A chaotic perturbation strategy is used to enhance the algorithm's ability to explore the global. Besides, an escape strategy is utilized to increase the diversity of the particle population. The performance of the ICPSO algorithm is verified by testing five benchmark functions, and the results show that the algorithm has good global exploration and convergence ability.

Keywords: Improved Chaotic particle swarm optimization · Chaotic mapping · Numerical optimization

1 Introduction

Intelligence algorithm is an important branch of artificial intelligence, and it defines design algorithms or problem-solving strategies inspired by the social behavior of insects. In swarm intelligence, a "swarm" is a collection of individuals, individuals with the same behavioral abilities, such as a bird colony or ant colony, or similar individuals with different responsibilities, such as detection in a bee colony. The bee leads the bee and follows the bee. The most representative swarm intelligence algorithms are the Particle Swarm Algorithm (PSO) which simulates the foraging behavior of birds; the artificial bee colony algorithm (ABC) simulates the behavior of bees in searching for food. And the artificial wolf colony algorithm (WOA) simulates the hunting behavior of wolves, drosophila algorithm (DA) simulates the food-seeking behavior of fruit flies [1].

The PSO algorithm was initially used to optimize the continuous function. Kennedy and Eberhart proposed the discrete binary PSO algorithm to solve discrete mixed optimization problems [2]. The main difference between this algorithm and the PSO algorithm lies in the difference in data processing between the two and in the use of update speed and position formulas. Since the inertial weight is one of the most important parameters in the PSO algorithm, many scholars continue to study its improvement methods [3]. Shi et al. proposed an improved particle algorithm for adaptive fuzzy adjustment of

L. Pan et al. (Eds.): BIC-TA 2022, CCIS 1801, pp. 158–164, 2023.
https://doi.org/10.1007/978-981-99-1549-1_13

inertia weight. Scholars such as P proposed three new non-linear strategies for inertia weights. Taherkhani M et al. proposed an adaptive inertial weight based on stability and determined the inertial weight of each particle under different sizes according to its performance and the distance of the best position. Considering the stability conditions and adaptive inertia weights, the acceleration parameters of PSO are adaptively determined [4–6].

The particle swarm algorithm optimizes the search through the cooperation between the particles and the guiding strategies generated by the particles themselves. Compared with evolutionary algorithms, particle swarm optimization has the following advantages: easy operation, global search strategy ability, and avoiding complex genetic processes. The unique memory part of the particle swarm algorithm makes it dynamically adjust the search strategy according to the current search situation [7]. Compared with evolutionary algorithms, the PSO algorithm is a more efficient parallel search algorithm. But there are some shortcomings, such as easy to fall into a local minimum, the parameter setting directly affects the search performance, and the theory is not perfect. Based on this, this paper proposes a chaos-based particle swarm optimization algorithm to improve the algorithm's performance.

2 Particle Swarm Algorithm

The PSO is also a population intelligence optimization algorithm that simulates the behavior of bird flocks. Since the algorithm has fewer parameters and can solve complex optimization problems more effectively, it has been applied to image processing, pattern recognition, and other fields. The PSO algorithm randomly generates an initial population and assigns a random position and velocity to each particle in the population. During the flight, the particle's position and velocity are dynamically adjusted by the flight experience of itself and its companions so that the whole population flies to the target position.

Assume that a particle swarm of n particles moves in the D-dimensional search space. The position of the i-th particle at the k-th iteration is expressed as $x_i^k = (x_{i1}^k, x_{i2}^k, \ldots, x_{iD}^k)$. The flying speed of particles is expressed by $v_i^k = (v_{i1}^k, v_{i2}^k, \ldots, v_{iD}^k)$. Each dimension of the extreme individual value of particle i in the $k + 1$ generation is updated by the following formula:

$$p_{id}^{k+1} = \begin{cases} x_{id}^{k+1} & F(x_{id}^{k+1} < F(p_{id}^{k+1})) \\ p_{id}^k & Otherwise \end{cases} \tag{1}$$

where $F(x)$ is the objective function.

For finding the minimum value, the optimal global solution is updated according to the following formula:

$$g_{id}^{k+1} = \arg\{\min F(x_{id}^{k+1})\} \tag{2}$$

The velocity and position of the particle i in the $k + 1$ iteration are updated according to the following formulas:

$$v_{id}^{k+1} = \omega v_{id}^k + c_1 r_1 (p_{id}^k - x_{id}^k) + c_2 r_2 (g_{id}^k - x_{id}^k) \tag{3}$$

$$x_{id}^{k+1} = x_{id}^k + v_{id}^{k+1} \tag{4}$$

In formula (3), ω represents the weight of inertia, c_1 and c_2 represent the learning factor (acceleration constant), and r_1 and r_2 represent random numbers in the range (0,1). Through the loop iterative calculation, the algorithm will find the optimal solution. The entire calculation process is shown in Algorithm 1.

Algorithm 1 Basic particle swarm algorithm process

Step 1 Initialize the particle swarm: initialize various parameters (learning factors c1 and c2, the maximum number of iterations K); random position and velocity.

Step 2 Evaluate particles: Calculate fitness values for all particles in the particle swarm according to fitness function.

Step 3 Update the best:

    Step 3.1 Update the individual best: Compare the particle fitness value with pbest, if it is better than pbest, replace the value with the current particle position.

    Step 3.2 Update the global best: Compare the particle fitness value with gbest, if it is better than gbest, replace the value with the current particle position.

Step 4 Update particles: According to formulas (3) and (4), the particle speed and orientation are updated.

Step 5 Terminate the loop: If the threshold of the number of iterations reaches the upper limit, stop the loop and output the optimal solution. Otherwise, enter Step 2.

---

3 Improved Chaos Particle Swarm Algorithm

3.1 Chaos Perturbation Strategy

The basic PSO algorithm lacks the mechanism to eliminate the local optimum and easily falls into the optimal local solution. In order to make the performance of PSO more efficient, this paper introduces the concept of chaotic perturbation so that the algorithm can search out for the local optimum, thus solving the shortcomings of the basic particle swarm algorithm, such as slow convergence speed and poor optimization accuracy.

According to the randomness and ergodicity of chaos theory, many researchers have thoroughly carried out chaotic particle swarm optimization (PSO), but their research on chaos theory can be summarized as follows: First, chaos theory is used in the initialization process of the population; The second is to apply chaos search to each iteration. Therefore, this paper proposes a particle swarm optimization algorithm based on chaos theory. The improved ideas of this algorithm are as follows: Firstly, chaos is used in initializing particle population; Then, when the optimal position of the population stagnates, chaos theory is used for mutation operation, which not only makes use of chaotic search iteration but also reduces the algorithm complexity. The chaos mapping formula used is the Logistic model regression equation as shown in Formula (5):

$$x_{id}^{k+1} = x_{id}^k \times \mu \times (1 - x_{id}^k) \tag{5}$$

The key idea of the chaos theory is as follows: Chaos theory is introduced to initialize the population. In the application of the particle swarm optimization algorithm, the process of population initialization is uncertain. Many particle groups may be distributed in the region far from the optimal solution, and the particles of these particle groups are not evenly distributed, but at this time, chaos population initialization can ensure that the particle group is evenly distributed. In the initialization stage, chaotic particle swarm optimization is used to make some particles with chaotic sequences to ensure the high mass of the particles in the initial population so as to find those particles that can be used as the initial particles of the population. In this way, not only the mass of the initial population can be guaranteed, but also the initial solutions can be uniformly distributed in the solution space. Because the optimal position of the population may converge with the local poles at the end of an iteration, the particle inertia weight and velocity at this time are small, which will easily lead to particle stagnation. The stagnation of particles will lead to local optima in the improved algorithm so that all the optima cannot be found. At this point, the state of the particle can be determined according to the parameter K (extreme value of stop times). In addition, chaos theory can be used to mutate the particle in the stagnant state to avoid the algorithm's local optimal trap.

3.2 Escape Strategy

Although chaotic perturbations can, to a certain extent, make the algorithm jump out of the search process local optimum, like the multi-peak Rastrigin function, there are many local minima in the search space. The introduction of chaotic perturbation to the PSO algorithm alone does not find the global optimum. The algorithm will be premature in the early stage of the search due to the decline of population diversity. The algorithm is premature in the early stage of the search due to the decrease in population diversity. Therefore, the algorithm can introduce an escape strategy to increase the population diversity when premature stagnation is detected in the search process. The escape strategy can be introduced to increase population diversity.

In the iterative process of particle swarm, if the global optimal solution pbest remains unchanged for M consecutive generations, it indicates that that the particle swarm has been or will be trapped in the local optimal solution according to the existing motion trajectory. Therefore, the constant pbest for M consecutive generations can be used to determine the algorithm's premature stagnation.

The PSO algorithm increases the number of iterations as the particles keep approaching pbest, and after pbest is invariant for M consecutive generations, all the particles achieve aggregation with pbest as the center. At this point, the particle survival density is too small, and the escape strategy will be used to find a new survival place and expand the particle search space. The escape strategy is actually a mutation operation to increase the diversity of particles so that the particles can jump out of the local extrema and enter other regions of the solution space to search. The escape strategy dramatically improves the convergence speed and the algorithm's accuracy. When a particle escapes, the process is characterized by Eq. (6), which is expressed mathematically as follows:

$$x_{id}^{k+1} = \text{rand} \times x_{id}^k \times (1 - x_{id}^k) \tag{6}$$

where rand denotes the random number within $[0,1]$.

The process of the proposed improved chaotic particle swarm optimization algorithm (ICPSO) is shown in Algorithm 2:

Algorithm 2 ICPSO algorithm

Step 1 Initialize the population based on chaos theory:

Randomly generate a D-dimensional vector $x1= (x11, x12, ..., x1D)$, D represents the dimension (solving optimization problems). According to Logistic mapping regression equation $xi+1, j=4xij(1-xij)$; Generate N x1, x2, xN-1, and put each component carrier into the variable value range of the optimization problem $zij=lj+(uj-lj)$ xi. Obtain the fitness values m of particles, determine m particles with better fitness values, these particles must come from the above N particles, and use these particles as the initial value. Generate m random velocities from the selected m particles, set kbest=0, which means the number of stops for optimization, and set Kbest to mean the maximum number of stagnation times for the optimal particle of the population the initialization process is completed.

Step 2 Evaluate particles: Calculate fitness valuesfor all particles in the particle swarm according to fitness function.

Step 3 Update the best:

Step 3.1 Update the individual best: Compare the particle fitness value with pbest, if it is better than pbest, replace the value with the current particle position.

Step 3.2 Update the global best: Compare the particle fitness value with gbest, if it is better than gbest, replace the value with the current particle position.

Step 4 Update the particles: Change the position and speed of the particles according to formulas (3) and (4).

Step 5 Stagnation judgment: If kbest>=Kbest, it is judged that the optimal particle pb of the population is in a stagnant state. At this time, follow the execution of the particle and set kbest=0; if the above condition is not met, that is, the particle is not in a stagnant state, then the particle is not aligned. Do any other operations.

Step5.1 Chaos mutation processing:

Randomly generate a vector $s0=(s01,s02,...,s0D)$, each component value of this vector is distributed in the interval (0,1), and its dimension is D. Update the speed according to formula (4) and control the speed in the interval [vmin, vmax], then use Logistic chaos and press formula (5) to generate s1, $s1j=4*s0j*(1-s0j)$, and then Each component of corresponds to the chaotic disturbance interval.

Step6 If pbest does not change in M consecutive evolutionary generations, the escape strategy is used.

Step 7 Loop termination: The optimal value will be output if the loop termination condition of the maximum number of iterations is reached. Otherwise, go to Step 2 to continue running.

---

4   Experimental Results and Discussion

4.1   Benchmark Function

The proposed algorithm is tested on five benchmark functions and compared with standard particle swarm optimization (PSO), adaptive particle swarm optimization (APSO) algorithms, and chaotic particle swarm optimization (CPSO) algorithms. The benchmark function is shown in Table 1 [8, 9].

Among them, F_1, F_2 are unimodal functions, and F3, F4, F5 are multimodal functions. All five functions are minimization problems.

Table 1. Benchmark functions

Benchmark functions	range	optimal
$F_1(x) = \sum_{i=1}^{n} x_i^2$	$[-50,50]$	0
$F_2(x) = \sum_{i=1}^{n-1} [100(x_{i+1} - x_i^2)^2 + (x_i - 1)]^2$	$[-20,20]$	0
$F_3(x) = \sum_{i=1}^{n} [x_i^2 - 10\cos(2\pi x_i) + 10n]$	$[-600,600]$	0
$F_4(x) = -20\exp\left(-0.2\sqrt{\frac{\sum_{i=1}^{n} x_i^2}{n}}\right) - \exp\left(\frac{1}{n}\sum_{i=1}^{n}\cos(2\pi x_i)\right) + 20 + e$	$[-100,100]$	0
$F_5 = \sum_{i=1}^{l} (10 + x_i^2 - 10\cos(2\pi x_i))$	$[-5.12,5.12]$	0

4.2 Experimental Results and Analysis

In the experiment, the population size is 20, the number of iterations is 100, and the problem dimension is 30; in order to reduce the error, the experiment takes an average of 50 experiments. In addition, the parameters of the other two comparison algorithms are set as follows, in PSO, $\omega = 0.75$, c1 = c2 = 2; in APSO, ω adaptive adjustment, c1 = c2 = 1.4; in the CPSO proposed in this paper. $\omega = 0.5$, c1 = c2 = 1.5; in the ICPSO proposed in this paper. $\omega = 0.4$, c1 = c2 = 1.2, the results are shown in Table 2 below:

Table 2. Test results

Algorithms	F1	F2	F3	F4	F5
PSO	3.56E+02	4.68E+03	2.65E+06	2.58E+01	1.36E+02
APSO	5.38E−03	6.35E−13	9.37E−03	2.50E−03	2.34E−01
CPSO	3.56E−05	7.26E−12	8.69E−04	2.39E−03	5.97E−02
ICPSO	**2.98E−09**	**2.16E−17**	**5.87E−10**	**5.54E−04**	**8.54E−04**

The experimental results show that the test results of the ICPSO algorithm proposed in this article are significantly better than other test results, and the search accuracy is improved by at least 100 times. It proves that the improvement proposed in this paper is effective.

5 Conclusion

This paper proposes an improved chaotic particle swarm algorithm, which uses a chaotic strategy to update the particle population and an escape strategy to generate the optimal individuals in the current population. The experimental results show that the ICPSO is significantly better than PSO, APSO and CPSO in terms of convergence speed and optimal solution quality.

Acknowledgment. This work is supported by the Guangdong Youth Characteristic Innovation Project (2021KQNCX120), the Natural Science Foundation of Guangdong Province of China (2020A1515010784), the Natural Science Project of Guangdong University of Science and Technology (GKY-2021KYYBK-20), the General Project of Science and Technology of Dongguan Social Development (20231800910352), the Natural Science Project of Guangdong University of Science and Technology (XJ2022003501), the University Distinguishing Innovation Project of Guangdong Provincial Department of Education (2021KTSCX149), and Key Project of Science and Technology of Dongguan Social Development (20211800905512).

References

1. Poli, R., Kennedy, J., Blackwell, T.: Particle swarm optimization. Swarm Intell. **1**(1), 33–57 (2007)
2. Bergh, F., Engelbrecht, A.: A cooperative approach to particle swarm optimization. IEEE Trans. Evol. Comput. **8**(3), 225–239 (2004)
3. Guedria, N.: Improved accelerated PSO algorithm for mechanical engineering optimization problems. Appl. Soft Comput. **40**, 455–467 (2016)
4. Bhandari, A., Singh, V., Kumar, A.: Cuckoo search algorithm and wind driven optimization-based study of satellite image segmentation for multilevel thresholding using Kapur's entropy. Expert Syst. Appl. **41**(7), 3538–3560 (2014)
5. Deng, W., Yao, R., Zhao, H., Yang, X., Li, G.: A novel intelligent diagnosis method using optimal LS-SVM with improved PSO algorithm. Soft. Comput. **23**(7), 2445–2462 (2017). https://doi.org/10.1007/s00500-017-2940-9
6. Bka, B., Ntc, D., Tttc, D.: Optimization of buckling load for laminated composite plates using adaptive Kriging-improved PSO: a novel hybrid intelligent method. Defence Technol. **17**(1), 15 (2021)
7. Wang, J., Gao, Y., Liu, W.: An improved routing schema with special clustering using PSO algorithm for heterogeneous wireless sensor network. Sensors **19**(3), 671 (2019)
8. Godio, A., Santilano, A.: On the optimization of electromagnetic geophysical data: application of the PSO algorithm. J. Appl. Geophys. **148**, 163–174 (2018)
9. Pace, F., Santilano, A., Godio, A.: A review of geophysical modeling based on particle swarm optimization. Surv. Geophys. **42**(3), 505–549 (2021)

Task Location Distribution Based Genetic Algorithm for UAV Mobile Crowd Sensing

Yang Huang[1]($\boxtimes$) (iD), Aimin Luo[1]($\boxtimes$), Mengmeng Zhang[1], Liang Bai[2], Yanjie Song[2] (iD), and Jiting Li[3]

[1] Science and Technology on Information Systems Engineering Laboratory, College of Systems Engineering, National University of Defense Technology, Changsha, China
huangyangnudt@163.com, amluo@nudt.edu.cn
[2] College of Systems Engineering, National University of Defense Technology, Changsha, China
[3] Chinese Academy of Military Science, Beijing, China

Abstract. The UAV mobile crowd sensing problem is a new research area due to the flexibility and low-cost advantage of UAVs. Current research rarely considers the task assignment and path planning problem simultaneously. In this paper, we describe the improved UAV mobile crowed sensing model that takes both the task assignment and path planning into consideration, meanwhile the model also considered the limit of UAV's power. In our paper, a task location distribution based genetic algorithm is proposed to solve the problem more efficiently. A series of instances involving different number of tasks and UAV bases is used in the paper. The results of the experiment indicate that our proposed method is efficient and can deal with large-scale problems. This research has resulted in a solution of the UAV mobile crowd sensing problem and can provide ideas to similar problems.

Keywords: UAV · Genetic Algorithm · Task Assignment · Mobile Crowed Sensing · Power Limit

1 Introduction

Mobile crowd sensing (MSC) is a new way to realize sensing coverage and data transmission by taking advantage of mobile devices. Traditional sensing relies on sufficient sensor node which the cost is high. Due to the low cost and flexibility of Uninhabited aerial vehicle (UAV), it has bright future in addressing the issue of MSC problem. This goal can be achieved by dispatching UAVs to fly over to the task position, then the UAVs can collect data by interconnecting their own sensor devices with sensors in task position.

The UAV based Mobile crowd sensing (UBMCS) problem is an autonomous task deployment problem. The conventional UAV task deployment usually consider the task assignment problem and the path planning problem unilaterally. However, UBMCS problem should consider those aspect at the same time. Due to the fact that the on-board battery capacity of UAVs is limited, the endurance capability of UAVs is hard to be

L. Pan et al. (Eds.): BIC-TA 2022, CCIS 1801, pp. 165–178, 2023.
https://doi.org/10.1007/978-981-99-1549-1_14

enlarged. The UBMCS problem is even more complicated because the energy efficiency of UAV should be considered. The general task deployment includes three phases: First the UAV should depart from the UAV base, then the UAV managed to fly to the assigned task position, finally when the UAV is about to run out of energy it will go back to the UAV base.

UBMCS problem have been receiving much attention due to the suitability of UAVs in the field of mobile crowd sensing. Several studied [1–6], for example, have been carried out on UBMCS problem.

In the area of UAV task deployment problem, Yang et al. [7] applied reinforcement learning method to solve the UAV swarm task assignment problem. Ejaz et al. [8] proposed an efficient task assignment method to optimize the path of UAV, this method is aiming at improving the performance of disaster management systems. Chang et al. [9] proposed an agent-based task planner to automatically decompose complex tasks into a series of interpretable subtasks under the constraints of resources, execution time, social rules and costs, the method applied deep reinforcement learning for UAV path planning. Li et al. [10] designed a scheduler based on aggregate flow. By assigning priority by calculating the urgency level of each AGFlow, this method can improve the average task completion rate. Tang et al. [11] combines improved particle swarm algorithm, optimized artificial potential algorithm, path detection switching mode and energy-based task scheduling mechanism, proposed a joint global and local path planning optimization method for UAV task scheduling for crowd air monitoring. Kurdi et al. [12] proposed a task assignment heuristic algorithm based on bacterial foraging behavior for solving the multi-UAV task assignment problem. Xu et al. [13] proposed a task assignment and sequencing method based on multi-objective shuffle frog-hopping algorithm and genetic algorithm for the optimization problem of multi-UAV plant protection operation. Chen et al. [14] proposed an intelligent task offloading algorithm for UAV edge computing networks to solve a large number of image or video processing computation problems. Chen et al. [15] proposes a systematic framework for solving the cooperative task assignment problem of multiple UAVs, the solution is solved by a directed graph method and the modified two-part wolfpack search algorithm.

There are also researches about some special cases in UAV task deployment problems. To cope with large-scale tasks, Rottondi et al. [16] propose two heuristic solutions for rapid response to emergencies, which draw our attention to focus on heuristic strategies. Hu et al. [17] questioned the problem of limited power for UAVs and studied the problem of cooperation between cars and UAVs, which considered power limitations. Ernest et al. [18] proposed a genetic algorithm based on a fuzzy logic system to model and solve the problems that a swarm of UAVs might encounter. Wang et al. [19] investigated the path planning problem of two UAVs performing a mission in cooperation, which considered the case of different initial energy of UAVs, and the authors designed and implemented an energy balancing path planning algorithm for multiple UAVs to solve the problem. Zhang et al. [20] studied the task planning problem considering both task assignment and route planning and wonders whether a hierarchical decision-making approach can be used for task planning.

The previous works shows that the task deployment problem which consider task assignment and path planning the same time still need to be solved more efficiently, meanwhile, the large-scale tasks scenarios place higher demands on algorithms.

To tackle those challenge, we proposed an improved genetic algorithm (GA) in this paper. We designed a heuristic rule for initial population generation in GA. The corresponding encoding method, evolutionary operators for this problem is proposed to improve the efficiency of solution.

The paper is organized as follows: we first give the mathematical model for UBMCS problem in Sect. 2, then introduce the proposed method in Sect. 3, experiment settings and results are demonstrated in Sect. 4.

2 Model

In this section, we first defined the problem, then denote the parameters and decision variables for our problem, followed by the objective function and constrains to construct the model.

2.1 Problem Defined

UAV mobile crowd sensing problem include UAV allocating problem and UAV path planning problem. UAV allocating problem need us to allocate UAV to tasks properly, where path planning is not in consideration. UAV path planning problem aims to arrange path for UAVs, where UAV allocating is a prerequisite. The UAV allocating problem is followed by the UAV path planning problem, therefor the UAV mobile crowd sensing problem should consider UAV allocating and path planning in order.

In this paper, we denote the set of UAV base as B, the number of UAV base is N_B, the geographic location for the m-th UAV base is (Lon_m, Lat_m). U_m represent the set of UAVs in the m-th UAV base. There are N_U UAVs in each UAV base, the cruising speed of the n-th UAV is v_n. The maximum energy of UAV is E_n. The energy expenditure per unit distance is e_n. T denotes the set of tasks. The task number is N_T, the location of the i-th task is denoted by (lon_i, lat_i). The UAVs sent out should obey the energy constrains and returns to the base successfully. All the tasks only need to be visited once. The goal of UBMCS is to minimize the task execution time and balance the task load of each UAV base.

2.2 Assumptions

Some assumptions are given for the proposed problem:

1) Assume that the UAV travels at a fixed speed.
2) Assume that the power of each UAV is the same.
3) Assume that there is no need to consider the turning arc of the UAV to reach the task point.
4) Assume that each task can be performed with only one UAV.

5) Assume that there is no need to stay at the task location and the task is regarded as completed once UAV arrives at the task location.
6) Assume that there exists several UAVs in the same UAV base.
7) Assume that the function of each UAV is the same.

2.3 Decision Variables and Parameters

The decision variables are as follows:

$z_{m,n}^{i,j}$ denotes that if the n-th UAV from the m-th UAV base execute the j-th task after executing the i-th task, there is $z_{m,n}^{i,j} = 1$, else we have $z_{m,n}^{i,j} = 0$.
$x_{m,n}^{i}$ denotes that if the i-th task is executed by the n-th UAV from the m-th UAV base, there is $x_{m,n}^{i} = 1$, else we have $x_{m,n}^{i} = 0$.

The parameters are as follows:

N_B denotes the UAV base number.
N_U denotes the UAV number in each UAV base.
N_T denotes the task number.
v_n denotes the cruising speed of the n-th UAV.
E_n denotes the energy of the n-th UAV.
e_n denoted the energy consumption of the n-th UAV per unit.
$d_{i,j}$ denotes the distance between the i-th task and the j-th task.
$t_{m,n}^{i,j}$ denotes the time that the time consumption for the n-th UAV from the m-th UAV base fly to the j-th task after executing the i-th task.

2.4 Objective Function

Objective function decides the evolution direction of our algorithm, proper objective function can improve the effectiveness of proposed evolution algorithm. The UBMCS problem aims to shorten the time span of the overall task execution, thus the finishing time of the last task is one of the objects to be optimized. Besides, we also need to balance the task load of each UAV base, therefor the standard deviation error for the number of UAV dispatched in each UAV base is another object to be optimized. The mathematic form of objective function is as follows:

$$F = \min\left(\alpha \cdot \max_{m \in B} FT_m + \beta \cdot TL\right) \quad (1)$$

$$TL = \sqrt{\sum_{m \in B}\left(\sum_{j \in T}\sum_{n \in U_m} z_{m,n}^{o,j} - \sum_{m \in B}\sum_{n \in U_m}\sum_{j \in T} z_{m,n}^{o,j}/N_B\right)^2 /N_B} \quad (2)$$

$$FT_m = \sum_{n \in U_m}\sum_{i \in T\cup\{o\}}\sum_{j \in T\cup\{o\}} z_{m,n}^{i,j} \cdot t_{m,n}^{i,j} \quad (3)$$

$$t_{m,n}^{i,j} = d_{i,j}/v_n, \forall m \in B, n \in U_m \tag{4}$$

$$\alpha + \beta = 1 \tag{5}$$

2.5 Constrains

The constrains mainly include task relevant constraints and UAV relevant constraints.

$$\sum_{i \in T} \sum_{m \in B} \sum_{n \in U_m} x_{m,n}^i = N_T \tag{6}$$

All tasks should be visited.

$$\sum_{m \in B} \sum_{n \in U_m} x_{m,n}^i \leq 1, \forall i \in T \tag{7}$$

Each task should be visited only once.

$$\sum_{j \in T} z_{m,n}^{o,j} = 1, \forall m \in B, n \in U_m \tag{8}$$

The UAV should depart from the UAV base.

$$\sum_{i \in T} z_{m,n}^{i,e} = 1, \forall m \in B, n \in U_m \tag{9}$$

The UAV should return to the UAV base.

$$\sum_{j \in T} z_{m,n}^{o,j} \cdot e_n \cdot d_{o,j} + \sum_{i \in T} \sum_{j \in T} z_{m,n}^{i,j} \cdot e_n \cdot d_{i,j} + \sum_{i \in T} z_{m,n}^{i,e} \cdot e_n \cdot d_{i,e} \leq E_n, \forall m \in B, n \in U_m$$
$$\tag{10}$$

The UAV should obey the energy constrains.

$$\sum_{i \in T} z_{m,n}^{i,j} = x_{m,n}^j, \forall m \in B, n \in U_m, j \in T \tag{11}$$

$$\sum_{i \in T \cup \{o\}} z_{m,n}^{i,j} = \sum_{k \in T \cup \{e\}} z_{m,n}^{j,k}, \forall m \in B, n \in U_m, j \in T \tag{12}$$

This denotes that the task assignment and the task flow obey the balance.

$$z_{m,n}^{i,j} \in \{0, 1\}, \forall m \in B, n \in U_m, i \in T, j \in T \tag{13}$$

$$x_{m,n}^i \in \{0, 1\}, \forall m \in B, n \in U_m, i \in T \tag{14}$$

The decision variable should be 0 or 1.

3 Method

We proposed an improved GA to search for the solution space more efficiently. In our arrange process, one of the UAV bases will be chosen, then one of UAV will be chosen to visit tasks in a specific order, when the UAV return to the UAV base, we will repeat the above process.

In this section, we first describe the encoding method for our problem, then introduce the heuristic rules for population generation, followed by the fitness evaluation method, finally we present evolutionary operators for our improved GA.

3.1 The Improved Genetic Algorithm Flowchart

The pseudocode is shown in Algorithm 1. The input of the algorithm contains the number of genetic generations N, population size N_P, UAV base set B, UAV set Um, crossover probability α, mutation probability β, task set T and crossover length L. The output is the UAV deployment solution.

Algorithm 1: Task Location Distribution based Genetic Algorithm (TLDGA)

Input: number of genetic generations N, population size N_P, UAV base set B, UAV set Um, crossover probability α, mutation probability β, task set T, crossover length L

Output: UAV deployment solution

1:      Initialization of algorithm parameters;

2:      $P_0 \leftarrow$ Initializing the population by the heuristic rule;

3:      $fit_{global\_best} \leftarrow$ Evaluate fitness and record the global best fitness;

4:      **For** $gen = 1 \rightarrow Gen$ **do**

5:      | **If** $rand() \leq \alpha$ **then**

6:      | | $P' \leftarrow Crossover(P, L, \alpha)$;

7:      | **End If**

8:      | **If** $rand() \leq \beta$ **then**

9:      | | $P' \leftarrow Mutation(P', \beta)$;

10:     | **End If**

11:     | $Fit(gen) \leftarrow$ Calculate the fitness function value of individuals in population;

12:     | $P' \leftarrow roulettewheel(P', Fit(gen), N_P)$;

13:     | $fit_{local\_best}, indi_{local\_best} \leftarrow$ Find the local best by $Fit(gen)$;

14:     | **If** $fit_{local\_best} > fit_{global\_best}$ **then**

15:     | | $fit_{global\_best}, indi_{global\_best} \leftarrow fit_{local\_best}, indi_{local\_best}$;

16:     | **End If**

17:     **End For**

Algorithm 2: Heuristic Rule Based Initial Population Generation (HRBIPG)

Input: population size N_P, task set T, population P, heuristic rule utilization index μ_1, heuristic rule utilization index μ_2, heuristic rule utilization index μ_3

Output: initial population P_0

1:    Set $\gamma \leftarrow 0, \Delta\gamma \leftarrow \mu$, $RT = [\ ]$;

2:    **For** $i = 1 \rightarrow N_P$ **do**

3:        **If** $i/(N_P + 1) \le \gamma + \Delta\gamma$ **then**

4:            $indi_i \leftarrow$ Select tasks using $HR(T, \mu_1)$;

5:            $RT \leftarrow$ The rest of tasks that have not been select by heuristic rule;

6:            **While** $RT \ne \varnothing$ **do**

7:                $task \leftarrow$ Randomly select a task from RT;

8:                $indi_i \leftarrow$ Select a gene position randomly and insert $task$;

9:                Remove $task$ from RT;

10:            **End While**

11:        **End If**

12:        **If** $i/(N_P + 1) \le \gamma + 2\Delta\gamma$ **then**

13:            $indi_i \leftarrow$ Select tasks using $HR(T, \mu_2)$;

14:            $RT \leftarrow$ The rest of tasks that have not been select by heuristic rule;

15:            **While** $RT \ne \varnothing$ **do**

16:                $task \leftarrow$ Randomly select a task from RT;

17. $indi_i \leftarrow$ Select a gene position randomly and insert $task$;

18:                Remove $task$ from RT;

19:            **End While**

20:        **End If**

21:        **If** $i/(N_P + 1) > \gamma + 2\Delta\gamma$ **then**

22:            $indi_i \leftarrow$ Select tasks using $HR(T, \mu_3)$;

23:            $RT \leftarrow$ The rest of tasks that have not been select by heuristic rule;

24:            **While** $RT \ne \varnothing$ **do**

25:                $task \leftarrow$ Randomly select a task from RT;

26:                $indi_i \leftarrow$ Select a gene position randomly and insert $task$;

27:                Remove $task$ from RT;

28:            **End While**

29:        **End If**

30:    **End For**

3.2 Encoding Method

The encoding is a representation of the solution space. Good encoding improves the efficiency of the algorithm for solving the problem. For this problem, the number of tasks to be performed is known, the chromosome encoded represent the task visitation priority order by UAVs.

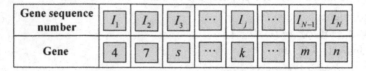

Gene sequence number	I_1	I_2	I_3	...	I_j	...	I_{N-1}	I_N
Gene	4	7	s	...	k	...	m	n

Fig. 1. Schematic diagram of individual encoding

An example of the encoding is show in Fig. 1, every gene in the chromosome represents a task, the sequence of gene represents the task priority order visited by UAVs. More specifically, the chromosome in the figure shows that task 4 has the highest priority order to be visited, task 7 has the second highest priority order to be visited, task n has the lowest priority to be visited.

Once the priority of task is provided, the task deployment scheme can be calculated according to the scheduling method. The idea of the scheduling method is arranging tasks to UAVs according to the task priority contained by the gene, we will arrange task to another UAV if current UAV is about to run out of energy. The UAV will visit a task close to itself in space. If the task represented in the next gene is not the surrounding task, the UAV will check the next gene until the task meet the conditions. If there is no task in the chromosome that followed meet the condition, then we arrange the first unvisited task in the chromosome to a new UAV. The UAV which is run out of energy will return to the base and replace the batteries, which can be the 'new UAV' that waiting for task assignments.

3.3 Population Generation Based on Heuristic Rules

Initial population is very important to evolution algorithm. Heuristic rules proposed here is based on the above ideas: For each task, calculate the average distance to each UAV base. Then attach priority to each UAV according to the average distance. Finally sort the tasks according to the priority.

According to the heuristic rule, the initial population is generated, the details are shown in Algorithm 2. As the algorithm table shows, the individual of initial population is generated by the combination of task sequences generated by heuristic rule and the rest of tasks inserted randomly. Meanwhile, the length of the task sequence generated by heuristic rule is different for different individuals. This method can avoid the circumstance that the gene sequences are similar to each other in the population which will lead to local optimal.

3.4 Evolutionary Operators

Crossover
The crossover operator is used to generate new individuals by swap genes in parent individual. Two kinds of crossover operators are introduced in our paper.

Crossover operators 1: The first crossover operator is gene fragment flipping operator, schematic diagram of gene fragment flipping operation is shown in Fig. 2. As the figure shows, a gene fragment contains m genes is selected randomly from the chromosome, then the gene fragment will be flipped and insert into the same place.

Crossover operators 2: The second crossover operator is gene fragment exchange operator; schematic diagram of gene fragment exchange operation is shown in Fig. 3. As the figure shows, two genes fragment contains m genes is selected randomly from the chromosome, then the genes fragment chosen will be swapped and insert into the location of the other gene.

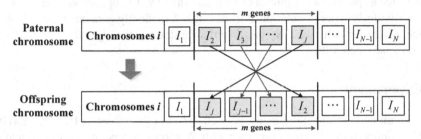

Fig. 2. Schematic diagram of gene fragment flipping operation

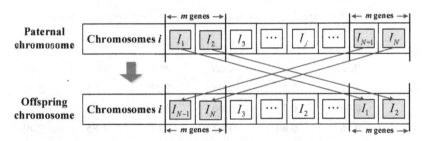

Fig. 3. Schematic diagram of gene fragment exchange operation

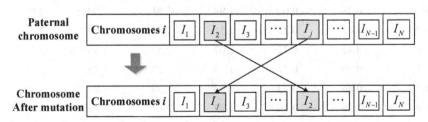

Fig. 4. Schematic diagram of gene recombination operation

We applied the above operators randomly in our proposed algorithm, this strategy can improve the diversity of individuals which is helpful to find a better solution.

Mutation

The mutation operator is gene recombination operator, schematic diagram of gene recombination operation is shown in Fig. 4. As the figure shows, two genes are selected randomly from the chromosome, then the genes will be swapped and insert into the location of the other gene.

Disturbance of Population

To avoid local optimal, we applied a disturbance strategy. The operation is replacing one of the individuals in our population at some stage. A randomly generated individual will replace the chosen individual. By this disturbance strategy, the algorithm can search the solution space more efficiently.

4   Experiment

The configuration of the experiments is Intel(R) Core (TM) i9-9980HK CPU, 2.4GHz, 32GB memory, on a desktop computer with Windows 10 operating system, and the coding environment is Matlab 2021a.

Table 1. The experiment results of our algorithm and contrast algorithms

Instance	TLDGA		GA		NS	
	Best	Ave.	Best	Ave.	Best	Ave.
500-1	**0.39**	**0.77**	1.75	2.27	1.71	2.11
500-2	**0.46**	**0.73**	1.90	2.34	1.80	2.29
500-3	**0.36**	**0.68**	1.82	2.61	1.85	2.49
500-4	**0.45**	**0.72**	1.92	2.34	1.72	2.30
500-5	**0.50**	**0.86**	1.83	2.38	1.93	2.66
500-6	**0.26**	**0.60**	1.98	2.81	1.93	2.82
500-7	**0.41**	**0.75**	1.75	2.31	1.77	2.41
500-8	**0.42**	**0.71**	2.07	2.58	1.89	2.44
800-1	**0.39**	**0.77**	1.75	2.27	1.71	2.11
800-2	**0.46**	**0.73**	1.90	2.34	1.80	2.29
800-3	**0.36**	**0.68**	1.82	2.61	1.85	2.49
800-4	**0.45**	**0.72**	1.92	2.34	1.72	2.30
800-5	**0.50**	**0.86**	1.83	2.38	1.93	2.66
800-6	**0.26**	**0.60**	1.98	2.81	1.93	2.82

(continued)

Table 1. (*continued*)

Instance	TLDGA		GA		NS	
	Best	Ave.	Best	Ave.	Best	Ave.
800-7	**0.41**	**0.75**	1.75	2.31	1.77	2.41
800-8	**0.42**	**0.71**	2.07	2.58	1.89	2.44

To test the efficiency of our proposed algorithm, we compared the proposed Task Location Distribution based Genetic Algorithm (TLDGA) with genetic algorithm (GA) and neighborhood search algorithm (NS). This experimental design was employed because those contract algorithm can solve the UBMCS problem from different evolutionary strategy.

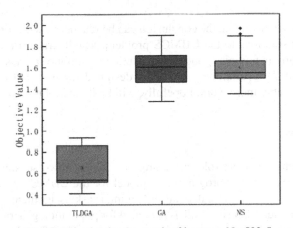

Fig. 5. Boxplot for the result of instance No.500-5

A series of scenarios are designed for the experiment which is randomly generated. The number of tasks in the instance is set to 500 and 800 respectively and the number of UAV base is set to 5 for all the instances. For each task scale, we provide 8 instances of tasks for experimental comparison.

Table 1 shows the results of simulation experiment, where we can find the solving result of our proposed TLDGA is excellent and our method has better convergence speed and stability. The result can support the conclusion that our method can handle the increase in the number of tasks, which can also search for a high-quality solution.

The boxplots of results for instance '500-5' and instance '800-6' are shown in Fig. 5 and Fig. 6. Results in the figure shows that the final objective function is much lower than the contract algorithms, meanwhile the result is stable enough which is important to our problem.

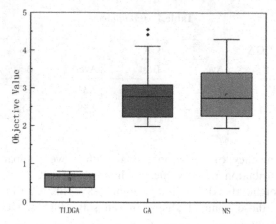

Fig. 6. Boxplot for the result of instance No.800-6

From the above discussion, the conclusion can be reached that our proposed method can obtain better solutions to the UBMCS problem. The heuristic rule for population generation is helpful in finding a good initial solution for mobile crowd sensing problem. It also shows that the evolutionary operators designed for is useful to deal with the problem. Considerably more work, hopefully, will be done in this area.

5   Conclusion

UAVs will play an increasing role in the area of mobile crowd sensing. In our paper, we proposed a mixed integer programming model for the UBMCS problem. Then we introduced an improved GA to calculate the optimal task assignment solution and the path planning scheme. A heuristic rule is designed for population generation and a serial of evolutionary operators are designed for the improved GA. The algorithm takes full advantage of the task position information to obtain a better initial population. Then the evolutionary operators such as crossover method, mutation operator and elitism operator will play an important role in the iterative process of the algorithm. The iterative process of evolutionary algorithms can be used to figure out the optimal solution of task assignment and path planning for UAVs. Our method has a strong applicability to solve similar problems as UBMCS.

Future research should consider the potential effects of dynamic environment more carefully, for example there can be emergent task in mobile crowd sensing process. It will be important that future research investigate algorithms for dynamic circumstances, new heuristic rules may be designed and a learning based method can be used to adapt to the need of dynamic environments.

References

1. Zhou, Z., et al.: When mobile crowd sensing meets UAV: energy-efficient task assignment and route planning. IEEE Trans. Commun. **66**(11), 5526–5538 (2018)
2. Edison, E., Shima, T.: Genetic algorithm for cooperative UAV task assignment and path optimization. In: AIAA Guidance, Navigation and Control Conference and Exhibit, p. 6317 (2008)
3. Shima, T., Rasmussen, S.J.: Sparks, A. G.: UAV cooperative multiple task assignments using genetic algorithms. In: Proceedings of the 2005, American Control Conference, pp. 2989–2994. IEEE (2005)
4. Sujit, P. B., Sinha, A., Ghose, D.: Multiple UAV task allocation using negotiation. In: Proceedings of the fifth international joint conference on Autonomous agents and multiagent systems, pp. 471–478 (2006)
5. Lin, J., Pan, L.: Multi-objective trajectory optimization with a cutting and padding encoding strategy for single-UAV-assisted mobile edge computing system. Swarm Evol. Comput. **75**, 101163 (2022)
6. Huang, P.Q., Wang, Y., Wang, K.Z.: Energy-efficient trajectory planning for a multi-UAV-assisted mobile edge computing system. Front. Inform. Technol. Electron. Eng. **21**(12), 1713–1725 (2020)
7. Yang, J., You, X., Wu, G., Hassan, M.M., Almogren, A., Guna, J.: Application of reinforcement learning in UAV cluster task scheduling. Futur. Gener. Comput. Syst. **95**, 140–148 (2019)
8. Ejaz, W., Ahmed, A., Mushtaq, A., Ibnkahla, M.: Energy-efficient task scheduling and physiological assessment in disaster management using UAV-assisted networks. Comput. Commun. **155**, 150–157 (2020)
9. Wang, C., Wu, L., Yan, C., Wang, Z., Long, H., Yu, C,: Coactive design of explainable agent based task planning and deep reinforcement learning for human-UAVs teamwork. Chin. J. Aeronaut. **33**(11), 2930–2945 (2020). https://doi.org/10.1016/j.cja.2020.05.001
10. Xiaohuan, L.I., et al.: An aggregate flow based scheduler in multi-task cooperated UAVs network. Chin. J. Aeronaut. **33**(11), 2989–2998 (2020)
11. Tang, Y., Miao, Y., Barnawi, A., Alzahrani, B., Alotaibi, R., Hwang, K.: A joint global and local path planning optimization for UAV task scheduling towards crowd air monitoring. Comput. Netw. **193**, 107913 (2021)
12. Kurdi, H., AlDaood, M.F., Al-Megren, S., Aloboud, E., Aldawood, A.S., Youcef-Toumi, K.: Adaptive task allocation for multi-UAV systems based on bacteria foraging behaviour. Appl. Soft Comput. **83**, 105643 (2019)
13. Xu, Y., Sun, Z., Xue, X., Gu, W., Peng, B.: A hybrid algorithm based on MOSFLA and GA for multi-UAVs plant protection task assignment and sequencing optimization. Appl. Soft Comput. **96**, 106623 (2020)
14. Chen, J., Chen, S., Luo, S., Wang, Q., Cao, B., Li, X.: An intelligent task offloading algorithm (iTOA) for UAV edge computing network. Dig. Commun. Netw. **6**(4), 433–443 (2020)
15. Chen, Y., Yang, D., Yu, J.: Multi-UAV task assignment with parameter and time-sensitive uncertainties using modified two-part wolf pack search algorithm. IEEE Trans. Aerosp. Electron. Syst. **54**(6), 2853–2872 (2018)
16. Rottondi, C., Malandrino, F., Bianco, A., Chiasserini, C.F., Stavrakakis, I.: Scheduling of emergency tasks for multiservice UAVs in post-disaster scenarios. Comput. Netw. **184**, 107644 (2021)
17. Hu, M., et al.: Joint routing and scheduling for vehicle-assisted multidrone surveillance. IEEE Internet Things J. **6**(2), 1781–1790 (2018)

178   Y. Huang et al.

18. Ernest, N., Cohen, K., Schumacher, C.: Collaborative tasking of UAVs using a genetic fuzzy approach. In: 51st AIAA Aerospace Sciences Meeting including the New Horizons Forum and Aerospace Exposition, p. 1032 (2013)
19. Wang, L., Zhang, X., Deng, P., Kang, J., Gao, Z., Liu, L.: An energy-balanced path planning algorithm for multiple ferrying UAVs based on GA. Int. J. Aerosp. Eng. **2020**, 1–15 (2020)
20. Zhang, L., Zhu, Y., Shi, X.: A hierarchical decision-making method with a fuzzy ant colony algorithm for mission planning of multiple UAVs. Information **11**(4), 226 (2020)

The Utilities of Evolutionary Multiobjective Optimization for Neural Architecture Search – An Empirical Perspective

Xukun Liu$^{(\boxtimes)}$ (ID)

Southern University of Science and Technology, Shenzhen, China
liuxk2019@mail.sustech.edu.cn

Abstract. Evolutionary algorithms have been widely used in neural architecture search (NAS) in recent years due to their flexible frameworks and promising performance. However, we noticed a lack of attention to algorithm selection, and single-objective algorithms were preferred despite the multiobjective nature of NAS, among prior arts. To explore the reasons behind this preference, we tested mainstream evolutionary algorithms on several standard NAS benchmarks, comparing single and multi-objective algorithms. Additionally, we validated whether the latest evolutionary multi-objective optimization (EMO) algorithms lead to improvement in NAS problems compared to classical EMO algorithms. Our experimental results provide empirical answers to these questions and guidance for the future development of evolutionary NAS algorithms.

Keywords: Evolutionary multiobjective optimization · neural architecture search · deep learning

1 Introduction

In recent years, researchers have proposed numerous new models to improve the performance of deep neural networks. However, these new models have significantly increased the number of parameters along with structural updates, making it more expensive to train and deploy these models. For example, the newly emerged GPT-3 [3] for natural language processing - with up to 175G parameters - makes it impossible to even train it in most hardware. For the smaller Transformer-Big model [31], even translating a sentence with only 30 words requires the execution of 13G FLOPS, which takes more than 20 s to run on less capable edge devices, such as Raspberry Pi. The emergence of this phenomenon has promoted the need for neural network optimization, and neural architecture search is one of the widely used methods. Compared to manually designed neural networks, neural architecture search tends to be simpler and more efficient, and enables trade-offs among multiple objectives, significantly reducing the labor cost of optimizing architectures while promoting the democratization of artificial intelligence (Table 1).

© The Author(s), under exclusive license to Springer Nature Singapore Pte Ltd. 2023
L. Pan et al. (Eds.): BIC-TA 2022, CCIS 1801, pp. 179–195, 2023.
https://doi.org/10.1007/978-981-99-1549-1_15

Table 1. Comparison of deep neural networks' parameter. "CV" denotes computer vision, and "NLP" denotes natural language processing.

DNN	Parameter	Task
ResNet-50 [16]	23.5M	CV
VGG-16 [27]	138M	CV
ViT-Base [12]	86M	CV
BERT-Base [9]	109M	NLP
GPT-3-Med [3]	350M	NLP
GPT-3-XL [3]	175G	NLP

Since neural architecture search is often considered a multi-objective problem, this has led to the widespread use of evolutionary algorithms in NAS problems. However, we also found some commonalities in recent work that are worth exploring. On the one hand, we notice that related papers in the field of NAS focus more on the search space design or parameter-sharing methods, while the search algorithms are often skimmed over and selected more casually. Considering the direct impact of search algorithms on the experimental results, failure to select a relatively suitable algorithm among the wide variety of search algorithms will greatly affect the experimental results and lead to a wrong estimation of the effectiveness of NAS algorithms. In other words, even if a NAS algorithm with better performance is selected, if an unsuitable search algorithm is chosen, it may obtain worse results than a NAS algorithm with poorer performance.

On the other hand, NAS, being a multi-objective problem, is often converted into a single-objective problem to solve. We are also hoping to explore the reasons behind this approach.

In the context of NAS, this paper is devoted to answering the following three research questions:

1. NAS is intrinsically a multi-objective problem. Even if the goal is to improve accuracy, can the diversity provided by adding extra helper objectives lead to better NAS problem-solving?
2. NAS is intrinsically a multi-objective problem. However, we may not always need a set of non-dominated architectures; instead, we most likely just need one or a few for deployment. Under this assumption, we would like to explore whether scalarization (i.e., the de-facto approach used by most existing multi-objective NAS works) or the population-based approach is more effective.
3. EMO has been a hot research area in EC for two decades, and a steady stream of new EMO algorithms with better performance on standard benchmark problems has been reported. We would like to revisit these EMO algorithms and see if these advancements on standard benchmark problems translate well to NAS.

The remainder of the paper is organized as follows.

2   Background

2.1   Evolutionary Multiobjective Optimization (EMO)

Generally, Multiobjective optimization problems (MOPs) refer to the optimization problems with multiple conflicting objectives, for example, which can be formulated as:

$$\min_{x} \quad F(\mathbf{x}) = \big(f_1(\boldsymbol{x}), f_2(\boldsymbol{x}), ..., f_M(\boldsymbol{x})\big),$$
$$\text{s.t.} \quad \boldsymbol{x} \in X, \quad F \in Y, \tag{1}$$

where $X \subset \mathbb{R}^D$ and $Y \subset \mathbb{R}^M$ are known as the *decision space* and the *objective space* respectively. Since $f_1(\boldsymbol{x}), \dots, f_M(\boldsymbol{x})$ are often conflicting with each other, there is no single solution that can achieve optima on all objectives simultaneously; instead, in multi-objective optimization, a Pareto dominance relationship is usually used to distinguish the quality of two different solutions, making the optimal solution for such an MOP often a set of solutions that trade-off between different objectives, known as the *Pareto optima*. Specifically, the images of the Pareto optima are known as the *Pareto set* (PS) and the *Pareto front* (PF) in the decision space and objective space respectively.

2.2   Neural Architecture Search (NAS)

Neural architecture search is a technique to automate the design of neural networks by searching for high-performance structures, which treats the architectural design of DNN models as an optimization problem. Currently, neural architecture search has been applied to several classical areas of artificial intelligence, such as CV [14], NLP [32], etc. In some tasks, NAS has met or exceeded human-designed networks and even discovered network structures that have never been proposed before.

The neural architecture search consists of three main components:

1. A search space of a human-specified DNN structure.
2. A pre-designed search algorithm.
3. An evaluator to estimate the performance of the structures (Fig. 1).

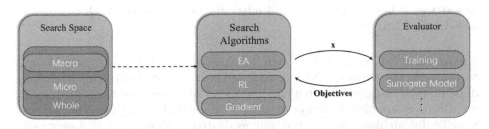

Fig. 1. Standard NAS Pipeline

Search Space: The architectural design of the DNN model can be decomposed into three parts:

1. The network skeleton (i.e., the depth and width of the network).
2. The design of layers (i.e., the type and arrangement of operators such as convolution, pooling, etc.).
3. The space contains the whole network.

Specifically, the micro-search space focuses on designing a modular computational block (also called a cell) that is iteratively stacked according to a pre-specified template to form a complete DNN architecture; the macro-search space focuses on designing the network framework, leaving the layers to mature designs. The complete network is a combination of micro and macro and is divided into homogeneous and heterogeneous. homogeneous is more like a micro search space, where the cell is repeatedly stacked according to the results of the macro search space. And heterogeneous implies the complete traversal of micro and macro search spaces.

Micro-search spaces limit the search to the internal configuration of layers, leading to a significant reduction in search space volume at the cost of inter-layer structural diversity, which is crucial for hardware efficiency. On the other hand, despite its convincing performance in hardware, the macro search space is often criticized for generating architecturally similar DNN models. While search against the complete network space addresses the diversity problem to some extent, the huge search space often implies a larger search overhead.

Search Algorithms: In general, existing NAS methods can be classified into reinforcement learning (RL) based methods, differential(or gradient) based methods (DARTS [21]), and evolutionary algorithms (EA). However, the first two approaches face the problem of excessive resource requirements. For example, the RL-based algorithm often requires hundreds of GPUs running for weeks to converge, and the excessive memory requirements of DARTS similarly limit the size of the search space. EA, on the other hand, strikes a better balance between resource consumption and effectiveness [23]. It makes NAS progressively better by treating it as a black-box discrete optimization problem, using genetic operations or heuristics to iterate over a batch of candidate solutions. Due to its modular framework, flexible coding, and population-based feature, EA has attracted increasing attention, leading to the emergence of a large number of EA-based NAS methods

Evaluator: For realistic-sized search spaces, thorough training from scratch to evaluate the accuracy (i.e., fitness) of the architecture is computationally difficult to achieve, which often consumes a lot of energy and time. As a result, architectures are almost always imperfectly evaluated during the NAS process. Specifically, architectures are typically evaluated in three ways: (i) proxy tasks (e.g., training and evaluating models using small-scale data sets); (ii) surrogate models (e.g., evaluating the performance of the network using predictors); and (iii) fixed training hyperparameters.

2.3   NAS Benchmarks

NASbench is a performance dataset of part or all of the structures in a given search space, often providing a variety of data on a given network structure for a specific task, including accuracy, latency, number of parameters, etc. The advent of NASBench solves the most overhead Evaluate step in the NAS algorithm so that users no longer need to train the network and can quickly evaluate the performance simply by querying. It can be said that NASBench not only provides a unified metric for NAS algorithms but also reduces the cost of developing related algorithms and promotes the democratization of this field.

The existing NASBench can be divided into two categories: (1) real test-based benchmark, NASBench's pioneering work NASBench101 [34], and the subsequent derivative NASBench201 [11], NATS [10], etc. are all in this category. (2) Benchmark based on the surrogate model, such as NASbench301 [26] and so on. Among these two types of benchmarks, the benchmark based on real data can provide the most accurate and various data, but it is limited by hardware devices and often comes from a smaller search space. The benchmark based on the surrogate model contains a larger search space, but the kind of information it can provide is often limited.

At this stage, NASBench has covered many fields including CV, ASR, GNN, NLP, etc., providing a variety of data such as energy consumption, hardware-related data, etc. (Table 2).

Table 2. NAS Benchmarks used in this work

Benchmark	Size	Type	Number of tasks
NAS-Bench-101 [34]	423,624	Image	1
NAS-Bench-201 [11]	15,625	Image	3
NATS-Bench [10]	32,768	Image	3
NAS-Bench-Macro [29]	6561	Image	1
NAS-Bench-ASR [24]	8242	ASR	1
NAS-Bench-Graph [25]	26206	Graph	9

Unlike the traditional test functions for multi-objective problems, NAS Benchmarks are often based on real data and have a relatively small number of objectives. Due to the complexity of neural networks, their metrics still cannot be effectively mapped by simple mathematical formulas. Meanwhile, the search space of NAS problems is non-continuous and unevenly distributed.

This makes it more irregular compared to manually designed problems, and the relationship between different objectives is also difficult to identify. This brings greater diversity and complexity to the NAS problem. Here we selected seven NAS datasets that have been fully traversed, comparing their distributions in the Y-space, the structure of the Pareto front, and the relative relationships between different objectives, and then obtained the following findings:

1. In terms of the distribution of the Y-space, the NAS problem shows a scattered and extremely uneven distribution. It is very similar to the complex multi-peaked problem.
2. In terms of the structure of the Pareto front, the NAS problem does not differ significantly from the traditional test function, but there are more inflection points.
3. For the relative relationship between individual objectives, the NAS problem greatly increases the complexity compared to the test function. For different objectives, the relative relationship between them cannot be described simply by positive or negative correlation.

Therefore, we believe that the diversity and complexity of NAS problems make it possible for algorithms that perform well on traditional test functions to perform poorly on NAS problems, and our experimental results support this idea as well (Fig. 2).

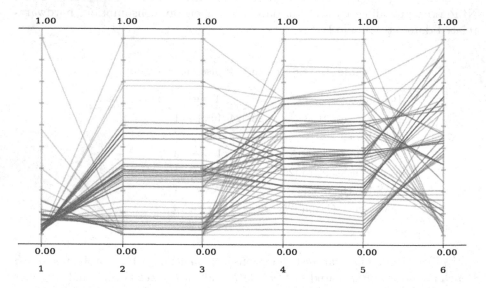

Fig. 2. Pareto front of a simple NAS problem defined by EVOXBench [22] (C10/MOP7) on NASBench-201. The six optimization objectives are 1: error, 2: number of parameters, 3: number of Flops, 4: latency on eyeriss, 5: energy used by eyeriss, 6: arithmetic intensity of eyeriss.

3   Scalarization vs. Population-Based Approaches

Although evolutionary multiobjective algorithms have shown notable utilities in neural architecture search, researchers in related studies have preferred to use scalarization to transform these multiobjective problems into single-objective problems to solve.

The first is the weighting method, which converts a multi-objective optimization problem into a single-objective optimization problem by assigning a weight to each objective and summing up the scalars for optimization. There are different variants of this method, such as the linear weighting method, the exponential weighting method, and so on.

$$\min_{x} \quad F(\mathbf{x}) = w_1 f_1(\boldsymbol{x}) + w_2 f_2(\boldsymbol{x}) + ... + w_M f_M(\boldsymbol{x}),$$
$$\min_{x} \quad F(\mathbf{x}) = f_1(\boldsymbol{x})^{w1} \times f_2(\boldsymbol{x})^{w2} \times ... \times f_M(\boldsymbol{x})^{w_M} \tag{2}$$

For example, MnasNet [30] uses this approach to trade off the accuracy of the target network with the latency and controls the extent to which the latency affects the network architecture by adjusting the hyperparameters.

$$\max_{x} \quad F(\mathbf{x}) = ACC(\boldsymbol{x}) \times [\frac{\boldsymbol{Latency(x)}}{T}]^{w} \tag{3}$$

The method is intuitive and easy to understand and operate. However, because the method requires manual setting of the weights of each item, i.e., there is a priori knowledge, it requires a lot of experiments and it is difficult to traverse the Pareto front, especially when the Pareto front is non-convex.

The second is the ϵ-constraint method. This method selects the most important sub-objective from the K objectives as the optimization objective and the rest of the sub-objectives as constraints. Each sub-objective is constrained by an upper bound ϵ_k.

$$\min_{x} \quad F(\mathbf{x}) = f_p(\boldsymbol{x}),$$
$$\text{s.t.} \quad f_k(\boldsymbol{x}) \leq \epsilon_k, k = 1, ..., M, k \neq p \tag{4}$$

This approach is also used in numerous articles including HAT [32]. In HAT, the hardware delay of the network is used as a constraint. This method is simple and is able to find the best possible target value for the main target while ensuring that the other sub-targets take the values allowed.

To compare the performance of this single-objective algorithm with that of the multi-objective algorithm, we experimentally compare the performance of the two in terms of diversity and performance of the final result, respectively.

To ensure fairness of the results, we use the same number of evaluations in the comparison and generate multiple sets of weights using the Riesz s-Energy [2] method to achieve as smooth a transition as possible. This allows even single-objective methods to converge to the true Pareto front surface by using different weights for multiple runs, which makes it possible to compare multi-objective algorithms with single-objective algorithms.

First, to demonstrate the effectiveness of our experimental method, i.e., that the single-objective algorithm can be made to converge to the true Pareto front by adjusting the weights for multiple runs. We generate the weights using the Riesz s-Energy method mentioned above, increasing the number of samples. Then, We compare the relationship between the true front and the Pareto front formed by the final solution. It can be seen that the Pareto front of the algorithm keeps converging to the true front as the number of samples increases. We have reason to believe that as the number of samples and the running time tend to infinitely, the frontier obtained by the algorithm will coincide with the true frontier.

In our experiments, we selected the most representative single-objective algorithms, GA [17], DE [28], and PSO [18], for comparison with the classical NSGA-II [7] algorithm. With the same number of evaluations guaranteed, we selected several sets of parameters for testing the single-objective algorithm and finally selected the best-performing parameters. The final experimental results are shown in Fig. 3.

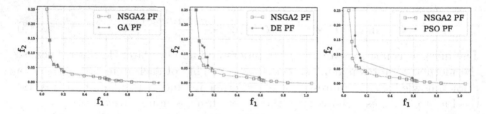

Fig. 3. Pareto front Single-objective algorithms and NSGA-II. The experiments are based on NAS-Bench-201, and the two optimization objectives are $f1$: error, $f2$: number of parameters.

In addition, we conducted replicated experiments on multiple benchmarks and compared the performance differences between the algorithms by measuring the metrics commonly used, HV [13] and IGD [6]. The results also demonstrated the superior performance of the multi-objective algorithms, with the NSGA-II algorithm outperforming the single-objective algorithm in the vast majority of metrics across all benchmarks (Table 3).

From the experimental results, it can be seen that among the three single-objective algorithms, GA performs the best on these problems and PSO performs the worst. However, even the best-performing single-objective algorithm still has a slight gap with the NSGA-II algorithm. Although the gap was small, we repeated the experiment several times and the gap persisted.

In the analysis of the experimental results, we found that even though we tried to make the single-objective algorithm's weights change as smooth as possible, the algorithm showed its insensitivity to the weights in all benchmarks. That is, changes in weights do not directly reflect in the algorithm's results,

Table 3. Performance on different search spaces.

Benchmark	Metrics	NSGA-II	GA	DE	PSO
NASBench101	HV	**0.949**	0.943	0.906	0.887
	IGD	**0.051**	0.067	0.142	0.169
NASBench201	HV	**0.837**	0.819	0.832	0.819
	IGD	**0.003**	0.051	0.047	0.055
NASBenchASR	HV	**0.956**	0.942	0.936	0.893
	IGD	**0.0**	0.527	0.197	0.205
NASBenchMacro	HV	**0.851**	0.844	0.846	0.837
	IGD	**0.023**	0.036	0.024	0.049
NATS	HV	**0.861**	0.828	0.836	0.836
	IGD	**0.022**	0.048	0.042	0.039

and often only affect the algorithm's preferences when the changes in weights accumulate to a certain level. This cumulative effect also leads to the generation of mutations, which in turn causes a decrease in the diversity of single-objective algorithms and is harmful to the subsequent analysis of the results.

In summary, we believe that the scalarization approach can only approximate the case where the Pareto front is convex, and if the Pareto front is non-convex, this approach cannot be equivalent to the multi-objective optimization problem, which can only be solved by directly dealing with the multi-objective algorithms. In addition, the solution of the multi-objective optimization problem can obtain a Pareto solution set, which contains a lot of information, for example, it can generate some interpretations of the model, analyze the correlation between multiple objectives, and so on. Therefore, in the choice of solution algorithm, in the field of NAS, the direct use of multi-objective algorithms will bring more advantages.

4 Revisit EMO Algorithm for NAS

In recent years, the field of multi-objective optimization was continually developing and many kinds of excellent algorithms have emerged. They have achieved notable results on numerous classical test problems. However, compared with the classical multi-objective algorithms, whether these algorithms can also get the corresponding improvement in solving the NAS problem has been less studied. Not only that, the existing multi-objective algorithms can be generally classified into three categories: dominance-based, indicator-based, and decomposition-based. Different classes of algorithms are often applied to different problems. Therefore, we also want to find out the most suitable type for solving NAS problems.

To evaluate the performance of the algorithm in a comprehensive view, we used 11 multi-objective tasks based on NASbench presented in EVOXBench, covering several problems in the mainstream NASBenchs: NASBench101, NAS-Bench201, NATS, and Darts. The problems are defined as shown in Table 4.

Table 4. Problems used in this work. D denotes number of decision variables, M denotes number of objectives, and M^* denotes number of true objectives.

Problem	Ω	D	M	Objectives	M^*
C-10/MOP1	NB101	26	2	f^e, f_1^c	2
C-10/MOP2	NB101	26	3	f^e, f_1^c, f_2^c	2
C-10/MOP3	NATS	5	3	f^e, f_1^c, f_2^c	2
C-10/MOP4	NATS	5	4	$f^e, f_1^c, f_2^c, f_1^{h1}$	2
C-10/MOP5	NB201	6	5	$f^e, f_1^c, f_2^c, f_1^{h1}, f_2^{h1}$	2
C-10/MOP6	NB201	6	6	$f^e, f_1^c, f_2^c, f_1^{h2}, f_2^{h2}, f_3^{h2}$	3
C-10/MOP9	DARTS	32	2	f^e, f_1^c	2
C-10/MOP10	DARTS	32	2	f^e, f_2^c	2
C-10/MOP11	DARTS	32	3	f^e, f_1^c, f_2^c	3
NASBenchASR/MOP1	NASBenchASR	9	2	f^e, f_1^c	2
NASBenchASR/MOP2	NASBenchASR	9	2	f^e, f_1^c, f_2^c	3
NASBenchMacro/MOP1	NASBenchMacro	8	2	f^e, f_1^c	2
NASBenchMacro/MOP2	NASBenchMacro	8	2	f^e, f_1^c, f_2^c	3

We chose these problems to test the algorithms because they are based on the classical NASBench and are very close to the actual problem while being diverse. The performance of the algorithms can be measured well in a variety of situations. Also, to make the test results as comprehensive as possible, we selected 14 algorithms to measure their performance in the above tasks. For the sake of narrative convenience, we select the most representative 6 algorithms to describe them (Table 5). This part of the results shows great similarity in trend with the full results, which also shows the generality of the conclusions we draw.

Tendency: In this section, we compare the trends in the performance of the algorithms as the number of targets increases. It can be seen that with a constant number of objectives, the performance of the algorithm does not change significantly with different search spaces. However, as the number of objectives increases, there is a sudden and dramatic decrease in several metrics of the algorithm (Fig. 4, Table 5). Among them, the dominance-based algorithm shows an earlier performance degradation compared to the other two algorithms, but its performance is particularly striking for a smaller number of targets. The dominance-based algorithm mostly ranks among the top in several statistical metrics.

In summary, the conclusions we draw from the experimental data can be summarized as follows.

Table 5. Performance on different search spaces.

Benchmark	Metrics	NSGA-II	SPEA2	Hype	IBEA	RVEa	MOEA/D
C10/MOP1	HV	0.913	**0.916**	0.835	0.867	0.886	0.904
	IGD	0.055	**0.051**	0.118	0.086	0.08	0.091
C10/MOP2	HV	0.897	**0.899**	0.803	0.817	0.84	0.858
	IGD	0.047	**0.046**	0.119	0.091	0.093	0.096
C10/MOP3	HV	0.804	**0.809**	0.793	0.805	0.788	0.782
	IGD	0.039	**0.037**	0.047	0.042	0.053	0.061
C10/MOP4	HV	0.743	0.748	0.744	**0.755**	0.727	0.717
	IGD	0.074	**0.071**	0.076	0.081	0.084	0.09
C10/MOP5	HV	0.824	**0.824**	0.823	0.822	0.819	0.813
	IGD	0.051	**0.049**	0.08	0.108	0.113	0.114
C10/MOP6	HV	0.112	0.114	**0.71**	0.708	0.702	0.113
	IGD	0.744	0.749	**0.105**	0.123	0.141	0.758
NASBenchASR/MOP1	HV	0.878	0.877	**0.878**	0.852	0.876	0.869
	IGD	0.013	**0.012**	0.165	0.369	0.013	0.527
NASBenchASR/MOP2	HV	1.032	**1.032**	1.032	0.779	-	1.027
	IGD	**0.086**	**0.086**	0.223	0.118	-	0.318
NASBenchMacro/MOP1	HV	**0.983**	0.983	0.970	0.982	0.981	0.913
	IGD	**0.007**	**0.007**	0.049	0.021	0.024	0.076
NASBenchMacro/MOP2	HV	0.880	**0.881**	0.830	0.876	0.859	0.824
	IGD	0.018	**0.012**	0.050	0.054	0.080	0.074
C10/MOP9	HV	0.948	**0.952**	0.949	0.948	0.921	0.876
	IGD	—	—	—	—	—	—
C10/MOP10	HV	0.95	**0.953**	0.943	0.945	0.904	0.882
	IGD	—	—	—	—	—	—
C10/MOP11	HV	0.934	**0.934**	0.926	0.932	0.886	0.824
	IGD	—	—	—	—	—	—

1. As the number of objectives increases, the algorithm performance decreases dramatically at 4 objectives. Fortunately, this is exactly the number of objectives that we are concerned with in most NAS problems. This finding demonstrates, to some extent, the advantage of multi-objective algorithms in solving NAS problems. That is, the algorithm remains stable and efficient under a common number of objectives.
2. With a small number of objectives, the dominance-based algorithm tops the performance on all problems. As the number of objectives gradually increases, the indicator-based and decomposition-based algorithms become having some advantages. However, the dominance-based algorithm still performs very well.

Diversity: In this section, we also try to compare the diversity of algorithmic results under multi-objective problems. In neural architecture search, on the one

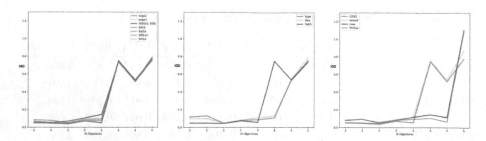

Fig. 4. IGD of different algorithms. The images from left to right are: dominance-based algorithms, indicator-based algorithms, and decomposition-based algorithms. The problems are: C10/MOP1, C10/MOP2, C10/MOP3, C10/MOP4, C10/MOP5, C10/MOP6, C10/MOP7, C10/MOP8

hand, a larger diversity increases considerably the freedom of subsequent manual screening, allowing the researcher to choose a model as close as possible to his or her vision. On the other hand, diversity implies an efficient traversal of the search space by the algorithm, which usually leads to a greater potential of the population.

For the diversity comparison, we selected the test set with the highest number of objectives to test the performance of the algorithm. This test set was chosen because as the number of targets increases, the algorithms will be forced to choose among conflicting objectives, making it more difficult to maintain diversity. The results obtained from the tests significantly reflect the differences in diversity among the algorithms. The highest diversity is found in the dominance-based algorithms, followed by the indicator-based algorithms, and finally the decomposition-based algorithms (Fig. 5).

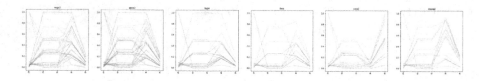

Fig. 5. Pareto front of six classical algorithms in C10/MOP6 (NASBench-201). The five optimization objectives are 1: error, 2: number of parameters, 3: number of Flops, 4: latency on edgegpu, and 5: energy used by edgegpu.

As can be seen from the figure, the dominance-based algorithm maintains a high diversity on almost all objectives. Such results have a great advantage compared to other algorithms. In recent years, evolutionary algorithms have continued to evolve at a high rate, and newer, more efficient algorithms have been proposed. Likewise, records of algorithm effectiveness are constantly being set. However, these newest algorithms are often proposed for specific problems or aim to address the shortcomings of existing algorithms. For example, the

RVEA [5] algorithm proposed in 2016 is aim to a better balance convergence and diversity. We wonder if these improvements will also work for neural architecture search.

In this section we used the same experimental approach as in the previous section, i.e., we tested up to 14 algorithms on the C10/MOP datasets and were surprised by the final results. Our experiments show that although these new algorithms have outperformed the old ones in several areas, the more classical old algorithms outperformed some of the new ones considerably on the NAS problem instead.

The best-performing algorithms are SPEA2 [37] and VaEA [33]. SPEA2 is a classical dominant algorithm proposed in the last century, while VaEA is a newer algorithm proposed in 2016, which considers both convergence and diversity, and improves the uniformity of the solution set. The two worst performing algorithms are MOEA/D [4], BiGE [20], both of which are decomposition-based algorithms, and BiGE, an algorithm that converts a multi-objective problem into a 2-objective problem, which outperforms the classical MOEA/D algorithm in most problems. Especially, when the number of objectives is small, the BiGE algorithm shows good strength compared with MOEA/D.

Here we provide the average ranking of the algorithms tested by us on all the problems (Table 6).

Table 6. Average ranking of algorithms.

Algorithms	IIV	IGD	Avg	Type
NSGA-II	4.73	3.27	4.00	Dominance-based
NSGA-III [8]	6.64	5.36	6.00	Dominance-based
NSGA-II-SDR	7.55	5.55	6.55	Dominance-based
BiGE	8.55	6.55	7.55	Dominance-based
KnEA [35]	6.73	4.73	5.73	Dominance-based
SPEA2	2.36	2.27	2.32	Dominance-based
NNIA [15]	5.82	3.09	4.46	Dominance-based
Hype [1]	6.64	4.82	5.73	Indicator-based
IBEA [36]	5.82	6.18	6.00	Indicator-based
VaEA	2.91	3.09	3.00	Indicator-based
GDE3 [19]	6.73	2.18	4.46	Decomposition-based
MOEA/D	11.0	8.73	9.87	Decomposition-based
RVEA	9.09	6.27	7.68	Decomposition-based
RVEAa [5]	6.46	4.09	5.28	Decomposition-based

It can be found that for the algorithms using decomposition and metrics, the newer algorithms have an advantage over the older ones, but for the dominance algorithm, the older algorithms have a better performance.

5   Conclusion

In the paper, we compare the differences between NAS search spaces and traditional multi-objective test problems, compare the advantages and disadvantages of Scalarization and Population-based Approaches, and also compare the performance of mainstream multi-objective evolutionary algorithms in solving NAS problems. This allows us to observe many patterns that have not been studied, which are important for guiding the design of algorithms for NAS problems and the selection of algorithms by researchers at this stage.

Search Space: First, the search space of neural networks is often more complex than the artificially designed search space. Since its indicators still cannot be effectively mapped by simple mathematical formulas, the relationships among the various objectives are more diverse and indeterminable. In addition, due to the discrete nature of its search space, NAS is more like a complex multi-peaked problem. The above two points greatly increase the difficulty of solving, so the algorithms that perform well on a specific test problem may not necessarily achieve excellent results on the NAS problem.

Scalarization and Population-Based Approach: Second, we compare two types of algorithms for solving NAS problems: the Scalarization approach, which converts a multi-objective problem into a single-objective problem, and the population-based multi-objective approach. We find that although Scalarization is simpler compared to the multi-objective problem, the population-based algorithm tends to perform better on the NAS problem. Moreover, the Pareto solution set generated by the population-based multi-objective algorithm is also beneficial for the subsequent analysis. Therefore, we believe that the population-based algorithm is more advantageous in the NAS problem.

Multi-objective Algorithms: Finally, we compare the performance between different algorithms. We found that the dominance-based algorithm performs much better in the NAS problem, especially when the number of targets is small. As the number of targets increases, the indicator-based algorithm and the decomposition-based algorithm perform better, but the dominance-based algorithm still has a good performance. In addition, the older dominance-based algorithms are more advantageous than the newest ones. This phenomenon is reversed for the other two types of algorithms, where the newer algorithms are more advantageous.

References

1. Bader, J., Zitzler, E.: HypE: an algorithm for fast hypervolume-based many-objective optimization. Evol. Comput. **19**(1), 45–76 (2011). https://doi.org/10.1162/EVCO_a_00009

2. Blank, J., Deb, K., Dhebar, Y., Bandaru, S., Seada, H.: Generating well-spaced points on a unit simplex for evolutionary many-objective optimization. IEEE Trans. Evol. Comput. **25**(1), 48–60 (2021). https://doi.org/10.1109/TEVC.2020.2992387
3. Brown, T.B., et al.: Language models are few-shot learners. CoRR abs/2005.14165 (2020). https://arxiv.org/abs/2005.14165
4. Cantú, V.H., Azzaro-Pantel, C., Ponsich, A.: Multi-objective evolutionary algorithm based on decomposition (MOEA/D) for optimal design of hydrogen supply chains. In: Pierucci, S., Manenti, F., Bozzano, G.L., Manca, D. (eds.) 30th European Symposium on Computer Aided Process Engineering, Computer Aided Chemical Engineering, vol. 48, pp. 883–888. Elsevier (2020). https://doi.org/10.1016/B978-0-12-823377-1.50148-8,https://www.sciencedirect.com/science/article/pii/B9780128233771501488
5. Cheng, R., Jin, Y., Olhofer, M., Sendhoff, B.: A reference vector guided evolutionary algorithm for many-objective optimization. IEEE Trans. Evol. Comput. **20**(5), 773–791 (2016). https://doi.org/10.1109/TEVC.2016.2519378
6. Coello Coello, C.A., Reyes Sierra, M.: A study of the parallelization of a coevolutionary multi-objective evolutionary algorithm. In: Monroy, R., Arroyo-Figueroa, G., Sucar, L.E., Sossa, H. (eds.) MICAI 2004. LNCS (LNAI), vol. 2972, pp. 688–697. Springer, Heidelberg (2004). https://doi.org/10.1007/978-3-540-24694-7_71
7. Deb, K., Pratap, A., Agarwal, S., Meyarivan, T.: A fast and elitist multiobjective genetic algorithm: NSGA-II. IEEE Trans. Evol. Comput. **6**(2), 182–197 (2002). https://doi.org/10.1109/4235.996017
8. Deb, K., Jain, H.: An evolutionary many-objective optimization algorithm using reference-point-based nondominated sorting approach, part i: Solving problems with box constraints. IEEE Trans. Evol. Comput. **18**(4), 577–601 (2014). https://doi.org/10.1109/TEVC.2013.2281535
9. Devlin, J., Chang, M.W., Lee, K., Toutanova, K.: BERT: pre-training of deep bidirectional transformers for language understanding (2018). https://doi.org/10.48550/ARXIV.1810.04805, https://arxiv.org/abs/1810.04805
10. Dong, X., Liu, L., Musial, K., Gabrys, B.: NATS-bench: benchmarking NAS algorithms for architecture topology and size. IEEE Trans. Pattern Anal. Mach. Intell. **44**(7), 3634–3646 (2021)
11. Dong, X., Yang, Y.: NAS-Bench-201: extending the scope of reproducible neural architecture search. In: Proceedings of International Conference Learning Representations (ICLR) (2020)
12. Dosovitskiy, A., et al.: An image is worth 16×16 words: transformers for image recognition at scale (2020). https://doi.org/10.48550/ARXIV.2010.11929, https://arxiv.org/abs/2010.11929
13. Fonseca, C., Paquete, L., Lopez-Ibanez, M.: An improved dimension-sweep algorithm for the hypervolume indicator. In: 2006 IEEE International Conference on Evolutionary Computation, pp. 1157–1163 (2006). https://doi.org/10.1109/CEC.2006.1688440
14. Gong, C., et al.: NASVit: neural architecture search for efficient vision transformers with gradient conflict aware supernet training. In: International Conference on Learning Representations (2022). https://openreview.net/forum?id=Qaw16njk6L
15. Gong, M., Jiao, L., Du, H., Bo, L.: Multiobjective immune algorithm with nondominated neighbor-based selection. Evol. Comput. **16**(2), 225–255 (2008). https://doi.org/10.1162/evco.2008.16.2.225

16. He, K., Zhang, X., Ren, S., Sun, J.: Deep residual learning for image recognition. In: 2016 IEEE Conference on Computer Vision and Pattern Recognition (CVPR), pp. 770–778 (2016). https://doi.org/10.1109/CVPR.2016.90
17. Holland, J.H.: Genetic algorithms. Scholarpedia **7**, 1482 (2012)
18. Kennedy, J., Eberhart, R.: Particle swarm optimization. In: Proceedings of ICNN 1995 - International Conference on Neural Networks, vol. 4, pp. 1942–1948 (1995). https://doi.org/10.1109/ICNN.1995.488968
19. Kukkonen, S., Lampinen, J.: Gde3: the third evolution step of generalized differential evolution. In: 2005 IEEE Congress on Evolutionary Computation, vol. 1, pp. 443–450 (2005). https://doi.org/10.1109/CEC.2005.1554717
20. Li, M., Yang, S., Liu, X.: Bi-goal evolution for many-objective optimization problems. Artif. Intell. **228**, 45–65 (2015). https://doi.org/10.1016/j.artint.2015.06.007, https://www.sciencedirect.com/science/article/pii/S0004370215000995
21. Liu, H., Simonyan, K., Yang, Y.: DARTS: differentiable architecture search. CoRR abs/1806.09055 (2018). https://arxiv.org/abs/1806.09055
22. Lu, Z., Cheng, R., Jin, Y., Tan, K.C., Deb, K.: Neural architecture search as multiobjective optimization benchmarks: problem formulation and performance assessment. arXiv e-prints arXiv:2208.04321 (2022)
23. Lu, Z., et al.: NSGA-NET: a multi-objective genetic algorithm for neural architecture search. CoRR abs/1810.03522 (2018). https://arxiv.org/abs/1810.03522
24. Mehrotra, A., et al.: NAS-bench-ASR: reproducible neural architecture search for speech recognition. In: Proceedings of International Conference Learning Representations (ICLR) (2021)
25. Qin, Y., Zhang, Z., Wang, X., Zhang, Z., Zhu, W.: NAS-bench-graph: benchmarking graph neural architecture search. arXiv preprint arXiv:2206.09166 (2022)
26. Siems, J., Zimmer, L., Zela, A., Lukasik, J., Keuper, M., Hutter, F.: NAS-bench-301 and the case for surrogate benchmarks for neural architecture search. CoRR abs/2008.09777 (2020). https://arxiv.org/abs/2008.09777
27. Simonyan, K., Zisserman, A.: Very deep convolutional networks for large-scale image recognition (2014). https://doi.org/10.48550/ARXIV.1409.1556, https://arxiv.org/abs/1409.1556
28. Storn, R., Price, K.V.: Differential evolution - a simple and efficient heuristic for global optimization over continuous spaces. J. Global Optim. **11**, 341–359 (1997)
29. Su, X., et al.: Prioritized architecture sampling with Monto-Carlo tree search. In: Proceedings of the IEEE Conference on Computer Vision Pattern Recognition (CVPR) (2021)
30. Tan, M., Chen, B., Pang, R., Vasudevan, V., Le, Q.V.: MnasNet: platform-aware neural architecture search for mobile. CoRR abs/1807.11626 (2018)
31. Vaswani, A., et al.: Attention is all you need (2017). https://doi.org/10.48550/ARXIV.1706.03762, https://arxiv.org/abs/1706.03762
32. Wang, H., et al.: HAT: hardware-aware transformers for efficient natural language processing. CoRR abs/2005.14187 (2020)
33. Xiang, Y., Zhou, Y., Li, M., Chen, Z.: A vector angle-based evolutionary algorithm for unconstrained many-objective optimization. Trans. Evol. Comput. **21**(1), 131–152 (2017). https://doi.org/10.1109/TEVC.2016.2587808
34. Ying, C., Klein, A., Christiansen, E., Real, E., Murphy, K., Hutter, F.: NAS-bench-101: towards reproducible neural architecture search. In: Proceedings of the International Conference on Machine Learning (ICML) (2019)
35. Zhang, X., Tian, Y., Jin, Y.: A knee point-driven evolutionary algorithm for many-objective optimization. IEEE Trans. Evol. Comput. **19**(6), 761–776 (2015). https://doi.org/10.1109/TEVC.2014.2378512

36. Zitzler, E., Künzli, S.: Indicator-based selection in multiobjective search. In: Yao, X., et al. (eds.) PPSN 2004. LNCS, vol. 3242, pp. 832–842. Springer, Heidelberg (2004). https://doi.org/10.1007/978-3-540-30217-9_84
37. Zitzler, E., Laumanns, M., Thiele, L.: SPEA 2: improving the strength pareto evolutionary algorithm (2001)

Research on Unmanned Ship Collision Avoidance Algorithm Based on Improved Particle Swarm Optimization Algorithm

Xiaoyu Liu[✉]

Zhongnan University of Economics and Law, Wuhan 430073, China
liuxiaoyu@zuel.edu.cn

Abstract. The local path planning of unmanned ships has high real-time require-ments, and the main research methods such as dynamic window method, random tree method, artificial potential field method, ant colony algorithm, particle swarm algorithm and other methods are difficult to be directly applied to path planning. Aiming at the local collision avoidance problem of unmanned ships, this paper proposes a local collision avoidance method for unmanned ships based on parti-cle swarm optimization algorithm. The algorithm introduces fuzzy mathematics theory to judge the collision risk of unmanned boats and obstacles, transforms the local obstacle avoidance problem of unmanned boats into the optimization problem of the speed and heading changes of small boats, and integrates maritime rules and unmanned boat dynamics. The constraint conditions further optimize the value range of the speed and heading, so that the optimal obstacle avoidance strategy can be calculated through the optimization of the particle swarm opti-mization algorithm. The simulation research simulates the complex environment at sea and selects typical encounter situations in the maritime rules for testing. The simulation results show that the proposed method can well adapt to obstacle avoidance situations such as ship encounters, overtaking, and cross encounters.

Keywords: particle swarm optimization · USV · local collision avoidance

1 Introduction

Local path planning is mainly aimed at avoiding unknown obstacles that affect navigation safety near the unmanned vehicle and has high real-time requirements. The main research methods of scholars at home and abroad include dynamic window method, random tree method, artificial potential field method, ant colony algorithm, particle swarm algorithm, etc.

Lu *et al.* proposed a path planning method based on a double-layer genetic algorithm [1]. Zheng *et al.* proposed a path planning method combining simulated annealing and particle swarms [2]. Lu Taizhi *et al.* used the particle swarm optimization algorithm to optimize the path in the visible view, and then smoothed the path [3]. Seder uses moving grids to model dynamic obstacles in the map and realizes local collision avoid-ance through prediction algorithms [4]. Tam proposed a collaborative algorithm that is compatible with maritime collision avoidance rules [5].

This paper adopts the local obstacle avoidance method of the UAV based on the particle swarm optimization algorithm. First, the obstacle avoidance parameters of the UAV are calculated by using the obstacle model, and then a comprehensive evaluation model for the risk degree of the UAV collision avoidance is constructed by combining the fuzzy theory. By establishing the mathematical model of the obstacle avoidance problem of the unmanned boat, the obstacle avoidance problem of the unmanned boat is transformed into the optimization problem of the heading angle and speed, and the particle swarm optimization algorithm is used to optimize the heading angle and speed. When dealing with the obstacle avoidance of ships at sea, the rules of collision avoidance at sea are combined to ensure that the unmanned vehicle can choose an effective obstacle avoidance strategy.

2 Construction of Collision Avoidance Model

2.1 Obstacle Model and Obstacle Avoidance Parameter Calculation

There are many kinds of obstacles encountered by USV in the process of performing tasks. When the real shape scanned by radar is used to model the environment, the processing time of the algorithm will be increased, and the real-time requirements of local collision avoidance will not be met, which may cause misunderstanding or collision accidents. In view of the obstacles encountered in the local collision avoidance process of USV are mostly small floating objects, ships, etc., even When the circular bounding box is used, it will not lose too much feasible space. As shown in Fig. 1, the outer edges of the USV and obstacles are all puffed to improve the safety of navigation.

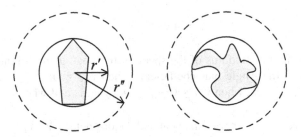

Fig. 1. Schematic of the safe expansion of the obstacle

When the USV meets the local obstacles, the motion state of them changes in real time, which is not conducive to the judgment of the dangerous situation. How to convert the relative motion of USV and obstacles into relative motion is the primary work for obstacle avoidance decision-making. Figure 2 shows the schematic diagram of the movement of the USV and the obstacles, the position of the USV is $P_o(x_o, y_o)$, the speed of the boat is $V_o(v_{xo}, v_{yo})$, the position of the obstacle is $P_s(x_s, y_s)$, the speed of the obstacle ship is $V_s(v_{xs}, v_{ys})$, and the relative orientation is φ. The following calculations are made:

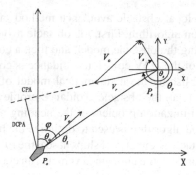

Fig. 2. Schematic diagram of boat and obstacle encounter

The velocity vector of the obstacle is $\overrightarrow{V_s}(V_s, \theta_s)$:

$$V_s = \sqrt{v_{xs}^2 + v_{ys}^2} \tag{1}$$

$$\theta_s = \arctan \frac{v_{xs}}{v_{ys}} + k_s \tag{2}$$

where, V_s represents the modulus of the obstacle velocity, θ_s represents the velocity direction, and k_s is the Angle bias coefficient. When $v_{xs} \geq 0$, $v_{ys} \geq 0$, $k_s = 0$, when $v_{xs} \geq 0$, $v_{ys} \leq 0$, $k_s = \pi$, when $v_{xs} \leq 0$, $v_{ys} \leq 0$, $k_s = \pi$, and when $v_{xs} \leq 0$, $v_{ys} \geq 0$, $k_s = 2\pi$.

$$V_o = \sqrt{v_{xo}^2 + v_{yo}^2} \tag{3}$$

$$\theta_o = \arctan \frac{v_{xo}}{v_{yo}} + k_o \tag{4}$$

where V_s represents the modulus of the speed of the boat, θ_o represents the direction of speed, and k_o is the Angle bias coefficient. When $v_{xo} \geq 0$, $v_{yo} \geq 0$, $k_o = 0$, when $v_{xo} \geq 0$, $v_{yo} \leq 0$, $k_o = \pi$, when $v_{xo} \leq 0$, $v_{yo} \leq 0$, $k_o = \pi$, and when $v_{xo} < 0$, $v_{yo} \geq 0$, $k_o = 2\pi$.

The relative velocity vector of the boat and the obstacle is $\overrightarrow{V_r}(V_r, \theta_r)$:

$$\overrightarrow{V_r} = \overrightarrow{V_s} - \overrightarrow{V_o} \tag{5}$$

$$\theta_s = \arctan \frac{v_{xs}}{v_{ys}} + k_s \quad \theta_o = \arctan \frac{v_{xo}}{v_{yo}} + k_o \tag{6}$$

where V_r denotes the modulus of the relative velocity, θ_r denotes the direction of the relative velocity, and k_r is the Angle bias coefficient. When $v_{xr} \geq 0$, $v_{yr} \geq 0$, $k_0 = 0$, when $v_{xr} \geq 0$, $v_{yr} \leq 0$, $k_r = \pi$, when $v_{xr} \leq 0$, $v_{yr} \leq 0$, $k_r = \pi$, and when $v_{xr} \leq 0$, $v_{yr} \geq 0$, $k_r = 2\pi$.

Distance between the obstacle and the boat L:

$$L = \sqrt{(x_o - x_s)^2 + (y_o - y_s)^2} \tag{7}$$

The position of the obstacle relative to the boat φ:

$$\varphi = \arctan \frac{x_o - x_s}{y_o - y_s} + k \tag{8}$$

where, k is the Angle bias coefficient. When $x_o - x_s \geq 0$ and $y_o - y_s \geq 0, k = 0$, when $x_o - x_s \geq 0$ and $y_o - y_s \leq 0, k = \pi$, when $x_o - x_s \leq 0$ and $y_o - y_s \leq 0, k = \pi$, when $x_o - x_s \leq 0$ and $y_o - y_s \geq 0, k = 2\pi$.

The relative position of the obstacle to the boat's course θ_t:

$$\theta_t = \varphi - \theta_o \tag{9}$$

The closest encounter distance between two objects keeping their current parametric motion $DCPA$:

$$DCPA = L * \sin(\theta_r - \varphi - \pi) \tag{10}$$

$DCPA$ directly reflects the danger degree between the USV and the obstacle in space, $0 < DCPA < d_{safe}$ indicates that the USV and the obstacle will certainly collide at some time in the future if they keep moving at the current speed, $0 < DCPA < d_{safe}$ indicates that the USV and the obstacle will be in danger at some time in the future, $DCPA > d_{safe}$ indicates that there is no danger of collision between the USV and the obstacle.

The shortest meeting time $TCPA$:

$$TCPA = \frac{L * \cos(\theta_r - \varphi - \pi)}{v_r} \tag{11}$$

$TCPA$ reflects the danger degree between the USV and the obstacle in time, $TCPA < 0$ indicates that the USV is moving away from the obstacle and there is no danger of collision avoidance in the future, $TCPA > 0$ indicates that there is a danger of collision between the USV and the obstacle, and the danger coefficient increases with the gradual decrease of $TCPA$.

2.2 Collision Avoidance Risk Assessment

Most of the existing obstacle avoidance Risk assessment models refer to the analysis method of Collision Risk Index (CRI) in the international Rules for Collision avoidance at sea and decide whether to take the obstacle avoidance behavior according to the value of CRI. According to the discussion of domestic and foreign researchers, there are five main factors that affect CRI: the shortest encounter distance DCPA, the shortest encounter time TCPA, the relative distance between USV and obstacles L, the relative orientation between USV and obstacles θ_t, and the velocity ratio between USV and obstacles K. Among them, DCPA and TCPA account for the largest proportion of influencing factors. In addition, the crew's driving level, navigation environment and meteorological conditions also affect the collision avoidance risk to a certain extent. In this paper, the fuzzy mathematics theory is used to describe CRI, and the fuzzy membership functions are established for DCPA and TCPA respectively, and the comprehensive evaluation model of CRI is constructed by synthetically processing them.

The membership function of DCPA with shortest encounter distance is shown in Eq. (12).

$$\lambda_{DCPA} = \begin{cases} 1, & DCPA \leq r \\ \frac{1}{2}\sin\left[\frac{\pi}{R-r}(DCPA - \frac{R+r}{2})\right], & r < DCPA < R \\ 0, & DCPA \geq R \end{cases} \tag{12}$$

where r represents the radius of the danger area and R represents the radius of the safety area. When the shortest encounter distance between the USV and the obstacle is less than r, it means that the USV will collide with the obstacle and relevant avoidance measures should be taken. When the shortest encounter distance between the USV and the obstacle is greater than R, it indicates that the USV will not collide with the obstacle, and it is not necessary to take relevant avoidance measures for the USV. When the shortest encounter distance between the USV and the obstacle is between the dangerous area and the safe area, the possibility of collision should be evaluated by the formula. The formulas for calculating r and R are given in Eqs. (13) and (14).

$$r = \begin{cases} 1.1 - 0.2\theta_t/180°, 0° \leq \theta_t \leq 112.5° \\ 1.0 - 0.4\theta_t/180°, 112.5° \leq \theta_t \leq 180° \\ 1.0 - 0.4(360° - \theta_t)/180°, 180° \leq \theta_t \leq 247.5° \\ 1.0 - 0.2(360° - \theta_t)/180°, 247.5° \leq \theta_t \leq 360° \end{cases} \tag{13}$$

$$R = 2r \tag{14}$$

The membership function of TCPA is shown in Eq. (15):

$$\lambda_{TCPA} = \begin{cases} 1 & TCPA \leq t \\ \left(\frac{T-TCPA}{T-t}\right)^2 & t < TCPA < T \\ 0 & TCPA \geq T \end{cases} \tag{15}$$

where, t and T represent the meeting time between the USV and the obstacle, which is calculated as follows:

$$t = \frac{\sqrt{D^2 - DCPA^2}}{V_r} \tag{16}$$

$$T = \frac{\sqrt{12^2 - DCPA^2}}{V_r} \tag{17}$$

where, t and T represent the encounter time between the USV and the obstacle, D represents the shortest distance needed for the USV to take emergency avoidance measures, and its value depends on the size of the USV's tonnage and handling performance.

The risk evaluation index λ of USV and obstacles is obtained by DCPA membership function and TCPA membership function:

$$\lambda = \lambda_{DCPA} \oplus \lambda_{TCPA} \tag{18}$$

where $\oplus$ is the composition operator, which can be discussed in three cases:

(1) If $\lambda_{DCPA} = 0$, then $\lambda = 0$;

(2) If $\lambda_{TCPA} = 0, \lambda_{DCPA} \neq 0$; then $\lambda = 0$;

(3) If $\lambda_{DCPA} \neq 0, \lambda_{TCPA} \neq 0$, then $\lambda = 0.5\lambda_{DCPA} + 0.5\lambda_{TCPA}$.

2.3 Mathematical Model of Local Obstacle Avoidance

In the process of unmanned vessel moving, obstacle avoidance can only be achieved by changing the rudder Angle and acceleration and deceleration. The key to collision avoidance is to find the appropriate change of rudder Angle and speed. As shown in Fig. 3, it shows the established obstacle avoidance model of USV. The speed of the USV is V_o, the speed of the obstacle is V_s, and the relative speed is V_r. The connection between the USV and the centerline of the obstacle is L, and the left and right tangent lines between the USV and the obstacle are L_l and L_r, respectively. The positive direction of x axis e_x is taken as the Angle reference direction. $\alpha = \angle(V_o, V_r)$, $\beta = \angle(V_r, L)$, $\varphi = \angle(L, e_x)$, $\mu_1 = \angle(L_l, L)$, $\mu_2 = \angle(L, L_r)$, in order to avoid collision with obstacles, it is necessary to ensure that the Angle β between the relative velocity V_r and L is an Angle other than $(\varphi - \mu_2, \varphi + \mu_1)$. The relative velocity is decomposed to obtain the velocity Δv_1 moving along the obstacle and the velocity Δv_2 moving away from it. During the obstacle avoidance analysis, it is assumed that the obstacle can maintain the existing motion speed until the end of obstacle avoidance.

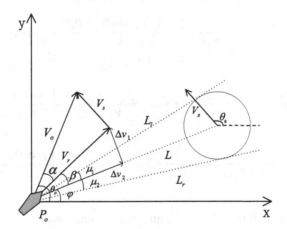

Fig. 3. Obstacle avoidance model of unmanned boat

The velocity along the obstacle obtained from the relative velocity decomposition Δv_1:

$$\Delta v_1 = V_o \sin(\theta_o - \varphi) - V_s \sin(\theta_s - \varphi) \qquad (19)$$

The velocity of the deviation from the obstacle motion obtained by the relative velocity decomposition Δv_2:

$$\Delta v_2 = V_o \cos(\theta_o - \varphi) - V_s \cos(\theta_s - \varphi) \qquad (20)$$

The tangent of the relative velocity to $L \tan \beta$:

$$\tan \beta = \frac{\Delta v_1}{\Delta v_2} = \frac{V_o \sin(\theta_o - \varphi) - V_s \sin(\theta_s - \varphi)}{V_o \cos(\theta_o - \varphi) - V_s \cos(\theta_s - \varphi)} \tag{21}$$

Let $\tan \beta$ be expressed as a function:

$$f(V_o, \theta_o, V_s, \theta_s) = \frac{V_o \sin(\theta_o - \varphi) - V_s \sin(\theta_s - \varphi)}{V_o \cos(\theta_o - \varphi) - V_s \cos(\theta_s - \varphi)} \tag{22}$$

$$d\beta = d(\arctan(f)) = \frac{1}{1+f^2} df \tag{23}$$

$$\frac{1}{1+f^2} = \frac{1}{1 + \frac{V_o \sin(\theta_o - \varphi) - V_s \sin(\theta_s - \varphi)}{V_o \cos(\theta_o - \varphi) - V_s \cos(\theta_s - \varphi)}} = \frac{(V_o \cos(\theta_o - \varphi) - V_s \cos(\theta_s - \varphi))^2}{V_o^2 + V_s^2 - 2V_o V_s \cos(\theta_o - \theta_s)} \tag{24}$$

$$df = df(V_o, \theta_o, V_s, \theta_s) = \frac{\partial f}{\partial V_o} dV_o + \frac{\partial f}{\partial \theta_o} d\theta_o + \frac{\partial f}{\partial V_s} dV_s + \frac{\partial f}{\partial \theta_s} d\theta_s \tag{25}$$

We know from the premise that: $dV_s = 0, d\theta_s = 0$, then:

$$\frac{\partial f}{\partial V_o} dV_o = \frac{V_s \sin(\theta_s - \theta_o)}{(V_o \cos(\theta_o - \varphi) - V_s \cos(\theta_s - \varphi))^2} dV_o \tag{26}$$

$$\frac{\partial f}{\partial \theta_o} d\theta_o = \frac{V_o^2 - V_o V_s \cos(\theta_o - \theta_s)}{(V_o \cos(\theta_o - \varphi) - V_s \cos(\theta_s - \varphi))^2} d\theta_o \tag{27}$$

$$df = \frac{V_s \sin(\theta_s - \theta_o)}{(V_o \cos(\theta_o - \varphi) - V_s \cos(\theta_s - \varphi))^2} dV_o + \frac{V_o^2 - V_o V_s \cos(\theta_o - \theta_s)}{(V_o \cos(\theta_o - \varphi) - V_s \cos(\theta_s - \varphi))^2} d\theta_o \tag{28}$$

$$d\beta = \frac{V_s \sin(\theta_s - \theta_o)}{V_o^2 + V_s^2 - 2V_o V_s \cos(\theta_o - \theta_s)} dV_o + \frac{V_o^2 - V_o V_s \cos(\theta_o - \theta_s)}{V_o^2 + V_s^2 - 2V_o V_s \cos(\theta_o - \theta_s)} d\theta_o \tag{29}$$

$$\Delta\beta = \frac{V_s \sin(\theta_s - \theta_o)}{V_o^2 + V_s^2 - 2V_o V_s \cos(\theta_o - \theta_s)} \Delta V_o + \frac{V_o^2 - V_o V_s \cos(\theta_o - \theta_s)}{V_o^2 + V_s^2 - 2V_o V_s \cos(\theta_o - \theta_s)} \Delta\theta_o \tag{30}$$

The vector triangle formed by V_s, V_o and V_r is denoted by:

$$\begin{cases} V_s \sin(\theta_s - \theta_o) = V_r \sin \alpha \\ V_o - V_s \cos(\theta_s - \theta_o) = V_r \cos \alpha \\ V_o^2 + V_s^2 - 2V_o V_s \cos(\theta_s - \theta_o) = V_r^2 \end{cases} \tag{31}$$

Then the change of the relative velocity $\Delta\beta$ between the USV and the obstacle can be expressed as:

$$\Delta\beta = \frac{-\sin \alpha}{V_r} \Delta V_o + \frac{V_s \cos \alpha}{V_r} \Delta\theta_o \tag{32}$$

In order to avoid collision with obstacles, the relative speed of the USV can be adjusted from two directions. The constraint conditions that the relative speed should be changed by the speed and heading Angle of the USV should be discussed separately:

When $\beta > 0$

$$\Delta\beta \geq \mu_1 - \beta \ or \ \Delta\beta \leq -(\beta + \mu_2) \tag{33}$$

When $\beta < 0 \beta < 0$

$$\Delta\beta \leq -(\beta + \mu_2) \ or \ \Delta\beta \geq \mu_1 - \beta \tag{34}$$

In the formula, $\Delta\beta < 0$ means that the USV will pass through the right side of the obstacle, $\Delta\beta > 0$ means that the USV will pass through the left side of the obstacle, $\beta > 0$ means that the relative velocity direction is on the left side of the line between the USV and the obstacle, and $\beta < 0$ means that the relative velocity direction is on the right side of the line between the USV and the obstacle. It can be seen that there is $\Delta\beta$ certain corresponding relationship between the change of the relative speed of the USV and the obstacles $\Delta\beta$, the change of the speed of the USV ΔV and the change of the heading Angle of the USV $\Delta\theta_o$. As long as the appropriate ΔV is determined, the obstacle avoidance of the USV can be realized.

2.4   The Constraints of Collision Avoidance Rules at Sea Are Considered

Unmanned boats sailing at sea will encounter various ships, and the "International Regulations for Preventing Collisions at Sea" is required to coordinate the actions of ships to avoid collisions. As shown in Fig. 4, it shows the encounter type diagram centered on own ship.

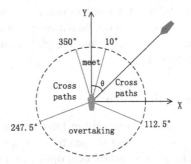

Fig. 4. The type of encounter defined by the maritime rules

3 USV Collision Avoidance Based on Improved Particle Swarm Optimization Algorithm

3.1 Standard Particle Swarm Optimization

Particle swarm optimization describes the D-dimensional problem to be solved as:

$$\begin{cases} \min f(x) = f(x_1, x_2, \cdots x_d) \\ s.t. x_{d\,min} \leq x_i \leq x_{d\,max}, i=1, 2, \cdots D \end{cases} \tag{35}$$

where $f(x)$ represents the particle fitness function, $X_i = [x_{i1}, x_{i2}, \cdots, x_{id}]$ represents the position of the number i particle, and $x_{d\,max}$ and $x_{d\,min}$ represent the upper and lower limits of the d-dimensional search space, respectively. The velocity of a particle is denoted by $V_i = [v_{i1}, v_{i2}, \cdots, v_{id}]$. During the iteration process, the particle updates its velocity and position by tracking the individual extreme value P_{best} of the particle and the global optimal value G_{best} of the population. The iterative formula is shown in Eqs. 36 and 37 as follows:

$$v_i^{k+1} = \omega v_i^k + c_1 \times r_1 \times (P_{best}^k - x_i^k) + c_2 \times r_2 \times (G_{best}^k - x_i^k) \tag{36}$$

$$x_i^{k+1} = x_i^k + v_i^{k+1} \tag{37}$$

where, v_i^{k+1} represents the velocity of particle i at the number $k + 1$ iteration, and ω is the inertia weight factor. c_1 and c_2 are the learning factors, which reflect the learning degree of the particle to the individual information and population information. If c_1 is too large and c_2 is too small, the particles tend to use their own experience to search, which is not conducive to global optimization. If c_1 is too small and c_2 is too large, the particles tend to accept the population experience and move toward the global extreme value, but there is a danger of falling into local optimum. r_1 and r_2 are random numbers between [0,1], which are used together with the learning factor, so that the particle can randomly use the individual extreme value and the global optimal extreme value and increase the randomness of the search.

3.2 Chaotic Perturbation Optimization

When the particle swarm optimization algorithm falls into the premature convergence state, the search process is corresponding to the chaotic trajectory traversal process to improve the ability of particles to jump out of local optimum. In the proposed algorithm, the population aggregation degree is judged according to the fitness variance. When the calculated fitness deviation is less than the given deviation, the particle with the best fitness value is selected to perform chaotic search. The fitness variance of the population is calculated as follows:

$$\sigma^2 = \sum_{i=1}^{n} \left(\frac{f_i - f_{av}}{f}\right)^2 \tag{38}$$

where f_i is the current fitness value of particle i, f_{av} is the current average fitness of the population, and f is the normalization factor.

$$f = \begin{cases} \max|f_i - f_{av}| & \max|f_i - f_{av}| > 1 \\ 1 & other \end{cases} \tag{39}$$

The introduction of chaos theory will definitely increase the complexity and operation time of the algorithm. In this paper, only a certain number of chaotic searches are carried out for the global optimal position. The specific operation steps are as follows:

1) At each iteration, the global optimal position $x_g = (x_{i1}, x_{i2}, ..., x_{iD})$ is selected as the initial chaotic quantity for chaos optimization.
2) Each dimensional component of decision variable x_g is mapped into chaotic variables in the range of [0,1] according to Eq. (38), and the corresponding chaotic variable $cx_d = (cx_{i1}, cx_{i2}, ..., cx_{id})$ is obtained.

$$cx_{id}^k = \frac{x_{id}^k - x_{\min,d}}{x_{\max,d} - x_{\min,d}}, d = 1, 2, ..., D \tag{40}$$

where $x_{\max,d}$ and $x_{\max,d}$ denote the upper and lower bounds of the d-dimensional search space, respectively.

3) Tent chaotic map formula is used to iterate K times for each dimension variable in the chaotic space ($K \leq 0.5n$, n is the number of particle swarm), and a group of chaotic sequences $CX = (cx^1, cx^2, ..., cx^K)$ is obtained.

$$cx_{id}^{k+1} = \begin{cases} cx_{id}^k/0.4 & (0 \leq cx_{id}^k \leq 0.4) \\ (1 - cx_{id}^k)/(1 - 0.4) & (0.4 < cx_{id}^k < 1) \end{cases} \tag{41}$$

4) The chaotic variables are mapped back to the decision space to generate a feasible solution sequence $X = (x^1, x^2, ..., x^K)$.

$$x_{id}^{k+1} = x_{\min,d} + cx_{id}^{k+1}(x_{\max,d} - x_{\min,d}) \tag{42}$$

5) The generated feasible solution is used to replace K particles with the worst fitness value in the particle swarm, evaluate the fitness value of each particle in the population, and update the global optimal position.

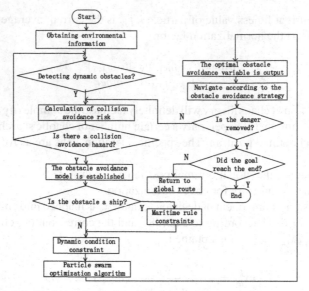

Fig. 5. Flow chart of dynamic obstacle avoidance for USV

According to the previous theoretical analysis, the key steps of local obstacle avoidance of USV are obtained. The flow chart of obstacle avoidance of USV is shown in Fig. 5.

4 Simulation Experiment

In order to verify the effectiveness of the proposed algorithm, this paper conducts simulation research on three typical states stipulated by maritime regulations and sea conditions with dynamic and static mixed obstacles. The blue track in the small picture is the sailing track of the unmanned boat, and the black track is the obstacle ship. Parameters such as starting point, moving speed, and moving direction can be set to simulate different encounter states.

(1) crossing ship

Suppose that the starting point of the obstacle vessel is (138.5, 1.75), and the starting point of the USV is (5, 5), and the two relative movements are in the situation of crossing and meeting. In this case, the obstacle vessel is coming from the starboard side, and the USV should turn right to avoid. The simulation schematic diagram is shown in Fig. 6, and the unmanned vehicle can complete the avoidance of cross encounters.

(2) Overtake

As shown in Fig. 7, the conflict situation of two ships pursuing is simulated, the starting point of the USV is set to (30, 50), the starting point of the obstacle ship is set to (100,115), and the initial speed is set to step 3 units in each cycle. The two ships are in the situation of pursuing conflict, and the USV should take the action of surpassing the left string. The simulation test verifies the correctness of the action of the USV.

(3) Meet head-on

The starting point of the unmanned boat is set as (30, 35) and the starting point of the obstructing ship is set as (160, 163). The two ships are in the collision situation. It is assumed that the obstructing ship does not take avoidance measures, and only the avoidance strategy of the unmanned boat is considered. As shown in Fig. 8, the unmanned boat detected the encountering vessel and took starboard avoidance measures.

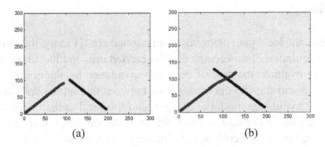

(a)   (b)

Fig. 6. Cross encounter situation

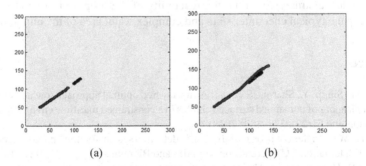

(a)   (b)

Fig. 7. Overtaking situation

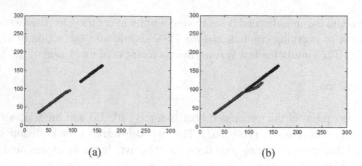

(a)   (b)

Fig. 8. Situation of meet head-on

5 Conclusion

This paper studies the local path planning of unmanned craft. Firstly, the dynamic obstacle avoidance calculation of unmanned craft is carried out, and the fuzzy mathematics theory is used to evaluate the risk of collision avoidance by integrating the shortest encounter distance and shortest encounter time between unmanned craft and obstacles. Then, an obstacle avoidance model is created, and the local collision avoidance problem is transformed into an optimization problem of the variation of speed and heading Angle of unmanned craft. The collision avoidance rules at sea and dynamics constraints of unmanned craft are introduced to limit the variation range of speed and heading Angle. Then particle swarm optimization algorithm is used to optimize the optimal obstacle avoidance strategy. Finally, the feasibility of the proposed method is verified by simulating the typical encounter state and complex water area of the unmanned craft.

References

1. Bibuli, M., Singh, Y., Sharma, S., et al.: A two layered optimal approach towards cooperative motion planning of unmanned surface vehicles in a constrained maritime environment. IFAC-Papers OnLine **51**(29), 378–383 (2018)
2. Wang, H., Wei, Z.: Stereovision based obstacle detection system for unmanned surface vehicle. In: IEEE International Conference on Robotics and Biomimetics, pp. 917–921. IEEE (2014)
3. Hansen, E., Huntsberger, T., Elkins, L.: Autonomous maritime navigation: developing autonomy skill sets for USVs. In: Unmanned Systems Technology VIII, vol. 6230, pp. 221–232. International Society for Optics and Photonics (2006)
4. Yan, R.-J., Pang, S., Sun, H.-B., Pang, Y.-J.: Development and missions of unmanned surface vehicle. J. Marine Sci. Appl. **9**(4), 451–457 (2010)
5. Seder, M., Petrovic, I.: Dynamic window based approach to mobile robot motion control in the presence of moving obstacles. In: IEEE International Conference on Robotics and Automation, pp. 1986–1991 (2007)
6. Tam, C., Bucknall, R.: Cooperative path planning algorithm for marine surface vessels. Ocean Eng. **57**, 25–33 (2013)
7. Feng, Y., Teng, G, F., Wang, A, X., et al.: Chaotic inertia weight in particle swarm optimization. In: 2nd International Conference on Innovative Computing, Information and Control, Kumamoto, Japan, p. 475. IEEE (2007)

8. Kuwata, Y., Wolf, M.T., Zarzhitsky, D., et al.: Safe maritime autonomous navigation with COLREGS, using velocity obstacles. IEEE J. Oceanic Eng. **39**(1), 110–119 (2014)
9. Johansen, T.A., Perez, T., Cristofaro, A.: Ship collision avoidance and COLREGS compliance using simulation-based control behavior selection with predictive hazard assessment. IEEE Trans. Intell. Transport. Syst. **17**(12), 3407–3422 (2016)
10. Naeem, W., Irwin, G.W., Yang, A.: COLREGs-based collision avoidance strategies for unmanned surface vehicles. Mechatronics **22**(6), 669–678 (2012)
11. Fiorini, P.: Motion planning in dynamic environments using velocity obstacles. Int. J. Robot. Res. **17**(7), 760–772 (1998)
12. Sun, J., Wu, X., Palade, V., Fang, W., Shi, Y.: Random drift particle swarm optimization algorithm: convergence analysis and parameter selection. Mach. Learn. **101**(1–3), 345–376 (2015)
13. Xu, X., Pan, W., Huang, Y., et al.: Dynamic collision avoidance algorithm for unmanned surface vehicles via layered artificial potential field with collision cone. J. Navig. **73**(6), 1306–1325 (2020)
14. Lin, J., He, C., Cheng, R.: Adaptive dropout for high-dimensional expensive multiobjective optimization. Complex Intell. Syst. **8**(1), 271–285 (2021)
15. Shabbir, F., Omenzetter, P.: Particle swarm optimization with sequential niche technique for dynamic finite element model updating. Comput.-Aided Civ. Infrastruct. Engineering **30**(5), 359–375 (2015)
16. Finaev, V.I., Medvedev, MYu., Pshikhopov, VKh., Pereverzev, V.A., Soloviev, V.V.: Unmanned powerboat motion terminal control in an environment with moving obstacles. Mekhatronika, Avtomatizatsiya, Upravlenie **22**(3), 145–154 (2021)

Solving Constrained Multi-objective Optimization Problems with Passive Archiving Mechanism

Huijuan Jia[1], Kai Zhang[1], and Chaonan Shen[1,2](✉)

[1] School of Computer Science and Technology, Wuhan University of Science and Technology, Wuhan 430065, China
449777215@qq.com
[2] Hubei Province Key Laboratory of Intelligent Information Processing and Real-Time Industrial System, Wuhan University of Science and Technology, Wuhan 430065, China

Abstract. During these years, many advanced constrained multi-objective evolutionary algorithms (CMOEAs) have been developed to solve constrained multi-objective optimization problems (CMOPs). However, some existing constrained multi-objective algorithms do not use an archiving mechanism, which leads to easy loss of the searched feasible solutions during the search process. In addition, even if some algorithms use an archiving mechanism, an unsuitable archiving update strategy can cause difficulty in balancing the diversity and convergence of the archive set. To address these challenges, a two-stage constrained multi-objective evolutionary algorithm based on a passive archiving mechanism (PA-CMOEA) is proposed in this paper. In the first stage, the constrained multi-objective problem is modeled as an unconstrained multi-objective problem, which allows the population to fully explore the decision space. The archive set is passively updated in a constraint-domination principle (CDP) criterion using population information. The goal of this stage is for the archive set to converge to the constrained Pareto front region of the problem with the help of population information. In the second stage, a local search is performed on the archive set. Specifically, the convergence and diversity of the archived solution set is improved while ensuring that the feasibility of the archived set is not worse. Finally, the algorithm outputs the archive set. To estimate the performance of the proposed algorithm, experiments are conducted on the LIRCMOP and CF benchmarks in comparison with some of the most popular algorithms. The experimental results demonstrate that the proposed algorithm has a competitive advantage compared with the comparative algorithms.

Keywords: Constrained Multi-objective Optimization · Two-stage · Passive Archiving

1 Introduction

Constraints are often present in practical engineering optimization problems. For example, in the problem of treating wastewater, in addition to energy consumption and penalty terms that can be treated as objective functions, there will be constraints such as attainment discharge limits for water quality indicators and upper and lower bounds for variables to be optimized [1]. In addition, such multi-objective optimization problems with constraints also exist in areas such as risk-constrained energy and reserve procurement [2], multi-objective path problems [3–6], and machine scheduling problems [7], which are collectively known to as constrained multi-objective optimization problems (CMOPs). In general, CMOPs can be defined as:

$$
\begin{aligned}
&\textit{minimine } F(x) = \min(f_1(x), ..., f_i(x), ..., f_M(x))^T \\
&\textit{subject to} \quad g_j(x) \leq 0, \ j = 1, ..., m \\
&\qquad\qquad h_j(x) = 0, \ j = m+1, ..., n \\
&\qquad\qquad x = (x_1, ..., x_D) \in \mathbb{S}
\end{aligned}
\tag{1}
$$

where x is a D-dimensional decision vector. $F(x)$ represents the M-dimensional objective function vector. m and $(n - m)$ denote the number of inequality and equation constraints, respectively.

The degree of constraint violation for the decision variable x is defined as:

$$
CV(x) = \sum_{i=1}^{m} \max\{g_i(x), 0\} + \sum_{j=m+1}^{n} |h_j(x)|
\tag{2}
$$

If $CV(x) = 0$ then it shows that x is a feasible solution, otherwise it is an infeasible solution.

For CMOPs, the Pareto set (PS) is composed of a set of non-dominated feasible optimal solutions, and the Pareto front (PF) is the objective vector corresponding to the PS. A good CMOEA needs to take into account three criteria simultaneously: (1) good convergence (2) good diversity (3) feasibility.

Although a great deal of CMOEAs have been proposed by researchers to solve CMOPs in recent decades, there are still many challenges in the constrained multi-objective domain that need to be addressed. For example, the existing constrained multi-objective algorithms with non-archiving mechanism are prone to miss potential infeasible or even feasible solutions when dealing with problems with inconsistent constrained and unconstrained Pareto front, resulting in wasting the number of evaluations or failure to find feasible solutions. In addition, even though some algorithms use archiving mechanism, inappropriate archiving update strategies can make it difficult to balance diversity and convergence of the archive set.

In view of the above problems, a two-stage constrained multi-objective evolutionary algorithm based on a passive archiving mechanism (PA-CMOEA) is proposed in this paper. During the first stage, the population *POP* approaches the unconstrained Pareto front quickly without considering the constraints, archive set is updated passively using information from the *POP*, and promising regions are preserved in archive set; The second stage performs a local search directly in the potential region, thus converging quickly to the constrained Pareto front. The archiving mechanism preserves the optimal set of solutions of the population at all times, avoiding missing feasible solutions or infeasible solutions with potential. The passive archiving mechanism does not destroy the nice diversity of the population itself and avoids unnecessary archive set updates.

The experimental part is a comparison of the algorithm performance with some efficient algorithms on the LIRCMOP [8] and CF [9] benchmarks, and the results verify that the algorithm proposed in this paper outperforms these comparative algorithms.

The remaining section provides an overview of the content of this paper. Section 2 develops a detailed analysis of the current work related to the algorithm. Section 3 provides a detailed description of the a two-stage constrained multi-objective evolutionary algorithm based on a passive archiving mechanism (PA-CMOEA). The experimental results of all the compared algorithms are given in Sect. 4 and the experimental results are discussed. Section 5 gives a summary of the whole paper.

2 Related Works

In the past period of time, researchers have proposed many CMOEAs with different strategies to address CMOPs, and in this paper, these algorithms are categorized by the presence or absence of archives as follows.

2.1 Evolutionary Algorithms Without Archiving Mechanism

NSGA-II-CDP [10], as the most typical representative of the initial dominance-based CMOEAs, combines the Deb constraint dominance principle (CDP) and the NSGA-II algorithm to solve CMOPs. Three rules are defined in this algorithm: given two solutions x_i and x_j, , if (1) both x_i and x_j are feasible solutions, the non-dominated solution is taken; (2) if both x_i and x_j are infeasible solutions, the solution with less constraint violation is taken; (3) if one is a feasible solution and one is infeasible solution, the feasible solution is taken. So feasible solutions are unsurpassable exist in this class of algorithms, which directly leads to the loss of potentially infeasible solutions and affects the convergence of the algorithm. Subsequent epsilon constraint processing (EC) [11], stochastic rank (SR) [12] and other algorithms slightly relaxed the constraint to accept infeasible solutions with certain probability. But these algorithms still use feasibility as the first selection criterion, although they relax the constraints. The population is difficult to cross large infeasible regions. In order to cross large infeasibility regions, researchers have successively proposed two-stage algorithms such as PPS [13] and TOP [14], where PPS approaches the unconstrained Pareto front (UPF, i.e., the Pareto optimal front obtained without considering any constraints) quickly in the first stage without considering constraints; and then considers constraints in the second stage to reach the final constrained Pareto front (CPF, i.e., the Pareto optimal front consisting of feasible non-dominated solution sets considering constraints). A relationship concept proposed in the literature [15] is used here to divide the relationship between UPF and CPF into four categories based on their distribution in the objective space: (Type-1) CPF is the same as UPF; (Type-2) CPF is part of UPF; (Type-3) CPF and UPF partially overlap; (Type-4) CPF and UPF do not overlap.

For the previous three types, the non-archived two-stage algorithm has been able to achieve better results, but in Type-4, since the first stage population focuses on finding the UPF, it will miss potential infeasible solutions or even feasible solutions, and when the area of the decision space corresponding to the UPF is too small, the CPF may not be found [16]. TOP aims to find feasible solutions in the first stage and the CPF in the second stage, and it is not easy to find a suitable transformation condition.

2.2 Evolutionary Algorithms with Archiving Mechanism

In addition to constrained multi-objective evolutionary algorithms without archiving mechanism, researchers have also proposed some algorithms with archiving mechanisms, such as CTAEA [17] proposed by Li et al. The convergence-focused archive and feasibility-focused archive exchange information bidirectionally to converge to the optimal front, but too close information exchange can hinder the direction of both archives. The dual populations in CCMO [18] evolve toward their respective goals, the process share information about the offspring. Later proposed MSCMO [19] to gradually increase constraints and EMCMO [16] to solve CMOPs using evolutionary multi-tasking. There is no doubt that high-quality information exchange can improve the performance of both populations, but the design of an update strategy for the archiving mechanism is not an easy task, and an inappropriate archive set update mechanism can easily lead to difficulty in balancing the diversity and convergence, which motivates us to design an effective archive set update mechanism to handle CMOPs.

3   Proposed Algorithm

For the purposes of this section., a two-stage constrained multi-objective evolutionary algorithm based on a passive archiving mechanism (PA-CMOEA) is proposed to solve CMOPs. In the first stage, the population *POP* approach the unconstrained Pareto front quickly without considering constraints, and archive set *Archive* is updated passively using the information of the *POP's* offspring; In the second stage, a local search is performed within the potential region obtained from the first stage search, and finally the Pareto front with good convergence and diversity.

3.1   Archiving Strategy

A number of existing algorithms often take the lead in searching the UPF of the problem in order to cross the large infeasible region in the target space. Taking LIRCMOP1 [8] as an example, as shown in Fig. 1(a), the population will quickly converge to the UPF and then spend numerous evaluations to pull back to the CPF of the problem. If the algorithm does not have enough number of evaluations or the area of the decision space corresponding to the UPF is too small, the population may not find a feasible solution. Then, if the algorithm introduces the population archiving mechanism in the search process, it will save the potential solutions in the search process. If the CDP criterion is used as the environment selection for archiving, the archived individuals are shown in Fig. 1(b), and obviously the individuals saved by the archiving mechanism are near the CPF of the problem. Then if the information of the archived individuals is effectively used, the algorithm is more likely to converge to the CPF of the problem. It follows that by using the information of archived individuals, the algorithm can save the number of evaluations of the algorithm by not having to go through the process of pulling the archive set from the UPF back to the CPF.

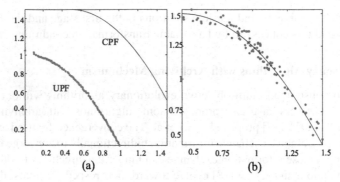

Fig. 1. (a) Distribution of individuals in the population (b) Distribution of individuals in the archive.

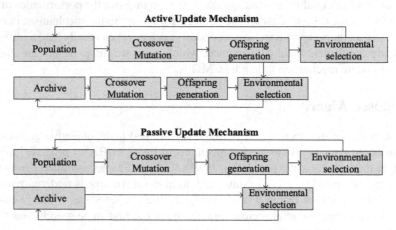

Fig. 2. Framework for active and passive update mechanisms.

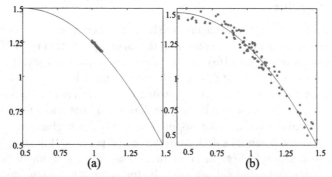

Fig. 3. (a) Archive set with active update mechanism. (b) Archive set with passive update mechanism.

3.2 Passive Update Mechanism

Although some algorithms now use archiving mechanism, the design of an archiving update mechanism is not an easy task, and an improper archiving update mechanism can lead to difficulties in balancing the convergence, diversity and feasibility of the archive set, so we propose a passive update mechanism for the archive set. Using this mechanism, the three major challenges of the CMOPs can be well balanced. To clarify the difference between the proposed passive update mechanism and the active update mechanism, the two mechanisms are shown in Fig. 2. The main difference between these two is in the red part in Fig. 2. Active update refers to the way in which the archive set actively updates the itself by generating offspring through evolutionary operators, while the passive update approach refers to the way in which the archive set can only be passively updated by information from the population, without actively evolving itself.

To elucidate the advantages of passive update, taking LIRCMOP1 as an illustrative case, Fig. 3 shows the distribution of the archive set on LIRCOMP1 using active and passive updates. The distribution of the archive set for the active update mechanism is shown in Fig. 3(a). The distribution of the archive set for the passive update mechanism is shown in Fig. 3(b). The common archive set use the mechanism of active update, where feasible parents can easily produce offspring satisfying the constraints through crossover and mutation, while the environment selection methods are all carried out in a constraint-first manner, which can result in some poorly diverse feasible solutions replacing potentially infeasible solutions that are well diverse and have a small degree of constraint violation. The active update approach tends to increase the convergence of the feasible solutions of the archive set, but leads to a sharp decrease in the diversity of the archive set, thus losing most of the Pareto front. In contrast, the archive set uses a passive update mechanism to overcome this drawback. Instead of relying on the information exchange within the archive set, the passive update only relies on the *POP* to passively update the archive set by means of CDP, and this update method can handle the priority relationship of convergence, diversity, and feasibility.

3.3 Design of Fitness Function

In order to consider both the convergence and diversity of the solution set, we design a new method for calculating the fitness function. The formula is as follows.

$$fitness\left(POP^{(i)}\right) = NUM\left(POP^{(i)}\right) + \frac{1}{1 + MED(POP^{(i)})} \tag{3}$$

where $NUM(\cdot)$ [20] indicates the number of individuals dominated by other individuals in the population, and a smaller $NUM(\cdot)$ value indicates better convergence of individuals. $MED(\cdot)$ [21] measures the diversity of individuals in the population, and a larger $MED(\cdot)$ value indicates better diversity of individuals. Since the value of $1/(1 + MED(\cdot))$ ranges from 0 to 1 and the value of $NUM(\cdot)$ is a non-negative integer. The formulation is designed to consider both diversity and convergence of the solution set, and the convergence of the solution set plays a dominant role. The smaller the value of fitness of an individual, the better the performance of the individual is indicated.

3.4   PA-CMOEA

This paragraph will explain the implementation process of the algorithm in detail. Firstly, in the first stage, because archive set *Archive* is updated passively, the population *POP* has to maintain a good diversity while converging rapidly. So the algorithm in reference [21], The usual method of environmental selection is to merge parents and offspring, and then get the next generation population by environmental selection. We call this environmental selection method as N-by-N selection method. However, in order to better explore the decision space, the environmental selection approach in this paper is a local selection between parents and their corresponding offspring. We call its method the One-by-One selection approach. The One-by-One approach preserves the maximum diversity of the population because selection is performed between only two individuals., as shown in Algorithm 1.

Algorithm 1 First stage of *PA-CMOEA*
1: Initialization POP, $t=0$, $Archive = POP$
2: while ($t <= 1/2$ $maxFE$){
3:    for $i = 1$: $Problem.N$
4:        $POP^{(i)}$ undergoes Gaussian mutation to generate $NewPOP^{(i)}$
5:        Add $NewPOP^{(i)}$ to $PopOff$
6:        Calculate the fitness value of $NewPOP^{(i)}$ with formula 3
7:        if $fitness(NewPOP^{(i)}) < fitness(POP^{(i)})$
8:            $POP^{(i)} = NewPOP^{(i)}$
9:        end if
10:   end for
11:   $Archive = EnvSelectionCDP([Archive, PopOff])$;
12:} end while

In the first line of the pseudocode, the *POP* and *Archive* are initialized. Each of the two stages of the algorithm accounts for 50% of the maximum number of evaluations. Lines 4 of the pseudocode generate the candidate individuals for *POP*, and this part uses Gaussian mutation operators to generate new solutions for the offspring. The fifth line of the pseudocode saves the candidate individuals for the subsequent *Archive*. Lines 7 to 9 of the algorithm use the idea of a binary tournament to select the next generation among the parent and the corresponding generated offspring. If the fitness value of the offspring is less than the corresponding parent, which indicates that the offspring outperforms the parent, then the offspring is selected to enter the next generation population. In the last line 11, the information of candidate individuals saved in the fifth row is used to do environmental selection to passively update the archive set, driving archive set closer to the CPF region of the problem and preparing for the second stage of local search.

At the end of the first stage, the individuals in *Archive* are already on or near the Pareto front, so the *POP* will no longer be needed to provide information in the second stage. Using LIRCMOP 1/5/9/11 as representatives of the four types of CPF and UPF relationships, respectively, to demonstrate the state of *Archive* at this point.

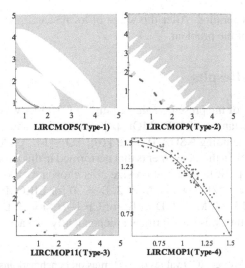

Fig. 4. The distribution of archive set after the first stage.

From the Fig. 4, it can be seen that for Type-1 and Type-3, *Archive* has basically reached the CPF, and for Type-2 and Type-4, Archive has not completely reached the CPF. Therefore, we improve the convergence and diversity of individuals while ensuring that the degree of individual constraint violation of *Archive* is not worse, so that the local search of *Archive* can be achieved and the quality of the solution set can be improved. Algorithm 2 of the second stage is shown in the pseudocode as above.

Algorithm 2 Second stage of *PA-CMOEA*
1: while ($t>1/2$ *maxFE*){
2:    for $i = 1$: *Problem. N*
3:       $Archive^{(i)}$ undergoes Gaussian mutation to generate $NewArchive^{(i)}$
4:       Calculating the CV of $Archive^{(i)}$ and $NewArchive^{(i)}$
5:          //CV is the degree of constraint violation
6:       if $(NewArchive_{CV}^{(i)}) \leq Archive_{CV}^{(i)}$)
7:          if $fitness(NewArchive^{(i)}) < fitness(Archive^{(i)})$
8:             $Archive^{(i)} = NewArchive^{(i)}$
9:          end if
10:    end if
11:   end for
12:} end while

The first 3 lines of Algorithm 2 are where *Archive* generates candidate solutions in the same way as in Algorithm 1. Line 4 calculates the degree of constraint violation for the *ith* individual and the candidate individuals in *Archive*. Lines 6–14 indicate that the fitness is compared under the premise that the degree of constraint violation is not worse, and if the fitness of the candidate individual is better, the candidate individual is selected as the next generation. After this stage of local search, *Archive* will gradually converge to the CPF of the problem.

4 Experimental Results

To verify the property of the presented algorithm, PA-CMOEA was next experimented on 24 benchmarks instances of LIRCMOP [8] and CF [9], and compared with five advanced CMOEAs including NSGA II-CDP [10], TOP [14], CTAEA [17], EMCMO [16], and MSCMO [19]. In the experiments that performed in this paper, the experimental platform is PlatEMO [22], the population size N is chosen to 100, and the maximum evaluations *maxFE* are 300000, where $M = 2$, $D = 30$ for LIRCMOP 1–12; $M = 3$, $D = 30$ for LIRCMOP 13–14; $M = 2$, $D = 10$ for CF 1–7; $M = 3$, $D = 10$ for CF 8–10. To be fair, each algorithm was run 30 times independently on each test problem.

Table 1. Average IGD values over 30 runs on benchmark instances.

Problem	NSGAII-CDP	ToP	CTAEA	EMCMO	MSCMO	PA-CMOEA
LIRCMOP1	2.7133e-1 (3.10e-2) -	3.0649e-1 (2.97e-2) -	1.6798e-1 (9.84e-2) -	2.0078e-1 (4.89e-2) -	2.0997e-2 (1.16e-2) =	1.9712e-2 (5.35e-4)
LIRCMOP2	2.4132e-1 (2.76e-2) -	2.7059e-1 (2.19e-2) -	1.2930e-1 (3.38e-2) -	1.7269e-1 (3.96e-2) -	9.3507e-3 (5.84e-3) -	1.5872e-2 (6.34e-3)
LIRCMOP3	3.0320e-1 (3.24e-2) -	3.4487e-1 (2.68e-2) -	2.4776e-1 (9.28e-2) -	1.9633e-1 (5.50e-2) -	2.8244e-2 (9.31e-3) -	2.2584e-2 (1.40e-2)
LIRCMOP4	2.8023e-1 (3.32e-2) -	3.1183e-1 (9.84e-3) -	2.4168e-1 (1.18e-1) -	2.2365e-1 (4.34e-2) -	4.4776e-2 (2.33e-2) -	1.8500e-2 (1.56e-2)
LIRCMOP5	1.1835e+0 (1.62e-1) -	1.1695e+0 (1.61e-2) -	1.0716e+0 (3.39e-1) -	2.5773e-1 (5.80e-2) -	7.0008e-3 (2.91e-4) +	8.7002e-3 (4.51e-4)
LIRCMOP6	1.3131e+0 (1.79e-1) -	1.1514e+0 (3.62e-1) -	1.3459e+0 (3.69e-4) -	2.6505e-1 (6.99e-2) -	2.7457e-1 (5.45e-1) -	9.3024e-3 (6.30e-4)
LIRCMOP7	5.9635e-1 (7.25e-1) -	9.8720e-1 (8.11e-1) -	2.1954e-1 (3.29e-1) -	1.1505e-1 (2.26e-2) -	8.7837e-3 (2.34e-2) =	8.6543e-3 (3.43e-4)
LIRCMOP8	9.0111e-1 (7.44e-1) -	1.1836e+0 (7.24e-1) -	7.3915e-1 (6.86e-1) -	1.8410e-1 (3.62e-2) -	7.2337e-3 (3.11e-4) +	9.5953e-3 (6.40e-4)
LIRCMOP9	8.1134e-1 (8.34e-2) -	4.5435e-1 (1.03e-1) -	4.6596e-1 (1.10e-1) -	3.4461e-1 (1.29e-1) -	1.1123e-1 (1.14e-1) -	1.2100e-2 (3.01e-3)
LIRCMOP10	7.2459e-1 (1.73e-1) -	3.6510e-1 (8.43e-2) -	2.4575e-1 (9.91e-2) -	8.9315e-2 (4.18e-2) -	6.4091e-3 (4.30e-4) +	7.5501e-3 (2.33e-4)
LIRCMOP11	7.2976e-1 (8.54e-2) -	3.9075e-1 (9.71e-2) -	1.8620e-1 (3.24e-2) -	4.4739e-2 (3.06e-2) -	2.8042e-3 (2.86e-4) -	2.2099e-3 (2.94e-4)
LIRCMOP12	5.2702e-1 (1.91e-1) -	2.8950e-1 (8.50e-2) -	1.7728e-1 (9.95e-2) -	1.4558e-1 (6.18e-2) -	1.3085e-1 (2.41e-3) -	2.7375e-3 (1.54e-5)
LIRCMOP13	1.3248e+0 (2.72e-3) -	1.2067e+0 (2.55e-1) -	1.0947e-1 (2.41e-3) +	9.1009e-2 (6.47e-4) +	1.1868e-1 (2.36e-3) -	1.4282e-1 (1.08e-2)
LIRCMOP14	1.2816e+0 (2.78e-3) -	1.2772e+0 (9.18e-2) -	1.1174e-1 (1.02e-3) +	9.6018e-2 (1.18e-3) +	1.0272e-1 (1.45e-3) +	1.2773e-1 (1.23e-2)
CF1	8.2981e-3 (1.40e-3) -	6.7662e-4 (3.84e-4) +	3.7717e-2 (7.00e-3) -	2.1750e-3 (3.14e-4) -	2.6405e-4 (1.91e-4) +	1.6422e-3 (2.18e-4)
CF2	2.5542e-2 (1.32e-2) -	3.8627e-3 (1.76e-4) -	2.2039e-2 (6.86e-3) -	1.7669e-2 (7.43e-3) -	3.3976e-3 (3.41e-5) -	3.1117e-3 (1.99e-4)
CF3	2.3440e-1 (8.81e-2) -	2.5680e-1 (1.59e-1) -	2.1681e-1 (9.51e-2) -	2.5185e-1 (9.26e-2) -	2.0050e-1 (7.29e-2) -	9.0548e-2 (2.45e-2)
CF4	7.6840e-2 (2.38e-2) -	2.8371e-2 (1.74e-2) -	7.5970e-2 (3.55e-2) -	7.0279e-2 (2.39e-2) -	3.3869e-2 (6.37e-3) -	1.0777e-2 (2.09e-3)
CF5	2.9784e-1 (9.44e-2) -	1.9839e-1 (7.94e-2) -	3.0418e-1 (1.42e-1) -	3.1082e-1 (1.24e-1) -	2.1312e-1 (8.57e-2) -	7.3487e-2 (7.48e-2)
CF6	9.3534e-2 (2.82e-2) -	5.1297e-2 (3.33e-2) -	9.1109e-2 (3.14e-2) -	4.7374e-2 (2.30e-2) -	2.0129e-2 (3.94e-3) +	2.5843e-2 (4.12e-3)
CF7	3.3590e-1 (1.31e-1) -	1.9714e-1 (1.02e-1) -	3.0330e-1 (1.52e-1) -	2.6062e-1 (9.85e-2) -	8.7469e-2 (3.85e-3) -	6.9761e-2 (3.39e-2)
CF8	3.6557e-1 (9.93e-2) -	6.6762e-1 (1.09e-1) -	2.1680e-1 (1.99e-2) -	2.2948e-1 (1.03e-1) -	3.5199e-1 (8.03e-2) -	1.0436e-1 (8.48e-3)
CF9	1.4916e-1 (2.96e-2) -	4.1241e-1 (1.07e-1) -	1.1296e-1 (2.08e-2) -	1.0511e-1 (3.27e-2) -	1.7061e-1 (3.41e-2) -	7.2324e-2 (5.15e-2)
CF10	NaN (NaN)	NaN (NaN)	3.3149e-1 (1.12e-1) -	3.8164e-1 (7.59e-2) -	4.6381e-1 (1.14e-1) -	1.7356e-1 (1.24e-1)
+/-/=	0/23/0	1/22/0	2/22/0	2/22/0	8/14/2	

Wilcoxon rank sum test at 0.05 significance level between PA-CMOEA and competing CMOEAs. "−" indicates that PA-CMOEA performs better than the other algorithms, "+" indicates that PA-CMOEA does not perform as well as the other algorithms, and " = " indicates that there is no comparability.

In the comparison of performance, the IGD [23] is used as a measure, and if the convergence and diversity of the solution is better, then it will have a smaller IGD value. The best average IGD value is shown on a gray background. As can be seen from

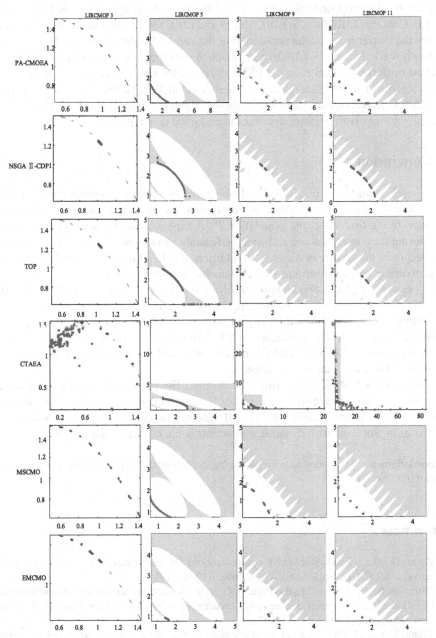

Fig. 5. Solutions obtained by PA-CMOEA, NSGA-II-CDP, TOP, CTAEA, MSCMO and EMCMO on LIRCMOP3, LIRCMOP5, LIRCMOP9 and LIRCMOP11 problems.

Table 1, the proposed PA-COMEA wins 16 instances on 24 CMOPs. MSCMO also has a significant advantage on 6 instances, where the recently proposed EMCMO [15] outperforms the other algorithms on LIRCMOP 13 and 14.

To further verify the advantages of this algorithm, Fig. 5 shows the contrast data acquired by the proposed PA-CMOEA and the other five algorithms on LIRCMOP. The LIRCMOP3, LIRCMOP5, LIRCMOP9 and LIRCMOP11 problems are selected to display the convergence of the population at the end of the algorithm search.

As shown in Fig. 5, PA-CMOEA achieves the best performance of the IGD metrics for most of the test instances in the LIRCMOP and CF benchmarks. The experimental results interpret that PA-CMOEA has the ability to handle CMOPs of various properties (including concave, convex, large infeasible domains and spherical Pareto fronts), finding optimal solutions with uniform distribution and good convergence.

5 Conclusion

In this paper, a two-stage constrained multi-objective evolutionary algorithm with passive archiving mechanism (PA-CMOEA) is proposed to solve the constrained multi-objective problem. In the first stage, the population *POP* explores the potential region without considering the constraints and archives the feasible and potentially infeasible solutions obtained from the search in *Archive*. In addition, the *MED* method is used to improve the probability of the algorithm to find feasible solutions and maintain the diversity of the population. The individuals in *Archive* are already on or near the CPF after the first stage, so the individuals in *Archive* in the second stage can reach the CPF by considering convergence and diversity in sequence without decreasing the degree of constraint violation. The performance of PA-CMOEA is compared with five advanced algorithms on the LIRCMOP and CF benchmarks, and the experimental results show that PA-CMOEA has a clear advantage.

Although PA-CMOEA performs well on the LIRCMOP and CF benchmarks, the constraints of these benchmarks are static. Research on multi-objective optimization problems with dynamic constraints is still scarce. Therefore, solving multi-objective optimization problems with dynamic constraints is the next research direction.

Acknowledgment. This work was supported by the National Natural Science Foundation of China (Grant No. 62176191).

References

1. Zhou, H., Qiao, J.: Multiobjective optimal control for wastewater treatment process using adaptive MOEA/D. Appl. Intell. **49**(3), 1098–1126 (2018)
2. Paterakis, N.G., Gibescu, M., Bakirtzis, A.G., et al.: A multi-objective optimization approach to risk-constrained energy and reserve procurement using demand response. IEEE Trans. Power Syst. **33**(4), 3940–3954 (2017)
3. Wang, J., Yuan, L., Zhang, Z., et al.: Multiobjective multiple neighborhood search algorithms for multiobjective fleet size and mix location-routing problem with time windows. IEEE Trans. Syst., Man, Cybern.: Syst. **51**(4), 2284–2298 (2019)

4. Wang, J., Ren, W., Zhang, Z., et al.: A hybrid multiobjective memetic algorithm for multi-objective periodic vehicle routing problem with time windows. IEEE Trans. Syst., Man, and Cybern.: Syst. **50**(11), 4732–4745 (2018)
5. Wang, J., Weng, T., Zhang, Q.: A two-stage multiobjective evolutionary algorithm for multi-objective multidepot vehicle routing problem with time windows. IEEE Trans. Cybern. **49**(7), 2467–2478 (2018)
6. Wang, J., Zhou, Y., Wang, Y., et al.: Multiobjective vehicle routing problems with simultaneous delivery and pickup and time windows: formulation, instances, and algorithms. IEEE Trans. Cybern. **46**(3), 582–594 (2015)
7. Zheng, X.L., Wang, L.: A collaborative multiobjective fruit fly optimization algorithm for the resource constrained unrelated parallel machine green scheduling problem. IEEE Trans. Syst., Man, and Cybern.: Syst. **48**(5), 790–800 (2016)
8. Fan, Z., et al.: An improved epsilon constraint-handling method in MOEA/D for CMOPs with large infeasible regions. Soft. Comput. **23**(23), 12491–12510 (2019)
9. Zhang, Q., Zhou, A., Zhao, S., et al.: Multiobjective optimization test instances for the CEC 2009 special session and competition. University of Essex, Colchester, UK and Nanyang technological University, Singapore, special session on performance assessment of multi-objective optimization algorithms, technical report, vol. 264, pp. 1–30 (2008)
10. Deb, K., Pratap, A., Agarwal, S., et al.: A fast and elitist multiobjective genetic algorithm: NSGA-II. IEEE Trans. Evol. Comput. **6**(2), 182–197 (2002)
11. Takahama, T., Sakai, S.: Constrained optimization by ε constrained particle swarm optimizer with ε-level control. Soft computing as transdisciplinary science and technology, pp. 1019–1029. Springer, Berlin, Heidelberg (2005)
12. Runarsson, T.P., Yao, X.: Stochastic ranking for constrained evolutionary optimization. IEEE Trans. Evol. Comput. **4**(3), 284–294 (2000)
13. Fan, Z., Li, W., Cai, X., et al.: Push and pull search for solving constrained multi-objective optimization problems. Swarm Evol. Comput. **44**, 665–679 (2019)
14. Liu, Z.Z., Wang, Y.: Handling constrained multiobjective optimization problems with constraints in both the decision and objective spaces. IEEE Trans. Evol. Comput. **23**(5), 870–884 (2019)
15. Ma, Z., Wang, Y.: Evolutionary constrained multiobjective optimization: Test suite construction and performance comparisons. IEEE Trans. Evol. Comput. **23**(6), 972–986 (2019)
16. Qiao, K., Yu, K., Qu, B., et al.: An evolutionary multitasking optimization framework for constrained multiobjective optimization problems. IEEE Trans. Evol. Comput. **26**(2), 263–277 (2022)
17. Li, K., Chen, R., Fu, G., et al.: Two-archive evolutionary algorithm for constrained multiobjective optimization. IEEE Trans. Evol. Comput. **23**(2), 303–315 (2018)
18. Tian, Y., Zhang, T., Xiao, J., et al.: A coevolutionary framework for constrained multiobjective optimization problems. IEEE Trans. Evol. Comput. **25**(1), 102–116 (2020)
19. Ma, H., Wei, H., Tian, Y., et al.: A multi-stage evolutionary algorithm for multi-objective optimization with complex constraints. Inf. Sci. **560**, 68–91 (2021)
20. Zitzler, E., Laumanns, M., Thiele, L.: SPEA2: Improving the strength Pareto evolutionary algorithm. TIK-report, vol. 103 (2001)
21. Zhang, K., Shen, C.N., He, J., et al.: Knee based multimodal multi-objective evolutionary algorithm for decision making. Inf. Sci. **544**, 39–55 (2021)
22. Tian, Y., Cheng, R., Zhang, X., et al.: PlatEMO: A MATLAB platform for evolutionary multi-objective optimization [educational forum]. IEEE Comput. Intell. Mag. **12**(4), 73–87 (2017)
23. Zitzler, E., Thiele, L., Laumanns, M., et al.: Performance assessment of multiobjective optimizers: an analysis and review. IEEE Trans. Evol. Comput. **7**(2), 117–132 (2003)

A Comparison of Large-Scale MOEAs with Informed Initialization for Voltage Transformer Ratio Error Estimation

Lianghao Li[1]([⊠]) [iD], Cheng He[2] [iD], and Hongbin Li[2]

[1] Key Laboratory of Image Information Processing and Intelligent Control of Education Ministry of China, School of Artificial Intelligence and Automation, Huazhong University of Science and Technology, Wuhan 430074, China
lianghaoli.hust@gmail.com
[2] School of Electrical and Electronic Engineering, Huazhong University of Science and Technology, Wuhan 430074, China
{chenghe_seee,lihongbin}@hust.edu.cn

Abstract. Large-scale multiobjective optimization problems (LSMOPs) exist widely in real-world applications. The large number of decision variables in LSMOP leads to a tremendous high-dimensional search space, which is still challenging for existing multiobjective evolutionary algorithms (MOEAs). The voltage transformer ratio error estimation (TREE) problem is a typical LSMOP in real-world applications. A number of large-scale multiobjective evolutionary algorithms (LSMOEAs) have been adopted for solving the TREE problem, but so far there is still considerable potential for improvement in their results. In this paper, we propose a informed initialization for solving the TREE problem. The physical properties of the TREE problem are used to construct an initial set of solutions with specific characteristics. The proposed initialization method is tested against the default initialization method on five TREE problems by being applied on six state-of-the-art LSMOEAs and a classical MOEA. The experimental results demonstrate that the initialization method has a significant effect on the performance of LSMOEAs in solving the TREE problems.

Keywords: Large-scale optimization · Multiobjective optimization · Population Initialization · Voltage transformer ratio error estimation

1 Introduction

Multiobjective optimization problems (MOPs) involve at least two objectives [32], whose mathematical formulation is

$$\begin{aligned} \text{minimize } & f_i(\mathbf{x}) \qquad i = 1, \dots, M \\ \text{subject to } & \mathbf{x} \in \Omega^D, \end{aligned} \qquad (1)$$

Supported by the National Natural Science Foundation of China (No. U20A20306, 61903178).

D is the number of decision variables, $\mathbf{x} = (x_1, \cdots, x_D)^T$ represents the decision vector, Ω^D represents the decision space, and M is the dimension of the objective space. MOPs arise in various practical applications, such as vehicle routing [2], production management [3], feature selection [21], auto-control [22], and voltage transformer ratio error estimation (TREE) [13]. In MOPs, there is typically a set of trade-off solutions that cannot be justified as superior or inferior, while in single-objective optimization there exists a global optimum. In the decision space, those trade-off solutions of an MOP form the Pareto-optimal set (PS), and they form the Pareto-optimal front (PF) in the objective space [27]. With their population-based characteristics, evolutionary algorithms (EAs) are naturally suitable for solving MOPs. During the past half-century, many multiobjective EAs (MOEAs) based on different solution updating strategies have been developed, including the elitist multiobjective genetic algorithm (NSGA-II) [7], the MOEA based on decomposition (MOEA/D) [30], the indicator-based EA (IBEA) [36], and others [9,10,20,33].

With more than 100 decision variables, solving large-scale MOPs (LSMOPs) poses significant difficulties to existing MOEAs [5]. As the amount of the decision variables increases in a linear fashion, the search space volume will grow exponentially, as will the search complexity, leading to performance degeneration for MOEAs. Although the research on solving large-scale single-objective optimization problems using EAs has gained significant development, the performance of current large-scale MOEAs is still not satisfying enough. Recently, large-scale multiobjective optimization has attracted growing interest, and several approaches with effective strategies for searching high-dimensional decision spaces efficiently have been proposed [19], which can be generally classified into five groups [11].

- **Cooperative co-evolution based approaches:** As one the most widely used approaches in large-scale optimization problems with a single objective, the cooperative co-evolution (CC) based framework handles a large number of variables using a divide-and-conquer manner [23,24]. The Cooperative co-evolution based algorithms divide the decision variables into a few non-overlapped groups, followed by cooperative optimization for each group. The widely used grouping methods in co-evolution The widely used grouping methods include random grouping, ordered grouping and differential grouping. CC-based approaches are also widely adopted in large-scale multiobjective optimization, such as CCGDE3 [1], MOEA/D-RDG [26], and CCLSM [17], where random grouping is used in CCGDE3, MOEA/D-RDG uses dynamic random grouping, and CCLSM uses differential grouping.
- **Decision variable analysis based approaches:** These algorithms are devoted to deal with LSMOPs by analyzing the characteristics and interrelationships of decision variables. Representative algorithms based on decision variable analysis mainly include MOEA/DVA [18] and LMEA [31]. Similar to coevolution, the algorithms based on variational analysis also group the decision variables, but they first conduct some analysis of the decision variables. Decision variable analysis can significantly improve the performance of

the algorithm on some variable separable problems, but it also consumes a large amount of computational resources, making it difficult to apply to solve multi-objective optimization problems with higher dimensions.

- **Problem reformulation based approaches:** These approaches reformulate the LSMOPs into problems with fewer decision variables and a simpler search space. For instance, transformation functions and decision variable grouping methods are used as dimension reduction approaches of the search space in the weighted optimization framework (WOF) [34]. In contrast, the large-scale multiobjective optimization framework (LSMOF) accelerates the searching of MOEAs in tackling LSMOPs by reformulating the search space and using direction vectors to steer the evolution process [15].
- **Offspring generation based approaches:** These algorithms deal with large-scale multi-objective optimization problems by designing novel offspring solution generation methods. When dealing with large-scale multi-objective optimization problems, suitable child solution generation methods can significantly improve the efficiency and performance of the algorithms, such as LMOCSO [29], DGEA [12], and FLEA [16].
- **Probability model based approaches:** This class of algorithms learns the probability distribution of the solution in a high-dimension space by learning the probability distribution and based on the probability distribution model. The algorithms using such strategies mainly include IM-MOEA [6], S^3-CMA-ES [4], and GMOEA [14]. IM-MOEA is base on Gaussian processes, S^3-CMA-ES is based on covariance matrix and evolutionary strategy, and GMOEA is based on generative adversarial network.

Under these approaches for tackling large numbers of decision variables, existing large-scale MOEAs have demonstrated their capability to deal with benchmark problems, such as ZDT [35], DTLZ [8], WFG [25], and LSMOP [5]. However, the performance of existing LSMOEAs on the real-world application based LSMOPs, such as TREE problem, is still unsatisfying.

2   Background

The ratio error of voltage transformer can be defined as:

$$RE = \frac{V_m - V_t}{V_t}, \tag{2}$$

where V_m denotes the voltage value obtained from the transformer measurement, V_t denotes the ground truth value of the voltage value which is usually unknown. Unlike existing artificial benchmark problems in the field of large-scale multiobjective optimization, TREE problems are extracted from real-world applications with data sampled from different substations in the power delivery system [13]. With enough sampled data, TREE problems transform conventional voltage transformer ratio error estimation tasks into large-scale multiobjective optimization. Since the data sampled from the power delivery system is irregular, the PS and PF of the TREE problems are also irregular, making it challenging to maintain diversity in both the decision and objective spaces for LSMOEAs. Figure 1 shows the irregular data sampled from a substation with 12 sets of VTs, where each VT set includes three primary voltage phases. The horizontal axis is the length of the measured sequential data, and the vertical axis shows the measured primary voltage value. The detailed formulations of two types of objectives are given in the follows.

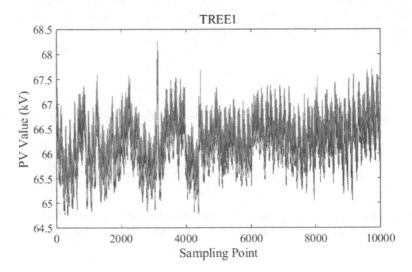

Fig. 1. The measured primary voltage values in a substation with 12 sets of VTs (each set contains three VTs, and thus a total number of 36 data sequences are involved.

Let us denote the sampled data from the ith VT is

$$\mathbf{d}^i = (d^i_{1,1}, \ldots, d^i_{1,T}, \ldots, d^i_{K,1}, \ldots, d^i_{K,T}), \tag{3}$$

where $d^i_{p,q}$ denotes the measured data from pth phase of the ith VT at time q, T denotes the number of sample time. If the problem involves the primary voltage values only, there will be three phases, *i.e.*, , $K = 3$; other wise $K = 6$. The decision variables are defined as the ground truth voltage values, which can be denoted as

$$\mathbf{x} = (x_{1,1}, \ldots, x_{1,T}, \ldots, x_{K,1}, \ldots, x_{K,T}). \tag{4}$$

Then the time-varying ratio error of the ith VT is

$$\mathbf{e}^i = (\frac{d^i_{1,1} - x_{1,1}}{x_{1,1}}, \ldots, \frac{d^i_{1,T} - x_{1,T}}{x_{1,T}}, \ldots, \frac{d^i_{K,1} - x_{K,1}}{x_{K,1}}, \ldots, \frac{d^i_{K,T} - x_{K,T}}{x_{K,T}}), \tag{5}$$

without loss of generality, it can be simplified as

$$\mathbf{e}^i = (e^i_{1,1}, \ldots, e^i_{1,T}, \ldots, e^i_{K,1}, \ldots, e^i_{K,T}). \tag{6}$$

The first order variance of the time-varying ratio error of the ith VT is

$$\boldsymbol{\Delta e}^i = (e^i_{1,2} - e^i_{1,1}, \ldots, e^i_{1,T} - e^i_{1,T-1}, \ldots, e^i_{K,2} - e^i_{K,1}, \ldots, e^i_{K,T} - e^i_{K,T-1}), \quad (7)$$

without loss of generality, it can be simplified as

$$\boldsymbol{\Delta e}^i = (\Delta e^i_{1,1}, \ldots, \Delta e^i_{1,T}, \ldots, \Delta e^i_{K,1}, \ldots, \Delta e^i_{K,T}). \quad (8)$$

In this case, we can form the two objective functions, one is to minimize the sum of time-varying ratio errors of all the VTs, and the other is to minimize the variation of time-varying ratio errors. We can have

$$f_1(\boldsymbol{x}, \boldsymbol{e}^1, \ldots, \boldsymbol{e}^P) = \sum_{j=1}^{T} \sum_{k=1}^{K} \sum_{i=1}^{P} e^i_{k,j}. \quad (9)$$

Once the variation of errors over time is considered, we can get the second objective function

$$f_2(\boldsymbol{x}, \boldsymbol{\Delta e}^1, \ldots, \boldsymbol{\Delta e}^P) = \sum_{i=1}^{P} \sqrt{std((\Delta e^i_{1,1}, \ldots, \Delta e^i_{K,T}))}, \quad (10)$$

where $std(*)$ denoting the standard deviation of vector $*$. Unlike objective f_1, the construction of f_2 is more complicated, involving different variable interaction relationships.

3   Empirical Studies

In this part, the proposed informed initialization for TREE is first introduced. Then, NSGA-II [7] five state-of-the-art LSMOEAs (DGEA [12], FLEA [16], LMOCSO [29], LSMOF [15], and WOF [34]) are chosen to compare on five TREE problems with random and informed initialization.

3.1   Informed Initialization

In TREE problems, the default initialization method is the same random initialization as in other problems, i.e., for each decision variable, the initial value is generated randomly between its upper and lower bounds. However, as defined previously, the decision variables of the TREE problems are the ground truth values of the time-series voltage. The ground truth voltage values can be calculated from the measured values with the ratio errors, which does not vary as drastically as the voltage values in general case.

Based on the above analysis, our proposed initialization method is to first randomly generate a fixed ratio error value for each individual in the initial population, and then calculate the initial values of the decision variables based on the ratio error values and the voltage measurement values. To be more specific, the default initialization method in TREE is to generate a random ratio error value for each sampling point and calculate the decision variable.

$$x_{k,t} = \frac{\bar{d}_{k,t}}{e_{k,t} + 1}$$
$$k \in 1, \cdots, K$$
$$t \in 1, \cdots, T,$$

(11)

where

$$\bar{d}_{k,t} = \frac{\sum_{i=1}^{P} d_{k,t}^i}{P},$$

(12)

P is the number of VTs that measure the same voltage values. Thus there are D different $e_{k,t} \sim \mathcal{U}(e_l, e_u)$ uniformly distributed between the lower boundary e_l and the upper boundary e_u.

In the proposed initialization method, only one random ratio error value will be generated for each phase.

$$x_{k,t} = \frac{\bar{d}_{k,t}}{e_k + 1}$$
$$k \in 1, \cdots, K$$
$$t \in 1, \cdots, T,$$

(13)

where K different $e_k \sim \mathcal{U}(e_l, e_u)$ are uniformly distributed between the lower boundary e_l and the upper boundary e_u.

3.2 Experimental Settings

The recommended parameter settings from the literature were used in the compared algorithms to ensure a fair comparison, as these settings have been reported to yield the best performance. All algorithms considered in the experiments are carried out on PlatEMO [28]. The experiments are executed on a PC with an AMD Ryzen™ 9 3950x CPU. Each algorithm is run 20 times independently on each test case.

NSGA-II [7] is embedded in DGEA, FLEA, LSMOF, and WOF. As recommended in the literature, the crossover probability (p_c) in SBX is set to 1.0, the mutation probability (p_m) in PM is set to $1/D$, the distribution index of crossover in SBX (n_c) and the distribution index of mutation in PM (n_m) are both set to 20. A population with 100 solutions are used for all the comparing algorithms on all test problems. A maximum time of function evaluations is adopted as the termination criterion for all compared algorithms, which is set to 10,000 for TREE1 to TREE5. The number of sample time T is set to 1000, thus the number of decision variables D is 3,000 for TREE1 and TREE2, 6,000 for TREE3, TREE4, and TREE5.

3.3 Results and Analysis

The HV results achieved by DGEA, FLEA, LMOCSO, LSMOF, WOF, and NSGA-II with random and informed initialization on 5 TREE problems are presented in Table 1. The best HV values on each test problem are bolded, and the "NaN" means the algorithm doesn't get any feasible solutions on the test problem with given time of function evaluations.

Table 1. The HV results achieved by DGEA, FLEA, LMOCSO, LSMOF, WOF, and NSGA-II with random and informed initialization on 5 TREE problems. The best HV values on each test problem are bolded.

Problem	D	Initialization	DGEA	FLEA	LMOCSO	LSMOF	WOF	NSGAII
TREE1	3000	Random	6.86e-1(1.48e-1)	7.98e-1(5.24e-2)	NaN	8.37e-1(2.58e-5)	7.90e-1(3.43e-2)	NaN
		Informed	8.33e-1(1.96e-2)	8.49e-1(3.03e-3)	8.24e-1(1.43e-3)	**8.50e-1(1.55e-4)**	8.44e-1(4.35e-3)	8.43e-1(1.96e-4)
TREE2	3000	Random	8.28e-1(2.68e-2)	8.11e-1(4.04e-2)	NaN	8.50e-1(1.50e-5)	7.85e-1(1.38e-1)	NaN
		Informed	8.54e-1(5.67e-4)	8.55e-1(4.28e-4)	8.52e-1(4.91e-4)	**8.56e-1(1.10e-4)**	8.54e-1(1.74e-3)	8.54e-1(1.39e-4)
TREE3	6000	Random	7.52e-1(1.73e-1)	NaN	NaN	**8.87e-1(8.02e-7)**	7.92e-1(3.24e-2)	NaN
		Informed	8.87e-1(8.95e-4)	8.87e-1(8.86e-7)	8.87e-1(2.06e-3)	8.87e-1(1.03e-6)	8.85e-1(9.44e-3)	**8.87e-1(2.24e-4)**
TREE4	6000	Random	3.55e-1(4.07e-1)	NaN	NaN	**9.64e-1(3.99e-5)**	4.38e-1(1.15e-1)	NaN
		Informed	9.64e-1(3.45e-5)	9.64e-1(6.31e-6)	9.53e-1(2.20e-2)	**9.64e-1(3.14e-5)**	9.64e-1(5.27e-4)	9.63e-1(2.33e-3)
TREE5	6000	Random	6.57e-1(2.45e-1)	9.21e-1(8.69e-3)	NaN	9.27e-1(1.59e-5)	7.83e-1(1.14e-1)	NaN
		Informed	9.31e-1(1.12e-4)	**9.35e-1(2.83e-3)**	9.21e-1(6.92e-3)	9.31e-1(1.25e-4)	9.27e-1(1.07e-2)	9.31e-1(3.00e-4)

As shown in Table 1, for almost all the comparing algorithms, the HV values on the problem with informed initialization are significantly better than those on the same problem with random initialization. Besides, with informed initialization, the performance gap between the comparing algorithms is significantly reduced. When using random initialization, LMOCSO and NSGA-II can not obtain any feasible solutions with given time of function evaluations, but they can achieve competitive HV results with informed initialization.

Figure 2 gives the curve of the mean value and the variance of HV obtained by each algorithm on TREE1-TREE5 with random and informed initialization over 20 independent runs. As shown in the figure, the initial population of algorithms with informed initialization can obtain better HV values than the well-converged population of the same algorithms with random initialization.

The above results indicate that for real-world application problems such as TREE, the initialization method may have a greater impact on the solution performance than the algorithm itself.

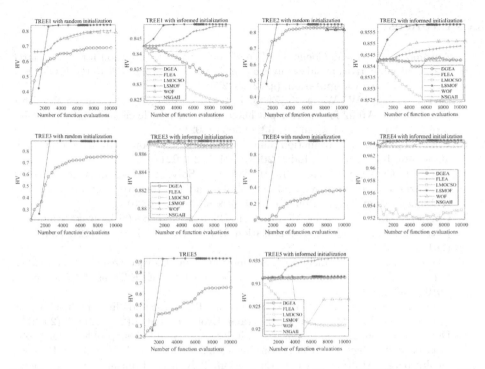

Fig. 2. The convergence profile (HV value) of each algorithm on TREE problems with random and informed initialization over 20 independent runs.

4   Conclusion

In this work, we investigate the effect of initialization method on LSMOEAs in solving TREE problems. Based on the realistic physical properties represented by the decision variables of the TREE problem, an informed initialization method is proposed. By comparing of NSGA-II and five representative LSMOEAs on TREE1 to TREE5 with different initialization methods, the effect of initialization method in solving TREE problems is examined. Experimental results have indicated the initialization method may have a greater impact on the solution performance than the algorithm itself. With informed initialization, the initial population of an MOEA can obtain better HV value than the well-converged population of the same MOEA on the same TREE problem.

In summary, for real-world application problems such as TREE, physical properties and prior information can be used to improve initialization methods or reduce the search space. A reasonable initialization method can have great potential to improve the performance of the LSMOEAs.

References

1. Antonio, L.M., Coello Coello, C.: Use of cooperative coevolution for solving large scale multiobjective optimization problems. In: Proceedings of 2013 IEEE Congress on Evolutionary Computation, pp. 2758–2765 (2013)
2. Cao, B., Zhang, W., Wang, X., Zhao, J., Gu, Y., Zhang, Y.: A memetic algorithm based on two_Arch2 for multi-depot heterogeneous-vehicle capacitated arc routing problem. Swarm Evol. Comput. **63**, 100864 (2021)
3. Chen, D., Zhao, X.: Production management of hybrid flow shop based on genetic algorithm. Int. J. Simul. Model. **20**(3), 571–582 (2021)
4. Chen, H., Cheng, R., Wen, J., Li, H., Weng, J.: Solving large-scale many-objective optimization problems by covariance matrix adaptation evolution strategy with scalable small subpopulations. Inf. Sci. **509**, 457–469 (2020)
5. Cheng, R., Jin, Y., Olhofer, M., Sendhoff, B.: Test problems for large-scale multiobjective and many-objective optimization. IEEE Trans. Cybern. **47**(12), 4108–4121 (2017)
6. Cheng, R., Jin, Y., Narukawa, K., Sendhoff, B.: A multiobjective evolutionary algorithm using gaussian process-based inverse modeling. IEEE Trans. Evol. Comput. **19**(6), 838–856 (2015)
7. Deb, K., Pratap, A., Agarwal, S., Meyarivan, T.: A fast and elitist multi-objective genetic algorithm: NSGA-II. IEEE Trans. Evol. Comput. **6**(2), 182–197 (2002)
8. Deb, K., Thiele, L., Laumanns, M., Zitzler, E.: Scalable test problems for evolutionary multiobjective optimization. In: Abraham, A., Jain, L., Goldberg, R. (eds.) Evolutionary Multiobjective Optimization. Advanced Information and Knowledge Processing, pp. 105–145. Springer, London (2005). https://doi.org/10.1007/1-84628-137-7_6
9. Gao, X., Liu, T., Tan, L., Song, S.: Multioperator search strategy for evolutionary multiobjective optimization. Swarm Evol. Comput. **71**, 101073 (2022)
10. González-Almagro, G., Rosales-Pérez, A., Luengo, J., Cano, J.R., García, S.: MEMEOA/DCC: multiobjective constrained clustering through decomposition-based memetic elitism. Swarm Evol. Comput. **66**, 100939 (2021)
11. He, C., Cheng, R.: Population sizing of evolutionary large-scale multiobjective optimization. In: Ishibuchi, H., et al. (eds.) EMO 2021. LNCS, vol. 12654, pp. 41–52. Springer, Cham (2021). https://doi.org/10.1007/978-3-030-72062-9_4
12. He, C., Cheng, R., Danial, Y.: Adaptive offspring generation for evolutionary large-scale multiobjective optimization. IEEE Trans. Syst. Man Cybern. Syst. **52**(2), 786–798 (2020)
13. He, C., Cheng, R., Zhang, C., Tian, Y., Chen, Q., Yao, X.: Evolutionary large-scale multiobjective optimization for ratio error estimation of voltage transformers. IEEE Trans. Evol. Comput. **24**(5), 868–881 (2020). https://doi.org/10.1109/TEVC.2020.2967501
14. He, C., Huang, S., Cheng, R., Tan, K.C., Jin, Y.: Evolutionary multiobjective optimization driven by generative adversarial networks (GANs). IEEE Trans. Cybern. **51**(6), 3129–3142 (2020)
15. He, C., et al.: Accelerating large-scale multiobjective optimization via problem reformulation. IEEE Trans. Evol. Comput. **23**(6), 949–961 (2019)
16. Li, L., He, C., Cheng, R., Li, H., Pan, L., Jin, Y.: A fast sampling based evolutionary algorithm for million-dimensional multiobjective optimization. Swarm Evol. Comput. **75**, 101181 (2022). https://doi.org/10.1016/j.swevo.2022.101181

17. Li, M., Wei, J.: A cooperative co-evolutionary algorithm for large-scale multi-objective optimization problems. In: Proceedings of 2018 Genetic and Evolutionary Computation Conference Companion, pp. 1716–1721 (2018)
18. Ma, X., et al.: A multiobjective evolutionary algorithm based on decision variable analyses for multi-objective optimization problems with large scale variables. IEEE Trans. Evol. Comput. **20**, 275–298 (2016)
19. Mahdavi, S., Shiri, M.E., Rahnamayan, S.: Metaheuristics in large-scale global continues optimization: a survey. Inf. Sci. **295**, 407–428 (2015)
20. Miguel Antonio, L., Coello Coello, C.A.: Coevolutionary multiobjective evolutionary algorithms: survey of the state-of-the-art. IEEE Trans. Evol. Comput. **22**(6), 851–865 (2018)
21. Nguyen, B.H., Xue, B., Andreae, P., Ishibuchi, H., Zhang, M.: Multiple reference points-based decomposition for multiobjective feature selection in classification: static and dynamic mechanisms. IEEE Trans. Evol. Comput. **24**(1), 170–184 (2020)
22. Patra, A.K., Nanda, A., Rout, B., Subudhi, D.K., Kar, S.K.: An automatic insulin infusion system based on the genetic algorithm FOPID control. In: Sharma, R., Mishra, M., Nayak, J., Naik, B., Pelusi, D. (eds.) Green Technology for Smart City and Society. LNNS, vol. 151, pp. 355–366. Springer, Singapore (2021). https://doi.org/10.1007/978-981-15-8218-9_30
23. Peng, X., Jin, Y., Wang, H.: Multimodal optimization enhanced cooperative coevolution for large-scale optimization. IEEE Trans. Cybern. **49**(9), 3507–3520 (2018)
24. Potter, M.A., De Jong, K.A.: A cooperative coevolutionary approach to function optimization. In: Davidor, Y., Schwefel, H.-P., Männer, R. (eds.) PPSN 1994. LNCS, vol. 866, pp. 249–257. Springer, Heidelberg (1994). https://doi.org/10.1007/3-540-58484-6_269
25. Huband, S., Hingston, P., Barone, L., While, L.: A review of multiobjective test problems and a scalable test problem toolkit. IEEE Trans. Evol. Comput. **10**(5), 477–506 (2006)
26. Song, A., Yang, Q., Chen, W.N., Zhang, J.: A random-based dynamic grouping strategy for large scale multi-objective optimization. In: Proceedings of 2016 IEEE Congress on Evolutionary Computation (CEC), pp. 468–475. IEEE (2016)
27. Tang, J., Liu, G., Pan, Q.: A review on representative swarm intelligence algorithms for solving optimization problems: applications and trends. IEEE/CAA J. Autom. Sin. **8**(10), 1627–1643 (2021)
28. Tian, Y., Cheng, R., Zhang, X., Jin, Y.: PlatEMO: a matlab platform for evolutionary multi-objective optimization [educational forum]. IEEE Comput. Intell. Mag. **12**(4), 73–87 (2017)
29. Tian, Y., Zheng, X., Zhang, X., Jin, Y.: Efficient large-scale multiobjective optimization based on a competitive swarm optimizer. IEEE Trans. Cybern. **50**(8), 3696–3708 (2020)
30. Zhang, Q., Li, H.: MOEA/D: a multiobjective evolutionary algorithm based on decomposition. IEEE Trans. Evol. Comput. **11**(6), 712–731 (2007)
31. Zhang, X., Tian, Y., Jin, Y., Cheng, R.: A decision variable clustering-based evolutionary algorithm for large-scale many-objective optimization. IEEE Trans. Evol. Comput. **22**, 97–112 (2016)
32. Zhou, A., Qu, B., Li, H., Zhao, S., Suganthan, P.N., Zhang, Q.: Multiobjective evolutionary algorithms: a survey of the state of the art. Swarm Evol. Comput. **1**(1), 32–49 (2011)
33. Zhou, S., Zhan, Z., Chen, Z., Kwong, S., Zhang, J.: A multi-objective ant colony system algorithm for airline crew rostering problem with fairness and satisfaction. IEEE Trans. Intell. Transp. Syst. **22**(11), 6784–6798 (2021)

34. Zille, H., Ishibuchi, H., Mostaghim, S., Nojima, Y.: A framework for large-scale multiobjective optimization based on problem transformation. IEEE Trans. Evol. Comput. **22**(2), 260–275 (2018)
35. Zitzler, E., Deb, K., Thiele, L.: Comparison of multiobjective evolutionary algorithms: empirical results. Evol. Comput. **8**(2), 173–195 (2000)
36. Zitzler, E., Künzli, S.: Indicator-based selection in multiobjective search. In: Yao, X., et al. (eds.) PPSN 2004. LNCS, vol. 3242, pp. 832–842. Springer, Heidelberg (2004). https://doi.org/10.1007/978-3-540-30217-9_84

Enabling Surrogate-Assisted Evolutionary Reinforcement Learning via Policy Embedding

Lan Tang[1], Xiaxi Li[2], Jinyuan Zhang[1], Guiying Li[1,3], Peng Yang[1,4(✉)], and Ke Tang[1,3]

[1] Guangdong Provincial Key Laboratory of Brain-Inspired Intelligent Computation, Department of Computer Science and Engineering, Southern University of Science and Technology, Shenzhen 518055, China
yangp@sustech.edu.cn
[2] Faculty of Engineering, Shenzhen MSU-BIT University, Shenzhen 518172, China
[3] Research Institute of Trustworthy Autonomous Systems, Southern University of Science and Technology, Shenzhen 518055, China
[4] Department of Statistics and Data Science, Southern University of Science and Technology, Shenzhen 518055, China

Abstract. Evolutionary Reinforcement Learning (ERL) that applying Evolutionary Algorithms (EAs) to optimize the weight parameters of Deep Neural Network (DNN) based policies has been widely regarded as an alternative to traditional reinforcement learning methods. However, the evaluation of the iteratively generated population usually requires a large amount of computational time and can be prohibitively expensive, which may potentially restrict the applicability of ERL. Surrogate is often used to reduce the computational burden of evaluation in EAs. Unfortunately, in ERL, each individual of policy usually represents millions of weights parameters of DNN. This high-dimensional representation of policy has introduced a great challenge to the application of surrogates into ERL to speed up training. This paper proposes a PE-SAERL Framework to at the first time enable surrogate-assisted evolutionary reinforcement learning via policy embedding (PE). Empirical results on 5 Atari games show that the proposed method can perform more efficiently than the four state-of-the-art algorithms. The training process is accelerated up to 7x on tested games, comparing to its counterpart without the surrogate and PE.

Keywords: Reinforcement learning · Evolutionary algorithms · Surrogates

L. Tang and X. Li—Contributed equally to this work.
This work was supported partly by the National Natural Science Foundation of China (Grants 62272210 and 62106099), partly by the Guangdong Provincial Key Laboratory (Grant 2020B121201001), partly by the Program for Guangdong Introducing Innovative and Entrepreneurial Teams (Grant 2017ZT07X386), and partly by the Stable Support Plan Program of Shenzhen Natural Science Fund (Grant 20200925154942002).

L. Pan et al. (Eds.): BIC-TA 2022, CCIS 1801, pp. 233–247, 2023.
https://doi.org/10.1007/978-981-99-1549-1_19

1 Introduction

In recent years, Evolutionary Algorithms (EAs) have been successfully applied to exploratively optimize the parameters of the Deep Neural Network (DNN) based policy for challenging reinforcement learning tasks, i.e., video games [17,29], Robotics [3], and computer vision [5]. The resultant Evolutionary Reinforcement Learning (ERL) [16] has been widely regarded as an alternative to traditional reinforcement learning methods [17], due to the ability of EAs on exploration, noise resistance, and parallel acceleration. In ERL, a diverse population of policies is generated by EA and is evaluated via interactions with the environment. The fitness value of each individual is considered as the total returns (e.g., cumulative rewards with discount factor 1.0) across the entire episode.

However, there is a notable issue with ERL [16]: the evaluation of the population usually requires a large amount of computational time and can be prohibitively expensive. First, ERL finds the optimal policy by iteratively searching in a population-based manner. This manner leads to the number of evaluation times of individual policies being numerous. Second, each real fitness evaluation can be computationally time-consuming due to the complex mechanism of the reinforcement learning simulators. These unfavorable characteristics can potentially restrict the application of ERL in real-world problems.

Actually, the computationally expensive problems have been extensively studied in the EA community for decades. Many Surrogate-Assisted Evolutionary Algorithms (SAEAs) have been proposed for expensive problems. In typical SAEAs, surrogate models are trained to replace the expensive fitness function for evaluation [22]. Based on this idea, this paper focus on how to apply surrogates into ERL to speed up training while keeping the ERL effective. And we call this framework as Surrogate-Assisted ERL (SAERL). In this paper, we argue that the existing SAEA approaches can hardly be directly applied to ERL. The reason is that each individual in ERL is usually a Deep Neural Network (DNN) based policy with millions of weights. This requires the surrogates accepting very high dimensional input vectors to represent the parameters of individual policies. This leads to three technical issues: First, the surrogate model can be complex and computationally expensive. Second, the training of the surrogate model can be very difficult due to the lack of specific training data. Third, the surrogate model may encounter the difficulty of distinguishing different weight vectors of policies in high-dimensional space. Therefore, the existing surrogate approaches may fail to accelerate the training of ERL. Few exceptions [6,20,24] employ the surrogates in the training phases, but they are not for high-dimensional DNN-based policies.

To eliminate the above issues, we first propose to project the surrogate model to work in a low dimensional space. A policy embedding (PE) module is proposed for the projection, and a new framework PE-SAERL is proposed. Specifically, before a high-dimensional policy vector is input into the surrogate model, the PE module first encodes the vector to a lower-dimensional space by preserving its major features. The existing surrogate models are thus able to be applied to the low-dimensional embedded input vectors for computationally cheap evaluation.

After the surrogate evaluation, the embedded policies are decoded to the original high-dimensional space for further processing of ERL, as shown in Fig. 1.

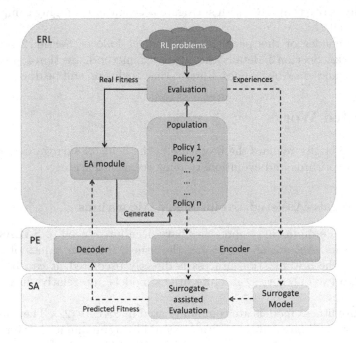

Fig. 1. The overview of the PE-SAERL framework.

The PE module is essentially a bidirectional mapper of both high-dimensional and low-dimensional spaces. Also note that, the focus of this paper is the first time discussing how to technically enable the SAERL. Hence, we are not focusing on sophisticated mappper designing. We restrict the complexity of the mapper used in this paper for better demonstrating that the PE module is useful for the SAERL framework.

Specifically, we choose the random embedding method [15] as the PE module here. This PE basically maps between the high-dimensional space and some randomly selected lower dimensions while is able to preserving the major features of the original space [15]. We believe other learning-based PE module can further improve the effectiveness of PE-SAERL, and this technically simple PE of random embedding can be viewed as an appropriate baseline for further investigations.

To instantiated the PE-SAERL for empirical studies, the EA and the surrogate model should be specified. In this work, we choose Negatively Correlated Search (NCS) [21] as the EA module and the Fuzzy Classification Pre-Selection (FCPS) [33] module as surrogate module. We should emphasize that generally any EA and its surrogate-assisted version can be adopted in the PE-SAERL framework for optimizing policies. Trade-offs will be considered between the computing complexity and the search effectiveness. As a result, we get an instantiated method called PE-FCPS-NCS. The empirical studies on 5 Atari games successfully verify that ERL with surrogate acceleration is highly competitive to traditional DRLs.

The remainder of this paper is organized as follows. Section 2 reviews the related works. Section 3 details the proposed method. Section 4 presents the empirical studies on Atari games. Finally, the conclusion will be drawn in Sect. 5.

2   Related Works

This section briefly reviews the literature in the fields of surrogate-assisted evolutionary algorithms and evolutionary reinforcement learning.

2.1   Surrogate-Assisted Evolutionary Algorithms

Many surrogate modelling approaches have been introduced to SAEAs over the past few years. According to whether the fitness function values of candidate solutions in the optimization process is directly predicted, it can be generally divided into two categories: absolute fitness models and relative fitness models [22].

Absolute fitness models are commonly used in SAEAs [22]. They aim at predicting new candidate solutions' fitness values by approximating the real fitness function, including Polynomial Regression (PR), Gaussian Process regression (a.k.a. Kriging model), Support Vector Regression (SVR), Radial Basis Function (RBF), Neural Network(NN) etc. Kriging [19,20] is a popular surrogate model in SAEAs because it can estimate the uncertainty as well as the fitness. SVR [11] is an effective surrogate model which has some practical applications, such as the optimization of railway wind barriers, but its training process will be expensive when the dimensionality of the problem is very high. RBF models [34] have also been used in SAEAs to solve many real-world problems.

Relative fitness models focus on providing the relative rank or preference of candidate solutions rather than their absolute fitness values [22]. Wang et al. in [14] proposed a hierarchical clustering algorithm by grouping the patient data into different numbers of clusters to reduce the amount of computation time. Subsequently, Wang et al. in [23] employed an artificial neural network as a classifier to predict the dominance relationship between candidate solutions and reference solutions. Zhou et al. in [32] used a fuzzy clustering method to extract the weight vectors, which works by constructing new objectives as linear combinations of the original Many-objective optimization problems. In addition, Zhang et al. in [31,33] applied a Fuzzy-KNN method to filter out 'unpromising' individuals before the actual evaluation.

2.2 Evolutionary Reinforcement Learning

ERL has been developed for years [26], and we noticed that ERL attracts increasing attentions recently with the DNN-based policies. As early as 1999, David et al. in [26] pointed out in the literature that evolutionary algorithm is a class of classical algorithms that search the space of policies for solving reinforcement learning tasks. Salimans et al. in [17] from OpenAI proposed to use a simplified natural evolution strategies to directly optimize the weights of policy neural networks, where the natural gradients was used to update a parameterized search distribution in the direction of higher expected fitness. It proved that Evolutionary Algorithm (EA) is competitive for policy neural networks search in deep RL [17]. Subsequently, Chrabaszcz et al. in [4] compared a simper and basic canonical ES algorithm with OpenAI ES further confirmes that the power of ES-based algorithms for policy search may rival that of the gradient-based algorithms. Recently, Yang et al. [29] proposed the CCNCS method, which uses a novel search method called Negatively Correlated Search [21,27,28] by explicitly measuring the diversity among different search processes to drive them to the regions with uncertainty. Another related work is a hybrid algorithm [9] combining gradient-based and derivative-free optimization.

Related to using surrogates in ERL, Wang et al. in [24] proposed Surrogate-assisted Controller (SC) applied on it to alleviate the computational burden of evaluation. Nevertheless, the core of SC is to replace the critic model, which aims to evaluate the action not the policy. In this regard, it does not follow the framework of SAEA and thus is quite different from the SAERL framework discussed in this work. In addition, there exists a few studies on applying surrogates to enhance EAs on RL tasks, such as Evolutionary Surrogate-assisted Prescription (ESP) [6], surrogate model-based Neuroevolution (SMB-NE) [20]. However, these methods do not take the DNN into account, and does not face the high-dimensional difficulties as is in this work.

As can be seen from the literature, ERL is a extremely promising method for solving RL tasks, but the evaluation process is prohibitively expensive. Using the surrogate model can solve the problems of low sampling efficiency and expensive evaluation in traditional SAEAs. However, considering that each individual in ERL usually involves millions of parameters and the commonly surrogate models (i.e., Kriging [19], kNN [31]) by calculating the genotype distance may have no way to extract effective features, the direct application of surrogates into ERL to speed up training may face inevitably failure. This article extends the surrogate model to ERL via policy embedding to speed up training.

3 Methodology

This section first introduces the proposed framework consisting of three main modules. Then we briefly describe random embedding employed by the PE module. Finally a concrete algorithm called PE-FCPS-NCS is described in detail.

3.1   PE-SAERL Framework

Intuitively, we intend to apply surrogates into ERL to speed up training by replacing a part of the real simulations with more computationally cheap surrogates. Importantly, the PE module is designed to address the three key technical issues faced by directly applying the surrogates to millions of dimensions, as mentioned previously. PE-SAERL Framework fully follows the PE-SAERL flowchart depicted in Fig. 1.

Specifically, the PE-SAERL Framework is roughly divided into 3 modules, which are ERL module, PE module and SA module.

- *ERL module*: ERL iteratively search for the optimal policy in a population-based manner with EAs. Each individual in the population is represented as a vector of all connection weights of the neural network policy. At the beginning of training, λ search processes are initialized in parallel. At each iteration, each of them separately preserves a individual x_i, and generates an offspring individual x_i' by crossover or variation operators. Eventually, under the effect of environmental selection, the individual with higher fitness value as new parent individual into the next generation, As depicted in Fig. 2.
- *PE module*: The PE module is essentially a bidirectional mapper of both high-dimensional and low-dimensional spaces. It generally consists of two steps, i.e., encoding and decoding. Encoding extracts a low-dimensional vector representation from a high-dimensional input sentence, and decoding generates a correct high-dimensional target translation from that representation.
- *SA module*: The SA module is used to speed up training by partially replacing expensive ground-truth fitness evaluations. The surrogate model uses all individuals in the population and their corresponding true fitness values as historical data to train the model, and then uses the trained model to evaluate candidate solutions and select the optimal solution y^* from them.

3.2   PE Module: Random Embedding

Random Embedding refers to a dimensional reduction technique with technically simple and desirable theoretical property, which project data into low-dimensional spaces with a randomly generated matrix [15,25,30]. The implicit assumption is that, for the SAERL framework, the fitness of candidate solutions are only affected by a few dimensions instead of all. This assumption is technically reasonable for the DNN-based policy since DNN often preserves redundent weights and can be effectively prunned [7,8,10].

Based on this assumption, given a high-dimensional function with low S-efficient dimension (i.e., $d_e \ll D$) and a random embedding matrix $A \in \mathbb{R}^{D \times d}$, for any $x \in \mathbb{R}^D$, there must exist $y \in \mathbb{R}^d$ such that $f(Ay) = f(x)$ with probability 1. That is, random embedding enables us to optimize the lower-dimensional function $g(y) = f(Ay)$ in $\mathbb{R}^d$ instead of optimizing the original high-dimensional $f(x)$ in $\mathbb{R}^D$, while the function value is still evaluated in the original solution

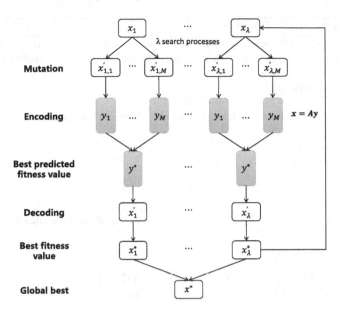

Fig. 2. The parallel architecture and main processes of the PE-SAERL framework.

space. For S-efficient dimension definition and provable completeness guarantee, please refer to [15].

Specifically, at encoding steps, a random embedding matrix $A \in \mathbb{R}^{D \times d}$ is generated drawn from $\mathcal{N}(0,1)$. For each solution x, there must be a unique y corresponding to it, according to $x = Ay$. On the embedding space, the surrogate preselects the best solution from M candidate solutions. The final candidate is then decoded to the original high-dimensional solution space.

3.3 PE-FCPS-NCS: A Concrete Algorithm

As mentioned above, the PE module provides the possibility for SA to be used in ERL. To instantiated the PE-SAERL for empirical studies, we choose Negatively Correlated Search (NCS) [21] as the EA module and the Fuzzy Classification Pre-Selection (FCPS) [33] as the surrogate module. The pseudo-code for the concrete algorithm is given in Algorithm 1.

Some components in Algorithm 1 are explained as follows.

* *Initialization*: A set of policies $\{x_1, x_2, \ldots, x_\lambda\}$ are initialized uniformly and first evaluated with respect to the RL simulator f in lines 1–2.
* *FCPS section*: In Line 5, M candidate policies are sampled by means of a Gaussian operator. In Line 8, the high-rank policies with the maximal membership degree belongs to "promising" class are chosen out. And one is randomly selected from the "promising" high-rank policies as the candidate y^*.

Algorithm 1. PE-FCPS-NCS

Input: Number of sub-process λ; RL simulator f; S-Effective embedding dimension d; Origin policy dimension D; Evaluation limitation max_steps; Number of candidate policies M.

Output: BestFound policy x^*.

1: Initialize λ policies $x_i \in \mathcal{R}^D$ uniformly, $i = 1, \cdots, \lambda$.
2: Evaluate the λ policies with respect to the objective function f.
3: **repeat**
4:    **for** $i = 1$ to λ **do**
5:       Sample candidate policies $\{x'_{i,1}, \cdots, x'_{i,M}\}$ from the distribution $p_i \sim \mathcal{N}(x_i, \Sigma_i)$;
6:       Generate a random matrices $A \in \mathcal{R}^{D \times d}$ with $\mathcal{N}(0, I)$;
7:       Encoding with $y_{i,j} = A^{-1} \cdot x'_{i,j}$ where $j = 1, \cdots, M$;
8:       Choose a policy y_i^* with maximal membership degree belongs to "promising" class by a fuzzy classifier model m;
9:       Let $x'_i = A \cdot y_i^*$;
10:      Evaluate $f(x'_i)$ as the qualities of x'_i;
11:      Calculate $d(p_i)$ and $d(p'_i)$, where $p'_i \sim \mathcal{N}(x'_i, \Sigma_i)$;
12:      **if** $f(x'_i) + \varphi \cdot d(p'_i) > f(x_i) + \varphi \cdot d(p_i)$ **then**
13:         $x_i = x'_i$; and $p_i = p'_i$;
14:      **end if**
15:      Update Σ_i according to the 1/5 successful rule;
16:    **end for**
17: **until** $steps\_passed \geq max\_steps$

---

* *PT section*: A Gaussian random matrix is generated in line 6 and applied in line 7 to encode the candidate policies from the original high-dimensional policy space into a low-dimensional effective subspace. Then, the policy space is decoded in line 9.
* *NCS section*: Considering both the quality and the diversity, i.e., $f(x'_i) + \varphi \cdot d(p'_i)$ and $f(x_i) + \varphi \cdot d(p_i)$, the following steps are carried out. In line 10, the candidate policy is evaluated with respect to f. And the diversity value of the current and candidate policies are also calculated as a basis for environmental selection in line 11. In Lines 12–14, the better one between the current and candidate policy is selected into the next generation. Finally, the step size is updated as described in line 15.
* *Stopping condition*: In Line 17, when the time budget has run out, i.e., the number of times the policy interacts with the environment exceeds the given maximum number max_steps, the best solution x^* ever found is output.

4   Experiments

In this experiment, we aim to answer the following questions: (1) Does the surrogate model really improve ERL compared to baselines? (2) Does the surrogate model really speed up training compared to original NCS? (3) Is the proposed method sensitive to parameters?

4.1   Experiments Settings

Environment. We apply the algorithm to a series of Atari 2600 games implemented in The Arcade Learning Environment (ALE) [1]. The Atari 2600 is a challenging RL testbed that provides agents with high-dimensional visual input and a diverse and interesting set of tasks, including obstacle avoidance (e.g. Freeway), shooting (e.g. Beamrider), two-player (e.g. DoubleDunk) and other types. These types of games provide agents with significantly different environmental settings and ways of interacting, thus enabling testing of RL methods with different tasks in maximizing long-term rewards. All these tasks are packaged according to the standard OpenAI Gym API [2] and are simulated friendly.

Baselines. Four state-of-the-art algorithms, namely A3C [12], PPO [18], CES [4], and NCS [21], are used as the major baselines and follow all original hyperparameter settings. Among them, PPO and A3C are popular gradient-based methods that use traditional backpropagation to train networks. CES is a recently proposed ERL that utilizes a canonical evolution strategy to directly optimize the weights of policy neural networks. Besides, to show how PE-SA promotes NCS, we directly apply NCS as a baseline to Atari games.

Performance Metric. The quality of the policy is measured by the testing score, i.e., the average score over 30 repetitions of the game without the frame limitations. More specifically, the testing score refers to the cumulative reward of the policy in an episode, and the episode refers to the time when there is no action in the random "noop" frame (sample in the interval 0 to 30) at the beginning of playing until the end signal is received, as mentioned by Machado et al. [1]. Technically, to prevent the agent from falling into a "dead situation", the frames of one episode is limited to a very large value (i.e. 100,000 frames).

Other Protocols. All the baselines share the policy network. We directly adopt the policy network proposed by Mnih et al. [13], which consists of three convolutional layers and two fully connected layers. The network architecture is shown in Table 1, which involves nearly 1.7 million connection weights that need to be optimized. As suggested by Mnih et al. [13], the raw observations are resized to 84×84, and the skipping frame is set to 4. The time budget is set to the consumed total game frames allowed in each training phase. For ERL methods (i.e., CES, NCS), The total game frames are set to 0.1B. For gradient-based methods (i.e., PPO), the time budget is set to 0.04B for fairness, as it works via back-propagation and the ratio of consumed frames on the same CPU-cored hardware is 2.5 [29]. Table 2 shows the hyper-parameters of PE-FCPS-NCS in this work, which are common across all environments.

4.2   Results and Analysis

Comparisons with Baselines. For each game, we train a neural network policy model and do 30 repeated testings of the trained model to obtain a average

Table 1. The network architecture of the policy.

	Input size	Output size	Kernel size	Stride	#filters	activation
Conv1	$4 \times 84 \times 84$	$32 \times 20 \times 20$	8×8	4	32	ReLU
Conv2	$32 \times 20 \times 204$	$64 \times 9 \times 9$	4×4	2	64	ReLU
Conv3	$64 \times 9 \times 9$	$64 \times 7 \times 7$	3×3	1	64	ReLU
Fc1	$64 \times 7 \times 7$	512	–	–	–	ReLU
Fc2	512	#Action	–	–	–	–

Table 2. The major hyperparameter settings of PE-FCPS-NCS. The hyperparameters consists of three parts: NCS [21], FCPS [33], and random embedding [15].

Parameter	Value	Remark
λ	6	The number of processes, that is, the number of populations
d	100	Effective subspace dimension of random embedding
$epoch$	5	The update interval of covariance matrix
r	1.2	Learning rate for covariance matrix
M	3	The number of candidate solutions
φ	1.0	The trade-off factor between quality and correlation
$[l, h]$	$[-0.1, 0.1]$	Search space boundaries in low-dimensional spaces

testing score as performance. To eliminate bias, we repeat the process three times, and the testing scores of each algorithm on 5 games are shown in Table 3. The Average row shows the average performance of the three executions where the best performance is marked in bold, and the Percent row indicates the average performance as a percentage of the best performance. It can be seen that PE-FCPS-NCS generally performs the best among all compared algorithms. In particular, in the Freeway and Bowling games, the best performances of baselines are only 65.23% and 69.49% compared to our proposed method. In the two-player game (DoubleDunk), a high score with a positive score is preferred, and a negative score implies agent failure. However, all algorithms score are negative, but relatively speaking, our algorithm performed best. Note that in the Alien game, PE-FCPS-NCS, while better than traditional gradient-based reinforcement learning algorithms (A3C, PPO) and evolution-based algorithms (CES), performs worse than the original NCS algorithm. In some specific tasks, the application of surrogate may encourage exploration in a worse direction.

Computational Time Analysis. Intuitively, we need to verify whether the surrogate model can speed up training. More specifically, we compare the train-

Table 3. The performance of PE-FCPS-NCS and four baselines on 5 Atari games. Priority is given to higher average scores in all games.

Game	Performance	A3C	PPO	CES	NCS	PE-FCPS-NCS
Time Budget		40M	40M	0.1B	0.1B	0.1B
Alien	Average	766.00	638.20	561.80	**1043.40**	825.68
	Percent	73.41%	61.17%	53.84%	100.00%	79.13%
BeamRider	Average	633.80	600.80	433.10	703.80	**724.80**
	Percent	87.44%	82.89%	59.75%	97.10%	100.00%
Bowling	Average	0	23.40	25	64.53	**92.87**
	Percent	0.00%	25.20%	26.92%	69.49%	100.00%
DoubleDunk	Average	-2	-3.1	-3.7	-1.3	**-0.60**
	Percent	54.78%	19.33%	0.00%	77.34%	100.00%
Freeway	Average	0.00	14.8	14.2	16.7	**22.69**
	Percent	0.00%	65.23%	62.58%	73.60%	100.00%

ing time consumed by NCS and PE-FCPS-NCS to reach a specified score. This score is set to the test performance achieved by the NCS method after training for 10 million game frames. The computational time for the PE-FCPS-NCS method to reach that score is then counted. Figure 3 shows the computational time costs (in minutes) of NCS and PE-FCPS-NCS on three games (i.e., Beam-Rider, Bowling, Freeway). The test performance achieved by the NCS method is 656, 34 and 14.9, respectively. From Fig. 3, it can be seen that the computational time cost of the surrogate-assisted NCS is quite superior to the original NCS. In particular, in the BeamRider game, the time consumption of the original NCS is nearly 7 times that of PE-FCPS-NCS. In addition, the insignificant difference in the Bowling game is mainly because the number of game frames consumed by each iteration is extremely large, which is nearly 4 times that of Freeway. Considering that some additional computational burden is introduced, e.g., the training of the surrogate model and the computation of the embedding, it suffices to show that PE-FCPS-NCS is promising not only for its solution performance but also in accelerated training.

Sensitivity Analysis. We vary the parameter value over a wide range and re-evaluate to perform parameter sensitivity analysis. More specifically, we choose the number of candidate policies involved in pre-selection in the surrogate model as the hyperparameter, and set it to four different values as [3, 5, 10, 100]. We also select three games (i.e., BeamRider, Bowling, Freeway) for sensitivity analysis. As shown in Fig. 4, the performance curves did not change drastically with the perturbation of the parameter, which means that PE-FCPS-NCS is usually not very sensitive to this important parameter.

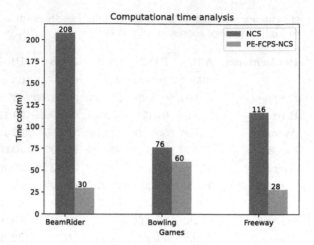

Fig. 3. The training time costs (in minutes) of NCS and PE-FCPS-NCS on three games (i.e., BeamRider, Bowling, Freeway) to reach a specified score that achieved by the NCS method after training for 10 million game frames.

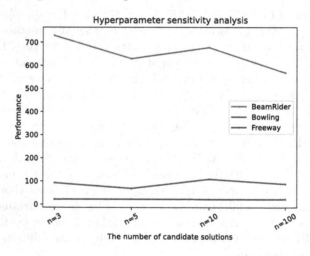

Fig. 4. The performance curves as the number of candidate policies vary in [3, 5, 10, 100].

5   Conclusion

This paper studies how to effectively enable surrogate-assisted evolutionary reinforcement learning to speed up training. First, To apply surrogate to preselect millions of connection weights of neural network policies, this paper employs the PE-SAERL Framework to scale up surrogate for large-scale search space. Next, we propose a concrete algorithm based on the framework, which called PE-FCPS-NCS. Then, empirical studies are conducted on 5 Atari games to verify the proposed method. Empirical results show that the proposed method can perform more efficiently than the four state-of-the-art algorithms (i.e., PPO, A3C, CES, NCS), while effectively accelerating training compared to the original NCS. Last, this paper studies the parameters sensitivity of the proposed method.

References

1. Bellemare, M.G., Naddaf, Y., Veness, J., Bowling, M.: The arcade learning environment: an evaluation platform for general agents. J. Artif. Intell. Res. **47**, 253–279 (2013)
2. Brockman, G., et al.: OpenAI gym. arXiv abs/1606.01540 (2016)
3. Castillo, G.A., Weng, B., Zhang, W., Hereid, A.: Robust feedback motion policy design using reinforcement learning on a 3D digit bipedal robot. In: Proceedings of the IEEE/RSJ International Conference on Intelligent Robots and Systems (IROS 2021), pp. 5136–5143. IEEE, Prague (2021)
4. Chrabaszcz, P., Loshchilov, I., Hutter, F.: Back to basics: benchmarking canonical evolution strategies for playing Atari. arXiv abs/1802.08842 (2018)
5. Chu, X., Zhang, B., Ma, H., Xu, R., Li, Q.: Fast, accurate and lightweight super-resolution with neural architecture search. In: Proceedings of the 25th International Conference on Pattern Recognition (ICPR 2020), pp. 59–64. IEEE, Milan (2020)
6. Francon, O., et al.: Effective reinforcement learning through evolutionary surrogate-assisted prescription. In: Proceedings of the 2020 Genetic and Evolutionary Computation Conference (GECCO 2020), pp. 814–822. Association for Computing Machinery, New York (2020)
7. Hong, W., Li, G., Liu, S., Yang, P., Tang, K.: Multi-objective evolutionary optimization for hardware-aware neural network pruning. Fundam. Res. (2022). https://doi.org/10.1016/j.fmre.2022.07.013
8. Hong, W., Yang, P., Wang, Y., Tang, K.: Multi-objective magnitude-based pruning for latency-aware deep neural network compression. In: Bäck, T., et al. (eds.) PPSN 2020. LNCS, vol. 12269, pp. 470–483. Springer, Cham (2020). https://doi.org/10.1007/978-3-030-58112-1_32
9. Khadka, S., Tumer, K.: Evolution-guided policy gradient in reinforcement learning. In: Advances in Neural Information Processing Systems (NeurIPS 2018), vol. 31. Curran Associates, Inc. (2018)
10. Li, G., Yang, P., Qian, C., Hong, R., Tang, K.: Stage-wise magnitude-based pruning for recurrent neural networks. IEEE Trans. Neural Netw. Learn. Syst. (2022). https://doi.org/10.1109/TNNLS.2022.3184730
11. Llorà, X., Sastry, K., Goldberg, D.E., Gupta, A., Lakshmi, L.: Combating user fatigue in iGAs: partial ordering, support vector machines, and synthetic fitness. In: Proceedings of the 7th Annual Conference on Genetic and Evolutionary Computation (GECCO 2005), pp. 1363–1370. Association for Computing Machinery, New York (2005)
12. Mnih, V., et al.: Asynchronous methods for deep reinforcement learning. In: Proceedings of the 33rd International Conference on Machine Learning (ICML 2016). Proceedings of Machine Learning Research, vol. 48, pp. 1928–1937. PMLR, New York (2016)
13. Mnih, V., et al.: Human-level control through deep reinforcement learning. Nature **518**(7540), 529–533 (2015)

14. Pan, L., He, C., Tian, Y., Wang, H., Zhang, X., Jin, Y.: A classification-based surrogate-assisted evolutionary algorithm for expensive many-objective optimization. IEEE Trans. Evol. Comput. **23**(1), 74–88 (2019)
15. Qian, H., Hu, Y.Q., Yu, Y.: Derivative-free optimization of high-dimensional non-convex functions by sequential random embeddings. In: Proceedings of the 25th International Joint Conference on Artificial Intelligence (IJCAI 2016), pp. 1946–1952. AAAI Press, New York (2016)
16. Qian, H., Yu, Y.: Derivative-free reinforcement learning: a review. Front. Comp. Sci. **15**(6), 156336 (2021)
17. Salimans, T., Ho, J., Chen, X., Sidor, S., Sutskever, I.: Evolution strategies as a scalable alternative to reinforcement learning. arXiv abs/1703.03864 (2017)
18. Schulman, J., Wolski, F., Dhariwal, P., Radford, A., Klimov, O.: Proximal policy optimization algorithms. arXiv abs/1707.06347 (2017)
19. Song, Z., Wang, H., He, C., Jin, Y.: A kriging-assisted two-archive evolutionary algorithm for expensive many-objective optimization. IEEE Trans. Evol. Comput. **25**(6), 1013–1027 (2021)
20. Stork, J., Zaefferer, M., Bartz-Beielstein, T., Eiben, A.E.: Surrogate models for enhancing the efficiency of neuroevolution in reinforcement learning. In: Proceedings of the Genetic and Evolutionary Computation Conference (GECCO 2019), pp. 934–942. Association for Computing Machinery, New York (2019)
21. Tang, K., Yang, P., Yao, X.: Negatively correlated search. IEEE J. Sel. Areas Commun. **34**(3), 542–550 (2016)
22. Tong, H., Huang, C., Minku, L.L., Yao, X.: Surrogate models in evolutionary single-objective optimization: a new taxonomy and experimental study. Inf. Sci. **562**, 414–437 (2021)
23. Wang, H., Jin, Y., Jansen, J.O.: Data-driven surrogate-assisted multiobjective evolutionary optimization of a trauma system. IEEE Trans. Evol. Comput. **20**(6), 939–952 (2016)
24. Wang, Y., Zhang, T., Chang, Y., Wang, X., Liang, B., Yuan, B.: A surrogate-assisted controller for expensive evolutionary reinforcement learning. Inf. Sci. (2022). https://doi.org/10.1016/j.ins.2022.10.134
25. Wang, Z., Zoghi, M., Hutter, F., Matheson, D., De Freitas, N.: Bayesian optimization in high dimensions via random embeddings. In: Proceedings of the 23th International Joint Conference on Artificial Intelligence (IJCAI 2013), pp. 1778–1784. AAAI Press, Beijing (2013)
26. Whiteson, S.: Evolutionary computation for reinforcement learning. In: Wiering, M., van Otterlo, M. (eds.) Reinforcement Learning. Adaptation, Learning, and Optimization, vol. 12, pp. 325–355. Springer, Heidelberg (2012). https://doi.org/10.1007/978-3-642-27645-3_10
27. Yang, P., Tang, K., Lozano, J.A.: Estimation of distribution algorithms based unmanned aerial vehicle path planner using a new coordinate system. In: Proceedings of the 2014 Congress on Evolutionary Computation (CEC 2014), pp. 1469–1476. IEEE, Beijing (2014)
28. Yang, P., Yang, Q., Tang, K., Yao, X.: Parallel exploration via negatively correlated search. Front. Comp. Sci. **15**(5), 155333 (2021)
29. Yang, P., Zhang, H., Yu, Y., Li, M., Tang, K.: Evolutionary reinforcement learning via cooperative coevolutionary negatively correlated search. Swarm Evol. Comput. **68**, 100974 (2022)
30. Yang, Q., Yang, P., Tang, K.: Parallel random embedding with negatively correlated search. In: Tan, Y., Shi, Y. (eds.) ICSI 2021. LNCS, vol. 12690, pp. 339–351. Springer, Cham (2021). https://doi.org/10.1007/978-3-030-78811-7_33

31. Zhang, J., Huang, J.X., Hu, Q.V.: Boosting evolutionary optimization via fuzzy-classification-assisted selection. Inf. Sci. **519**, 423–438 (2020)
32. Zhou, A., Wang, Y., Zhang, J.: Objective extraction via fuzzy clustering in evolutionary many-objective optimization. Inf. Sci. **509**, 343–355 (2020)
33. Zhou, A., Zhang, J., Sun, J., Zhang, G.: Fuzzy-classification assisted solution preselection in evolutionary optimization. In: Proceedings of the AAAI Conference on Artificial Intelligence, vol. 33, no. 01, pp. 2403–2410 (2019)
34. Østergård, T., Jensen, R.L., Maagaard, S.E.: A comparison of six metamodeling techniques applied to building performance simulations. Appl. Energy **211**, 89–103 (2018)

An Improved MPCA Algorithm with Weight Matrix Based on Many-Objective Optimization

Jianrou Huang, Jingbo Zhang, Qian Wang, and Xingjuan Cai[✉]

Complex System and Computational Intelligent Laboratory, Taiyuan University of Science and Technology, Taiyuan 030024, China
xingjuancai@163.com

Abstract. Data compression and dimensionality reduction play an essential role in machine learning. In recent years, due to the relatively high dimensionality of data, dimensionality reduction based on vectors cannot be well processed and will destroy the structure of high-dimensional data. Therefore, research on dimensionality reduction based on tensors is increasing daily. This paper mainly studies the extended multi-linear principal component analysis (MPCA) of principal component analysis (PCA) in the tensor space. This paper proposes a MPCA dimension reduction algorithm based on the weight matrix, followed by the construction of a multi-objective model. An IP-NSGA-III is proposed to solve the weight matrix. Experimental results of hyperspectral image classification show that, compared with different algorithms, the MPCA algorithm based on the weight matrix proposed in this paper has a better dimension-reduction effect.

Keywords: Dimensionality Reduction · Tensor · MPCA · Many-objective Optimization

1 Introduction

Nowadays, in the 21st century, when computer technology is developing more rapidly, the demands of human beings are increasing for the data of daily life and work. As data grows in complexity and dimension, subsequent computing tasks can be hugely burdensome and cause dimensional disasters. In order to get the data, we need from big data efficiently, data reduction [1, 2] and feature extraction [3, 4] have a lot of potential applications in machine learning and other fields. In general, the data we get has different amounts of information in different dimensions, some of which contain a lot of useful information, but some of which do not. Dimensionality reduction mainly uses a linear or nonlinear mapping mechanism to map high-dimensional data to low-dimensional space and keep as much information in the data as possible. And it not only avoids the dimensional disaster of high-dimensional data but also saves the storage space of high-dimensional data and improves the data transmission efficiency in different application scenarios. A common dimension reduction algorithm used in machine learning [5] is principal component analysis (PCA) [6, 7]. PCA is a kind of mature data dimension reduction algorithm, which can achieve ideal results for some specific data

L. Pan et al. (Eds.): BIC-TA 2022, CCIS 1801, pp. 248–262, 2023.
https://doi.org/10.1007/978-981-99-1549-1_20

sets or applications. But PCA dimensionality reduction is mainly through processing high-dimensional data into vector form and then carrying on the dimension reduction, which can destroy the data between the different dimensions of internal structure and spatial structure, greatly reducing the dimension reduction effect.

As a multi-dimensional matrix, tensors [8] have the characteristics of preserving space structure and data internal relations. Tensor is a type of generalization based on vector and matrix, and tensor-based dimensionality reduction techniques are becoming more mature. For linear discriminant analysis (LDA) [9, 10] to the tensor space, Yan et al. [11] put forward multilinear discriminant analysis (MDA) [12], but not convergence, and the linear discriminant analysis variables are highly affected by the parameters, making them unable to accurately determine the dimensions of each subspace. They need to list the exhaustive method, but the tensor subspace dimension may be high, so this approach is not feasible [13]. Lu et al. [14] generalized PCA to a high-dimensional tensor space and proposed a multilinear principal component analysis (MPCA). The dimensionality reduction principle of MPCA is shown in Fig. 1. PCA is mapping high-dimensional vectors to low-dimensional vectors, MPCA is through a set of multiple linear projections to data dimension reduction. It is multiplied by a projection matrix in each mode for input tensor, as shown in Fig. 1, and it does not need to be converted to a vector but directly to the high-dimensional tensor projection to low-dimensional tensor, which not only records the value of each element but also retains the correlation between each element. Using MPCA to reduce the dimensionality of high-dimensional data can retain its internal structure, making more features retained after dimensionality reduction, and the structural features of the data remain intact [15, 16].

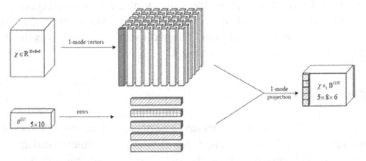

Fig. 1. Dimension reduction principle of MPCA.

The MPCA reduces the dimension along all directions of the tensor at the same time, and the difference of the tensor after the dimension reduction reaches the maximum among various data samples. However, the traditional MPCA method does not have a good algorithm to determine the dimension of the projection matrix in all directions of the tensor, and the contribution of the feature of each column in the projection matrix also has a great influence on the dimension reduction effect. In this paper, the projection matrix is optimized by using the many-objective optimization algorithm. The main contributions of this work are as follows:

- The eigenvectors of each dimension are treated equally, and there is no role in distinguishing the eigenvectors in the projection matrix. Therefore, this paper proposes multiplying the projection matrix of each MPCA mode by a weight matrix.
- Considering various factors, this paper establishes a new many-objective MPCA model to optimize the projection matrix by compression ratio, reconstruction error, difference value, and similarity.
- In the traditional NSGA-III algorithm, population competition is not too fierce during the evolution process, and it is difficult to evolve individuals with better adaptability. Therefore, this paper improves the population of the NSGA-III algorithm and divides it into two sub-populations to enhance individual diversity to get better individuals.

The rest of the paper is arranged as follows. The second section describes the related work of this paper. In Sect. 3, the problems in this paper are briefly described, the proposed improvement strategies are explained in detail, and the four objective functions are mainly introduced. The solution algorithm for improved MPCA is given in the following section. In the fifth section, the validity of the proposed objectives and the efficiency of the proposed method are proved by simulation experiments, and a series of comparative experiments are carried out. The sixth section summarizes the main work of this paper and puts forward the future work.

2 Related Works

A tensor is a generalization of vectors and matrices, and an n-order tensor is an n-dimensional vector [17]. As an intuitive expression of multi-dimensional arrays, tensors can not only record data information but also preserve the structural connections between data. However, when faced with tensor data, it is usually necessary to convert the tensor into a vector form to be suitable for various learning algorithms. In order to reduce the loss of structure, we introduced MPCA. As an extended application of PCA on tensor data, MPCA adopts matrix or high-order tensor form to obtain effective features, and this can not only avoid dimensional disasters but also reflect the advantages of keeping the basic structure information of original data when tensor data is directly treated as objects.

The dimension reduction method of tensors has been preliminarily explored by scholars in recent years. For the dimensionality reduction problem of second-order tensor data of image matrices, that is, non-vector form, Yang et al. [18] proposed two-dimensional principal component analysis (2DPCA), which constructs the image covariance matrix through the two-dimensional matrix to avoid the dimension growth in the vectorization process. But the dimensionality reduction process mainly focuses on the second module and ignores the column-column dependence, which leads to poor dimensionality reduction effect. For this reason, Zhang et al. [19] took into account the internal structure between the two modes of the second-order image matrix and reduced the dimension of both modes of the image matrix. Considering the correlation between images and the computational cost, Zhou et al. [20] proposed a fast-tensioning quantum space analysis algorithm based on second order orthogonality. Based on the above research content, a second order tensor dimensionality reduction method was proposed [21, 22]. All of these are based on the matrix form of second order tensor data research. It is a simple form of higher-order tensor, which is also a special case of higher-order tensor. The study of the second order tensor is much keener on the study of the tensor general pattern.

For the research of dimensionality reduction based on the general tensor model, some progress has been made. Lu et al. proposed a multi-linear principal component analysis algorithm. Lai et al. [23] proposed multi-linear sparse principal component analysis (MSPCA) for tensor data by introducing the concept of sparsity in sparse principal component analysis. Merola et al. [24] proposed an algorithm framework (SIMPCA) to calculate sparse components by rotating pivot elements. In this algorithm, the real sparse components are calculated by projecting the rotated principal component onto the sample variable subset. In order to make the prior information of the data can be used to guide the calculation of the subspace and realize the automatic selection and processing of the subspace variables of each mode of tensor data, Filisbino et al. [25] introduced the spatial weight into the MPCA framework. Correspondingly, Ouamane et al. [26] applied MPCA to face recognition of mixed 2D and 3D image data, which improved the recognition efficiency. Wu et al. [27] proposed a multi-linear principal component analysis network (MPCANet) based on PCANet tensor extension by utilizing the spatial relations within the image structure, which has achieved the purpose of improving the extraction and classification of advanced semantic features of multidimensional images. In addition, in order to enhance the processing ability of the algorithm to dynamically update data, Han et al. [28] proposed an online multi-linear principal component analysis algorithm, which reduced the dimensionality reduction time by making use of historical data and continuously meeting the newly increased data requirements.

However, there are few studies on the dimensions of the projection matrix and the eigenvectors that make up the projection matrix in the MPCA projection process. This paper mainly uses many-objective optimization algorithms [29, 30] to solve the dimensions of the projection matrix and improve the components of the projection matrix. The many-objective optimization algorithm refers to a multi-objective problem with more than three objectives. Based on the multi-objective algorithm NSGA-II [31], a reference point method is proposed, which emphasizes that the population members are non-dominated and the distance from the provided reference point is very close [31, 32]. Many-objective optimization algorithms can comprehensively consider multiple factors of the problem. In recent years, many-objective optimization algorithms and their improvements [33] have been applied in many scenarios, such as wireless sensor location [34, 35], malicious code detection [36], and so on.

3 The Improvement of MPCA Algorithm

3.1 Problem Description

According to the principle of dimensionality reduction of the MPCA algorithm, the projection matrix plays a vital role in the dimensionality reduction process, and it can keep the important information in the data. But the projection matrix is made up of eigenvectors corresponding to the first P largest eigenvalues by expanding each pattern of the tensor to obtain the eigenvalues of its covariance matrix, and then sorting them according to the size of the eigenvalues. Thus, it can be seen that eigenvectors corresponding to large eigenvalues and those corresponding to small eigenvalues have different influences in the dimensionality reduction process. However, the traditional MPCA projection matrix does not differentiate them. Further research and improvement were carried out in this paper in order to distinguish the influence of eigenvectors corresponding to different eigenvalues on the MPCA dimension reduction process.

3.2 Weighted MPCA

In this paper, an improved MPCA algorithm based on weight matrix is proposed to solve the problem that there is no distinction between the eigenvectors that constitute the projection matrix in the MPCA algorithm. The proposed algorithm preserves the tensor space structure and improves the dimensionality reduction effect by adding the weight matrix. This chapter first randomly initializes a set of weights, and then, based on the hyperspectral dimensionality reduction process, four target models are constructed. Using the improved algorithm to optimize the weights of the NSGA-III, the weights are obtained by solving the diagonalization and then multiplied by the corresponding projection matrix, respectively, to implement the weights, so that we can effectively distinguish between different characteristic vectors in the process of dimension reduction performance of different roles. It is guaranteed that the eigenvectors corresponding to large eigenvalues will not annihilate the influence brought by the eigenvectors corresponding to small eigenvalues when they play their roles. The dimensionality reduction formula of the weighted improved MPCA algorithm based on NSGA-III is shown in formula 1.

$$y_m = \tilde{\chi}_{m(n)} X_1 (W_{(1)} \times U^{(1)T}) X_2 \left(W_{(2)} \times U^{(2)T} \right) \cdots X_N \left(W_{(N)} \times U^{(N)T} \right),$$
$$m = 1, 2, \cdots, M \tag{1}$$

$$W_{(n)} = diag \left(w_{n1}, w_{n2}, \cdots, w_{nP_{(n)}} \right) \tag{2}$$

where $W_{(n)}$ is the weight matrix obtained by diagonalization of weight.

3.3 Objective Functions

This paper proposes four objective functions for the MPCA dimension reduction process, which are described as follows:

Compression Ratio (CR). In this paper, the compression ratio is defined as the ratio of the product of the order dimension of the output tensor after dimensionality reduction of the original input tensor of MPCA and the original input tensor. The lower the dimension is, the easier it is to store and use in machine learning. In other words, the smaller the compression ratio, the better. The equation is shown in Eq. 3.

$$obj1 : CR = \frac{P_1 \times P_2 \times \cdots \times P_N}{I_1 \times I_2 \times \cdots \times I_N} \tag{3}$$

where I_1 represents the first order's dimension of input tensor χ, I_2 represents the second order's dimension of input tensor χ, I_N represents the N-th order's dimension of input tensor χ, so $I_1 \times I_2 \times \cdots \times I_N$ represents the dimension of the input tensor χ. In a similar way, $P_1 \times P_2 \times \cdots \times P_N$ represents the dimension of the output tensor γ after dimension reduction by MPCA.

Reconstruction Error (RE). Reconstruction error refers to the difference between the low-order tensor after dimensionality reduction and the original high-order tensor. The smaller the reconstruction error is, the less data information is lost in the dimensionality reduction process [37].

$$obj1 . RE = \| \chi' - \chi \|_F^2 \tag{4}$$

where χ represents the higher-order tensor of the input. χ' refers to the higher order tensor data obtained after reconstruction is based on the lower dimension tensor after dimension reduction γ.

Similarity (Sim). Similarity compares how similar two matrices are. Generally speaking, the smaller the distance, the greater the similarity. In this paper, the Euclidean distance is used to calculate the similarity. The formula of distance function is shown in Eq. 5.

$$obj3 : Sim = \sqrt{\sum_{i=1}^{k}(x_i - y_i)^2} \tag{5}$$

where x_i represents the i-th property of the input tensor χ, y_i represents the i-th property of the output tensor γ.

Difference Value (DV). The difference value refers to the difference value between the lower order tensor after dimensionality reduction and the higher order tensor in the original input. The smaller the difference is, the less information is lost in the process of dimension reduction, and the better the dimension reduction effect is. In this paper, peak signal-to-noise ratio (PSNR) was used to calculate the difference before and after feature extraction. The specific formula is as follows:

$$obj4 : DV = 1/PSNR \tag{6}$$

$$PSNR = 10 \times log_{10}\left(\frac{255^2}{MSE}\right) \tag{7}$$

$$MSE = \frac{1}{m}\sum_{i=1}^{m}\left((x_i - \widehat{y}_i)^2\right) \tag{8}$$

where PSNR is peak signal-to-noise ratio, which is often used as an index to evaluate images, and MSE is the mean value.

Finally, the many-objective model problem in the dimensionality reduction process can be described in Eq. 9.

$$F = [min(CR), min(RE), min(Sim), min(DV)] \tag{9}$$

4 The Solving Algorithm of Weighted MPCA

4.1 PIP-NSGA-III Algorithm

The NSGA-III algorithm adds a reference point mechanism on the basis of NSGA-II, and its main idea is to introduce an elite strategy to retain excellent individuals for the next generation and avoid the loss of excellent individuals. The method based on reference points is used to select excellent individuals so that the algorithm can better deal with the many-objective function optimization problems and improve the global scarch ability of the algorithm to a certain extent. At the same time, there are many improved algorithms for NSGA-III, such as the improvement of the crossover selection operator [38, 39]. Since the initial population has a great impact on the results of the algorithm, we proposed to divide the initial population into two sub-populations, and independently optimize and update them with different crossover operator strategies, respectively. The specific process is shown in Fig. 2. The improved algorithm for NSGA-III is named IP-NSGA-III. The specific realization process of the optimized weight matrix of the IP-NSGA-III algorithm is shown as follows:

Step 1: Initialize a weight population W and a reference point (the initial weight population is processed by the method shown in Fig. 2 of this section, and the overall population size is set to N).
Step 2: The weight population is selected, crossed, and mutated to produce the offspring population, Q.
Step 3: The weight population of the offspring Q and the parent W was combined to obtain 2N individuals, and the population was set as P. Individual fitness values were calculated by formulas 3, 4, 5, and 6.

Step 4: Fast non-dominated sorting, which divides the solutions into different non-dominated solution sets.

Step 5: Suppose a new set of empty solutions is placed one by one in the new empty population, starting from the first layer of the non-dominated solution set, until the number of solutions in the new set is greater than or equal to N.

Step 6: The new parent population is selected from the new solution set obtained in step 5.

Step 7: Keep iterating until the termination condition is met. The final solution set obtained is the optimal weight required in this section.

In the whole algorithm, the selection of solutions is based on the reference point method.

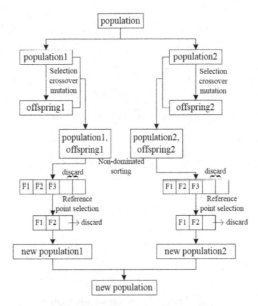

Fig. 2. IP-NSGA-III algorithm.

4.2 MPCA Based on Weight Matrix by IP-NSGA-III

The core of the MPCA algorithm is to solve and determine the projection matrix of each order [39, 40]. The projection matrix directly affects how much information the low-dimensional tensor can preserve in the original high-dimensional tensor after dimension reduction. The process of MPCA to obtain the projection matrix is mainly composed of the eigenvectors corresponding to the first P maximum eigenvalues obtained by solving the covariance matrix of the N-mode expansion of the tensor, and the projection space directly affects how much the low-order tensor retains the information of the original high-order tensor after dimensionality reduction. However, each mode tensor contains information and digital meaning is different. For hyperspectral images used in this paper, the first two modes represent the data of space dimension, and the third mode is the spectral dimension information of the hyperspectral image, so the choice of each order projection space should also be different, not the same for each mode. The method proposed in this paper is to improve the tensor projection matrix for each mode, which can ensure that the eigenvectors with large eigenvalues play a major role in the projection process, while the eigenvectors with small eigenvalues do not overwhelm the role played by them. However, the traditional MPCA treats each eigenvector equally and ignores the role played by eigenvectors with small eigenvalues. In this paper, a weighting strategy is implemented for eigenvectors corresponding to eigenvalues; that is, the projection matrix of each pattern is multiplied by a weight matrix. The improved MPCA flow chart is shown in Fig. 3.

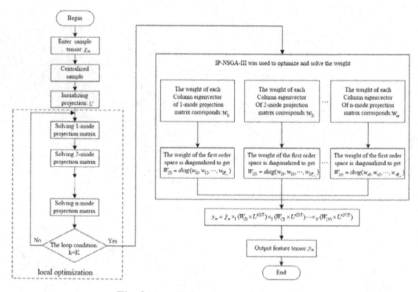

Fig. 3. MPCA based on weight matrix.

The specific solving process of MPCA based on the weight matrix mentioned in this paper is shown in Algorithm 1.

Algorithm 1: MPCA based on weighted matrix

Begin

Input: N-order tensor: $\{\chi_m \in R^{I_1 \times I_2 \times \cdots I_N}, m = 1, 2, \cdots, M\}$.

1. **(Centralized):** Centralized input sample $\tilde{\chi}_m = \chi_m - \bar{\chi}, \bar{\chi} = \frac{1}{M}\sum_{m=1}^{M} \chi_m$, $m = 1, 2, \cdots, M$, M is the sample number.

2. **(Initialization):** Construct the covariance matrix on each of χ_m. modes $\varphi^{(n)} = \sum_{m=1}^{M} \tilde{\chi}_{m(n)} \cdot \tilde{\chi}_{m(n)}^T$ $(n = 1, 2, \cdots, N)$; Calculate the eigenvalues and eigenvectors, the eigenvectors corresponding to the maximum eigenvalues of the previous $P_{(n)}$ are selected to form the projection matrix $U^{(n)}$ $(n = 1, 2, \cdots, N)$ for initialization.

3. **(Local optimization):** Calculate

$y_m = \tilde{\chi}_{m(n)} \times_1 U^{(1)T} \times_2 U^{(2)T} \cdots \times_N U^{(N)T}$, $m = 1, 2, \cdots, M$ and $\psi_{y_0} = \sum_{m=1}^{M} \parallel y_m \parallel^2$.

For k = 1: K (K is the number of iterations)

    For n = 1: N (N is the order of the input tensor sample)

    $U_{\varphi^{(n)}} = U^{(N)} \otimes U^{(N-1)} \otimes \cdots \otimes U^{(n+1)} \otimes U^{(n-1)} \otimes \cdots \otimes U^{(2)} \otimes U^{(1)}$;

Calculate $\varphi^{(n)} = \sum_{m=1}^{M} \tilde{\chi}_{m(n)} U_{\varphi^{(n)}} U_{\varphi^{(n)}}^T \tilde{\chi}_{m(n)}^T$;

Select the eigenvectors corresponding to the first $P_{(n)}$ maximum eigenvalues of $\varphi^{(n)}$ to update $U^{(n)}$ $(n = 1, 2, \cdots, N)$;

Calculate $\{y_m, m = 1, 2, \cdots, M\}$ and ψ_{y_k};

If k = K, break and go Step 4.

And get the optimal projection matrix $U^{(n)}$ $(n = 1, 2, \cdots, N)$.

4. **(Weighted):** The eigenvectors of each column of the optimal projection matrix obtained in step 3 are multiplied by a weight w_{ni} $(i = 1, 2, \cdots, P_{(n)})$, which is the optimal weight selected by the many-objective optimization algorithm under the four objectives (formula 3, 4, 5, 6).

5. **(Diagonalization):** Diagonalizing the weights obtained in step 4. $W_{(n)} = diag(w_{n1}, w_{n2}, \cdots, w_{nP_{(n)}})$.

    Calculate $y_m = \tilde{\chi}_{m(n)} \times_1 (W_{(1)} \times U^{(1)T}) \times_2 (W_{(2)} \times U^{(2)T}) \cdots \times_N (W_{(N)} \times U^{(N)T})$, $m = 1, 2, \cdots, M$.

Output: Low order characteristic tensor $\{y_m \in R^{P_1 \times P_2 \times \cdots \times P_N}, m = 1, 2, \cdots, M\}$.

End

---

5 Experimental Studies

In this section, the experiments are mainly through the vector-based PCA, the traditional MPCA, and the improved MPCA in this paper for classification experiment comparison analysis, in order to prove the effect of the improved MPCA algorithm. In addition, through the comparison of different many-objective optimization algorithms, it is concluded that the IP-NSGA-III algorithm has better convergence, and the obtained solutions have better diversity. All experiments are simulated on MATLAB 2018a in this paper.

5.1 Data Sets

In order to prove the effectiveness of the proposed model and the effect of dimension reduction, three different data sets of hyperspectral images were used in this experiment: the Salinas image dataset, the Pavia University image dataset, and the Indian Pines image dataset. The parameters of the dataset are shown in Table 1 (Fig. 4).

Table 1. Data set parameters.

Parameters	Salinas	Pavia University	Indian Pines
Number of categories	16	9	16
Number of wavelengths	204	103	200
Image size	512217	610340	145145
Number of samples	54129	42776	10249

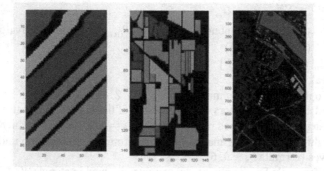

Fig. 4. Hyperspectral images.

5.2 Evaluation Indicators

In order to objectively evaluate the effectiveness of the model and the effect of MPCA dimension reduction, this paper uses the evaluation indexes of classification to conduct experimental analysis, which are: overall classification accuracy (OA) and Kappa coefficient.

Overall classification accuracy (OA): the ratio of the correct data points to the total data points after hyperspectral image classification. The formula is as follows:

$$OA = \frac{\sum_{i=1}^{C} M_{ii}}{N} \tag{10}$$

$$N = \sum_{i=1}^{C} \sum_{j=1}^{C} M_{ij} \tag{11}$$

where N represents the total number of samples, M represents the confusion matrix obtained according to the comparison between the classified data and the real data, which can be expressed as $C \times C$, where C represents the number of categories in the hyperspectral image, and M_{ij} represents the number of data samples in class j that are identified as data samples in class i.

Kappa coefficient: Since OA only considers the diagonal elements in M and does not consider other elements, sometimes it cannot accurately indicate the classification performance. Therefore, Kappa coefficient, which takes into account all elements in M, is used as another evaluation index for the experiment in this paper. Its mathematical expression is shown in Formula 12.

$$\text{Kappa} = \frac{N \sum_{i=1}^{C} M_{ii} - \sum_{i=1}^{C} \left(\sum_{j=1}^{C} M_{ij} \sum_{j=1}^{C} M_{ji} \right)}{N^2 - \sum_{i=1}^{C} \left(\sum_{j=1}^{C} M_{ij} \sum_{j=1}^{C} M_{ji} \right)} \tag{12}$$

5.3 Comparison with Other Many-Objective Optimization Algorithms

In order to verify the performance of the algorithm proposed in this paper, experimental comparison results between the IP-NSGA-III and other many-objective optimization algorithms in this paper are given. The three many-objective optimization algorithms used in this section are, respectively, NSGA-III, RVEA, and KnEA, and the experimental results are shown in Table 2.

Table 2. Performance comparison of different many-objective optimization algorithms.

Algorithm	GD	IGD
IP-NSGA-III	1.0008e-02	1.3875e-01
NSGA-III	3.5101e-02	1.5448e-01
RVEA	6.6195e-01	1.7934e+00
KnEA	6.1189e-02	3.6489e-01

As can be seen from Table 2, the GD value of the IP-NSGA-III algorithm in this paper is significantly lower than that of the other three algorithms, and the NSGA-III algorithm is lower than that of the RVEA and KNEA algorithms. The smaller the GD, the better, which indicates that the solution set obtained by the NSGA-III algorithm has a better effect and is closer to the optimal solution, while the IP-NSGA-III algorithm has better experimental results than the original NSGA-III algorithm. Obviously, the order of IGD from small to large is IP-NSGA-III, NSGA-III, KnEA, and RVEA, and the smaller the IGD, the better the convergence and diversity of the algorithm, that is, the better the effect of the IP-NSGA-III algorithm. To sum up, the IP-NSGA-III algorithm in this paper has better performance to obtain the optimal solution. In Table 3, the popular evaluation indicators GD and IGD in the optimization field are compared. To sum up, our algorithm has a good convergence index and diversity index. This is because, in this paper, we decompose the traditional single population into two populations. Selective crossover and mutation operations are performed in the two populations, respectively, and then the two populations are combined for non-dominated sorting. After the matching selection between the two populations, the information belonging to their own populations has been generated separately. The information of the two populations can be effectively coupled and interacted with in the joint non-dominated sorting, reflecting the character-istics of collaborative optimization. Although this approach seems simple, often simple strategies can produce good results. Table 4 shows the information loss of PCA, MPCA, and weighted MPCA. The calculation of RE adopts Eq. (4). The RE value is normalized to [0, 1], and the closer to 0, the smaller the information loss.

Table 3. Comparison of RE.

Data sets		PCA	MPCA	Weighted MPCA
Indian Pines	RE	0.50	0.46	0.37
Pavia University	RE	0.73	0.62	0.55
Salinas	RE	0.47	0.44	0.34

It can be seen from Table 4 that the proposed algorithm has the smallest RE on the three datasets. Therefore, it has less information loss. In the subsequent classification experiments, the proposed algorithm also showed better performance.

6   Conclusion

In this paper, the projection matrix of MPCA is analyzed, and a new algorithm for solv-ing the projection matrix of MPCA is proposed. By analyzing the different proportions of eigenvectors corresponding to the magnitude eigenvalues in the projection matrix, a weighted matrix is proposed to be used to multiply the projection matrix for improve-ment, and four different objective functions are established, which are solved by using the IP-NSGA-III algorithm in this paper. Through the experimental analysis of three hyperspectral images under PCA, MPCA, and weighted MPCA, the results show that the algorithm proposed in this paper has a better dimension reduction effect. In future work, we will consider the redundancy of the projection matrix. We will perform a dimensionality reduction on the projection matrix, and then use the MPCA algorithm to reduce the dimensionality of the tensor data.

Acknowledgment. This work is supported by the National Natural Science Foundation of China under Grant No.61806138; Natural Science Foundation of Shanxi Province under Grant No. 201801D121127; Science and Technology Development Foundation of the Central Guiding Local under Grant No. YDZJSX2021A038.

References

1. Ireneusz, C., Jdrzejowicz.: An approach to data reduction and integrated machine classification. New Generation Comput. **28**(1), 21–40 (2010)
2. Liu, Y., Yi, X., Rong, C., Zhai, Z., Gu, J.: Feature extraction based on information gain and sequential pattern for English question classification. IET Softw. **12**(6), 520–526 (2018)
3. Zhang, W., Kang, P., Fang, X., Teng, L., Han, N.: Joint sparse representation and locality preserving projection for feature extraction. Int. J. Mach. Learn. Cybern. **10**(7), 1731–1745 (2018). https://doi.org/10.1007/s13042-018-0849-y
4. Chuang, M., Zhang, H., Wang, X.: Machine learning for big data analytics in plants. Trends Plant Sci. **19**(12), 798–808 (2014)
5. Rodarmel, C., Shan, J.: Principal component analysis for hyperspectral image classification. Eng. Surv. Map. **62** (2017)
6. Li, L., Zhao, J., Wang, C., Yan, C.: Comprehensive evaluation of robotic global performance based on modified principal component analysis. Int. J. Adv. Robot. Syst. **17**(4) (2020)
7. Jiao, A., Li, C., Li, Y.: SDB-tensors and SQB-tensors. Linear & Multilinear Algebra (2017)
8. Riffenburgh, R.H., Clunies-Ross, C.W.: Linear discriminant analysis. Chicago **3**(6), 27–33 (2013)
9. Bandos, T.V., Bruzzone, L., Camps-Valls, G.: Classification of hyperspectral images with regularized linear discriminant analysis. IEEE Trans. Geosci. Remote Sens. **47**(3), 862–873 (2009)
10. Yan, S., Xu, D., Yang, Q., Zhang, L., Zhang, H.J.: Multilinear discriminant analysis for face recognition. IEEE Trans. Image Process. **16**, 212–220 (2006)
11. Li, Q., Schonfeld, D.: Multilinear discriminant analysis for higher-order tensor data classification. IEEE Trans. Pattern Anal. Mach. Intell. **36**(12), 2524–2537 (2014)
12. Mabrouk, M.S., Afify, H.M., Marzouk, S.Y.: 3D reconstruction of structural magnetic resonance neuroimaging based on computer aided detection. Int. J. Bio-Insp. Comput. **17**(3), 174–181 (2021)
13. Lu, H., Plataniotis, K.N., Venetsanopoulos, A.N.: MPCA: multilinear principal component analysis of tensor objects. IEEE Trans. Neural Netw. **19**(1), 18–39 (2008)
14. Agarwal, S., Ranjan, P.: TTPA: a two tiers PSO architecture for dimensionality reduction. Int. J. Bio-Inspired Comput. **13**(2), 119–130 (2019)
15. Handa, H.: Neuroevolution with manifold learning for playing Mario. Int. J. Bio-Inspired Comput. **4**(1), 14–26 (2012)
16. Chen, Y., Haber, E., Yamamoto, K., Georgiou, T.T., Tannenbaum, A.: An efficient algorithm for matrix-valued and vector-valued optimal mass transport. J. Sci. Comput. **77**(1), 79–100 (2018)
17. Jian, Y., David, Z., Frangi, A.F., Jing-Yu, Y.: Two-dimensional PCA: a new approach to appearance-based face representation and recognition. IEEE Trans. Pattern Anal. Mach. Intell. **26**(1), 131–137 (2004)
18. Zhang, Z.D.: 2D2PCA: 2-Directional 2-Dimensional PCA for Efficient Face Representation and Recognition (2005)
19. Zhou, Y., Bao, L., Lin, Y.: Fast second-order orthogonal tensor subspace analysis for face recognition. J. Appl. Math. 1–11(2013)
20. Ming, L., Yuan, B.: 2D-LDA: a statistical linear discriminant analysis for image matrix. Pattern Recogn. Lett. **26**(5), 527–532 (2005)
21. Chen, Y., Lai, Z., Wen, J.: Nuclear norm based two-dimensional sparse principal component analysis. Int. J. Wavelets Multiresolution Inf. Process. (2018)
22. Lai, Z., Xu, Y., Chen, Q., Yang, J., Zhang, D.: Multilinear sparse principal component analysis. IEEE Trans. Neural Netw. Learn. Syst. **25**(10), 1942–1950 (2014)

23. Merola, G.M.: SIMPCA: a framework for rotating and sparsifying principal components. J. Appl. Stat. **47**(8), 1325–1353 (2020)
24. Filisbino, T.A., Giraldi, G.A., Thomaz, C.: Ranking tensor subspaces in weighted multilinear principal component analysis. Int. J. Pattern Recog. Artif. Intell. **31**(7), 1751003.1–1751003.35 (2017)
25. Ouamane, A., Chouchane, A., Boutellaa, E., Belahcene, M., Bourennane, S., Hadid, A.: Efficient tensor-based 2D+3D face verification. IEEE Trans. Inf. Forensics Secur. **12**(11), 2751 (2017)
26. Zeng, R., Wu, J., Shao, Z., Senhadji, L., Shu, H.: Multilinear principal component analysis network for tensor object classification. IEEE Access (2014)
27. Han, L., Wu, Z., Zeng, K., Yang, X.: Online multilinear principal component analysis. Neurocomputing (2017)
28. Huang, J., Huang, X., Ma, Y., Liu, Y.: On a high-dimensional objective genetic algorithm and its nonlinear dynamic properties. Commun. Nonlinear Sci. Numer. Simul. **16**(9), 3825–3834 (2011)
29. Du, R., Santi, P., Xiao, M., Vasilakos, A.V., Fischione, C.: The sensable city: a survey on the deployment and management for smart city monitoring. IEEE Commun. Surv. Tutor. **21**(2), 1533–1560 (2019)
30. Deb, K., Jain, H.: An evolutionary many-objective optimization algorithm using reference-point based nondominated sorting approach. Part II: handling constraints and extending to an adaptive approach. IEEE Trans. Evol. Comput. **18**(4), 602–622 (2014)
31. Deb, K., Jain, H.: An evolutionary many-objective optimization algorithm using reference-point based nondominated sorting approach, Part I: solving problems with box constraints. IEEE Trans. Evol. Comput. **18**(4), 577–601 (2014)
32. Cui, Z., Zhang, M., Wang, H., Cai, X., Zhang, W.: A hybrid many-objective cuckoo search algorithm. Soft. Comput. **23**(21), 10681–10697 (2019). https://doi.org/10.1007/s00500-019-04004-4
33. Wang, P., Huang, J., Cui, Z., Xie, L., Chen, J.: A Gaussian error correction multi-objective positioning model with NSGA-II. Concurr. Comput. Pract. Exp. **32**(5), (2020)
34. He, C., et al.: Accelerating large-scale multiobjective optimization via problem reformulation. IEEE Trans. Evol. Comput. **23**(6), 949–961 (2019)
35. Cai, X., Wang, P., Du, L., Cui, Z., Zhang, W., Chen, J.: Multi-Objective three-dimensional DV-Hop localization algorithm with NSGA-II. IEEE Sens. J. **19**(21), 10003–10015 (2019)
36. Cui, Z., Du, L., Wang, P., Cai, X., Zhang, W.: Malicious code detection based on CNNs and multi-objective algorithm. J. Parallel Distrib. Comput. **129**(Jul), 50–58 (2019)
37. Zhao, J., Qiang, W., Ji, G., Zhou, X.: 3D reconstruction of pulmonary nodules in PET-CT image sequences based on a novel 3D region growing method combined with ACO. Int. J. Bio-Inspired Comput. **11**(1), 54–59 (2018)
38. Cui, Z., Chang, Y., Zhang, J., Cai, X., Zhang, W.: Improved NSGA-III with selection-and-elimination operator. Swarm Evol. Comput. **49** (2019)
39. Kang, Q., Wang, K., Huang, B., An, J.: Kernel optimisation for KPCA based on Gaussianity estimation. Int. J. Bio-Inspired Comput. **6**(2), 91–107 (2014)

Machine Learning and Deep Learning

Research on Helmet Wearing Detection Based on Improved YOLOv4 Algorithm

Wei Zhao[1,2]([✉]) and Jing Wang[2]

[1] Nanjing University of Aeronautics and Astronautics, Nanjing, China
18936685616@189.cn
[2] Jiangsu Automation Research Institute, Lianyungang, China

Abstract. In view of the difficulty and low precision of the traditional method in the detection of helmet wearing, this paper proposes an improved YOLOv4 algorithm model. First, use the k-means clustering algorithm to readjust the parameters of the prior frame to improve the matching degree between the prior frame and the target, and then add a layer of feature layer to process the features, improve the network structure of YOLOv4, and improve the accuracy of the target through feature fusion. On this basis, the Pyramid Split Attention (PSA) model is introduced to further process multi-scale feature information and improve detection accuracy. The experimental results show that the improved YOLOv4 algorithm has an average accuracy of 95.20% on the helmet target detection data set, which is 1.64% higher than the original YOLOv4 algorithm, and meets the accuracy of the helmet wearing detection task.

Keywords: Target detection · Helmet detection · YOLOv4 · Pyramid segmentation attention

1 Introduction

In recent years, with the country's vigorous development of basic economic construction, the issue of production safety has become the primary issue that we need to urgently study and break through. Although the country and governments at all levels are constantly emphasizing production safety, there are still serious hidden dangers in many construction sites. Safety helmets have a variety of protective capabilities, such as flame retardancy, electrical insulation, puncture resistance, impact resistance, etc. The impact resistance of safety helmets can disperse the impact force when an accident occurs, thereby effectively reducing the impact on the head and neck. Therefore, wearing a helmet can effectively protect the head. Although through safety education and training, the safety awareness of workers can be enhanced and the probability of safety accidents can be reduced to a certain extent, but the danger is unpredictable. Safety accidents caused by not wearing safety helmets.

According to relevant research statistics, 47.3% of people with head injuries on construction sites were injured by falling objects and impacts of objects because they did not wear safety helmets [1]. Therefore, it is necessary to monitor whether workers wear safety helmets. Management is very important.

With the development of science and technology, monitoring equipment is installed in most construction sites, and the personnel in charge of monitoring and management monitor and process in real time to realize the real-time monitoring of the helmet wearing status. However, the current helmet detection system still has the following problems:

(1) Low detection accuracy: The current helmet detection algorithm is divided into traditional target detection algorithm and deep learning detection algorithm. However, due to the complex construction scene, partial occlusion between detection targets, on-site weather changes, lighting and other influences, the detection accuracy of helmet wearing status is low.
(2) High detection cost: Deploying cameras on the entire construction site has a large scale and high economic cost. Moreover, the work on the construction site is busy and complicated. If special personnel are assigned to monitor it, it will greatly consume manpower, material resources, and financial resources. Moreover, when the testing workers are not at work, it is easy to miss the inspection. Once there is an omission, it will cause immeasurable consequences.

Target detection is a machine vision method based on deep learning, the purpose is to accurately find the desired target from a given image and give specific location information. In the process of continuous research and updating, target detection can be roughly divided into two categories, one is a two-stage (Two-stage) detection algorithm, and the other is a one-stage (One-stage) detection algorithm. Two-stage representative algorithms include R-CNN [2], Fast R-CNN [3], Faster R-CNN [4], etc. The detection process of this type of algorithm is divided into two steps. First, the candidate area is selected, and then the Select an area to classify and locate. This type of detection method has high detection accuracy, but the model structure is complex, the calculation parameters are more, and the real-time detection performance is poor. One-stage algorithms, such as SSD [5], YOLO series [6–9], etc. These algorithms directly complete the selection and classification of candidate frames through the network structure, saving computing costs and improving the speed of detection. The feature of real-time detection is more advantageous in actual engineering needs.

Literature [10] is based on the YOLOv4 algorithm, and introduces a depth separable convolution module for indoor scene target detection, replacing the original 3 × 3 convolution layer in the model, reducing model parameters, and improving detection speed and accuracy. Literature [11], based on YOLOv4-tiny, proposes bottom-up multi-scale fusion, combines low-level information to enrich the feature level of the network, improves feature utilization, and thus improves detection accuracy.

Based on the target detection algorithm YOLOv4, this paper introduces the Pyramid Split Attention module (PSA) on the basis of the original YOLOv4 algorithm and strengthens the network depth and Shallow feature fusion and other methods can further enhance the detection performance of the algorithm and improve the detection accuracy of helmets.

2 YOLOv4 Target Detection Algorithm

The YOLOv4 algorithm is an improved version of the YOLOv3 algorithm. Although most of the YOLOv4 algorithm still uses the network structure of YOLOv3, YOLOv4 has made great breakthroughs in the speed and accuracy of target detection.

The backbone network Darknet53 of the YOLOv3 model is composed of a series of residual network structures. On this basis, the YOLOv4 model has made two improvements:

(1) Change the LeakyReLU activation function to the Mish activation function. Its low cost and its smoothness, non-monotonicity, upper unbounded, and lower bounded characteristics improve its performance compared with other commonly used functions such as ReLU and Swish. The formula (1) of the Mish activation function and Fig. 1 are as follows:

$$\text{Mish} = x \times \tanh\left(\ln\left(1 + e^x\right)\right) \tag{1}$$

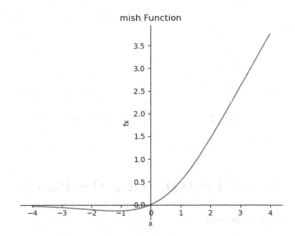

Fig. 1. Mish activation function image

(2) Drawing on the idea of splitting in the CSPnet structure, the residual block in Res-block_body is split into two parts, one part still performs the stacking operation in the original network structure, and the other part is connected to the end after a small amount of processing, which can Enhance the learning ability of the network to maintain accuracy while reducing weight. And it also has a good effect on reducing computing bottlenecks and memory costs.

The structure of the YOLOv4 network model is shown in Fig. 2. Taking the image input size of 416 × 416 as an example, the Darknet-53 network extracts features of different scales from the input image through the convolution and pooling of each network layer, and then uses the YOLO detector to extract the feature maps of different scales output by the feature extraction network. Fusion is carried out, and the result prediction is finally performed on three different output dimensions, and finally the location and category information of the target is obtained.

Among them, the YOLO detection layer generates prediction results at three different scales of 13 × 13, 26 × 26 and 52 × 52, which are suitable for the detection of large-scale targets, medium-scale targets and small-scale targets respectively.

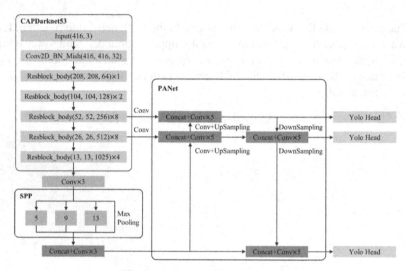

Fig. 2. YOLOv4 network structure

3 Improved YOLOv4 Target Detection Algorithm

3.1 Prior Frame Re-clustering

The prior frame used by the original network of YOLOv4 is generated using a clustering algorithm on the COCO public dataset. The COCO data set is mostly based on natural scenes, and there are 80 target categories. There are only two types of helmet detection in this paper, so the prior frame parameters in the original network cannot meet the actual needs of helmet detection. Box size has been redesigned.

In this paper, the k-means clustering algorithm is used to re-cluster on the helmet dataset to increase the matching degree between the prior box and the actual target box as much as possible, thereby further improving the detection accuracy. After clustering, 9 prior boxes are obtained. Among them (5,10), (6,13), (7,15) are used for 52 × 52 feature map output detection, (9,19), (13,24), (19,32) are used for 26 × 26 Feature map output detection and (27,47), (43,74), (81,140) are used for 13 × 13 feature map output detection, these three different feature map outputs correspond to small, medium, and large helmet target detection respectively.

3.2 Pyramid Segmentation Attention

In recent years, the idea of attention mechanism has been widely used in many machine vision fields, such as object detection, image classification, semantic segmentation and so on. Existing studies have shown that the performance of the network can be improved by embedding channel attention, spatial attention or both in the network. For example, attention mechanisms such as SENet [12], BAM [13], and CBAM [14] have all brought considerable performance improvements.

There are still two key problems with this type of attention mechanism. On the one hand, how to efficiently extract the spatial information of the feature map and enrich the feature space. Another aspect is how to improve long-range channel dependencies instead of just getting local information. Correspondingly, some scholars have proposed methods based on multi-scale feature expression and cross-channel information interaction, such as PyConv [15], Res2Net [16] and HS-ResNet [17]. However, the above method also increases the complexity of the model and the computational cost of the network is relatively high.

To solve these problems, this paper adopts the Pyramid Split Attention (PSA) [18]. This model mainly introduces multi-scale ideas based on channel attention. The PSA module can handle the spatial information of multi-scale input feature maps and can effectively establish long-range dependencies among multi-scale channel attention. As shown in Fig. 3, the PSA module is mainly realized through four steps.

First, use the Squeeze and Concat (SPC) module to segment the channel. The SPC module structure diagram is shown in Fig. 4, and then perform multi-scale feature extraction for the spatial information on each channel feature map; secondly, use the SEWeight module to extract The channel attention of feature maps of different scales is used to obtain the channel attention vectors at different scales; again, Softmax is used to perform feature recalibration on the multi-scale channel attention vectors to obtain new attention weights after multi-scale channel interaction. Finally, the recalibrated weights and the corresponding feature maps are element-wise dot-multiplied, and the output is a multi-scale feature information attention-weighted feature map. The multi-scale information expression ability of the feature map is richer.

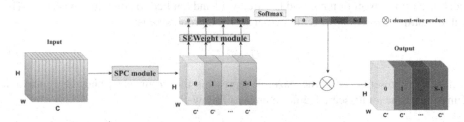

Fig. 3. Pyramid Split Attention (PSA) structure

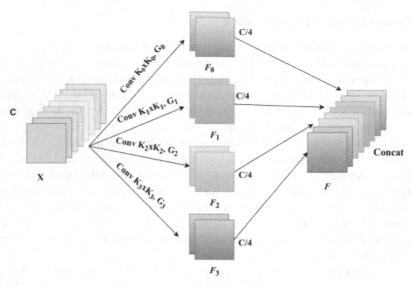

Fig. 4. Squeeze and Concat (SPC) structure

SPC structure is the key to multi-scale feature extraction in the PSA module. This structure extracts the spatial information of the feature map by using multiple branches. The input channel dimension of each branch is C, so that more abundant input tensor positions can be obtained. Information and process it in parallel at multiple scales. This results in a feature map containing a single type of kernel. At the same time, using multi-scale convolution kernels in the pyramid structure can produce different spatial resolutions and depths. By compressing the channel dimension of the input tensor, spatial information at different scales on each channel feature map can be efficiently extracted. Each feature map of different scales has a common channel dimension $C' = \frac{C}{S}$, $i = 0, 1, \cdots , S - 1$, and each branch is able to learn multi-scale spatial information independently.

To handle input tensors at different kernel scales without increasing computational cost, a group convolution method is introduced and applied to convolution kernels. The relationship between multi-scale kernel size and group size is:

$$G = 2^{\frac{K-1}{2}} \tag{2}$$

where K represents the kernel size, G represents the group size, and the generation function of the multi-scale feature map is as follows:

$$F_i = Conv(k_i \times k_i, G_i)(X) \quad i = 0, 1, 2 \cdots S - 1 \tag{3}$$

where the size of the i-th core, the size $k_i = 2 \times (i + 1) + 1$ of the i-th group is

$$G_i = 2^{\frac{k_i-1}{2}} \tag{4}$$

Represents feature maps of different scales. The entire multi-scale preprocessed feature map can be obtained by cascading:

$$F = Cat([F_0, F_1, \cdots, F_{S-1}]) \tag{5}$$

where $F \in R^{C \times H \times W}$ is the obtained multi-scale feature map. The attention weight vectors of different scales are obtained by extracting the channel attention weight information from the multi-scale preprocessing feature map. Expressed as:

$$Z_i = SEWeight(F_i), \quad i = 0, 1, 2 \cdots S - 1 \tag{6}$$

where $Z_i \in R^{C' \times 1 \times 1}$ is the attention weight. The SEWeight module is used to obtain attention weights from input feature maps of different scales. With such an approach, the PSA module can fuse different scales of contextual information and generate better pixel-wise attention for high-level feature maps. Further, on the premise of not destroying the original channel attention vector, the interaction of attention information is realized, and the dimension vector is fused. Thus, the concatenation of the entire multi-scale channel attention vector is obtained as

$$Z = Z_0 \oplus Z_1 \oplus \cdots \oplus Z_{S-1} \tag{7}$$

where $\oplus$ is the concat operator, which Z_i is the attention value from F, and Z is the multi-scale attention weight vector.

3.3 Enhanced Feature Fusion

The location environment where the wearer of the helmet is often relatively complex, including building materials, mechanical equipment, etc. For such a complex environment, the size of the helmet is relatively small, so the extraction of low-level features of the image is enhanced in the network structure, which can improve the performance of the network model.

In the YOLOv4 network, there are three feature layers of 13×13, 26×26, and 52×52 in different scales. Through multiple convolution, upsampling, and downsampling feature fusion, the feature layer is finally output. There are certain limitations in the detection of hard hats with complex scenes and small targets. Therefore, on the basis of the original network structure of YOLOv4, a 104×104 feature output layer is added, and through feature fusion, the network's ability to detect small targets is improved.

The improved network structure is shown in Fig. 5. The shallow features are more sensitive to the location information of the construction personnel target than the deep features, and are more accurate for the personnel positioning in the construction image, and pay more attention to the global information. The sensitivity of layer features improves the detection performance of small targets.

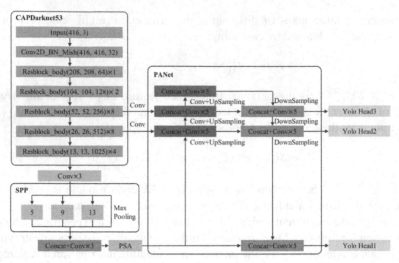

Fig. 5. Improved YOLOv4 network model

4 Experimental Results and Analysis

4.1 Dataset

The data set used in this article comes from the online open source Safety Helmet Wearing-Dataset (SHWD), which is mainly obtained through the method of crawlers. After data cleaning, 7581 images are obtained, including 9044 images wearing safety helmets. Bounding box (positive class), and 111514 bounding boxes without helmets (negative class). In addition, part of the negative data in this data set comes from the SCUT-HEAD data set, which is used to judge people who do not wear hard hats. Before training, the data set is divided into training set, validation set and test set in a ratio of 7:1:2.

4.2 Experimental Platform and Configuration

The operating system version of the experimental machine in this paper is windows 10, the CPU model is AMD Ryzen 7 5800H, the GPU model is GeForce RTX 3060, the memory size is 6 GB, and the memory size is 16 GB. All models are based on Pytorch 1.6 and use cuda 11.0 to accelerate GPU.

In the analysis of experimental results, Mean Average Precision (mAP) is used as the evaluation index of model detection accuracy, and the calculation formula is as follows.

$$Precision = \frac{TP}{TP + FP} \tag{8}$$

$$Recall = \frac{TP}{TP + FN} \tag{9}$$

$$AP = \int_0^1 PRdr \tag{10}$$

$$mAP = \frac{\sum\limits_{1}^{n} AP}{n} \qquad (11)$$

In the above formula, TP refers to true examples, FP refers to false positive examples, FN refers to false negative examples, P refers to Precision (precision rate), R refers to Recall (recall rate), n is the number of categories, and the helmet used in this article The number of categories in the dataset is $n = 2$.

4.3 Experimental Process and Result Analysis

Since the features of the backbone feature extraction network are universal, in order to speed up the model training and prevent the weight from being destroyed in the early stage of training, the experiment in this paper adopts the method of freezing the training, first freezing the training of this part of the weight, and putting more resources in the training The network parameters in the latter part, so that the time and resource utilization can be greatly improved.

The training model uses the Adam optimizer, the initial learning rate of the frozen training is set to 0.001, the Epoch is set to 100, and the Batch_size is set to 4. Table 1 shows the comparative detection results of different algorithms.

Table 1. Detection results of different object algorithms

Algorithm	AP/%		mAP/%
	Hat	Person	
YOLOv3	92.49%	83.69%	88.09%
YOLOv4	95.43%	91.69%	93.56%
YOLOv4-improve	96.92%	93.48%	95.20%

Through experiments on the hard hat dataset SHWD, the results show that compared with YOLOv4, YOLOv4-improve has a mAP increase of 1.64%, and the AP on the hat and person categories has increased by 1.49% and 1.79%, respectively. It can be seen that the algorithm YOLOv4-improve in this paper has a better detection effect than YOLOv4.

5 Conclusion

The problem of low detection accuracy of personnel wearing helmets in complex construction scenarios, a helmet detection model based on the improved YOLOv4 algorithm is proposed, and a 104×104 feature output layer is added to the backbone feature extraction network of YOLOv4. Through multiple convolutions and sampling, feature fusion is enhanced, thereby improving the network's ability to detect small targets; a PSA module is added behind the SPP module to enhance the ability to process spatial information of multi-scale feature maps; k-means clustering is used The algorithm re-clusters the prior frame to further improve the detection accuracy. Although the method in this paper has certain improvements in performance and accuracy, further research is needed on how to reduce the number of parameters of the algorithm and make the model lighter.

References

1. Hao, Z., Wei, Y.: Analysis of standard statistical characteristics of 448 construction accidents. China Standardization (2), 245–247, 249 (2017)
2. Girshick, R., Donahue, J., Darrell, T., Malik, J.: Rich feature hierarchies for accurate object detection and semantic segmentation. In: IEEE Conference on Computer Vision and Pattern Recognition (CVPR). IEEE (2014)
3. Girshick, R.: Fast R-CNN. In: IEEE International Conference on Computer Vision, pp. 1440–1448. IEEE (2015)
4. Ren, S., He, K., Girshick, R., Sun, J.: Faster r-cnn: towards real-time object detection with region proposal networks. In: Advances in Neural Information Processing Systems, 28 (2015)
5. Liu, W., Anguelov, D., Erhan, D., et al.: SSD: single shot multibox detector. In: European Conference on Computer Vision, pp. 21–37 (2016)
6. Bochkovskiy, A., Wang, C.Y., Liao, H.Y.M.: Yolov4: optimal speed and accuracy of object detection. In: IEEE Conference on Computer Vision and Pattern Recognition CVPR (2020)
7. Redmon, J., Farhadi, A.: Yolov3: an incremental improvement. In: IEEE Conference on Computer Vision and Pattern Recognition (CVPR) (2018)
8. Redmon, J., Farhadi, A.: Yolo9000: better, faster, stronger. In: IEEE Conference on Computer Vision and Pattern Recognition (CVPR) (2016)
9. Redmon, J., Divvala, S., Girshick, R., Farhadi, A.: You only look once: unified, real-time object detection. In: IEEE Conference on Computer Vision and Pattern Recognition (CVPR) (2016)
10. Li, W., Yang, C., Jiang, L., Zhao, Y.: Indoor scene target detection based on improved YOLOv4 algorithm. Prog. Laser Optoelectron. 59(18), 1815003 (2022)
11. Wang, B., Le, H., Li, W., Zhang, M.: Improved YOLO lightweight network mask detection algorithm. Comput. Eng. Appl. 57(08), 62–69 (2021)
12. Hu, J., et al.: Squeeze-and-excitation networks. IEEE Trans. Pattern Anal. Mach. Intell. 42(8), 2011–2023 (2020)
13. Park, J., Woo, S., Lee, J.Y., Kweon, I.S.: Bam: bottleneck attention module. In British Machine Vision Conference (BMVC) (2018)
14. Woo, S., Park, J., Lee, J.-Y., Kweon, I.S.: CBAM: convolutional block attention module. In: Ferrari, V., Hebert, M., Sminchisescu, C., Weiss, Y. (eds.) ECCV 2018. LNCS, vol. 11211, pp. 3–19. Springer, Cham (2018). https://doi.org/10.1007/978-3-030-01234-2_1
15. Duta, I.C., Liu, L., Zhu, F., Shao, L.: Pyramidal convolution: rethinking convolutional neural networks for visual recognition. arXiv preprint arXiv:2006.11538 (2020)
16. Gao, S.H., Cheng, M.M., Zhao, K., Zhang, X.Y., Yang, M.H., Torr, P.: Res2net: a new multi-scale backbone architecture. IEEE Trans. Pattern Anal. Mach. Intell. 43(2), 652–662 (2021)
17. Yuan, P., et al.: Hs-resnet: hierarchical-split block on convolutional neural network. arXiv preprint arXiv:2010.07621 (2020)
18. Zhang, H., Zu, K., Lu, J., et al.: EPSANet: an efficient pyramid split attention block on convolutional neural network. In: Proceedings of IEEE Conference on Computer Vision and Pattern Recognition. Piscataway. arXiv (2021)

Research on Target Detection Algorithm of Unmanned Surface Vehicle Based on Deep Learning

Fan Huang[1,2], Yong Chen[1(✉)], Xinlong Pan[3], Haipeng Wang[3], and Heng Fang[4]

[1] Wuhan Institute of Digital Engineering, Wuhan 430205, China
chen_dadi@163.com
[2] Shanghai Jiao Tong University, Shanghai 200240, China
[3] Naval Aviation University, Yantai 264001, China
[4] Wuhan University of Technology, Wuhan 430070, China

Abstract. The typical YOLO algorithm, SSD algorithm and RCNN model all have the problem of low detection accuracy or slow detection speed in surface target detection, and cannot be applied to surface unmanned ships that require high real-time and precision. In response to the above problems, this paper designs a maritime target detection model based on multi-scale features, uses four basic modules to build a backbone network for feature extraction, optimizes the YOLOv3 model, and designs a maritime target detection model Net-Y; Based on the analysis of the application scenario requirements and characteristics of unmanned vehicle target detection, a special data set for maritime target detection was constructed; Net-Y and Net-S were respectively trained by using a loss function based on the intersection and union ratio. Experimental results show that the proposed model has improved performance compared with other sea surface object detection algorithms.

Keywords: target recognition · deep learning · YOLO v3 · unmanned surface vehicle

1 Introduction

Maritime target detection is to find out a variety of targets from an image and mark their exact positions. The image is generally a visible light visual image. In 2002, the U.S. unmanned boat took the lead in using various sensors such as cameras to detect ships. Liang Xiumei established a sea target image library, used the Mean-Shift algorithm to extract the features of the target, and realized the detection and recognition function of various water targets [1]. Sergiy et al. used binarization and filtering algorithms to achieve target detection [2]. The above-mentioned methods are all traditional image processing, which need to calculate the image pixel by pixel, and are greatly affected by the external environment, so they cannot be applied to surface unmanned ships that require high real-time and precision.

The deep learning target detection algorithm is mainly divided into two categories: the first stage and the second stage. The first stage is typically represented by the YOLO algorithm and the SSD algorithm, and the second stage is represented by the RCNN series. Many scholars at home and abroad combined the application scenarios of unmanned boats and proposed Various improvement schemes. Fu et al. built a new data set for ship targets at sea and trained the Faster-RCNN model [3]. Zou et al. added the ResNet structure to the Faster-RCNN network [4]. Nie et al. improved the YOLOv3 network structure [5], Wang et al. used the YOLOv3 algorithm of deep learning to realize ship target detection at sea [6], He et al. trained the SSD model to realize the target detection function [7], Li et al. adopted the idea of saliency detection improves the detection accuracy of object locations [8].

To sum up, for the maritime target detection technology of unmanned boats, especially small USV, visible light images are mostly used as data sources, and deep learning algorithms are used to realize recognition and detection functions. Aiming at the problems of low accuracy and poor real-time performance of existing models, this paper designs a maritime target detection model based on multi-scale features. The application requirements of unmanned boats and the structure of classic networks are analyzed, and four basic modules are used to build a backbone network for feature extraction. The YOLO v3 model is optimized, and the maritime target detection model Net-Y is designed, which improves the real-time and accuracy of surface target detection.

2   Algorithm Design of Maritime Target Detection

2.1   Basic Module Design of Maritime Target Detection Network

The ResNet model designs two basic modules based on the idea of skip connections, called ResBlock, as shown in Fig. 1. The two structures have different functions. The former is used for network superposition and feature extraction, and the latter can be used for downsampling instead of the MaxPooling layer.

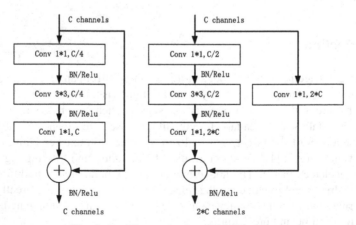

Fig. 1. ResNet two basic modules ResBlock

The Dilated Block module is different from the five downsampling layers of the commonly used classification model. This model has four downsampling layers in total, that is, the size of the network output feature map is 1/16 of the input. As shown in Fig. 2, the Dilated Block module has two forms, both of which can increase the feeling of the network. Generally, the number of channels on the left side does not change, and the number of channels on the feature map can be changed on the right side.

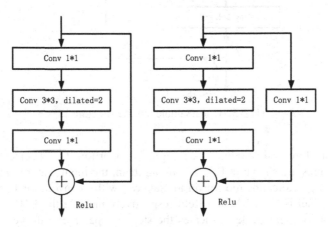

Fig. 2. Dilated Schematic diagram of Block module

At present, in most network models, the MaxPooling layer is generally used for downsampling to reduce the data dimension. The calculation method of the pooling layer is relatively simple, and useful information may be filtered out. The pooling layer can be regarded as a special convolutional layer. Using a convolution with a step size of 2 can not only achieve the same function as the pooling layer, but also extract features during the downsampling process to reduce information loss. The convolutional layer with a step size of 2 is shown in Fig. 3.

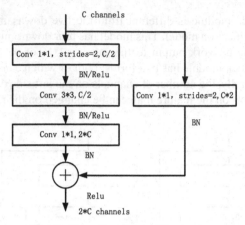

Fig. 3. DownSample module structure

In summary, four basic modules are designed for maritime target detection, and their specific structures are shown in Fig. 4. Among them, the input feature map channel is C, Conv1 ∗ 1, C/2 indicating the convolutional layer with a convolution kernel of 1, and the output channel is C/2, BN and Relu respectively indicate the BN layer and Relu activation function, and strides indicates the step size parameter in the convolutional layer.

2.2 Maritime Object Detection Network Design

Figure 5 shows the designed target detection network, denoted as Net-Y, where the numbers in each module, such as 64 and 128, indicate the number of channels of the output feature map of the module, and ResBlock × 2 indicates that ResBlock is repeated twice. The left half is the BackBone network with feature extraction function, all of which are composed of convolutional layers, with a total of 71 layers and a network parameter of 10M. The target position prediction part on the right side uses the design idea in YOLOv3, uses the upsampling layer for the deep advanced semantic features, increases the size of the feature map, and then fuses it with the shallow large-size feature map, make the shallow feature maps responsible for predicting small objects more informative. In the network design, a normalization layer (BN) is added after each convolutional layer. The BN layer normalizes the input data of each layer, and then can pull the output value from the saturation area of the activation function to the unsaturation area, effectively solve the problem of gradient disappearance and gradient explosion during training, and improve the training speed and convergence speed of the network.

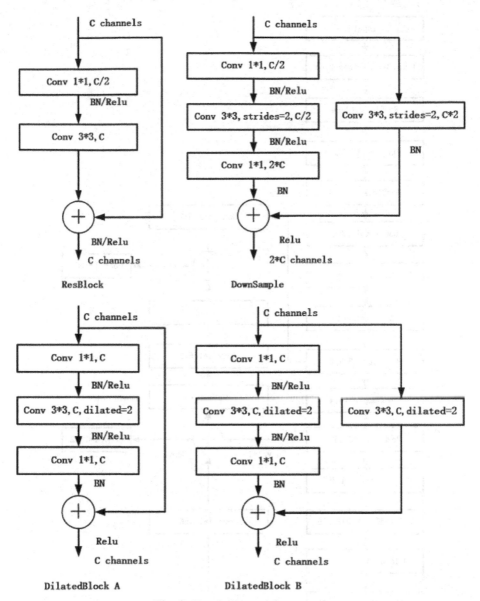

Fig. 4. Four basic module structure

Due to the existence of multiple size pre-selection boxes, an object is often detected by multiple prediction boxes. Therefore, it is necessary to select a prediction box that is most likely to have a target and the most accurate prediction box from these prediction boxes. Generally, a non-maximum value suppression method is used. [9] and its improved algorithm, the effect diagram of NMS algorithm is shown in Fig. 6.

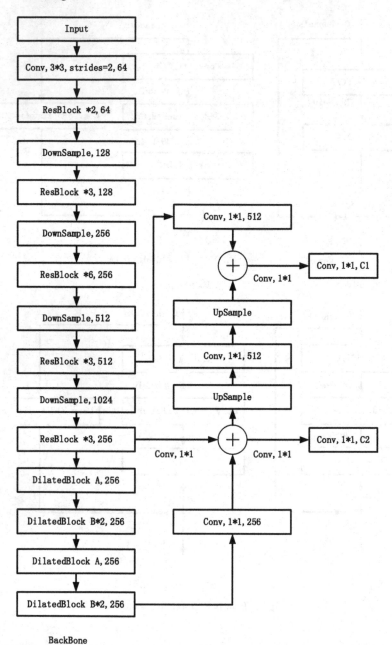

Fig. 5. Net-Y network structure

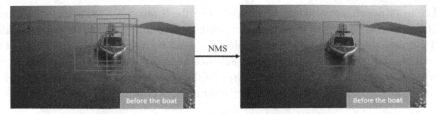

Fig. 6. Schematic diagram of the effect of the NMS algorithm

3 Maritime Object Detection Dataset Construction

3.1 Dataset Screening and Labeling

The designed target detection network is used for the unmanned vehicle perception system to identify obstacles encountered during navigation, and the background of the data set picture used for the model is the lake or sea. For this scenario, there is currently no dedicated large-scale public dataset, so the dataset used for training the model needs to be constructed and labeled by itself, and a dedicated dataset is constructed by labeling suitable images selected from the public dataset and self-collected images, including ImageNet [10], MSCOCO2017 [11], Marvel Dataset [12], SeaShips [13] and many other public datasets.

The data set used in this article is extracted from the above-mentioned multiple data sets, and the data in line with the navigation process of the unmanned boat is selected, which is close to the real scene. Select pictures from the four datasets and the data pictures collected by yourself to build a dataset, and use the tool Labelme to label the filtered pictures, and finally get the dataset.

The constructed data set is used for maritime target recognition and detection. According to actual application scenarios (lakes and offshore waters), the targets in the data set are divided into three categories, namely Ship (including various types of ships and boats), Ship (including various types of ships and boats), Island Hill (including reefs) and floating objects Other (including unknown category targets). In the subsequent planning and decision-making module, it is only necessary to determine that it is a ship, and there is no need to pay attention to the specific type of ship. There are many types of ships with different appearances, and it is difficult to subdivide various ships. Therefore, in order to improve the accuracy of the target detection model, all ship types are classified into ship categories. This will not affect the function of the planning decision-making module.

3.2 Data Augmentation

The size of the data volume greatly affects the performance of the deep learning model. However, the data volume of the compiled maritime target dataset is too small for deep learning, and the diversity of collected data may not be enough. Therefore, it is necessary to use data augmentation (Data Augmentation) to expand the data volume and increase the diversity of data. Therefore, the image is randomly flipped left and right, brightness and contrast are changed, sharpened and other methods are enhanced to expand the data set. Considering the sea application scene, there is often fog, so Gaussian noise is added to the image to simulate fog. In particular, it should be noted that for image flipping, the annotation information needs to be adjusted accordingly. As shown in Fig. 7, use various methods to enhance the comparison of the obtained data effects, and normalize the pictures to $[-1, 1]$ between when performing model training.

Fig. 7. Kinds of image enhancement methods

4   Target Detection Model Design

The target detection task actually includes two sub-tasks of classification and target positioning. When designing the loss function of the target detection task, it is also divided into two parts, as shown in the following formula. Many target detection model loss functions include classification loss, confidence loss, and localization loss. In this paper, confidence loss is classified into classification loss.

$$Loss = L_{class} + L_{loc} \tag{1}$$

For the classification subtask, the output of the model is the probability value that the predicted box belongs to each category. Not only does the correct category need to have the highest probability value, but the smaller the probability values of other categories, the better. Cross-entropy can measure this relationship very well, but the idea of pre-selection boxes is often used, and there is a problem of sample imbalance, which will make the performance of the trained model lower.

Kaiming He proposed a Korean-style Focal Loss [14] based on cross-entropy improvement in Fast - RCNN, and the calculation formula is as follows. Among them, α and γ are constants, and the two weights are effectively positive and negative, respectively, and the problem of unbalanced difficult and easy samples. The typical value is 0.25, which $\alpha\,\gamma$ is 2

$$FL = \begin{cases} -\alpha(1-p)^{\gamma} \, log(p), & if \ y = 1 \\ -(1-\alpha)p^{\gamma} \, log(1-p), & if \ y = 0 \end{cases} \tag{2}$$

Focal Loss loss function has been widely recognized and verified since it was proposed, and it also has good results in practical applications. Therefore, Focal Loss is selected as the loss function of the classification subtask.

For the positioning subtask, when designing the positioning loss function, three factors need to be considered at the same time: the overlapping area of the prediction box and the label box, the distance between the center point and the aspect ratio. IOU Loss, GIOU, DIOU, and CIOU are loss functions for these three factors. Among them, CIOU loss [15] considers the above three aspects at the same time, the evaluation of the model is more accurate, and the convergence speed is also greatly improved.

The CIOU loss function is calculated as follows:

$$R_{CIOU} = \frac{d^2}{c^2} + \alpha \upsilon \tag{3}$$

$$\alpha = \frac{\upsilon}{(1 - IOU) + \upsilon} \tag{4}$$

$$\upsilon = \frac{4}{\pi^2}(arctan\frac{w^{gt}}{h^{gt}} - arctan\frac{w}{h})^2 \tag{5}$$

$$L_{CIOU} = 1 - IOU + R_{CIOU} \tag{6}$$

CIOU Loss considers the three most important factors in target positioning at the same time, which greatly improves the positioning accuracy of the model, and is very suitable as a loss function for positioning tasks.

To sum up, the loss function of the sea target detection model of the surface unmanned vehicle is as follows. Assuming that there are k categories, the size of the output feature map of the model on a certain scale is W × H, and the number of pre-selected boxes set for each pixel of the feature map is B, then the output prediction box on this scale is M = B × W × H 1, $\hat{p}_{cij}$ indicating ij the probability that the prediction box output by the network p_{cij} belongs to the category, Indicates c the probability that the $\hat{y}_{ij}$ actual preselected box ij belongs to the category, c indicates the confidence of the predicted box output by the network, indicates the ij confidence of the y_{ij} actual preselected box ij, α and γ is the parameter of Focal Loss, whose value is $\alpha = 0.25, \gamma = 2$. Indicates I_{ij}^{obj} whether there is a target in the predicted box, ij If it exists, it is 1, otherwise it is 0. It means that I_{ij}^{noobj} the prediction frame has no target, and the calculation is performed outside the target prediction $I_{ij}^{obj} = 0$ frame, satisfying the condition that the maximum IOU is less than the given threshold (Iou loss thresh) is considered as having no target, and I_{ij}^{noobj} it is 1, otherwise it is 0. For prediction frames of different sizes, the tolerance for offset is different. The same deviation has little effect on the prediction frame with a large size but may have a great impact on the prediction frame with a small size. Therefore, in order to reduce the sensitivity to the size of the prediction frame, a scale factor is added to the prediction frame positioning loss item Scale$_{ij}$, and its value is the same as that of the preselected frame The size is related to the calculation method as follows, that is, the ratio of the area of the preselection box anchor to the area of the input image.

$$Loss = \sum_{i=0}^{M}\sum_{j=0}^{B} I_{ij}^{obj}Scale_{ij}L_{CIOUij} + \sum_{i=0}^{M}\sum_{j=0}^{B} I_{ij}^{obj}\sum_{c=0}^{k-1}(p_{cij}log(\hat{p}_{cij}))$$

$$- (1 - p_{cij})\log(1 - \widehat{p}_{cij})) + \sum_{i=0}^{M}\sum_{j=0}^{B}I_{ij}^{obj}(-\alpha(1 - \widehat{p}_{ij})^{\gamma}\log(\widehat{y}_{ij}))$$

$$+ \sum_{i=0}^{M}\sum_{j=0}^{B}I_{ij}^{noobj}(-(1 - \alpha)\widehat{p}_{ij}^{\gamma}\log(1 - \widehat{y}_{ij})) \tag{7}$$

$$\text{Scale}_{ij} = 2.0 - \frac{\text{anchor\_w}_{ij} \times \text{anchor\_h}_{ij}}{\text{image\_w} \times \text{image\_h}} \tag{8}$$

5 Model Training and Result Analysis

Through the hyperparameters, when the parameters are updated, the calculation is performed according to multiple gradients, which can avoid fluctuations caused by outliers and make the learning process more stable. The parameter update formulas are as formulas (9) and (10), m is a hyperparameter, generally set to 0.9, and the initial value v is 0. The Adam algorithm [16] is called adaptive momentum gradient descent, which combines the ideas of momentum GD and RMSprop algorithm and adjusts the learning rate parameters when doing weighted average. The calculation method is as follows: (11), (12), (13) and (14), β_1 and β_2 are hyperparameters, generally with values of 0.9 and 0.999.

$$v = m * v + \nabla J(\theta_n) \tag{9}$$

$$\theta_{n+1} = \theta_n - \eta * v \tag{10}$$

$$v = \beta_1 * v + (1 - \beta_1) * \nabla J(\theta_n) \tag{11}$$

$$s = \beta_2 * s + (1 - \beta_s) * \nabla J(\theta_n)^2 \tag{12}$$

$$v_{cor} = \frac{v}{1 - \beta_1^n}, \, s_{cor} = \frac{s}{1 - \beta_2^n} \tag{13}$$

$$\theta_{n+1} = \theta_n - \eta * \frac{v_{cor}}{\sqrt{s_{cor}}} \tag{14}$$

Build Net-Y network and miniaturized model Net-S based on Keras and Tensorflow [17] and use the loss function designed in the previous section. The size of the input original image of the model is 720×1280 to perform target detection on two scales of 1/16 and 1/32, and the size of the feature map output by the two scales is 45×80 sum 23×40. The above two networks are trained separately, and the effects of the two networks are compared.

When training Net-Y, the Adam algorithm is selected as the optimization method, the learning rate is 0.0001, and it decays exponentially, the decay coefficient is 0.95, and the final learning rate is 1×10^{-5}. In order to avoid over-fitting, the L2 method is used for regularization. The optimization coefficient is 4×10^{-5}. Two NVIDIA 1080 T i GPUs are used for parallel computing during training, and the achievable Batch Size is 32. After training for 600,000 steps, the model reaches convergence. For Net-S network, BatchSize can be set to 6 4, and other parameters are consistent with Net-Y.

Figure 8 shows the variation curves of the classification subtask cross-entropy loss, learning rate, model total loss and recall rate with the number of training steps during the training process of the model Net-Y. In sub-graph A and sub-graph C, both the cross-entropy loss and the total loss of the model decrease rapidly. As the number of training steps increases, the convergence speed decreases slowly, and the loss tends to be stable. Subfigure B shows the change in the learning rate, which decays exponentially. Subgraph D shows the change curve of the model recall rate. As the training progresses, the recall rate continues to rise, and finally reaches nearly 0.9.

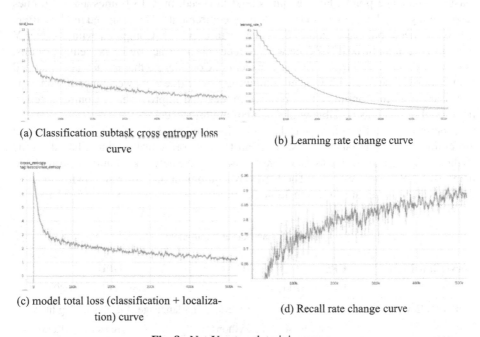

(a) Classification subtask cross entropy loss curve

(b) Learning rate change curve

(c) model total loss (classification + localization) curve

(d) Recall rate change curve

Fig. 8. Net-Y network training curve

The trained network models Net-Y and Net-S are verified on the verification set, and the above-mentioned mAP index and model inference speed FPS (i.e. frame per second) are used to evaluate the model performance. The selected usage scenarios are compared with similar algorithms in this paper, as shown in Table 1. It should be noted that the mAP value is calculated when the IOU threshold is 0.5, and the speed is the result of testing on a single NVIDIA GTX 1080Ti.

Table 1. Performance comparison of various models (the larger the detection accuracy and speed value, the better the model)

model name	Detection accuracy mAP	Detection speed FPS	model source
Net-Y	0.83 2	13.4	This article
Net-S	0.81 4	17.8	This article
Fast-Det	0.82 0	< 1	Literature [8]
CFE-YOLO	0.748	29.8	Literature [6]
SSD-Faster-RCNN	0.752 _	2 3	Literature [18]

It can be seen from the table that the maritime target detection models Net-Y and Net-S designed based on YOLOv3 and DetNet have a greater improvement in detection accuracy than the CFE-YOLO model and the SSD-Faster-RCNN model, improving at least 6 percentage points, but lags far behind in speed, only 13 frames and 17 frames respectively, which cannot meet the real-time requirements. The designed model suffers a loss in speed, but has a good performance in accuracy. Compared with the Net-Y network, the miniaturized Net-S network reduces the accuracy by 1.8 percentage points, but increases the speed by 4.4 frames per second. Considering that the actual unmanned boat chooses a device with weaker computing power due to power consumption, it is necessary to further optimize the proposed model, and improve the reasoning speed of the model as much as possible while maintaining high accuracy.

The training and verification tasks of the designed maritime target detection model are completed on the laboratory server, and the network construction is based on the Python language using Keras and Tensorflow1.2. The network training is completed using two GPUs on the server. The hardware equipment and software environment of the laboratory server are shown in Table 2.

Table 2. Lab Server Parameters

server hardware	CPU	RAM	GPU
	Intel i7–6700	64G	NVIDIA 1080Ti x2
Software Environment	operating system	Programming language	deep learning library
	Ubuntu 1604	Pythons	Tensorflow, Keras

Surface unmanned boats generally cannot provide power for computing centers for a long time, so unmanned boats choose computing platforms with lower power consumption. The surface unmanned vehicle used in the test uses NVIDIA GTX 1070 as the computing platform, and the power consumption is only 150 W, and the computing power is also reduced. Integrate the model algorithm into the unmanned vehicle autopilot software, and use the TensorRT deployment tool to further optimize the structure of the model, which is denoted as Net-ST, so that the model can run better on low-performance computing platforms. The unmanned vehicle autopilot software environment is as follows: Table 3 shows.

Table 3. Unmanned boat autopilot software environment

operating system	Programming Tools and Languages	database	dependent library
Ubuntu 1604	QT5.9 C++	redis MOOSDB	TensorRT CUDA/cuDNN

On NVIDIA GTX 1070, including CUDA, cuDNN, Keras and Tensorflow. The software environment is basically the same as that of the laboratory. After converting the trained model file into a static file and testing it in the unmanned ship software, the model reasoning speed has been greatly reduced, which cannot meet the real-time operation requirements. In order to further improve the model reasoning speed and meet the real-time requirements of the application scenario, the TensorRT tool is used to quantify the model.

Figure 9 shows the detection effect of the Net-ST model on the verification set, where the red, purple and blue boxes represent the targets of the S hip, Hill and Other categories, respectively. It can be seen from the figure that the model has good detection results for various types of targets in an environment with good visibility. In severe weather such as heavy fog, rain and snow, there are phenomena such as inaccurate detection positions and false detection. This is also unavoidable. But in general, it has good detection ability and accuracy for various targets at sea.

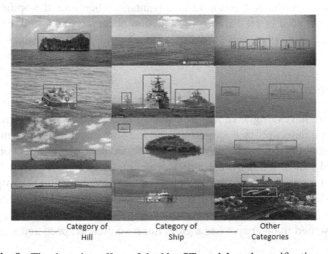

Fig. 9. The detection effect of the Net-ST model on the verification set

The following table shows the comparison of the inference speed and accuracy of the two models on the verification data set before and after using the TensorRT tool to optimize the designed Net-S model, and the performance comparison with other maritime target detection models. The Net-ST model is tested in On NVIDIA GTX 1070, other models tested on NVIDIA GTX 1080Ti.

Table 4. Lab Server Parameters

model name	Detection accuracy mAP	Detection speed FPS	model source
Net-Y	0.832	13.4	This article
Net-S	0.81 4	17.8	This article
Net-ST	0.795	3 5.1	This article
Fast-Det	0.82 0	<1	Literature [8]
CFE-YOLO	0.748	29.8	Literature [6]
SSD-Faster-RCNN	0.752	23	Literature [18]

It can be seen from Table 4 that the Net-ST model, which has been quantized and compressed by the TensorRT tool, has greatly improved the inference speed compared with the Net-S model, and meets the real-time requirements of the unmanned boat, even in the high-speed sailing state, it basically meets the requirements, while the accuracy drops by 2 percentage points. In practical applications, the performance loss of the model after optimization is within an acceptable range, while the inference speed has been greatly improved. Compared with other models, Net-ST is higher than the CFE-YOLO model and SSD-Faster-RCNN model in terms of sea target detection accuracy, and the speed is slightly improved; the detection accuracy is slightly lower than the Fast-Det model, but in terms of speed have an advantage. Therefore, the optimized Net-ST model has achieved a better balance in detection accuracy and speed, and is suitable for mobile computing platforms for small surface unmanned boats, which can meet the real-time requirements of unmanned boats, and achieved a high detection accuracy.

6 Conclusion

In this paper, we have designed a real-time maritime target detection algorithm for small USV, which can detect obstacle targets around the unmanned boat accurately. First, we analyzed the structure of the classic network, designed the backbone network for feature extraction and Net-Y for maritime target detection. Secondly, YOLO v3 structure is optimized and improved by using the idea of multi-scale detection to design the target detection model. Thirdly, we constructed a maritime target data set to train and test the target detection model. The simulation results show that the maritime target detection algorithm designed in this paper has a great improvement in speed and accuracy.

Acknowledgement. This work was supported partly by the National Nature Science Foundation of China (62076249), partly by the Key Research and Development Plan of Shandong Province (2020CXGC010701, 2020LYS11), and partly by the Natural Science Foundation of Shandong Province (ZR2020MF154).

References

1. Liang, X.: Research on feature extraction and recognition technology of target image in unmanned vehicle vision system. Harbin Engineering University, Harbin (2013)
2. Fefilatyev, S., Goldgof, D., Shreve, M., Lembke, C.: Detection and tracking of ships in open sea with rapidly moving buoy-mounted camera system. Ocean Eng. **54**(6), 1–12 (2012)
3. Fu, H., Li, Y., Wang, Y., et al.: Maritime ship targets recognition with deep learning. In: Proceedings of the 37th Chinese Control Conference (CCC), pp. 9297–9302. IEEE (2018)
4. Zou, J., Yuan, W., Yu, M.: Maritime target detection of intelligent ship based on faster R-CNN. In: Proceedings of the 2019 Chinese Automation Congress (CAC), pp. 4113–4117. IEEE (2019)
5. Nie, X., Yang, M., Liu, R. W.: Deep neural network-based robust ship detection under different weather conditions. In: Proceedings of the 2019 IEEE Intelligent Transportation Systems Conference (ITSC), pp. 47–52. IEEE (2019)
6. Wang, Y., Ning, X., Leng, B., et al.: Ship detection based on deep learning. In: Proceedings of the 2019 IEEE International Conference on Mechatronics and Automation (ICMA), pp. 275–279. IEEE (2019)
7. He, W., Xie, S., Liu, X., et al.: A novel image recognition algorithm of target identification for unmanned surface vehicles based on deep learning. J. Intell. Fuzzy Syst. **37**(4), 4437–4447 (2019)
8. Li, C., Cao, Z., Xiao, Y., et al.: Fast object detection from unmanned surface vehicles via objectness and saliency. In: Proceedings of the 2015 Chinese Automation Congress (CAC), pp. 500–505. IEEE (2015)
9. Li, Z., Peng, C., Yu, G., et al.: DetNet: a backbone network for object detection. arXiv preprint arXiv:1804.06215 (2018)
10. Deng, J., Dong, W., Socher, R., et al.: ImageNet: a large-scale hierarchical image database. In: Proceedings of the IEEE Conference on Computer Vision and Pattern Recognition (CVPR), pp. 248–255. IEEE (2009)
11. Lin, T.-Y., et al.: Microsoft COCO: common objects in context. In: Fleet, D., Pajdla, T., Schiele, B., Tuytelaars, T. (eds.) ECCV 2014. LNCS, vol. 8693, pp. 740–755. Springer, Cham (2014). https://doi.org/10.1007/978-3-319-10602-1_48
12. Gundogdu, E., Solmaz, B., Yücesoy, V., Koç, A.: MARVEL: a large-scale image dataset for maritime vessels. In: Lai, S.H., Lepetit, V., Nishino, K., Sato, Y. (eds.) ACCV 2016. LNCS, vol. 10115, pp. 165–180. Springer, Cham (2016). https://doi.org/10.1007/978-3-319-54193-8_11
13. Shao, Z., Wu, W., Wang, Z., et al.: SeaShips: a large-scale precisely annotated dataset for ship detection. IEEE Trans. Multimed. **20**(10), 2593–2604 (2018)
14. Lin, T., Goyal, P., Girshick, R., et al.: Focal loss for dense object detection. In: Proceedings of the IEEE International Conference on Computer Vision (CVPR), pp. 2980–2988. IEEE (2017)
15. Zheng, Z., Wang, P., Liu, W., et al.: Distance-IoU loss: faster and better learning for bounding box regression. arXiv preprint arXiv:1911.08287 (2019)
16. Kingma, D. P., Ba, J.: Adam: a method for stochastic optimization. arXiv preprint arXiv:1412.6980 (2014)
17. Abadi, M., Barham, P., Chen, J., et al.: TensorFlow: a system for large-scale machine learning. In: Proceedings of the 12th Symposium on Operating Systems Design and Implementation (OSDI), pp. 265–283. USENIX (2016)
18. Song, X., Jiang, P., Zhu, H.: Research on unmanned vessel surface object detection based on fusion of SSD and faster-RCNN. In: Proceedings of the 2019 Chinese Automation Congress (CAC), pp. 3784–3788. IEEE (2019)

A Review on Bio-inspired Fluid Mechanics via Deep Reinforcement Learning

Jianxiong Wang[1], Zhangze Jiang[2,3(✉)], Yi Yang[2,4], and Wulong Hu[2(✉)]

[1] Naval Research Institute, Beijing 100094, China
[2] Green & Smart River-Sea-Going Ship, Cruise and Yacht Research Center, Wuhan University of Technology, Wuhan 430070, China
1142811013@qq.com, wulong.hu@whut.edu.cn
[3] Hubei Key Laboratory of Theory and Application of Advanced Materials Mechanics, School of Science, Wuhan University of Technology, Wuhan 430070, China
[4] School of Naval Architecture, Ocean and Energy Power Engineering, Wuhan University of Technology, Wuhan 430063, China

Abstract. As a result of advancements in artificial intelligence and big data technology, the field of study on bio-inspired fluid mechanics is shifting toward data-driven methodologies. Deep reinforcement learning (DRL) has recently been adopted across a wide range of bio-inspired fluid mechanics domains due to its potential to solve problem-solving issues that were previously unsolvable due to a combination of non-linearity and high dimensionality. This study discusses the most recent advances in bio-inspired fluid mechanics using deep reinforcement learning. First, a timeline of bio-inspired fluid mechanics evolution is provided. Several popular deep reinforcement learning techniques are introduced in bio-inspired fluid mechanics, along with a detailed discussion of the benefits and drawbacks of each technique. Furthermore, data-driven bio-inspired fluid mechanical advancements are presented. Finally, several new challenges are proposed based on recent progress. The goal of this publication is to give researchers interested in tackling new problems with data-driven fluid mechanics a better understanding of DRL capabilities as well as cutting-edge applications in data-driven fluid mechanics.

Keywords: Fluid Mechanics · Deep Reinforcement Learning · Bio-inspired

1 Introduction

The origins of bio-inspired fluid mechanics can be traced back to a 1510 sketch of birds soaring in the wind by Leonardo da Vinci [38]. Because of the development of bionic robots, bionic fluid mechanics has advanced significantly in recent decades: people learn from birds, fish, and plankton how to optimize and control their shape and motion, allowing them to use unstable fluid forces to achieve agile propulsion, efficient migration, and other operations [11, 14, 20, 43, 44]. However, due to the high dimensionality, nonlinearity, and multi-scale nature of the flows, the cost of repeating a large number of numerical calculations is significant, limiting real-time optimization and control.

© The Author(s), under exclusive license to Springer Nature Singapore Pte Ltd. 2023
L. Pan et al. (Eds.): BIC-TA 2022, CCIS 1801, pp. 290–304, 2023.
https://doi.org/10.1007/978-981-99-1549-1_23

With the rise of deep neural networks and reinforcement learning in recent years, the system has been able to interact with the environment through trial and error, as well as use deep neural networks to extract information and process high-dimensional data. Deep reinforcement learning is the result of combining deep and reinforcement learning (DRL). Volodymyr Mnih's 2015 study in Nature established DRL as a research hotspot [29]. Nowadays, DRL has reached or even surpassed the human level in many fields, including: using DRL for induction, summary, and even reasoning; using DRL to assist vehicle control; and using DRL to assist vehicle control. DRL was taught to play games and went on to defeat the world champion in Go and the Champion Team in DOTA II [6, 10, 16, 42].

DRL methods have been introduced into bio-inspired fluid mechanics in this context. DRL in this paper will be based on existing research achievements in bio-inspired fluid mechanics introduce the basic concepts and methods of existing depth of intensive study, and then from three application scenarios, the review of the existing DRL in bio-inspired fluid mechanics to the current situation of the application of flow control, including effective strategy of the fish school, flow navigation by microswimmers, and bio-inspired structural optimization. Finally, the future of DRL applications in bio-inspired fluid mechanics is predicted.

2 Deep Reinforcement Learning

Deep reinforcement learning (DRL) is a machine learning method that combines deep learning and reinforcement learning. DRL solves complex decision-making problems by using neural networks as function approximators for reinforcement learning. This section introduces the fundamental concepts of DRL, divides DRL into value-based algorithms and policy-based algorithms, and lists some of them. The benefits and drawbacks of these algorithms are then discussed.

2.1 Reinforcement Learning

Reinforcement learning is a type of unsupervised machine learning [46]. Its core function is to simulate the interaction of agents with the environment and to receive rewards and observation data from the environment (see Fig. 1). An agent's goal is to take the best action in a given scenario in order to maximize the return. The following formula can be used to calculate the cumulative return along the trajectory:

$$R(\tau) = \sum_{t=0}^{T} \gamma^t r_t \tag{1}$$

where r_t is the reward obtained by the agent at the time t, τ is the trajectory of the agent's movement, and γ is the weighting coefficient introduced to define the importance of obtaining rewards at different time points. Markov processes are commonly used to describe reinforcement learning tasks.

Markov decision process can be represented by a tuple composed of (S, A, P, R) [5, 17], where S, A, R respectively represent the set of non-terminated states, all actions,

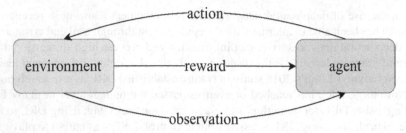

Fig. 1. Iteration between environment and agent.

and all returns; P represents the state transition probability, for example, $P(s' \mid s, a)$ is the probability that states transfers to state s' under action a. The goal of the system is to find a strategy $\pi(s)$ to maximize the expected reward.

2.2 Artificial Neurons and Neural Networks

An artificial neuron is the fundamental unit of a neural network. It has one or more inputs (x_1, x_2, etc.). Each input has a weight, and each neuron has a bias w. The weighted sum result is then fed into the activation function σ, which calculates the neuron's output (see Fig. 2). The following formula can be used to express the output z:

$$z = \sigma\left(\sum w_i x_i + b\right) \tag{2}$$

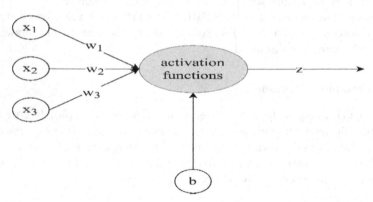

Fig. 2. Artificial neuron, where x_1, x_2 and x_3 are the three inputs, w is the weight, b is the bias, and z is the output.

A neural network is composed of multiple neurons connected with each other. If each neuron in one layer is connected to the neuron in the next layer, such a neural network is called a full connection layer (see Fig. 3). The learning process of the neural network includes adjusting all offsets and weights of the network to reduce the value of a loss function that represents the prediction quality of the network. This iteration is usually realized by gradient descent method, in which the gradient of loss function relative to weight and deviation is estimated by using the back propagation algorithm. The specific parameter update method and back propagation algorithm can be referred to Bucci MA et al. [8].

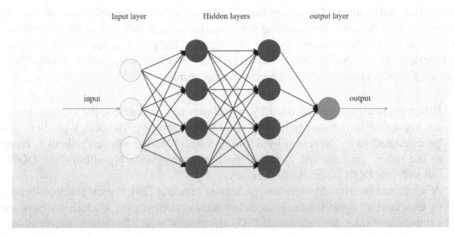

Fig. 3. Fully connected neural network with three inputs, one output and two hidden layers, each hidden layer contains four artificial neurons.

2.3 DRL Algorithm

DRL algorithms are classified as model-based or model-free. Model-based methods enable an agent to learn a model that describes how the environment works based on its observations, and then use this model to make decisions. It will not be covered in this article because it is rarely used. Model-free methods, which are currently the most popular, allow agents to interact with their surroundings immediately. Model-free methods are further subdivided into value-based and policy-based methods.

Value-Based Methods. The value-based methods use the artificial neural network to estimate the value estimate Q or the action value function A, specifically including the DQN algorithm and a series of improved algorithms:

1) DQN: The DQN algorithm uses the artificial neural network to generate a $S^+ \times A \to R$ [39], so that it can provide the possible value estimation Q of each action under the given input state. The Q grid uses the gradient descent method commonly used by the neural network to update the grid parameters and optimize it by the Bellman optimal equation.
2) DDQN: when π_θ is derived by systematically selecting the role of the highest q value, which may overestimate the q value, leading to suboptimal decision-making. Two networks are used by Van H et al. [48]: one is used to select the best action, and the other is used to estimate the target value. DDQN has a more stable strategy than DQN.

3) DQN based on experience replay: Van H et al. [48] and Schaul T et al. [41] create an experience replay area, which can store previously learned experiences in the experience replay area. This improvement also reduces the correlation between experiences.

Policy-Based Methods. The policy-based methods optimize the policy function π_θ ($a \mid s$) by gradient descent. Compared to the value-based methods, policy-based methods can be used to handle continuous states and actions. The policy-based methods mainly include deep deterministic policy gradient (DDPG), proximal policy optimization (PPO), actor-critic (A2C) and its relative improved algorithms:

1) When the strategy is determined by neural network parameters, the strategy gradient algorithm improves similarly to the DQN algorithm. The policy gradient can be evaluated when expressed by defining a loss function whose gradient is equal to the policy gradient and applying the back propagation algorithm [24]. DDPG outperforms DQN in terms of efficiency.
2) A3C: Based on asynchronous reinforcement learning [28], it uses multiple threads to execute the agent's actions asynchronously, removes the correlation between samples, and integrates almost all DRL algorithms, which has strong universality.

3 Application

In this section, several applications combining DRL and bio-inspired fluid mechanics found in the literature are presented in detail. For each case, the algorithms and experimental results used in each case are also introduced.

3.1 Effective Swimming Strategy of the Fish

The swimming strategy of fish is a classic field of bio-inspired fluid mechanics and with the development of data-driven technology, research such as [19, 25, 33, 49, 52, 57] has merged.

Mattia Gazzola et al. [16] first used the one-step-Q-learning algorithm as the DRL algorithm, swimmers in a viscous incompressible flow are simulated with a remeshed vortex method coupled with Brinkman penalization and projection approach. Let single or multiple swimmer complete tasks such as parallel swimming and V-shaped swimming through RL algorithm.

Verma et al. [33] studied the swimming strategy of two swimmers in viscous incompressible flow. Among them, the first swimmer is the leader who moves with the specified behavior, and the second swimmer is the follower whose movement strategy is determined by the DQN algorithm. The agent observes the displacement Δx and Δy in both directions from the environment, the included angle θ between the navigator and the follower (see Fig. 4), and the moment of action and the last two actions taken by the swimmer. The reward is defined as $r_t = 1 - 2 |\Delta y|/L$, where L is the length of the swimmer. When the follower deviates laterally from the leader's path, the follower will be punished. Finally, when two fish are too far away, the current event is terminated by any reward of $r_t = -1$ to artificially limit the state space. The research found that under the optimal behavior strategy of agent discovery, the swimming efficiency was 20% higher than that of a single swimmer.

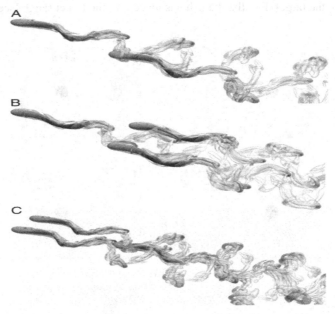

Fig. 4 Efficient coordinated swimming of one leader and one follower(A), one leader and two followers(B), two leaders and one follower(C) in 3D [49].

Verma et al. [49] further studied the swimming strategies of multiple swimmers in two-dimensional and three-dimensional flows based on synchronisation of two swimmers. The LSTM (long short-term memory network) layer recurrent neural network is added to the DQN model. Past observations, on the other hand, contain information about future changes (i.e., the process is no longer Markov). The results show that, when compared to multiple individual swimmers, appropriate use of the wake generated by other swimmers can result in collective energy conservation. Swimmers use DNSs (direct numerical simulations) of the Navier-Stokes (NS) equations. The findings show that, when compared to multiple individual swimmers, appropriate use of vortices generated by other swimmers can achieve collective energy conservation (see Fig. 5). Long-term memory cells are also required for capturing the unpredictability of the two-way interactions between the fish and the vortical flow field.

Unlike [25, 33, 49, 52, 57], Yan L et al. [56] studies the problem of obstacle avoidance by individual self-propelled fish under intelligent control. In this research, an AC agent learn to avoid the obstacle and reach the target. The moving mesh based on radial basis function and overset grid technology is taken to achieve a wide range of maneuvering. The reward function is defined as $r = 1/(1 + R_t)$, where R_t is the distance from agent to the target. Simultaneously, a penalty is imposed: when the agent hits an obstacle, it receives a 10 bonus; when it goes out of bounds, it receives a 100 reward; but if it reaches the target, it receives an additional $+100$ bonus. The agent begins by cruising toward the target point. Then, as the distance between the fish and the obstacle decreases, it changes direction gradually to avoid colliding with the obstacle. When the fish's center of mass crosses the horizontal coordinate of the obstacle center, it changes course and moves toward the target. Finally, the fish has made it to the target range (see Fig. 6).

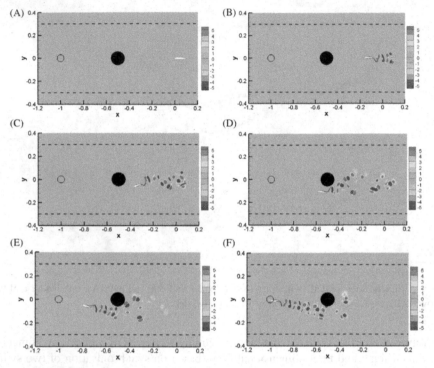

Fig. 5. The vorticity contours during motion (filled circle: obstacle; hollow circle: target; dashed black line: boundary) at $t = 0.0$ (A); $t = 1.4$ (B); $t = 2.8$ (C); $t = 4.2$ (D); $t = 5.6$ (E); $t = 7.0$ (F) [56].

3.2   Flow Navigation of Microswimmers

Compared with other bio-inspired fluid mechanics, the influence of microswimmers on the flow field is usually ignored [1, 2, 7, 9, 12, 13, 18, 19]. Based on the background of vertical migration of plankton, Amoudruz L et al. [2] used actor critic method to compare simple strategy swimmers, RL swimmers who do not perceive flow information, RL swimmers who perceive eddy currents, and RL swimmers who perceive velocity fields. The reward function is defined as:

$$r_n = -\Delta t + 10[\frac{\|X_{n-1} - X_{\text{target}}\|}{U_{\text{swim}}} - \frac{\|X_n - X_{\text{target}}\|}{U_{\text{swim}}}] + \text{bonus} \qquad (3)$$

where X_n is the coordinate of the swimmer and X_{target} is the coordinate of the target point. U_{swim} is the speed, and bonus is the reward. The research shows that velocity RL swimmers have the highest success rate in point-to-point path planning, followed by vorticity RL swimmers, and naive swimmers have the lowest success rate.

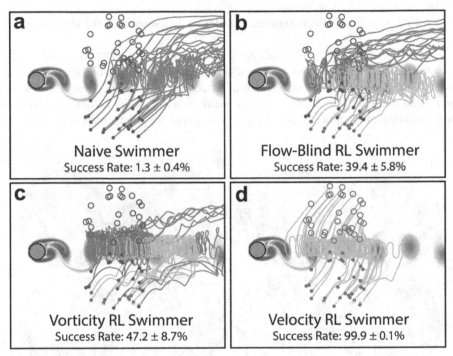

Fig. 6. Naive swimmers (a), flow-bind RL swimmers(b), vorticity RL swimmers (c), velocity RL swimmers(d) average success rate and path of success (green) and failure (red) on point-to-point problems [19].

Microswimmers must decide which direction to swim in the presence of a complex three-dimensional underlying flow with chaotic trajectories, according to Gustavsson K et al. [19]. One-step Q-learning method is used to train the agents. This research shows how smart microswimmers can learn approximately optimal swimming policies to escape fluid traps and to efficiently swim upwards even in the presence of a complex three-dimensional underlying flow with chaotic trajectories.

3.3 Motion Control of the Bio-robot

The motion control of the bio-robot is a field of bionic control. It is also the widest used bionic method in bionic robots. It can be divided into bionic control based on numerical simulation [27, 30, 32, 37] and experiment [45, 58].

Hydrodynamic invisibility refers to the swimming process in which swimmers weaken their hydrodynamic signals to hide themselves. Mehdi Mirzakhanloo et al. [27]

used the Q learning method to study the active invisibility of micro swimmers in the Stokes flow. In the study, Mirzakhanloo et al. transformed the learning objectives of two invisibles into two special tasks:

1) Learn how to optimally capture (randomly moving) intruders by forming an ideal stealth arrangement around them.
2) Learn how to follow the designated intruder while maintaining the ideal layout.

The information obtained by each agent includes the normalized distance $\zeta = |x' - x|/L_s$ and the relative direction θ. In the tracking phase, θ is the deviation angle perceived by each agent relative to the designated intruder's swimming direction. The research shows that the stealth intelligent body can maintain a high stealth efficiency under the fluid mediated interaction and can also maintain a specific arrangement (Fig. 7).

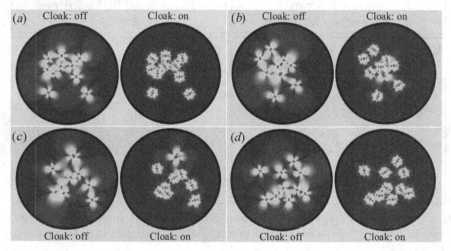

Fig. 7. Comparison of agents before and after starting stealth at different time points [27].

In Novati G et al. [32], the optimal glide strategy for two-dimensional ellipses is studied. The author considers two different criteria: minimum energy expenditure or fastest time of arrival. The author explores the gliding strategies of the RL agents that aim to minimize either time-to-target or energy expenditure by varying the aspect ratio β and density ratio ρ^* of the falling ellipse. These two optimization objectives may be seen as corresponding to the biologic scenarios of foraging and escaping from predators. The research shows that in both cases (energy and time) the gliding trajectories are smooth. In addition, model-free reinforcement learning has been shown to achieve more robust glide with lower computational costs compared to model-based optimal control strategies (Fig. 8).

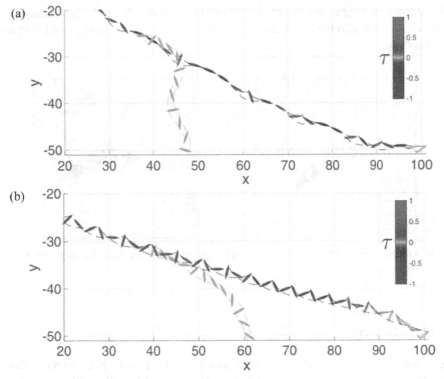

Fig. 8. Motion path of the ($\beta = 0.1$, $\rho^* - 200$) (a) and ($\beta = 0.1$, $\rho^* = 100$) (b) [32].

3.4 Shape Optimization

Shape optimization is a field of bionic structure. The goal is to optimize the shape to meet certain criteria such as drag reduction. Many studies now use supervised neural networks combined with gradient-based and gradient-free methods for shape optimization.

Based on two studies that use reinforcement learning to optimize shapes [21, 22] and inspired by studies [3] that use deep learning to solve optimization problems. Viquerat J et al. [50] used DRL to optimize the shape which is determined by a Bessel curve:

$$\begin{cases} r = r_{\max} \max(|p|, r_{\min}), \\ \theta = \frac{\pi}{n}\left(i + \frac{q}{2}\right), \\ x = r\cos(\theta) \\ y = r\sin(\theta) \\ e = \frac{1}{2}(1 + s) \end{cases} \tag{4}$$

where (p, q, s) is provided by neural network, a transformed triplet (x, y, e) is obtained that generates the position x, y and local curvature e of the ith point. The PPO agent get drag and lift though CFD and maximize the lift-to-drag ratio C_l/C_d by optimizing the shape. In addition, the author adds area penalization to limit the area of agent. As shown in Fig. 9, the agent can design airfoil-like shapes maximizing the lift-to-drag ratio at

Reynolds numbers of about a few hundred and area penalization induces a reduction of approximately 30% in the lift-to-drag ratio compared to the optimal shape without area penalization.

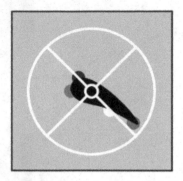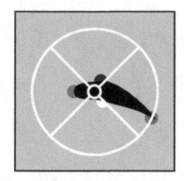

Fig. 9. Optimal shapes obtained with(left) and without(right) area penalization using 4 points with 3 free points [50].

4   Conclusion

This paper summarizes the existing research of data-driven bio-inspired fluid mechanics and classifies it according to application scenarios. In addition, the DRL method used in each example, the design of reward and the conclusions drawn are summarized. Compared with traditional flow control methods, DRL method has good efficiency, performance, and robustness. At the same time, the parallel computing capability of DRL library can save expensive CFD computing resources. In addition, transfer learning can also be used to save computing time.

Up to now, the application of DRL in the flow control of unmanned clusters is still in its infancy, and it still faces the following new challenges:

1) Most of the studies in this paper take the velocity field and some simple parameters such as coordinates and shapes as observations. How to obtain more flow field characteristics and apply them to the control strategies of agents.
2) The research on DRL and experiment coupling is very rare, and the ability to transfer from numerical simulation to physical test remains to be explored. This is also a big challenge for DRL.
3) The large number of repeated calculations of DRL puts forward higher requirements on the accuracy and efficiency of CFD. To solve this problem, researchers have proposed parallel computing and distributed computing methods [35, 50], With the development of GPU, GPU acceleration has become a feasible solution [25].
4) The exploration of the neural network layer is not enough: most current studies choose the fully connected layer as the neural network layer of DRL instead of other networks. Only Verma S et al. [49] and Yu H et al. [57] use the LSTM, and their results also show that the LSTM shows better performance. Therefore, the type

and structure of neural network layer should be considered comprehensively in the research.

5) Influenced by early open-source projects [35, 55], the majority of research employs the Proximal Policy Optimization (PPO) and q-learning algorithms. However, in the literature [15], PPO, TPRO, and A2C algorithms are compared, and the results show that SAC outperforms the other two in cylinder control strategy. It is unclear whether SAC superior to other algorithms, but it is clear that researchers must select the appropriate algorithm for each situation. Reinforcement learning has been criticized for a long time for its randomness [4]. Hence, it's suggested to open the code and datasets. The known open-source literature and some of its information are summarized in Table 1.

In practical applications, the application of data-driven bio-inspired fluid mechanics will encounter many complex problems, which need researchers to explore.

Table 1. Open-source literature.

Reference	DRL	CFD
Novati G et al. [33]	Smarties [31]	Custom solver
Verma et al. [49]	Smarties	Custom solver
Liu W et al. [25]	Pytorch [34]	LBM [23]
Gunnarson P et al. [18]	Smarties	Custom solver
Novati G et al. [32]	Smarties	Custom solver
Viquerat J et al. [50]	Tensorforce [40]	Fenics [26]
Rabault J et al. [35]	Stable-baselines [36]	Fenics
Darshan T et al. [47]	Pytorch	Openfoam [53]
Wang Q et al. [51]	Tianshou [54]	Openfoam

References

1. Alageshan, J.K., Verma, A.K., Bec, J., et al.: Machine learning strategies for path-planning microswimmers in turbulent flows. Phys. Rev. E **101**(4), 043110 (2020)
2. Amoudruz, L., Koumoutsakos, P.: Independent control and path planning of microswimmers with a uniform magnetic field. Adv. Intell. Syst. **4**(3), 2100183 (2022)
3. Andrychowicz, M., Denil, M., Gomez, S., et al.: Learning to learn by gradient descent by gradient descent. In: Advances in Neural Information Processing Systems, vol. 29 (2016)
4. Bajorath, J., Coley, C.W., Landon, M.R., et al.: Reproducibility, reusability, and community efforts in artificial intelligence research. Artif. Intell. Life Sci. **1**, 100002 (2021)
5. Bellman, R.: A Markovian decision process. J. Math. Mech. 679–84 (1957)
6. Berner, C., Brockman, G., Chan, B., et al.: Dota 2 with large scale deep reinforcement learning. arXiv preprint arXiv:191206680 (2019)

7. Biferale, L., Bonaccorso, F., Buzzicotti, M., et al.: Zermelo's problem: optimal point-to-point navigation in 2D turbulent flows using reinforcement learning. Chaos: Interdisc. J. Nonlinear Sci. **29**(10), 103138 (2019)

8. Bucci, M.A., Semeraro, O., et al.: Control of chaotic systems by deep reinforcement learning. Proc. R. Soc. A **475**(2231), 20190351 (2019)

9. Buzzicotti, M., Biferale, L., Bonaccorso, F., Clark di Leoni, P., Gustavsson, K.: Optimal control of point-to-point navigation in turbulent time dependent flows using reinforcement learning. In: Baldoni, M., Bandini, S. (eds.) AIxIA 2020. LNCS (LNAI), vol. 12414, pp. 223–234. Springer, Cham (2021). https://doi.org/10.1007/978-3-030-77091-4_14

10. Catarau-Cotutiu, C., Mondragon, E., Alonso, E.: AIGenC: AI generalisation via creativity. arXiv preprint arXiv:220509738 (2022)

11. Clark, I.A., Daly, C.A., Devenport, W., et al.: Bio-inspired canopies for the reduction of roughness noise. J. Sound Vib. **385**, 33–54 (2016)

12. Colabrese, S., Gustavsson, K., Celani, A., et al.: Smart inertial particles. Phys. Rev. Fluids **3**(8), 084301 (2018)

13. Colabrese, S., Gustavsson, K., et al.: Flow navigation by smart microswimmers via reinforcement learning. Phys. Rev. Lett. **118**(15), 158004 (2017)

14. Costa, D., Palmieri, G., Palpacelli, M.-C., et al.: Design of a bio-inspired autonomous underwater robot. J. Intell. Rob. Syst. **91**(2), 181–192 (2018)

15. Garnier, P., Viquerat, J., Rabault, J., et al.: A review on deep reinforcement learning for fluid mechanics. Comput. Fluids **225**, 104973 (2021)

16. Gazzola, M., Hejazialhosseini, B., Koumoutsakos, P.: Reinforcement learning and wavelet adapted vortex methods for simulations of self-propelled swimmers. SIAM J. Sci. Comput. **36**(3), B622–B639 (2014)

17. Goodfellow, I., Bengio, Y., Courville, A.: Deep learning (adaptive computation and machine learning series). Cambridge Massachusetts, pp. 321–59 (2017)

18. Gunnarson, P., Mandralis, I., Novati, G., et al.: Learning efficient navigation in vortical flow fields. Nat. Commun. **12**(1), 1–7 (2021)

19. Gustavsson, K., Biferale, L., Celani, A., Colabrese, S.: Finding efficient swimming strategies in a three-dimensional chaotic flow by reinforcement learning. Eur. Phys. J. E **40**(12), 1–6 (2017). https://doi.org/10.1140/epje/i2017-11602-9

20. Lagor, F.D., DeVries, L.D., Waychoff, K., et al.: Bio-inspired flow sensing and control: autonomous rheotaxis using distributed pressure measurements. J. Unmanned Syst. Technol. **1**(3), 78–88 (2013)

21. Lampton, A., Niksch, A., Valasek, J. (eds.): Morphing airfoils with four morphing parameters. In: AIAA Guidance, Navigation and Control Conference and Exhibit (2008)

22. Lampton, A., Niksch, A., Valasek, J.: Reinforcement learning of a morphing airfoil-policy and discrete learning analysis. J. Aerosp. Comput. Inf. Commun. **7**(8), 241–260 (2010)

23. Li, W., Chen, Y., Desbrun, M., et al.: Fast and scalable turbulent flow simulation with two-way coupling. ACM Trans. Graph. **39**(4), Article no. 47 (2020)

24. Lin, L.-J.: Reinforcement Learning for Robots Using Neural Networks. Carnegie Mellon University (1992)

25. Liu, W., Bai, K., He, X., et al.: FishGym: a high-performance physics-based simulation framework for underwater robot learning. arXiv preprint arXiv:220601683 (2022)

26. Logg, A., Wells, G.N.: DOLFIN: automated finite element computing. ACM Trans. Math. Softw. (TOMS) **37**(2), 1–28 (2010)

27. Mirzakhanloo, M., Esmaeilzadeh, S., Alam, M-R.: Active cloaking in Stokes flows via reinforcement learning. J. Fluid Mech. **903** (2020)

28. Mnih, V., Badia, A.P., Mirza, M., et al. (eds.): Asynchronous methods for deep reinforcement learning. In: International Conference on Machine Learning. PMLR (2016)

29. Mnih, V., Kavukcuoglu, K., Silver, D., et al.: Human-level control through deep reinforcement learning. Nature **518**(7540), 529–533 (2015)
30. Nair, N.J., Goza, A.: Bio-inspired variable-stiffness flaps for hybrid flow control, tuned via reinforcement learning. arXiv preprint arXiv:221010270 (2022)
31. Novati, G., Koumoutsakos, P. (eds.): Remember and forget for experience replay. In: International Conference on Machine Learning. PMLR (2019)
32. Novati, G., Mahadevan, L., Koumoutsakos, P.: Controlled gliding and perching through deep-reinforcement-learning. Phys. Rev. Fluids **4**(9), 093902 (2019)
33. Novati, G., Verma, S., Alexeev, D., et al.: Synchronisation through learning for two self-propelled swimmers. Bioinspiration Biomimetics **12**(3), 036001 (2017)
34. Paszke, A., Gross, S., Massa, F., et al.: Pytorch: an imperative style, high-performance deep learning library. In: Advances in Neural Information Processing Systems, vol. 32 (2019)
35. Rabault, J., Ren, F., Zhang, W., Tang, H., Xu, H.: Deep reinforcement learning in fluid mechanics: a promising method for both active flow control and shape optimization. J. Hydrodyn. **32**(2), 234–246 (2020). https://doi.org/10.1007/s42241-020-0028-y
36. Raffin, A., Hill, A., Ernestus, M., et al.: Stable baselines3 (2019)
37. Reddy, G., Celani, A., Sejnowski, T.J., et al.: Learning to soar in turbulent environments. Proc. Natl. Acad. Sci. **113**(33), E4877–E4884 (2016)
38. Rival, D.E.: Biological and Bio-inspired Fluid Dynamics: Theory and Application. Springer, Heidelberg (2022)
39. Ruder, S.: An overview of gradient descent optimization algorithms. arXiv preprint arXiv: 160904747 (2016)
40. Schaarschmidt, M., Kuhnle, A., Ellis, B., et al.: LIFT: reinforcement learning in computer systems by learning from demonstrations. arXiv preprint (2018)
41. Schaul, T., Quan, J., Antonoglou, I., et al.: Prioritized experience replay. arXiv preprint arXiv: 151105952 (2015)
42. Silver, D., Huang, A., Maddison, C.J., et al.: Mastering the game of Go with deep neural networks and tree search. Nature **529**(7587), 484–489 (2016)
43. Sun, T., Chen, G., Yang, S., et al.: Design and optimization of a bio-inspired hull shape for AUV by surrogate model technology. Eng. Appl. Comput. Fluid Mech. **15**(1), 1057–1074 (2021)
44. Takizawa, K., Tezduyar, T.E., Kostov, N.: Sequentially-coupled space–time FSI analysis of bio-inspired flapping-wing aerodynamics of an MAV. Comput. Mech. **54**(2), 213–233 (2014). https://doi.org/10.1007/s00466-014-0980-x
45. Tedrake, R., Jackowski, Z., Cory, R., et al. (eds.): Learning to fly like a bird. In: 14th International Symposium on Robotics Research, Lucerne, Switzerland (2009)
46. Thrun, S., Littman, M.L.: Reinforcement learning: an introduction. AI Mag. **21**(1), 103 (2000)
47. Thummar, D.: Active flow control in simulations of fluid flows based on deep reinforcement learning. Zenodo (2021)
48. Van Hasselt, H., Guez, A., Silver, D. (eds.): Deep reinforcement learning with double q-learning. In: Proceedings of the AAAI Conference on Artificial Intelligence (2016)
49. Verma, S., Novati, G., Koumoutsakos, P.: Efficient collective swimming by harnessing vortices through deep reinforcement learning. Proc. Natl. Acad. Sci. **115**(23), 5849–5854 (2018)
50. Viquerat, J., Rabault, J., Kuhnle, A., et al.: Direct shape optimization through deep reinforcement learning. J. Comput. Phys. **428**, 110080 (2021)
51. Wang, Q., Yan, L., Hu, G., et al.: DRLinFluids–an open-source python platform of coupling deep reinforcement learning and OpenFOAM. arXiv preprint arXiv:220512699 (2022)
52. Weber, P., Wälchli, D., Zeqiri, M., et al.: Remember and forget experience replay for multi-agent reinforcement learning. arXiv preprint arXiv:220313319 (2022)
53. Weller, H.G., Tabor, G., Jasak, H., et al.: A tensorial approach to computational continuum mechanics using object-oriented techniques. Comput. Phys. **12**(6), 620–631 (1998)

54. Weng, J., Chen, H., Yan, D., et al.: Tianshou: a highly modularized deep reinforcement learning library. arXiv preprint arXiv:210714171 (2021)
55. Xu, H., Zhang, W., Deng, J., Rabault, J.: Active flow control with rotating cylinders by an artificial neural network trained by deep reinforcement learning. J. Hydrodyn. **32**(2), 254–258 (2020). https://doi.org/10.1007/s42241-020-0027-z
56. Yan, L., Chang, X., Wang, N., et al.: Learning how to avoid obstacles: a numerical investigation for maneuvering of self-propelled fish based on deep reinforcement learning. Int. J. Numer. Meth. Fluids **93**(10), 3073–3091 (2021)
57. Yu, H., Liu, B., Wang, C., et al.: Deep-reinforcement-learning-based self-organization of freely undulatory swimmers. Phys. Rev. E **105**(4), 045105 (2022)
58. Zhang, T., Tian, R., Wang, C., et al.: Path-following control of fish-like robots: a deep reinforcement learning approach. IFAC-Papers OnLine **53**(2), 8163–8168 (2020)

Memristor-Based Neural Network Circuit of Operant Conditioning with Overshadowing

Yuanpeng Xu, Ronghang Liao, and Junwei Sun[✉]

School of Electrical and Information Engineering,
Zhengzhou University of Light Industry, Zhengzhou 450002, Henan, China
junweisun@yeah.net

Abstract. Overshadowing in operant conditioning is laid less attention in recent years. A memristor-based neural network circuit of operant conditioning with overshadowing is proposed in this paper, which realizes the overshadowing effect of two kinds of rewards in strengthening animal behaviors. The circuit consists of a button module, a reward module, an inhibition module and a synapse module. The button module realizes the effect of buttons on dogs. The reward module can realize the effect of rewards to operant conditioning. The inhibition module realizes the inhibition effect between two rewards. The synapse module reflects the variation of synaptic strength in the operant conditioning. The circuit designed in this work can realize the overshadowing effect in operant conditioning, improve the study of operant conditioning on reward, and provides more references for neural networks of operant conditioning with further development.

Keywords: Memristor · Neural network · Associative memory · Overshadowing · Operant conditioning

1 Introduction

Associative memory is divided into conditional association and operant conditioning [1,2]. Since the 1930s, American behaviorist psychologist Skinner has created operant conditioning on the basis of classical conditioning [3,4]. Operant conditioning occurs when an association is made between a particular behavior and a consequence for that behavior and reinforcement that closely follows a

This work was supported in part by the National Natural Science Foundation of China under Grant 62276239 and 62272424, in part by the Joint Funds of the National Natural Science Foundation of China under Grant U1804262, in part by Henan Province University Science and Technology Innovation Talent Support Plan under Grant 20HASTIT027, in part by Zhongyuan Thousand Talents Program under Grant 204200510003, in part by Zhongyuan Talents Program under Grant ZYY-CYU202012154, and in part by Henan Natural Science Foundation-Outstanding Youth Foundation under Grant 222300420095.

L. Pan et al. (Eds.): BIC-TA 2022, CCIS 1801, pp. 305–315, 2023.
https://doi.org/10.1007/978-981-99-1549-1_24

behavior will reinforce the behavior [5,6]. In practical application, the principle of operant conditioning has been widely used in teaching and learning [7,8]. Operant conditioning is a type of associative learning process and is one of the hot research fields in artificial neural network [9–11]. However, the neural network circuits which can realize operant conditioning most have high consumption [12,13]. The appearance of memristor brings possibility to neural network circuits with low consumption.

The memristor was first proposed by Professor Chua in 1971 [13]. It was first made in 2008 by Hewlett-Packard Labs [14]. It has excellent non-volatility and non-linear characteristics [15–17]. Memristor values will change when voltages are applied in different directions and sizes. Currently, memristors are mainly used in chaotic circuits, digital logic circuits and neural network circuits. In this paper, an operant conditioning neural network circuit with overshadowing based on memristors is proposed.

At present, most of the researches on operant conditioning are about the reward of contingency, immediacy, magnitude and so on [18–20], but the research on the overshadowing of operant conditioning is very few. However, in operant conditioning, reinforcement is the key factor in determining whether an animal can learn. The overshadowing effect in operant conditioning is each reward will reduce the control exerted by the other when there are two rewards training [21]. It is difficult to use one of the two rewards to promote the certain behavior of animals after using two kinds of rewards.

The rest of the paper is organized as follows. Section 2 presents the memristor models. Modules design of circuit is shown in Section 3. Section 4 presents experiment design and simulation analysis. The conclusion is drawn in Sect. 5.

2   Memristor Model

Memristor models have developed rapidly and a growing number of models have been proposed successively in recent decades. The threshold memristor model used to mimic the AgInSbTe(AIST)-based memristor is selected in this work. The model is described as

$$M(t) = R_{on}\frac{w(t)}{D} + R_{off}(1 - \frac{w(t)}{D}) \tag{1}$$

where $M(t)$ is the memristance, R_{off} denotes the resistance at highly doped, $w(t)$ is the width of the high doped region, D is the full thickness of memristive material. R_{on} is the resistance at low doped. The derivative of the state variable $w(t)$ is

$$\frac{dw(t)}{dt} = \begin{cases} \mu_v \frac{R_{ON}}{D}\frac{i_{off}}{i(t)-i_0}f(w(t)), & v(t) > V_{T+} > 0 \\ 0, & V_{T-} \leqslant v(t) \leqslant V_{T+} \\ \mu_v \frac{R_{ON}}{D}\frac{i(t)}{i_{on}}f(w(t)), & v(t) < V_{T-} < 0 \end{cases} \tag{2}$$

where μ_v stands for the ionic mobility, i_0, i_{on} and i_{off} are constants. V_{T+} and V_{T-} are positive and negative threshold voltages, respectively. $f(w(t))$ is given as a window function

$$f(w(t)) = 1 - (\frac{w(t)}{D} - sgn(-i))^{2p} \tag{3}$$

These parameter settings of memristors in the circuit are shown in Table 1. The memristance will be reduced if the positive voltage applied to the memristor is greater than V_{T+}. The memristance will be increased if the nagative voltage applied to the memristor is less than V_{T-}. The greater the positive voltage is, the slower $w(t)$ increases, and the slower the memristance decreases. All the simulation processes in this paper are implemented by PSPICE.

Table 1. Parameters of memristors

Parameters	M_1	M_2	$M_3 \& M_5$	$M_4 \& M_6$	$M_7 \& M_8$	M_9
$D(nm)$	3	3	3	3	3	3
$R_{init}(\Omega)$	300	$1k$	200	200	200	$1.5k$
$R_{OFF}(\Omega)$	$1k$	$1k$	800	800	800	$1.5k$
$R_{ON}(\Omega)$	300	500	200	200	200	50
$V_{T+}(v)$	15	0.1	0.2	2	0.1	3
$V_{T-}(v)$	-1	-0.1	-1	-1	-0.01	-0.01
$\mu_v(m^2s^{-1}\Omega^{-1})$	$1.6e-15$	$1.6e-18$	$1.6e-18$	$1.6e-18$	$1.6e-18$	$1.6e-18$
$i_{on}(A)$	1	1	1	1	1	1
$i_{off}(A)$	$1e-5$	$1e-5$	$1e-5$	$1e-5$	$1e-5$	$1e-5$
$i_0(A)$	$1e-3$	$1e-3$	$1e-3$	$1e-3$	$1e-3$	$1e-3$
p	10	10	10	10	10	10

3   Modules Design of Circuit

3.1   Button Module

The button module is shown in Fig. 1 to simulate the behavior of a dog pressing a button. In the illustration of operant conditioning with overshadowing, the dog will get eggs or meat when it occasionally presses the button. The button module is designed for changes in the neural network when the button is pressed. The synaptic strength will not increase when rewards are not given after pressing the button. When the button is pressed, N_1 generates a 10 v pulse with the width of 10 s and is sent to a feedback amplifier module. The connection mode of M_1 is reverse connection. The memristance of M_1 is 300 Ω when the applied voltage does not exceed its positive threshold voltage. In the opposite case, the memristance of M_1 will rise from 300 Ω to $1\,k\,\Omega$ in a short time to prevent excessive input voltage from damaging the circuit network. NMOS T_1 uses M2SK530 model (threshold voltage is about $2.3\,v$) and PMOS T_2 uses M2SJ136 model (threshold voltage is about $-2\,v$) in JPWRMOS library.

$V_{OP1} < 0.1v$, D_1 produces a high level. T_1 is opened and T_2 is closed. S_1 is closed. V_2 is applied to M_2 and R_2 by port 1 of SUM_1. M_2 and R_2 form a voltage divider to control the applied voltage on R_2. The applied voltage of R_2 can be expressed by

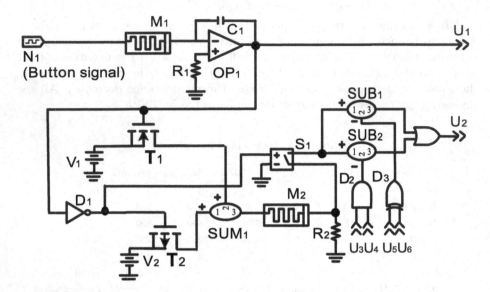

Fig. 1. Button module. N_1 is button signal. R_1 and R_2 are resistors. M_1 and M_2 are memristors. OP_1 is a precision operational amplifier. T_1 is NMOS and T_2 is PMOS. D_1 is a NOT gate. D_2 is an AND gate. D_3 is an Exclusive-OR gate. SUM_1 is a sum component. SUB_1 and SUB_2 are subtract components. S_1 is the voltage-controlled switch.

$$V_{R2} = \begin{cases} (V_2 * R_2) / (M_2 + R_2), & V_{OP1} < 0 \\ 0, & 0 \leqslant V_{OP1} \leqslant 1 \\ (V_1 * R_2) / (M_2 + R_2), & V_{OP1} > 1 \end{cases} \qquad (4)$$

The reverse integral operation circuit increases the synaptic weight, while the forgetting voltage decreases the synaptic weight. The initial memristance of M_2 is $1\,k\,\Omega$. V_{R2} is sent to the synaptic neuron as the forgetting voltage. U_1 and U_2 are voltages to synapse module. U_3 and U_4 are output voltages from the reward module. U_5 and U_6 are output voltages from the inhibition module.

3.2   Reward Module

Reward is positive reinforcement in operant conditioning. Positive reinforcement will promote the occurrence of a certain behavior. Reward will reinforce the behavior of pressing button until the dog learns to press the button. However, the behavior will not be strengthened if the reward does not follow the behavior. The reward module is designed to reinforce the synaptic weight between button neuron and output neuron in the neural network. The reward module is shown in Fig. 2. Both rewards are set to be equally attractive to dogs in this paper. The synaptic strength achieved by either of the two rewards is the same in the training. N_2 produces a $1.5\,v$ pulse with the width of $15\,s$ when the egg signal occurs. N_3 produces a $1.5\,v$ pulse with the width of $15\,s$ when the meat signal

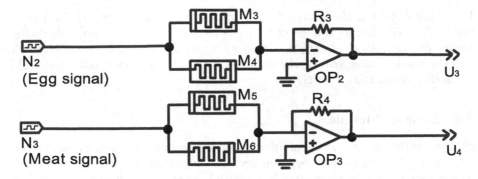

Fig. 2. Reward module. N_2 is egg signal. N_3 is meat signal. R_3 and R_4 are resistors. M_3, M_4, M_5 and M_6 are memristors. OP_2 and OP_3 are precision operational amplifiers.

occurs. The connection mode of M_3 is the opposite of M_4. When N_2 produces a positive voltage, the memristance of M_3 does not change and the memristance of M_4 begins to rise rapidly. The output voltage of OP_2 will decrease with the memristance of M_4 increases. U_3 is the output voltage of OP_2. U_4 is the output voltage of OP_3.

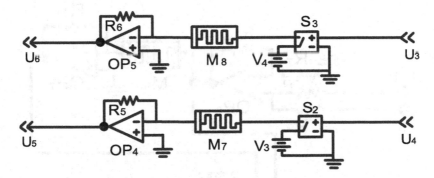

Fig. 3. Inhibition module. R_5 and R_6 are resistors. M_7 and M_8 are memristors. OP_4 and OP_5 are precision operational amplifiers. S_2 and S_3 are the voltage-controlled switches.

3.3 Inhibition Module

The overshadowing effect in operant conditioning is each reward will reduce the control exerted by the other. The inhibition module is the key module to realize the overshadowing effect in operant conditioning. The module is designed to release inhibition voltages between two rewards. Inhibition voltages are present when every reward is trained singlely. The inhibition module is shown in Fig. 3, which can realize the interaction between rewards. $U_3 > 0.1v$, S_3 is closed and

V_4 is exerted to M_8. $U_4 > 0.1v$, S_2 is closed and V_3 is exerted to M_7. V_3 and V_4 are $-1.5\,v$. The output voltage of OP_4 is $-(R_5/M_7) * V_3$. The lower M_7 is, the higher output voltage of OP_4 is. The output voltage of OP_5 is $-(R_6/M_8) * V_4$ and the voltage change rule is the same as OP_4. U_6 is the output voltage of OP_5. U_5 is the output voltage of OP_4.

3.4   Synapse Module

Synapse is the basic structure of information transmission in the nervous system. The strength of connections between neurons is determined by synaptic weights. The synapse module is shown in Fig. 4, which is designed to reflect the strength of the connection between the dog and button. Output voltages from other modules are given to synapse module to change the synaptic weight. The initial value of the synaptic weight of the neuron is $W = R_{11}/M_9$. The memristance of M_9 is regulated by the voltage of V_{OP6}. If V_{OP6} is greater than the positive threshold voltage of M_9, the memristance of M_9 will gradually decrease. The greater the voltage is, the slower the memristance of M_9 decreases. An opposite case occurs when V_{OP6} is lower than the negative threshold of M_9. ABM is a math module which is $V_{ABM} = -V_{IN2}/V_{IN1}$. OP_8 is as to a comparator.

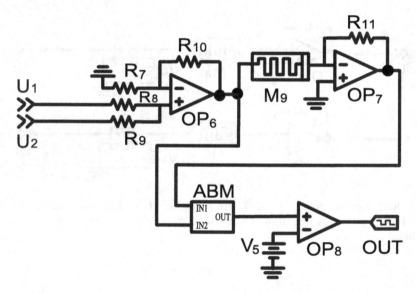

Fig. 4. Synapse module. R_7, R_8, R_9 R_{10} and R_{11} are resistors. M_9 is a memristor. OP_6, OP_7 and OP_8 are precision operational amplifiers.

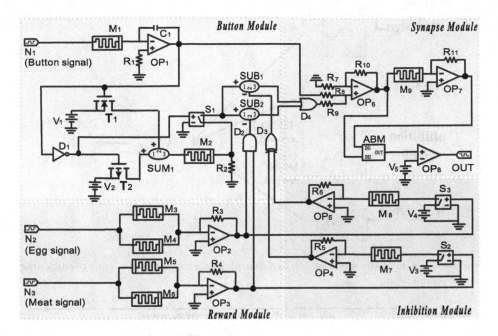

Fig. 5. Memristor-based neural network circuit of operant conditioning with overshadowing.

3.5 Complete Circuit

The memristor-based neural network circuit of operant conditioning with overshadowing is shown in Fig. 5, including button module, reward module, inhibition module and synapse module. The button module simulates the dog occasionally presses the button. The purpose of the reward module is to reinforce the behavior of pressing button. The inhibition module is to inhibit the control exerted by the other reward. The synapse module is to reflect the strength of the connection between the dog and button.

4 Experiment Design and Simulation Analysis

4.1 Experiment Introduction

As shown in Fig. 6, the memristor-based neural network circuit of operant conditioning with overshadowing contains five neurons ($N_1 - N_5$). N_1, N_2, N_3, N_4 and N_5 represent button neurons, egg neurons, meat neurons, inhibitory interneurons and output neurons, respectively. The neurons are interconnected by memristive synapses. The connection strength between neurons is represented by the synaptic weight W_{15}. The greater the synaptic weight, the stronger the connection strength between the dog and button. The red dotted arrows represent the

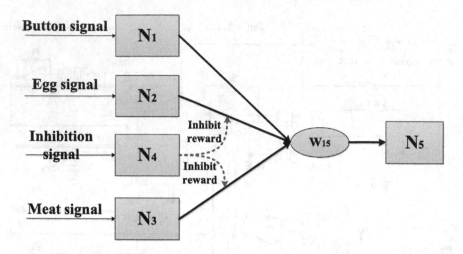

Fig. 6. Diagram of the memristor-based neural network circuit of operant conditioning with overshadowing.

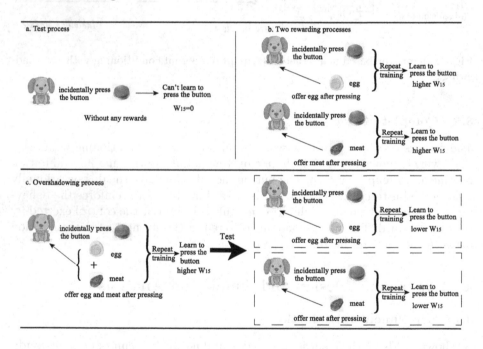

Fig. 7. Process of operant conditioning with overshadowing.

inhibitory signal inputs. Egg signal or meat signal performs associative learning singly, neuron N_4 will be activated and generate inhibitory signals.

The experiment process of operant conditioning with overshadowing is shown in Fig. 7. The dog will get eggs or meat when it occasionally presses the button in the cage. In process a, nothing happens when the dog presses the button occasionally. The dog can't learn to press the button. In process b, reward is given immediately when the dog presses the button, and then the training is repeated several times. The dog can learn to press the button after repeating trainings. In process c, two rewards are given simultaneously when the dog presses the button, and then the training is repeated several times. The dog will learn to press the button after repeating trainings. The value of W_{15} is higher. Then giving eggs or meat singly to the dog when it presses the button. The dog can learn to press the button but have lower W_{15}. This phenomenon shows that each reward will reduce the control exerted by the other and demonstrates the overshadowing effect in operant conditioning.

4.2 Simulation Results and Analyses

The simulation results of the memristor-based neural network circuit of operant conditioning with overshadowing in SPICE are shown in Fig. 8. At the beginning, N_1 generates a voltage of $10\,v$ with the width of $10\,s$ and period is $65\,s$. N_2 and N_3 generate a voltage of $1.5\,v$ with the width of $15\,s$ and period is $65\,s$. V_{S1} is the reinforcing voltage and it varies with the strength of the reward voltage. V_{OUT1} is the output voltage by two rewards. V_{OUT2} is the output voltage by a single reward. V_{OUT1} descends faster than V_{OUT2} because V_{OUT1} represents stronger synaptic weight. V_{OUT2} has lower synaptic weight because of the overshadowing effect in operant conditioning. The blue line represents the experiment results of two rewards, and the red line represents the experiment results of a single reward.

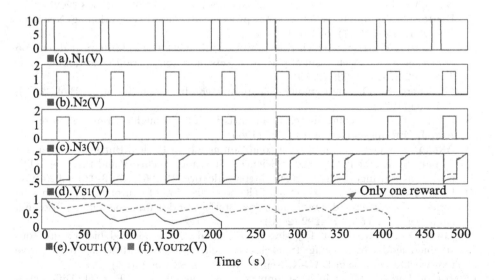

Fig. 8. The simulation result of operant conditioning with overshadowing. (a) Button signal. (b) Egg signal. (c) Meat signal. (d) The output voltage of S_1. (e) The output voltage of synapse module.

Y. Xu et al.

5 Conclusion

In order to improve the study of operant conditioning, the effect of positive reinforcement in operant conditioning on animal behavior is studied. In this work, a memristor-based neural network circuit of operant conditioning with overshadowing is constructed by designing the button module, reward module, inhibition module and synapse module. The circuit uses two rewards to reinforce behaviors, and SPICE simulation proves that the circuit can realize the overshadowing effect in operant conditioning. The memory characteristics of memristor and associative memory circuit are combined in this work, which broadens the application range of the circuit, and can build the memristive circuit flexibly according to different requirements. This work provides a reference for the development of operant conditioning with overshadowing and hopes to realize the blocking phenomenon because of the overshadowing effect in the future research.

References

1. Raaijmakers, J.G., Shiffrin, R.M.: Search of associative memory. Psychol. Rev. **88**(2), 93 (1981)
2. Reijmers, L.G., Perkins, B.L., Matsuo, N., et al.: Localization of a stable neural correlate of associative memory. Science **317**(5842), 1230–1233 (2007)
3. D'Souza, D., Avati, A.: Memory and learning: basic concepts. In: Thomas, K.A., Kureethara, J.V., Bhattacharyya, S. (eds.) Neuro-Systemic Applications in Learning, pp. 227–240. Springer, Cham (2021). https://doi.org/10.1007/978-3-030-72400-9_11
4. Nery, D., Palottini, F., Farina, W.M.: Classical olfactory conditioning promotes long-term memory and improves odor-cued flight orientation in the South American native bumblebee *Bombus pauloensis*. Curr. Zool. **67**(5), 561–563 (2021)
5. Bouton, M.E., Moody, E.W.: Memory processes in classical conditioning. Neurosci. Biobehav. Rev. **28**(7), 663–674 (2004)
6. Adamczyk, W.M., Wiercioch-Kuzianik, K., Bajcar, E.A., et al.: Rewarded placebo analgesia: a new mechanism of placebo effects based on operant conditioning. Eur. J. Pain **23**(5), 923–935 (2019)
7. Grossberg, S.: On the dynamics of operant conditioning. J. Theor. Biol. **33**(2), 225–255 (1971)
8. Hewett, F.M.: Teaching speech to an autistic child through operant conditioning. Am. J. Orthopsychiatry **35**(5), 927 (1965)
9. Akpan, B.: Classical and operant conditioning—Ivan Pavlov; Burrhus Skinner. In: Akpan, B., Kennedy, T.J. (eds.) Science Education in Theory and Practice. STE, pp. 71–84. Springer, Cham (2020). https://doi.org/10.1007/978-3-030-43620-9_6
10. Ito, H., Fujiki, S., Mori, Y., et al.: Self-reorganization of neuronal activation patterns in the cortex under brain-machine interface and neural operant conditioning. Neurosci. Res. **156**, 279–292 (2020)
11. Maffei, G., Santos-Pata, D., Marcos, E., et al.: An embodied biologically constrained model of foraging: from classical and operant conditioning to adaptive real-world behavior in DAC-X. Neural Netw. **72**, 88–108 (2015)
12. Jurado-Parras, M.T., Sánchez-Campusano, R., Castellanos, N.P., et al.: Differential contribution of hippocampal circuits to appetitive and consummatory behaviors during operant conditioning of behaving mice. J. Neurosci. **33**(6), 2293–2304 (2013)

13. Tanner, M.K., Davis, J.K.P., Jaime, J., et al.: Duration-and sex-dependent neural circuit control of voluntary physical activity. Psychopharmacology **239**(11), 3697–3709 (2022)
14. Strukov, D.B., Snider, G.S., Stewart, D.R., Williams, R.S.: The missing memristor found. Nature **453**(7191), 80–83 (2008)
15. Itoh, M., Chua, L.O.: Memristor oscillators. Int. J. Bifurcat. Chaos **18**(11), 3183–3206 (2008)
16. Li, B., Shan, Y., Hu, M., et al.: Memristor-based approximated computation. In: International Symposium on Low Power Electronics and Design (ISLPED), pp. 242–247, IEEE (2013)
17. Corinto, F., Forti, M.: Memristor circuits: bifurcations without parameters. IEEE Trans. Circuits Syst. I Regul. Pap. **64**(6), 1540–1551 (2017)
18. Sun, J., Han, J., Wang, Y., et al.: Memristor-based neural network circuit of operant conditioning accorded with biological feature. IEEE Trans. Circuits Syst. I Regul. Pap. **69**(11), 4475–4486 (2022)
19. Sun, J., Wang, Y., Liu, P., et al.: Memristor neural network circuit based on operant conditioning with immediacy and satiety. IEEE Trans. Biomed. Circuits Syst. (2022)
20. Yang, C., Wang, X., Chen, Z., et al.: Memristive circuit implementation of operant cascaded with classical conditioning. IEEE Trans. Biomed. Circuits Syst. **16**(5), 926–938 (2022)
21. Miles, C.G., Jenkins, H.M.: Overshadowing in operant conditioning as a function of discriminability. Learn. Motiv. **4**(1), 11–27 (1973)

Continual Learning with a Memory
of Non-similar Samples

Qilang Min[1,2], Juanjuan He[1,2(✉)], Liuyan Yang[1,2], and Yue Fu[1,2]

[1] College of Computer Science and Technology, Wuhan University of Science and Technology,
Wuhan 430081, China
{minqilang,hejuanjuan}@wust.edu.cn
[2] Hubei Province Key Laboratory of Intelligent Information Processing and Real-Time
Industrial System, Wuhan, China

Abstract. The replay method is an effective strategy to address the problem of
catastrophic forgetting, a key challenge for continuous learning. However, the
sample sets obtained by replay-based methods generally suffer from local data
information deficiencies. This problem leads to an imbalance in the plasticity-
stability of the model on older tasks. This paper proposes a novel method, called
non-similar sample storage (NSS). Non-similar refers to the Euclidean distance
for feature vectors of different samples being far. NSS extracts the feature vectors
of the samples and then calculates the similarity of the feature vectors for each
sample after the current task training. Samples that contribute less to the model
classification effect among similar samples are iteratively deleted, and the subset
of low-similarity samples is retained. Moreover, NSS reserves 30% of the storage
space for saving samples near the center of the sample set. Low-similarity samples
stored by NSS get larger losses during the replay process, leading to lower training
effectiveness of the current task. This paper introduces a knowledge distillation
strategy to solve this problem. A variable parameter was used to balance the
classification loss of the new task with the distillation loss of the old task (NSS-
D). Experimental results in CIFAR10 and imbalanced CIFAR10 show that NSS
maximizes the data's global information and can retain the model's ability to
recognize old tasks better. Compared with classical algorithms, NSS-D performs
better on CIFAR100 (48.8%) and ImageNet-200 (36.7%).

Keywords: Catastrophic Forgetting · Data Replay · Similarity

1 Introduction

Catastrophic forgetting [1, 2] is one of the critical challenges facing continuous learning. In many application scenarios, one needs to learn a series of tasks without access
to historical data, called continual learning (CL). Under continual learning [3, 4], the
parameters used to fit the old task classification model are trained again to fit the new
task classification model when learning a new task, causing catastrophic forgetting.
This classical problem keeps the neural network model from continually learning new
knowledge without forgetting the old knowledge. The model cannot continually learn.

Various advanced algorithms have been proposed to solve the problem of catastrophic forgetting. The literature [5] divides these algorithms into three main categories: Replay methods [6, 7], Regularization-based methods [8, 9], and Parameter isolation methods [10, 11]. Replay methods [12] expect to achieve a joint training-like effect by replaying old data or pseudo-data while learning a new task; Regularization-based methods [13] consolidate old knowledge by introducing an additional Regularization term when training a new task; Parameter isolation methods [14] dedicates different model parameters to each task to prevent forgetting. Replay methods and Parameter isolation methods are similar ideas, which use extra memory to store samples or network parameters and reduce memory requirements by various advanced techniques of compressing storage. These methods yield better results, particularly Parameter isolation methods achieve incredibly high performance. However, the memory requirement increases linearly with the number of tasks. Regularization-based strategies overcome this problem. Nevertheless, the soft penalty introduced is insufficient to restrict the optimization process to stay in the feasible region of previous tasks [15], which may sometimes lead to an increase in forgetting the previous task [5]. Replay methods generally perform better than Regularization-based.

In this paper, we focus on continuous learning through data replay. Although many approaches to data replay have made significant contributions, they generally suffer from local information deficits due to the uneven distribution of sample subsets. This makes the model plasticity-stability [16] imbalanced on old tasks. Our goal is to preserve the global information of the old data to the maximum extent. To this end, this paper proposes Non-Similar Sample Storage (NSS). Non-similar refers to the similarity of data feature vectors being low. We use the Euclidean distance for the similarity measure [17]. The larger the value, the more non-similar the sample is. NSS extracts the feature vectors of the samples and then calculates the similarity of the feature vectors for each sample after the current task training. Iteratively, the samples that contribute less to the classification effect of the model among similar samples are deleted to obtain a subset of non-similar samples. Here contribution value we measure the average distance from samples to other samples. Meanwhile, we reserve 30% of the storage space for storing samples near the center of the sample set. This allows for maximum preservation of the global information of the old data while maintaining the stability of the old data distribution. The experiments on CIFAR10 illustrate that NSS is efficient and reliable and retains the model's ability to recognize old tasks better. We further test the general adaptability of NSS to imbalanced datasets at imbalanced [18, 19] CIFAR10. Low-similarity samples stored in the NSS get a large loss during the replay process, resulting in reduced training effectiveness for the current task. This paper introduces the knowledge distillation strategy [20, 21] to solve this problem. A variable parameter is used to balance the classification loss of the new task with the distillation loss of the old task (NSS-D). Experimental results on CIFAR100 and ImageNet-200 show that the proposed algorithm performs better than most classical methods.

In a word, our main contributions concern:

- This paper proposes a novel data replay-based approach that maximally preserves the global information of old data, called non-similar sample storage (NSS).

- This paper introduces a custom variable parameter to balance the distillation loss of the old task and the classification loss of the new task.
- The general adaptation of current classical storage strategies to NSS on CIFAR10 and imbalanced CIFAR10 is tested. NSS-D performs better on CIFAR100 and ImageNet-200 compared to the most classical approach.

2 Related Work

A series of different ways of storage strategies have arisen with data replay methods. They contribute significantly to solving the problem of catastrophic forgetting. However, the problem that sample subsets cannot represent the global information of the old data is prevalent in these methods. This leads them to forget information about the missing data during training.

Herding. This method is a strategy proposed by iCaRL [6]. Herding preserves the nearest neighbor samples from the center of the sample and removes the edge samples. The process iteratively selects a subset of samples for each class by calculating the distance between each sample and class mean and chooses the subset closest to the class mean in the learned feature space. The subset of samples obtained by the Herding strategy is highly sequential, and the earlier the samples are stored, the closer they are to the center of the training samples. Therefore, when some samples need to be removed to make room for new ones, the ones added later in the same class will be selected first.

Distance and Entropy. They are proposed in RWalk [22]; they preserve the data calculated in a specific way and lose the data of the opposite strategy. Distance selects samples by calculating the distance between the sample and the decision boundary. For a given sample, the inverse distance from the sample to the class mean is calculated, approximating the distance from the sample to the decision boundary instead, meaning that the smaller the distance, the farther it is from the decision boundary. The entropy strategy calculates the entropy of the SoftMax output and selects the samples with higher entropy. Since the SoftMax layer has been removed from our network, we artificially add a SoftMax output that does not affect the current network when calculating the entropy. In [23], the inverse form of distance and entropy is proposed. It is worth noting that inverse distance is particularly similar to herding, but differs in that inverse distance iteratively selects the single sample closest to the class mean, while herding selects the overall sample set closest to the class mean.

Random. This sampling strategy is the simplest way to select samples to add to memory. The reservoir sampling algorithm generates random numbers from all training samples to ensure that each sample has the same probability of being randomly selected. It maintains the distribution of the original data as much as possible. However, due to the random uncertainty, it is highly uncertain whether the remaining samples still represent the characteristics of the original data of that class as the samples are gradually reduced.

Preserving the data's global information maintains the model's ability to recognize old data. The model will overfit the sample subsets during training, thus forgetting the information expressed by the missing samples. Herding stores, the overall sample subsets closest to the sample center, forgetting the information of the edge data during training. Distance keeps the decision boundary data and forgets the information about the data at the center of the sample. Entropy stores the samples with higher entropy and forgets the information of samples with lower entropy. Random retains the global information as much as possible, but its random uncertainty leads to a rapid decline in model robustness as the buffer size decreases. NSS calculates the similarity between samples and retains the samples contributing more to the model classification effect among similar samples. Finally, the subset of low-similarity samples is obtained. This sample subset can maximally replace the old data. The problem of missing local information on the old data is solved. Moreover, NSS reserves 30% of the storage space for saving the samples near the center of the sample set. It can maximally preserve the global information of the old data while maintaining the stability of the old data distribution.

3 Non-similar Sample Storage and Distillation Loss

We aim to preserve the global information of the old data using sample subsets to reduce the forgetting of the information of the old data that is not preserved during the training process. In addition, it is necessary to balance the mutual interference of old and new data during training. In this section, we describe our algorithm in detail.

3.1 Non-similar Sample Storage

To solve the problem of missing local information due to uneven subsets of samples, we propose a memory management strategy, called non-similar sample storage (NSS). NSS maintains the diversity of samples by calculating the similarity between samples and iteratively culling similar samples. Finally, we obtain a subset of samples with low similarity. Simultaneously, we reserve 30% of the storage space for saving the center samples in the original sample set. The obtained subset of samples retains the global information of the old data to the maximum extent under the fixed storage capacity.

We argue that the knowledge obtained by the model from similar samples is similar. Similar samples get similar outputs through the model, and the presence of multiple similar samples contributes less to the classification effect of the model. Diverse samples are more representative of the global information of the data. Diverse samples also produce larger losses in the training process, making the network parameter distribution biased toward the class to which they belong. In conclusion, we believe that only a small number of similar samples should be retained that contribute more to the classification effect of the model.

Algorithm 1 Non-similar Sample Storage

input $X_s, \ldots, X_t$ // training examples
input M // memory size
require θ // current model parameters
require $P = (P_1, \ldots, P_{s-1})$ // current exemplar sets
  $\theta \leftarrow Train(X_s, \ldots, X_t; P, \theta)$
  $m \leftarrow 0.7 * M/(t*2)$ // number of exemplars per class
  for $y = 1, \ldots, s-1$ **do**
    $P_y \leftarrow delete\_sub - sample(P_y, m)$
  end for
  for $y = s, \ldots, t$ **do**
    $f \leftarrow Extract\_features(X_s, \ldots, X_t, \theta)$
    $P_y \leftarrow Similarity\_selection(X_y, m, f)$
    $P_y \leftarrow Add\_data(X_y, m)$
  end for
  $P \leftarrow (P_1, \ldots, P_t)$ // new exemplar sets

To obtain different samples, i.e., diverse samples, NSS calculates the similarity of each sample to the other samples. Here, we use Euclidean distance to calculate the similarity between samples.

Algorithm 1 summarizes our proposed sample storage update strategy. Specifically, we introduce $P_y \leftarrow similarty\_selecton(X_y, m, f)$. We train the current network model and save the current model parameters at the end of training. Using the saved model parameters θ we perform feature extraction to obtain the feature vector of all current data:

$$F_i = (x_{i1}, x_{i2}, \ldots, x_{ip})^T \tag{1}$$

where p is the feature dimension. Calculating the Euclidean distance d from each data to other data of the same class:

$$d(F_i, F_j) = \sqrt{\sum_{k=1}^{p} (x_{i1} - x_{j1})^2} \tag{2}$$

Finding the two data with the lowest similarity X_i, X_j. Calculating the sum of the distances from X_i and X_i to other similar samples respectively and iteratively remove the minimum value:

$$delete \leftarrow \min(d_{sum}^i = \sum_{s \neq i,j} \sqrt{\sum_{k=1}^{p} (x_{ik} - x_{sk})^2}, d_{sum}^j = \sum_{s \neq i,j} \sqrt{\sum_{k=1}^{p} (x_{jk} - x_{sk})^2}) \tag{3}$$

Record the order of deleting samples and save the last x samples deleted. When new data needs to be stored, continue deleting samples to make room for new ones. In addition, we set aside 30% of the storage space to save the heart samples in the sample set, citing the herding sampling strategy to obtain samples.

3.2 Distillation Loss

Let us briefly review the knowledge distillation strategy [20]. Knowledge distillation uses the Teacher-Student model, where the teacher is the exporter of "knowledge" and the student is the recipient of "knowledge". The process of knowledge distillation is divided into two stages:

- Training "Teacher model": abbreviated as Net-T, it is characterized by a relatively complex model and can be integrated by several separately trained models. The only requirement is that for each input X, the output Y, where Y is mapped by SoftMax, corresponds to the probability value of the corresponding category.
- Training "Student model": abbreviated as Net-S, it is a single model with a small number of parameters and a relatively simple model structure. Similarly, input X can output Y, and Y can also output the probability value corresponding to the corresponding category after SoftMax mapping.

Following the knowledge distillation strategy proposed by LWF in 2017 iCaRL extended knowledge distillation to the form of distillation based on old data. The model trained on the old data as Net-T and the model trained on the new data as Net-S. During the training process, the classification loss of the new data is expressed in combination with the distillation loss of the old data as:

$$\ell(\Theta) = - \sum_{(x_i,y_i)\in D} \left[\sum_{y=s}^{t} \delta_{y=y_i} \log g_y(x_i) + \delta_{y\neq y_i} \log(1 - g_y(x_i)) \right.$$
$$\left. + \sum_{y=1}^{s-1} q_i^y \log g_y(x_i) + (1 - q_i^y) \log(1 - g_y(x_i)) \right] \tag{4}$$

The notation of Formula 4 can be found in iCaRL [6]. Formula 4 can be written formally as follows:

$$L(\theta) = \alpha L_{classification}(\theta) + \beta L_{distillation}(\theta) \tag{5}$$

where α and β are reconciling factors. With the continuous learning setting, tasks arrive in order. Let the new task be T_1, and the old task is T_2:

$$P(y_1 \cdots y_k | f_\theta(x)) = P(T_1, T_2 | f_\theta(x))$$
$$= P(T_1 | f_\theta(x)) \cdot P(T_2 | f_\theta(x)) \tag{6}$$

where $y_1 \cdots y_k$ is all classes (k) of T_1 and T_2. $f_\theta(x)$ is the output obtained by passing the data x of each class through the parametric model. In maximum likelihood inference, we maximize the log-likelihood of the model.

$$\log P(T_1 | f_\theta(X_1)) \propto - \sum_{i=1}^{n} \|y_i - f_\theta(x_i)\|^2 \tag{7}$$

where x_i is all the data of y_i. A regularization item can usually be introduced to retain old knowledge for the model.

$$\log P(T_1|f_\theta(X_1)) \propto -\sum_{i=1}^{n} \|y_i - f_\theta(x_i)\|^2 + \lambda\|w\|_2^2 \tag{8}$$

where w is the weight. Here we do not discuss the form of λ, and L2 regularization [24] explains it. We use the old data instead of λw_2^2:

$$\log P(T_1, T_2|f_\theta(X)) \propto -\sum_{(x_i,y_i)\in T_1} \|y_i - f_{\theta_1}(x_i)\|^2 - \sum_{(x_i,y_i)\in T_2} \|y_i - f_{\theta_2}(x_i)\|^2 \tag{9}$$

To balance the effect of both on the variation of the model parameters, we add some weights:

$$\log P(T_1, T_2|f_\theta(X)) \propto -\frac{\lambda_1\|\Delta\theta_2\|_2^2}{\|\Delta\theta_1\|_2^2 + \|\Delta\theta_2\|_2^2 + \xi} \sum_{(x_i,y_i)\in T_1} \|y_i - f_{\theta_1}(x_i)\|^2$$
$$-\frac{\lambda_2\|\Delta\theta_1\|_2^2}{\|\Delta\theta_1\|_2^2 + \|\Delta\theta_2\|_2^2 + \xi} \sum_{(x_i,y_i)\in T_2} \|y_i - f_{\theta_2}(x_i)\|^2 \tag{10}$$

where $\Delta\theta_i$ is the variation of the model parameters obtained from the data of the training task T_i relative to the old model parameters, and ξ is a minimal value to prevent division by 0 in the calculation process. Let $\alpha = \frac{\|\Delta\theta_2\|_2^2}{\|\Delta\theta_1\|_2^2 + \|\Delta\theta_2\|_2^2 + \xi}$:

$$\log P(T_1, T_2|f_\theta(X)) \propto -\lambda_1\alpha \sum_{(x_i,y_i)\in T_1} \|y_i - f_{\theta 1}(x_i)\|^2$$
$$-\lambda_2(1 - \alpha) \sum_{(x_i,y_i)\in T_2} \|y_i - f_{\theta 2}(x_i)\|^2 \tag{11}$$

$$L(\theta) = \lambda_1\alpha \sum_{(x_i,y_i)\in T_1} \|y_i - f_{\theta 1}(x_i)\|^2 + \lambda_2(1 - \alpha) \sum_{(x_i,y_i)\in T_2} \|y_i - f_{\theta 2}(x_i)\|^2 \tag{12}$$
$$= \lambda_1\alpha L1 + \lambda_2(1 - \alpha)L2$$

where λ_1 and λ_2 are super parameters, L_1 and L_2 are the loss of T_1 and T_2 respectively. Smaller weights are given to the losses corresponding to samples with large variations in model parameters during the training process so that the loss imbalance effect can be reduced.

4 Experiments

This section compares NSS with multiple sampling strategies on the CIFAR10 and imbalanced CIFAR10 datasets. It is used to test the effectiveness of the NSS algorithm and its adaptability to unbalanced datasets. In addition, we compare NSS-D with the current baseline of the most classical continuous learning methods on CIFAR100 and ImageNet-200.

4.1 Effectiveness and General Adaptability of NSS

Datasets and Metrics. We use CIFAR10 to configure CL task setups for evaluations. We randomly divide CIFAR10 into five tasks to generate a CL task setup. Each task includes two classes of data. We use a popular metric in the literature, Last Accuracy (A5), "Last" refers to the value measured after all tasks are learned, and we denote it with the number "5" here because CIFAR10 has five tasks.

$$Acc = \frac{1}{t} \sum\nolimits_{i=1}^{t} a_{t,i} \tag{13}$$

where $a_{t,i}$ is the test accuracy of task i after training task t.

To create data imbalance, we proportionally cull the number of samples in a class for each task. We refer to the majority of samples in each task as positive samples and the minority of samples as negative samples. r is the sample imbalance ratio and is used here as $R = positive_{num}/negative_{num}$, R = 2, 5, 10. we use the mean F1-score [25] of all classes to evaluate the performance of different sampling algorithms.

$$F1 - score = \frac{1}{n} \sum_{i=1}^{n} \frac{2 \cdot precision_i \cdot recall_i}{precision_i + recall_i} \tag{14}$$

Baselines. We compare our proposed NSS with the standard Sampling strategy including random, herding, distance, inverse distance, entropy, and inverse entropy. All experiments were performed in the same backbone, modifying only the model's sampling strategy and memory size (K).

Table 1. Average accuracy of different sampling strategies on CIFAR10

	K = 200	K = 500	K = 1000
Entropy	57.94 ± 2.88	63.42 ± 1.49	67.93 ± 2.48
Inverse entropy	49.91 ± 4.27	59.00 ± 2.76	63.98 ± 0.79
Herding	59.20 ± 1.50	65.73 ± 3.56	69.96 ± 1.98
Distance	50.77 ± 3.06	59.82 ± 2.65	60.71 ± 2.55
Inverse distance	**59.80 ± 2.91**	65.46 ± 2.03	69.00 ± 3.92
Random	47.74 ± 1.34	58.84 ± 1.76	68.07 ± 2.05
NSS	59.37 ± 2.12	**67.11 ± 1.53**	**73.12 ± 1.31**

Experimental results in Table 1 and Fig. 1 show that NSS can better preserve the global information of samples and solve the problem of missing information of unsaved samples. Herding and inverse distance are similar in that they both preserve the samples closest to the center of the sample set. In contrast, distance keeps the boundary samples. It can be inferred from the experimental results that not all samples are equally important. Sample-centered samples contribute more to model accuracy than boundary samples for the current training model and CIFAR10 datasets. Entropy uses the maximum entropy algorithm, and it can be concluded from the experimental results that entropy performs better than inverse entropy on CIFAR10. Random samples have better results when the memory size K is larger. Random samples can get better results when the cache capacity K is larger. However, when K is small, the performance of the random algorithm decreases significantly.

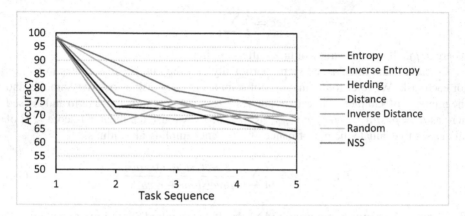

Fig. 1. The average accuracy of all tasks after training task T on CIFAR10 (K = 1000).

Table 2. Mean F1-score for different sampling strategies on imbalanced CIFAR10 (K = 500)

	R = 2	R = 5	R = 10
Entropy	0.61 ± 0.02	0.53 ± 0.02	0.51 ± 0.01
Inverse entropy	0.52 ± 0.04	0.49 ± 0.03	0.45 ± 0.04
Herding	0.63 ± 0.02	0.54 ± 0.05	0.54 ± 0.03
Distance	0.51 ± 0.03	0.46 ± 0.01	0.41 ± 0.02
Inverse distance	0.62 ± 0.01	0.57 ± 0.02	0.52 ± 0.02
Random	0.58 ± 0.02	0.57 ± 0.02	0.55 ± 0.04
NSS	**0.64 ± 0.02**	**0.61 ± 0.02**	**0.57 ± 0.03**

Experimental results in Table 2 show roughly the same conclusions as in Table 1. All methods are universally adaptable on imbalanced CIFAR10. Surprisingly, random gradually show an advantage with increasing R. Random's equal-probability storage strategy makes it store a subset of samples that also roughly exhibit a similar proportional relationship with R values. We believe that an unbalanced subset of samples will lead to significantly lower prediction results due to the low number of negative samples in the training process. However, it can be inferred from the experimental results of random that the number of positive samples dominates and masks the poor classification effect of negative samples. This can be verified by outputting the F1-score of each category.

4.2 Experimental Setup of NSS-D

Datasets and Metrics. We use CIFAR100 and ImageNet-200 to configure CL task setups for evaluations. We randomly divide CIFAR100 and ImageNet-200 into 10 tasks to generate a CL task setup. Thus, each task has 10 and 20 major classes for CIFAR100 and ImageNet-200 datasets. Also, we use Last Accuracy and Last Forgetting as the evaluation metric.

$$F = \frac{1}{t} \sum_{i=1}^{t} \frac{(a_{t,i} - a_{i,i})}{a_{i,i}} \tag{15}$$

where the notation for $a_{t,i}$ can be found in Formula 13.

Baselines. NSS-D is compared with three Knowledge Distillation-based methods (iCaRL, LWF), one regularization-based method (SI), and four rehearsal-based methods (ER, GSS, FDR, HAL). We further provide a lower bound for finetuning without any forgetting countermeasures, and an upper bound given by the joint training of all tasks (JOINT). Table 4 reports the average precision and forgetting rate at the end of all tasks. The average results were calculated for ten runs, each involving a different initialization.

Experimental results in Table 3 illustrate that our method yields higher accuracy and less forgetting than other methods. This is attributed to the fact that we maximally retain the global information of the old data while preserving the central sample. As well as using reasonable parameters to balance the effect of different losses on the variation of model parameters during backpropagation. When the memory size is increased from 500 to 5000, ER performance improves significantly. This is because ER is a replay method based on random policy, and ER performance depends on the size of the cache capacity and the number of various types of data in the cache. iCaRL resists forgetting to some extent using the knowledge distillation strategy. FDR does not perform well resisting partial forgetting, the same as ER, and the increase in cache capacity also brings a performance gain to FDR. Due to the excessive training time of GSS and HAL, we only conduct experiments on CIFAR100. We also observe that HAL and GSS show unsatisfactory results. LwF and SI do not use replay data. They aim at parameter regularization, which avoids the dependence on old data, but is generally less effective than replay-based methods, as we have concluded from other papers.

5 Ablation Experiment

To further understand the working mechanism of NSS and NSSD. We conducted additional experiments on CIFAR10 and CIFAR100. We analyze why our method improves over plain finetuning-based training, which differs in two aspects: by using a non-similar sample strategy and distillation loss. We, therefore, set up three hybrid setups: the first using neither NSS nor distillation loss, the second using NSS but not distillation loss, and the third using both NSS and distillation loss.

Table 4 explains the performance of our method gains simultaneously from NSS with distillation loss. Their synergy results in a better performance of the model NSS retains the global information of the old data to the maximum extent while retaining the central sample. And distillation loss can better overcome the problem of transitional fitting of sample subsets to the model.

Table 3. Comparison with some current classical methods on CIFAR100 and ImageNet-200

Memory	Method	CIFAR100		ImageNet-200	
		Accuracy	Forgetting	Accuracy	Forgetting
____	Joint	70.28 ± 0.81	0	59.99 ± 0.19	0
	Finetuning	9.01 ± 0.28	1	7.92 ± 0.26	1
	SI [26]	19.48 ± 0.17	0.35 ± 0.02	6.58 ± 0.31	0.87 ± 0.01
	LwF [27]	19.61 ± 0.05	0.41 ± 0.02	8.46 ± 0.22	0.36 ± 0.01
500	ER [12]	21.21 ± 1.55	0.82 ± 0.02	9.99 ± 0.29	0.92 ± 0.03
	iCaRL [6]	19.84 ± 0.66	0.34 ± 0.01	17.38 ± 1.53	0.47 ± 0.01
	FDR [28]	22.21 ± 0.29	0.78 ± 0.01	10.54 ± 0.21	0.97 ± 0.01
	GSS [29]	13.53 ± 1.54	0.95 ± 0.00	____	____
	HAL [30]	11.55 ± 0.73	0.90 ± 0.01	____	____
	NSS-D	**27.32 ± 2.35**	**0.62 ± 0.01**	**20.45 ± 1.21**	**0.82 ± 0.02**
5000	ER	43.61 ± 0.30	0.45 ± 0.02	27.40 ± 0.31	0.72 ± 0.03
	iCaRL	41.93 ± 0.21	0.27 ± 0.00	14.08 ± 1.92	0.36 ± 0.01
	FDR	30.34 ± 1.28	0.72 ± 0.01	28.97 ± 0.41	0.70 ± 0.01
	GSS	17.51 ± 0.38	0.90 ± 0.00	____	____
	HAL	18.69 ± 0.82	0.69 ± 0.03	____	____
	NSS-D	**48.80 ± 0.55**	**0.29 ± 0.03**	**36.73 ± 0.64**	**0.47 ± 0.02**

Table 4. Average accuracy on CIFAR10 and CIFAR100 for different settings

	CIFAR10		CIFAR100	
	K = 500	K = 1000	K = 500	K = 5000
Finetuning	19.62 ± 0.05	19.62 ± 0.05	9.01 ± 0.28	9.01 ± 0.28
NSS	67.11 ± 1.53	73.12 ± 1.31	21.04 ± 1.71	45.65 ± 0.82
NSS-D	69.25 ± 0.81	77.38 ± 1.10	27.32 ± 2.35	48.80 ± 0.55

6 Conclusions

This paper proposes a novel sampling strategy to solve the problem that the subset of stored samples does not represent the global information of old data. The model overfits the stored samples during the training process, which makes the local information of the old data missing and leads to the reduced recognition ability of the model for the old task.

NSS extracts the feature vectors of the samples, then calculates the feature vectors' similarity for each sample after the current task training, and keeps the non-similar sample sets. Meanwhile, NSS reserves 30% of the storage space for storing the samples near the center of the sample set. This can preserve the global information of old data as much as possible while maintaining the stability of the model. In addition, this paper introduces a knowledge distillation strategy for the problem of a large loss of dissimilar samples in the training process. A variable parameter is introduced to balance the classification loss of the new task with the distillation loss of the old task. Experimental results on CIFAR10 and imbalanced CIFAR10 demonstrate the effectiveness and generalizability of NSS. In addition, NSS-D achieves higher accuracy and a lower forgetting rate on CIFAR100 and ImageNet-200.

Acknowledgments. This work was supported by the National Natural Science Foundation of China under Grant 62272355, 61702383, and 62176191.

References

1. French, R.M.: Catastrophic forgetting in connectionist networks. Trends Cogn. Sci. **3**(4), 128–135 (1999)
2. Kemker, R., McClure, M., Abitino, A., Hayes, T., Kanan, C.: Measuring catastrophic forgetting in neural networks. In: Proceedings of the AAAI Conference on Artificial Intelligence, vol. 32, No. 1 (2018)
3. Aljundi, R., Babiloni, F., Elhoseiny, M., Rohrbach, M., Tuytelaars, T.: Memory aware synapses. learning what (not) to forget. In: Ferrari, V., Hebert, M., Sminchisescu, C., Weiss, Y. (eds.) ECCV 2018. LNCS, vol. 11207, pp. 144–161. Springer, Cham (2018). https://doi.org/10.1007/978-3-030-01219-9_9
4. Van de Ven, G.M., Tolias, A.S.: Three scenarios for continual learning. arXiv preprint arXiv: 1904.07734 (2019)
5. De Lange, M., et al.: A continual learning survey: defying forgetting in classification tasks. IEEE Trans. Pattern Anal. Mach. Intell. **44**(7), 3366–3385 (2021)
6. Rebuffi, S.A., Kolesnikov, A., Sperl, G., Lampert, C.H.: iCaRL: incremental classifier and representation learning. In: Proceedings of the IEEE Conference on Computer Vision and Pattern Recognition, pp. 2001–2010 (2017)
7. Isele, D., Cosgun, A.: Selective experience replay for lifelong learning. In: Proceedings of the AAAI Conference on Artificial Intelligence, vol. 32, No. 1 (2018)
8. Tang, S., Su, P., Chen, D., Ouyang, W.: Gradient regularized contrastive learning for continual domain adaptation. In: Proceedings of the AAAI Conference on Artificial Intelligence, vol. 35, No. 3, pp. 2665–2673 (2021)
9. Kirkpatrick, J., et al.: Overcoming catastrophic forgetting in neural networks. Proc. Natl. Acad. Sci. **114**(13), 3521–3526 (2017)
10. Rusu, A.A., et al.: Progressive neural networks. arXiv preprint arXiv:1606.04671 (2016)
11. Aljundi, R., Chakravarty, P., Tuytelaars, T.: Expert gate: lifelong learning with a network of experts. In: Proceedings of the IEEE Conference on Computer Vision and Pattern Recognition, pp. 3366–3375 (2017)
12. Rolnick, D., Ahuja, A., Schwarz, J., Lillicrap, T., Wayne, G.: Experience replay for continual learning. Adv. Neural Inf. Process. Syst. **32** (2019)

13. Rannen, A., Aljundi, R., Blaschko, M.B., Tuytelaars, T.: Encoder based lifelong learning. In: Proceedings of the IEEE International Conference on Computer Vision, pp. 1320–1328 (2017)
14. Fernando, C., et al.: Pathnet: evolution channels gradient descent in super neural networks. arXiv preprint arXiv:1701.08734 (2017)
15. Farquhar, S., Gal, Y.: Towards robust evaluations of continual learning. arXiv preprint arXiv: 1805.09733 (2018)
16. Takesian, A.E., Hensch, T.K.: Balancing plasticity/stability across brain development. Prog. Brain Res. **207**, 3–34 (2013)
17. Lin, Y.S., Jiang, J.Y., Lee, S.J.: A similarity measure for text classification and clustering. IEEE Trans. Knowl. Data Eng. **26**(7), 1575–1590 (2013)
18. He, H., Garcia, E.A.: Learning from imbalanced data. IEEE Trans. Knowl. Data Eng. **21**(9), 1263–1284 (2009)
19. Sun, Y., Wong, A.K., Kamel, M.S.: Classification of imbalanced data: a review. Int. J. Pattern Recognit. Artif. Intell. **23**(04), 687–719 (2009)
20. Gou, J., Yu, B., Maybank, S.J., Tao, D.: Knowledge distillation: a survey. Int. J. Comput. Vis. **129**(6), 1789–1819 (2021)
21. Kim, Y., Rush, A.M.: Sequence-level knowledge distillation. arXiv preprint arXiv:1606.07947 (2016)
22. Chaudhry, A., Dokania, P.K., Ajanthan, T., Torr, P.H.S.: Riemannian walk for incremental learning: understanding forgetting and intransigence. In: Ferrari, V., Hebert, M., Sminchisescu, C., Weiss, Y. (eds.) ECCV 2018. LNCS, vol. 11215, pp. 556–572. Springer, Cham (2018). https://doi.org/10.1007/978-3-030-01252-6_33
23. Masana, M., Liu, X., Twardowski, B., Menta, M., Bagdanov, A.D., van de Weijer, J.: Class-incremental learning: survey and performance evaluation on image classification. IEEE Trans. Pattern Anal. Mach. Intell. (2022)
24. Cortes, C., Mohri, M., Rostamizadeh, A.: L2 regularization for learning kernels. arXiv preprint arXiv:1205.2653 (2012)
25. Chicco, D., Jurman, G.: The advantages of the Matthews correlation coefficient (MCC) over F1 score and accuracy in binary classification evaluation. BMC Genomics **21**(1), 1–13 (2020)
26. Zenke, F., Poole, B., Ganguli, S.: Continual learning through synaptic intelligence. In: International Conference on Machine Learning, pp. 3987–3995. PMLR (2017)
27. Li, Z., Hoiem, D.: Learning without forgetting. IEEE Trans. Pattern Anal. Mach. Intell. **40**(12), 2935–2947 (2017)
28. Benjamin, A.S., Rolnick, D., Kording, K.: Measuring and regularizing networks in function space. arXiv preprint arXiv:1805.08289 (2018)
29. Aljundi, R., Lin, M., Goujaud, B., Bengio, Y.: Gradient based sample selection for online continual learning. Adv. Neural Inf. Process. Syst. **32** (2019)
30. Chaudhry, A., Gordo, A., Dokania, P., Torr, P., Lopez-Paz, D.: Using hindsight to anchor past knowledge in continual learning. In: Proceedings of the AAAI Conference on Artificial Intelligence, vol. 35, No. 8, pp. 6993–7001 (2021)

Research on Soil Moisture Prediction Based on LSTM-Transformer Model

Tao Zhou[1,2] (ID), Yuanxin He[1,2], Liang Luo[1,2](✉), and Shengchen Ji[1,2]

[1] Wuhan University of Technology, Wuhan 430000, China
luoliang@whut.edu.cn
[2] Ministry of Education Key Laboratory of High Performance Ship Technology,
Wuhan 430000, China

Abstract. Soil moisture is one of the basic climate variables of the global climate observation system, and the prediction of soil moisture is of great significance for agricultural yield assessment, flood and drought prediction, and soil and water conservation. Aiming at the complexity of soil moisture influencing factors and their time-varying time series characteristics, we propose a Transformer model that introduces LSTM, which uses the sequential modeling capability of LSTM to extract contextual information for each data, and plays the role of position coding in the LSTM-Transformer model, and the multi-head attention mechanism in the model can highlight important features by weighting, so as to effectively process time series data. Taking soil moisture, soil evaporation, vegetation index, runoff and climate data at different depths of Xilin Gol grassland in Inner Mongolia from 2012 to 2022 as input variables, soil moisture at different depths from 2022 to 2023 was predicted, and the model prediction performance was compared with the traditional long short-term memory neural network (LSTM) and bidirectional long short-term memory neural network (BiLSTM) through the three statistical indicators of MAE, MAPE and RMSE. The LSTM-Transformer model has better performance for prediction of soil moisture at different depths. The prediction of soil moisture has great guiding significance for timely grasping grassland soil moisture and adopting proactive agricultural production.

Keywords: Soil moisture · Forecast · Time series · LSTM-Transformer

1 Introduction

Soil Wet (SW), as an important indicator to quantify the wetness and dryness of surface soil, has an important impact on ecology [1–3]. Soil moisture determines the water supply of plants, and too high or too low can adversely affect crops [4]. When the soil water content is too low, it is difficult for plant roots to absorb enough water from the soil to compensate for transpiration consumption. The lack of sufficient water affects the photosynthesis of crops, resulting in reduced crop quality and yield, or even withering and death. When the soil water content is too high, it will limit the soil aeration, which will not only affect the respiration and growth of crop roots, but also hinder the growth and development of the aboveground part of the crop. With 25% of the world's

L. Pan et al. (Eds.): BIC-TA 2022, CCIS 1801, pp. 329–342, 2023.
https://doi.org/10.1007/978-981-99-1549-1_26

land area, grasslands are an important terrestrial component and one of the planet's largest reservoirs of carbon. Grassland can regulate climate, conserve water sources, reduce soil erosion, improve ecological environment, maintain biodiversity and other functions, which makes grassland have special strategic significance in ecological security, agricultural development, people's livelihood and other aspects [5, 6]. Based on the important influence of soil moisture content on the growth and development of grassland plants, it is of great significance to obtain grassland soil moisture data. At present, soil moisture data is mainly obtained through two ways: sensor on-site measurement and satellite remote sensing technology [7]. Measuring soil moisture content using sensors in the field can obtain more accurate results, but this method is time-consuming and labor-intensive [8]. Remote sensing satellites can obtain soil moisture data on a large area of the earth because they observe the earth's surface from space. However, satellite remote sensing technology has certain limitations in measuring depth, which makes it difficult to obtain soil moisture data in the root zone [9].

In order to guide the production activities in the grassland area, it is necessary to make reasonable and effective predictions of future soil moisture data and their change trends. At present, scholars at home and abroad mainly have the following methods [10] to predict soil moisture: Practical experience, water balance, soil dynamics [11], time series [12–14], remote sensing detection [15–17] and neural network prediction method. In 2017, Oliviu and colleagues used data mining techniques to mine meteorological data from weather stations and establish a soil moisture prediction system to dynamically predict soil moisture in real time on the second day. In actual tests, the system can get accurate prediction results and operate stably. In 2018, Shikha [18] compared mainstream machine learning algorithms such as multiple linear regression and recurrent neural networks and used these algorithms to predict soil moisture values on days 1, 2, and 7. According to the obtained regression coefficient index, multiple linear regression is more accurate in soil moisture prediction. In 2019, Gursimran Singh [19] used smart irrigation as a framework to predict the future soil moisture in fields. The framework uses ML technology and is driven by the Internet of Things to realize the optimization of irrigation water in this area. In the same year, Cai Yu [20] established a soil moisture prediction model based on the regression network, which is capable of deep learning. He predicted soil moisture in an area of Beijing and verified the effectiveness of the model. In 2021, Meng Chu [21] used the BAS-BP neural network to verify and test the average soil moisture at a depth of 40cm in a certain place in Changchun City. The prediction model used in this test has the characteristics of fast convergence speed and high prediction accuracy, and can accurately predict the changes of soil moisture in the next five days after using weather forecast data. In 2022, Nguyen Thu Thuy and colleagues [22] combined sensors with advanced machine learning models to create a new and low-cost method for soil moisture prediction. Compared with random forests and support vector machines, this method has higher prediction accuracy. In the same year, Li Qingliang [23] proposed a novel encoder-decoder model and used it to predict soil moisture. The model is a deep learning model with residual learning that enhances forecasting capabilities with intermediate time series data.

Inner Mongolia Xilin Gol Grassland, located in the Xilin River Basin of Inner Mongolia, is one of the four major grasslands in China. It is not only an important animal husbandry production base in China, but also an important green ecological guardian in China because of its special role in reducing sandstorms and bad weather. In addition, the Xilin Gol grassland is one of the typical areas for studying the response mechanism of ecosystems to human disturbance and global climate change.

Soil moisture prediction in Xilin Gol grassland in Inner Mongolia will provide a strong scientific basis for the prevention of grassland flood and drought disasters, ecological environmental protection and sustainable social and economic development.

2 Overview of Study Areas and Datasets

2.1 Study Area

Located in the eastern central part of Inner Mongolia Autonomous Region, Xilin Gol Grassland in Inner Mongolia is one of the four major grasslands in China, with geographical coordinates between $110°50' - 119°58'$ east longitude and $41°30' - 46°45'$ north latitude. The grassland covers an area of 179,600 square kilometers, with a slope from southeast to northwest, and an altitude of 800 m–1200 m. The area has the characteristics of wind, drought and cold, and its annual precipitation is 150 mm–400 mm from northwest to southeast. On the whole, Xilin Gol grassland in Inner Mongolia is a representative and typical grassland in temperate grassland, with complex types, relatively well-preserved and rich biodiversity, and is an important ecological guardian in Beijing-Tianjin and northern China.

2.2 Data Set

The soil and climate data of the Xilin Gol grassland are monitored and provided by specialized institutions, including soil moisture (kg/m^2), soil evaporation (W/m^2), vegetation index (*NDVI*), runoff (m^3/s), altitude (m), average temperature (°C), precipitation (mm) and average wind speed (knots).

Soil Moisture. Soil moisture (*SW*), also known as soil moisture content, when expressed in absolute soil water content, means that 100 g of dried soil contains several grams of water. Soil moisture is mainly affected by meteorological factors (such as local precipitation), vegetation conditions and other factors. The soil moisture data used in this paper include soil moisture at depths of 10 cm, 40 cm, 100 and 200 cm, which is intended to consider the soil moisture requirements of different crops in different growth cycles. The data collected spans 2012 to 2022 and is measured monthly.

Soil Evaporation. Soil Evaporation (*SE*) is the process by which water in soil vaporizes into the atmosphere in units of W/m^2 As an important part of the hydrological cycle, soil evaporation process will have a profound impact on the change of soil water content, which in turn affects the growth and development of plants [24–28]. The movement of various forms of water is mainly limited by soil moisture content and meteorological factors, such as regional precipitation, temperature changes, and land surface conditions.

The soil evaporation data used in this study spans the period from 2012 to 2022 and is measured monthly.

NDVI. As an important indicator for macroscopic monitoring of vegetation, the normalized vegetation index *NDVI* can characterize land surface vegetation cover and vegetation growth. *NDVI* is the most commonly used vegetation index reflecting plant chlorophyll activity, related to the proportion of radiation absorbed by photosynthesis to total radiation, *NDVI* is calculated as:

$$NDVI = \frac{NIR-R}{NIR+R} \tag{1}$$

where *NIR* is the reflectance in the near-infrared band, *R* is the reflectivity of the red band.

After the normalization process, the *NDVI* value ranges from -1 to 1, and there are three cases of value: negative value, 0 value and positive value, respectively. 1) When the *NDVI* is less than 0, it means that visible light can be reflected at a high level; 2) When *NDVI* is equal to 0, it means that *NIR* and *R* are approximately equal; 3) When the *NDVI* is greater than 0, the vegetation here is considered to be sufficient to cover the surface.

Runoff. Runoff (*R*) is mainly affected by rainfall and vegetation cover, and the water cycle in the area changes due to its changes. Changes in runoff have been found to correlate well with temperature and precipitation [29], while other studies have shown that greater vegetation cover is associated with lower runoff [30]. The runoff data used in this paper covers the period from 2012 to 2022 and is collected monthly.

Climate Data. The climate data used in this paper were all measured at the same site (site number: 54102099999), which has an altitude of 1004.0 m. The measured data includes altitude, average temperature, precipitation, and average wind speed. Data is collected monthly, covering the period from 2012 to 2022.

3 Model Building

The soil moisture prediction dataset is a typical multivariate time series data, which is more consistent with the core structure of the Transformer model, but because the Transformer model has no memory function when processing time series, we propose an LSTM-Transformer model that introduces LSTM to enhance its sequential modeling ability and improve its sensitivity to time series forecasting, and the model is mainly composed of three parts: input layer, encoding-decoding layer and output layer (see Fig. 1).

Output layer

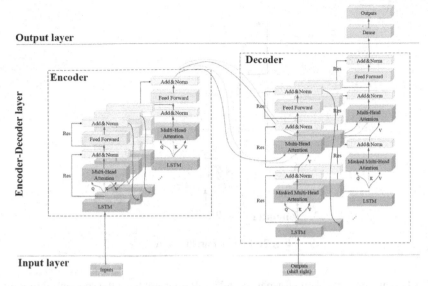

Fig. 1. LSTM-Transformer model architecture diagram.

When the model works, the variables such as soil evaporation, vegetation index, and runoff are first imported into the model through the input layer, and the data is encoded by the encoder, then the long and short-term information of the time series is first captured through the LSTM layer in the encoder. The importance of different features is calculated through the multi-head attention mechanism and the Add & Norm layer, connected by the residuals. Features weighted by importance enter the feed forward layer and the Add & Norm layer to further consider the importance of different features. A matrix of encoded information for all data is then obtained and output from the encoder. Then the data output from the encoder enters the multi-head attention mechanism in the decoder, at the same time, the output result of the decoder enters the Mask multi-head attention mechanism after LSTM encoding, then enters the output layer after decoding by the decoder, finally obtains the prediction result after processing by the full connection layer.

3.1 Encoder Architecture

The encoder structure consists of the LSTM layer, the multi-head attention layer, the Add & Norm layer, the residuals and the feed forward layer (see Fig. 2). The input variable input from the input layer to the encoder, first through the LSTM layer. LSTM layers capture the long and short-term information of the time series. Then the input variable enters into the multi-head attention mechanism, we intend to use eight heads. The variable information is weighted to highlight its important characteristics, and then transported to the next network in the form of parallel, the data is added and normalized, and its residuals are calculated, and then enter the forward propagation layer, and the data is added and normalized again, and its residuals are calculated, and finally the encoder outputs the results.

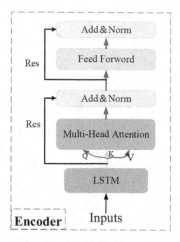

Fig. 2. Encoder architecture.

LSTM Layer. Since the Transformer model itself does not have the sequential modeling capability of time series, the introduction of the LSTM layer (see Fig. 3) allows the model to better learn the long and short distance dependencies of the data [31–33]. In LSTM, the information flow is transmitted through a mechanism called "cell state" [34, 35], and the information of "cell state" is removed or increased through the four gate structures of forget gates, input gates, update gates, and output gates, and the learning of long-term and short-term relationships between time series data is realized, and the specific functions of the four gating are described below.

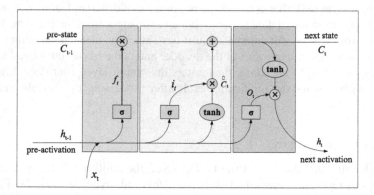

Fig. 3. LSTM layer architecture diagram.

Forget gate: Forget gate uses the sigmoid function (σ) to decide what information is discarded and what to keep. The previous hidden state (h_{t-1}) and the current input (x_t) serve as inputs to this gated. When the output of the forget gate is False, the network forgets the previous information; When output is True, the network retains all information in the cell state C_{t-1}, such as Eq. (2).

$$f_t = \sigma\left(W_f\left[h_{t-1}, x_t\right] + b_f\right) \tag{2}$$

Input gate: The sigmoid layer of the input gate determines which values the network will update, and the tanh activation layer generates a new candidate value $\tilde{C}t$, which is calculated as (3), (4).

$$i_t = \sigma\left(W_i\left[h_{t-1}, x_t\right] + b_i\right) \tag{3}$$

$$\tilde{C}t = \tanh\left(W_c\left[h_{t-1}, x_t\right] + b_i\right) \tag{4}$$

Update gate: In the update gate, the new cell state C_t is equal to the sum of the results of cell state C_{t-1} multiplied by C_{t-1} and i_t multiplied by $\tilde{C}_t$ at time t-1 with the Eq. (5).

$$C_t = f_t C_{t-1} + i_t \tilde{C}_t \tag{5}$$

Output gate: The responsibility of the output gate is to solve the number of unit state outputs at the current moment, first determine the cell state to be retained through the sigmoid layer, and then process the cell state through the tanh activation function, and multiply the processed cell state with the output of the sigmoid layer to obtain the final output such as Eq. (6), (7).

$$o_t = \sigma\left(W_o\left[h_{t-1}, x_t\right] + b_o\right) \tag{6}$$

$$h_t = o_t \tanh C_t \tag{7}$$

Multi-head Attention. In the LSTM-Transformer model, the multi-head attention mechanism is the core. Its structure is below (see Fig. 4).

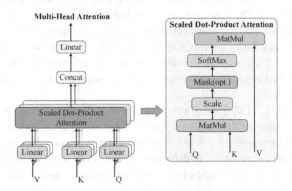

Fig. 4. Scaled Dot-Product Attention structure.

The input of the multi-head attention mechanism has three components, namely Q (query), K (key), and V (value), and the three vectors are divided into multiple input heads and passed to Scaled Dot-Product Attention through the linear connection layer, and Scaled Dot-Product Attention receives three sets of input data (Q, K, V), and calculates its output using Eq. (8).

$$\text{Attention}(Q, K, V) = \text{softmax}\left(\frac{QK^T}{\sqrt{dk}}\right)V \tag{8}$$

Among them, d_k is the column number of Q, K matrix.

Add & Norm Layer. The Add & Norm layer consists of two parts, Add and Norm, As shown in Eq. (9).

$$\text{Outputs} = \begin{cases} Normalize(X + Multi - HeadAttention(x)) \\ Normalize(X + FeedForward(x)) \end{cases} \tag{9}$$

In the formula, the input of Multi-Head Attention or Feed Forward is represented by X, and the output of Multi-Head Attention or Feed Forward is represented by MultiHeadAttention(X) and FeedForward(X).

Add refers to the sum of the input of Multi-Head Attention or Feed Forward with the output of Multi-Head Attention or Feed Forward, which is a residual join. This approach provides a solution to the multi-layer network training problem, and these features make the network focus on differences in the current state. Since the input of each layer of neurons can be converted to the same mean and variance by Normalization, Normalization can be used to accelerate the process of network convergence.

3.2 Decoder Architecture

The decoder consists of the LSTM layer, the masked multi-head attention mechanism, the Add & Norm layer, the multi-head attention mechanism, and the feed forward layer (see Fig. 5). One input to the decoder is the output of the encoder, and the other input is the combination of the output of the decoder and the LSTM decoding layer. The decoder contains two multi-head attention mechanisms, one is the multi-head attention mechanism, which accepts the output from the decoder as input, the input is the K vector and the V vector, and the other input of this attention mechanism is the output of the Masked multi-head attention mechanism. The second attention mechanism is masked long attention, and its input is a combination of the output of the previous decoder and the LSTM decoding.

In summary, the LSTM-Transformer model uses the data of the multivariate time series of soil moisture as the input of the encoder when forecasting and models the location of the data through the LSTM. The decoding process is autoregressive, and the prediction value of soil moisture at different depths is obtained after repeated multiple times by entering the decoder prediction result and encoder output into the decoder.

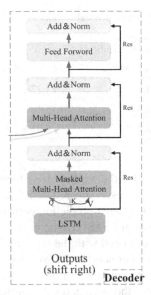

Fig. 5. Decoder Architecture.

4   Results and Analysis

After the model is established, it is necessary to evaluate and analyze its prediction effect, and one way to evaluate the LSTM-Transformer model is to test and verify it with the traditional LSTM and BiLSTM models. The model prediction performance index uses Mean Absolute Percentage Error Eq. (10), Mean Absolute Error Eq. (11) and Root Mean Square Error Eq. (12).

$$n_{MAPE} = \frac{1}{n} \sum_{i=1}^{n} \left| \frac{\hat{y}_i - y_i}{y_i} \right| \times 100\% \tag{10}$$

$$n_{MAE} = \frac{1}{n} \sum_{i=1}^{n} |\hat{y}_i - y_i| \tag{11}$$

$$n_{RMSE} = \sqrt{\frac{1}{n} \sum_{i=1}^{n} (\hat{y}_i - y_i)^2} \tag{12}$$

where $\hat{y}_i$. is the predicted soil moisture value; y_i is the true soil moisture value; n is the soil moisture statistics to be treated.

Table 1 below shows the values of soil moisture prediction performance indicators for each model.

Table 1. The predictive performance index values of the model for soil moisture.

Model	Predicted indicators		
	MAE (MW)	MAPE (%)	RMSE (MW)
LSTM	724.7	13.27%	859.7
BiLSTM	342.4	8.46%	583.2
LSTM-Transformer	335.3	2.01%	484.1

It can be seen from Table 1 that the prediction performance of LSTM-Transformer model for soil moisture is the strongest compared with that of BiLSTM and LSTM models, followed by BiLSTM and the worst prediction performance of LSTM model. The MAE, MAPE and RMSE indexes of LSTM-Transformer model were 335.3 MW, 2.01% and 484.1 MW respectively. Compared with the other two models, MAE decreased by 389.4 MW and 7.1 MW, MAPE decreased by 11.26% and 6.45%, and RMSE decreased by 375.6 MW and 99.1 MW respectively. Based on soil moisture data from 2012 to March 2022, this paper compares the predicted value of the model with the actual value to visualize the prediction effect of the LSTM-Transformer model (see Fig. 6).

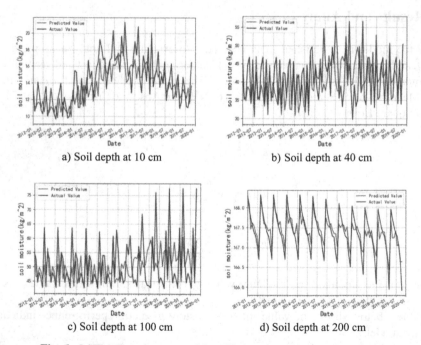

a) Soil depth at 10 cm   b) Soil depth at 40 cm

c) Soil depth at 100 cm   d) Soil depth at 200 cm

Fig. 6. LSTM-Transformer model predictions compared to true values.

Table 2 below shows the model's predictions of soil moisture at different depths from April 2022 to December 2023.

Table 2. Prediction results of soil moisture.

year	month	Humidity at 10 cm depth (Kg/m^2)	Humidity at 40 cm depth (Kg/m^2)	Humidity at 100 cm depth (Kg/m^2)	Humidity at 200 cm depth (Kg/m^2)
2022	04	11.79	35.79	46.15	167.46
	05	11.41	41.29	69.39	166.03
	06	14.32	47.98	73.39	166.40
	07	13.99	37.47	54.71	166.59
	08	14.42	38.24	57.97	166.66
	09	16.98	43.82	57.48	166.51
	10	21.42	59.49	75.36	166.14
	11	18.88	51.72	66.28	167.21
	12	19.30	52.78	75.64	166.34
2023	01	16.55	51.73	74.47	166.53
	02	12.59	43.08	58.13	167.07
	03	12.48	47.42	75.41	165.97
	04	13.99	37.47	54.71	166.59
	05	14.42	38.24	57.97	166.66
	06	16.98	43.82	57.48	166.51
	07	21.42	59.49	75.36	166.14
	08	18.88	51.72	66.28	167.21
	09	19.30	52.78	75.64	166.34
	10	16.55	51.73	74.47	166.53
	11	12.59	43.08	58.13	167.07
	12	12.48	47.42	75.41	165.97

5 Conclusion

In this paper, a Transformer model introduced into LSTM is presented to predict soil moisture values at different depths at future moments. The model used various data collected from 2012 to 2022, including soil moisture values at different depths, soil evaporation, vegetation index, runoff, and climate data as input variables, and predicted soil moisture values at different depths from April 2022 to December 2023.In order to compare with the traditional LSTM and BiLSTM, the MAE, MAPE and RMSE values of each model were measured to comprehensively evaluate the prediction performance of the LSTM-Transformer model. At last, the results show that the three index values of the LSTM-Transformer model are the lowest and the prediction error is the smallest. In summary, this method has the following advantages and significance:

1) By introducing LSTM into the Transformer, the excellent prediction ability of the Transformer and the sequential modeling ability of LSTM are combined, so that the LSTM-Transformer model can learn the internal laws of the time series and the long-distance dependence between the sequences effectively, and realize more accurate prediction of soil moisture data with time series attributes.
2) The multi-head attention mechanism in the LSTM-Transformer model can not only dynamically weight different input features to highlight important features, but also input feature extraction network in parallel, which promotes the distributed training of the model and improves the training speed of the model.
3) The LSTM-Transformer model presented in this paper comprehensively considers a variety of factors affecting soil moisture, and compares the predicted performance of the traditional model with the three statistical indicators of MAE, MAPE and RMSE. Thus, the validity of the model is verified.

The prediction of soil moisture is of great significance for crop yield estimation, hydrological model construction, and water resources management, and the LSTM-Transformer model proposed in this paper can make an effective prediction of soil moisture at different depths in the future. Since soil moisture is affected by a combination of factors, in future work, researchers can consider more influencing factors and extract more in-depth information from them to achieve more accurate predictions of soil moisture.

References

1. Souissi, R., et al.: Integrating process-related information into an artificial neural network for root-zone soil moisture prediction. Hydrol. Earth Syst. Sci. **26**(12), 3263–3297 (2022)
2. Filipović, N., Brdar, S., Mimić, G., Marko, O., Crnojević, V.: Regional soil moisture prediction system based on long short-term memory network. Biosys. Eng. **213**(2022), 30–38 (2022)
3. Niu, H., Meng, F., Yue, H., Yang, L., Dong, J., Zhang, X.: Soil moisture prediction in peri-urban Beijing, China: gene expression programming algorithm. Intell. Autom. Soft Comput. **28**(1), 93–106 (2021)
4. Bodo, G.: Plant mass and yield of broccoli as affected by soil moisture. HortScience **41**(1), 113–118 (2006)
5. Beautiful grassland, the wealth of mankind. Green China **3**(12) (2022)
6. Geng, G.: Protect beautiful grassland in accordence with the law. Green China **13**, 20–23 (2022)
7. Yang, Y., Liu, H., Wan, X., Cui, J., Zhang, F., Cai, T.: Research on soil moisture and temperature prediction based on environmental temperature and humidity. Modern Electron. Technol. **45**(18), 159–165 (2022)
8. Chan, Y.J., Carr, A.R., Roy, S., Washburn, C.M., Neihart, N.M., Reuel, N.F.: Positionally-independent and extended read range resonant sensors applied to deep soil moisture monitoring. Sens. Actuators: A. Phys. **333**, 113227 (2022)
9. Yuan, L., Fang, X., Guo, X., Ynag, L., Zhang, X., Ren, L.: Calculation of root zone soil moisture using MIV-BP neural networks. Sci. Technol. Eng. **22**(17), 6911–6919 (2022)
10. Xu, X., Yi, S., Huang, C.: Soil moisture content prediction situation review. J. Agric. Mechanization Res. **35**(07), 11–15 (2013)
11. Zhang, G., Fei, Y., Wang, H., Lian, Y.: Specific characteristics of soil hydrodynamic field state and its application to irrigational infiltration. J. Hydraul. Eng. **41**(09), 1032–1037 (2010)

12. Liu, H., Wu, W., Wei, C., Xie, D.: Soil water dynamics simulation by autoregression models. Mt. Res. **01**, 121–125 (2004)

13. Deng, J., Chen, X., Fang, K., Du, Z.: Prediction of chaotic soil moisture time series based on artificial neural network. Bull. Soil Water Conserv. **28**(06), 82–85 (2008)

14. Wang, P., Sun, W.: Comparison study on NDVI and LST based drought monitoring approaches. J. Beijing Normal Univ. (Nat. Sci.) **43**(03), 319–323 (2007)

15. Lee, K., Anagnostou, E.N.: A combined passive/active microwave remote sensing approach for surface variable retrieval using Tropical Rainfall Measuring Mission observations. Remote Sens. Environ. **92**(1), 112–125 (2004)

16. Jackson, T.J., Chen, D.: Vegetation water content mapping using Land sat data derived normalized difference water index for corn and soybeans. Remote Sens. Environ. **92**(4), 225–236 (2004)

17. Zhang, Y., Wang, J., Bao, Y.: Soil moisture retrieval from multi-resource remotely sensed images over a wheat area. Adv. Water Sci. **21**(02), 222–228 (2010)

18. Shikha, P., Animes, S., Sitanshu, S.: Soil moisture prediction using machine learning. In: 2018 2nd International Conference on Inventive Communication and Computational Technologies (ICICCT), pp. 1–6. IEEE (2018)

19. Gursimran, S., Deepak, S., Amarendra, G., Sugandha, S., Shukla, A., Satish, K.: Machine learning based soil moisture prediction for internet of things based smart irrigation system. In: 2019 5th International Conference on Signal Processing, Computing and Control (ISPCC), pp. 175–180. IEEE (2019)

20. Yu, C., Zheng, W., Zhang, X., Zhang, Z., Xue, X.: Research on soil moisture prediction model based on deep learning **14**(4), (2019)

21. Meng, C.: Research on Field Irrigation Method Based on Soil Moisture Prediction. Jilin Agricultural University (2021)

22. Nguyen, T., et al.: A low-cost approach for soil moisture prediction using multi-sensor data and machine learning algorithm. Sci. Total Environ. **833**, 155 (2022)

23. Li, Q., Li, Z., Wei, S., Wan, X., Li, L., Yu, F.: Improving soil moisture prediction using a novel encoder-decoder model with residual learning. Comput. Electron. Agric. **195**, 106816 (2022)

24. Ma, J., Feng, K., Li, W., Hao, L., Li, Y., Gao, H.: Using water surface evaporation to estimate soil surface evaporation in arid regions in central ningxia. J. Irrig. Drainage **39**(10), 35–41 (2020)

25. Bai, W., et al.: Effect of super absorbent polymer on vertical infiltration characteristics of soil water. Trans. Chin. Soc. Agric. Eng. **25**(02), 18–23 (2009)

26. Dani, O., Lehmann, P., Shahraeeni, E.: Advances in soil evaporation physics: a review. Vadose Zone J. **12**(4), 1–16 (2013)

27. Liu, P., Xia, Y., Shang, M.: Estimation methods of phreatic evaporation for different textures in bare soil area. Trans. Chin. Soc. Agric. Eng. **36**(01), 148–153 (2020)

28. Jiang, Y., Tang, R., Jiang, X., Li, Z., Gao, C.: Estimation of soil evapotranspiration and vegetation evapotranspiration using two trapezoidal models based on MODIS data. Geophys. Res. Atmos. **124**(14), 7647–7664 (2019)

29. Zhang, G., Wang, X., Guo, M.: The spatial and temporal structure of runoff variation and the climate background in the Yellow River basin during the past 60 years. J. Arid Land Res. Environ. **27**(7), 91–95 (2013)

30. Li, K., Yao, W., Xiao, P.: Advances in research on the effects of vegetation on soil infiltration and surface runoff processes. Soil Water Conserv. China **2017**(3), 27–30 (2017)

31. Kadyan, V., Dua, M., Dhiman, P.: Enhancing accuracy of long contextual dependencies for Punjabi speech recognition system using deep LSTM. Int. J. Speech Technol. **24**(2), 517–527 (2021). https://doi.org/10.1007/s10772-021-09814-2

32. Jeena, K., Abdul, N.: An enhanced Tree-LSTM architecture for sentence semantic modeling using typed dependencies. Inf. Process. Manag. **57**(6), 102362 (2020)
33. Apeksha, S., Deepika, N., Simone, A.: Performance evaluation of deep neural networks applied to speech recognition: RNN, LSTM and GRU. J. Artif. Intell. Soft Comput. Res. **9**(4), 235–245 (2019)
34. Zhang, Z., Zhou, J., Ma, G., Zeng, T.: Kashi district mumps prediction model based on LSTM neural network. Mod. Electron. Tech. **45**(19), 127–132 (2022)
35. Guo, H., Feng, X.: CSI gesture recognition algorithm based on Bi-LSTM. Comput. Eng. Des. **43**(09), 2614–2621 (2022)

Graph Contrastive Learning with Intrinsic Augmentations

Dengdi Sun[1,3] , Mingxin Cao[2] , Zhuanlian Ding[4(✉)] , and Bin Luo[2]

[1] Key Laboratory of Intelligent Computing and Signal Processing (ICSP),
Ministry of Education, School of Artificial Intelligence,
Anhui University, Hefei 230601, China
[2] Anhui Provincial Key Laboratory of Multimodal Cognitive Computing,
School of Computer Science and Technology, Anhui University, Hefei 230601, China
[3] Institute of Artificial Intelligence, Hefei Comprehensive National Science Center,
Hefei 230026, China
[4] School of Internet, Anhui University, Hefei 230039, China
dingzhuanlian@163.com

Abstract. Graph contrastive learning has become an important app-roach for learning unsupervised representations of graphs, with the key idea of maximizing the consistency of representations in both augmented views through data augmentation. Existing graph contrastive learning models concentrate on topology enhancement of the graph structure by simply removing/adding edges between nodes randomly, which may not only destroy the important structure of the graph but also generate meaningless graphs, leading to a significant degradation of the contrastive learning performance. In addition, current research is too focused on minimizing losses, while ignoring the intrinsic factors that can affect the quality of node representations, which is detrimental to the training of models and the generation of high-quality node representations. To address these issues, we propose intrinsic augmented graph contrastive learning, named IAG, which consists of two components: 1) In the topology augmentation part, we propose a novel topology augmentation strategy based on potential connections in the feature space, which complements the traditional topology augmentation by allowing different graphs to obtain augmentation strategies more suitable for their own characteristics to advance the traditional graph contrastive learning. 2) We explored the effect of temperature coefficient in the loss function on the quality of the final representation and proposed dynamic temperature with penalty terms, which helps to generate high-quality node representations. Finally, we conducted extensive node classification experiments on 8 real-world datasets. The experimental results show that our proposed method is highly competitive with the existing state-of-the-art baselines and even surpasses some supervised methods.

Keywords: Graph neural networks · Contrastive learning · Graph representation learning · Unsupervised learning

© The Author(s), under exclusive license to Springer Nature Singapore Pte Ltd. 2023
L. Pan et al. (Eds.): BIC-TA 2022, CCIS 1801, pp. 343–357, 2023.
https://doi.org/10.1007/978-981-99-1549-1_27

1 Introduction

Graph representation learning is an important approach to analyze graph data, which is dedicated to embedding non-Euclidean graph data into a low-dimensional vector space, and such low-dimensional vectors that retain underlying structural and attributive information can be better used as input for downstream tasks [4,6,14]. Although current graph neural networks (GNNs) have made significant progress in graph representation learning, their application to more common unlabeled data is hindered by the fact that most of their models need to be trained on human-labeled [5]. Therefore, in order to facilitate the development of GNNs for practical applications, it is important to develop graph representation learning that does not require labels.

In recent years, InfoMax [9] based contrastive learning methods have achieved great success in computer vision [1]. These contrastive learning methods seek to maximize the Mutual Information (MI) between the input (i.e., images) and its representations (i.e., image embeddings) by contrasting positive pairs with negative-sampled counterparts. Inspired by these contrastive learning methods, Deep Graph InfoMax (DGI) [21] combines GNN and InfoMax to propose an alternative objective based on maximizing MI in the graph domain. Specifically, DGI first performs a simple random disruption of node features to obtain an augmented graph, and then employs GNN to learn node embeddings and obtain graph embeddings through a readout function, with the final goal of maximizing the MI between node embeddings and graph embeddings. Based on this, MVGRL [3] introduces the concept of graph multi-view contrastive learning, where the input graph is augmented by a graph diffusion kernel [8] and Personalized PageRank [15] that compares the node representations and global embeddings of two augmented views from the input graph. GMI [16] and GRACE [30] extend the concept of MI maximization to contrast the representation of a node with its raw information (e.g., node features) or neighboring representations in different views.

Despite the great success of the above approaches, as with visual representation learning, graph augmentation, a key component of graph contrastive learning, has seen limited progress [25]. Graph augmentation includes graph topology augmentation and node attribute augmentation, and studies have shown that effective graph augmentation methods can greatly improve the performance of graph contrastive learning [29]. However, in topology augmentation, most existing methods apply the same augmentation strategy to all graphs in a given scenario, such as GRACE [30] and CCA-SSG [28] which perform two random edge removing on the original graph to obtain two topology augmented views, and although this method of making the graph sparse may lead to improvement in contrast learning performance, randomly removing valid edges in complex graphs [10] may destroy some important structures. Other methods [27] use randomly adding edges between nodes, which does not break the network structure but may generate meaningless graphs, and thus both methods bring uncertain results. Taking a real-world network of commodity co-purchase relationships [19] as an example, what is represented by the established connection is that two commodities are often purchased together, and this relationship is strongly connected. Randomly removing edges may lead to the deletion of edges containing strong connections, while randomly adding edges may lead to the formation of strong connections between meaningless commodities, all of which obviously have a large impact on the final node representations. Since this randomness can lead to great uncertainty, does a method exist to guide the generation of meaningful topology augmentation?

Furthermore, the node representations generated by contrastive learning are used as input for downstream tasks, and therefore the quality of the generated node representations greatly affects the accuracy of the final downstream tasks. The intrinsic factors affecting the alignment and uniformity of the final representations have been analyzed in the visual domain [23], while the current graph contrastive learning method only focuses on the loss reduction during training, ignoring the influence of internal factors in the loss on the final representation quality, which not only leads to the overfitting of the model but also is not conducive to the generation of high-quality node representations.

To solve the above problems, we propose a graph contrastive learning method using intrinsic augmentation called IAG, specifically, we first propose a novel graph topology augmentation strategy, which is to obtain the potential underlying relationships of graph topology by exploring the relationships in the node feature space, and later expand these potential underlying relationships to the original graph to form the augmented graph. This augmentation strategy not only reflects the intrinsic patterns of the original graph, but also complements the traditional topological augmentation, allowing different graphs to obtain augmentation strategies more suitable for their own characteristics to advance the traditional graph contrastive learning. In addition, we investigate the effect of temperature coefficient on node representations in the loss function and propose dynamic temperature with penalty term, which helps to generate high-quality node representations with compact intra-cluster and separated inter-cluster.

2 Related Works

The goal of graph representation learning is to transform graph-available information, such as graph topology and high-dimensional complex node features, into low-dimensional fine-grained vectors. Traditional methods have the disadvantages of high complexity and inability to make full use of graph information. After traditional methods, graph neural networks have achieved great success in node representation [6,20], which effectively use graph structure and node attributes to update node representation through the message passing process, but are limited by the need for node labels and cannot be widely used for real-world data.

In recent years, graph contrastive learning without labels and with excellent performance has received a lot of attention in self-supervised graph learning [11]. The basic idea is to maximize the consistency of representations between different augmented perspectives. For example, inspired by InfoMax, DGI [21] has maximized the mutual information between global and local representations as a learning goal. MVGRL [3] diffuses the graph into multiple structural views and learns by comparing the differences between views. GMI [16] proposes two comparison goals that directly measure the MI between inputs and representations of nodes and edges, respectively. GRACE [30] uses random edge removal and random feature masking generates different views for comparison learning. Considering that random edge deletion may bring meaningless graphs, GCA [31] adopts adaptive augmentation by considering graph topology and semantics.

3   The Proposed Method

In this section, we begin with a brief overview of the model, then present our topology augmentation method, and finally analyze the effect of temperature coefficient on node representation and propose a dynamic temperature based on penalty term.

3.1   Overview

A undirected attributed graph can be represented by $\mathbf{G} = (\mathbf{V}, \mathbf{X})$, where $\mathbf{V} = \{v_1, v_2, \cdots, v_N\}$ is the set of nodes with $|\mathbf{V}| = N$, $\mathbf{X} \in \mathbb{R}^{N \times F}$ is the feature matrix, each row of which corresponds the F-dimensional attributes of one node. The graph structure can be represented by the adjacency matrix $\mathbf{A} = [a_{ij}] \in \mathbb{R}^{N \times N}$, where $a_{ij} = 1$ if an edge exists between the node v_i and v_j, and vice versa.

In our model, the given node label information is not used in training, instead, the goal of learning is to improve the consistency of node representations of the two augmented views. Specifically, for the above graphs, the original graph $\mathbf{G}$ is first topologically augmented twice (see Sect. 3.2), followed by attribute augmentation (feature random mask) for each of the two graphs, and the two new graphs generated are noted as $\tilde{\mathbf{G}}_1$ and $\tilde{\mathbf{G}}_2$, respectively, and then a shared GNN is used as the encoder for $\tilde{\mathbf{G}}_1$ and $\tilde{\mathbf{G}}_2$, with the goal of learning from $\tilde{\mathbf{G}}_1$ and $\tilde{\mathbf{G}}_2$ to the low-dimensional embeddings $\mathbf{Z}'$ and $\mathbf{Z}''$, and during the training process, the goal of contrastive learning is to encourage the GNN encoder to minimize the difference between node representations $\mathbf{Z}'$ and $\mathbf{Z}''$. To obtain high-quality node representations, we use dynamic temperature based on penalty term in training (see Sect. 3.3). After the training is completed, we can use the topology and node features of the original graph as the input of the trained GNN, and its output $\mathbf{Z} = f(\mathbf{X}, \mathbf{A})$ is the desired low-dimensional embedding. These representations can be used in downstream tasks, such as node classification and node clustering.

3.2   Graph View Generation

In this section, we first introduce the traditional graph augmentation strategy and analyze its drawbacks, and then propose a complementary graph topology augmentation approach.

Removing Edges (RE). Graph topology augmentation is an important step in generating graph contrastive views. Most current methods [28,30] use graph topology augmentation by randomly removing edges, i.e., first sample a random mask matrix $\tilde{\mathcal{R}} \in \{0,1\}^{N \times N}$ which obeys the Bernoulli distribution $\tilde{\mathcal{R}}_{ij} \sim \mathcal{B}(1 - p_r)$, p_r is the probability of each edge being removed, the enhanced graph topology becomes:

$$\tilde{\mathbf{A}} = \mathbf{A} \circ \tilde{\mathcal{R}} \tag{1}$$

where $\circ$ is Hadamard product.

Many current methods for augmenting graphs use randomly removing edges for both views [28,30], the limitations of this idea are obvious. For some sparse graphs and graphs with strong structural correlation, if both augmented graphs remove the same strongly connected edges, this important information is equivalent to disappearing and possibly adding isolated nodes, the resulting node embedding will obviously affect the accuracy of downstream tasks. For such graphs, the most important thing may be how to effectively expand the graph based on the existing information instead of removing edges randomly, for which we propose a strategy of topology augmentation based on the potential connections in the feature space.

Topology Expansion with Feature Space (TE). It is found that in some networks with strong correlations among features of the same class, the feature space information has a greater impact on the final node embedding [24]. Even, in these networks, the potential underlying topology brought by the feature space information is more important than the original topology. Therefore, when expanding the graph topology, we consider discovering the potential underlying links generated in the feature space, and augmenting the graph structure by fusing the original topology with the underlying topology from the feature space. In this way, deeper information can be learned on the basis of taking full advantage of the feature space information. Specifically, we first discover the potential topology of the feature space by cosine similarity, which is calculated as:

$$\mathcal{S}_{ij} = \frac{x_i \cdot x_j}{\| x_i \| \cdot \| x_j \|} \tag{2}$$

We consider the top k nodes that are most similar to the central node i as having potential links to the central node i, and the set is denoted as P_i:

$$P_i = topk(\mathcal{S}_i) \tag{3}$$

Afterwards, using potential links and the original graph topology merged as an augmentation of the original graph:

$$A'_{ij} = \begin{cases} A_{ij}, & if \quad A_{ij} = 1 \\ 1, & if \quad A_{ij} = 0 \quad and \quad j \in P_i \end{cases} \tag{4}$$

At this point, we can obtain a new graph $\tilde{\mathbf{G}}$, which is equivalent to an expansion of the original graph. In this paper, we use a combination of RE and TE for topology augmentation, i.e., randomly removing edges and using feature space for topology expansion to generate $\tilde{\mathbf{G}}_1$ and $\tilde{\mathbf{G}}_2$ respectively from the original graph.

3.3   Dynamic Temperature with Penalty Item

Next, we will explore the intrinsic factors that affect node representation. First look at the common contrastive learning loss, whose goal is to maximize the consistency of representation in both augmented views. It can be formulated as:

$$
\mathcal{L}_i^{u,v} = -\log \frac{\exp\left(\frac{1}{\tau} \cdot \theta\left(u_i, v_i\right)\right)}{\underbrace{\exp\left(\frac{1}{\tau} \cdot \theta\left(u_i, v_i\right)\right)}_{\text{positive pair}} + \underbrace{\sum_{k \neq i} \exp\left(\frac{1}{\tau} \cdot \theta\left(u_i, v_k\right)\right)}_{\text{inter-view negative pairs}} + \underbrace{\sum_{k \neq i} \exp\left(\frac{1}{\tau} \cdot \theta\left(u_i, u_k\right)\right)}_{\text{intra-view negative pairs}}}
$$
(5)

where u_i and v_i respectively represent the representation of node i in the two augmented views, τ is a temperature coefficient. The similarity function is defined as $\theta(u, v) = sim(h(u), h(v))$, where $sim(\cdot)$ and $h(\cdot)$ are respectively cosine similarity and non-linear projection transformation (i.e., the two-layer MLP). Since two views are symmetric, so the overall objective to be minimized is then defined as the average over all positive pairs:

$$
\mathcal{L} = \frac{1}{2N} \sum_{i=1}^{N} \left(\mathcal{L}_i^{u,v} + \mathcal{L}_i^{v,u}\right)
$$
(6)

The goal of Eq. 5 is that with the progress of training, the similarity of the same nodes from different perspectives will gradually increase, while the similarity between different nodes from the same perspective and different nodes from different perspectives will become smaller and smaller. In essence, it is InfoNCE loss. [22] has fully demonstrated how temperature coefficient τ affects the final representation through the gradient analysis and experiment of the loss function. In a nutshell, the function of the temperature coefficient is to regulate the degree of attention paid to the hard negative samples: the larger the temperature coefficient is, the less attention paid to the more hard negative samples; However, the smaller the temperature coefficient is, the more attention will be paid to the hard negative sample which is very similar to the sample, and a larger gradient will be given to the hard negative sample to separate from the positive sample.

Most of the existing studies set a fixed temperature coefficient to the model and afterwards focus only on the decrease of loss during training, ignoring the effect of temperature coefficient on the final representation quality. Suppose we set a smaller fixed temperature coefficient, which would then force the model to focus on the hard negative samples and obtain a relatively uniform distribution, but the hard negative samples tend to be more similar to the positive samples and are likely to be potentially positive (e.g., different nodes in a class), which

(a) Small τ (b) Medium τ (c) Large τ

Fig. 1. T-SNE visualization of the embedding distribution. The model is trained on Cora. From Fig. (a) to Fig. (c), a smaller temperature coefficient is, less tolerant to difficult negative nodes and generates a more uniform distribution of node representations. As the temperature coefficient increases, the obtained representations gradually tend to local clustering and global separation.

would overly force the positive samples to be separated from the hard negative samples if training continues, destroying the quality of the node representations (the learned semantic structure). If we do not set a fixed temperature coefficient, but make the temperature coefficient slowly increase during training, its penalty for all negative samples will gradually converge, and the model will pay more attention to those simple negative samples (i.e., different classes) and move those simple negative samples as far away from the current sample as possible, thus achieving local clustering, which is obviously what we hope to get.

An easier way to achieve this local clustering effect is to know in advance which difficult negative samples are potential positive samples, that would undoubtedly make training easier, but in practice we do not know, so we want to use dynamic temperature to control the quality of the final node representation, and we consider that the trend of temperature coefficient is not the same but depends on the results of training, specifically, we design the penalty function for temperature coefficient as:

$$\tau_t = \tau_{t-1} + \beta \left(\underbrace{\frac{\sum_1^N sim(z_i^t, z_i^t)}{\sum_{j \neq i} sim(z_i^t, z_j^t) + \sum_{j' \neq i} sim(z_i^t, z_{j'}^t)}}_{\text{penalty term}} \right) \tag{7}$$

where $sim(\cdot)$ is cosine similarity and t denotes the epoch of training. We can find that the numerator in penalty term gets larger and the denominator gets smaller as the training progresses. In order to prevent the temperature from rising too fast, we use β control, and too large temperature coefficient is not conducive to the training of the model. Therefore, when the temperature coefficient rises to a certain range, the temperature coefficient will be fixed and recorded as τ_{max}. We experimentally verified the validity of our idea. Figure 1 shows how the node representations change with the training process; see Sect. 4.5 for more detailed results.

4   Experiments

In this section, we present the experimental dataset used, the experimental setup, Evaluation protocol and Baseline Methods, respectively, and finally we perform the node classification and ablation experiments.

4.1   Datasets

In this section, we present the experimental dataset we used. Based on previous papers [30,31], we used 8 datasets including Cora [18], Citeseer [18], Pubmed [18], DBLP [26], Wiki-CS [13], Amazon-Computers [19], Amazon-Photo [19], and Coauthor-CS [19], to evaluate the performance. Some statistics of these datasets are shown in Table 1.

Table 1. Dataset statistics.

Dataset	Nodes	Edges	Features	Classes
Cora	2708	5429	1433	7
Citeseer	3327	4732	3703	6
Pubmed	19717	44338	500	3
DBLP	17716	105734	1639	4
Wiki-CS	11701	216123	300	10
Amazon-Computers	13752	245861	767	10
Amazon-Photo	7650	119081	745	8
Coauthor-CS	18333	81894	6805	15

4.2   Experimental Setup and Evaluation Protocol

In our experiments, we performed attribute augmentation and topological augmentation on the original graph, where attribute augmentation was performed using feature randomly masking, and topological augmentation was performed using randomly removing edges for one of the two views and topological augmentation based on feature space for the other view, called RE-TE. In order to verify the effectiveness of the scheme, we set up two additional enhancement schemes in the ablation experiment, the first scheme: both views perform randomly removing edges, named RE-RE, and the second scheme: one of the views uses the original graph and the other view is topologically enhanced by feature space, named Raw-TE. Either way, we can get two new graphs $\tilde{\mathbf{G}}_1$ and $\tilde{\mathbf{G}}_2$, and then use the classic two-layer GCN [6] model as the encoder, the form is:

$$\mathbf{Z}' = softmax\left(\hat{\mathbf{A}}'ReLU\left(\hat{\mathbf{A}}'\mathbf{X}'W^0\right)W^1\right) \tag{8}$$

where $\hat{\mathbf{A}} = \hat{\mathbf{D}}^{-\frac{1}{2}}(\mathbf{A} + \mathbf{I})\hat{\mathbf{D}}^{-\frac{1}{2}}$, $\hat{\mathbf{D}}$ is the degree matrix with self-loops, $W^0(W^1)$ are first-layer (second-layer) neural network parameters, respectively. And we used dynamic temperature during the training.

To evaluate the performance of the model, we use the same linear evaluation scheme as in the literature [21]. specifically, the model is first trained in an unsupervised manner, after which the obtained node representations are trained and tested using a logistic regression model, and different training-validation-test partitions are used in order to evaluate the quality of the representations in a comprehensive manner. To ensure the fairness of the results, we trained the model 20 times and reported the average accuracy on each dataset.

4.3 Baseline Methods

To validate the effectiveness of IAG, we compare it with the following state-of-the-art competitors, including supervised GCN [6] and GAT [20], logistic regression classification using raw features, walk-based deepwalk [17] and node2vec [2], traditional graph autoencoders GAE [7] and VGAE [7], and more recently, deep graph information maximization DGI [21], GMI [16], GRACE [30] and GCA [31], multi-view representation learning MVGRL [3], for all baselines, we reproduce their results and report their performance according to their official implementations.

4.4 Results of Node Classification

In the node classification task, the structure augmentation scheme we use is RE-TE and dynamic temperature, its hyperparameter settings are shown in Table 5. We first conduct experiments in four common citation networks (Cora, Citeseer, Pubmed and DBLP), and the experimental results are shown in Table 2, where available data for each method during the training phase is shown in the second column, $\mathbf{X}$, $\mathbf{A}$, $\mathbf{Y}$ correspond to node features, the adjacency matrix, and labels respectively, the highest performance of models is highlighted in boldface. From the table, we can observe that IAG has the best classification accuracy on all four datasets, and this result can be attributed to two key components we propose: (1) Unlike the traditional randomly removing edges, we propose a topological enhancement based on feature space expansion graph topology, which can strengthen the relationship between potentially strongly connected nodes in certain graphs. (2) To further obtain high-quality node representations, we adopt a dynamic temperature guided by training, which is more conducive to distinguish different node classes.

To validate the generality of our model, we conducted the same experiments on four other large datasets with more classes (Wiki-CS, Amazon-Computers, Amazon-Photo, Coauthor-CS), and the experimental results are shown in Table 3. From the table we can see that our classification results are still optimal, which further validates the effectiveness of our proposed component. The specific effects produced by the two components are given in Sect. 4.5.

Table 2. Node classification results on citation networks.

Methods	Training Data	Cora	Citeseer	Pubmed	DBLP
Supervised GAT	**X, A, Y**	82.30	71.42	84.62	81.91
Supervised GCN	**X, A, Y**	82.82	72.00	84.91	82.75
Raw features	**X**	64.81	64.63	84.84	71.65
node2vec	**A**	74.86	52.30	80.31	78.84
DeepWalk	**A**	75.75	50.54	80.52	75.90
GAE	**X, A**	76.91	60.62	82.90	81.23
VGAE	**X, A**	78.90	61.25	83.02	81.71
DGI	**X, A**	82.66	68.85	85.34	83.27
GRACE	**X, A**	83.32	71.31	85.30	84.24
GCA	**X, A**	84.14	72.32	86.16	84.10
IAG	**X, A**	**86.10**	**73.64**	**86.10**	**85.35**

Table 3. Node classification results on Wiki, Amazon and Coauthor.

Method	Training Data	Wiki-cs	Amazon-comput	Amazon-photo	Coauthor-cs
Supervised GCN	**X, A, Y**	77.19	86.51	92.42	93.03
Supervised GAT	**X, A, Y**	77.65	86.93	92.56	92.31
Raw features	**X**	71.98	73.81	78.53	90.37
node2vec	**A**	71.79	84.39	89.67	85.08
DeepWalk	**A**	74.35	85.68	89.44	84.61
DeepWalkX	**X, A**	77.21	86.28	90.05	87.70
GAE	**X, A**	70.15	85.27	91.62	90.01
VGAE	**X, A**	75.63	86.37	92.20	92.11
DGI	**X, A**	75.35	83.95	91.61	92.15
GMI	**X, A**	74.85	82.21	90.68	OOM
MVGRL	**X, A**	77.52	87.52	91.74	92.11
GRACE	**X, A**	78.00	87.40	91.59	92.62
GCA	**X, A**	78.35	87.85	92.53	93.10
IAG	**X, A**	**81.80**	**88.70**	**92.65**	**93.50**

4.5 Ablation Study

In this section, we explore the role of the proposed two components, feature space based topology augmentation and dynamic temperature by penalty term, respectively.

Effect of Feature Space Topology Augmentation. To investigate the impact of our proposed feature space-based topological enhancement (i.e., TE), we first fix the temperature coefficient and later perform topological augmentation using RE-RE, Raw-TE and RE-TE, respectively, and the results are shown in Table 4, where the first column represents the topological augmentation method used and the second column represents whether dynamic temperature (i.e., DT) is used. From the first and third rows of the table we can see that Raw-TE achieves better results compared to RE-RE on all but Pubmed and Amazon-Computers, which indicates that removing edges randomly for both views at the same time may indeed lead to the loss of some important connections in the graph, when it is more effective to use the original graph and the feature-based topology enhancement; RE -TE also achieves better results on most of the datasets compared to Raw-TE, which indicates that the combination of RE and TE provides a richer perspective compared to the original graph, which further facilitates the training of comparative learning.

Table 4. Effectiveness evaluation of TE and DT.

Aug	DT	Cora	Citeseer	Pubmed	DBLP	Wiki-cs	Amazon-c	Amazon-p	Coauthor-cs
RE-RE	✗	83.32	71.31	85.30	84.24	78.00	87.40	91.59	92.62
RE-RE	✓	84.21	72.93	85.62	84.54	78.62	87.76	92.63	92.95
Raw-TE	✗	85.30	72.45	85.21	84.62	79.56	87.35	92.52	93.31
Raw-TE	✓	85.90	72.75	86.10	85.23	79.72	87.51	92.83	93.42
RE-TE	✗	85.72	73.11	85.55	84.52	81.04	88.00	92.33	93.12
RE-TE	✓	86.10	73.64	86.10	85.35	81.80	88.70	92.65	93.50

Effect of Dynamic Temperature. We conducted experiments on the dynamic temperature on each of the three topology augmentations, and the experimental results results are the second, fourth and sixth rows of Table 4. From the results in the table we can see that either RE, TE or RT are improved after the addition of dynamic temperature, which also proves the effectiveness of dynamic temperature. And to verify that the representation of the nodes in the graph tends to converge within clusters and separate between clusters as the temperature coefficient keeps changing, we visualize the training process of Cora in two dimensions by using the t-SNE [12] algorithm, whose results are shown in

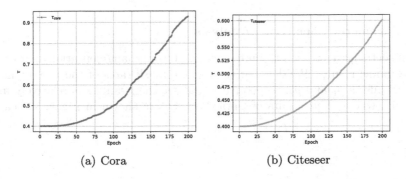

(a) Cora   (b) Citeseer

Fig. 2. Temperature coefficient changes during training.

Fig. 3, from left to right epochs of 10-40-90-150-200. From the figure we can see that initially we get relatively uniform representations (e.g., epoch 10 and 40), because at this time the main focus is on difficult negative samples, and the goal of training is to separate nodes with higher similarity to positive samples, and as the training proceeds with increasing temperature coefficient, our focus on difficult negative samples decreases, and at this time the tendency is to separate samples with lower similarity (i.e., samples of different classes) are separated, producing a clustering-like effect, such as epoch 90–200. In Fig. 2, we visualized the trend of temperature coefficient change on the Cora and Citeseer datasets. It can be seen from the figure that the slope of the curve gradually increased, which also means that with the increasing similarity between the positive samples (i.e., the molecule of the penalty term in Eq. 7), the model was required to pay less attention to the hard negative samples, which was consistent with our expectation.

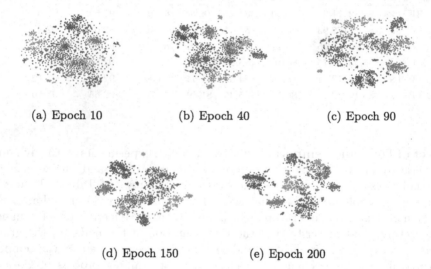

(a) Epoch 10   (b) Epoch 40   (c) Epoch 90

(d) Epoch 150   (e) Epoch 200

Fig. 3. Visualization results presented on Cora using t-SNE as training is iterated.

4.6 Parameter Sensitivity Analysis

In this section, we studied the influence of attribute augmentations parameters on the model results. We denoted the feature mask rates of the two augmentations graphs as p_1 and p_2, respectively, and set their ranges as relatively meaningful 0.1 to 0.6. The results of Cora data set are shown in Fig. 4. We can see

Table 5. The hyperparameters of IAG are set, where p_t represents the probability of RE, p_1 and p_2 represent the probability of graph $\tilde{G}_1$ and $\tilde{G}_2$ feature masks respectively, β controls the penalty term, and k represents topk.

Dataset	p_t	p_1	p_2	β	k	τ	τ_{max}	Learning rate	Training epochs	Hidden dimension	Activation function
Cora	0.2	0.4	0.1	0.1	9	0.4	1.0	0.0005	200	128	ReLU
Citeseer	0.3	0.5	0.5	0.001	9	0.4	1.0	0.001	200	256	PReLU
Pubmed	0.1	0.2	0.1	0.2	2	0.6	1.0	0.001	600	256	ReLU
DBLP	0.1	0.1	0.5	0.001	7	0.8	1.0	0.001	1000	256	ReLU
Wiki-CS	0.1	0.1	0.1	0.002	2	0.5	1.0	0.01	3000	256	RReLU
Amazon-Computers	0.2	0.1	0.1	0.001	9	0.2	0.5	0.01	2000	128	RReLU
Amazon-Photo	0.1	0.1	0.1	0.005	7	0.5	1.0	0.1	2000	256	ReLU
Coauthor-CS	0.1	0.1	0.1	0.01	8	0.4	0.6	0.8 0.0005	1000	256	RReLU

that when parameters change, the point classification results are relatively stable, which indicates that our model is not sensitive to these probabilities and proves the robustness of hyperparameter tuning.

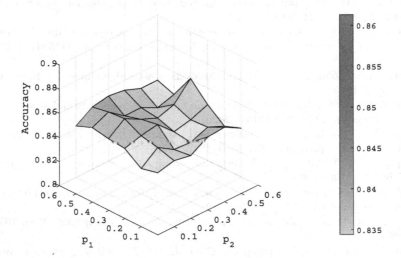

Fig. 4. Performance of IAG with different feature mask hyperparameters in node classification of Cora dataset in terms of accuracy.

5   Conclusion

In this paper, we propose graph contrastive learning based on intrinsic augmentation, named IAG, which consists of two components: 1) In the topology augmentation part, we propose a novel topology augmentation strategy based on potential connections in the feature space, which complements traditional topology augmentation by allowing different graphs to obtain augmentation strategies that are more suitable for their own characteristics to advance traditional graph contrast learning. 2) We explore the effect of temperature coefficient in the loss function on the quality of the final representation and propose a dynamic temperature with penalty term, which helps to generate high-quality node representations. The effectiveness of the method is verified by comparing its experimental results with various existing algorithms.

Acknowledgement. This study is funded in part by the National Natural Science Foundation of China (No. 61906002, 62076005, U20A20398), the Natural Science Foundation of Anhui Province (2008085QF306, 2008085MF191, 2008085UD07), and the University Synergy Innovation Program of Anhui Province, China (GXXT-2021-002).

References

1. Bachman, P., Hjelm, R.D., Buchwalter, W.: Learning representations by maximizing mutual information across views. Adv. Neural Inf. Process. Syst. **32** (2019)
2. Grover, A., Leskovec, J.: node2vec: scalable feature learning for networks. In: Proceedings of the 22nd International Conference on Knowledge Discovery and Data Mining, pp. 855–864 (2016)
3. Hassani, K., Khasahmadi, A.H.: Contrastive multi-view representation learning on graphs. In: International Conference on Machine Learning, pp. 4116–4126. PMLR (2020)
4. Horn, R.A., Johnson, C.R.: Matrix Analysis. Cambridge University Press, Cambridge (2012)
5. Jin, W., et al.: Self-supervised learning on graphs: deep insights and new direction. arXiv preprint arXiv:2006.10141 (2020)
6. Kipf, T.N., Welling, M.: Semi-supervised classification with graph convolutional networks. arXiv preprint arXiv:1609.02907 (2016)
7. Kipf, T.N., Welling, M.: Variational graph auto-encoders. arXiv preprint arXiv:1611.07308 (2016)
8. Klicpera, J., Weißenberger, S., Günnemann, S.: Diffusion improves graph learning. arXiv preprint arXiv:1911.05485 (2019)
9. Linsker, R.: Self-organization in a perceptual network. Computer **21**(3), 105–117 (1988)
10. Liu, N., Tan, Q., Li, Y., Yang, H., Zhou, J., Hu, X.: Is a single vector enough? Exploring node polysemy for network embedding. In: Proceedings of the 25th International Conference on Knowledge Discovery & Data Mining, pp. 932–940 (2019)
11. Liu, X., et al.: Self-supervised learning: generative or contrastive. IEEE Trans. Knowl. Data Eng. **35**, 857–876 (2021)
12. Van der Maaten, L., Hinton, G.: Visualizing data using t-SNE. J. Mach. Learn. Res. **9**(11) (2008)
13. Mernyei, P., Cangea, C.: Wiki-CS: a Wikipedia-based benchmark for graph neural networks. arXiv preprint arXiv:2007.02901 (2020)
14. Neyshabur, B., Khadem, A., Hashemifar, S., Arab, S.S.: NETAL: a new graph-based method for global alignment of protein-protein interaction networks. Bioinformatics **29**(13), 1654–1662 (2013)
15. Page, L., Brin, S., Motwani, R., Winograd, T.: The PageRank citation ranking: bringing order to the web. Technical report, Stanford InfoLab (1999)
16. Peng, Z., et al.: Graph representation learning via graphical mutual information maximization. In: Proceedings of The Web Conference 2020, pp. 259–270 (2020)

17. Perozzi, B., Al-Rfou, R., Skiena, S.: DeepWalk: online learning of social representations. In: Proceedings of the 20th International Conference on Knowledge Discovery and Data Mining, pp. 701–710 (2014)
18. Sen, P., Namata, G., Bilgic, M., Getoor, L., Galligher, B., Eliassi-Rad, T.: Collective classification in network data. AI Mag. 29(3), 93–93 (2008)
19. Shchur, O., Mumme, M., Bojchevski, A., Günnemann, S.: Pitfalls of graph neural network evaluation. arXiv preprint arXiv:1811.05868 (2018)
20. Veličković, P., Cucurull, G., Casanova, A., Romero, A., Lio, P., Bengio, Y.: Graph attention networks. arXiv preprint arXiv:1710.10903 (2017)
21. Velickovic, P., Fedus, W., Hamilton, W.L., Liò, P., Bengio, Y., Hjelm, R.D.: Deep graph infomax. In: ICLR (Poster), vol. 2, no. 3, p. 4 (2019)
22. Wang, F., Liu, H.: Understanding the behaviour of contrastive loss. In: Proceedings of the IEEE/CVF Conference on Computer Vision and Pattern Recognition, pp. 2495–2504 (2021)
23. Wang, T., Isola, P.: Understanding contrastive representation learning through alignment and uniformity on the hypersphere. In: International Conference on Machine Learning, pp. 9929–9939. PMLR (2020)
24. Wang, X., Zhu, M., Bo, D., Cui, P., Shi, C., Pei, J.: AM-GCN: adaptive multi-channel graph convolutional networks. In: Proceedings of the 26th International Conference on Knowledge Discovery & Data Mining, pp. 1243–1253 (2020)
25. Wu, M., Zhuang, C., Mosse, M., Yamins, D., Goodman, N.: On mutual information in contrastive learning for visual representations. arXiv preprint arXiv:2005.13149 (2020)
26. Yang, J., Leskovec, J.: Defining and evaluating network communities based on ground-truth. In: Proceedings of the ACM SIGKDD Workshop on Mining Data Semantics, pp. 1–8 (2012)
27. You, Y., Chen, T., Sui, Y., Chen, T., Wang, Z., Shen, Y.: Graph contrastive learning with augmentations. Adv. Neural. Inf. Process. Syst. 33, 5812–5823 (2020)
28. Zhang, H., Wu, Q., Yan, J., Wipf, D., Yu, P.S.: From canonical correlation analysis to self-supervised graph neural networks. Adv. Neural. Inf. Process. Syst. 34, 76–89 (2021)
29. Zhu, Y., Xu, Y., Liu, Q., Wu, S.: An empirical study of graph contrastive learning. arXiv preprint arXiv:2109.01116 (2021)
30. Zhu, Y., Xu, Y., Yu, F., Liu, Q., Wu, S., Wang, L.: Deep graph contrastive representation learning. arXiv preprint arXiv:2006.04131 (2020)
31. Zhu, Y., Xu, Y., Yu, F., Liu, Q., Wu, S., Wang, L.: Graph contrastive learning with adaptive augmentation. In: Proceedings of the Web Conference 2021, pp. 2069–2080 (2021)

Research on Global Collision Avoidance Algorithm for Unmanned Ship Based on Improved Artificial Potential Field Algorithm

Yue You[1]([✉]), Ke Chen[1], Yang Zhang[1], Jingxiang Feng[2], and Yushen Huang[2]

[1] Naval Research Institute, Beijing 100161, China
youyue_nudt@126.com
[2] Jiangsu Automation Research Institute, Lianyungang 222006, Jiangsu, China

Abstract. Collision avoidance technology is very important in the research of unmanned ship path planning. Aiming at the problem that the existing global path planning algorithm of unmanned vehicle is easy to fall into the local optimal solution and the target point is unreachable, a path planning algorithm based on improved artificial potential field was designed. According to the route requirement of USVS, the power function of target distance change is introduced to improve the repulsive field model. The dangerous target points are screened by using the target point judgment method based on the safe arrival radius. Finally, by dynamically adjusting the artificial potential field coefficient, the unmanned vehicle can jump out of the local optimal trap. Based on the theoretical research, the simulation experiment of path planning is designed. The simulation results show that the algorithm can jump out of the local minimum point trap, and the path planned by the algorithm can successfully reach the target point under the condition of multiple obstacles.

Keywords: artificial potential field method · USV · global collision avoidance · environment model

1 Introduction

The goal of global path planning for unmanned vehicles is to plan a navigation path that avoids obstacles and provide a reference for the safe navigation of unmanned vehicles. The premise of the algorithm is that the starting position, arrival position and obstacle information of the USV are known. The advantage of the global planning method is that it can find the optimal route in the current environment and solve the problem of unmanned vehicle track tracking. At present, many USV collision avoidance methods based on global road planning have been proposed.

With the development of modern intelligent optimization theory, algorithms such as genetic algorithm, ant colony algorithm and particle swarm algorithm are gradually applied to the research of collision avoidance in the field of unmanned boats. The genetic algorithm has the advantages of good parallelism and is not easy to fall into the local optimal solution, so it is a good choice to solve the global planning problem of the

L. Pan et al. (Eds.): BIC-TA 2022, CCIS 1801, pp. 358–369, 2023.
https://doi.org/10.1007/978-981-99-1549-1_28

unmanned vehicle. Liu et al. proposed a path planning method based on evolutionary genetic algorithm and hierarchical strategy to solve the problem of long planning time due to the large navigation environment of unmanned vehicles. This method can avoid planning from falling into local optimal solution and improve planning efficiency [1]. On this basis, compared with the genetic algorithm, the ant colony algorithm has memory. Although it increases some computational overhead, the solution efficiency is improved. Chang et al. proposed an unmanned vehicle collision avoidance method based on the improved ant colony algorithm to solve the problem of insufficient search ability and stagnation of the algorithm. The selection probability is changed by increasing the weight of the direction angle to realize the optimization of the global path. Compared with the traditional algorithm, the path planning effect of this method is smoother, and the search speed is faster [2, 3]. The particle swarm optimization algorithm has the advantages of low time complexity and fast solution speed and has been widely used in the global planning research of unmanned vehicles. Shriyam et al. proposed an unmanned vehicle path planning method based on particle swarm optimization combined with simulated annealing algorithm to solve the problem that traditional planning algorithms tend to fall into local minimum points. The simulation results verify that the method has a high degree of convergence and is easy to jump out of the local optimal trap [4, 5].

In this paper, the artificial potential field method is applied to the unmanned boat platform, and a path planning algorithm based on the improved artificial potential field is designed to solve the problems of local optimal traps and unreachable targets in the algorithm and realize the path planning of the unmanned boat. The planning function is used as the premise constraint for the local collision avoidance of the unmanned vehicle.

2 USV Collision Avoidance Based on Improved Artificial Potential Field Method

2.1 Improvement of Artificial Potential Field Method

It is necessary to consider the problem that the movement and manipulation characteristics of the unmanned vehicle platform are different from those of mobile robots [9]. In particular, the traditional artificial potential field method has the problem that it is easy to fall into the local optimal solution trap. Once the unmanned vehicle falls into the local minimum trap, the course and speed calculated by the algorithm will be in a state of rapid change and oscillation. The change of the position at any time will cause the change of the relative distance from the obstacle, which will lead to a large change in the strength of the repulsive force field, that is, the strength of the repulsive force field changes too fast. In the actual navigation of the unmanned vehicle, the rapid change of the potential field may cause the heading to change too fast, which will cause frequent steering [10–12].

(1) **Improvement of the problem of unreachable target points**

According to the expression of the classic artificial potential field repulsion function, it can be concluded that the obstacle repulsion is inversely proportional to the distance. When there are obstacles around the target point, the repulsion of the unmanned vehicle will become very large. At this time, the distance between the USV and the target point is very close, and the gravitational force is relatively small. In this case, it is difficult for the unmanned boat to reach the target point. The phenomenon shown is that the unmanned boat will circle around the target point and obstacles, and the course and rudder angle of the unmanned boat will change greatly. For the surface unmanned vehicle platform, frequent changes in navigation and rudder operations at high speeds may cause problems such as hull rollover and navigation system instability. So in this case it is very dangerous. The unreachable problem of the target point of the artificial potential field method is shown in Fig. 1.

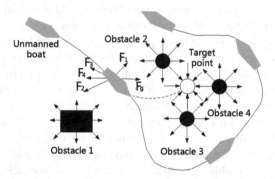

Fig. 1. The problem of unreachable target points in the artificial potential field method

In Fig. 1, $F_1 - F_4$ the repulsive forces generated by obstacles 1–4 are represented respectively, and F_g the gravitational force generated by the target point is represented. The 4 black dots represent 4 obstacles, the white dots represent the target point that the USV is going to, the dotted line is the ideal path planned by the artificial potential field algorithm, and the solid line is the collision avoidance path actually planned by the algorithm. It can be seen that the USV did not follow the optimal path to reach the target point, but circled around the target point and three obstacles along the actually planned collision avoidance path, and could never reach the target point.

Aiming at the problem of target unreachability, literature [13] adopts the method of introducing a new artificial potential field repulsion function to solve this problem. The specific implementation method is to change the repulsion field function of the classical artificial potential field, and add the distance influence factor between the target point and the object to the obstacle repulsion function on the basis of the original repulsion field. When the distance between the target point and the obstacle is relatively close, the USV is very close to the obstacle and the target point. Although the repulsive force of obstacles is very large, the repulsive force function is affected by the distance factor, which can drag the repulsive force field to a certain extent, and the final output repulsive force of the algorithm will decrease, making the unmanned vehicle continue to approach the target point. The improved repulsion field function expression is as follows:

$$U_r^*(q) = \begin{cases} \frac{1}{2}\eta\left(\frac{1}{d(q-q_{obs})} - \frac{1}{\rho_{max}-\rho_{min}}\right)^2 d^k(q-q_{goal}), & d(q-q_{obs}) \le \rho_0 \\ 0, & d(q-q_{obs}) > \rho_0 \end{cases} \quad (1)$$

where $U_r^*(q)$ is the improved repulsion potential field function, η is the gain factor of q the repulsion potential field, is the current position of the USV, is q_{obs} the position of the obstacle, is $d(q - q_{obs})$ the distance ρ_0 between the USV and the obstacle, and is the influence radius of the obstacle, is k the distance influence factor of the repulsive potential field function.

(2) **Improvement of local minimum point problem**

During the voyage of the unmanned boat, if there is a certain point in the artificial potential field, the gradient value of the potential field here is zero, and the gravitational force and repulsive force at this point are exactly equal in magnitude and opposite in direction. When the unmanned boat sails to this point, it is easy to fall into the local optimal solution or oscillation, so that the unmanned boat cannot move forward to the target point. The heading output by the algorithm will change frequently, causing frequent steering problems, which may lead to problems such as hull rollover and navigation instability. In this case, unmanned boats are very dangerous to sail. Therefore, in order to ensure the safety of the USV in the process of collision avoidance, it should avoid falling into the local optimal solution. The local minima of the artificial potential field method are shown in Fig. 2.

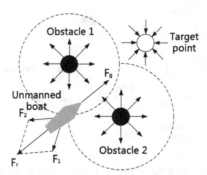

Fig. 2. The situation of local minimum point problem of artificial potential field method

In Fig. 2, respectively represent F_1, F_2 the repulsive force generated by F_r obstacles 1–2, represent the combined repulsive force of obstacles, and represent F_g the attractive force generated by the target point. The black dots represent obstacles, and the white dots represent the target points that the USV is going to. The dotted line represents the repulsive influence range of obstacles. A case of the local minimum point problem is when the unmanned boat, the obstacle and the target point are on a straight line, that is $F_g = F_r$, the gravitational force of the target point is equal to the combined repulsion force of the obstacle, and the direction is opposite. The resultant force of the boat here is exactly zero; another situation is that under the action of the attraction of the target point and the repulsion of obstacles, a circular potential field is generated, which causes the unmanned boat to fall into a circular potential field when it sails to this point. In this case, the output value of the algorithm will fall into a local optimal solution at the next moment, and the course change of the unmanned boat will cause a large oscillation, which will cause the unmanned boat to be unable to move forward and never reach the target point.

On this basis, a method to dynamically adjust the artificial potential field coefficient is proposed to solve this problem. When the USV falls into a local optimal trap, the resultant force of the artificial potential field is zero, and the magnitude of the repulsive force is adjusted by changing the coefficient of the potential field so that the resultant force is not zero, thus breaking the local balance between the attraction and repulsion and jumping out of the local minimum value points. This method is divided into methods of increasing the gravitational potential field coefficient and reducing the repulsive potential field coefficient, that is, increasing the gravitational force of the target point, or reducing the repulsive force of obstacles, changing the total potential field strength, so that the unmanned vehicle moves towards the target point. And limit the heading angle change between the next step and the previous step within 45°. The method of dynamically adjusting artificial potential field coefficients is shown in Fig. 3.

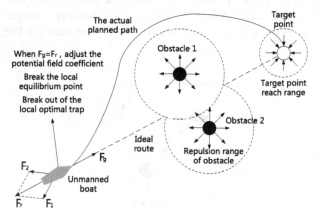

Fig. 3. Schematic diagram of the solution to the problem of 8 local minimum points

Other hand, it is very critical to accurately judge the situation where the USV falls into a local minimum point. This paper adds two judgment conditions for local minimum points. When the following two conditions are met, it is judged that the USV has fallen into the local optimal solution trap.

(a)   Before calculating the next position, that is, before outputting the heading at the next moment, if the unmanned boat does not reach the target point, that is, the distance between the unmanned boat and the target point is not zero.
(b)   for five consecutive times is less than the step size, then when the above two conditions are met, it is judged that the unmanned boat has fallen into the trap of a local minimum point.

When the path of the USV falls into the local optimal solution trap, the magnitude and direction of the resultant force are adjusted by dynamically adjusting the coefficient of the artificial potential field. First increase the gravitational field coefficient to increase the gravitational force of the target point, and then reduce the obstacle repulsion by reducing the repulsive field coefficient. If after adjusting the parameters, it is still unable to jump out of the trap of local optimal solution, then use the method of adding new target points to solve it. At 5 steps above/below the midpoint of the connecting line between the USV and the target point, plan a new target point. Through the gravitational traction of the new target point, the unmanned vehicle can jump out of the local optimal trap. If the planned target point is a dangerous target point, that is, there are obstacles around, then continue to plan new target points on this point until it jumps out of the local optimal trap.

2.2   Algorithm Design of Unmanned Vessel Path Planning

In the actual navigation of the unmanned vehicle, if the potential field changes too fast, it may cause the heading to change too fast, resulting in frequent steering. In this case, the planned path is unreasonable, so the repulsion function of the artificial potential field needs to be improved.

The improved expression of the repulsion field function is as follows:

$$U_r^*(q) = \begin{cases} (\frac{1}{6})^{n-1} \eta \frac{(\rho_{max} - d(q - q_{obs}))^n}{\rho_{max} - \rho_{min}}, & d(q - q_{obs}) \leq \rho_{max} - \rho_{min} \\ 0, & d(q - q_{obs}) > \rho_{max} - \rho_{min} \end{cases} \tag{2}$$

where $U_r^*(q)$ is the improved repulsion field function, η is the gain factor of the repulsion field, q is the current position of the unmanned boat, q_{obs} is the position of the obstacle, $d(q - q_{obs})$ is the distance from the unmanned boat to the obstacle, ρ_{max} is the maximum influence radius of the obstacle, and ρ_{min} is the obstacle. The influence radius of the dead zone is the influence radius of the obstacle $\rho_{max} - \rho_{min}$, and n is the control parameter of the repulsion field function. The larger n is, the faster the function converges, and the smaller n is, the more severe the path oscillation is.

In order to solve the target unreachable problem of the algorithm, the factor of the relative position between the target point and the USV is introduced into the improved repulsion field function, and the improved repulsion field function is multiplied by the formula $(q - q_{goal})^n$. At this time, the repulsion field will also be affected by the gravitational field. When there is an obstacle near the target point, the repulsion force of the unmanned ship at the target point will be zero, thereby solving the problem of the target being unreachable and obtaining a further improved repulsion field. The expression of the function is as follows:

$$U_r^*(q) = \begin{cases} (\frac{1}{6})^{n-1} \eta d(q - q_{goal})^n \frac{(\rho_{max} - d(q - q_{obs}))^n}{\rho_{max} - \rho_{min}}, & d(q - q_{obs}) \leq \rho_{max} - \rho_{min} \\ 0, & d(q - q_{obs}) > \rho_{max} - \rho_{min} \end{cases} \quad (3)$$

where q is the current position of the USV, and q_{goal} is the position of the target point, $(q - q_{goal})^n$ representing the distance factor between the USV and the target point, and the meanings of other expressions are the same as above. The negative gradient of the improved repulsion field is calculated, and the expression of the final improved repulsion function is as follows:

$$F_r^*(q) = -\nabla U_r^*(q) = \begin{cases} (\frac{1}{6})^{n-1} n\eta d(q - q_{goal})^n \frac{(\rho_{max} - d(q - q_{obs}))^{n-1}}{\rho_{max} - \rho_{min}} \nabla d(q - q_{obs}), & d(q - q_{obs}) \leq \rho_{max} - \rho_{min} \\ 0, & d(q - q_{obs}) > \rho_{max} - \rho_{min} \end{cases} \quad (4)$$

where $F_r^*(q)$ is the improved repulsive force function, $-\nabla U_r^*(q)$ is the negative gradient function of the repulsive force field to the distance, and $\nabla d(q - q_{obs})$ is the gradient value of the distance between the obstacle and the unmanned vehicle, and the meanings of other expressions are the same as above. The unmanned boat advances under the gravitational force of the target point and will also be affected by the repulsive force of obstacles during navigation, and the direction of repulsive force is opposite to the direction of obstacles. Under the joint force of the artificial potential field, the unmanned boat finally reaches the target point.

Next, improve the gravitational field function. This part mainly considers the problem of excessive gravity of the target point. According to the improvement method of this paper, the gravitational field function is corrected for the problem of encountering obstacles, and the limit threshold of the influence range of the target point is added to avoid excessive gravity caused by excessive distance. The expression of the improved gravitational field function is as follows:

$$U_g^*(q) = \begin{cases} \frac{1}{2}\xi d^2(q - q_{goal}), & d(q - q_{goal}) \leq \rho_{goal}^* \\ \xi\rho_{goal}^* d(q - q_{goal}) - \frac{1}{2}\xi(\rho_{goal}^*)^2, & d(q - q_{goal}) > \rho_{goal}^* \end{cases} \quad (5)$$

where $U_g^*(q)$ is the improved gravitational field function, $d(q - q_{goal})$ is the distance from the USV to the target point, ξ is the gain factor of the gravitational field, and ρ_{goal}^* is the limiting threshold of the influence range of the target point, and the meanings of other expressions are the same as above. The gravitational force generated by the target point is constrained by setting the distance threshold range, and the negative gradient is calculated for the improved gravitational field function. The expression of the final improved gravitational function is as follows:

$$F_g^*(q) = -\nabla U_g^*(q) = \begin{cases} \xi d(q - q_{goal})\nabla d(q_{goal} - q), & d(q - q_{goal}) \leq \rho_{goal}^* \\ \xi\rho_{goal}^* \nabla d(q_{goal} - q), & d(q - q_{goal}) > \rho_{goal}^* \end{cases} \quad (6)$$

where $F_g^*(q)$ is the improved gravitational function, $-\nabla U_g^*(q)$ is the negative gradient function of the gravitational field to the distance, and $\nabla d(q - q_{goal})$ is the gradient value of the distance between the USV and the target point, and the meanings of other expressions are the same as above. The resultant force potential field is obtained by vector synthesis of the improved repulsive force and gravitational field, and the resultant force is obtained by calculating the negative gradient of the distance. The resultant force direction of the artificial potential field is the motion direction of the USV in path planning. The expression of the resultant force function is as follows:

$$F_h^*(q) =$$

$$\begin{cases} \sqrt{(\xi d_g \nabla d_g)^2 + g_o^*(q)^2 + 2\xi d_g \nabla d_g g_o^*(q) \cos(\pi - \frac{\theta}{2})}, & d_o \le \rho_o \wedge d_g \le \rho_g^* \\ -\xi \rho_g^* \nabla d_g, & d_o > \rho_o \wedge d_g > \rho_g^* \end{cases} \quad (7)$$

where $F_h^*(q)$ is the resultant force function of the improved artificial potential field, $g_o^*(q)$ expresses the expression of the improved repulsive force function, $g_o^*(q) = n\eta d_g^n(\rho_{max} - d_o)^{n-1}\nabla d_o/6^{n-1}\rho_o$, d_g and d_o represent the distances from the unmanned boat to the target point and obstacles respectively, and ∇d_g is the gradient value of the distance η between the unmanned boat and the target point, The sum ξ is the gain factor of the repulsive force and the gravitational potential field, θ is the angle between the directions of the gravitational force and the repulsive force, ρ_o represents the influence radius of each obstacle, and ρ_g^* is the limiting threshold of the influence range of the target point, and the meanings of other expressions are the same as above. The $\overrightarrow{F}_h^*(q)$ direction of the resultant force is the course of the unmanned boat φ_h, and the magnitude of the $\overrightarrow{F}_h^*(q)$ resultant force is the speed of the unmanned boat v_s. According to the course and speed, the position of the unmanned boat at the next moment can be calculated as:

$$\begin{cases} x_t = \lambda \cdot v_s \cos\varphi_h + x_0 \\ y_t = \lambda \cdot v_s \sin\varphi_h + y_0 \end{cases} \quad (8)$$

where (x_0, y_0) is the coordinate point of the current position of the unmanned boat, (x_t, y_t) is the position coordinate of the unmanned boat at the next moment, λ is the moving step of the unmanned boat, φ_h is the course angle of the unmanned boat, and v_s is the speed value of the unmanned boat.

3 Simulation

Firstly, the simulation of the single static obstacle scene is carried out. The initial navigation parameters of the unmanned vehicle are the heading $\varphi = 90°$ (the direction of the y-axis is true north) 0°, the speed, $v = 10$ kn the 0° azimuth of the obstacle at the heading angle of the unmanned vessel, and the distance 2 km. The simulation results are shown in Table 1.

The starting point of the USV is the (0, 10) location in the figure, and the ending point is the (30, 10) location. Obstacles are red dots, and the blue curve is the trajectory of the

Table 1. Simulation results of a single static obstacle scene

Test No.	number of obstacles	D CPA (m)	T CPA (s)	Minimum safe encounter distance (m)	The shortest distance from the target point (m)	avoid Amplitude (°)
1	1	136.589	60.265	58.721	8.652	21.46

unmanned boat from the starting point to the ending point for collision avoidance. From the above table, it can be seen that the collision avoidance method of the unmanned boat is turning right, the shortest safe encounter distance is 58.721 m, the avoidance range is 21.46°, and the target point is successfully reached. The simulation trajectory of a single static obstacle scene is shown in Fig. 4.

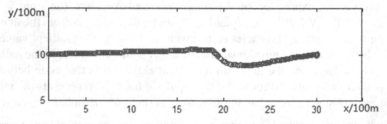

Fig. 4. Simulation trajectory of a single static obstacle scene (Color figure online)

Next, simulate multiple static obstacle scenarios. The initial navigation parameters of the unmanned boat are, heading $\varphi = 45°$, speed $v = 10\,\text{kn}$, and the number of obstacles. There are a total of 18 obstacles, of which 9 are fixed obstacles, and 9 are randomly generated obstacles during the navigation of the unmanned boat. Within the test area. The simulation results of using the original artificial potential field method and the improved algorithm in this paper are shown in Table 2.

Table 2. Simulation results of three static obstacle scenarios

Test No.	number of obstacles	D CPA (m)	T CPA (s)	Minimum safe encounter distance (m)	The shortest distance from the target point (m)	avoid Amplitude (°)
1	18	147.924	66.782	62.136	11.384	60.92
2	18	145.846	65.165	68.429	7.465	12.06

The starting point of the USV is the (0, 0) location, and the ending point is the (30, 16) location. Obstacles are red dots, and the blue curve is the trajectory of the unmanned boat from the starting point to the end point of collision avoidance. Experiments 1 and 2 are the simulation results of the original artificial potential field method and the improved calculation respectively. It can be seen from the above table that the shortest safe encounter distance using the original artificial potential field method is 62.136 m, and the avoidance range is 60.92°. The shortest safe encounter distance using the improved algorithm is 68.429 m, and the avoidance range is 12.06°. The simulation trajectories of multiple static obstacle scenarios using the traditional artificial potential field method and the improved path planning algorithm are shown in Fig. 5.

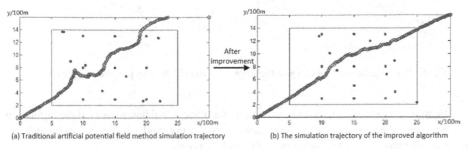

(a) Traditional artificial potential field method simulation trajectory   (b) The simulation trajectory of the improved algorithm

Fig. 5. Simulation trajectories of more than 15 static obstacle scenarios (Color figure online)

From Fig. 5(a), it can be seen that when the traditional artificial potential field method is used, the planned path has oscillations, and the avoidance of obstacles is too large, which will lead to frequent steering of the unmanned vehicle in the process of collision avoidance problem, so the collision avoidance action is unqualified. From Fig. 5(b), it can be seen that the improved algorithm in this paper has improved the problem of path oscillation, and the smoothness of the path has been improved. The unmanned boat can avoid a small range of navigation by following this path, which enhances navigation. Stability. By adjusting the control parameters of the algorithm, the planned USV trajectory is optimized. The conditions of course $\varphi = 45°$, speed $v = 10\,\text{kn}$. The simulation results after algorithm parameter optimization are shown in Table 3.

Table 3. Simulation results after optimization of algorithm parameters

Test No.	number of obstacles	D CPA (m)	T CPA (s)	Minimum safe encounter distance (m)	The shortest distance from the target point (m)	avoid Amplitude (°)
1	18	160.113	68.233	67.254	10.996	10.77
2	18	158.463	67.797	69.798	8.034	7.34

From the table above, it can be seen that Tests 1 and 2 are the simulation results before and after algorithm parameter optimization. The shortest safe encounter distance of Test 1 is 67.254 m, and the avoidance range is 10.77°. The shortest safe encounter distance of Test 2 after algorithm parameter optimization is The encounter distance is 69.798 m, and the avoidance range is 7.34°. The simulation trajectories of multiple static obstacle scenarios before and after algorithm parameter optimization are shown in Fig. 6.

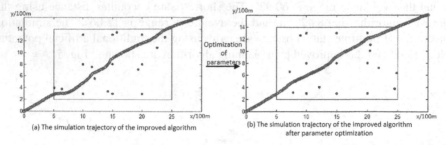

Fig. 6. The simulation trajectory of the improved algorithm after parameter optimization

From Fig. 6(a) and (b), it can be seen that after the control parameters of the algorithm are adjusted, the smoothness of the planned path is improved, obstacles can be avoided by a small margin, and the stability of the path is good.

4   Conclusion

In this paper, a path planning algorithm for USV based on improved artificial potential field is designed. Based on the analysis of the classic artificial potential field model according to the route requirements of the unmanned ship, a method of dynamically adjusting the coefficient of the artificial potential field is proposed to improve the repulsive and gravitational potential field functions, so as to solve the problem of easily falling into the trap of local minimum points; Based on the target point judgment method of the safe arrival radius, the dangerous target points are screened, and the relative position factor between the target point and the unmanned ship is introduced into the repulsion field function to solve the problem of target point inaccessibility; after simulation comparison, the improved algorithm Compared with the classic artificial potential field method, the time to jump out of the local optimal trap is shortened, the path oscillation is reduced, the smoothness of the path is improved, and the planned path can reach the target point when there are obstacles near the target point. Target. Finally, the simulation results of USV path planning are shown.

References

1. Zhuang, J., Wan, L., Liao, Y., et al.: Research on global path planning of unmanned surface vehicle based on electronic chart. Comput. Sci. **38**(9), 211–214 (2011)
2. Oh, H., Niu, H., Tsourdos, A., et al.: Development of collision avoidance algorithms for the C-Enduro USV. IFAC Proc. Vol. **47**(3), 12174–12181 (2014)
3. Chen, C., Tang, J.: Path planning and design of unmanned surface vehicle based on visible graphics method. China Shipbuild. **54**(1), 129–135 (2013)
4. Liu, J.: Research on collision avoidance system of unmanned vessel based on evolutionary genetic algorithm. Dalian Maritime University (2015)

5. Song, C.H.: Global path planning method for USV system based on improved ant colony algorithm. Appl. Mech. Mater. **568**, 785–788 (2014)
6. Shang, M., Zhu, Z., Zhou, T.: Research on intelligent collision avoidance method for unmanned surface vehicle based on improved ant colony algorithm. Ship Eng. **38**(9), 6–9 (2016)
7. Shriyam, S., Shah, B.C., Gupta, S.K.: Decomposition of collaborative surveillance tasks for execution in marine environments by a team of unmanned surface vehicles. J. Mech. Robot. **10**(2), 7–25 (2018)
8. Laval, B., Bird, J.S., Helland, P.D.: An autonomous underwater vehicle for the study of small lakes. J. Atmos. Oceanic Tech. **17**(1), 69–76 (2000)
9. Luo, G., Zhang, H., Wang, H., et al.: Application of improved artificial potential field method in robot path planning. Comput. Eng. Design **32**(4), 279–281 (2011)
10. Song, J.: Research on robot obstacle avoidance based on artificial potential field method. Shenyang University of Technology (2017)
11. Song, A., Su, B., Dong, C., et al.: A two-level dynamic obstacle avoidance algorithm for unmanned surface vehicles. Ocean Eng. **170**(15), 351–360 (2018)
12. Fujimori, A., Teramoto, M., Nikiforuk, P.N., et al.: Cooperative collision avoidance between multiple mobile robots. J. Robot. Syst. **17**(7), 347–363 (2000)
13. Wu, P., Xie, S., Liu, H., et al.: Autonomous obstacle a voidance of an unmanned s surface vehicle based on cooperative manoeuvring. Industr. Rob.: Int. J. **44**(1), 64–74 (2017)
14. Yin, G.: Research on path planning of mobile robots based on improved artificial potential field method. Tianjin University of Technology (2017)
15. Wang, X., Yadav, V.: Cooperative USV formation flying with obstacle/collision avoidance. IEEE Trans. Control Syst. Technol. **15**(4), 672–679 (2007)
16. Moldovan, E., Tatu, S.O., Gaman, T., et al.: A new 94-GHz six-port collision-avoidance radar sensor. IEEE Trans. Microw. Theory Tech. **52**(3), 751–759 (2004)

OCET: One-Dimensional Convolution Embedding Transformer for Stock Trend Prediction

Peng Yang[1,2], Lang Fu[2], Jian Zhang[4], and Guiying Li[2,3](✉)

[1] Department of Statistics and Data Science, Southern University of Science
and Technology, Shenzhen 518055, China
ligy@sustech.edu.cn
[2] Guangdong Provincial Key Laboratory of Brain-Inspired Intelligent Computation,
Department of Computer Science and Engineering, Southern University of Science
and Technology, Shenzhen 518055, China
[3] Research Institute of Trustworthy Autonomous Systems, Southern University
of Science and Technology, Shenzhen 518055, China
[4] Shenzhen Securities Information Co., Ltd., Shenzhen, China

Abstract. Due to the strong data fitting ability of deep learning, the use of deep learning for quantitative trading has gradually sprung up in recent years. As a classical problem of quantitative trading, Stock Trend Prediction (STP) mainly predicts the movement of stock price in the future through the historical price information to better guide quantitative trading. In recent years, some deep learning work has made great progress in STP by effectively grasping long-term timing information. However, as a kind of real-time series data, short-term timing information is also very important, because stock trading is high-frequency and price fluctuates violently. And with the popularity of Transformer, there is a lack of an effective combination of feature extraction and Transformer in STP tasks. To make better use of short term information, we propose One-dimensional Convolution Embedding (OCE). Simultaneously, we introduce effective feature extraction with Transformer into STP problem to extract feature information and capture long-term timing information. By combining OCE and Transformer organically, we propose a noval STP prediction model, One-dimensional Convolution Embedding Transformer (OCET), to capture long-term and short-term time series information. Finally, OCET achieves a highest accuracy up to 0.927 in public benchmark FI-2010 When reasoning speed is twice that of SOTA models and a highest accuracy of 0.426 in HKGSAS-2020. Empirical results on these two datasets show that our OCET is significantly superior than other algorithms in STP tasks. Code are available at https://github.com/langgege-cqu/OCET.

This work was supported partly by the National Natural Science Foundation of China (Grants 62272210), partly by the Guangdong Provincial Key Laboratory (Grant 2020B121201001), partly by the Program for Guangdong Introducing Innovative and Entrepreneurial Teams (Grant 2017ZT07X386), and partly by the Stable Support Plan Program of Shenzhen Natural Science Fund (Grant 20200925154942002).

Keywords: Stock · Limit-Order-Books · Trend · Transformer

1   Introduction

Nowadays, more than half of the stock market transactions are completed through Limit Order Books (LOBs) [21] data which records the information of stock trading in the financial market. In the electronic stock market, price change movement of stocks are of great significance to the stability and circulation of the financial market. Stock Trend Prediction (STP) is to automatically predict the movement direction of stock price in the future. It is a very challenging task, because the stock data itself is chaotic, volatile, high-frequency and sensitive data [1–3]. Meanwhile, market experts can use the values of time series, such as past trading prices and quantities, to construct STP tasks, try to infer price trends mathematically, and can be used to provide useful instructions for future price movements, roughly infer the moving direction of the stock market, and promote the circulation and stability of the financial market [8]. Here, we give a specific example of STP task. As shown in Fig. 1, this picture describes the price change of bitcoin from 19:15 to 20:30 on March 30, 2021 in America. It needs to predict the future price trend (the trend of the yellow arrow) based on the past stock price information (the part surrounded by the black rectangular box).

Starting from the characteristics of stock data, the best model structure in recent years is defined by the paradigm of Convolutional Neural Network (CNN) [15] + Recurrent Neural Networks (RNN) [24] (CRNN), such as DeepLOB [32], DeepFolio [23], MTDNN [17], etc. Such a network extracts the data characteristics of each stock timeline through a CNN network, and aggregates the information of different timelines through a RNN network, so as to predict the movement of stock price. Such networks only aggregate long-term timing information at the last layer of the network, ignoring the short-term timing information of stock attachment time at each specific time point. Moreover, the feature extraction layer of these models is a CNN with deep network layers, which will affect the network speed and cause information redundancy to a certain extent.

In this paper, we propose a noval neural network model called One-dimensional Convolution Embedding Transformer (OCET), whose structure is composed of CNN and Transformer neural networks, for STP tasks. Based on CNN structure, we specially designed OCE module to capture short-term information according to the high-frequency characteristics of stock data. Simultaneously, in order to better extract feature information and grasp the long-term information, we introduced Transformer module with effective CNN module. By combining the two efficiently, OCET can make better use of long-term information and short-term information at the same time, so as to better improve the solution effect of STP.

It should be noted that the stock data itself is high-frequency and volatile, and there is some location information in the short time interval of adjacent transactions. Inspired by SENet [12] and ECA-Net [22], this paper propose

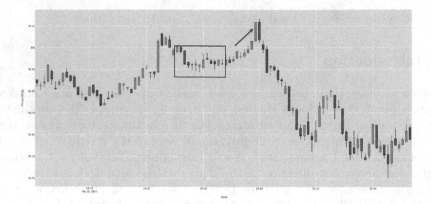

Fig. 1. A picture of Bitcoin price trend, the horizontal axis is the time and the vertical axis is the transaction price, green box chart represents rise and the red box chart represents fall. (Color figure online)

a learnable convolutional relative Positional Embedding (PE) without changing the dimension and order of stock data, called One-dimensional Convolution Embedding (OCE). Our OCE module is essentially the attention of each feature to its adjacent timeline. It will automatically and selectively learn short-term timing information. To aggregate long-term timing information, we use Transformer to aggregate long-term timing information. Transformer [28] was first proposed in 2017 for machine translation tasks. In the past two years, Transformer has made great achievements in the fields of computer vision and natural language processing. On the STP task, the introduction of Transformer is relatively few. In this regard, OCET can be viewed as an example for other researchers to study the application of Transformer in STP task. With the combination of OCE and Transformer, our model achieves an average accuracy of 0.895 in FI-2010 [20] and an average accuracy of 0.416 in HKGSAS-2020 [13].

The rest of the paper is as follows. Section 2 introduces related works. Section 3 describes the the problem of STP. Section 4 describes the architecture of OCET in detail. Section 5 introduces dataset, experiments. Section 6 is the conclusion and future works.

2   Related Work

Traditional statistical methods usually assume that the time series studied of STP are generated by parametric processes [6]. However, it is generally agreed that the performance of stock returns is more complex and usually highly nonlinear [5]. Machine learning technology can capture this arbitrary nonlinear relationship, and there is little or no need for a priori knowledge about the input data [4]. With the popularity of deep learning in various fields, in 2017 tsantekidis [27] first used Support Vector Machine (SVM) and CNN to analyze LOB data to predict the trend of intermediate price change, and achieved better results than manual empirical statistical methods at that time. Because CNN can only extract local information and cannot capture long-term timing information, tsantekidis and others used Long Short Term Memory (LSTM) [9] network to capture long-term timing information and predict the price trend of LOB in the same year, and achieved better results than before. In 2018, kanniainen [26] designed a time channel attention network and used a novel data normalization method to predict stock price trend, which obtained better results.

Later, in order to combine the ability of CNN to extract local information and the ability of RNN to extract long-term timing information, researchers gradually combined CNN and RNN to build a new neural network, CRNN, to predict STP tasks. In 2019, Zhang [32] combined the characteristics of Inception [25] and LSTM to design DeepLOB, which improves the index of stock price trend prediction (accuracy, f1-score, etc.) where the network parameters are greatly reduced. In 2020, sangadiev [23] based on DeepLOB, combined DeepLOB with ResNet [10], a classic network in the field of computer vision, and proposed DeepFolio to improve the network accuracy and ensure that the network can evolve to a deeper level. In 2020, Hailong Huang [17] and others used Multi-scale Deep Neural Network (MTDNN) to predict the trend of LOB intermediate price, and made a new breakthrough by making the network similar to an integrated operation. In 2021, Zhang [31] proposed DeepLOBAttention, which combined the attention mechanism with DeepLOB on the basis of the previous DeepLOB network, and used their attention to improve the network performance to better than before.

In particular, our OCET uses the structure of efficient CNN + Transformer to break the previous design of CRNN, capture long-term and short-term timing information, and better complete STP tasks.

3   Task Formulation

In this section, we will describe two different problems of STP tasks. One is the mid-price prediction task, which is relatively simple and has sparked a lot of research, which is also our main research task; The other is volume-weighted average price predict task, which is proposed with the proposal of HKGSAS-2020 dataset, which is relatively complex.

3.1   Mid-Price Predict

Mid-price predict task uses Level-10 bid orders and ask orders to predict the movement of mid-price. Input is defined $X = [\ x_1,\ x_2, \cdots,\ x_t, \cdots,\ x_{100}]^T \in \mathbb{R}^{100 \times 40}$, where $x_t = [\ p_a^i(t),\ v_a^i(t),\ p_b^i(t),\ v_a^i(t)]_{i=1}^{n=10}$, t is window size, which denote t-th time from now to the past. a and b denote ask orders and bid orders. p^i and v^i denote the price and volume size at i-th level of stock data. $i = 1$ denotes the ask price is minimum and bid price is highest, which means that the priority of price limit transaction is the highest. We create labels through the future movement of mid-price:

$$p_t = \frac{p_a^1(t) + p_b^1(t)}{2} \tag{1}$$

Due to the high frequency and confusion of stock data, researchers [27] generally use percentage change of mean future mid-price, Δp, to predict the long price trend of stocks, Δp is as follows,

$$\Delta p = \frac{m_t(k) - p_t}{p_t} \tag{2}$$

where

$$m_t(k) = \frac{1}{k} \sum_{j=1}^{k} p_{t+j} \tag{3}$$

$m_t(k)$ is mean future mid-price, k is the prediction horizon. Label y is calculated according to the size of Δp, as follows,

$$y = \begin{cases} 1, & \alpha \leq \Delta p \\ 0, & -\alpha < \Delta p < \alpha \\ -1, & \Delta p \leq -\alpha \end{cases} \tag{4}$$

where 1, 0, −1 represent up, stationary, down moving trend of mid-prices. Since the mid-price has the highest priority, the moving direction of the mid-price is generally used to replace the moving direction of the stock price. α is an artificially designed threshold. In this scenario, the STP task is a time series 3 classification task.

3.2   Volume-Weighted Average Price Predict

Volume-Weighted Average Price Prediction (VWAPP) task is to predict the quantified Volume-Weighted Average Price (VWAP), which is proposed from the HKGSAS-2020 dataset. VWAP is defined as,

$$vwap = \frac{\sum_{i=1}^{n} p_i \times v_i}{\sum_{i=1}^{n} v_i} \tag{5}$$

where i represents the i-th price and quantity of the transaction at this time. VWAPP calculates changes of logarithmic VWAP and then discretize it into five different quantized sizes – 10%, 20%, 40%, 20%, 10%, respectively represent labels of −2, −1, 0, 1, 2, as follows,

$$\ln \frac{vwap_t}{vwap_0} \subseteq [2, -1, 0, 1, 2] \tag{6}$$

where t represents a horizon of t seconds in the future. In this scenario, the STP task is a time series 5 classification task.

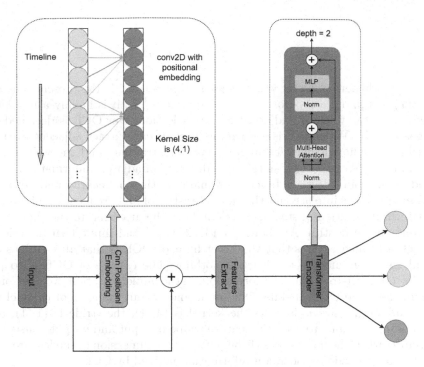

Fig. 2. Illustration architecture of OCET. Raw stock feature captures short-term timing information through Cnn Positional Embedding (OCE), aggregates features through effective Feature Extract, and finally captures long-term timing information through Transformer to complete prediction.

4   Model

The architecture of OCET is illustrated in Fig. 2, which consists of the following three main modules: an convolutional positional embedding layer, feature extract layers and a Transformer encoder layer. Convolutional positional embedding layer is used to capture short-term timing information through a residual connection. Features extract layers are used to extract features and make the feature information of stock more abstract and aggregated. Finally, when capturing long-term timing information, we use Transformer instead of RNN structure, so that the information at different times can be fully. Transformer module also makes up for the disadvantage that RNN can not be calculated in parallel due to its own structure. It greatly improves the prediction accuracy and speed of the whole model.

4.1   One-Dimensional Convolution Embedding

We provide a new learnable positional embedding for stock data, called One-dimensional Convolution Embedding (OCE). Its mathematical description is as follows,

$$oce(x_i) = \sigma(\sum_{j=0}^{l} w_j x_{i+j}) \tag{7}$$

where l is a hyperparameter which is short time span, i is the present moment and $\sigma(x)$ is sigmoid activation function: $\sigma(x) = \frac{1}{1+e^{-x}}$, it is mainly a nonlinear change. w is the weight of adjacent feature selection. Our OCE is designed for two reasons: (1) We hope it is a learnable part. It can extract adjacent features according to different data features during data training, and the weight w is updatable; (2) We want to maintain this diversity of adjacent information, so we introduce sigmoid activation function to maintain the nonlinearity and diversity. OCE mainly takes into account the characteristics of high frequency and obvious short-term fluctuation of stocks, and completes the attention to the short-term adjacent characteristics. As shown in Fig. 2, OCE and input form a residual connection, which ensures that the lower limit of OCE is that all w values are 0, and the OCE module has learned nothing. The essence of OCE is to pay attention to short-term timing information. In the implementation, we use Con2d to implement the OCE module. Our input and output numbers of channel are all 1, and weight size w is 4, so the kernel is (4, 1), the stride is (1, 1), and the padding is 'same' to keep the feature shape of input and output consistent. Therefore, when the data passes through OCE, the dimension of the feature does not change, just add the location information noticed by OCE.

4.2   Rest of the Model

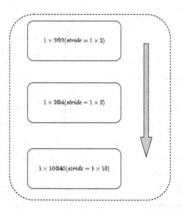

Fig. 3. The architecture of CNN Feature Extract

Features Extract. Features Extract(FE) module is a module that abstracts and aggregates features of original data. We think it is an expandable part according to actual needs. For Mid-Prices Predict task, we provide a CNN FE module that does not change the size of feature space, as shown in Fig. 3. The entire FE module contains three convolution layers, each of which is designed according to the characteristics of LOB data, because LOB data has 40 charac- teristics at each time, ranging from 1 to 10 levels. The first layer of convolution aims to aggregate the information of each internal action transaction of ask (price, volume) and bid (price, volume) respectively; The second layer of con- volution aims to distribute and aggregate the information of ask and bid, and each quotation priority; The last convolution is to aggregate the information of 10 quotations; The whole FE is progressive layer by layer, and the feature extraction is simple and efficient. The number of channels will expand as much as its feature map is reduced. FE module completes the transformation of tensor shape [1, 100, 40] to [40, 100, 1]. For WVAPP task, because the data feature space of this task is relatively complex, in order to keep the size of the feature space unchanged, we directly use two-layer transformer encoder for FE. The set- tings are as follows: depth is 2, head is 4, dim of head is 32, dim of MLP is 248. The shape of the data input to the FE is consistent with the output, it is [50, 124].

Transformer Encoder. Finally, we use a two layers Transformer to aggregate long-term timing information. As shown in Fig. 1, Transformer is mainly com- posed of Norm, Multi-Head Attention and MLP. Its core is Multi-Head Atten- tion, and its mathematical expression is as follows,

$$MHA(x) = [SA_1(X), SA_2(X), \cdots, SA_i(X)]Z_{linear} \tag{8}$$

where input is $X \in \mathbb{R}^{L \times D}$, i is i-th head, Z_{linear} means that after multi head attention, the concatenated features of multiple heads are projected into the original feature space, $SA(X)$ is

$$SA(X) = softmax(\frac{qk^T}{\sqrt{D_h}})v \tag{9}$$

where $[q, k, v] = [xw^q, xw^k, xw^v]$, represents the aggregate information of each head after linear projection, $D_h = D/i$, which represents a scaling, in order to prevent the gradient from disappearing due to the value of softmax being too average. We think that making qk^T this kind of matrix calculation is a cosine, which is equivalent to paying attention to different times. It will notice the long- term timing information, that is, the information of our window size. Similar to the cls token of Bert [7], we directly take the first token for classification. We believe that the features extracted by the FE module are abstract enough. In order to prevent further over fitting, our transformer has only two layers to aggregate the long-term timing information. The parameters of the actual experimental Transformer have two different settings. For Mid-Prices Predict task, the settings of Tranformer are as follows: depth is 2, head is 4, dim of head is 25, dim of MLP is 200. For WVAPP task, the setting of Transformer are: depth is 2, head is 4, dim of head is 32, dim of MLP is 248.

5 Experiments

We have done sufficient experiments on two public well-known dataset FI-2010 and HKGSAS-2020. The main purposes of the experiment are: (1) Verify the ability of OCE to capture short-term information; (2) Solution performance of OCET for STP problem.

5.1 Datasets and Settings

FI-2010 is the first public dataset for high frequency stock Limit Order Books. It consists of 4 million time series events about five stocks for 10 consecutive days from the Nasdaq Nordic stock market. It has 144-D feature vector at every moment. We follow previous researchers [32], using the data of the first 7 days for training and the data of the next 3 days for testing. We also use the threshold of $\alpha = 0.002$ for labels to predict the prediction horizons of $k = 10$, 50 and 100. The input window size is 100, that is, we use the information of the past 100 moments to predict the change trend of future prices. The input feature is a vector of 40-D, which is the information of stock level 1 to 10. We train our model with Adam [14] optimizer, the learning rate is set to 0.001, and weight decay 0.0005 is used for L2 [19] regularization. The epoch of training is 100.

HKGSAS-2020. HKGSAS-2020 records the LOB information of 20 important stocks in Shenzhen Stock Exchange of China, with a time span from June 2020 to September 2020. It has 124-D feature vector at every moment, including size, price, order quantity and weighted average of order staleness. It also provides a more difficult VWAPP task dealing with the volatile stock trading environment. We directly follow the research work of HKGSAS-2020, train with the data from June 2020 to August 2020, and then test with the data in September 2020. Here, we maintain and FI-2010 select multiple prediction horizons for prediction, and the prediction lines we select are $k = 5$, 60, 240. The prediction horizons here is calculated in seconds, because the electronic stock trading frequency in Europe is much faster than that in China. The input window size is 50, and the optimizer is also Adam with the learning rate 0.0001, weight decay 0.0001. The epoch of training is 20.

5.2 Experiments on the FI-2010 Dataset

The performance of the overall model is listed in Table 1. Some models have only one prediction line and cannot be reproduced. For these models, we directly quote the results of the original paper. We also use accuracy, precision, recall and f1-score to evaluate the prediction results of the model as previous researchers. From the three different prediction horizons of $k = 10$, 50 and 100, it can be seen that our model is much better than the previous model, and the results of these three different prediction horizons further show that our model has strong robustness, rather than over fitting a prediction horizon. It can be found that with the

Table 1. Experiment results for FI-2010

Model	Year	Accuracy %	Precision %	Recall %	F1 %
Prediction horizon k = 10					
B(TABL) [26]	2018	78.91	68.04	68.81	71.21
C(TABL) [26]	2018	84.70	76.95	78.44	72.84
DeepLOB [32]	2019	83.15	82.74	83.15	81.55
DeepFolio [23]	2020	83.21	82.57	83.21	81.95
DeepLOBAttention [31]	2021	84.07	83.39	84.07	83.22
OCET(Ours)	2022	**87.75**	**88.07**	**87.75**	**86.94**
Prediction horizon k = 50					
B(TABL) [26]	2018	75.58	74.58	73.09	73.64
C(TABL) [26]	2018	79.87	79.05	77.04	78.44
DeepLOB [32]	2019	79.83	79.71	79.83	79.69
BL-GAM-RHN-7 [18]	2019	82.00	81.45	80.43	80.88
MTDNN [17]	2020	81.12	–	–	81.05
DeepFolio [23]	2020	80.09	79.94	80.09	79.95
DeepLOBAttention [31]	2021	79.18	79.11	79.18	79.10
OCET(Ours)	2022	**88.05**	**88.15**	**88.05**	**88.00**
Prediction horizon k = 100					
DeepLOB [32]	2019	80.12	80.53	80.12	80.11
DeepFolio [23]	2020	80.41	80.38	80.41	80.39
DeepLOBAttention [31]	2021	78.72	78.81	78.72	78.75
OCET(Ours)	2022	**92.74**	**92.76**	**92.74**	**92.75**

gradual increase of k value, the precision gap of our OCET model is larger than that of the best CRNN model, and it is improved more. It can also be well explained that Transformer can simultaneously process the characteristics from each time in the past, instead of gradually transmitting information backward at each time like RNN, so that Transformer can timely handle the volatility of LOB high-frequency data.

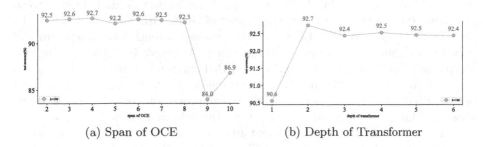

(a) Span of OCE   (b) Depth of Transformer

Fig. 4. Hyperparameter of OCE

Ablation Experiment. We found that the SOTA models in the past few years are the structure of CNN + RNN, CRNN paradigm. They only use RNN to notice the long-term time series information at the last layer. From 2019 to

Table 2. Ablation experiment for OCE

Model	Accuracy %	F1 %	Parameters	Forward (ms)
Prediction horizon k = 10				
DeepLOB	83.15	81.55	143907	0.052
DeepLOB(+OCE)	**83.87(0.72↑)**	**82.75(1.20↑)**	143912	0.055
DeepFolio	83.21	81.95	59443	0.046
DeepFolio(+OCE)	**84.25(1.04↑)**	**82.84(0.91↑)**	59448	0.048
DeepLOBAttention	84.07	83.22	177315	0.063
DeepLOBAttention(+OCE)	**84.09(0.02↑)**	**83.28(0.06↑)**	177320	0.054
OCET(-OCE)	87.61	86.81	163661	0.027
OCET(+OCE)	**87.75(0.14↑)**	**86.94(0.13↑)**	163666	**0.026**
Prediction horizon k = 50				
DeepLOB	79.83	79.69	–	0.052
DeepLOB(+OCE)	**80.14(0.31↑)**	**79.95(0.26↑)**	–	0.053
DeepFolio	80.09	79.95	–	0.045
DeepFolio(+OCE)	**80.51(0.42↑)**	**80.31(0.36↑)**	–	0.047
DeepLOBAttention	79.18	79.10	–	0.056
DeepLOBAttention(+OCE)	**79.56(0.38↑)**	**79.34(0.24↑)**	–	0.057
OCET(-OCE)	87.88	87.79	–	0.029
OCET(+OCE)	**88.05(0.17↑)**	**88.00(0.21↑)**	–	**0.027**
Prediction horizon k = 100				
DeepLOB	80.12	80.11	–	0.055
DeepLOB(+OCE)	**80.44(0.32↑)**	**80.45(0.34↑)**	–	0.055
DeepFolio	80.41	80.39	–	0.045
DeepFolio(+OCE)	**81.82(1.41↑)**	**81.83(1.44↑)**	–	0.047
DeepLOBAttention	78.72	78.75	–	0.054
DeepLOBAttention(+OCE)	**79.86(1.14↑)**	**79.89(1.14↑)**	–	0.056
OCET(-OCE)	92.37	92.37	–	**0.024**
OCET(+OCE)	**92.74(0.37↑)**	**92.75(0.38↑)**	–	0.027

2021, we selected CRNN models with strong representativeness, high accuracy and reproducibility, namely deepLOB, DeepFolio and DeepLOBAttention. We finally add our OCE module in the best CRNN models to this kind of model to do ablation experiments to test the performance of OCE. After adding OCE module, DeepLOB, DeepFolio, DeepLOBAttention can pay attention to short-term and long-term timing information at the same time. As shown in Table 2, DeepLOB, DeepFolio, DeepLOBAttention all exceed their original accuracy and F1-score with adding OCE on the prediction horizons $k = 10$, 50 and 100. Especially on the prediction horizon of $k = 100$, the improvement of each model is the largest. This is because our OCE notices the information of $k = 4$ adjacent time intervals, which is quite different from $k = 100$, so that OCE will notice the fluctuation of stock price in real time and learn the fluctuation information selectively. The OCE module itself has only five parameters, which will hardly affect the reasoning speed of the model.

We also conducted the ablation experiment of hyperparameter, mainly including l (span of OCE) and depths of transformer. As shown in Fig. 4, We mainly select the prediction line when $k = 100$. It can be seen from Fig. 4(a) that when $l = 4$, OCET gets the maximum score. And the smaller l is, the more stable its score will be. This shows that short-term fluctuations will affect the price. The larger the l is, the greater the l will be, and it will notice that some long-term information conflicts with the transformer information, resulting in a decline in accuracy. Figure 4(b) also revealed that the larger the depts of transformer, the higher the score of OCET. It also shows that the role of transformer here is to aggregate long-term information rather than feature extraction. Therefore, when the depts of transformer became larger, the score of OCET did not become higher, which was consistent with the experimental results. As shown in Fig. 5, after adding our simple and effective CNN FE, score of OCET has been greatly improved at all times of different k, which also shows the efficiency of the three layers of CNN FE.

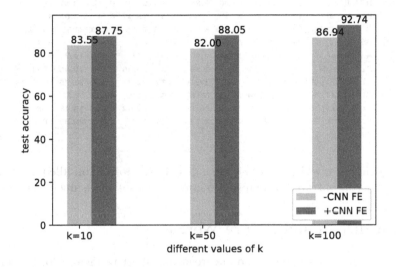

Fig. 5. Ablation experiment of CNN FE

5.3   Experiments on the HKGSAS-2020 Dataset

To verify the generalization of OCET in more complex WVAPP STP tasks, we tested our OCET model on the HKGSAS-2020 dataset and compared it with the two SOTA models of HKGSAS-2020 itself. The experimental results are shown in Table 3. The prediction horizon we selected is $k = 5, 60, 240$. Overall, The accuracy of OCET in the three prediction horizon is the highest. All three models perform best in [0.3, 0.7] because the number of labels in this interval accounts for 40% of the samples. And all models performed poorly in the edge areas [0, 0.1], [0.9, 1.0]. We analyze the reasons for the low accuracy: (1) the WAVPP task is more difficult. It represents the moving trend of the weighted average price of the overall stock. It is not only related to the price of the transaction, but also related to the number of transactions. The volatility is too large; (2) The input feature dimension of the model itself has 124 dimensions, which may be difficult for the model itself to learn and easy to fall into local optimization. (3) The price distribution of wvap in the marginal area [0, 0.1], [0.9, 1.0] is

Table 3. Experiment for HKGSAS-2020

Model	LSTM			CNN-LSTM			OCET		
Horizon	k=5	k=60	k=240	k=5	k=60	k=240	k=5	k=60	k=240
Accuracy %	42.29	41.17	40.50	41.91	41.09	37.18	**42.69**	**41.69**	**40.51**
Precison (P) %	**41.85**	37.02	35.91	40.48	37.50	32.56	38.82	**38.72**	**35.93**
Recall (R) %	42.29	41.17	40.50	41.91	41.09	37.18	**42.69**	**41.69**	**40.51**
F1 %	**34.09**	30.72	30.49	33.62	30.12	**31.91**	32.62	**34.35**	29.77
P [0, 0.1] %	27.00	34.53	**27.96**	**27.08**	32.53	20.57	16.76	**36.50**	27.71
P [0.1, 0.3] %	33.71	34.72	**35.42**	**34.83**	35.92	27.85	31.92	**36.26**	33.97
P [0.3, 0.7] %	41.06	42.38	42.19	40.82	42.03	**43.38**	41.07	**43.36**	41.99
P [0.7, 0.9] %	**60.59**	31.53	31.17	55.45	33.66	26.91	57.77	**34.84**	**33.25**
P [0.9, 1.0] %	**28.59**	33.68	**29.23**	26.31	**35.20**	22.05	19.17	35.12	29.20
R [0, 0.1] %	2.20	9.22	11.06	**2.89**	9.43	**21.35**	0.14	**13.17**	11.09
R [0.1, 0.3] %	**14.74**	11.09	8.02	11.84	7.76	**18.47**	6.74	**15.53**	8.60
R [0.3, 0.7] %	90.58	90.56	89.26	89.93	**91.78**	72.28	**94.21**	84.58	**90.54**
R [0.7, 0.9] %	25.77	2.50	5.96	27.00	3.61	**8.24**	**29.75**	**9.95**	3.84
R [0.9, 1.0] %	2.27	13.01	**8.93**	**2.60**	11.54	7.86	0.67	**14.35**	6.91
F1 [0, 0.1] %	4.06	14.56	15.85	**5.22**	14.62	**20.95**	0.27	**19.35**	15.84
F1 [0.1, 0.3] %	**20.51**	16.81	13.08	17.68	12.76	**22.21**	11.13	**21.75**	13.72
F1 [0.3, 0.7] %	56.50	**57.74**	57.30	56.15	57.65	54.22	**57.14**	57.33	**57.37**
F1 [0.7, 0.9] %	36.17	4.64	10.00	36.32	6.52	**12.62**	**39.28**	15.48	6.88
F1 [0.9, 1.0] %	4.21	18.77	**13.68**	**4.74**	17.38	11.59	1.30	**20.38**	11.18

relatively discrete, while the wvap price distribution of the middle part [0.3, 0.7] is relatively concentrated, which is also in line with the modeling of the problem.

6   Conclusion and Future Works

This paper proposes a model for aggregating short-term and long-term timing information at the same time, OCET, for STP tasks. OCET is composed of OCE and Transformer, where OCE is our novel design to capture short-term timing information. Our motivation is how to efficiently aggregate long-term and short-term timing information of LOB data and efficiently extract LOB features. Under the combined application of these two modules, OCET achievements state of the art performance on the benchmark FI-2010 and HKGSAS dataset. Simultaneously, we also verify the application of our newly proposed location coding (OCE) in network structures such as CRNN. It only needs five parameters to improve the prediction accuracy of CRNN networks. In the future, we will focus on the deployment of STP task prediction model in the mobile terminal [11,16] to assist stock investors in decision-making [29,30] and trading.

References

1. Ahn, H.J., Cai, J., Hamao, Y., Ho, R.Y.: The components of the bid-ask spread in a limit-order market: evidence from the Tokyo stock exchange. J. Empir. Financ. **9**(4), 399–430 (2002)

2. Aitken, M.J., Berkman, H., Mak, D.: The use of undisclosed limit orders on the Australian stock exchange. J. Bank. Financ. **25**(8), 1589–1603 (2001)

3. Anagnostidis, P., Papachristou, G., Thomaidis, N.S.: Liquidity commonality in order-driven trading: evidence from the Athens stock exchange. Appl. Econ. **48**(22), 2007–2021 (2016)

4. Atsalakis, G.S., Valavanis, K.P.: Surveying stock market forecasting techniques-part II: soft computing methods. Expert Syst. Appl. **36**(3), 5932–5941 (2009)

5. Cao, Q., Leggio, K.B., Schniederjans, M.J.: A comparison between Fama and French's model and artificial neural networks in predicting the Chinese stock market. Comput. Oper. Res. **32**(10), 2499–2512 (2005)

6. Cavalcante, R.C., Brasileiro, R.C., Souza, V.L., Nobrega, J.P., Oliveira, A.L.: Computational intelligence and financial markets: a survey and future directions. Expert Syst. Appl. **55**, 194–211 (2016)

7. Devlin, J., Chang, M.W., Lee, K., Toutanova, K.: BERT: pre-training of deep bidirectional transformers for language understanding. arXiv preprint arXiv:1810.04805 (2018)

8. Gould, M.D., Porter, M.A., Williams, S., McDonald, M., Fenn, D.J., Howison, S.D.: Limit order books. Quantit. Financ. **13**(11), 1709–1742 (2013)

9. Greff, K., Srivastava, R.K., Koutník, J., Steunebrink, B.R., Schmidhuber, J.: LSTM: a search space odyssey. IEEE Trans. Neural Netw. Learn. Syst. **28**(10), 2222–2232 (2016)

10. He, K., Zhang, X., Ren, S., Sun, J.: Deep residual learning for image recognition. In: Proceedings of the IEEE Conference on Computer Vision and Pattern Recognition, pp. 770–778 (2016)

11. Hong, W., Li, G., Liu, S., Yang, P., Tang, K.: Multi-objective evolutionary optimization for hardware-aware neural network pruning. Fundam. Res. (2022). https://doi.org/10.1016/j.fmre.2022.07.013

12. Hu, J., Shen, L., Sun, G.: Squeeze-and-excitation networks. In: Proceedings of the IEEE Conference on Computer Vision and Pattern Recognition, pp. 7132–7141 (2018)

13. Huang, C., Ge, W., Chou, H., Du, X.: Benchmark dataset for short-term market prediction of limit order book in china markets. J. Financ. Data Sci. **3**(4), 171–183 (2021)

14. Kingma, D.P., Ba, J.: Adam: a method for stochastic optimization. arXiv preprint arXiv:1412.6980 (2014)

15. LeCun, Y., Bengio, Y., et al.: Convolutional networks for images, speech, and time series. Handb. Brain Theory Neural Netw. **3361**(10), 1995 (1995)

16. Li, G., Yang, P., Qian, C., Hong, R., Tang, K.: Stage-wise magnitude-based pruning for recurrent neural networks. IEEE Trans. Neural Netw. Learn. Syst. (2022). https://doi.org/10.1109/TNNLS.2022.3184730

17. Liu, G., et al.: Multi-scale two-way deep neural network for stock trend prediction. In: IJCAI, pp. 4555–4561 (2020)

18. Luo, W., Yu, F.: Recurrent highway networks with grouped auxiliary memory. IEEE Access **7**, 182037–182049 (2019)

19. Marquardt, D.W., Snee, R.D.: Ridge regression in practice. Am. Stat. **29**(1), 3–20 (1975)

20. Ntakaris, A., Magris, M., Kanniainen, J., Gabbouj, M., Iosifidis, A.: Benchmark dataset for mid-price forecasting of limit order book data with machine learning methods. J. Forecast. **37**(8), 852–866 (2018)

21. Parlour, C.A., Seppi, D.J.: Limit order markets: a survey. Handb. Financ. Intermediat. Bank. **5**, 63–95 (2008)

22. Qilong, W., Banggu, W., Pengfei, Z., Peihua, L., Wangmeng, Z., Qinghua, H.: ECA-net: efficient channel attention for deep convolutional neural networks 2020 ieee. In: CVF Conference on Computer Vision and Pattern Recognition (CVPR) (2020)
23. Sangadiev, A., et al.: DeepFolio: convolutional neural networks for portfolios with limit order book data. arXiv preprint arXiv:2008.12152 (2020)
24. Schuster, M., Paliwal, K.K.: Bidirectional recurrent neural networks. IEEE Trans. Signal Process. 45(11), 2673–2681 (1997)
25. Szegedy, C., Vanhoucke, V., Ioffe, S., Shlens, J., Wojna, Z.: Rethinking the inception architecture for computer vision. In: Proceedings of the IEEE Conference on Computer Vision and Pattern Recognition, pp. 2818–2826 (2016)
26. Tran, D.T., Iosifidis, A., Kanniainen, J., Gabbouj, M.: Temporal attention-augmented bilinear network for financial time-series data analysis. IEEE Trans. Neural Netw. Learn. Syst. 30(5), 1407–1418 (2018)
27. Tsantekidis, A., Passalis, N., Tefas, A., Kanniainen, J., Gabbouj, M., Iosifidis, A.: Forecasting stock prices from the limit order book using convolutional neural networks. In: 2017 IEEE 19th Conference on Business Informatics (CBI), vol. 1, pp. 7–12. IEEE (2017)
28. Vaswani, A., et al.: Attention is all you need. In: Advances in Neural Information Processing Systems, pp. 5998–6008 (2017)
29. Yang, P., Yang, Q., Tang, K., Yao, X.: Parallel exploration via negatively correlated search. Front. Comp. Sci. 15(5), 1–13 (2021)
30. Yang, P., Zhang, H., Yu, Y., Li, M., Tang, K.: Evolutionary reinforcement learning via cooperative coevolutionary negatively correlated search. Swarm Evol. Comput. 68, 100974 (2022)
31. Zhang, Z., Zohren, S.: Multi-horizon forecasting for limit order books: novel deep learning approaches and hardware acceleration using intelligent processing units. arXiv preprint arXiv:2105.10430 (2021)
32. Zhang, Z., Zohren, S., Roberts, S.: DeepLOB: deep convolutional neural networks for limit order books. IEEE Trans. Signal Process. 67(11), 3001–3012 (2019)

A Rapid and Precise Spiking Neural Network for Image Recognition

Cheng Zhu and Chuandong Li[✉]

College of Electronic and Information Engineering, Southwest University,
Chongqing 400715, China
cdli@swu.edu.cn

Abstract. Spiking neural network is a new information processing method that combines energy conservation and biologic plausibility. With the development of neuromorphic hardware, it is attracting more and more attention. However, in terms of solving practical problems, spiking neural networks still lack mature training methods. To improve the performance and accuracy of Spiking Neural Networks (SNN), in this paper, we propose a two-layer convolutional adaptive encoder and a supervised training algorithm based on surrogate gradient method, which is expected to overcome the non-differentiable characteristics of the activation function and increase the training speed. In comparison with the previous results of image recognition on the neuromorphic dataset DVS128 Gesture and the non-neuromorphic dataset Fashion-MNIST, the presented method leads to higher accuracy with the significantly shorter time. Our work promotes the application of SNN in real life, especially in the field of image recognition.

Keywords: Spiking neural network · Image recognition · DVS128 Gesture dataset

1 Introduction

Spiking neural network (SNN) is a new kind of artificial neural network that has emerged in recent years. Because SNN is inspired by how the brain works, it is highly biologic plausible. SNN is stateful neural network, so it works well in processing time-series data because of its spatiotemporal correlation property. Moreover, because of the event-driven nature of SNN and the binary nature of activation function's output, SNN can be deployed on neuromorphic hardware with significant power conservation such as SpiNNaker [6], TrueNorth [10], and Tianji [12]. With the development of neuromorphic hardware, more work is now being done to study the application of SNN in real life.

Similar to training recurrent neural network (RNN), training SNN is complex due to the spatiotemporal dependence of its state. Moreover, because of the binary characteristics of the activation function's output, training SNN has the difficulty of non-differentiable activation function compared to training RNN.

L. Pan et al. (Eds.): BIC-TA 2022, CCIS 1801, pp. 385–393, 2023.
https://doi.org/10.1007/978-981-99-1549-1_30

The known work has proved the feasibility of training SNN, but the low accuracy and the long simulation time of the obtained model hinders the use of SNN in solving practical problems, so it is very valuable to overcome this difficulty.

In order to input images into SNN, we need to use an encoding layer to convert the pixel value of image into spikes. Poisson encoding is most commonly used. The rule of Poisson encoding is that the probability of spike emission is determined by pixel value [1]. However, the encoding scheme requires long simulation time to encode the pixel value precisely, resulting in long training and inference time. Weighted phase encoding is a encoding scheme based on binary representation. It converts the input data to binary numbers and emit spikes sequentially from the high bit to the low bit of binary numbers [7]. Compared to Poisson coding, weighted phase coding transmits more information per time step and therefore requires fewer simulation time steps. However, the encoding layer using weighted phase encoding takes more time to encode and the first hidden layer also takes more time to decode. We took inspiration from recent work, Wu et al. [16] proposed the idea of using part of network as encoder. In this paper, we propose a new encoding scheme that uses the first two layers of convolutional SNN as adaptive convolutional encoder, through which the input data can be encoded using shorter simulation time. The scheme significantly reduces the training time and inference time of SNN.

In terms of training algorithm, there is currently no mature training method for SNN. According to whether label is used for training, the training algorithm can be divided into supervised learning algorithm and unsupervised learning algorithm.

The main method of unsupervised learning algorithm is STDP learning algorithm, which uses the relative time of pre-synaptic spike and post-synaptic spike to adjust synaptic weight [9]. STDP learning algorithm mainly considers the local optimization of the neural network. although it can increase the global variables to regulate the whole network, it still lacks a mechanism for the overall optimization of the neural network, so the accuracy of SNN trained by the algorithm is relatively low.

Supervised learning algorithm can be divided into direct training algorithm and indirect training algorithm, and indirect training algorithm is also called transformation algorithm, which obtain trained SNN by converting trained artificial neural network (ANN) into SNN [2]. At present, the transformation algorithm is the most successful method for training large-scale precise SNN, but there are some problems in the transformation algorithm. The SNN obtained by the transformation algorithm can't adjust the weight directly. When the weight needs to be adjusted, we need to retrain ANN and convert the trained ANN to SNN, so the algorithm is rigid. Moreover, when converting ANN to SNN, we require sufficient simulation steps to reduce the loss of convertion, so more simulation steps are required for the inference of the converted SNN, which is not suitable for tasks that require fast response [13].

In terms of direct training algorithm, Neftci et al. [11] proposed an idea of using a surrogate gradient function to replace non-differentiable step function in SNN and then using the derivative of this surrogate gradient function to optimize the weight of SNN when performing gradient descent. Lee et al. [8] proposed gradient estimation algorithm, which estimate the gradient of LIF activation function according to the update and activation characteristics of LIF activation function and uses the estimated value for gradient descent. However, these algorithms are not well optimized and require too long simulation time. We propose a new SNN training algorithm, which selects the appropriate surrogate gradient function and function smoothness when training SNN, and combines surrogate gradient algorithm with Adam algorithm commonly used in ANN. By using the training algorithm and above-mentioned double-layer convolutional adaptive encoder, SNN takes a shorter time to reach the fitting state and achieves slightly higher classification accuracy on datasets Fashion-MNIST and DVS128 Gesture. This work explores direct training algorithm for SNN, promoting the application of SNN in the field of image recognition.

2 Proposed Method

In this section, we propose a double-layer convolutional adaptive encoder to reduce simulation steps when encoding. Then we propose a new SNN training algorithm, which combines surrogate gradient algorithm with Adam algorithm to improve the performance of the model.

2.1 Encoding and Decoding Scheme

Poisson encoding is a type of rate encoding. The spike emission frequency of the encoding layer using rate encoding is proportional to image pixel value. However, rate encoding requires sufficient simulation steps to encode image information precisely. In order to solve the problem, we adopt a new coding scheme that reduces the simulation steps of encoding without affecting the performance of the model.

We propose an adaptive encoding scheme that uses the first two layers of SNN as encoder. The encoder can receive spike signal directly from neuromorphic datasets or receive real value signal from non-neuromorphic datasets and convert it into spikes. The encoder can continuously adjust to encode picture information during training. Because it does not adopt a fixed encoding mode like rate encoding, the accuracy of the encoder is not very dependent on the length of simulation step.

As Fig. 1 shows, our proposed adaptive encoder contains two convolutional layers, ReLU activation function and LIF neuron. We normalize the signal before they forward propagate to the LIF neuron. Compared with the work of Wu et al. [16], we use a deeper network structure. Due to the increase of layers and the use of traditional ANN network structure, our encoder can better extract the features of the image and the energy consumption of each time step during encoding is increased to a certain extent. However, the SNN using this encoder needs significantly fewer time steps, so the total energy consumption during training and inference is greatly reduced.

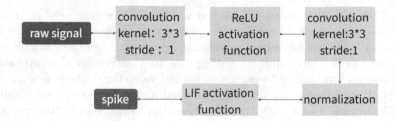

Fig. 1. Description of the adaptive encoder

2.2   Neuron Model Iterative Update Scheme

LIF model is the most commonly used model to describe the update of neuronal membrane potential and spike emission. Its update formula is

$$\tau\frac{du}{dt} = -u + I, u < V_{th} \tag{1}$$

$$fire\ a\ spike\ \&\ u = V - V_{th}, u \geq V_{th} \tag{2}$$

where τ is the time constant, u is the membrane potential, I is the total current transmitted by presynaptic neurons and V_{th} is the threshold at which spike is emitted. Equation (1) describes that when membrane potential is less than threshold, neuron receives the total current transmitted by the presynaptic neurons and update membrane potential. Equation (2) describes that when membrane potential reaches the threshold, the neuron fires a spike and resets membrane potential. There are two types of reset methods: hard method and soft method. Hard method resets the membrane potential to assigned potential regardless of current membrane potential. Because hard method reduces the selectivity of neurons, we use soft method, which subtracts the threshold from the membrane potential

The update of membrane potential in the above formula is in continuous domain. To simulate LIF neurons in deep learning framework, we need to convert Eq. (1) into discrete, iterative form. Therefore, we use the difference equation to approximate Eq. (1) and obtain the iterative expression

$$u^{t+1} = (1 - \frac{dt}{\tau})u^t + \frac{dt}{\tau}\sum_j W_j o^{t+1}(j) \tag{3}$$

where j denotes the index of the presynaptic neuron, W_j represents the synaptic weight and $o(j)$ denotes the spike emitted by presynaptic neuron at time $t + 1$, which has a value of 1 or 0. We simplify $(1 - \frac{dt}{\tau})$ as the decay factor k_τ and incorporate $\frac{dt}{\tau}I$ into the weight and get

$$u^{t+1} = k_\tau u^t + \sum_j W_j o^{t+1}(j) \tag{4}$$

Then we incorporate the fire and reset mechanism into Eq. (4) to obtain the final neuronal membrane potential update formula

$$u^{t+1}(i) = k_r u^t(i) + \sum_j W_{ij} o(j)^{t+1} - V_{th} \Theta(u^t(i) - V_{th})$$ (5)

where i denotes the index of neuron in the current layer and W_{ij} denotes the weight of synapse connecting neuron j in the upper layer and neuron i in the current layer. $\Theta(x)$ is a step function that has a value of 0 at $x < 0$ and 1 in other cases. It can be seen in Eq. (5) that the membrane potential at time $t + 1$ updates on the basis of the membrane potential at time t.

2.3   Training Scheme

We convert the label y into label vector Y using one hot encoding. Then we calculate the error between Y and the actual output spike frequency vector to get our loss function

$$L = \frac{1}{2}(Y - \frac{1}{T}\sum_{t=1}^{T} o^t)^2$$ (6)

where o^t denotes the spike vector emitted by the output layer at time t, and T denotes the length of simulation step. When training SNN, we use the same operation as ANN. We backpropagate error and calculate gradient for weight update. Backpropagation is the reverse operation of forward propagation. We obtain the error of each layer

$$L^n = \frac{1}{2}\lambda(Y - \frac{1}{T}\sum_{t=1}^{T} \Theta(u^t - V_{th}))^2$$ (7)

where λ denotes the inverse weight tensor to backpropagate to the n layer and L^n denotes the error of the n layer. After backpropagating error, we need to calculate the gradient of each layer. We use e to denote the error between label vector Y and the actual output spike frequency vector. We use the chain derivation rule to get the gradient

$$\frac{\partial L^n}{\partial W^n} = \frac{\partial L^n}{\partial \sum_{t=1}^{T} \Theta(u^t - V_{th})} \sum_{t=1}^{T} (\Theta^{'}(u^t - V_{th}) \frac{\partial u^t}{\partial W^n})$$
$$= -\frac{\lambda}{T} e \sum_{t=1}^{T} (\Theta^{'}(u^t - V_{th}) \frac{\partial u^t}{\partial W^n})$$ (8)

We can see $-\frac{\lambda}{T}e$ as constant. we set the decay factor k_{tau} to 1. According to Eq. (5), we expand $\frac{\partial u^t}{\partial W^n}$ along the temporal dimension

$$\frac{\partial u^t}{\partial W^n} = \frac{\partial u^{t-1}}{\partial W^n} + O^{t,n} - \Theta^{'}(u^{t-1}) \frac{\partial u^{t-1}}{\partial W^n}$$ (9)

$$\frac{\partial u^1}{\partial W^n} = O^{1,n} \tag{10}$$

where $O^{t,n}$ denotes the spike tensor received by n layer at time t. We can use the saved spike tensor to sequentially calculate the partial derivative of the membrane potential relative to the weight at each time step from $t = 1$ and finally get $\frac{\partial u^t}{\partial W^n}$. However, step function $\Theta(x)$ of Eqs. (8) and (9) is not derivable, so we use the derivative of a function that approximates the step function to replace $\Theta'(x)$ to participate in operation. Arctangent function has the characteristic of changing sharply around 0 and slowly in other range, so we choose arctangent function as the surrogate gradient function. Specifically, the arctangent function we choose is

$$sg(x) = \frac{1}{\pi}arctan(\frac{\pi}{2}\alpha x) + \frac{1}{2} \tag{11}$$

where we use parameter α to control the smoothness of $sg(x)$. With α larger, the derivative of $sg(x)$ around 0 is larger, $sg(x)$ is more similar to step function (see Fig. 2). However, when α increases to a certain extent, the derivative of sg(x) around 0 is too large, which may cause gradient explosion, affecting the accuracy and fitting speed of the model. Adam algorithm can make each weight parameter update more balanced, which can effectively alleviate the influence of gradient explosion. Therefore, in order to approximate step function and alleviate the influence of gradient explosion during training, we combine surrogate gradient algorithm with Adam algorithm to train SNN. Specifically, the gradient is calculated using the surrogate gradient algorithm. Then we use Adam algorithm and gradient to update the weight parameters.

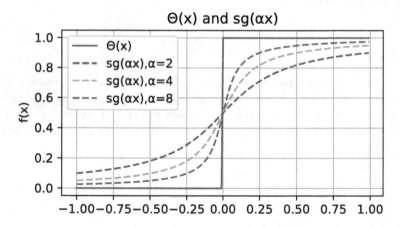

Fig. 2. Comparison of the curves of arctangent function $sg(x)$ and step function $\Theta(x)$

3   Experiment

We test neuromorphic dataset (DVS128 Gesture) and non-neuromorphic dataset (Fashion-MNIST) from the aspects of training and inference time and classification accuracy. For simple dataset Fashion-MNIST, we use a network of two convolutional layers and two fully-connected layers. For the multi-channel complex dataset DVS128 Gesture, we use a network of five convolutional layers and two fully-connected layers to better extract image features. After each convolution layer we use an average-pooling layer. The average-pooling layer is more suitable for SNN. If we use max-pooling layer, its output will only be 1 and 0, which will lead to insufficient selectivity of information.

3.1 Training and Inference Time

SNN requires enough simulation steps to simulate neural dynamics, encode and decode data, so the length of simulation step is important. For SNN, past work often require 100 or even 1000 steps to achieve good performance [14]. Using our adaptive encoder, SNN requires significantly less simulation steps. As shown in Fig. 3, the increase of simulation steps will lead to the performance's improvement of the model. However, when simulation steps reaches 4, the increase of simulation steps has few influence on the model's performance, so we choose 4 as the length of simulation steps when training SNN and inferencing. Therefore, the use of our adaptive encoder can accelerate the training and inference of SNN.

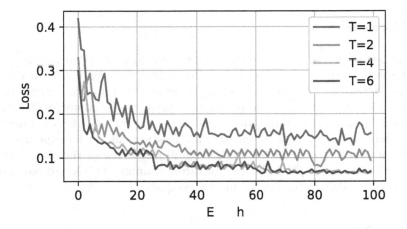

Fig. 3. The influence of simulation step on Fashion-MNIST

3.2 Classification Accuracy

Table 1 summarizes the latest results of existing SNN direct training methods on the non-neuromorphic dataset Fashion-MNIST and neuromorphic dataset DVS128 Gesture. Our model achieves 93.72% accuracy on the Fashion-MNIST dataset, outperforming the previous best result of 92.13%. DVS128 Gesture dataset is a neuromorphic dataset that consists of several sequences of events captured by a DVS camera, each corresponding to a gesture. Our model achieves an accuracy of 95.74%, which is better than the previous best result of 95.54%. Moreover, our model only needs 67 epochs to achieve the highest accuracy because of using Adam algorithm.

Table 1. Comparison with existing results on datasets

Model	Method	Accuracy Fashion-MNIST	Accuracy DVS128 Gesture
[15]	Spike-based BP	–	93.64%
[5]	Spike-based BP	–	95.54%
[18]	Spike-based BP	92.13%	–
[3]	Spike-based BP	92.07%	–
[17]	Spike-based BP	–	92.01%
[4]	Spike-based BP	–	93.40%
Our model	Spike-based BP	**93.72%**	**95.74%**

4   Conclusion

In this paper, we propose a double-layer convolutional adaptive encoder. It requires only a few simulation steps to accurately extract features of image information. Moreover, it can be used as part of network to improve the performance of the model. In addition, we derive the membrane potential update equation of LIF neurons and then optimize the direct training algorithm of SNN. Finally, we select Fashion-MNIST dataset and DVS128 Gesture dataset for experiment. We achieve greater accuracy on them using significantly shorter time than previous work. Our work could point the way for direct training of SNN and promote the application of SNN in the field of image recognition.

References

1. Adrian, E.D., Zotterman, Y.: The impulses produced by sensory nerve endings. J. Physiol. **61**(4), 465–483 (1926)
2. Cao, Y., Chen, Y., Khosla, D.: Spiking deep convolutional neural networks for energy-efficient object recognition. Int. J. Comput. Vis. **113**(1), 54–66 (2015). https://doi.org/10.1007/s11263-014-0788-3
3. Cheng, X., Hao, Y., Xu, J., Xu, B.: LISNN: improving spiking neural networks with lateral interactions for robust object recognition. In: IJCAI, pp. 1519–1525 (2020)
4. He, W., et al.: Comparing SNNs and RNNs on neuromorphic vision datasets: similarities and differences. Neural Netw. **132**, 108–120 (2020)

5. Kaiser, J., Mostafa, H., Neftci, E.: Synaptic plasticity dynamics for deep continuous local learning (DECOLLE). Front. Neurosci. **14**, 424 (2020)
6. Khan, M.M., et al.: SpiNNaker: mapping neural networks onto a massively-parallel chip multiprocessor. In: 2008 IEEE International Joint Conference on Neural Networks (IEEE World Congress on Computational Intelligence), pp. 2849–2856. IEEE (2008)
7. Kim, J., Kim, H., Huh, S., Lee, J., Choi, K.: Deep neural networks with weighted spikes. Neurocomputing **311**, 373–386 (2018)
8. Lee, C., Sarwar, S.S., Panda, P., Srinivasan, G., Roy, K.: Enabling spike-based backpropagation for training deep neural network architectures. Front. Neurosci. **14**, 119 (2020)
9. Masquelier, T., Thorpe, S.J.: Learning to recognize objects using waves of spikes and spike timing-dependent plasticity. In: The 2010 International Joint Conference on Neural Networks (IJCNN), pp. 1–8. IEEE (2010)
10. Merolla, P.A., et al.: A million spiking-neuron integrated circuit with a scalable communication network and interface. Science **345**(6197), 668–673 (2014)
11. Neftci, E.O., Mostafa, H., Zenke, F.: Surrogate gradient learning in spiking neural networks: bringing the power of gradient-based optimization to spiking neural networks. IEEE Sign. Process. Mag. **36**(6), 51–63 (2019)
12. Pei, J., et al.: Towards artificial general intelligence with hybrid Tianjic chip architecture. Nature **572**(7767), 106–111 (2019)
13. Rueckauer, B., Lungu, I.A., Hu, Y., Pfeiffer, M., Liu, S.C.: Conversion of continuous-valued deep networks to efficient event-driven networks for image classification. Front. Neurosci. **11**, 682 (2017)
14. Sengupta, A., Ye, Y., Wang, R., Liu, C., Roy, K.: Going deeper in spiking neural networks: VGG and residual architectures. Front. Neurosci. **13**, 95 (2019)
15. Shrestha, S.B., Orchard, G.: Slayer: spike layer error reassignment in time. In: Advances in Neural Information Processing Systems, vol. 31 (2018)
16. Wu, Y., Deng, L., Li, G., Zhu, J., Xie, Y., Shi, L.: Direct training for spiking neural networks: faster, larger, better. In: Proceedings of the AAAI Conference on Artificial Intelligence, vol. 33, pp. 1311–1318 (2019)
17. Xing, Y., Di Caterina, G., Soraghan, J.: A new spiking convolutional recurrent neural network (SCRNN) with applications to event-based hand gesture recognition. Front. Neurosci. **14**, 1143 (2020)
18. Zhang, W., Li, P.: Spike-train level backpropagation for training deep recurrent spiking neural networks. In: Advances in Neural Information Processing Systems, vol. 32 (2019)

Research on Improved Image Dehazing Algorithm Based on Dark Channel Prior

Xinlong Pan[1](✉), Haipeng Wang[1](✉), Yu Liu[1], Ziwei Zhao[2], and Fan Huang[3,4]

[1] Naval Aviation University, Yantai 264001, China
airadar@126.com, whp5691@126.com
[2] Wuhan University of Technology, Wuhan 430070, China
[3] Wuhan Institute of Digital Engineering, Wuhan 430205, China
[4] Shanghai Jiao Tong University, Shanghai 200240, China

Abstract. The dark channel prior dehazing algorithm can clear the foggy images to different degrees. However, the algorithm still has some deficiencies, such as the halo phenomenon at the edge of the image in the area of sudden change in depth of field; inaccurate estimation of transmittance, resulting in color shift in the image; inaccurate estimation of atmospheric light value, resulting in darker images after defogging, etc. Therefore, it is necessary to improve the dark channel prior algorithm. This paper improves the dehazing algorithm on the basis of in-depth study of the dark channel prior algorithm, and proposes a new dehazing algorithm based on the atmospheric scattering model. The simulation results show that the improved algorithm in this paper can effectively suppress the halo and color distortion in the abrupt depth of field area, and the obtained defogged image has rich detail information, clearer image, and moderate brightness. The algorithm has improved in objective parameters such as average gradient, structural similarity, peak signal-to-noise ratio and information entropy.

Keywords: image dehazing · dark channel prior · image restoration · transmittance optimization

1 Introduction

Image defogging technology is to use certain algorithms or means to reduce or even eliminate the fog in the image. Image defogging technology has always been an important research content in the field of image processing. The current mainstream image defogging algorithms can be divided into three categories [1]: one is based on image enhancement, the other is based on image restoration, and the third is based on deep learning.

The prior-based image defogging method is based on the atmospheric scattering model and uses certain prior conditions to solve the parameters in the atmospheric scattering model, mainly the solution of atmospheric light value and transmittance, and then deduces the haze-free image. Dark channel prior algorithm in outdoor fog-free images, except for the sky area, in most cases, a channel with a very low pixel value can be found in the R, G, and B channels, that is, the dark channel. The coarse transmittance is estimated according to the dark channel prior theory, and the first 0.1% of the pixels are taken from the dark channel image according to the brightness, and the maximum value of the corresponding pixel is found in the original foggy image as the atmospheric light value, so as to obtain the Fog image. However, the dark channel prior algorithm will cause a serious Halo effect in the sudden change of field depth, and the transmission rate estimation of the sky area is also inaccurate.

Aiming at the deficiencies in the dark channel prior algorithm, this paper improves the dark channel prior algorithm: first, the atmospheric light value is improved, and the Otsu algorithm is used to initially segment the image, and the white noise in the segmentation is improved through morphological knowledge. Calculate the atmospheric light value in the sky area. Secondly, to improve the transmittance, a linear model is established according to the relationship between the foggy image and the fog-free image, the adaptive parameters are introduced to constrain the rate of linear transformation, and the initial transmittance is obtained by logarithmic transformation of the linear mapping relationship. Then the depth-of-field formula is introduced to compensate the transmittance refined by guided filtering, and then the optimized transmittance is obtained through the weighted L1 norm context regularization algorithm. Finally, contrast enhancement processing is carried out on the obtained restored image with histogram equalization algorithm.

2 Atmospheric Scattering Model

The atmospheric scattering model in this paper can be divided into three parts: incident light attenuation model, ambient light imaging model and fog image degradation model.

2.1 Incident Light Attenuation Model

The incident light attenuation model describes the attenuation process of light from the object to the imaging device.

Suppose there is a space, the material distribution of this space is the same as that of the atmosphere. Assuming its thickness is dx, when a beam of parallel light with intensity $E(x, \lambda)$ passes through the space, the energy change $dE(x, \lambda)$ of this beam of parallel light is expressed as:

$$\frac{dE(x, \lambda)}{E(x, \lambda)} = -\beta(\lambda)dx \tag{1}$$

In formula (1), λ represents the wavelength of light; β is a function of λ, , also known as the atmospheric scattering coefficient. Assuming that the particles in the atmosphere are evenly distributed, the definite integral is used to deal with both sides of the formula (1) in the interval from $x = 0$ to $x = d$, and after simplified processing, it can be obtained:

$$E_d(d, \lambda) = E_0(\lambda)e^{-\beta(\lambda)d} \tag{2}$$

Equation (2) is the expression of the incident light attenuation model, where $E_0(\lambda)$ represents the light intensity of the object at $x = 0$. As the depth of field d gradually increases, the energy $E_d(d, \lambda)$ of the incident light gradually decays.

2.2 Incident Light Attenuation Model

The ambient light imaging model refers to the process in which ambient light such as sunlight and ground reflected light participate in object imaging. These ambient lights increase the ability to attenuate. Due to the large amount of ambient light, the three channel values of R, G, and B increase, which eventually leads to problems such as blurring and color distortion of the acquired image.

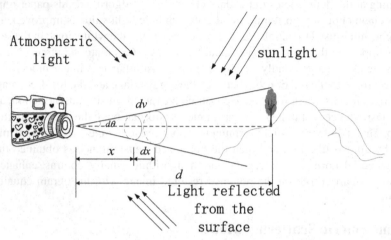

Fig. 1. Schematic diagram of ambient light imaging model

As shown in Fig. 1, assuming that the ambient light in the imaging scene is in a constant state, let the angle between the camera and the edge of the object and the horizontal direction be $d\theta$, the distance between the object and the camera be d, and the unit at x from the camera The microelement part of the volume is dv, and the microelement dv at the distance x from the object to the camera is used as the ambient light point source, then within a unit volume, the obtained light intensity microelement dI can be expressed as:

$$dI(x, \lambda) = dv \cdot \beta(\lambda) = \beta(\lambda) \cdot d\theta \cdot x^2 \cdot dx \qquad (3)$$

According to the law of incident light attenuation, it can be obtained that the ambient light enters the propagation path and passes through the loss of suspended particles in the atmosphere, and the light intensity dE reaching the camera is as follows:

$$dE(x, \lambda) = \frac{dI(x, \lambda)e^{-\beta(\lambda)}}{x^2} \qquad (4)$$

The definite integral of formula (4) in the interval from $x = 0$ to $x = d$, considering the incident attenuation, can obtain the energy relationship of the ambient light imaging model under the ambient light intensity E_q:

$$E_q(d, \lambda) = E_\infty(\lambda)(1 - e^{-\beta(\lambda)d}) \qquad (5)$$

2.3 Image Degradation Model in Foggy Weather

Due to the influence of severe weather such as smog and system-related parameters, image degradation can be understood as the attenuation of each pixel in a clear image to varying degrees. Therefore, the atmospheric scattering model can be approximately regarded as the superposition of the incident light attenuation model and the ambient light imaging model, and its mathematical expression is:

$$E(d, \lambda) = E_0(\lambda)e^{-\beta(\lambda)d} + E_\infty(\lambda)(1 - e^{-\beta(\lambda)d}) \tag{6}$$

For the convenience of calculation, it is generally considered that the pixel gray value of a certain point in the image is approximately proportional to the radiation intensity received by the point, then:

$$E(d, \lambda) = I(x), E_0(\lambda) = J(x), E_\infty(\lambda) = A \tag{7}$$

$$t(x) = e^{-\beta d(x)} \tag{8}$$

where $x = (i, j)$ is a two-dimensional vector, which represents the coordinate position of each point in the image, I represents the foggy image, J represents the image after defogging, $t(x)(0 \le t(x) \le 1)$ is the transmittance distribution function, and A is the atmospheric light value. $t(x)J(x)$ is the attenuation term and $(1 - t(x))A$ is the ambient light term.

Simplifying the processing of formula (7), the radiation intensity obtained after dehazing can be expressed as:

$$J(x) = \frac{I(x) - A}{t(x)} + A \tag{9}$$

3 Improved Dehazing Algorithm Based on Dark Channel Prior

3.1 Dark Channel Prior Transmittance and Atmospheric Light Value Estimation

Assuming that the transmittance $t(x)$ in each minimum filtering window is constant, assuming $\tilde{t}(x)$, and A (global atmospheric composition) is a known quantity, we can obtain [2]:

$$\min_{y\in\Omega(x)} (\min_C (\frac{I^C(y)}{A^C})) = \tilde{t}(x) \min_{y\in\Omega(x)} (\min_C (\frac{J^C(y)}{A^C})) + 1 - \tilde{t}(x) \tag{10}$$

In fog-free weather, there are also some particles in the air, which is a slight fog. In order to produce the texture of depth of field, it is necessary to retain a certain amount of fog. Therefore, assuming a retention factor ρ, in general, $\rho = 0.95$ can be taken. Then formula (10) can be rewritten as:

$$\tilde{t}(x) = 1 - \rho \cdot \min_{y\in\Omega(x)} (\min_C (\frac{I^C(y)}{A^C})) \tag{11}$$

For the selection of the atmospheric light A value, combined with the ambient light imaging model, the area with the densest fog concentration is generally selected. In fact, the A value can also be obtained from the dark channel map, and the first 0.1% of the pixels are taken according to the brightness. Find the corresponding point with the highest brightness in the original fog image, and the pixel value of this point is used as the A value. Combined with formula (9), the formula for restoring the image can be obtained as:

$$J(x) = \frac{I(x) - A}{\max(\tilde{t}(x), t_d)} + A \tag{12}$$

where $t_d = 0.1$ is a threshold set for the transmittance, which can restrain the phenomenon that the restored image is too bright due to too small transmittance, resulting in serious distortion of the restored image color.

3.2 Optimization of Atmospheric Light Value Selection Method

The atmospheric light value model uses the Otsu algorithm principle [3], and the specific analysis process is shown in the figure below (Fig. 2):

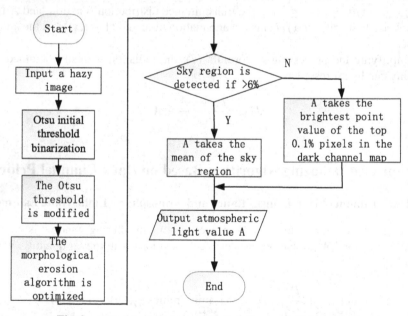

Fig. 2. Atmospheric light value calculation module flow chart

Sky Area Judgment

If there is a sky area, calculate the atmospheric light value in the sky area, if there is no sky area, then use the method of calculating the atmospheric light value by He et al. The number of pixels defining the sky area is n_{sky}, the total number of pixels of the foggy image is MN, and the ratio of the sky area to the entire image is p_{sky}, then:

$$p_{sky} = \frac{n_{sky}}{MN} \tag{13}$$

At that time $p_{sky} \geq 6\%$, it is determined that there is a sky area. At that time $p_{sky} < 6\%$, it was determined that there was no sky area.

Atmospheric Light Value Calculation

After judging that there is a sky area, find the values in the dark channel map correspond-ing to the sky area, and calculate the average of these values as the value of atmospheric light. The calculation formula is as follows:

$$A_1 = \frac{\sum\limits_{(i,j)\in sky} I_{dark}(i,j)}{|n_{sky}|} \tag{14}$$

where $I_{dark}(i,j)$ is the dark channel information map of the foggy image.

After many experiments and tests, this method has a good correction effect on the situation where the atmospheric light value is too large or too small, and the picture will not appear too white or the color is too saturated, and the restored image is more natural.

When the sky area is not detected, the method used to calculate the atmospheric light value is as follows:

(1) In the dark channel map, take the top 0.1% of the pixels according to the brightness,
(2) Correspond the pixel position found in (1) to the original fog image and find the value of the corresponding point with the highest brightness as the A value.

3.3 Transmittance Method Improvements

Firstly, by analyzing the foggy image and the fogless image, a linear model is used, and its logarithmic transformation is performed to obtain the initial transmittance. The initial transmittance is refined by guided filtering algorithm to obtain the refined transmittance. Then, the depth-of-field formula is introduced to compensate the thinned transmittance. The transmittance was optimized using the weighted norm A regularization algorithm. Figure 3 below is the flow chart of transmittance optimization:

Transmittance Estimation Based on Logarithmic Adaptation

Transmittance is also one of the important parameters in image restoration, and the accuracy of transmittance estimation has a direct impact on the quality of defogged images. The transmittance function is:

$$t(i,j) = \frac{A - I(i,j)}{A - J(i,j)} \tag{15}$$

where $I(i,j)$ and $J(i,j)$ represent the gray value of the foggy image and the haze-free image at the C middle coordinate point (i,j) respectively.

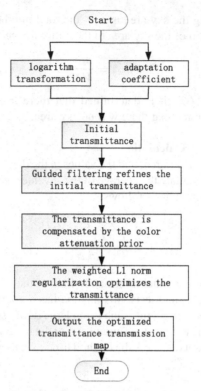

Fig. 3. Transmittance Optimization Flowchart

According to the dark channel prior theory, after the minimum filtering process on the transmittance function, we can get:

$$t(i,j) = \frac{\min\limits_{C \in \{R,G,B\}} A - \min\limits_{C \in \{R,G,B\}} I^C(i,j)}{\min\limits_{C \in \{R,G,B\}} A - \min\limits_{C \in \{R,G,B\}} J^C(i,j)} \quad (16)$$

Assuming that the atmospheric light A^C is known to be a constant, it will be abbreviated as $J^{dark}(i,j)$, then it can be obtained as:

$$t(i,j) = \frac{A^C - I^{dark}(i,j)}{A^C - J^{dark}(i,j)} \quad (17)$$

In the existing single image defogging algorithm based on linear mapping [4], the impact of ambient light on imaging is closely related to the depth of field distance, and the farther the depth of field distance is, the greater the impact will be.

Therefore, using the segmented area of the quadratic function to approximate it, the above formula can be modified as:

$$\min\limits_{C \in \{R,G,B\}} J^C(i,j) = \frac{\min\limits_{C \in \{R,G,B\}} I^C(i,j) - RGB_{\min}}{RGB_{\max} - RGB_{\min}} \cdot *I^{dark}(i,j) \quad (18)$$

where RGB_{max} and RGB_{min} are the maximum and minimum values of $\min_{C \in \{R,G,B\}} I^C(i,j)$, respectively.

Increase the luminance value with obvious difference in fog concentration in the dark channel, and perform logarithmic transformation processing on formula (18), it can be obtained:

$$\min_{C \in \{R,G,B\}} J^C(i,j) = \log(\delta + \frac{\min\limits_{C \in \{R,G,B\}} I^C(i,j) - RGB_{min}}{RGB_{max} - RGB_{min}} . * I^{dark}(i,j)) \qquad (19)$$

The logarithmic transmittance function can be obtained:

$$t_2(i,j) = \frac{\left| A^C - \min\limits_{C \in \{R,G,B\}} I^C(i,j) \right|}{\left| A^C - \log(\frac{\min\limits_{C \in \{R,G,B\}} I^C(i,j) - RGB_{min}}{RGB_{max} - RGB_{min}} . * \min\limits_{C \in \{R,G,B\}} I^C(i,j) + \delta) \right|} \qquad (20)$$

Transmittance Compensation Based on Color Attenuation Prior Algorithm
For most outdoor foggy images, the difference between fog density and brightness and saturation is positively correlated in the same area. Assuming that there is a linear relationship between the fog density, the difference between brightness and saturation, the specific mathematical expression is as follows:

$$d(x) \propto c(x) \propto v(x) - s(x) \qquad (21)$$

where x is the pixel point, $d(x)$ is the depth of field, $c(x)$ is the concentration of the fog, $v(x) - s(x)$ is the difference between the brightness and saturation of the image, and $\propto$ is a positive correlation sign.

The above law is the prior law of color attenuation, and its corresponding depth of field model, the formula is as follows:

$$d(x) = \theta_0 + \theta_1 v(x) + \theta_2 s(x) + \varepsilon(x) \qquad (22)$$

where $v(x)$ is the bright spot of the pixel, $s(x)$ is the saturation of the pixel, and $\theta_0, \theta_1, \theta_2$ are the linear coefficients. The function $\varepsilon(x)$ represents the random variable from which the model produces random errors, using a Gaussian density function with mean 0 and variable $\sigma^2 (\varepsilon(x) \sim N(0, \sigma^2))$.

According to the CAP principle, the depth of field formula can be introduced to compensate the transmittance. Its mathematical expression is:

$$t_b(x,y) = \max(t_3(x), q_i) \qquad (23)$$

Context Regularization Based on Weighted L1 Norm
Pixels in local facets of an image generally have similar depth values, deriving a piecewise transfer from boundary constraints. However, this creates patches in areas where the depth of field changes dramatically, leading to noticeable halo artifacts in the defogged

results. The way to solve this problem is to introduce a weighting function $W(x, y)$ on the constraints, namely:

$$W(x, y)(t_b(y) - t_b(x)) \approx 0 \tag{24}$$

where x and y are the pixel points of two adjacent positions in the transmittance map.

Through the feature statistics of a large number of haze images, it is found that the adjacent pixels on the edge of the image depth have similar color grayscales. Therefore, we can use the difference between local adjacent pixels to construct a weighting function. Below is the squared difference between vectors based on two adjacent pixels.

$$W(x, y) = e^{-\frac{\|I(x)-I(y)\|^2}{2\sigma^2}} \tag{25}$$

where σ is the standard deviation, which $I(x) - I(y)$ is the difference in gray value between two adjacent pixels in the local window.

Another formula for brightness difference based on adjacent pixels is as follows:

$$W(x, y) = \left(|\ell(x) - \ell(y)|^\phi + \zeta\right)^{-1} \tag{26}$$

where $\ell(x)$ is the logarithmic bright channel of the hazy image, and the exponent ϕ is the sensitivity of the sum of two adjacent pixels, ζ is a small constant (usually 0.0001) to prevent division by zero.

Based on the variable splitting method to optimize the above formula, the basic idea is to introduce several auxiliary variables, decompose them into a series of simple sub-problems, and construct a new transmittance target optimization function as:

$$\gamma \|t_b - t_z\|_2^2 + \sum_{j \in \eta} \|W_j \circ u_j\|_1 + \tau(\sum_{j \in \eta} \|u_j - (D_j \otimes t_b)\|_2^2) \tag{27}$$

Among them, $u_j(j \in \eta)$ is the auxiliary function introduced, and τ is the weight.

Through the 2-dimensional Fast Fourier Transform (FFT) algorithm and assuming a circular boundary condition, the optimal transmittance t^* can be directly calculated:

$$t^* = F^{-1}\left(\frac{\frac{\tau}{\gamma}F(t_z) + \sum_{j \in \eta} \overline{F(D_j)} \circ F(u_j)}{\frac{\tau}{\gamma} + \sum_{j \in \eta} \overline{F(D_j)} \circ F(D_j)}\right) \tag{28}$$

Among them, the class $F(\bullet)$ is the Fourier transform, which $F^{-1}(\bullet)$ is the inverse Fourier transform, which $(\bullet)$ means conjugation, and "∘" means element-wise multiplication.

3.4 Image Restoration

According to the description in Sect. 3.2, the optimal transmittance obtained through L1 regularization constrained iterative solution of the compensated transmittance and the atmospheric light value obtained based on the Otsu algorithm described in Sect. 3.1 are abbreviated in this paper as, by substituting the formula (9) A clear and fog-free image can be obtained, and a corrected image can be obtained.

$$J(x, y) = \frac{I(x, y) - A_1}{t^*(x, y)} + A_1 \tag{29}$$

4 Image Restoration Experiment and Result Analysis in Foggy Weather

4.1 Algorithm Process

The overall flowchart of the algorithm in this paper is shown in Fig. 4 below:

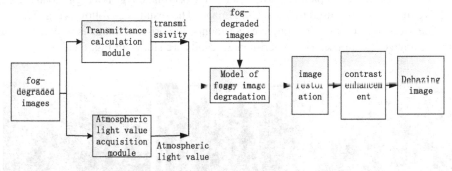

Fig. 4. The overall flow chart of the algorithm in this paper

The main steps of the algorithm are described as follows:

(1) Get a foggy image.
(2) The transmittance obtaining module and the atmospheric light value obtaining module obtain the atmospheric light value and the transmittance respectively.
(3) Substitute the foggy image, the calculated atmospheric light value and transmittance into the foggy image degradation model to obtain the restored image.
(4) Contrast enhancement processing is performed on the restored image with histogram equalization algorithm to obtain the final dehazed image.

4.2 Experimental Results and Analysis

Algorithms such as He algorithm, Meng algorithm, Zhu algorithm, Berman algorithm and the improved algorithm are selected for comparative experiments.

Original Image and Algorithm Meng Algorithm

Zu hu Algorithm   Berman Algorithm   Algorithm

Fig. 5. Defog comparison chart of road fog image

The first group of comparative experiments: the original image is a foggy road image, and five algorithms are used to dehaze the image. The result is as follows (Fig. 5):

The second group of comparative experiments: the original image is selected as the foggy image of the house. The experimental results are shown in Fig. 6 below:

Original Image and Algorithm Meng Algorithm

Zhu Algorithm   Berman Algorithm   Algorithm

Fig. 6. Defog comparison chart of houses in foggy days

The third group of comparative experiments: the original image used in this experiment is the ginkgo tree fog image (Fig. 7).

Original Image and Algorithm Meng Algorithm

Zhu Algorithm   Berman Algorithm   Algorithm

Fig. 7. Defog comparison chart of houses in foggy days

Table 1. Objective evaluation results

Dehaze image	algorithm	AG	SSIM	PSNR	EN
First group	He Algorithm	8.5661	0.4453	57.4749	6.7301
	Meng algorithm	8.4818	0.7240	59.6940	6.8776
	Zhu algorithm	6.0281	0.71206	60.0981	7.5663
	Berman Algorithm	10.4880	0.6462	59.2993	7.1226
	Algorithm	8.7614	0.7309	64.4671	7.9326
Second Group	He Algorithm	4.8137	0.6817	60.1483	6.5308
	Meng algorithm	5.4477	0.9141	64.9243	6.5664
	Zhu algorithm	2.7155	0.7261	61.7485	6.3269
	Berman algorithm	7.1307	0.7945	63.4003	7.3606
	Algorithm	9.6705	0.7454	65.1649	7.6837
The third group	He Algorithm	8.5661	0.4453	57.4749	6.7301
	Meng algorithm	8.4818	0.7240	59.6940	6.8776
	Zhu algorithm	6.0281	0.71206	60.0981	7.5663
	Berman Algorithm	10.4880	0.6462	59.2993	7.1226
	Algorithm	8.7614	0.7309	64.4671	7.9326

As AG, SSIM, EN, and PSNR are used to objectively evaluate the image after defogging. The following Table 1 shows the index data measured in each group of experimental result graphs.

From the data indicators, we can see that the average gradient of the algorithm in this paper is slightly lower than the average gradient of the Meng algorithm in the experiment. The SSIM, PSNR and E N of the algorithm in this paper are all higher than those of the Meng algorithm. In contrast, although the AG of the algorithm in this paper is slightly lower than that of the Meng algorithm, it provides the highest SSIM, P SNR and E N. It shows that the color of the algorithm in this paper is the most realistic, that is, the degree of image distortion is the lowest, and the detailed information of the image after defogging is rich, that is, the image is clearer.

5 Conclusion

After studying a large number of relevant literatures in the field of image dehazing, this paper decides to focus on the image dehazing algorithm based on the atmospheric scattering model. After deeply understanding the theoretical knowledge of the dark channel prior algorithm and the shortcomings of the dark channel prior algorithm, a new image defogging algorithm is proposed. It is verified by comparative experiments that the algorithm in this paper has no halo and color shift in the area of sudden depth of field, and the visual effect is more real and natural. Moreover, the defogged image obtained by the algorithm in this paper has rich detail information, clearer image, and has certain advantages in the degree of distortion. The defogged effect is good, and it is more in line with people's visual aesthetics.

Acknowledgement. This work was supported by the National Nature Science Foundation of China (62076249), Key Research and Development Plan of Shandong Province (2020CXGC010701, 2020LYS11), Natural Science Foundation of Shandong Province (ZR2020MF154).

References

1. Gonzalez, R.C., Woods, R.E.: Digital Image Processing, 3rd edn. Electronic Industry Press, Beijing (2011)
2. Bai, Z., Wang, B.: Air Particulate Matter Pollution and Prevention. Chemical Industry Press, Beijing (2011)
3. Wang, D., Zhang, T.: Review and analysis of image dehazing algorithms. J. Graph. **154**(06), 861–870 (2020)
4. Tarel, J.P., Hautière, N.: Fast visibility restoration from a single color or gray level image. In: IEEE International Conference on Computer Vision. IEEE (2010)
5. Fattal, R.: Single image dehazing. ACM Trans. Graph. **27**(3), 1–9 (2008)
6. Tan, R.T.: Visibility in bad weather from a single image. In: 2008 IEEE Computer Society Conference on Computer Vision and Pattern Recognition (CVPR), Anchorage, Alaska, USA. IEEE (2008)
7. He, K., Sun, J., Tang, X.: Single image haze removal using dark channel prior. IEEE Trans. Pattern Anal. Mach. Intell. **33**(12), 2341–2353 (2011)
8. Harald, K.: Theorieder horizontalen sichtweite: Kontrast und Sichtweite. Keim & Nemnich (1925)

9. Mccartney, E.J.: Scattering phenomena. Optics of the atmosphere: scattering by molecules and particles. Science **196**, 1084–1085 (1977)
10. Nayar, S.K., Narasimhan, S.G.: Vision in bad weather. In: Proceedings of the Seventh IEEE International Conference on Computer Vision. IEEE (2002)
11. Shen, Y., Wang, Q.: Sky region detection in a single image for autonomous ground robot navigation. Int. J. Adv. Rob. Syst. **10**(4), 1 (2013)
12. Li, X., Zhou, L.: Open-set voiceprint recognition adaptive threshold calculation method based on Otsu algorithm and deep learning. J. Jilin Univ. (Nat. Sci. Ed.) **59**(04), 909–914 (2021)
13. Li, X., Niu, H., Zhong, H.: Image defogging method based on mean standard deviation and weighted transmittance. J. Railway Sci. Eng. **17**(11), 2938–2945 (2020)
14. Zhao, X., Huang, F.: Image enhancement based on dual-channel prior and illumination map-guided filtering. Adv. Lasers Optoelectron. **58**(8), 45–54 (2021)
15. Wang, P., Zhang, Y., Bao, F., et al.: Optimal dehazing method based on foggy image degradation model. Chin. J. Image Graph. **23**(4), 12 (2018)
16. Cai, B., Xu, X., Jia, K., et al.: DehazeNet: an end-to-end system for single image haze removal. IEEE Trans. Image Process. **25**(11), 5187–5198 (2016)
17. Ren, W., Liu, S., Zhang, H., Pan, J., Cao, X., Yang, M.-H.: Single image dehazing via multi-scale convolutional neural networks. In: Leibe, B., Matas, J., Sebe, N., Welling, M. (eds.) ECCV 2016. LNCS, vol. 9906, pp. 154–169. Springer, Cham (2016). https://doi.org/10.1007/978-3-319-46475-6_10
18. Rashid, H., Zafar, N., Iqbal, M.J., et al.: Single image dehazing using CNN. Procedia Comput. Sci. **147**, 124–130 (2019)
19. He, Z., Patel, V.M.: Densely connected pyramid dehazing network. In: IEEE/CVF Conference on Computer Vision and Pattern Recognition. IEEE (2018)

Reinforcement Learning Based Vertical Scaling for Hybrid Deployment in Cloud Computing

Jianqi Cao[1] , Guiying Li[1,2(✉)] , and Peng Yang[1,3]

[1] Guangdong Provincial Key Laboratory of Brain-Inspired Intelligent Computation, Department of Computer Science and Engineering, Southern University of Science and Technology, Shenzhen 518055, China
ligy@sustech.edu.cn
[2] Research Institute of Trustworthy Autonomous Systems, Southern University of Science and Technology, Shenzhen 518055, China
[3] Department of Statistics and Data Science, Southern University of Science and Technology, Shenzhen 518055, China

Abstract. To maximize the CPU utilization of the server, offline tasks are usually deployed to the same server where the online service is running. Considering the necessity to ensure the service quality of online services, it is common practice to isolate the resources of online services. How to set the resource quota for online services not only affects the service quality of online services, but also affects the number and the stability of offline tasks that can be run on the server. Traditional rule-based methods or prediction-based methods will cause over-provision and fail to consider the stability of offline tasks, which often cannot achieve stability and efficiency. In this paper, reinforcement learning is proposed for the first time to solve the hybrid deployment of online services and offline tasks and dynamically adjust the resource quota of online services more effectively. Compared with the original state of the server, our proposed method reduces CPU idleness rate by 35.32% and increases CPU resource utilization rate by 3.84%.

Keywords: Cloud Computing · Reinforcement Learning · Vertical Scaling

1 Introduction

The hybrid deployment of online and offline tasks has become an increasingly important problem in cloud computing. With hybrid deployment, users can

This work was supported partly by the National Natural Science Foundation of China (Grants 62272210), partly by the Guangdong Provincial Key Laboratory (Grant 2020B121201001), partly by the Program for Guangdong Introducing Innovative and Entrepreneurial Teams (Grant 2017ZT07X386), and partly by the Stable Support Plan Program of Shenzhen Natural Science Fund (Grant 20200925154942002).

deploy offline tasks on the same host as online services. The hybrid deployment can improve server CPU usage and prevent users from renting cloud hosts when online services are idle. The traditional method to solve the problem of mixing online services and offline tasks is to plan the resource quota of online services in advance and schedule offline tasks based on the resource quota. However, the planning approach does not properly reserve resources for online services. Because users access online services at different times of the day, the load on online services fluctuates according to the volume of traffic [1]. The fixed resource reservation method may cause inaccurate resource estimation or waste of resources. In recent research, there are several methods to dynamically adjust resource quotas by predicting the load of online services [7–10,12,13]. However, these methods only consider the load of online services, and do not consider the stability of offline tasks. Because the resource quota of online tasks frequently changes, offline tasks are frequently scheduled and evicted from the server.

This paper proposes an effective vertical scaling method, under the premise of ensuring the quality of service (QoS) of online services, reduces the CPU resource idleness rate as much as possible, and improves the utilization rate. Dynamic online vertical scaling is essentially a Markov Decision Process (MDP), in which the agent will constantly adjust the resource quota of online tasks according to the time series and get rewards according to the server status. This is the first time, we model this problem as a reinforcement learning problem, regard the CPU load of online service and the running state of offline task as the state, and regard the adjustment of resource quota of online task as the action. The goal of reinforcement learning is to search for the optimal strategy and output the optimal resource quota for online tasks.

We use the dataset provided by Aliyun in 2021 for analysis [2]. The data set includes 12 h of run-time data for 1300 servers, with 1300 microservices corresponding to 42 min of run-time data for 90,000 containers. Based on the analysis of 1300 nodes and 90,000 containers, the average CPU usage of all servers is 64%, and the average CPU usage of all containers is 18%. Figure 1 shows the typical CPU usage of the servers (up) and the online service (down).

We used Q-learning method and verified them on the aliyun dataset. Empirical results show that the reinforcement learning method we proposed can reduce the idleness rate of CPU resources in the server by 35.32% and improve the resource utilization rate of the server by 3.84%.

This paper is summarized as follows. The second part shows the common methods to solve the hybrid deployment problem of the online services and offline tasks. The third part introduces how to model this problem as a reinforcement learning problem, and gives the implementation of Q-learning algorithm. The fourth part is the experimental part of this paper, through the comparison of other methods to prove the effectiveness of the proposed algorithm. The fifth part summarizes the work of this paper and discusses the future work.

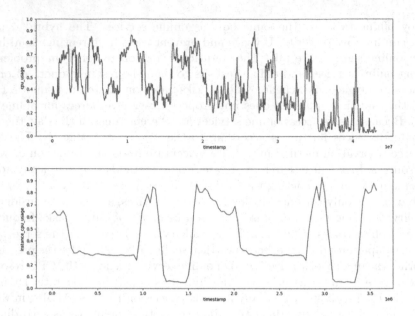

Fig. 1. Typical CPU usage of the servers (up) and the online service (down).

2   Related Work

In cloud computing, there have been many public cloud vendors and scholars devote to study the hybrid deployment problem of resource allocation, online service and offline task. Aliyun has released Koordinater [3], a hybrid workloads orchestration for the Kubernetes platform. The product first sets several levels of water level, when the online service CPU usage reaches a certain level, the corresponding level of vertical scaling will be triggered. But the settings of the horizontal line and the degree of expansion process depends on manual setting. Google Cloud [4] also provides vertical scaling on the Kubernetes platform. When the amount of resources used by a container reaches a certain level, the container expansion is triggered. However, the container cannot be shrunk and this vertical scaling strategy is not recommended for use in a production environment. [5] and [6] use rule-based method to vertically scale containers and virtual machines.

Rattihalli et al. [13] designed an automatic expansion system based on container resource utilization, i.e. RUBAS, which realized uninterrupted vertical automatic expansion and shrinkage of containers and resource usage estimation. Alam et al. [7] proposed a reliability-based resource allocation strategy for cloud environments, which minimizes cost and maximizes user application reliability at the same time. ZHOU et al. [15] proposed a container quota optimization algorithm based on LSTM and GRNN to reduce resource allocation rate by predicting the load of online tasks, but did not discuss the hybrid deployment scenario. Buchaca et al. [9] used perceptrons to predict the stage of container operation, and then dynamically modified the container quota based on policies. Berral et al. [8] used historical resource usage to predict future resource usage. The prediction methods of different time windows are also provided. In [10–12,14,20], appropriate resources are allocated to the application by predicting the future load. In addition to reinforcement learning, which can optimize decision-making, Yang et al. [19] proposed more efficient evolutionary reinforcement learning. Evolutionary learning is another way to solve problems. Although some works [16–18] proposed specifically designed evolutionary algorithms with efficiency optimized, it still takes a huge amount of time to solve the problems raised in this paper. In the above mentioned work, the stability and scheduling status of offline tasks are not considered as influencing factors to adjust resource quotas for online services.

3 Methodology

This paper models the hybrid deployment problem of online service and offline task as reinforcement learning problem and uses Q-learning algorithm to solve this problem. Figure 2 shows the process by which an agent interacts with environment. The following details how the agent interacts with the environment. 1) the agent in reinforcement learning first observes the current server environment, including the CPU usage of online service, the CPU usage of offline service, and the remaining CPU capacity of the server; 2) the agent selects an appropriate action based on the current environment, such as increasing or decreasing the resource quota of the online service; 3) the environment will change the current state (e.g., schedule and evict offline tasks) according to the actions made by the agent and the agent will get the reward value according to the change of the state. Reinforcement learning is to repeat the above process continuously until the agent can output the required result. In this part, we will introduce the definition of environment and reward function in Q-learning algorithm respectively, and the specific implementation of the Q-learning algorithm will be introduced in detail.

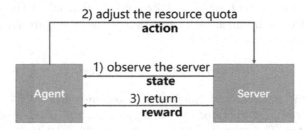

Fig. 2. Reinforcement learning include two part: agent and server. Agent observes the server and takes an action. Server is changed and returns a reward to agent.

Table 1. The attributes of server and offline tasks.

attribute	symbol	explain
server.duration	T	running time of server
server.current-cpu-usage	SC	server CPU usage at a time
server.current-cpu-usage-line	SL	server CPU quota at a time
server.online-task-cpu-usage	ONC	online task CPU usage
offline-task.num	OFN	the number of offline task
offline-task.cpu-usage	OFC	CPU usage of a offline task
offline-task.duration	OFT	running time of a offline task
offline-task.completed	OFCOM	completed offline task
offline-task.running	OFRUN	running offline task
offline-task.readied	OFRED	readied offline task

3.1 Environment Definition

A server includes the definitions in Table 1. The CPU usage of the online service is also an attribute of the server because it monopolizes the resources of the entire server in its initial state. Several offline tasks act as a resource pool with the definitions in Table 1. Given a moment, each offline task corresponds to a state.

3.2 Reward Function

In this paper, Q-learning algorithm is used to realize the dynamic adjustment of online task resource quota. Equation 1 shows the overall training process of Q-learning algorithm. The action $a0$ is selected by ϵ-greedy strategy. However, the ϵ-greedy parameter is set to 0 in this paper, because the agent of reinforcement learning needs to obtain the optimal decision in the training process, rather than the diversity of decisions. The agent first selects action $a0$ with the highest value of $s0$ in the current state and executes action $a0$ in the environment. Subsequently, the environment will transition to the next state $s1$, and the agent will get different degrees of reward r.

$$Q(s0, a0) \leftarrow Q(s0, a0) + \alpha(r + \gamma * \max Q(s1, a1) - Q(s0, a0)) \qquad (1)$$

The state of the environment consists of three indicators: CPU usage of the online service, resource quota of the online service, and number of offline tasks in different states. The agent has five actions A: $-20, -10, 0, 10, 20$. The agent can adjust the resource quota of online services. So, Q table is a $|T|*|SC|*|A|$ table. T indicates the running time of the online service, SC indicates the CPU usage of the online service, and A indicates the number of agent actions.

When designing the reward Eq. 2 of Q-learning algorithm, this paper mainly considers two indicators: the gap between CPU quota and real CPU usage, and the state of offline task.

Algorithm 1: Q-learning-based vertical scaling

Input: env:{server, online service, offline tasks}
Output: SL, OFCOM
1 **for** *i = 1 to episodes* **do**
2   reset env
3   **for** *t = 1 to NT* **do**
4   1. agent pick a action according to current state;
5   2. change to next state and return reward:
       `// state of online service`
6   **if** *SL in a suitable gap* **then**
7 | get a positive reward
8   **end**
       `// state of offline tasks`
9   **if** *OFCOM* **then**
10 | get a positive reward
11   **end**
       `// state of offline tasks`
12   **if** *evict OFRUN* **then**
13 | get a negative reward
14   **end**
15   3. Q-learning update according to reward;
16   **end**
17 **end**

$$reward = -|SL - ONC| + (|OFCOM| - |OFRED|) \qquad (2)$$

For the first indicator($-|SL - ONC|$), the online service gets a low reward when its CPU quota is far from the actual usage, and gets a high reward when the quota and the actual usage come closer. However, when the resource quota is in the appropriate range, a higher reward will be obtained. And when the resource quota does not change, it will get an additional reward. The resource quota for online services does not change frequently. Otherwise, offline tasks may be affected. For example, the offline task will be evicted due to small changes, resulting in the instability of the offline task. For the second indicator, when the offline task is scheduled or completed, the agent will get a higher reward. When an offline task is evicted, it gets a lower reward. Based on the above two indicators, the agent can get timely feedback when adjusting the resource quota of online tasks.

3.3 Q-Learning Design

Algorithm 1 shows the procedure of Q-learning interact with environment. We take the CPU usage of the online service, offline task, and node as input, and the resource quota of the online service (SL) and the completed offline task

(OFCOM) as output. The execution of the algorithm consists of two layers of loops, the outer loop is to perform episodes training for a certain environment, and the inner loop is a complete training process. A complete training process includes: 1) the agent chooses an optimal action according to the CPU usage of the online service and offline task on the current server; 2) the server changes the state according to the action (e.g., schedule and evict offline tasks) and return rewards; 3) Q-learning updates according to the rewards obtained.

4   Experiments

4.1   Experiments Setup

In this paper, two problems are mainly verified through experiments: 1) the agent trained by Q-learning algorithm is used to dynamically adjust the resource quota of online services to reduce the CPU resource idleness rate (Eq. 3) of the server; 2) improve the CPU utilization rate (Eq. 4) of the server by scheduling and executing offline tasks based on dynamically adjusted resource quotas. Figure 3 visually shows the meaning of the two indicators. In this experiment, we use the data set provided by aliyun and implement the Q-learning algorithm in the simulation environment.

$$idleness\ rate = \frac{1}{|T|} \sum_{t=1}^{T} \frac{SL_t - ONC_t}{100} \tag{3}$$

$$utilization\ rate = \frac{\sum_{t=1}^{T} ONC_t + \sum_{i=1}^{|OFCOM|} OFT_i * OFC_i}{T * 100} \tag{4}$$

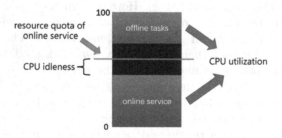

Fig. 3. CPU idleness refers to the portion allocated for online service but is not being used. CPU utilization refers to the total usage of online service and offline tasks.

In addition, this experiment also implements the unified reservation method and rule-based reservation method to compare with the agent trained by Q-learning algorithm. In addition to data sets and evaluation metrics, this experiment also considers how to set the specification and scheduling mode of offline tasks. Because the scheduling mode of offline task will affect the training process of different algorithms, different scheduling order will cause the algorithm cannot converge. By analyzing the aliyun dataset, we consider using several offline tasks with different sizes and the same scheduling order. At each point of time, the scheduler will prioritize scheduling large-size offline tasks. When the load of the current server is greater than the server capacity, the scheduler will evict the minimum offline tasks.

4.2   Algorithm Setup

This paper implements three ways of adjusting resource quota: unified resource reservation, rule-based resource reservation method and Q-learning method. The following describes the settings for each of the three methods.

Unified Resource Reservation. The unified resource reservation method collects the CPU usage of online services during a historical period. During this time, the highest CPU usage is captured for the online service. The unified reservation method uses the highest CPU usage as the resource quota of the online service and does not change dynamically. In this experiment, we will use the highest CPU usage of the online service plus 10% as the resource quota.

Rule-Based Resource Reservation. In the rule-based resource reservation mode, operator need to observe online services and set rules based on the historical CPU usage of the microservice. For example, if the CPU usage of an online service reaches a certain threshold, the system increases the resource quota. In this experiment, we determine whether to increase the resource quota based on the CPU usage at each moment, and we need to reserve a certain interval with the real CPU usage. In this experiment, we set the interval to 10%, because reserving an appropriate interval can cope with online service load requests.

Q-Learning Algorithm. The third method is to dynamically adjust the resource quota of online service by the agent trained by Q-learning. In the algorithm, two parameters need to be adjusted: Q-learning algorithm parameters and action space size. For algorithm parameters, after many experiments, we set the parameters in Eq. 1 as: $\epsilon = 0$, $\alpha = 0.1$, $\gamma = 0.9$. The ϵ parameter is used to allow the algorithm to explore richer strategies. However, for the problem proposed in this paper, we expect to find the optimal scheduling decision in a small action space without frequently trying other decisions, so we set ϵ to 0. The α parameter is the learning rate of the algorithm and represents how many steps are required for each update of the algorithm. The parameter γ is the algorithm's discount factor, indicating how much a reward in the future state affects the value of the present state. For the parameter of action, we set the action space to five discrete values: -20, -10, 0, 10, 20. The purpose of this is that we need to trade-off the amount of action space and the breadth of adjustment. Too large action space will slow down the algorithm training speed, and too small adjustment range will lead to frequent changes in the online service resource quota.

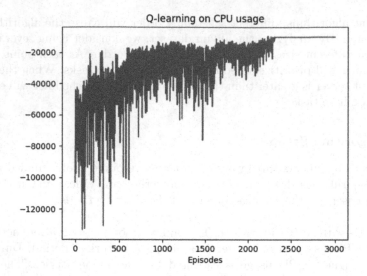

Fig. 4. The convergence process of the Q-learning in our experiments. The algorithm iterates and gets higher rewards

4.3 Experiment Results and Analysis

In our experiments, the Q-learning algorithm reaches convergence after 2500 iterations. Figure 4 shows the convergence process of Q-learning algorithm. The algorithm gains higher rewards as it iterates.

Figure 5 visually shows the resource quotas adjusted by each of the three algorithms on a certain server, and the offline tasks can be scheduled for execution on that server. Intuitively, the agent trained by Q-learning can reduce more CPU idleness rate and improve CPU utilization rate. Through the execution of three algorithms on the aliyun dataset, Table 2 shows that Q-learning algorithm has better performance than the other two methods. Q-learning algorithm reduces the CPU idleness rate by 35.32% and increased the CPU utilization rate by 3.84%.

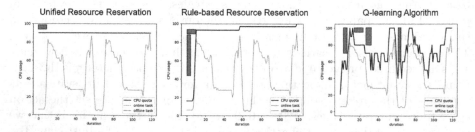

Fig. 5. The efficiency of the three methods under the same online service load. The red rectangle represents the offline tasks scheduled to this server (Color figure online).

Table 2. Performance of three methods on the same data set.

method	CPU idleness rate	CPU utilization rate
origin	81.15%	18.84%
Unified Resource Reservation	46.39%	20.31%
Rule-based Resource Reservation	**44.41%**	20.44%
Q-learning Algorithm	45.83%	**22.68%**

In this experiment, the agent trained by Q-learning algorithm can improve the CPU utilization rate of server. However, the agent did not have a good effect on CPU idleness rates. This is because the algorithm also considers the offline task scheduling in the training process. The agent does not adjust the resource quota of online services too much without affecting the QoS. Therefore, the agent does not perform well on this metric. In the scenario where online services and offline tasks are hybrid deployed, the CPU utilization rate directly reflects the efficiency of the algorithm. The Q-learning algorithm proposed in this paper has achieved good results indeed.

5   Conclusion

In this paper, Q-learning algorithm is proposed for the first time to solve the hybrid deployment problem of online service and offline task. Compared with other methods that rely on human expertise and fixed rules, Q-learning can consider the stability of both online services and offline tasks. Through the experiment on the open source data set, Q-learning method reduces the overall CPU idleness rate of 35.32% and improves the CPU utilization rate of 3.84%.

In addition, the Q-learning algorithm used in this paper is running in a simulated environment. The next step will be to extend it on real container platforms, and as an extension of the platform. The resource quota on the container platform will be reasonably organized by our algorithm.

References

1. Guo, J., et al.: Who limits the resource efficiency of my datacenter: an analysis of alibaba datacenter traces. In: Proceedings of the International Symposium on Quality of Service, pp. 1–10 (2019)
2. Luo, S., et al.: Characterizing microservice dependency and performance: Alibaba trace analysis. In: Proceedings of the ACM Symposium on Cloud Computing, pp. 412–426 (2021)
3. Koordinator. https://koordinator.sh/. Accessed 4 Oct 2022
4. Google Cloud. https://cloud.google.com/kubernetes-engine/docs/concepts/verticalpodautoscaler. Accessed 4 Oct 2022
5. Hoenisch, P., Weber, I., Schulte, S., Zhu, L., Fekete, A.: Four-fold auto-scaling on a contemporary deployment platform using docker containers. In: Barros, A., Grigori, D., Narendra, N.C., Dam, H.K. (eds.) ICSOC 2015. LNCS, vol. 9435, pp. 316–323. Springer, Heidelberg (2015). https://doi.org/10.1007/978-3-662-48616-0_20

6. Turowski, M., Lenk, A.: Vertical scaling capability of openstack. In: Toumani, F., et al. (eds.) ICSOC 2014. LNCS, vol. 8954, pp. 351–362. Springer, Cham (2015). https://doi.org/10.1007/978-3-319-22885-3_30

7. Alam, A.B., Zulkernine, M., Haque, A.: A reliability-based resource allocation approach for cloud computing. In: 2017 IEEE 7th International Symposium on Cloud and Service Computing (SC2), pp. 249–252. IEEE (2017)

8. Berral, J.L., Buchaca, D., Herron, C., Wang, C., Youssef, A.: Theta-scan: leveraging behavior-driven forecasting for vertical auto-scaling in container cloud. In: 2021 IEEE 14th International Conference on Cloud Computing (CLOUD), pp. 404–409. IEEE (2021)

9. Buchaca, D., Berral, J.L., Wang, C., Youssef, A.: Proactive container auto-scaling for cloud native machine learning services. In: 2020 IEEE 13th International Conference on Cloud Computing (CLOUD), pp. 475–479. IEEE (2020)

10. Chen, X., Wang, H., Ma, Y., Zheng, X., Guo, L.: Self-adaptive resource allocation for cloud-based software services based on iterative QoS prediction model. Futur. Gener. Comput. Syst. **105**, 287–296 (2020)

11. Cheng, Y.L., Lin, C.C., Liu, P., Wu, J.J.: High resource utilization auto-scaling algorithms for heterogeneous container configurations. In: 2017 IEEE 23rd International Conference on Parallel and Distributed Systems (ICPADS), pp. 143–150. IEEE (2017)

12. Li, C., Tang, J., Luo, Y.: Elastic edge cloud resource management based on horizontal and vertical scaling. J. Supercomput. **76**(10), 7707–7732 (2020). https://doi.org/10.1007/s11227-020-03192-3

13. Rattihalli, G., Govindaraju, M., Lu, H., Tiwari, D.: Exploring potential for non-disruptive vertical auto scaling and resource estimation in kubernetes. In: 2019 IEEE 12th International Conference on Cloud Computing (CLOUD), pp. 33–40. IEEE (2019)

14. Roy, N., Dubey, A., Gokhale, A.: Efficient autoscaling in the cloud using predictive models for workload forecasting. In: 2011 IEEE 4th International Conference on Cloud Computing, pp. 500–507. IEEE (2011)

15. Zhou, H.C., Bai, H., Cai, Z.G., Cai, L., Gu, J., Tang, Z.M.: Container quota optimization algorithm based on GRNN and LSTM. Acta Electonica Sinica **50**(2), 366 (2022)

16. Yang, P., Tang, K., Yao, X.: Turning high-dimensional optimization into computationally expensive optimization. IEEE Trans. Evol. Comput. **22**(1), 143–156 (2017)

17. Yang, P., Yang, Q., Tang, K., Yao, X.: Parallel exploration via negatively correlated search. Front. Comp. Sci. **15**(5), 1–13 (2021). https://doi.org/10.1007/s11704-020-0431-0

18. Yang, P., Tang, K., Yao, X.: A parallel divide-and-conquer-based evolutionary algorithm for large-scale optimization. IEEE Access **7**, 163105–163118 (2019)

19. Yang, P., Zhang, H., Yu, Y., Li, M., Tang, K.: Evolutionary reinforcement learning via cooperative coevolutionary negatively correlated search. Swarm Evol. Comput. **68**, 100974 (2022)

20. Yang, J., Liu, C., Shang, Y., Mao, Z., Chen, J.: Workload predicting-based automatic scaling in service clouds. In: 2013 IEEE Sixth International Conference on Cloud Computing, pp. 810–815. IEEE (2013)

Swing-Up and Balance Control of Cart-Pole Based on Reinforcement Learning DDPG

Jie Liu[✉], Xiangtao Zhuan, and Chuang Lu

Wuhan University, Bayi Road No. 299, Wuhan 430072, China
791428282@qq.com

Abstract. As a typical strong artificial intelligence method, reinforcement learning has been applied to real control tasks. Cart-pole system is an ideal controlled object, which is often used to verify the feasibility of control theory. In order to explore the effect of continuous reinforcement learning in the control of physical systems, this paper proposes a full control method for the swing up and stabilization of the cart-pole based on reinforcement learning DDPG. By interacting the cart-pole model with the main body of reinforcement learning, the Actor-Critic framework and the deterministic gradient algorithm DDPG are used to complete the learning of the cart-pole and realize the whole process of swing-up and balance control. Finally, the stability and effectiveness of reinforcement learning for the control of cart-pole are verified by simulation and experiment.

Keywords: Reinforcement learning · DDPG · Cart-pole system

1 First Section

The cart-pole system is a typical high-order, multi-variable, severely unstable and strongly coupled nonlinear system. It is an ideal controlled object in the study of control theory and an experimental platform to verify the feasibility of control theory and methods [1]. At the same time, the research of cart-pole also has an important engineering background. From any control problem with the center of gravity at the top and the fulcrum at the bottom seen in daily life to the stability of spacecraft and various servo platforms, it is very similar to the control of cart-pole. Therefore, it is widely used in the stability control of offshore drilling platforms, satellite launchers, aircraft control, chemical process control and robot walking [2, 3].

Due to the effect of its own gravity, when the cart-pole system is in a static state, the pole is in a vertical downward state. The process of making the pole automatically from the vertical state to the vertical upward state is the swing-up control of the cart-pole system, and the swing-up control process is required to be fast. Keeping the pole upright is the balance control of the cart-pole system. The balance control requires that the angle change of pole and the displacement change of cart should be as small as possible, that is, high precision is required.

In the current research on cart-pole, most of the research results mainly focus on the balance control, because the balance control of the cart-pole system can linearize the

L. Pan et al. (Eds.): BIC-TA 2022, CCIS 1801, pp. 419–429, 2023.
https://doi.org/10.1007/978-981-99-1549-1_33

model near the equilibrium point, and then a variety of mature linear control theories can be used to control the pole. However, the automatic swing-up control of the cart-pole cannot be linearized due to the large-scale movement of the pole, so it cannot be controlled by the traditional control theory, and it also has strong nonlinear and chaotic motion, which is quite difficult to control. It is a challenging research topic, and many scholars at home and abroad are also working on this research [4].

The Bang-Bang control method based on time optimal control adopts minimum value to solve the time optimal control problem. Its advantages are easy to realize and fast response, and its disadvantages are strong limitations, rough control and poor control effect [5, 6]. The energy control algorithm mainly controls the energy of the pole in the cart-pole system to achieve swing-up control. It can swing up quickly. Although it can achieve good control effect, it has difficult requirements for the accuracy of object parameters [7]. The fuzzy control algorithm does not need a specific object model, it only needs to obtain appropriate control rules to achieve good control effect. However, it also has the disadvantages of large amount of calculation and large amount of experimental data [8].

In recent years, reinforcement learning, as an intelligent learning algorithm, has gradually attracted the attention of various experts and scholars. Through its unique "trial and error" mechanism, it continuously tries optional behaviors in various states, and uses the given reward mechanism to maximize the reward that can be obtained in the current state, so as to continuously explore the optimal behavior through self-learning to achieve the control effect. To solve the nonlinear, unstable and highly complex problem of cart-pole, reinforcement learning, a method with robustness, simple algorithm and strong generality, is extremely suitable.

As early as 2004, reinforcement learning has been applied to cart-pole. Barto proposed to use ASE and ACE neurons to train neurons with reinforcement learning method, and successfully realized the balance control of cart-pole under small disturbance near the balance position, but did not solve the swing-up control of cart-pole [9]. Later, some scholars used SAC [10], Q-learning [11], DQN and other methods to realize the swing-up control of cart-pole.

Among various reinforcement learning algorithms, Q-learning only expresses the value function in the form of a table, and DQN constructs the value function in a neural network method but is not suitable for continuous state control. Reference [12] decomposes the cart-pole into two subtasks, the swing-up control is solved by Q-learning, and the balance control of the cart-pole is solved by Actor-Critic algorithm. Reference [13] verifies the effectiveness of DQN on the swing-up and balance control of cart-pole respectively. Such methods of decomposing tasks are relatively independent, but the two learning processes will increase the computational complexity of the system and the control accuracy is poor [14].

DDPG is a deep reinforcement learning algorithm. It uses the experience pool playback technology and target network technology in the DQN algorithm to improve the convergence and stability of the algorithm. On the basis of the deterministic gradient algorithm, it uses deep function approximation and learning strategies to make it can be applied in high-dimensional, continuous action spaces.

In this paper, a full control method for swing-up and balance control of cart-pole based on reinforcement learning DDPG is proposed. The torque applied to the cart is determined by observing the angle difference between the pole and the balance position. The cart-pole model is integrated into the reinforcement learning environment to interact and transfer information with the reinforcement learning subject. The reinforcement learning subject builds the corresponding neural network based on the Actor-Critic framework and updates it with the deterministic gradient algorithm DDPG to complete the learning of the left and right swing of the cart during the swing track of the pole, so as to realize the full control process of the swing-up and balance control of cart-pole.

2   Model of Cart-Pole

2.1   Mathematical Model of the Cart-Pole

The cart-pole system discussed in this paper is abstracted as a cart and a homogeneous pole, as shown in Fig. 1.

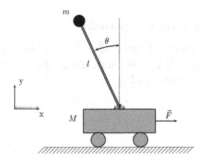

Fig. 1. Mathematical model of the cart-pole

The model parameters of the cart-pole system are as follows: the mass of the cart $M = 1$ kg, the mass of pole $m = 1$ kg, the friction coefficient of cart $b = 0.1$ N/(m $*$ sec), the moment of inertia of pole $I = 0.0034$ kg $*$ m^2, the length of pole $l = 0.5$ m. Ignoring the secondary friction such as air flow, the force analysis of the cart and pole shows that the dynamic equation of the cart-pole system is:

$$(M + m)\ddot{x} + b\dot{x} + ml\dot{\theta}\cos\theta - ml\dot{\theta}^2\sin\theta = F \tag{1}$$

$$(I + ml^2)\ddot{\theta} + mgl\sin\theta = -ml\ddot{x}\cos\theta \tag{2}$$

Since these two equations are nonlinear, the exact linearization method based on differential geometry is adopted to approximate linearization, and the following results are obtained:

$$(I + ml^2)\ddot{\theta} - mgl\theta = ml\ddot{x} \tag{3}$$

$$(M + m)\ddot{x} + b\dot{x} - ml\ddot{\theta} = u \tag{4}$$

where the control quantity u is the force exerted on the cart. For the motion equation, the control force is selected as the system input. The cart-pole system has four outputs: the displacement of the cart x, the speed of the cart $\dot{x}$, the angle of the pole θ, and the angular speed of the pole $\dot{\theta}$. Suppose the output of the system is y, the state space expression of the system can be obtained as:

$$\begin{bmatrix} \dot{x} \\ \ddot{x} \\ \dot{\theta} \\ \ddot{\theta} \end{bmatrix} = \begin{bmatrix} 0 & 1 & 0 & 0 \\ 0 & \frac{-(I+ml^2)b}{I(M+m)+Mml^2} & \frac{m^2gl^2}{I(M+m)+Mml^2} & 0 \\ 0 & 0 & 0 & 1 \\ 0 & \frac{-mlb}{I(M+m)+Mml^2} & \frac{mgl(M+m)}{I(M+m)+Mml^2} & 0 \end{bmatrix} \begin{bmatrix} x \\ \dot{x} \\ \theta \\ \dot{\theta} \end{bmatrix} + \begin{bmatrix} 0 \\ \frac{I+ml^2}{I(M+m)+Mml^2} \\ 0 \\ \frac{ml}{I(M+m)+Mml^2} \end{bmatrix} u \tag{5}$$

$$y = \begin{bmatrix} x \\ \theta \end{bmatrix} = \begin{bmatrix} 1 & 0 & 0 & 0 \\ 0 & 0 & 1 & 0 \end{bmatrix} \begin{bmatrix} x \\ \dot{x} \\ \theta \\ \dot{\theta} \end{bmatrix} + \begin{bmatrix} 0 \\ 0 \end{bmatrix} u \tag{6}$$

2.2   Swing-Up Control of Cart-Pole System

The essence of swing-up control of cart-pole is the process from the vertical downward state of the pole to the vertical upward state under the action of external force. The swing-up control process is shown in Fig. 2:

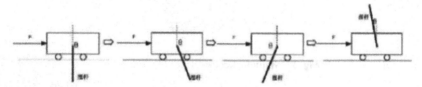

Fig. 2. Swing-up control process of cart-pole

Generally, due to the restriction of the length of the guide rail, a one-way force F cannot swing the pole from the vertical downward position to the vertical upward position. Therefore, the swing-up process can be divided into the following steps:

1) Under external force F, the cart moves horizontally along the guide rail from the static state. Due to inertia, the pole swings from the vertical downward position ($\theta = 0$) to the fourth quadrant position ($0 < \theta < \pi/2$).
2) Apply force -F in the opposite direction of the force F, and the pole swings from the fourth quadrant position ($0 < \theta < \pi/2$) to the third quadrant position ($-\pi/2 < \theta < 0$).
3) Apply a force in opposite direction of force -F, and the pole swings from the third quadrant position ($-\pi/2 < \theta < 0$) to the first quadrant position ($\pi/2 < \theta < \pi$).
4) Repeat steps 2) and 3) until the pole reaches the vertical upward position $[-2°, 2°]$. Complete the swing-up control process.

2.3 Balance Control of Cart-Pole System

The essence of the balance control of the cart-pole is to stabilize the pole within a small angle deviation (generally $[-12°, 12°]$), through the left and right swing of the cart within a specified displacement, keep the pole does not tip over, and make the angle deviation as close to 0 as possible. The schematic diagram of balance control is as follows (Fig. 3):

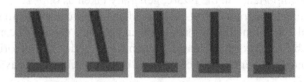

Fig. 3. Balance control process of cart-pole

At first, PID controller was used to control the balance control of cart-pole, but single PID could not control the angle of pole and the displacement of cart at the same time, so double loop PID could be used to effectively control the cart-pole system. When PID controller is used to control cart-pole, trial and error method is usually used for parameter tuning. This method has randomness and uncertainty and is time-consuming and difficult to improve accuracy. Some scholars have studied PID parameter tuning based on improved genetic algorithm and improved particle swarm optimization algorithm, which has significantly improved compared with the traditional trial and error method. The balance control algorithm based on particle swarm optimization algorithm has simple steps, fast search speed and high efficiency, but its performance is poor, the coding of network weights and the selection of genetic operators are sometimes troublesome, so it is easy to fall into local optimization. Traditional fuzzy control can realize fixed-point control, but it can only ensure local stability. In contrast, reinforcement learning can effectively reduce the amount of experimental training and obtain better control effect. For example, Soft Action-Critic (SAC) method and Q-learning algorithm have been applied to the balance control of cart-pole.

3 Reinforcement Learning of Cart-Pole System

3.1 Cart-Pole Model in Reinforcement Learning

Firstly, the control problem of the cart-pole system by reinforcement learning is described as a Markov decision process, which is solved by the continuous reinforcement learning method DDPG. In the environment of cart-pole system, both state $S(s_t \in S)$ and action $A(a_t \in A)$ are continuous, and the state transition equation is $T : T(s, a, s')$, a represents under the action, the probability of transition from state s to state s'; reward function is $R : R(s, a, s')$, a indicates under the action, the reward value of transition from state s to state s'. Starting from the initial state s_0, the agent takes actions a_t according to the DDPG strategy algorithm to obtain rewards $r(a_t, s_t)$ and transfer to the new state s_{t+1}. The purpose of reinforcement learning is to learn the best control strategy π^*

through interaction with the environment and trial and error, so that the agent can get the maximum expected total return from the environment, that is

$$\pi^* = \arg_\pi \max E_\pi \left[\sum_{t=0}^{h} \gamma^t R(t) \right] \tag{7}$$

where, h is the length of the decision-making sequence, γ is the discount coefficient, and indicates the importance to the future, generally taken as $0.8-1$.

The control block diagram of cart-pole based on reinforcement learning is shown in Fig. 4. In the full control method for swing-up and balance control of cart-pole, DDPG algorithm used by reinforcement learning agent is a model free algorithm based on deterministic policy gradient, which can be applied to continuous behavior space.

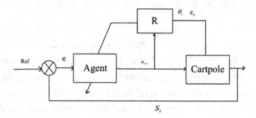

Fig. 4. Control block diagram of cart-pole based on Reinforcement Learning

In this paper, based on Actor-Critic framework, DDPG is used to update its policy gradient. The learning process is as follows:

Initialize Critic network $Q(s, a|\theta^Q)$ and Actor network $\mu(s|\theta^\mu)$ with random weights θ^Q and θ^μ.

Use the weight assignment $\theta^{Q'} \leftarrow \theta^Q$, $\theta^{\mu'} \leftarrow \theta^\mu$, to initialize the Critic target network Q' and the Actor target network μ'.

Select the action according to the current strategy of adding white noise N_t

$$a_t = \mu(s_t|\theta^\mu) + N_t \tag{8}$$

Execute action a_t and join the next status s_{t+1} and get reward r_t.

$$r_t = -0.1(5\theta_t^2 + x_t^2 + 0.05\mu_{t-1}^2) - 100B \tag{9}$$

where μ_{t-1} is the control force of the previous time step. B is a sign indicating whether the cart exceeds the specified range. If the cart is out of bounds, it is 1, and if it is not out of bounds, it is 0.

Store a set of data (s_t, a_t, r_t, s_{t+1}) in R. A small number of N groups of data (s_i, a_i, r_i, s_{i+1}) are randomly selected from R for learning. Set the target value of the value function as follows, where the discount factor γ is 0.99.

$$y_i = r_i + \gamma Q'(s_{i+1}, \mu'(s_{i+1}|\theta^{\mu'})|\theta^{Q'}) \tag{10}$$

Update the Critic network by minimizing the loss function, where $N = 128$:

$$L = \frac{1}{N} \sum_i (y_i - Q(s_i, a_i|\theta^Q))^2 \tag{11}$$

Update the Actor network with the policy gradient obtained from the sampled data, where $N = 128$:

$$\nabla_{\theta^\mu} \mu|_{s_i} \approx \frac{1}{N} \sum_i \nabla_a Q(s, a|\theta^Q)|_{s=s_i, a=\mu(s_i)} \nabla_{\theta^\mu} \mu(s|\theta^\mu)|_{s=s_i} \tag{12}$$

Update target network regularly, and the update parameter is $\tau = 0.001(1e-03)$:

$$\theta^{Q'} \leftarrow \tau\theta^Q + (1-\tau)\theta^{Q'} \tag{13}$$

$$\theta^{\mu'} \leftarrow \tau\theta^\mu + (1-\tau)\theta^{\mu'} \tag{14}$$

3.2 Simulation of Cart-Pole Based on Reinforcement Learning

In this paper, the model building and simulation of the cart-pole system are carried out in the Simulink environment of MATLAB. The cart-pole system is modeled by Simscape Multibodies. The specific environment of reinforcement learning is described as a pole connected to the cart without a drive joint, moving along a friction-free track. The control process is that after a force F is applied to the cart, the system will transfer from state s_0 to state s_1, and the environment will return to the system a reward r. The training goal of reinforcement learning is to use the minimum control force, start from the vertical downward position of the pole, and apply external forces in different directions to the cart to make the cart move left and right, swing the pole left and right, quickly reach the vertical upward position and always keep the upright, while maximizing the accumulated reward.

The state space of the cart-pole system environment is $s = [x, \dot{x}, \theta, \dot{\theta}]$, which is composed of the position offset of the cart relative to the central position x, the speed of cart $\dot{x}$, the displacement angle of the pole relative to the vertical upward position θ, and the angular speed of the pole $\dot{\theta}$. In this environment, if the upward position of the stabilizer bar is radian 2π (or 0π) and the downward suspension position is radian π, the initial state of the system is $[0\ 0\ \pi\ 0]$ and the position range of the cart in the direction x is $[-3.5\ 3.5]$, if the range is exceeded, the training of this set will be terminated. The action space of the system environment is μ, which means adding a right or left force to the cart; control force μ is the system input, and its interval is $[-15N\ 15N]$. The state transition is calculated by the system dynamics equation.

In the experiment of DDPG learning and training cart-pole model, set the Critic learning rate as 0.001 and the Actor learning rate as 0.0005. The sampling time in the training process is 0.02 s, the single round training time is 25 s, and the maximum number of training steps is 1250 steps. Stop training when the average reward of the system is greater than −400. The changes of loss value and reward value during the training of cart-pole model are shown in Fig. 5. From the change trend of reward value curve and

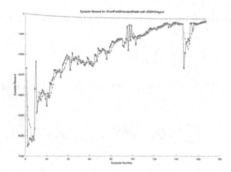

Fig. 5. Learning and training process of cart-pole model

strategy loss value curve, it can be seen that the reward value has met the requirements when the training process reaches about 170 rounds, that is, the cart-pole system has reached a stable state at this time.

The action sequence after the training of the cart-pole system is shown in Fig. 6. The action sequence is, the trolley moves forward to the right, driving the swing bar to swing to the left, then the trolley moves forward to the left, the swing bar swings to the right under the effect of inertia, and the swing bar reaches the vertical upward state after several rounds of repetition.

Fig. 6. Action sequence of cart-pole model completed by DDPG training

During the full control method for swing-up and balance control of cart-pole after DDPG training, the changes of the four state quantities of the system are shown in Fig. 7. As can be seen from the figure, it can be seen that driven by the cart, the pole continuously swings and accumulates momentum from the vertical downward position (π). After 1.2 s (60step) control, it finally enters the swing up successful angle area (0π), and successfully realizes the swing-up control of the cart-pole. The validity of DDPG algorithm for swing-up control of cart-pole is verified. It can also be seen from Fig. 7 that the position change range of the cart is $[-0.5, 0.2]$, and the angle change range of the pole is within 0.05 rad, meeting the stability target.

To verify the universal applicability of the agents trained by DDPG, change the pole mass to 0.042 kg, and the changes of the four state quantities of the system are shown in Fig. 8. As can be seen from the figure, it can be seen that driven by the cart, the pole continuously swings and accumulates momentum from the vertical downward position (π). After 2.2 s (110step) control, it finally enters the swing up successful angle area (0π), and successfully realizes the swing-up control of the cart-pole. The validity of DDPG algorithm for swing-up control of cart-pole is verified. It can also be seen from Fig. 8 that the position change range of the cart is $[-0.4, 0.1]$, and the angle change range of the pole is within 0.05 rad, which satisfies the stability target.

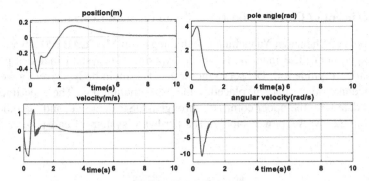

Fig. 7. Change curve of cart position and pole angle of cart-pole model trained with DDPG

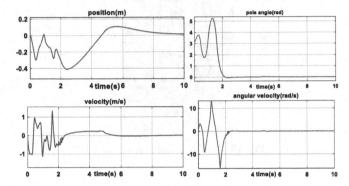

Fig. 8. Change curve of cart position and pole angle after changing pole mass

3.3 Comparative Simulation

The simulation of using dual PID controller to control the cart cart-pole is shown in Fig. 9. The inner loop PID parameters are $K_p = 190$, $K_i = 0$, $K_d = 20$ and the outer loop PID parameters are $K_p = 0.1$, $K_i = 0$, $K_d = 0.1$. It can be seen from the Fig. 9 that the pole recovers to be stable and basically remains unchanged at 1.4 s.

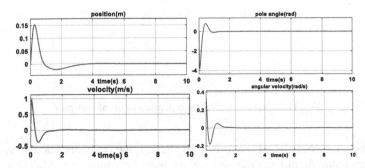

Fig. 9. Response curve of cart-pole system based on double PID control

428   J. Liu et al.

3.4   Experiment of Cart-Pole Based on Reinforcement Learning

The Cart-pole of Shenzhen Yuanchuang technology and MATLAB r2021b are selected for this experiment. The main body of the experimental system includes pole, cart, portable support, guide rail, DC servo motor etc. The main body, driver, power supply and data acquisition card are all placed in the experimental box. The experimental box exchanges data with the upper computer through a USB data line, and the other line is connected to 220 V AC power supply. The hardware system of cart-pole is shown in Fig. 10:

Fig. 10. Cart-pole hardware system

After careful adjustment of the controller, the model parameters of the cart-pole system are as follows: mass of the cart $M = 1\,\text{kg}$, mass of pole $m = 0.0426\,\text{kg}$, gravitational acceleration $g = 9.81\,\text{m/s}^2$, length of pole $l = 0.25\,\text{m}$. Matlab parameter setting is consistent with simulation. The pole angle change (a) and displacement change (b) of the cart-pole balance control are shown in Fig. 11:

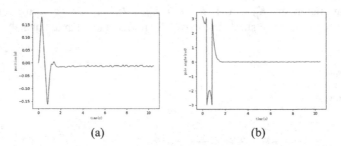

(a)   (b)

Fig. 11. Change of angle and displacement of cart-pole

It can be seen from Fig. 11 that driven by the cart, the pole continuously swings and accumulates momentum from the vertical downward position (π). After 2 s, it finally enters the swing up successful angle area (0π) and remains stable,The cart shaking error is between 4 cm and the angle error of the swing bar is between ($-0.02\pi, 0.02\pi$), which verifies the experimental effectiveness of DDPG algorithm for the swing-up and balance control of cart-pole.

4 Conclusion

In this paper, the cart-pole is taken as the research object, and the optimal control problem in the full control for swing-up and balance control is studied by reinforcement learning method. The environment and return function required by reinforcement learning in the control process of cart-pole are established, which solves the problem that the traditional control method cannot control the whole process of swing-up and balance control at the same time. Compared with the traditional algorithm, it is proved that reinforcement learning has better control accuracy and stability. The algorithm is used to complete the full control of the cart-pole hardware experimental platform, which verifies the stability and effectiveness of the proposed method. The application of reinforcement learning in practical engineering has a preliminary practice and has a good application prospect.

References

1. Li, X., Kallepalli, P., Mollik, T., et al.: The pendulum adaptive frequency oscillator. Mech. Syst. Signal Process. **179**, 109361 (2022)
2. Kong, F., Li, Z., Zhang, J., et al.: Quadcopter-research on control algorithm of inverted pendulum system. J. Intell. Sci. Technol. **1**(2), 140–144 (2019)
3. Chang, L.: Positioning, navigation and planning of biped robot in complex environment. Harbin Institute of Technology, Harbin (2021)
4. Zhang, L., Li, C.: Research and Simulation of control algorithm for swing up of Cart-pole. Industr. Control Comput. **21**(12), 40–42 (2008)
5. Wang, Y., Zhao, Y.: Swing-up and balance of inverted pendulum system based on bang-bang control. Mech. Electron. **8**, 16–18 (2004)
6. Hou, X., Yu, H., Chen, C.: The bang-bang-adjust control algorithm and simulation during the swing-up process of circular rail inverted pendulum. In: 2013 25th Chinese Control and Decision Conference, Harbin, pp. 373–378 (2013)
7. Gao, H., Cao, L., Cao, Y.: Research on swing-up control strategy of inverted pendulum system based on energy control. J. Shandong Univ. Sci. Technol. **35**(6), 95–100 (2016)
8. Wang, G., Xu, Z., Wang, L.: Research on adaptive fuzzy control system of inverted pendulum. Numer. Control Technol. **39**(7), 4–6 (2021)
9. Barto, A.G., Sutton, R.S., Anderson, C.W.: Neuronlike adaptive elements that can solve difficult learning control problems. IEEE Trans. Syst. Man Cybern. **13**(5), 834–846 (1983)
10. Chen, L., Jia, W.: Comparative analysis on the application of continuous reinforcement learning and PID control: taking the first-order inverted pendulum system as an example. Industr. Control Comput. **34**(10), 20–22 (2021)
11. Araújo, J.P., Figueiredo, M.A., Botto, M.A.: Control with adaptive Q-learning: a comparison for two classical control problems. Eng. Appl. Artif. Intell. **112**, 104797 (2022)
12. Zhang, R., Chen, W.: Whole process control of swing-up and balance of inverted pendulum based on reinforcement learning. Syst. Eng. Electron. Technol. **26**(1), 72–76 (2004)
13. Yang, W.: Research on inverted pendulum control algorithm based on reinforcement learning. Xi'an University of technology (2019)
14. Mao, W.: Research on the application of reinforcement learning in the swing up and balance control of inverted pendulum. Xi'an University of technology (2018)

Yolov5 Outdoor Dynamic Object Detection Based on Multi-scale Feature Fusion

Chengfeng Yu[1]($^{\boxtimes}$), Ken Chen[1], and Lin Jiang[1,2]

[1] School of Computer Science and Technology, Wuhan University of Science and Technology,
Wuhan 430081, China
353488210@qq.com

[2] Institute of Robotics and Intelligent Systems, Wuhan University of Science and Technology,
Wuhan 430081, China

Abstract. Outdoor object target detection is a very popular research task. 7 categories of data that fit the outdoor scenario were selected from the VOC2012 dataset as the current dataset for this study. In order to improve the detection accuracy of outdoor objects while keeping the network in good real-time, we propose a new yolov5 outdoor object detection network structure based on multi-scale feature fusion, which fuses shallow and deep features at different scales so that the fused feature layer has rich semantic information while enhancing the positioning information. The MAP of its proposed network on the VOC2012 dataset improved by 1.5% compared to Yolov5, and the FPS reached 61.51, which is better than the original Yolov5 method.

Keywords: Outdoor · Object detection · Yolov5 · Multi-scale · Feature Fusion

1 Introduction

Outdoor dynamic object detection is a front-end technology in artificial intelligence, automatic driving, robot obstacle avoidance and other directions, the improvement of its detection accuracy and real-time performance is also extremely important for the development of the entire industry. The rise of deep learning convolutional neural networks has also greatly promoted the research and development of related industries. Object Detection is an important application of convolutional neural networks, whose main task is to find out the interested objects in images and output the location and category information of detected objects. Therefore, improving the accuracy of target detection is one of the core issues in the field of computer vision.

Convolutional neural network (CNN) has been widely used in the field of computer vision and has been continuously developed. For example, VGG [1], GooLeNet [2], ResNet [3], Mobilenet [4], FCN [5] and U-net [6] are all classic CNN models. It is mainly used in the field of object detection or image segmentation. At present, the mainstream target detection algorithms are mainly divided into two categories: one-stage detection algorithm and two-stage detection algorithm. The former is represented by RCNN [7]

L. Pan et al. (Eds.): BIC-TA 2022, CCIS 1801, pp. 430–441, 2023.
https://doi.org/10.1007/978-981-99-1549-1_34

series, which mainly includes FastR-cnn [8], FasterR-cnn [9], R-FCN [10] and LibraR-CNN [11], while the latter is represented by Yolo series algorithms [12–15] and SSD [16].The two-stage algorithm has the advantages of fast speed and easy deployment, so the object of study of this paper is the two-stage algorithm.

For convolutional neural networks, different network depths correspond to different levels of semantic characteristics, shallow characteristics have high resolution, and carry more details of the characteristics, while deep web has low resolution, and carries more semantic characteristics. Different levels of characteristic maps correspond to characteristics which use different targets, so the theoretical development of feature fusion is conducive to improve the detection effect of network to targets.

Early object detection algorithms, such as SPP-Net, Fast R-CNN, Faster R-CNN, etc., all used the last layer of feature map in the network to externally connect the detection head for target detection. If the output size of the last layer of the feature map is 1/32 of the original image size and the original image size of the object is less than 32 * 32, it will lead to the inability to effectively detect objects, that is, it is impossible to use a single-scale feature map to represent objects of different scales. To solve such problems, Feature Pyramid Network (FPN) [17] came into being and proposed a method of feature fusion of different depths, that is, feature map of each resolution and the up-sample low-resolution features were spliced in the dimension of the channel, which enhanced the features of different levels and formed feature pyramids to represent objects of different sizes, thus improving the detection effect of small targets. With the development of deep learning feature fusion theory, feature fusion networks such as CEM (Context Enhancement Module) [18], PANet [19], Balanced Feature Pyramid [20], NAS-FPN [21] and BiFPN [22] have emerged continuously, further improving the efficiency of neural network feature fusion. PANet creates a bottom-up path enhancement on the basis of FPN, which is used to shorten the information path, and improve the feature pyramid structure by using the accurate positioning signals stored in the low-level features. The HRNet [23] model, by gradually adding low-resolution feature graph sub-networks in parallel to the main network of high-resolution feature graph, realizes multi-scale fusion and feature extraction among different subnetworks, but its computation is complex and its real-time performance is poor.

This paper makes improvements on the basis of PANet and proposes a new feature fusion method, which fuses high-resolution features in Backbone into Path Aggregation Network (PAN) [19], further strengthening the feature information connection between Backbone and Neck, in order to reduce the loss of detail information in the process of feature extraction. The improved PANet network structure is added to the Yolov5m [24] network structure to detect outdoor dynamic objects, which makes the network structure more efficient use of feature information with different resolutions, thus improving the target detection accuracy of the whole network.

2 Structure and Principle of Yolov5

The Yolov5 detection network consists of five versions: Yolov5n, Yolov5s, Yolov5m, Yolov5l and Yolov5x. Among them, Yolov5n is specially designed for embedded devices, with less parameters and calculation, but its detection accuracy will also decrease. The five structures control the depth and width of the network by the parameter depth multiple and width multiple respectively. The deeper the network is, the higher the detection accuracy is and the slower the detection speed is. Because the image itself of VOC2012 [25] data set contains more information, picture scenes are abundant, and the network has low requirements on hardware devices, the network structure of Yolov5m is adopted in this paper.

The network structure of Yolov5 is mainly composed of CSPDarkNet53 [26], SPPF, FPN and PAN. It uses CSPDarknet53 (cross-stage local network) as the Backbone, PANet (Path aggregation network) composed of FPN and PAN as the Neck, and the Head of YOLOv3 at same time.

In the Backbone part, Yolov5 adopts Conv, C3 and SPPF structures in the Backbone. Conv module is the combination of Convolution, Batch Normalization [27] and SiLu [28] activation functions. C3 is a residuals structure, as shown in Fig. 1. And Fig. 2 is the schematic diagram of Bottlenect module in C3.

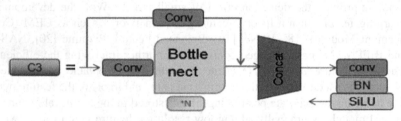

Fig. 1. Schematic diagram of C3 module

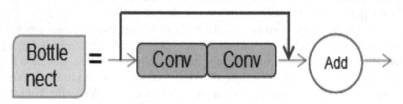

Fig. 2. Schematic diagram of Bottlenect module

SPPF module is improved on the basis of SPP [29]. It realizes the fusion of local features and global features by SPPF module, enriches the expression ability of feature map and enlarging the receptive field. SPPF adopts 5×5 maximum pooling method, while pools three times, and multi-scale feature fusion is carried out after obtaining different levels of features. Padding the input image, so SPPF will not change the size of the image. The SPPF module is shown in Fig. 3.

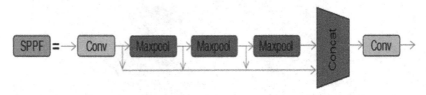

Fig. 3. SPPF structure diagram

In the Neck section, Yolov5 adopts the structure of PANet, which contains one FPN structure and two PAN structures. FPN is the top-down transfer and fusion of high-level semantic information through up-sample. PAN is the bottom-up feature pyramid. FPN layer conveys strong semantic features from top to bottom, while PAN conveys strong positioning features from bottom to top. FPN and PAN mainly play the role of divide-and-conquer. Feature maps of different scales are used to represent objects of different sizes. At the same time, the function of feature fusion is taken into account. However, with the deepening of the network, it will also lead to the loss of object features and reduce the accuracy of object detection by the model.

Head is used for the final detection part. The three C3 modules at the tail of Neck are external connected with detection heads for target detection, which can predict on three different scales to realize target detection of objects of different scales. It applies the anchor boxes to the feature graphs and generates the final output vector with class probability, object score, and enclosing box.

3   Multi-scale Feature Fusion

In general, when using the network to detect objects, the shallow network has a high resolution, learning the detailed features of the picture, while the deep network has a low resolution, learning more semantic features. In the early CNN network (as shown in Fig. 3(a), only used the last feature layer with more semantic information to exter-nally connected detection head for target detection, while the importance of its location features was ignored, resulting in the decline of the accuracy of target detection and unfriendly to small objects. Figure 3(b) is improved on the basis of Figure a network. The feature pyramids are formed by using different scale features extracted from images, and the generated feature maps of different scales are predicted respectively. The bottom feature map can identify large-scale targets well, but it is easy to miss the detection of small targets. The semantic information of the feature map of shallow large scale is less, although small targets can be framed, small targets are easily classified by mistakes.

Figure 3(c) uses the image pyramid to generate the feature pyramid, the image scales are resized in different proportions, and then the feature layer of each scale is extracted separately for prediction, which is equivalent to multi-scale transformation of the target, so that the network can better learn the information of different scales of the same target and improve the detection accuracy, but it will consume a lot of time, so it is not applicable.

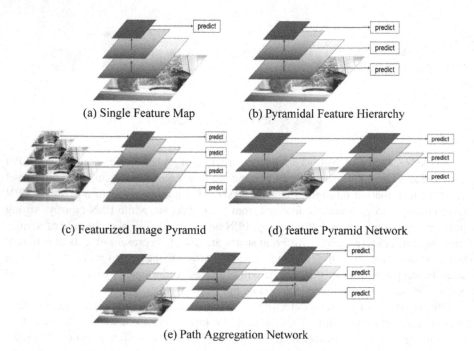

(a) Single Feature Map   (b) Pyramidal Feature Hierarchy

(c) Featurized Image Pyramid   (d) feature Pyramid Network

(e) Path Aggregation Network

Fig. 4. The structures of different Networks

With the development of deep learning feature fusion theory, the emergence of FPN has solved the related problems of the above-mentioned networks. Based on the previous network structure, up-sample and horizontal connection are added for feature fusion. The process from top to bottom is to up-sample the high-level features that carry more semantic information, and then connect the features to the previous features horizontally for feature fusion. Therefore, the semantic information and location information of each layer's feature map are enhanced, and calculation mount remains basically unchanged, thus improving the accuracy of target detection. Figure 3(e) is improved on the basis of FPN. On the original basis, down-sample and horizontal connection are added. Although FPN transmits semantic information of high-level features back to low-level features, but it ignores that the location information of high-level features is weak, so PANet transmits the rich location information of low-level features to high-level features, which further improves the target detection effect. However, with the deepening of the network structure, it will also cause the loss of some object features, resulting in missing and false detection of object.

4 Yolov5 Multi-scale Feature Fusion

4.1 Improved Network Structure

This article mainly aims at the two types of problems of target detection: location and recognition, this paper proposes an ideal of multi-scale feature fusion, which fuses different feature information carried by different scale feature maps, so make the network better fuse the detailed features of high-resolution targets and effectively improve the accuracy of outdoor dynamic object detection.

This paper is improved on the basis of Yolov5m target detection network, as shown in Fig. 4. The difference between it and Yolov5m is that in this paper the high-resolution features in Backbone are directly added to PAN for feature fusion after down-sample, and superimposed in the channel direction, that is to say, features of different scales in Neck and Backbone are fused to increase the spatial information exchange between Backbone and Neck, thus reducing the loss of spatial information in the process of feature extraction. By fusing the location information provided by the bottom-level detail features with the context information provided by the high-level semantic information, in the case of adding a few additional parameters, the detection accuracy of the target can be further improved, and the real-time performance can be maintained. Its improved structure is shown in Fig. 3.

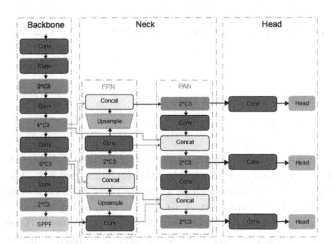

Fig. 5. Schematic diagram of Yolov5m network structure after improvement

4.2 The Structure of the Improved PANet Model

Traditional network (CNN) usually uses a single-line structure from shallow to deep to extract image features. For small objects, their semantic information appears in the shallow characteristic figure, but along with the network gradually deepened, the details information of small objects may disappear completely, thus causing the decrease of the accuracy of target detection, especially the decrease of the accuracy of small target detection is more serious.

As shown in Fig. 4, in order to reduce the information loss of some large-scale feature maps in the down-sample process of the network, we convolve the bottom features in the Backbone with 3 * 3 and step size of 2, to make the length and width of the feature map become half of the original size, unify the feature scale, keep the depth unchanged, and keep the fused feature sizes be same. Through Concat operation, the features after convolution operation are stacked with those in the original PAN in the channel direction, and then the detailed features at the bottom layer are transferred to the high-level features in Neck. The combination of the position information at the bottom layer and the semantic information at the high level avoids the feature loss caused by the deepening of the network and improves the robustness and accuracy of the network. The combination of the position information at the bottom layer and the semantic information at the high level avoids the feature loss caused by the deepening of the network and improves the robustness and accuracy of the network.

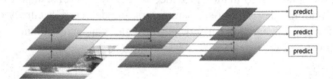

Fig. 6. Schematic diagram of improved PANet network structure

5 Experimental Results and Analysis

5.1 Data Set

The data set of outdoor dynamic objects used in this paper is VOC2012. Seven categories which are suitable for outdoor scenes are selected from 20 categories in VOC2012 data set as experimental data in this paper, including people, bicycles, cars, motorcycles, buses, cats, and dogs. The data samples are RGB images with different sizes of resolution. Although there are only 7 types of target samples, each sample has different shapes and sizes, and the image scenes are rich. Moreover, RGB images itself contain relatively rich information, which increases the representation ability of the model.

After image processing of the original VOC2012 data set, the number of samples is 12,868. The training set, verification set, and test set are randomly selected according to the ratio of 6: 2: 2, that is, 7720 training sets, 2574 verification sets and 2574 testing machines each. After the images are processed, the corresponding defect category and location information are recorded in xml format. After data processing, the number of various targets in the training set was detected as follows: 10,568 people, 500 bicycles, 1510 cars, 476 motorcycles, 402 buses, 783 cats and 977 dogs.

5.2 Data Enhancement

In this paper, Mosaic and image rotation, translation, scaling, clipping, Mixup [30], horizontal and upside-down flipping are used to enhance the data set, so as to enhance the generalization ability in network training.

At the input end, Yolov5 adopts Mosaic data enhancement method, that is, images are spliced by random scaling, random cropping, and random arrangement. Simply put, four training pictures are scaled and spliced into one picture. Mosaic is conducive to improving the detection of network for small targets. MixUp uses different weight coefficients to weight sum the two images and adjusts the labels according to the weight coefficients to increase the generalization ability of the model.

5.3 Experimental Environment and Sample Training

This paper uses local server to build the deep learning environment required by model training. The hardware environment of the test platform is Intel(R) CPU i9-10900x, 64GB running memory, NVIDIA GeForce RTX 3090 graphics cards, and 24GB display card memory. The software environment is: Ubuntu18.04 operating system, pytorch1.8.1 deep learning framework, CUDA11.1, using GPU to accelerate computing.

During the training of the network model, the initial weights of the feature extraction network are adopted pre-trained Yolov5m weight on the ImageNet to carry out the transfer learning. The optimization algorithm is SGD optimizer, aiming at large data sets, the training speed is faster. Specific parameters are set as follows: the whole training process is divided into 200 epochs, the number of images (batch size) read each time is 64, the initial learning rate is 0.0032, the learning rate drops once every five epochs of training, and the weight attenuation coefficient is set as 0.00036; The momentum is set as 0.843. The changes of target frame loss, category loss and target loss value of the training set are shown in Fig. 5. It can be seen from Fig. 5 that when the training approaches 180 epochs, the loss curve tends to be horizontal, and the model converges.

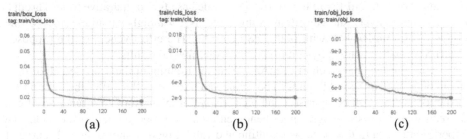

Fig. 7. Schematic diagram of proposed algorithm bounding box loss (a), class loss (b) and object loss (c)

5.4 Experimental Results and Analysis

This paper selected the classical two-stage algorithm (Faster-CNN) and the one-stage algorithm (Yolov5m, SSD) for comparison. The main evaluation indexes of detection accuracy use Average Precision (AP) and mean Average precision (mAP), which are commonly used in target detection models. The evaluation index of detection speed uses FPS (Frame Per Second) to evaluate the real-time performance of network monitoring.

From the comparison of network parameters in Table 1, it can be seen that the video memory occupation of the network image test in this paper is 1941MB, which only increases a small number of parameters compared with Yolov5m. Meanwhile, the model size and calculation amount are also kept within a controllable range. The hardware requirements of network reasoning are not high, and it can be deployed in practical applications.

Table 1. Comparison of network parameters

Method	Memery/MB	The model size/MB	Params	GFLOPS
Faster R-CNN	2765	330.5	41530000	--
SSD	2141	202.2	25270000	88.56
Yolov5m	1923	43.4	21467516	48
Yolov5m-FPN	1865	28.7	14186172	40.5
Yolov5m-BiFPN	1919	42.5	21024636	48.5
Proposed algorithm	1941	46.8	23200444	50.6

As can be seen from Table 2, no matter which kind of target, the AP value of the proposed algorithm is higher than that of FaSTER-CNN, SSD and Yolov5m, etc. The mAP of the seven types of defects reaches 83.3%, which is 1.5% higher than that of Yolov5m and still ensures a high detection speed. The FPS of the algorithm in this paper reaches 61.65, which is higher than the requirement of real-time detection of 25FPS and meets the requirement of real-time detection. At the same time, we use the YOLOV5-FPN and Yolov5m-BiFPN in the Yolov5 [24] code base for comparative tests. Through the test data analysis, the algorithm in this paper is 1.9% higher than the mAP of the highest Yolov5m-BiFPN, with both FPS being 61. The effectiveness of the proposed algorithm can be obtained by comparing the experimental results.

Figure 6 is the change chart of mAP during the training of the proposed algorithm. It can be seen that the mAP tends to be stable and has good effect after the training for 180 epochs. Through comparison and analysis with the curve of loss function, it can be seen that there is no overfitting phenomenon in the proposed network model. The experimental data in this paper is scientific and valid. Figure 7 shows the P_R curve of the proposed algorithm in the test set. The horizontal axis is recall and the vertical axis is precision. Recall refers to the percentage of the number of Class A objects correctly predicted in the total number of Class A objects; Precision refers to the percentage of the number of Class A objects correctly predicted in the total number of Class A objects predicted. The area below the P_R curve reflects the quality of the model. As can be seen from Fig. 7, while Recall value increases, Precision value remains at a high level and then gradually decreases. The model is more friendly to buses, cats, and dogs. In general,

Table 2. Comparison of detection performance of different algorithms

Method	FPS	Person's AP@0.5	Bicycle's AP@0.5	Car's AP@0.5	Motorbike's AP@0.5	Bus's AP@0.5	Cat's AP@0.5	Dog's AP@0.5	mAP @0.5
Faster R-CNN	11.93	0.798	0.644	0.658	0.731	0.762	0.944	0.880	0.774
SSD	35.77	0.675	0.397	0.464	0.430	0.591	0.850	0.699	0.587
Yolov5m	62.42	0.825	0.706	0.741	0.776	0.9	0.9	0.875	0.818
Yolov5m-fpn	70.82	0.780	0.626	0.657	0.619	0.861	0.851	0.752	0.735
Yolov5m-bifpn	61.92	0.822	0.688	0.736	0.792	0.893	0.907	0.857	0.814
Proposed algorithm	61.65	0.828	0.758	0.761	0.78	0.905	0.915	0.883	0.833

P_R curve declines in a balanced and orderly manner, which proves the effectiveness of the model (Figs. 8 and 9).

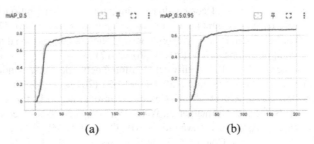

(a)   (b)

Fig. 8. Algorithm in this paper mAP@0.5 (a), mAP@0.95 (b) Variation diagram

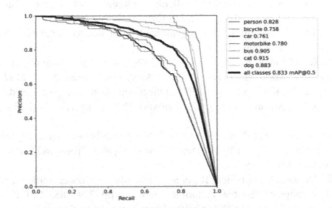

Fig. 9. Algorithm in this paper P_ R curve

6 Conclusions

In this paper, a new Yolov5 outdoor dynamic object detection network on multi-scale feature fusion is proposed. That is, in Yolov5m, high-resolution features of Backbone are fused with low-resolution features in Neck after down-sample, and shallow features and deep features are fused. It avoids the loss of some features with the deepening of the network, enriches the position information and semantic information of the tail feature layer, and improves the accuracy of target detection. The experimental results show that the improved Yolov5m has higher detection accuracy and speed for all seven outdoor objects, and AP values of all categories have been improved. In VOC2012 data set, the mAP of our network is increased by 1.5% compared with Yolov5, and FPS reaches 61.51. The improved structure of Yolov5m is proved to be effective and scientific.

References

1. Simonyan, K., Zisserman, A.: Very deep convolutional networks for large-scale image recognition. In: International Conference on Learning Representations, pp. 1–14 (2015)
2. Szegedy, C., Wei, L., Jia, Y., et al.: Going deeper with convolutions. In: 2015 IEEE Conference on Computer Vision and Pattern Recognition (CVPR), pp. 1–9 (2015)
3. He, K., Zhang, X., Ren, S., et al.: Deep residual learning for image recognition. In: Proceedings of the IEEE Conference on Computer Vision and Pattern Recognition, pp. 770–778 (2016)
4. Howard, G., Zhu, M., Chen, B., et al.: Mobilenets: effificient convolutional neural networks for mobile vision applications, arXiv:1704.04861 (2017)
5. Long, J., Shelhamer, E., Darrell, T.: Fully convolutional networks for semantic segmentation. IEEE Trans. Pattern Anal. Mach. Intell. **39**(4), 640–651 (2015)
6. Ronneberger, O., Fischer, P., Brox, T.: U-Net: convolutional networks for biomedical image segmentation. In: Navab, N., Hornegger, J., Wells, W., Frangi, A. (eds.) Medical Image Computing and Computer-Assisted Intervention – MICCAI 2015. MICCAI 2015. LNCS, vol. 9351, pp. 234–241. Springer, Cham (2015). https://doi.org/10.1007/978-3-319-24574-4_28
7. Girshick, R., Donahue, J., Darrell, T., et al.: Rich feature hierarchies for accurate object detection and semantic segmentation. In: IEEE Conference on Computer Vision and Pattern Recognition, pp. 580–587, IEEE, Columbus (2014)
8. Girshick, R.: Fast R-CNN. In: IEEE International Conference on Computer Visio. Santigago, pp. 1440–1448, IEEE, Chile (2015)
9. Ren, S., He, K., Girshick, R., et al.: Faster R-CNN: towards real-time object detection with region proposal networks. IEEE Trans. Pattern Anal. Mach. Intell. **39**(6), 1137–1149 (2017)
10. Dai, J.F., Li, Y., He, K.M., Sun, J.: R-FCN: object detection via region-based fully convolutional networks. In: Advances in Neural Information Processing Systems (NIPS), pp. 379–387 (2016)
11. Pang, J.M., Chen, K., Shi, J.P., et al.: Libra R-CNN: towards balanced learning for object detection. In: Proceedings of the IEEE Conference on Computer Vision and Pattern Recognition (CVPR), pp. 821–830 (2019)
12. Redmon, J., Divvala, S., Girshick, R., et al.: You only look once: unified, real-time object detection. In: Computer Vision & Pattern Recognition, pp. 779–788. IEEE (2016)
13. Redmon, J., Farhadi, A.: YOLO9000: better, faster, stronger. In: Computer Vision and Pattern Recognition (CVPR), pp. 6517–6525. IEEE (2017)
14. Redmon, J., Farhadi, A.: Yolov3: an incremental improvement. arXiv (2018)
15. Bochkovskiy, A., Wang, C.Y., Liao, H.: YOLOv4: optimal speed and accuracy of object detection, arXiv preprint arXiv, 10934 (2020)
16. Liu, W., Anguelov, D., Erhan, D., et al.: SSD: single shot MultiBox detector. In: Leibe, B., Matas, J., Sebe, N., Welling, M. (eds.) Computer Vision–ECCV 2016. ECCV 2016. LNCS, vol. 9905, pp. 21–37. Springer, Cham (2016). https://doi.org/10.1007/978-3-319-46448-0_2

17. Lin, T Y., Dollar, P., Girshick, R B., et al.: Feature pyramid networks for object detection. In: CVPR (2017)
18. Qin, Z., Ming, Z., Zhang, Z.N., et al.: ThunderNet: towards real-time generic object detection. In: CVPR (2019)
19. Liu, S., Qi, L., Qin, H.F., et al.: Path aggregation network for instance segmentation. In: Proceedings of the IEEE Conference on Computer Vision and Pattern Recognition (CVPR), pp. 8759–8768. IEEE (2018)
20. Zhang, T., Zhang, X., Shi, J., et al.: Balanced feature pyramid network for ship detection in synthetic aperture radar images. In: 2020 IEEE Radar Conference (RadarConf20). IEEE (2020)
21. Ghaisi, G., Lin, T.Y., Le, Q.V., et al.: NAS-FPN: Learning Scalable Feature Pyramid Architecture for Object Detection. arXiv preprint arXiv:1904.07392 (2019)
22. Tan, M.X., Pang, R.M., Le, Q.V.: EfficientDet: scalable and efficient object detection. In: Proceedings of the IEEE Conference on Computer Vision and Pattern Recognition (CVPR) (2020)
23. Sun, K., Xiao, B., Liu, D., et al.: Deep High-Resolution Representation Learning for Human Pose Estimation. arXiv e-prints (2019)
24. Jocher, G., Stoken, A., Borovec, J., et al.: Ultralytics Yolov5: v6.0–Yolov5-P6 1280 models, AWS, Supervise.ly and YouTube integrations (2021)
25. Shetty, S.: Application of Convolutional Neural Network for Image Classification on Pascal VOC Challenge 2012 dataset. arXiv preprint arXiv:1607.03785 (2016)
26. Wang, C.Y., Liao, H.Y., Wu, Y.H., et al.: CSPNet: a new backbone that can enhance learning capability of CNN. In: Proceedings of the IEEE Conference on Computer Vision and Pattern Recognition Workshop (CVPR) (2020)
27. Sergey, I., Christian, S.: Batch normalization:Accelerating deep network training by reducing internal covariate shift. arXiv preprint arXiv:1502.03167 (2015)
28. Elfwing, S., Uchibe, E., Doya, K.: Sigmoid-weighted linear units for neural network function approximation in reinforcement learning. Neural Netw. **107**, 3–11 (2018)
29. He, K., Zhang, X., Ren, S., Sun, J.: Spatial pyramid pooling in deep convolutional networks for visual recognition. In: Fleet, D., Pajdla, T., Schiele, B., Tuytelaars, T. (eds.) ECCV 2014. LNCS, vol. 8691, pp. 346–361. Springer, Cham (2014). https://doi.org/10.1007/978-3-319-10578-9_23
30. Zhang, H.Y., Cisse, M., Dauphin, Y.N., et al.: MixUp: beyond empirical risk minimization. arXiv preprint arXiv:1710.09412 (2017)

Intelligent Control and Simulation

Multi-UUV Formation Cooperative Full-Area Coverage Search Method

Shuo Zhang[✉]

China Ship Development and Design Center, Wuhan 430064, China
zhangshuo1t@163.com

Abstract. At present, multi-UUV formation cooperative full-area coverage search method multi-sampling "Z" shaped search path. The traditional "Z" shaped search path has the problem that all UUVs need to turn outside the search area during the search process, which greatly reduces the efficiency of the full-area coverage search. In this paper, an improved "Z-shaped" search path is proposed to solve this problem. Under the premise of ensuring complete coverage of the region, the method adaptively adjusts according to specific tasks, and selects the appropriate formation in real time according to the shape of the region, thus reducing the overall search path length of the UUV formation. Based on the improved Z-shaped search path, the sustainability of the search algorithm after formation reconstruction caused by faults is analyzed. The simulation results show that the improved algorithm can effectively improve the search efficiency of underwater vehicle formation compared with the traditional "Z" shaped search path.

Keywords: UUV · Collaborative search · Coverage search · Search path

1 Introduction

Unmanned underwater vehicle (UUV), as the backbone to promote the development of marine industry, is the focus of attention of all countries in both military and civilian fields. With the complexity of marine operations, compared with the single UUV operation method, the multi-UUV collaborative method meets the needs of the times and will be the trend leading the future of marine operation tasks. However, in the actual execution of tasks, due to the complexity of the task environment, the diversity of task requirements, the differences among members of the multi-UUV system, and the particularity of the underwater communication environment, the collaborative operation between multiple UUVs is an extremely difficult task. Complex and challenging process [1–3]. With the continuous development of multi-UUV collaborative technology, the use of multi-UUV cooperative operations to perform tasks such as detection and search, patrol and reconnaissance in specific underwater areas is not only the frontier field of the current military, but also an important manifestation of the value of marine scientific research. The problem of underwater multi-UUV cooperative search has always been a hot topic in the field of UUV research. This paper will mainly focus on the problem of multi-UUV cooperative search for multiple targets.

L. Pan et al. (Eds.): BIC-TA 2022, CCIS 1801, pp. 445–456, 2023.
https://doi.org/10.1007/978-981-99-1549-1_35

The problem of target search is a hot spot in academia and industry. With the development of UUV technology, the problem of multi-UUV cooperative target search has also received more and more attention. Most of the research on multi-UUV cooperative search focuses on the allocation of search resources and the planning of search paths [4]. In the optimal allocation of search resources, Rajnarayan and others applied the collaboration theory to multiple agent search problems and proposed an updated formula for multi-agent search problems, through the decentralized cooperation theory, the cooperative optimality between multi-agents is transformed into the global optimality of the entire cooperative system. On this basis, the centralized collaboration strategy and distributed collaboration strategy are deduced. Thi and others studied a hierarchical online search planning model, the main idea of which is to divide the search space into multiple subtask spaces, and then in the subtask space Quadratic programming search path. This search method by dividing the search space is efficient and accurate. Sayyaadi et al. studied the problem of multi-agent task allocation ratio. The goal of task allocation is to make each agent have equal tasks and capacity ratio. This paper mainly uses an asynchronous, distributed and scalable continuous time domain algorithm to achieve the balanced consistency of the task load of each agent in the group. Healay et al. studied the complete coverage problem in the process of multi-UUV cooperative search. In order to ensure complete coverage of the search area even if a UUV fails when searching in formation mode, an effective coordination strategy is designed. When UUV searches in an unknown environment information mode, it mainly relies on changing formation to adapt to the dynamic changes of the environment. Ferrant proposed an autonomous search strategy. This method is to add tags to the unknown environment by the robot, and indirectly communicate with other robots to solve the problem of poor communication environment. In the case of collaborative search problems among robots, it can also coordinate the movement of multiple robots in environments with insufficient prior information and different topological terrains. Cai and other distributed organizations based on the particle swarm optimization model abstracts the search process for the optimal solution in the solution space to an unknown environment The search process of the target is used to realize the collaborative search of multiple robots. However, this type of method needs to record the path points passed by the robot. When the search area is large and the number of robots is large, the storage capacity of the system is required to be high. Therefore, it is not suitable for large-scale search tasks [5–8].

Affected by the complexity of the underwater environment, the general cooperative search method cannot play an effective role in the underwater environment. At the same time, the time-varying characteristics of UUV hydrodynamic parameters and the complexity of the underwater communication environment also increase the number of UUVs in underwater execution. The difficulty of the search task. In a search mission, the number of UUVs performing the search mission usually depends on factors such as the size of the search area, the speed of the UUV, and its endurance. Due to the complex underwater communication environment, when the search area is a large area of the vast sea, to meet the communication requirements, multiple UUVs are required to be equipped with powerful communication equipment. In this way, it is not economical, but also violates the original design intention of the multi-UUV system. Therefore, in the face of a search task with a large group size, by organizing multiple UUV formations, each formation is equipped with a UUV with strong communication capabilities to achieve communication between the formations, thereby reducing the number of UUVs due to large task areas. Communication pressure [9–13]. This paper will take the UUV formation as the research object and discuss the target search method in the form of formation. By analyzing the formation search problem, a formation cooperative search method that can not only meet the area coverage requirements but also improve the overall search efficiency is designed.

2 Multi-UUV Formation Cooperative Search Problem Description

UUV formation, as a typical multi-UUV collaborative operation method, adopts a master-slave network structure to perform tasks. When UUVs perform search tasks in formation, they usually search according to a certain search method under a specific formation. At present, when searching for static targets scattered in a large area of water, most studies mainly regard it as a coverage path planning problem, and most of the solutions come from the path planning technology of ground robots [14]. Considering the impact of factors such as voyage time and turning radius on the search efficiency when UUV conducts underwater search tasks, it is necessary to select an appropriate search method in combination with UUV's own constraints. Usually, the UUV formation uses the "Z" search algorithm to complete the coverage search of the mission area [15]. When using the "Z" search method to search the mission area, due to the limitation of the UUV turning radius, it is generally necessary for the entire formation to sail outside the search area to make a turn, and then enter the search area again, to ensure full coverage of the search area. But this undoubtedly increases the search path and reduces the efficiency of formation search. To enable the UUV formation to better complete the area coverage search task, it is necessary to seek an efficient search strategy [16]. First, describe the UUV formation search problem:

Assuming that the search environment is a rectangular sea area $M \times N$, the UUV formation is composed of isomorphic UUVs n_v, the detection radius of each UUV is R_s, the minimum turning radius is R_v, and the detection radius is greater than the minimum turning radius, i.e., $R_s > R_v$. The UUV formation sails at a constant speed v_u. To ensure that the UUV formation does not miss the target when covering the search area. Follow these principles when analyzing search methods:

(1) The number of UUVs, $n_v < N/2R_s$ that is, the UUV formation must make turns to complete the search of the entire target area.
(2) The UUV starts from the left side of the search boundary.
(3) The UUV formation only starts searching from the top or bottom of the boundary.

Under the above principles, the research seeks out a formation search strategy to complete the coverage search of the mission area.

3 Multi-UUV Formation Coverage Search Method

3.1 Improved "Z" Shape Multi-UUV Formation Coverage Search Path

Based on the UUV formation is that the sonar array composed of sonar detection equipment carried by each member can be adaptively adjusted according to the specific task situation. When covering the area, it is necessary to select the appropriate formation according to the shape of the area [17]. The traditional "Z" search path usually needs to turn outside the search area, which reduces the efficiency of formation search. Assume that the search area is a rectangular area ABCD as shown in Fig. 1. Taking two UUVs as an example, record them as U_1 and U_2 respectively. In the figure, M and N are the width and length of the search area, respectively. R_s and R_v are respectively the detection radius of the UUV; $a_1, b_1, \ldots, j_1$ is the navigation path point of U_1, and $a_2, b_2, \ldots, j_2$ is the navigation path point of U_2. The description of the improved "Z" search path is as follows:

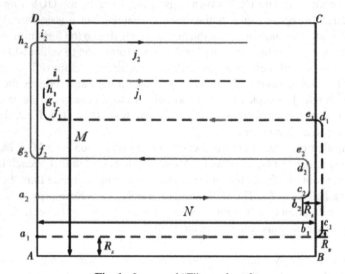

Fig. 1. Improved "Z" search path

(1) The improved UUV formation search algorithm can achieve complete coverage of the mission area. When n_v UUVs line up in tandem and sail from the lower boundary point A of the search area to the other boundary, the area before and after the turn has been fully covered and searched. During the turning process, U_1 starts to turn after passing the right boundary, so all areas within the U_1 search width between AB in Fig. 2 are searched.

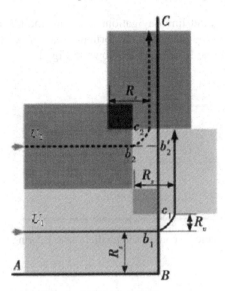

Fig. 2. UUV turning diagram

(2) Assume that when the UUV formation completes the coverage search of the search
area, all UUVs just stop on the boundary of BC, n_v and the distance between the
first UUV and point C is exactly R_s. If the UUV formation adjusts its direction $2k$
times in the entire coverage path, then under the traditional "Z" search method, the
distance traveled by each UUV in the UUV formation is:

$$L_c = (2k + 1)L + 2k\pi R_v + 4kn_v R_s - 4kR_v \tag{1}$$

Similarly, it can be calculated that the sailing distance of each UUV after
improvement is:

$$L_G = 2(k + 1)L + 2k\pi R_v + 4kn_v R_s - 4kR_v - 2k(n_v - 1)(R_s - R_v) \tag{2}$$

Because $n_v > 1$, $R_s > R_v$, obviously $L_C > L_G$, the search is performed according
to the improved "Z" shape search method, the total path length of each UUV navigation
is reduced, and the overall search path is also reduced accordingly.

3.2 Formation Reconstruction of Multi-UUV Formation Search in Unknown Environment

Due to the unpredictability of the underwater environment, the UUV formation may
experience UUV failure during the search process, that is, a certain UUV in the formation cannot continue to perform the search task. If the remaining UUVs want to better
complete the task, the search formation needs to be carried out. Adjustment.

In the process of straight-line navigation, there are mainly two types of failures, namely, a UUV failure in the middle of the formation, as shown in Fig. 3, and a failure of a UUV on one side of the formation, as shown in Fig. 4.

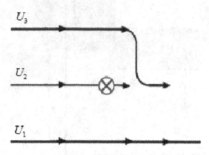

Fig. 3. Formation reconfiguration failure case 1 during line search

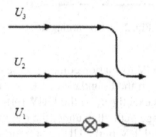

Fig. 4. Formation reconstruction failure case 2 during line search

When a UUV fails during the straight-line search phase, in order to ensure the continuation of the coverage search, the remaining UUVs in the formation approach the area not searched by the faulty UUV in the manner of "turn - straight - turn" with the minimum turning radius R_v. In Fig. 3, U_2 breaks down, U_1 continues to sail according to the original route, and U_3 approaches U_2 to complete the formation reconstruction. At this time, U_3 lags behind U_1 by

$$L_{a13} = (\pi - 4)R_v + 2R_s \tag{3}$$

Through simple analysis, if the traditional "Z" search method is followed, the distance between the two UUVs is the smallest during the turning process, and $(5 - \pi)R_v$ collisions are very likely to occur. In Fig. 4, U_1 fails, at this time U_2 approaches U_1, and U_3 approaches U_2. There will be no distance difference between UUVs after the formation is reconstructed. Therefore, there is little difference in the formation reconstruction process between the traditional "Z" shape and the improved "Z" shape formation.

The situation of failure during turning is shown in Fig. 5, Fig. 6, and Fig. 7.

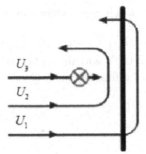

Fig. 5. Formation reconstruction failure case 1 during turn search

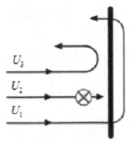

Fig. 6. Formation reconstruction failure case 2 during turn search

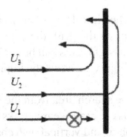

Fig. 7. Formation reconstruction failure case 3 during turn search

In Fig. 5, after U_3 fails, U_1 and U_2 follow the original path of "turn left-go straight", and when they go straight to the point after the U_3 turn is completed, U_2 "turn left - go straight" with the minimum turning radius and enter in the original U_3 navigation trajectory, similarly, U_1 searches along the original planned trajectory of U_2 in the same way. So far, formation reconstruction is completed. When the situation in Fig. 6 occurs, only U_3 needs to travel more distance $R_s - R_v$, and other changes do not need to be made. In this case, after the formation reconstruction is completed, the distance between the two UUVs is $2R_s$, when searching according to the traditional "Z" shape, the positions of U_1 and U_3 coincide at the moment of the next turn, that is, the two UUVs There was a collision. In the improved "Z" shape, the turning trajectories of each UUV are different, so they will not appear on the same channel, which can effectively avoid the occurrence of collisions between UUVs.

In addition, when there are many UUVs and multiple UUVs fail, it is difficult to adjust the distance between each UUV with the traditional "Z" search method, while the improved method can adjust the turning trajectory of each UUV, Thereby shortening the gap between each UUV.

4 Simulation Analysis

To verify the effectiveness of the improved "Z" shape search method, a simulation verification is carried out here. Each initial condition is set as:

$$M \times N = 1500\,\text{m} \times 1200\,\text{m} \tag{4}$$

$$v_u = 4\,\text{m/s} \tag{5}$$

$$R_v = 20\,\text{m} \tag{6}$$

$$R_s = 50\,\text{m} \tag{7}$$

$$n_v = 4 \tag{8}$$

Divide the sea area to be searched into 15×12 grids, and each grid is assigned a value of 0. When the UUV passes through the grid, the value of the grid becomes 1, so that the growth of the sum of all grid assignments can be analyzed to test the effectiveness of the improved search algorithm. Figure 8 shows the improved "Z" search path proposed in this paper.

The UUV formation enters the search area from the bottom left of the search area in a tandem formation, and performs coverage search tasks from left to right. Figures 9 and 10 are the diagrams of the horizontal and vertical path changes of each UUV in the UUV formation under the improved "Z" search method, respectively. Combining the search path diagram and the path change diagram in the horizontal and vertical directions, it can be seen that the total length of the search path of each UUV is equal in the entire search task.

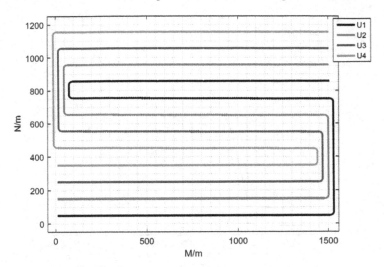

Fig. 8. Improve the "Z" font search path

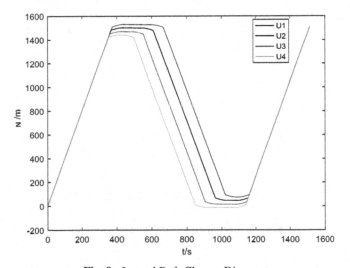

Fig. 9. Lateral Path Change Diagram

In order to verify the effectiveness of the improved algorithm, the task of covering the area is completed under the same conditions according to the traditional "Z" search method, and the search efficiency of the two is compared, as shown in Fig. 11.

It can be seen from the figure that both the traditional "Z" search and the improved "Z" search can complete the search with full coverage of the search area. Since the search strategies and search paths in the initial stage of the two methods are the same, there is no difference in the search effect in the initial stage. As the search task progresses, the advantages of the improved search algorithm gradually emerge. At about 300s, the UUV formation began to turn, and the search efficiency of the improved search algorithm was

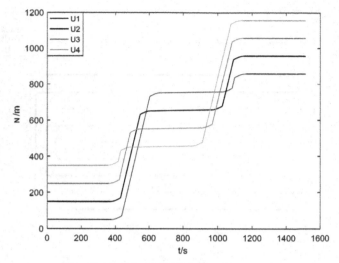

Fig. 10. Longitudinal Path Change Diagram

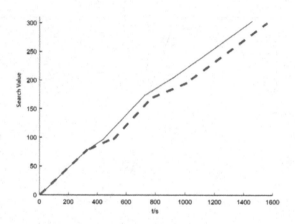

Fig. 11. Search efficiency comparison chart

gradually better than that of the traditional "Z" search. This is due to the improved search algorithm that reduces the area that UUV formations search repeatedly during turns. From the simulation results, it can be seen that the improved "Zigzag" search algorithm takes less time to complete the coverage search than the traditional advanced "Zigzag" search algorithm, which significantly improves the efficiency of the UUV formation to perform search tasks. The simulation results are consistent with the theoretical analysis.

5 Conclusion

This paper studies the multi-UUV cooperative search method with UUV formation as the cooperative method. First, the UUV formation coverage search problem is described according to the traditional "Zigzag" search method. Secondly, the traditional "Z" search algorithm is improved, and at the same time, the UUV formation reconstruction strategy is given, and the improved "Z" search algorithm is proved to improve the search efficiency and formation reconstruction through theoretical analysis. Effectiveness. Finally, a simulation experiment is designed to compare the search efficiency of the traditional "Z" font and the improved "Z" font. The simulation results are consistent with the theoretical analysis, which proves the effectiveness of the algorithm proposed in this paper.

References

1. Duan, J., Jiang, Z.: Joint scheduling optimization of a short-term hydrothermal power system based on an elite collaborative search algorithm. Energies **15**(13), 4633 (2022)
2. Zheng, D., Liu, C., Huang, L.: Spatio-temporal coverage planning algorithm for multi-UAV and multi-sensor cooperative search. J. Phys: Conf. Ser. **2187**(1), 012047 (2022)
3. Li, L., Zhang, X., Yue, W., Liu, Z.: Cooperative search for dynamic targets by multiple UAVs with communication data losses. ISA Trans. **114**, 230–241 (2021)
4. Hou, K., Yang, Y., Yang, X., Lai, J.: Distributed cooperative search algorithm with task assignment and receding horizon predictive control for multiple unmanned aerial vehicles. IEEE Access **9**, 6122–6136 (2021)
5. Sun, Z., Yen, G.G., Wu, J., Ren, H., An, H., Yang, J.: Mission planning for energy-efficient passive UAV radar imaging system based on substage division collaborative search. IEEE Trans. Cybern. **53**, 275–288 (2021)
6. He, C., et al.: Accelerating large-scale multi-objective optimization via problem reformulation. IEEE Trans. Evol. Comput. **23**(6), 949–961 (2019)
7. Lin, J., He, C., Cheng, R.: Adaptive dropout for high-dimensional expensive multiobjective optimization. Complex Intell. Syst. **8**(1), 271–285 (2021). https://doi.org/10.1007/s40747-021-00362-5
8. Liang, H., Fu, Y., Kang, F., Gao, J., Qiang, N.: A behavior-driven coordination control framework for target hunting by UUV intelligent swarm. IEEE Access **8**, 4838–4859 (2020)
9. Liu, Y., Wang, M., Su, Z., et al.: Multi-AUVs cooperative target search based on autonomous cooperative search learning algorithm. J. Marine Sci. Eng. **8**(11), 843 (2020)
10. Ni, J., Tang, G., Mo, Z., Cao, W., Yang, S.X.: An improved potential game theory based method for Multi-UAV cooperative search. IEEE Access **8**, 47787–47796 (2020)
11. Bo, L., et al.: Multi-UAV collaborative search and strike based on reinforcement learning. J. Phys. Conf. Ser. **1651**(1), 012115 (2020)
12. Ma, X.W., Chen, Y.L., Bai, G.Q., Sha, Y.B., Liu, J.: Multi-autonomous underwater vehicles collaboratively search for intelligent targets in an unknown environment in the presence of interception. Proc. Inst. Mech. Eng. Part C: J. Mech. Eng. Sci. **235**(9), 1539–1554 (2021)
13. Shao, Q., Jia, M., Xu, C., Zhu, Y.: Multi-helicopter collaborative search and rescue operation research based on decision-making. J. Supercomput. **76**(5), 3231–3251 (2018). https://doi.org/10.1007/s11227-018-2555-7
14. Huang, P.Q., Wang, Y., Wang, K., Liu, Z.Z.: A bilevel optimization approach for joint offloading decision and resource allocation in cooperative mobile edge computing. IEEE Trans. Cybern. **50**(10), 4228–4241 (2019)
15. Wenjun, D., et al.: Investigation on optimal path for submarine search by an unmanned underwater vehicle. Comput. Electr. Eng. **79**(C), 106468 (2019)

456   S. Zhang

16. Ziyang, Z., et al.: Distributed intelligent self-organized mission planning of multi-UAV for dynamic targets cooperative search-attack. Chin. J. Aeronaut. **32**(12), 2706–2716 (2019)
17. John, G.B., Thomas, A.W.: A ROC-based approach for developing optimal strategies in UUV search planning. IEEE J. Oceanic Eng. **43**(4), 843–855 (2018)

Multi-scale Simulations of Flow in the Upper Open Area of a Luxury Cruise Ship

Zhongqiu Yan[1], Liujun Zhang[2,3], Weiguo Wu[2(✉)], Zhenguo Song[4(✉)], Heng Luo[4], Yi Yang[2], and Wulong Hu[2]

[1] Jiangsu Institute of Automation, Lianyungang 222061, China
[2] Green & Smart River-Sea-Going Ship, Cruise and Yacht Research Center, Wuhan University of Technology, Wuhan 430070, China
`mailjt@163.com`
[3] Hubei Key Laboratory of Theory and Application of Advanced Materials Mechanics, School of Science, Wuhan University of Technology, Wuhan 430070, China
[4] China Ship Development and Design Center, Wuhan 430000, China
`570706684@qq.com`

Abstract. Luxury cruise ships have complicated superstructures, which are greatly affected by wind loads. At present, the research on the external flow field of luxury cruise ships mainly adopts the method of directly calculating the whole ship model. Due to the limited computing resources, this method is difficult to accurately describe the local complex flow field structure. This paper proposes a multi-scale simulation method based on the entire ship model of the luxury cruise ship and the refined model of the superstructure. Firstly, the flow field around the cruise ship is obtained through the whole ship simulation, and the local flow field data is extracted; then the local flow field data is imported into the local refined model of the cruise ship as the speed entrance boundary condition; finally, the refined simulation analysis of the cruise ship is carried out. In this paper, a multi-scale simulation method is used to analyze the characteristics of the flow field around the superstructure of the cruise ship by combining pressure, flow velocity and vorticity. The results show that the local simulation can obtain a finer flow field structure than the whole ship simulation, and the result obtained by using multi-scale simulation is more accurate than the result obtained by directly using the natural wind speed profile velocity entrance. This article provides a new idea for wind field simulation of luxury cruise ships and other ship types with complex superstructures.

Keywords: Luxury Cruise Ship · Complex Superstructure · Multi-scale Simulation

1 Introduction

As the "pearl on the crown" of shipbuilding industry, luxury cruise ship has attracted people's attention in recent years, and is a key field that has not achieved a major breakthrough among Chinese high-tech ships. Luxury cruise ships have both transportation and entertainment functions, aiming to provide tourists with the most comfortable travel and vacation experience. Therefore, in addition to safety requirements, they also have high requirements for passenger comfort. Due to the large number of leisure and entertainment areas and facilities, the superstructure of luxury cruise ships is very complicated, resulting in the surrounding flow field is often more complex than that of other types of ships [1], and the calculation of wind loads is more difficult. The wind field characteristics in the open area of the cruise ship will greatly affect the comfort of passengers, including the direct impact of wind on passengers, wind noise, wind influence on chimney smoke emission, etc. Therefore, the optimization design of the cruise ship should be based on the simulation analysis of the flow field around the cruise superstructure and the wind loads calculation [2, 3].

Wind loads research methods are generally divided into three categories: wind tunnel test, empirical formula and CFD method [4]. Wind tunnel test results have high reliability, but it is difficult to describe the details of the flow field, and the cost is high, so it is generally only used for the verification of structural design. The empirical formula has great convenience and wide application in engineering application, but it is not strong pertinence and cannot fully consider the complicated flow problems. CFD method has strong applicability, can simulate various complex environmental conditions and obtain results quickly, with high accuracy and high-cost performance. In the design of ship cruise ship, the three methods have their advantages and disadvantages and are all used, but CFD method is the most widely used method in the study of ship wind loads because of its obvious advantages.

When using CFD method to study ships with complicated superstructure, scholars often choose to simplify the ship geometric model to save computing resources. For example, Gao Ye has studied and analyzed the deck vortex structure of a simplified aircraft carrier model. Based on the simplified ship model, Liu Changmeng simulated and analyzed the characteristics of the deck flow field by CFD method. Zhang Yahui used the empirical formula and CFD method to evaluate the wind load of the simplified container crane carrier model. Without considering the characteristics of local flow field, the simplified ship geometry model can greatly improve the computational efficiency. However, the upper deck of a luxury cruise ship has the function of recreation and entertainment, which has a high requirement on the comfort of tourists. Therefore, it is necessary to carry out detailed simulation and analysis on the wind load, surrounding flow field and noise of each local structure. If the cruise ship geometric model is simplified, the simulated local environmental features will be distorted, and the refined cruise ship geometric model simulation will face a huge amount of computing costs. Therefore, a multi-scale CFD simulation method is needed. On the one hand, the characteristics of the flow field around the hull can be obtained through efficient whole-ship simulation, on the other hand, the details of the diverting field can be obtained through local simulation. Some scholars in the field of ships have started to study the local model. Zhang Xiuyuan analyzed the characteristics of the flow field around the mast of an aircraft carrier by studying the uniform flow. Based on the independent tank model and natural gradient wind speed, Bi Lili studied the leakage and diffusion laws of LNG storage tanks. Huang Shaoxiong simulated flue gas diffusion based on the chimney model and uniform incoming flow. However, these methods study the local structure apart from the whole ship, and do not consider that the local structure will be affected by other superstructures of the whole ship [5–9]. The inlet condition of the local structure is not a simple uniform flow or natural gradient wind speed, but a part of the flow field of the whole ship.

In this paper, a method based on multi-scale simulation is proposed to establish a local refined model. The local velocity field is extracted from the whole ship velocity field and used as the velocity inlet boundary condition of the local model. The CFD method is used to compare the simulation results of the whole ship and the local model to verify the accuracy of the local calculation results and the rationality of this method.

2   Multi-scale Simulations

2.1   Introduction to Methods and Models

In this paper, the CFD method is used to simulate the flow field of the whole ship and obtain the information of the flow field of the whole ship. Then, the velocity data at the local structure inlet profile A (Fig. 1) is extracted and imported into the local model as the velocity inlet for calculation. Since the distribution of the whole ship's flow field changes periodically, which will lead to periodic changes of the profile velocity with space and time, this paper programmed the velocity data into the form of v(x, y, z, t). In Fig. 1, the two straight lines I and II are the intersection lines between the central section of the tourist-intensive area and the superstructure of the ship, which need to be studied with emphasis.

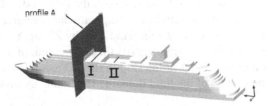

Fig. 1. Inlet profile of the whole ship model and local model

The Reynolds number of the whole ship model and the local model is greater than 10^5, which belongs to the turbulent state. At present, turbulence numerical simulation methods are mainly divided into three categories: direct numerical simulation (DNS), large eddy simulation (LES) and Reynolds-averaged Navier-Stokes (RANS). Direct numerical simulation method uses numerical methods to directly solve the N-S equation without any turbulence model. It requires very high computational resources to solve and calculate in the temporal and spatial scale grid of turbulence. Currently, it is generally used for low Reynolds number flows. Based on the self-similarity theory, the LES method uses sub-grid-scale to simulate small-scale vortexes, and calculates the velocity fields of large-scale vortexes and small-scale vortexes separately. The calculation accuracy is high, but it is difficult to calculate in the near-wall area and requires a high-quality computational grid. RANS method makes statistical average of the control equation, so that it does not need to calculate the turbulent fluctuation of various scales, but only needs to calculate the mean motion, reducing the calculation amount, and the calculation accuracy is high, so it is the most widely used method at present. In this paper, RANS method is used, which requires turbulence model to seal the equation. Studies have shown that among several common RANS turbulence models, the calculated results of realizable k-ε turbulence model are close to the experimental results. Therefore, this paper adopts realizable k-ε turbulence model and its turbulence equation is as follows:

$$\frac{\partial(\rho k)}{\partial t} + \frac{\partial(\rho k u_i)}{\partial x_i} = \frac{\partial}{\partial x_j}\left[\left(\mu + \frac{\mu_t}{\sigma_k}\right)\frac{\partial k}{\partial x_i}\right] + G_k - \rho\varepsilon \tag{1}$$

$$\frac{\partial(\rho\varepsilon)}{\partial t} + \frac{\partial(\rho\varepsilon u_i)}{\partial x_i} = \frac{\partial}{\partial x_j}\left[\left(\mu + \frac{\mu_t}{\sigma_k}\right)\frac{\partial\varepsilon}{\partial x_i}\right] + \rho C_1 E\varepsilon - \rho C_2\frac{\varepsilon^2}{k + \sqrt{v\varepsilon}} \tag{2}$$

Because the first-order upwind scheme cannot accurately capture the flow field structure parameters such as the recirculation zone length, the back pressure coefficient and the separated angle, and the central difference will lead to serious numerical dissipation in the region with low grid resolution, the second-order upwind scheme is adopted for the convection term dispersion. In incompressible flows, the SIMPLEC algorithm will obtain better convergence results and its convergence rate is faster, so the pressure-velocity coupling equation is solved by the SIMPLEC algorithm.

2.2 Hull Model Above the Waterline

This paper builds a geometric model based on the Carnival Vista cruise ship. Since only the effect of air on the cruise ship is discussed, only the hull model above the waterline needs to be built. Referring to the setting of computational domain and boundary conditions in the numerical simulation of ship wind load by Zhang Ye, the computational domain in this paper is 5 times the length and 5 times the width of the ship in the x and y directions, and 3 times the height of the ship in the z direction, and the cruise model is at the bottom in the middle of the fluid domain. The boundary conditions are shown in Table 1.

Table 1. Fluid computational domain boundary settings

Wall	boundary conditions
inlet	Velocity inlet
outlet	pressure outlet
Surface of bilateral boundary	symmetry
Upper and lower boundary	slip interface
The internal boundary	no slip interface

The structural grid division was adopted. The grid division of the hull surface was shown in Fig. 2, and the grid refinement was carried out on the bow and other complex areas. The total number of grids was 3.97 million.

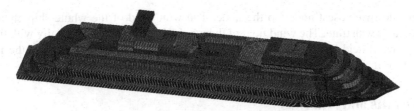

Fig. 2. The grid division of the hull surface

Wind speed with natural gradient is selected at the entrance of velocity calculation for the whole ship [3]. The frictional resistance generated by waves near the sea will reduce wind energy near the sea, thus reducing wind speed near the sea. In the atmospheric boundary layer, wind speed VW changes with the height h above the sea, and its natural velocity gradient is usually expressed as follows:

$$V_W = V_0 \left(\frac{h}{h_0} \right)^{\alpha} \tag{3}$$

V_0 is the wind speed at h_0 above the sea level, and h_0 is usually 10 m. α is the wind speed profile index, usually between 0.11 and 0.14. In this paper, V_0 is 10 m/s, α is 0.125 [10], and the wind direction is 0°, that is, the inlet wind direction is x negative direction.

2.3   Local Numerical Simulation

The water park area on the top deck of the cruise ship was selected as the fine local model. The length of the calculation domain in the x direction is equal to the size of the model, and the size of the model in other directions is 5 times. Structural grid division is adopted. The grid division of water park is shown in Fig. 3.

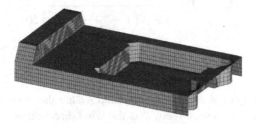

Fig. 3. Surface mesh of water park region

The inlet wind speed of the local model of the water park comes from the wind field data of the whole ship. The velocity information on the grid node of profile A in Fig. 1 is programmed to correspond one-to-one to the grid node of the inlet profile of the local model. Because the size of the whole ship grid is different from that of the local model grid, the grid nodes cannot be completely one-to-one corresponded. For the uncorresponded grid nodes, spatial linear interpolation is carried out according to the data of the two closest nodes to the node. The wind field of the whole ship changes periodically with time. The wind field of the whole ship changes periodically with time, and the wind field data in one period of profile A is extracted and used as the inlet velocity of the local model.

3 Results and Analysis

3.1 Method Validation

Based on aerodynamic theory and wind tunnel test data, researchers have established various empirical formulas for calculating ship wind loads. Jiang Wei took the basic wind pressure as the calculation basis, considered the effects of height and structure size, and proposed the empirical formula for calculating ship wind load according to the wind area of ships as follows:

$$F_{xw} = 73.6 \times 10^{-5} A_{xw} V_x^2 \zeta_1 \zeta_2 \qquad (4)$$

$$F_{yw} = 73.6 \times 10^{-5} A_{yw} V_y^2 \zeta_1 \zeta_2 \qquad (5)$$

F_{xw}, F_{yw} is the x and y component of the wind load on the ship, A_{xw}, A_{yw} is the wind area of the ship in the x and y directions respectively, V_x, V_y are the x and y component of the wind speed respectively, ζ_1 is the wind pressure non-uniform reduction coefficient, ζ_2 is the wind pressure height change correction coefficient.

Blendermann systematically studied ship's wind load by taking ship's shape, super-structure distribution and ship's wind area as independent variables, carried out a series of wind tunnel tests on various types of ship models, obtained a large number of experimental data, derived ship's wind load coefficient from it, and put forward the following empirical formula:

$$C_x = -D_{AF} \frac{\cos \theta}{1 - \frac{\delta}{2}\left(1 - \frac{D_{AF}}{D_t}\frac{A_F}{A_L}\right)\sin^2 2\theta} \qquad (6)$$

$$C_y = -D_t \frac{\sin \theta}{1 - \frac{\delta}{2}\left(1 - \frac{D_{AF}}{D_t}\frac{A_F}{A_L}\right)\sin^2 2\theta} \qquad (7)$$

D_{AF}, D_t, δ are empirical coefficients, A_F and A_L are the wind affected area of the ship in x and y directions respectively, θ is the wind direction angle. C_x and C_y are x and y direction coefficients respectively, defined as follows:

$$C_x = \frac{2F_x}{\rho u^2 A_x} \qquad (8)$$

$$C_y = \frac{2F_y}{\rho u^2 A_y} \qquad (9)$$

ρ is the air density, u is the average wind speed experienced by the superstructure, A_x and A_y are the x and y direction wind area above the surface of the hull.

Fujiwara Toshifumi assumed that the wind pressure acting on the hull was composed of mainstream resistance, cross-fluid drag, lift force and induced drag, proposed the composition segregation wind load derivation method, and summarized the following empirical formula through a large number of wind tunnel experiments [12]:

$$C_X(\psi) = C_{VD}\cos^2\psi + (C_{LF}\sin\psi\cos\psi)\sin\psi$$
$$+ \sin\left(\psi - \frac{\pi}{2}\right)\left(C_{XID}\sin^2\psi\cos\psi\right)\cos\psi \qquad (10)$$

$$C_Y(\psi) = C_{CR}\sin^2\psi + (C_{LF}\sin\psi\cos\psi)\cos\psi$$
$$+ \left(C_{YID}\sin^2\psi\cos^2\psi\right)\sin\psi \qquad (11)$$

C_{VD}, C_{LF}, C_{XID}, C_{CR}, and C_{YID} are all empirical coefficients, whose values are related to ship shape and distribution of superstructure, and ψ is wind direction angle.

The comparison between CFD simulation results and empirical formulas in this paper is shown in Table 2. At $0°$ wind direction angle, the difference between CFD simulation results and the empirical formulas of Blendermann and Fujiwara is 5% and 23.8% respectively, but the difference between CFD simulation results and the empirical formulas of Jiangwei is 176.2%. This is because Jiangwei formula only considers the wind area of the ship in the x and y directions without considering the shape and superstructure form of the ship, which will have a large error when applied to the cruise ship with complex superstructure.

Table 2. Comparison of results between CFD method and empirical formula

	CFD	Jiangwei	Blendermann	Fujiwara
C_x	−0.42	−1.16	−0.4	−0.52
C_y	−1.29e-4	0	0	0

3.2 Analysis of Wind Field Characteristics of the Whole Ship

Simulate the entire cruise ship according to the model and method in Sect. 1.2. Figure 4(a) shows the vorticity distribution of the whole ship, and Fig. 4(b) and Fig. 4(c) respectively show the partial streamlines and wind pressure distribution of the water park on the whole ship. Figure 4(a) shows that the whole ship's vortex shedding is mainly composed of bow separation vortices, hull trailing vortices, hull edge separation vortices and superstructure trailing vortices, among which the chimney trailing vortices are the most obvious. However, in the water park, there is only a small amount of vortex shedding, and no vortex shedding are found in low-lying areas.

It can be seen from Fig. 4(b) that the platform part of the water park has relatively obvious downwash wind and vortex flow, while the low-lying part only has observed downwash wind without obvious vortex structure. In Fig. 4(c), the wind load of the water park basically presents a symmetrical distribution, and the platform part presents a strip distribution along the y direction, with the maximum positive pressure area appearing in the low-lying part.

Through the simulation of the whole cruise ship, the velocity of profile A changing with the time period is obtained, and the velocity distribution at a certain time within the period is extracted, as shown in Fig. 4(d). The wind speed distribution in Fig. 4(d) is obviously different from the natural gradient distribution, especially the wind speed distribution near the hull is more complex than the natural gradient distribution, because the wind close to the hull surface will be affected by the hull structure.

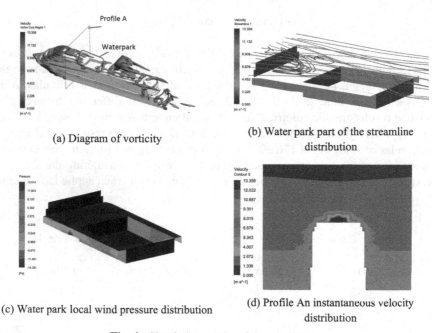

(a) Diagram of vorticity

(b) Water park part of the streamline distribution

(c) Water park local wind pressure distribution

(d) Profile An instantaneous velocity distribution

Fig. 4. Simulation results of the whole ship

3.3   Verification and Analysis of Numerical Simulation of Local Wind Field

According to the wind speed distribution of the water park entrance profile obtained from the whole ship numerical simulation in the last section, the data were extracted and calculated as the boundary conditions of the velocity inlet of the local model. The following results were obtained.

Figure 5 compares the calculation results of the whole ship and the local ones, and analyzes the law of wind speed v changing with the y coordinate in two lines I and II (Fig. 1). It can be found that the change law of the whole ship and the local calculation is very similar, and the amplitude is also similar, which further verifies the accuracy of the multi-scale numerical simulation method.

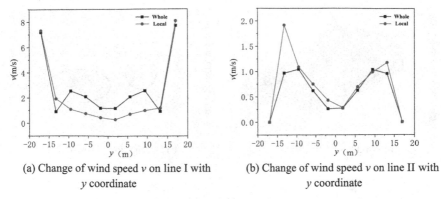

(a) Change of wind speed v on line I with
y coordinate

(b) Change of wind speed v on line II with
y coordinate

Fig. 5. Comparison of whole ship and local simulation results

Figure 6(a) shows the wind load distribution of the water park. By comparison with Fig. 4(c) and Fig. 6(a), the maximum positive pressure and negative pressure of simulated wind load of the whole ship are 13.814 Pa and 14.291 Pa, and the maximum positive pressure and negative pressure of simulated wind load in local areas are 10.241 Pa and 12.060 Pa respectively. The maximum positive pressure in the local area is 25.87% smaller than that in the whole ship simulation, and the maximum negative pressure is 15.08% smaller than that in the whole ship simulation. The wind load distribution rules of the two are similar.

The flow lines of the water park are shown in Fig. 6(b). It can be seen that there are obvious vortices in places 1, 2 and 3 in the figure, and there are obvious downwashing phenomena in the three places. The vortices in places 2 and 3 are larger, and the flow lines are distributed disordered. By comparing Fig. 4(a) and 6(b) (the streamline density Settings in the two figures are the same), it is not difficult to find that the simulation results of the local model describe the streamline more clearly and specifically, and can reflect more details. The vortices in Fig. 4(a) are not well reflected, because the details in the whole ship model are simplified and the mesh size is relatively large.

Figure 6(c) is the vorticity diagram of the water park. The vortex shedding mainly consist of the top trailing vorteices and low-lying vortices, which are basically distributed symmetrically. The water park has two structures with reduced height, which will generate airflow separation and backflow after collision with the wall, resulting in the emergence of low pressure area, and then the generation of vortex. By comparing Fig. 4(a) and Fig. 6(c), it can be found that in Fig. 4(a), only separated vortices on both sides of the ship can be observed in the water park, while other vortices of different forms and sizes are completely absent. However, the multi-scale analysis method can clearly describe the vortices of different forms and sizes.

Through the above comparative analysis, it can be seen that a separate simulation of the local model can improve the accuracy of the results, and more detailed changes of the local flow field can be observed.

If the local model is separated from the whole ship flow field and formula (3) is used as its velocity inlet to simulate separately, the wind load distribution as shown in Fig. 7 will be obtained. By comparing Fig. 6(a) and Fig. 7, it can be found that the

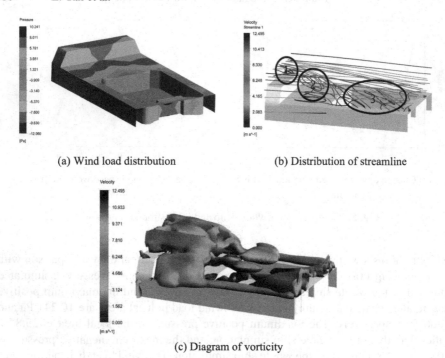

(a) Wind load distribution   (b) Distribution of streamline

(c) Diagram of vorticity

Fig. 6. Local calculation results of water park based on the whole ship flow field

amplitude and distribution of wind loads obtained by the natural gradient wind inlet for the local model are far different from those obtained by the multi-scale simulation method mentioned above, indicating that the whole ship flow field has a great influence on the local wind field and wind load.

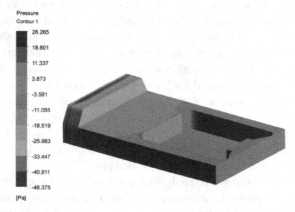

Fig. 7. Wind load distribution of water park under natural wind velocity profile

4 Conclusion

(1) A multi-scale numerical simulation method is proposed in this paper. The whole ship flow field information is programmed to input the velocity boundary conditions of the local model, which greatly improves the accuracy of the numerical simulation of the local model, and also makes up for the shortage that the whole ship calculation cannot accurately describe the flow field details. The problem that the simplified processing of the whole ship model will lead to the detail distortion of the local model flow field is solved. The multi-scale numerical simulation method meets the requirements of saving computation and improving accuracy at the same time and provides a new idea for the study of wind field characteristics in open area of luxury cruise ships.

(2) The wind field and wind load in the local area are affected by the whole ship structure and other areas, so the numerical simulation of the local area model needs to be based on the flow field information of the whole ship, rather than completely isolating the local area model for simulation.

References

1. Szymonski, M.: Some effects of wind on ship's manoeuvrability. TransNav Int. J. Mar. Navig. Saf. Sea Transp. **13**(3), 623–626 (2019)
2. Huang, L., Wen, Y., Geng, X., et al.: Integrating multi-source maritime information to estimate ship exhaust emissions under wind, wave and current conditions. Transp. Res. Part D Transp. Environ. **59**, 148–159 (2018)
3. Wang, P., Wang, F., Chen, Z.: Investigation on aerodynamic performance of luxury cruise ship. Ocean Eng. **213**, 107790 (2020)
4. Wang, P., Wang, F., Chen, Z., et al.: Aerodynamic optimization of a luxury cruise ship based ona many-objective optimization system. Ocean Eng. **236**, 109438 (2021)
5. Li, Z., Qi, S., Hu, T., et al.: Error source analysis of true wind for sailing ships. J. Ocean Technol. **38**(02), 78–84 (2019)
6. You, D., Sun, Y.: Application of robust neural networks in nonlinear ship model motion control. Ship Sci. Technol. **44**(12), 161–164 (2022)
7. Liu, W., Huang, Y., Wang, X.M., et al.: Numerical simulation and test of ship model collapse under waves. China Navig. **44**(01), 21–26 (2021)
8. Chen, G., Wang, W., Huo, C.: Identification of ship dynamics model based on Gaussian process regression. Ship Sci. Technol. **44**(19), 1–5 (2022)
9. Xu, J.: Numerical simulation study on the influence of wind and current conditions on ship navigation. China Water Transp. **42**(05), 122–124 (2021)
10. Kaneda, Y., Yamanoto, Y.: Velocity gradient statistics in turbulent shear flow: an extension of Kolmogorov's local equilibrium theory. J. Fluid Mech. **929**, 101017 (2021)
11. Blendermann, W.: Parameter identification of wind loads on ships. J. Wind Eng. Ind. Aerodyn. **51**(3), 339–351 (1994)
12. Kitamura, F., Ueno, M., Fujiwara, T., et al.: Estimation of above water structural parameters and wind loads on ships. Ships Offshore Struct. **12**(8), 1100–1108 (2017)

Research on Course Control Algorithm of Unmanned Craft Based on Model Predictive Control

Wei Wu[1](✉), Xuemei Qin[2], Jianhua Qin[3], Bing Song[1], and Xingbang Chen[1]

[1] China Ship Development and Design Center, Wuhan, Hubei, China
601600752@qq.com
[2] Wuhan Geomatics Institute, Wuhan, Hubei, China
[3] Wuhan Geotechnical Engineering and Surveying Co., Ltd., Wuhan, Hubei, China

Abstract. Course control is one of the key technologies for autonomous navigation of surface unmanned ship. In order to realize course control of surface unmanned ship, the main problems need to be solved include the steering constraint of surface unmanned ship and the disturbance of wind and waves during navigation. Based on the analysis of these two problems, this paper designs a course controller of surface unmanned ship based on predictive function control. The controller adopts the idea of prediction before control. In the process of solving the optimal control quantity online, the influence of above constraints and interference on heading control is fully considered, so as to realize timely compensation of interference. Then, the course controller based on predictive function control is simulated and compared with the course controller based on traditional PID control to verify the effectiveness of the controller.

Keywords: Model prediction · USV · Course control · Wind and wave disturbance

1 Introduction

With the characteristics of low risk, high maneuverability, deployability, and environmental adaptability, unmanned ships have important military and civilian values. At present, unmanned ships have been gradually applied in many fields. For example, our country is now facing the problem of water pollution, and the problem of governance and supervision is relatively difficult [1]. The use of unmanned ships can continuously sample and detect the water quality in the water environment [2], unmanned ships can expand the measurement range and improve the efficiency of surveying and mapping in channel mapping [3], unmanned ships can also apply high-intensity, low-security tasks such as water search and rescue [4].

Regarding the course control algorithm of unmanned boats, Minorsky [5] proposed that the traditional PID control algorithm has excellent robustness, but the algorithm cannot be effectively controlled in severely disturbed waters. Experts and scholars at home and abroad have proposed various improvements to the traditional PID course

L. Pan et al. (Eds.): BIC-TA 2022, CCIS 1801, pp. 468–483, 2023.
https://doi.org/10.1007/978-981-99-1549-1_37

control algorithm. Methods. For example, improve the parameter setting of traditional PID control and integrate other advanced algorithm technologies [6]. Yang Juncheng [7] used the improved particle swarm optimization algorithm to optimize the parameters of the traditional PID algorithm. Bai Yiming [8] proposed an efficient control algorithm based on dynamic surface control technology. Wen Jingsong [9] proposed to introduce ADRC into heading controller. Hu Junxiang [10] introduced linear active disturbance rejection technology.

The surface unmanned ship itself has steering constraints, and there are disturbances such as wind, wave and current during its navigation. In view of the above problems, combined with the motion model of the unmanned ship, this paper uses the model predictive control algorithm to design the heading controller of the unmanned ship to realize the control of the unmanned ship. Effective control of heading [11].

2   Construction of Mathematical Model of Unmanned Boat Control

2.1   Coordinate System

In order to describe the motion of the unmanned ship in the six degrees of freedom [12, 13], the unmanned ship coordinate system as shown in Fig. 1 should be established, including the fixed coordinate system $o - x_1 y_1 z_1$ (briefly referred to as "fixed system") and the moving coordinate system $G - xyz$ (briefly referred to as "moving system"). "Fixed system" refers to the inertial coordinate system fixed to the Earth. "Moving system" refers to the attached coordinate system following the movement of the hull. The origin is set at G, the center of gravity of the hull of the unmanned ship. Gx is the horizontal direction of the unmanned ship, Gy is the cross section direction of unmanned ship, and Gz is the longitudinal and middle section direction of the unmanned ship. The positive direction is right - handed.

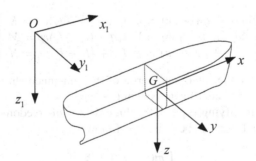

Fig. 1. Unmanned ship coordinate system

The motion parameters involved in the attached coordinate system $G - xyz$ and the inertial coordinate system $o - x_1 y_1 z_1$ are shown in Table 1. Where X, Y and Z are the projection of external forces on the attached coordinate system $G - xyz$. K, M, N are the projection of the unmanned ship moment on the attached coordinate system $G - xyz$, u, v, and w are the projection of the unmanned ship's velocity on the attached coordinate system $G - xyz$, p, q, and r are the projection of the angular velocity of the unmanned ship on the attached coordinate system $G - xyz$, (x, y, z) is the space coordinate of the unmanned ship's center of gravity in the inertial coordinate system $o - x_1y_1z_1$; ϕ, θ, and ψ are the three Euler angles of the unmanned ship in the inertial coordinate system $o - x_1y_1z_1$, where ϕ is the roll Angle, θ is the trim Angle, and ψ is the azimuth Angle.

Table 1. Motion parameters of unmanned ship

Degree of freedom	Forces and moments	Velocity and angular velocity	Position and Euler Angle
surge	X	u	x
Transverse oscillation	Y	v	y
heaving	Z	w	z
roll	K	p	ϕ
pitch	M	q	θ
Bow wave	N	r	ψ

The sailing condition of the ship on the ocean can be expressed by the velocity vector u, v, w and the angular velocity vector p, q, r in the "moving system", as well as the derivative x, y, z of the position vector in the fixed system and the derivative ϕ, θ and ψ of the Euler angular attitude vector.

2.2 Linear Mathematical Model

The six-degree-of-freedom motion equation of an unmanned ship is as follows:

$$\left\{ \begin{array}{l} m(u + qw - rv) = X \quad I_z p + (I_x - I_y)qr = K \\ m(v + ru - pw) = Y \quad I_y q + (I_x - I_z)rp = M \\ m(w + pv - qu) = Z \quad I_z r + (I_y - I_z)pq = N \end{array} \right\} \tag{1}$$

where I_x, I_y and I_z are the rotational inertia of the unmanned ship with respect to the three axes of the attached coordinate system $G - xyz$.

Therefore, by simplifying Eq. (1), the three-degree-of-freedom motion equation of the surface unmanned ship can be written as follows:

$$\left\{ \begin{array}{l} m(u - rv) = X \\ m(v + ru) = Y \\ I_z r = N \end{array} \right. \tag{2}$$

When conducting the research on the course stability and maneuvering motion of the unmanned ship under the small rudder Angle, the motion parameter Δu is very small. Therefore, it meets the following requirements:

$$ur = (u_0 + \Delta u)r = u_0 r + \Delta u r \approx u_0 r \tag{3}$$

In addition, the motion parameters v, r and δ are also very small, so the quantities above the second order of motion parameters v, r and δ in Eq. (2) can be ignored, and then combined with Eq. (3), the linearized result of Eq. (6) can be written:

$$\begin{cases} mu = X \\ m(v + u_0 r) = Y \\ I_z r = N \end{cases} \tag{4}$$

Before linearization of formula (2), it is agreed that the motion parameter Δu is very small, so it is generally believed that $u = u_0$. Therefore, the first equation in Formula (4) is only used to determine the speed u_0.

The linear equation of the maneuvering motion of the unmanned ship can be written:

$$\begin{cases} (m + \lambda_{22})v - Y_v v + (m + \lambda_{11})u_0 r - Y_r r = Y_\delta \delta \\ (I_z + \lambda_{66})r - N_v v - N_r r = N_\delta \delta \end{cases} \tag{5}$$

When studying the heading control of unmanned ship, this paper pays more attention to the response speed of unmanned ship's bow Angle to steering. Therefore, after v elimination operation of the unmanned ship's steering linear Eq. (5), the linear equation of unmanned ship's bow rolling response can be derived:

$$T_1 T_2 r + (T_1 + T_2)r + r = K\delta + KT_3 \delta \tag{6}$$

The expression of relevant parameters is as follows:

$$\begin{cases} T_1 T_2 = (m + \lambda_{22})(I_z + \lambda_{66}) \big/ C \\ T_1 + T_2 = [(m + \lambda_{22})N_r + (I_z + \lambda_{66})Y_v] \big/ C \\ K = Y_\delta N_v - N_\delta Y_v \big/ C \\ T_3 = \frac{(m+\lambda_{22})N_\delta}{Y_\delta N_v - N_\delta Y_v} \\ C = Y_v N_r + N_v[(m + \lambda_{11})u_0 - Y_r] \end{cases} \tag{7}$$

When steering is not very frequent, formula (7) can be approximated as:

$$Tr + r = K\delta \tag{8}$$

Formula (8) is called the first order Nonmoto equation. Among them, K and T have distinct physical significance. K stands for the rotary index of unmanned ship, and T stands for the rudder index of unmanned ship. The values of K and T can be obtained from the Z-shaped test of the unmanned ship. Equation (8) fully expresses the relationship between the bow Angle and the rudder Angle of the unmanned ship, which facilitates the analysis and study of the heading control of the unmanned ship when steering with a small rudder Angle.

2.3 Nonlinear Mathematical Model

The first-order field equation expressed by formula (8) is no longer applicable to the study of ship handling characteristics with large rudder Angle motion or in an unstable state. Birch and Nomoto have done a lot of research in this regard and improved formula (6). Some nonlinear terms were added to the formula, and the following two commonly used nonlinear maneuvering response equations were proposed.

$$T_1 T_2 r + (T_1 + T_2)r + r + \alpha r^3 = K\delta + KT_3\delta \tag{9}$$

$$T_1 T_2 r + (T_1 + T_2)r + r + KH(r) = K\delta + KT_3\delta \tag{10}$$

where α is the coefficient and $H(r)$ is the nonlinear function of r. α and $H(r)$ can be obtained from a spiral test of an unmanned ship.

Similarly, for the first-order linear field equation, the nonlinear manipulation response equation can be expressed as:

$$Tr + r + \alpha r^3 = K\delta \tag{11}$$

where α is the coefficient.

3 Course Control of Unmanned Craft Based on Model Predictive Control

3.1 Classical Model Predictive Control Algorithm

The basic principle of MPC is shown in Fig. 2. At time k, the optimal predictive control input sequence $\hat{y}(k + i)$ within the predictive step size P is solved according to the measured output $y(h)$ and reference track y_d at the current time, so as to minimize the objective function defined by the predictive output and reference track. The first control input in the optimal predictive control input sequence is taken as the system control input at time k. When it is entered into the system, The calculation process of time k is repeated at time $k + 1$, so as to obtain the optimal system input at time $k + 1$. This rolling optimization strategy can ensure that the input of each step is the optimal value calculated based on the current state. The control performance of MPC is related to modeling accuracy, prediction step size and optimal solving efficiency of objective function. In the control process, appropriate simplified model, appropriate control parameters and objective function should be selected according to specific requirements to meet the needs of different control systems.

The predictive control method is based on solving the optimal control problem at any sampling distance. The algorithm can be expressed by the following main principles:

1) Use recognizable models to predict the future output of the system.
2) The control signal is calculated by optimizing the objective function, which is usually a function of prediction error and control efficiency.
3) Using the optimized signal to implement the control.
4) Repeat the above three steps.

The basic structure of the main ideas based on model predictive control can be shown in Fig. 3. The cost function of predictive control is usually the function of prediction

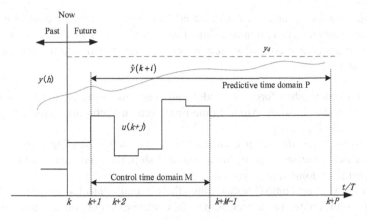

Fig. 2. MPC basic schematic diagram

error. In most cases, the change of control signal is also included in the cost function. By minimizing the cost function, the prediction error and the change of control signal can be optimized. The cost function can be single or multi-step, that is, changes in the prediction error and control signal may only be considered in the cost function at some future time, or their sum may be included in the cost function at several future times.

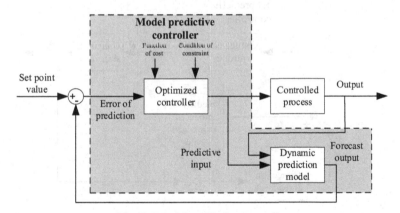

Fig. 3. Model predictive controller

3.2   Course Controller Based on Predictive Function Control

The overall block diagram of course control system of surface unmanned ship based on predictive function control (PFC) is shown in Fig. 4. Wherein, the input signal of the heading control system is the target heading Angle ψ_d of the unmanned ship, the output signal of the heading control system is the actual heading Angle ψ of the unmanned ship, and the input signal of the heading controller based on predictive function control is the target heading Angle ψ_d and the actual heading Angle ψ of the unmanned ship.

The output signal of course controller based on predictive function control is the input rudder Angle control δ of unmanned ship. The control process of the predictive function controller in the course control system of surface unmanned ship mainly includes the following steps:

1) The unmanned ship forecast model calculates the forecast output value of the unmanned ship heading Angle in the time domain according to the current rudder Angle control quantity.
2) According to the deviation between the current actual heading Angle output and the predicted output value of the unmanned ship, the predicted output value in the forecast time domain is corrected.
3) The optimization controller calculates the optimal control sequence of the rudder Angle control of the unmanned ship in the control time domain according to the optimization performance index function and the target course of the current unmanned ship, and determines the rudder Angle control signal applied to the unmanned ship at the current moment according to the optimal control sequence of the rudder Angle control of the unmanned ship.
4) A rolling optimization strategy was adopted to continue the above calculation process (as shown in Fig. 4), and steps 1–3) were repeated.

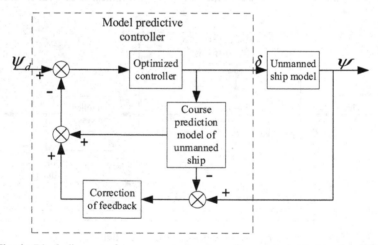

Fig. 4. Block diagram of course control system of unmanned ship based on PFC

(1) **Prediction model**

The predictive model of course control system of unmanned ship refers to the model which can correctly describe the relationship between input and output of course control system. The Nonmoto equation, which can describe the maneuverability of unmanned ship, is used as the prediction model in the course control research of unmanned ship. The Nonmoto equation can be expressed as:

$$T\dot{\psi} + \psi = K\delta \tag{12}$$

where K is the cyclicity index of unmanned ship, and T is the followability index of unmanned ship. δ is the rudder Angle of the unmanned ship; ψ is the bow Angle of the unmanned ship.

Taking the sampling time as T_s and discretizing Eq. (12), the difference equation of the unmanned ship prediction model can be expressed as:

$$\psi(k) = e^{-\frac{T_s}{T}}\psi(k-1) + K\left(1 - e^{-\frac{T_s}{T}}\right)\delta(k-1) \tag{13}$$

In Eq. (12), the cyclicity index K of the unmanned ship and the followership index T of the unmanned ship can be obtained by conducting the Z-shaped maneuverability test of $10°/10°$ on the existing experimental ship. The basic parameters of the experimental ship in this paper are shown in Table 2.

Table 2. Main parameters of unmanned ship

Parameter	Length/m	Draft/m	Full load displacement/T	Maximum speed/kn	Cruising speed/kn
Value	8.075	0.6	3.2	12	6

The Z-shape $(10°/10°)$ test of the experimental ship was carried out on the calm lake surface. The experimental data of the unmanned ship were fitted, and the values of K and T parameters of the experimental ship were obtained as follows: $K = 0.19, T = 1.91$.

Unmanned ships are prone to wind and wave disturbances when sailing at sea. These disturbances f_a are equivalent to rudder Angle disturbances and added into the prediction model, that is, formula (12) is written as:

$$T\dot{\psi} + \psi = K(\delta + f_a) \tag{14}$$

(2) **Correction of feedback**

The feedback correction in the course controller of unmanned ship based on predictive function control (PFC) is mainly to correct the course prediction output value in the forecast time domain given by the forecast model of unmanned ship, so that the predicted value is closer to the actual output value. Here, the error between the predicted output heading Angle $\psi_m(k)$ of the unmanned ship prediction model known at time k and the actual output heading Angle $\psi(k)$ of the system is adopted to correct the forecast output heading Angle $\hat{\psi}(k+i)$ in the future prediction time domain of the system. The error expression is:

$$e(k+i) = \psi(k) - \psi_m(k) \tag{15}$$

After correction, the $\hat{\psi}(k+i)$ expression of the output heading Angle of the unmanned ship prediction model in the time domain of future prediction of the system is as follows:

$$\hat{\psi}(k+i) = \psi_m(k+i) + e(k+i) \tag{16}$$

(3) **Optimization of rolling**

The course controller based on predictive function control obtains the optimal rudder Angle input corresponding to each moment by means of rolling optimization. The criterion of the rudder Angle control quantity is the tracking effect of the controller on the preset course. In addition, the selection of rudder Angle control quantity should not only consider the course maximization tracking, but also consider the change of rudder Angle in line with the actual situation of unmanned ship steering. Therefore, the standard function J, as shown in Formula (17), is constructed to calculate the weighted average of the square of the course tracking error in the prediction time domain and the square of the change increment of the rudder Angle control quantity in the control time domain. The optimal rudder Angle input control quantity of the unmanned ship is obtained by minimizing the standard function J. The importance of course error and rudder Angle increment in the optimal rudder Angle control in the control time domain is reflected by adjusting the weighting factor in the standard function. Under normal circumstances, under the condition that the rudder Angle constraint is satisfied, more attention is paid to the course control effect. Therefore, in formula (17), the course error weighting factor a_i should be set larger than the rudder Angle control error weighting factor b_i.

$$J = \frac{1}{2} \left\{ \sum_{i=1}^{H_p} a_i [\psi(k+i) - \psi_r(k+i)]^2 \right. $$
$$\left. + \sum_{i=1}^{H_c} b_i [\delta(k+i-1) - \delta(k+i-2)]^2 \right\} \tag{17}$$

where ψ is the actual output heading Angle of the heading controller, ψ_r is the reference heading Angle of the unmanned ship, H_p is the prediction time domain of the predictive function control, H_c is the control time domain of the predictive function control, and satisfies $H_c \leq H_p$, a_i is the weighting factor of the heading error, and b_i is the incremental weighting factor of the rudder Angle control.

In addition, according to the control characteristics of the unmanned ship itself, the model predictive course controller should consider the rudder Angle constraint when it outputs the rudder Angle control quantity of the unmanned ship. The rudder Angle constraint of the unmanned ship model used in this paper is:

$$-35° \leq \delta \leq 35° \tag{18}$$

Therefore, the problem of obtaining the optimal rudder Angle control quantity at time k in rolling optimization can be transformed into the local optimal control problem expressed in Formula (19). The first control value of the optimal control sequence obtained by formula (19) at time k is applied to the course control system of unmanned ship as the optimal rudder Angle control quantity at time k to realize the course control of unmanned ship.

$$\min J = \frac{1}{2} \left\{ \sum_{i=1}^{H_p} a_i [\psi(k+i) - \psi_r(k+i)]^2 \right.$$

$$+ \sum_{i=1}^{H_c} b_i [\delta(k+i-1) - \delta(k+i-2)]^2 \Big\}$$

$$s.t. \quad \delta_{min} \leq \delta \leq \delta_{max} \tag{19}$$

4 Simulation Experiment and Analysis

The course control system is simulated in Matlab simulation environment to verify the feasibility of the controller. The system simulation mainly involves two aspects: course keeping and course tracking. The algorithm designed in this paper is compared with the PID control algorithm, and the specific parameters are set as follows:

- The initial parameters of the unmanned ship are set as: speed v = 1 m/s, heading Angle $\psi = 0°$.
- PID Heading controller parameter setting: $K_p = 0.5$, $K_i = 0.00005$, $K_d = 0.9$.
- Parameter setting of MPC heading controller: predict $H_p = 10$ in time domain. Control time domain $H_c = 2$. In objective function J, course error weighting factor $a_i = 1$, $i = 1, 2, \ldots, H_p$, rudder Angle control increment weighting factor $b_i = 1$, $i = 1, 2, \ldots, H_c$.

4.1 Course Keeping Effect

Under the ideal condition without interference, course keeping effect of two unmanned ship course controllers is simulated. Given the constant input of target heading Angle $\psi_d = 10°$, the resulting heading control effect is shown in Fig. 5, and the change of rudder Angle control quantity is shown in Fig. 6.

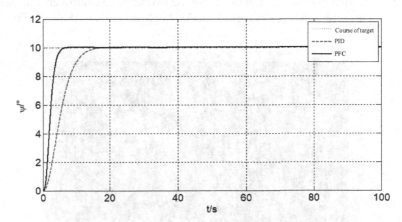

Fig. 5. Course keeping effect diagram (no interference)

As can be seen from the simulation results in Fig. 5 and Fig. 6, under the ideal condition without interference, both the MPC-based heading controller and the PID-based heading controller can reach the given target heading Angle without overshooting.

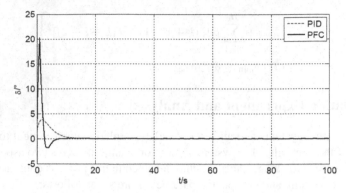

Fig. 6. Course hold rudder Angle control (no interference)

The response speed of the MPC-based heading controller is faster than that of the PID-based heading controller, but its rudder Angle control varies greatly.

In course control simulation with external interference, the wind and wave interference term are replaced by a white noise multiplied by the second-order wave function [14]. The mean value of the white noise signal in the interference signal is 2 and the power spectral density is 0.5. The expression of the second-order wave function is shown in Formula (20), and the parameters are as follows: gain constant $K_\omega = 0.42$, dominant wave frequency $\omega_0 = 0.606$, damping coefficient $\zeta = 0.3$.

$$h(s) = \frac{K_\omega s}{s^2 + 2\zeta\omega_0 s + \omega_0^2} \tag{20}$$

The interference signals in the simulation experiment are shown in Fig. 7. In the case of external interference, the control effects of the two heading controllers are shown in Fig. 8, and the corresponding changes of rudder Angle control are shown in Fig. 9.

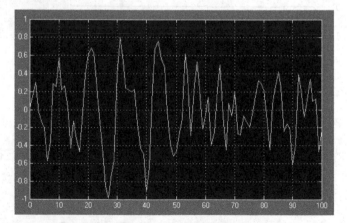

Fig. 7. Wind and waves interfere with the signal

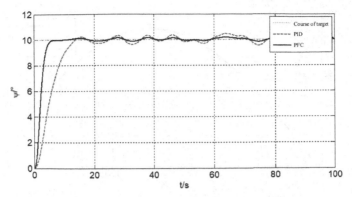

Fig. 8. Course keeping effect (presence of interference)

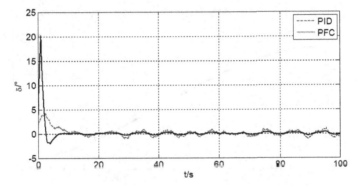

Fig. 9. Course hold rudder Angle control (with interference)

According to the simulation results shown in Fig. 8 and Fig. 9, it can be seen that the control effect of predictive function control is better than that of classical PID control when there is external wind and wave interference in the course control system of unmanned ship [15, 16]. As can be seen from the course tracking curve in Fig. 8, the course control system based on predictive function control can track the target course faster, and the course jitter is smaller when tracking the target course. As can be seen from the curve of rudder Angle control variation in Fig. 9, the output rudder Angle control variation of the course controller based on predictive function control is smoother.

4.2 Course Tracking Effect

Under the ideal condition of no interference, course tracking effect of two course control systems is simulated. In the course tracking simulation experiment, the target course Angle of the unmanned ship is given as a sinusoidal signal with amplitude of 0.02 rad/s and frequency. Course tracking effects of the two controllers obtained by the simulation experiment are shown in Fig. 10, and the variation curve of rudder Angle control quantity is shown in Fig. 11.

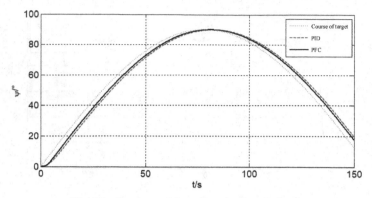

Fig. 10. Course tracking effect (no interference)

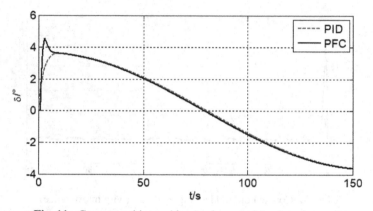

Fig. 11. Course tracking rudder Angle control (no interference)

In order to further analyze the control effects of the PID-based heading controller and the MPC-based heading controller, the average heading tracking errors of the two controllers are shown in Table 3. As can be seen from Fig. 10, Fig. 11 and Table 3, the MPC-based heading controller can achieve stable heading tracking performance, and the tracking speed and tracking accuracy are higher than the traditional PID heading controller.

Table 3. Course tracking error

Controller	Mean tracking error
PID	4.9180°
MPC	3.4953°

In the presence of external interference, the target heading Angle input of the unmanned ship is given as a sinusoidal signal with amplitude of 0.02 rad/s, and the heading tracking effect simulation of two heading controllers is carried out. The interference signal still adopts the wind and wave interference described in formula (24). The obtained course tracking effect is shown in Fig. 12, and the change of rudder Angle control is shown in Fig. 13.

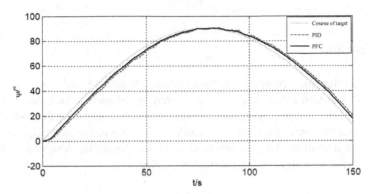

Fig. 12. Course tracking effect (with interference)

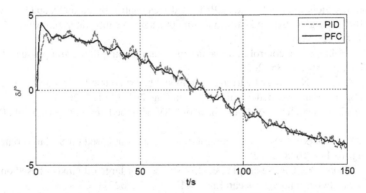

Fig. 13. Course tracking rudder Angle control (with interference)

In order to further analyze the control effect of the heading controller based on PID and the heading controller based on MPC, the heading tracking errors of the two controllers are shown in Table 4. As can be seen from Fig. 12, Fig. 13 and Table 4, the MPC-based heading controller can also achieve stable heading tracking performance in the case of interference, and the tracking speed and tracking accuracy are higher than the traditional PID heading controller.

Table 4. Course tracking error (with interference)

Controller	Mean tracking error
PID	4.9384°
MPC	3.5001°

5 Conclusion

Based on the algorithm principle of model predictive control and the requirements of course control of unmanned ship, a course controller of surface unmanned ship based on model predictive control is proposed in this paper. By comparing the course keeping and course tracking with the course controller of surface unmanned ship based on the classical PID algorithm, it is verified that the course controller designed in this paper has better course keeping and course tracking effects.

References

1. Sivaraj, S., Rajendran, S., Prasad, L.P.: Data driven control based on Deep Q-Network algorithm for heading control and path following of a ship in calm water and waves. Ocean Eng. **259**, 111802 (2022)
2. Hao, Y., et al.: Course control of a manta robot based on amplitude and phase differences. J. Mar. Sci. Eng. **10**(2), 285 (2022)
3. Zhang, Z., et al.: Nonlinear hydrodynamic model based course control and roll stabilization by taking rudder and propeller actions. Ocean Eng. **263**, 112377 (2022)
4. Li, Y., et al.: Research on heading control of USV with the lateral thruster. Math. Probl. Eng. (2022)
5. He, Y., et al.: Depth and heading control of a manta robot based on S-plane control. J. Mar. Sci. Eng. **10**(11), 1698 (2022)
6. Gao, S., Zhang, X.: Course keeping control strategy for large oil tankers based on nonlinear feedback of swish function. Ocean Eng. **244**, 110385 (2022)
7. Fujiwara, T., Brotas, M., Chiappe, M.E.: Walking strides direct rapid and flexible recruitment of visual circuits for course control in Drosophila. Neuron **110**(13), 2124–2138 (2022)
8. Zhang, P., Chen, Q., He, P.: Path following of underactuated vehicles via integral line of sight guidance and fixed-time heading control. IET Cyber-Syst. Robot. **4**(1), 51–60 (2022)
9. Luo, Z., et al.: Research on a course control strategy for unmanned surface vessel. In: Journal of Physics: Conference Series, vol. 1948, no. 1, p. 012106 (2021)
10. Liu, Z.: Adaptive extended state observer based heading control for surface ships associated with sideslip compensation. Appl. Ocean Res. **110**, 102605 (2021)
11. Dong, Y., et al.: Predictive course control and guidance of autonomous unmanned sailboat based on efficient sampled Gaussian process. J. Mar. Sci. Eng. **9**(12), 1420 (2021)
12. Lin, J., Pan, L.: Multiobjective trajectory optimization with a cutting and padding encoding strategy for single-UAV-assisted mobile edge computing system. Swarm Evol. Comput. **75**, 101163 (2022)
13. Mirzaei, M., Taghvaei, H.: Heading control of a novel finless high-speed supercavitating vehicle with an internal oscillating pendulum. J. Vib. Control **27**(15–16), 1765–1777 (2021)

14. Dong, X., Wang, H.: Research on the influence of yaw control on wind turbine performance under wake effect. In: IOP Conference Series: Earth and Environmental Science, vol. 651, no. 2, p. 022004 (2021)
15. Cepeda-Gomez, R., et al.: Switched adaptive scheme for heading control in ships. IFAC PapersOnLine **54**(16), 378–383 (2021)
16. Mu, D., Wang, G., Fan, Y.: A novel model switching course control for unmanned surface vehicle with modeling error and external disturbance. IEEE Access **9**, 84712–84723 (2021)

Multi-UUV Underwater Target Cooperative Detection Task Planning and Assignment

Yongzhou Lu[1(✉)], Hao Zhou[2], Heng Fang[2], and Ziwei Zhao[2]

[1] Naval Academy of Armament, Beijing 100161, China
lulufalcon@163.com
[2] Wuhan University of Technology, Wuhan 430070, China

Abstract. Multi-UUV collaboration is critical for ocean exploration missions. Aiming at the problem of underwater wreck coverage detection, this paper designs a task allocation method based on the biological balance principle and a path planning strategy for underwater target UUV area coverage detection suitable for near-seabed environment. Voronori principle is used to divide the core detection area where the crash target wreckage is located, and then the UUV task load is balanced by the biological balance principle. According to the actual situation analysis of underwater vehicle coverage detection in near-seabed environment, combined with the intelligent unit reliability q function, the path planning strategy of underwater vehicle area coverage detection underwater target is designed. Simulation results show that the proposed algorithm can achieve multi-UUV underwater target detection task region allocation and achieve the consistency of load balancing among UUV tasks.

Keywords: Cooperative detection · UUV · Task Planning

1 Introduction

With the improvement of underwater task requirements, the operating range and functional load types of a single UUV are difficult to meet the complex requirements of ocean exploration tasks. The multi-UUV cooperative operation method can solve this problem well. The multi-UUV cooperative system has various characteristics such as time, space, and function distribution [1], which can effectively improve operation efficiency. It is of great significance and can be applied to information search and detection of underwater targets.

The multi-UUV cooperative detection task, due to the complex combination relationship between the target and the unmanned system, there will be a variety of combinations between the task target and the unmanned system, which makes the task allocation problem more complicated and requires more Only a large amount of calculation can make the system reach the state of optimal assignment of tasks [2]. At present, the underwater communication capability and stability of multi-UUV systems are difficult to meet the above requirements. In 2016, Sahar proposed a market-based clustering method to reduce the computational complexity of the system. This method first grouped task targets into clusters according to their respective information, and then assigned the grouped clusters to Individuals in the cluster, thereby reducing the task time of the entire system [3]. Murugappan et al. proposed a K-means clustering allocation method that added the system task load balancing allocation factor. This method exists the task allocation results are greatly affected by the initial state of the system [4]. Due to the high autonomy of the individual in the distributed multi-robot system, the individual follow-up task planning can be realized through the information interaction between the individuals, so as to complete the task goal of the entire cluster, because there are many uncertain factors in the operation process of a single robot, so the optimization principle in the traditional control method is difficult to realize for this distributed multi-robot autonomous planning system. There are many biological communities in the natural environment [5]. Individuals in these biological communities can well complete the allocation of living resources and work tasks in the community. Some researchers have applied the relationship between biological communities to the task allocation of multi-robots, Research on motion control and other issues. Mistumoto et al. proposed a multi-robot population control model inspired by the biological immune network. This model is not affected by the control of individual individuals. Individuals in the cluster can properly adjust their tasks according to the incentive and inhibition information they receive. Change, so that the task distribution of the entire system is balanced [6]. Parker proposed a behavior-based multi-robot distributed collaborative relationship model for heterogeneous multi-robot task assignment methods because the previous research focused on improving the efficiency of the system and often ignored the needs of individual robots for fault tolerance and adaptability. ALLICANCE, this collaborative model incorporates task-oriented action selection mechanisms into behavior-based systems, thereby enabling multi-robot distributed systems to improve the efficiency of the overall multi-robot system while maintaining the required fault-tolerant and adaptable characteristics [7]. However, since the communication method between robots in this research uses implicit communication, the processing results for the task assignment problem of the system with high complexity are not ideal [8]. Aiming at the problem that the effect of information interaction is not ideal due to the complexity of robot information interaction in a multi-robot system, Dahl et al. from the Robot Intelligence Laboratory of the University of Wales, inspired by the ubiquitous resource allocation process in nature, proposed a method based on vacancy chain (Vacancy chain scheduling, VCS) task allocation method, as a resource allocation method, vacancy chain is very common in nature and human society. It means that individuals in the system can evaluate and exchange system resources. This method also considers Influenced by the information interaction between robots in a multi-robot system, the method can achieve better results in a cooperative operation system with priority transportation [9].

From the above research status at home and abroad, it can be seen that the task allocation of multi-robot systems is very important to the research of cluster systems, and the real-time performance, reliability, communication capabilities, robustness and other indicators of task allocation methods are the focus of research [10]. The research content of this paper is the task allocation problem in the process of multi-UUV search and exploration of underwater targets, and the efficiency of the multi-UUV system to complete the search and exploration operation is used as the system evaluation index [11].

2 Problem Analysis and Modeling

In order to confirm the specific location information of the underwater crash target, it is necessary to conduct underwater imaging scanning operations on the core area where the crash target obtained in Sect. 3 is obtained through UUV to confirm the specific underwater situation of the crash target. Firstly, it is necessary to model the UUV detection area and divide it into several small areas reasonably. The Voronoi diagram partition theory [12] is introduced to provide a theoretical basis for the division of detection regions. Voronoi diagrams are widely used to solve problems related to region division. Considering the UUV clusters as generators one by one, the detection area of the UUV is the Voronoi polygon corresponding to the generator. Each UUV only needs to perform detection operations in its assigned task area. This division makes every point in the detection area. The distance to its corresponding UUV is the shortest. However, since the position of the UUV in the core area is randomly set, the area size of the detection area allocated to each UUV is inconsistent, and the probability of the existence of a crash target in the detection area of different UUVs is also different. In order to make the task distribution of the entire UUV cluster more reasonable, the Voronoi diagram should be partitioned on the core and then redistributed according to the area size of each Voronoi polygon and the possibility of underwater crash targets in the area [13].

Assuming that the number of UUVs in S in the task area is n, then for these n UUVs, each UUV can be represented by the corresponding generator p, which $P = \{p_1, p_2, \cdots p_n\}$ is the set of all generators, and q is a point in the task area S, then the generator The Voronoi polygon corresponding to p can be written as:

$$V(p_i) = \{q \in S \mid \|q - p_i\| \leq \|q - p_j\|, \forall j \neq i\} \tag{1}$$

The above formula represents the Voronoi polygon $V(p_i)$ corresponding to the generator p_i, which $V(p) = \{V(p_1), V(p_2), \cdots, V(p_n)\}$ can be expressed as a set of all generators.

When UUV conducts underwater detection operations, it is necessary to judge the importance of the area based on environmental information, so as to determine the importance of UUV detection operations in this area. A series of basic functions are used to describe the environmental information of the underwater task area. as follows:

$$K = [a_1 * K_1, \ldots, a_n * K_n]^T \tag{2}$$

$$K_j = \frac{1}{\sigma_j \sqrt{2\pi}} \exp\left(-\frac{(q - \mu_j)^2}{2\sigma_j^2}\right) \tag{3}$$

In the above formula, q is a random point in the core area, a_n is a weighting coefficient obtained from environmental information by UUV, σ is the standard deviation representing the concentration of the basis function, μ is set as the mean value of the basis function, K is known. The environmental information is simulated by Matlab. Since the acoustic signal of the underwater target is emitted from the black box, the protruding part in the middle is the coordinate position of the black box found. As a part of the crashed target, the black box is surrounded by the wreckage of the wreckage of the crashed target has the highest probability, and the outward probability decreases successively, so the underwater environment information is simulated as shown in Fig. 1.

Fig. 1. Schematic diagram of environmental information

Let any point in the area be r, there are m target points in the area, and the environmental information is given by K Indicates that the importance index of this position is expressed as:

$$E(r) = \frac{1}{2\pi} \sum_{i=1}^{m} \exp\left[-\frac{1}{2}(r - r_i)^T K_i (r - r_i)\right] \tag{4}$$

In the formula, r_i is the point with a large importance index, which is determined by the known environmental information, and K_i is a diagonal matrix, and its value represents r_i the existence intensity. Then for UUV, the task load of this area is determined by the probability of the existence of crashed targets in this area, that is, the sum of the importance indicators of each position in this area is expressed as:

$$T_i = \int_{r \in V(p_i)} E(r) dr \tag{5}$$

Among them $V(p_i)$ is the multi-deformation of the core detection area divided by the Voronoi principle. Let the number of vertices of the Voronoi polygon be Ntr_i. According to mathematical knowledge, the polygon can be divided into Ntr_i triangles, and the area of $j \in \{1, 2, \cdots, Ntr_i - 1\}$ each triangle is s_j. Bring formula (4) into formula (5), we have:

$$T(i) = \frac{1}{2\pi} \sum_{j=1}^{Nt_i} \int_{s_j} \sum_{i=1}^{m} \exp\left[-\frac{1}{2}(r - r_i)^T K_i (r - r_i)\right] dr \tag{6}$$

The average problem of multi-UUV underwater detection tasks is solved by assigning the task load of UUV, that is, the assigned task load of each UUV conforms to the following relationship:

$$T(1) = T(2) = \cdots = T(n) \tag{7}$$

3 Multi-UUV Detection Task Assignment Algorithm Based on Biological Balance

Due to the application of the Voronoi principle to partition the core detection area, the task area assigned by a single UUV has different area sizes, and the importance of targets at different positions in the core detection area is also different, making the task load distribution of a single UUV in the cluster different. This paper mainly solves the problem of uneven task load distribution of multi-UUV full-area coverage detection of underwater targets. The research content of this paper is the design of multi-UUV autonomous task assignment method based on the principle of balance. The principle of balance is a very common phenomenon in nature. Individuals in nature will be affected by the constraints of natural resources such as food, water, and living space in the environment. A state of survival will be formed during this period, and this balance state formed by multiple constraints enables the organisms in the ecosystem to survive, and makes the ecosystem maintain a positive and virtuous cycle state [14]. The UUV operating environment is regarded as the natural environment, and UUV corresponds to the biological individual in the natural environment. Multi-UUV distributed collaborative operation is equivalent to the competition and cooperation relationship between individuals in the natural environment, and the detection task load is used as an allocatable resource in nature. Therefore, this section will use the idea of natural balance in the follow-up research to model and analyze the task of multi-UUV detection of underwater targets, so that the multi-UUV system can achieve autonomous task assignment.

When multiple UUVs work together, each UUV is in a relationship of mutual cooperation and mutual competition. When the detection task is allocated, in order to maximize the search efficiency of the UUV system, this paper realizes the autonomous task allocation of multiple UUV systems through a task allocation strategy based on the principle of balance. The principle of balance is based on the phenomenon that multiple biological individuals in nature maintain a balanced state in a fixed environment area and combined with the characteristics that UUV clusters need to conduct full-coverage detection of the entire task area when conducting underwater target search and detection operations. The task allocation of the entire UUV cluster is regarded as a dynamic process determined by the detection task factors. First of all, UUV corresponds to the biological individual in the natural environment. The stronger the biological individual's own ability in nature, the stronger its competitive ability, which corresponds to the performance of UUV. Secondly, individual organisms in nature have their own living territories, which can correspond to the task area of UUV. In nature, the survival resources required for biological survival are limited, and individual organisms need to obtain survival resources through competition, and the resources correspond to the underwater accident targets in the core detection area, which can be regarded as the UUV in formula (6) task load. In nature, the resources allocated to organisms are often determined by their own abilities. The greater the ability, the more survival resources they can obtain. Therefore, the relationship between the survival resources of the environment in which the individual organisms live and Relationship to survivability:

$$r = R/A \tag{8}$$

Among them, R is the total resources in the biological natural area, A is the sum of the survivability of all biological individuals in the area, and r is the ratio between the two. We know that in the natural environment, the higher the r value of an individual organism, it indicates that it occupies more resources, and its control over the living area is low. At this time, the organism with a low r value will actively occupy the r value because of survival needs the living resource of the individual with high value is equal until the r value of the two is equal, and the two reach an equilibrium state. In the UUV cluster, R and A can be regarded as the UUV task load and UUV execution capability. This ratio is the task-ability ratio of a single UUV in a multi-UUV system. The multi-UUV task allocation result is adjusted by the ratio r between UUVs.

Next, analyze the above allocation method, and prove that the balance principle conforms to the optimal principle when resource allocation is carried out by this method. Let S represent the total resource in the environment, which is the resource S_i occupied by the individual, as follows:

$$S = \sum_{i=1}^{n} s_i \tag{9}$$

In general, the relationship between UUV's ability to execute tasks and resources can be described by the following nonlinear function expression:

$$J_i(s_i) = A_i(1 - \exp(-a_i s_i)) \tag{10}$$

where A_i is the maximum execution capacity, $a_i \geq 0$ and is a constant. For normalization J_i, we get:

$$\overline{J}_i(s_i) = 1 - \exp(-a_i s_i) \tag{11}$$

For the entire UUV cluster, the overall execution capability can be written as:

$$\tilde{J} = \sum_{i=1}^{n} \tilde{J}_i = \sum_{i=1}^{n} 1 - \exp(-a_i s_i) \tag{12}$$

It can be concluded that when the execution capability $\tilde{J}$ is at the maximum value, the resources allocated to a single UUV S_i satisfy the following equations:

$$\begin{cases} s_1 + s_2 + \cdots + s_n = S \\ 2s_2 + s_3 + \cdots + s_n = S \\ \vdots \\ s_1 + s_2 + \cdots + 2s_n = S \end{cases} \tag{13}$$

Solve Eqs. (12), get:

$$s_1 = s_2 = \cdots = s_n = \frac{S}{n} \tag{14}$$

Equation (13) is the optimal solution of the execution capability J of the UUV swarm system. It can be known from mathematical knowledge that the balance degree of a group of numbers can be described by the variance, then:

$$
\begin{cases}
E_i(t) = \sum_{i=1}^{n} J_i(t)/n \\
D_i(t) = \sum_{i=1}^{n} (J_i(t) - E_i(t))^2
\end{cases}
\tag{15}
$$

where $E_i(t)$ is the standard deviation $D_i(t)$ of the system, and is the variance of the system; it can be seen from formula (15) that when $\tilde{J}_i = \tilde{J}/n$, $D_i(t) = 0$ is just in equilibrium.

4 Simulation Experiment

In the above content, the research on the task load distribution method for multi-UUV detection of underwater targets is completed. Next, the movement path of UUVs in the underwater detection process is analyzed. In the process of performing multi-UUV detection operations, the first thing that needs to be satisfied is to perform full-coverage detection operations on the entire detection area, to detect every corner as much as possible, and to avoid repeated detection operations in the same area; The operation area is close to the seabed, and it is necessary to consider the impact of the complex environment on the seabed on the safety of UUV navigation. Therefore, the UUV should be able to perform simple obstacle avoidance actions against obstacles and actively stay away from the area where obstacles are located [15–17]. This paper proposes a multi-UUV regional collaborative detection method based on smart cells, which enables UUVs to quickly complete the coverage detection operation and reduce the repetition rate of regional detection.

Define the cell's confidence function assignment, which is determined by the current state of the cell and the UUV heading change value. The current state of a ρ_i smart cell p_i is represented by a function:

$$
\rho_i = \begin{cases}
1, & \text{The cell is not explored} \\
1 - \lambda/2, & \text{Have been detected} \\
-\infty, & \text{There are obstacles}
\end{cases}
\tag{16}
$$

Before the job starts, the current state of each cell is the initial value $\rho_i = 1$. In order to improve the efficiency of the UUV in the detection job, it is necessary to reduce the number of UUV detections on the same cell and reduce the repetition rate of the path. Therefore, when UUV The depreciation operation will be performed on the current cell ρ_i, as shown in formula (16), the value of the obstacle in the cell ρ_i is set to negative infinity, and the aircraft is prohibited from passing through this area. In order to reduce the energy consumption of the UUV during navigation, the detection route of the UUV should be kept as straight as possible, and the number and angle of turning should be reduced, so the ρ_i value of the turning cost function is assigned to the cell:

$$
\omega_i = 1 - \frac{\Delta\theta_i}{\pi}
\tag{17}
$$

Among them, the steering cost that needs to be paid $\Delta\theta_i$ for the UUV to go to the cell.

When the UUV performs detection operations in the rasterized mission area, it finds the optimal cell around the UUV through the path planning strategy, and uses the cell as the target coordinate of the UUV at the next moment, so as to cover the entire mission area probing. Based on the above analysis of the reliability function of the smart cell, it is assumed that:

$$F_i = \max\left\{F_j = \rho_j + \beta\omega_j, j = 1, 2, \cdots, k\right\} \tag{18}$$

Among them, F_i is the reliability function value of the UUV cell at the next moment, β is the weighting coefficient, and j is the number of cells centered on the UUV. The reliability function is designed so that the UUV will give priority to straight driving and reduce the number of turns of the UUV, to reduce energy consumption. When the UUV state is inactive, it is necessary to increase the selected cell range and increase the j value so that the UUV can continue to go to other cell areas.

The improved cell-based area coverage strategy design algorithm first obtains the seabed environment information through the UUV sonar equipment to assign values to the current state of the gridded cells, and then introduces the steering cost function during UUV navigation, which reduces the number of UUV steering times and reduces Energy consumption during the UUV detection process; optimize the design for the non-action state during the UUV operation, and reduce the path repetition rate during the UUV detection operation.

5 Simulation Analysis

The task area is set as a square area with a side length of 2000 m. $O(x_0, y_0)$ The probability of the target wreckage at the coordinates of the black box is 1. Concentric circles are generated with the black box as the center, and the probability of the existence of points on the concentric circles is equal. Therefore, the probability of the existence of the target at each position point in each task area can be obtained according to the distance from the position point in the task area to the black box. The calculation formula is:

$$P_{(x,y)} = \mu\sqrt{(x - x_0)^2 + (y - y_0)^2} \tag{19}$$

where $P_{(x,y)}$ is the probability value of the existence of the target at the point (x, y) in the area; μ is the target existence probability coefficient, which is determined by the location $O(x_0, y_0)$ and the size of the core detection area.

When detecting the wreckage of the wrecked target in the core detection area, four UUVs are used to form a multi-UUV system to detect the mission area. The task load description for UUV can be regarded as the superimposed value of the target existence probability value of all position points in the task area assigned by UUV. The initial position is set to be randomly assigned, and the ability to perform tasks of the UUV in the system is directly given as (Table 1):

According to the multi-UUV task allocation method designed above, the UUV task load is first allocated according to the initial position of the UUV, and then based on the

Table 1. The ability to perform tasks of the UUV

UUV	1	2	3	4
x (m)	111	408	1057	1786
y (m)	1571	208	1115	484
Capability	681	920	864	819

principle of biological balance, the task load allocated to the UUV is calculated twice according to the performance value of each UUV load in the system. The distribution makes the task load distribution of each UUV in the UUV system reach the optimal state. Next, a simulation test is carried out on the multi-UUV task allocation algorithm based on the balance principle to verify the feasibility of the algorithm. The simulation results are shown in Fig. 2 and Fig. 3.

Figure 3 shows the UUV survival coefficient values in the process of multi-UUV system task load distribution. In the above analysis, the survival coefficient of UUV will eventually tend to be the survival coefficient of the whole system. It can be seen from Fig. 3 that in the initial state, the initial value of UUV1 is the highest, and the initial value of UUV4 is the lowest. After a period of time, the survival coefficient of all UUVs the coefficient values have reached an equilibrium state.

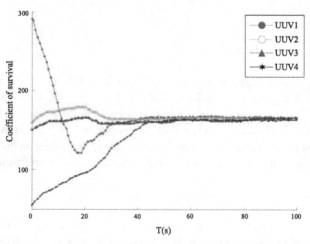

Fig. 2. The change curve of the survival coefficient of each UUV in the process of task assignment

Figure 4 shows the size of the task area allocated by UUV during the task load allocation process of the multi-UUV system. It can be seen from the above figure that the task-ability ratio of individuals in the multi-UUV system has reached consistency, and the task areas corresponding to the individuals have also been allocated. Therefore, the above simulation test verifies the feasibility and stability of the task allocation algorithm based on the principle of biological balance. The initial allocation of UUV task areas is realized through the Voronoi diagram. In order to make the multi-UUV system task

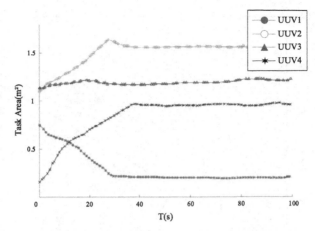

Fig. 3. The change curve of UUV task area in the process of task allocation

allocation process more intuitive and easier to understand, here is the change process of the task area during the task allocation process of the system, as shown in Fig. 4 and Fig. 5 shown.

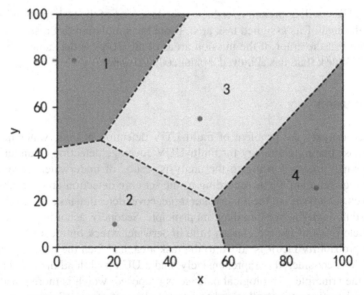

Fig. 4. Initial allocation results of multiple UUV task areas

Figure 4 shows the initial allocation results of multiple UUV task areas. The four different colored punctuation points in the figure represent the positions of UUVs. The color depth of the task area indicates the ratio of UUVs current task load to task performance. The darker the color, the higher the degree of UUV task overload, its task performance cannot withstand the sum of task loads in this area. At this time, the UUVs

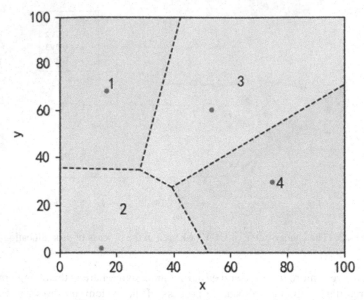

Fig. 5. UUV task area final allocation results

around the UUV will approach it, reducing the task load of the task overload UUV and reducing the area of its assigned task area. From the simulation test results in Fig. 5, it can be seen that the color of the mission areas of all UUVs is the same, which means that the UUV task load has achieved a balanced distribution.

6   Conclusion

This paper analyzes the problem of multi-UUV detection of underwater targets, and designs a task planning strategy for multi-UUV coverage detection of underwater targets in a small area according to the analysis results of underwater target coverage detection requirements. First, according to the set core detection area, combined with the actual characteristics of the underwater target cover detection task, the UUV task area is allocated through the Voronoi diagram principle. Secondly, according to the distribution characteristics of the wreckage of the underwater wreck on the seabed, the target existence probability factor is added to the task area, and then the load capacity factor of the UUV is considered comprehensively, and a UUV task load distribution method based on the principle of biological balance is proposed, which is more reasonable for UUV. The detection area is allocated to improve the detection efficiency of the whole system. Finally, a coverage detection method based on improved cells is designed, so that the UUV can perform coverage detection operations in the assigned task area, which solves the problem of path point selection in the detection process, reduces the energy consumption of UUV detection operations, and improves the path repetition rate. Low, high area coverage. A simulation test was carried out for the UUV task load distribution problem. The test results show that the algorithm proposed in this paper can realize the task area distribution of multi-UUV underwater target detection operations, and realize the task load balance consistency of each UUV.

References

1. Xu, J., Zhou, X., Chen, X., et al.: Multi-scale adaptive corner detection and feature matching algorithm for UUV task target image. In: IEEE International Conference on Mechatronics and Automation, pp. 1104–1109, IEEE (2017)
2. Chen, Y.L., Ma, X.W., Bai, G.Q., et al.: Multi-autonomous underwater vehicle formation control and cluster search using a fusion control strategy at complex underwater environment. Ocean Eng. **216**(7), 108048 (2020)
3. Zhang, G.P., Liu, Z., Tian, X.D.: Searching and simulation of underwater cooperative mine detection based on UUV group. Ordnance Industry Automation (2007)
4. Zhu, D., Liu, Y., Sun, B.: Task assignment and path planning of a multi-AUV system based on a Glasius bio-inspired self-organising map algorithm. J. Navig. **71**(2), 482–496 (2018)
5. Deng, Y.: Task allocation and path planning for acoustic networks of AUVs. Florida Atlantic University (2010)
6. Du, X., Guo, Q., Li, H., et al.: Multi-UAVs cooperative task assignment and path planning scheme. J. Phys: Conf. Ser. **1856**(1), 012016 (2021)
7. Lin, J., Pan, L.: Multiobjective trajectory optimization with a cutting and padding encoding strategy for single-UAV-assisted mobile edge computing system. Swarm Evol. Comput. **75**, 101163 (2022)
8. Deng, Y., Beaujean, P.J., An, E., et al.: Task allocation and path planning for collaborative autonomous underwater vehicles operating through an underwater acoustic network. J. Robot. 483095.1–483095.15 (2013)
9. Mahmoud Zadeh, S., Powers, D.M.W., Sammut, K., Yazdani, A.M.: A novel versatile architecture for autonomous underwater vehicle's motion planning and task assignment. Soft. Comput. **22**(5), 1687–1710 (2016). https://doi.org/10.1007/s00500-016-2433-2
10. Eun, Y., Bang, H.: Cooperative task assignment and path planning of multiple UAVs using genetic algorithm. In: AIAA Infotech@Aerospace 2007 Conference and Exhibit (2007)
11. Chen, X., Liu, Y.: Cooperative task assignment for multi-UAV attack mobile targets. In: 2019 Chinese Automation Congress (CAC) (2019)
12. Ghafoor, H., Noh, Y.: An overview of next-generation underwater target detection and tracking: an integrated underwater architecture. IEEE Access **99**, 1 (2019)
13. Xu, J., Du, X., Li, J., et al.: Kernel two-dimensional nonnegative matrix factorization: a new method to target detection for UUV vision system. Complexity (2020)
14. Liu, C., Wang, H., Yingmin, G.U., et al.: UUV path planning method based on QPSO. In: Global Oceans 2020: Singapore - U.S. Gulf Coast (2020)
15. Li, Z., Li, S., Wen, S.: Automatic detection of an underwater target based on UUV. J. Harbin Eng. Univ. (2017)
16. Yang, F., Chakraborty, N.: Multirobot simultaneous path planning and task assignment on graphs with stochastic costs. In: 2019 International Symposium on Multi-Robot and Multi-Agent Systems (MRS) (2019)
17. Zhao, M., Li, T., Su, X.H., Zhao, L.L., Zhang, Y.H.: A survey on key issues of cooperative task planning for 3D multi-UAVs system. Intell. Comput. Appl. **6**(1), 31–35 (2016)

Multi-UUV Cooperative Navigation and Positioning Algorithm Under Communication Delay

Junjun Wang[✉]

Marine Equipment Major Project Center, Beijing 100841, China
z18336265177@163.com

Abstract. Aiming at the communication delay problem in the process of multi-UUV cooperative navigation and positioning, this paper takes the single pilot master-slave structure as the research object, analyzes the basic principles of multi-UUV cooperative navigation and positioning, and establishes the UUV kinematics model and measurement model. On the basis of the standard Kalman filter algorithm, considering the constraints of underwater acoustic communication, the EKF algorithm is improved based on the principle of state compensation, and a DEKF algorithm is proposed. A simulation experiment was carried out on the DEKF algorithm, and the EKF was compared with the dead reckoning. The simulation results showed that the average value of the EKF positioning error was smaller than that of the dead reckoning, which could effectively correct the UUV's own positioning error. Comparing EKF and DEKF, DEKF can effectively compensate the system state estimation error under communication delay, and the positioning accuracy and stability are improved compared with the EKF algorithm.

Keywords: UUV · Delay of communication · Navigation and positioning algorithm

1 Introduction

When the vehicle cooperates to perform tasks underwater, high-precision navigation and positioning information is crucial to the efficient cooperation between multiple UUV systems. However, due to the particularity of the underwater environment, GPS positioning on land cannot be used when UUVs operate underwater, especially after reaching a certain depth. Moreover, traditional underwater acoustic positioning and geomagnetic navigation also have many defects. Different UUVs observe each other based on underwater acoustic ranging and realize the collaborative correction of their own navigation and positioning errors through information fusion. The use of multi-UUV cooperative navigation and positioning can overcome some of the defects of traditional positioning methods, and has the characteristics of relatively low cost and good system robustness, which can improve the overall navigation and positioning performance of the system, thereby enhancing the collaborative operation capability of underwater multi-UUV systems, has important research value. However, due to the particularity of the underwater

L. Pan et al. (Eds.): BIC-TA 2022, CCIS 1801, pp. 496–506, 2023.
https://doi.org/10.1007/978-981-99-1549-1_39

environment, the multi-UUV cooperative navigation and positioning system based on the underwater acoustic communication network has unique technical difficulties, such as modeling and algorithm, error compensation, etc., which greatly limit the positioning performance of UUV. Therefore, it is of great significance to explore more efficient collaborative navigation and positioning methods.

At present, the problem of multi-UUV underwater cooperative navigation and positioning is still a major technical challenge for UUV. Generally speaking, the research on cooperative navigation and positioning began at the end of last century. Relevant experts, scholars, and research institutions at home and abroad have carried out relevant analysis of cooperative navigation algorithms for multi-UUVs in the past two decades, and have also conducted related research on collaborative navigation network error compensation and formation configuration design methods. The research team led by American Roumeliotis has done some research work on the basic theory of coordinated navigation and positioning, such as Engel R et al. analyzed error compensation and modeling. Roumeliotis et al. is oriented towards centralized positioning, and uses distributed ideas to establish a decentralized framework, which can theoretically be applied to formations with a large number of UUVs [1]. In the GREX project of the European Union, the researchers of the project team proposed a method for UUV to use only these two methods for UUV cooperative navigation and positioning without underwater acoustic positioning based on distance measurement and dead reckoning [2]. Based on geosynchronous satellites, IEC Communication Company has conducted research on the method of distributed cooperative navigation and positioning [3]. Italian Garello and other research teams analyzed the performance of cooperative navigation and positioning in some specific scenarios and the impact on the average capture time of terminals and compared a variety of common different algorithms [4]. MIT Bahr and others in the United States studied the UUV cooperative navigation and positioning method under the consideration of using the surface unmanned vehicle and not using the surface unmanned vehicle [5]. Khalaji et al. aimed at the reconstruction and formation control in the cooperative navigation and positioning system and proposed a multi-UUV formation control method based on the master-slave structure and Lyapunov theory [6–8].

From the perspective of core algorithms, multi-UUV cooperative navigation and positioning algorithms can be roughly divided into the following three types: navigation and positioning algorithms based on graph theory, navigation and positioning algorithms based on Bayesian estimation, and cooperative navigation and positioning algorithms based on optimization theory. Specifically, the multi-UUV cooperative navigation and positioning problem is modeled based on graph theory [9–13]. First, a measurement graph model is established, so that the navigation status at all historical moments corresponds to the node, and the sensor measurement corresponds to the edge; Bayesian estimation involves Cooperative navigation and positioning algorithm based on Kalman filter, cooperative navigation and positioning algorithm based on particle filter and cooperative navigation and positioning algorithm based on information filter, algorithm based on optimization theory There are various ways to realize the algorithm through optimization technology [14–16]. As mentioned above, the specific implementation of multi-UUV cooperative navigation and positioning related algorithms is also varied. Each algorithm has its own characteristics and is not perfect. There are defects to varying degrees, and there is still room for improvement in algorithm performance.

This paper first deals with the typical nonlinear model for the multi-UUV cooperative navigation and positioning problem and analyzes the optimal estimation problem of the stochastic nonlinear system, and then designs the cooperative navigation and positioning algorithm based on EKF. Then, considering the constraints of underwater acoustic communication, an improved DEKF algorithm based on state compensation is designed to solve the communication delay problem in the process of multi-UUV cooperative navigation and positioning.

2   Multi-UUV Co-location Measurement Modeling

In the master-slave UUV cooperative navigation and positioning system, the relative displacement vector obtained from the dead reckoning of the UUV at adjacent sampling moments is called the moving vector [17]. Similar to the displacement vector, the moving vector plays an important role in the specific measurement and modeling process combined with the cooperative navigation and positioning mechanism in this section.

Records (x_k^M, y_k^M, z_k^M), (x_k^S, y_k^S, z_k^S) represent t_k the coordinate positions of the master and slave UUVs at the time respectively. The master UUV broadcasts its own position information and sends out the acoustic pulse signal. The relative distance between UUVs is recorded as l_k, and the three-dimensional positional relationship is converted to a two-dimensional plane according to their respective depths, as shown in Fig. 1.

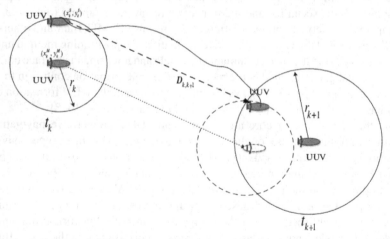

Fig. 1. Schematic diagram of single navigator cooperative navigation and positioning

The circle and the radius of the UUV at t_k time,

$$r_k = \sqrt{l_k^2 - |z_k^M - z_k^S|^2} \tag{1}$$

The equation of the circle is:

$$(x_k^S - x_k^M)^2 + (y_k^S - y_k^M)^2 = r_k^2 \tag{2}$$

At the moment t_{k+1}, the master UUV also sends the acoustic pulse signal to each slave UUV and broadcasts its own position and calculates the relative distance information between the master and the slave l_{k+1}, then the slave UUV is located on the circle with the center of the circle and the radius of the circle at the moment t_k, where,

$$r_{k+1} = \sqrt{l_{k+1}^2 - \left| z_{k+1}^M - z_{k+1}^S \right|^2} \tag{3}$$

The equation of the circle is:

$$(x_{k+1}^S - x_{k+1}^M)^2 + (y_{k+1}^S - y_{k+1}^M)^2 = r_{k+1}^2 \tag{4}$$

UUV $D_{k,k+1}$ can use its own dead reckoning to calculate the movement t_k vector to t_{k+1} the moment, remember $D_{k,k+1} = (Dx_{k,k+1}^S, Dy_{k,k+1}^S)^T$, there is

$$\begin{cases} x_{k+1}^S = x_k^S + Dx_{k,k+1}^S \\ y_{k+1}^S = y_k^S + Dy_{k,k+1}^S \end{cases} \tag{5}$$

According to the moving vector, the plane relative positions of the t_{k+1} master and slave UUV at the moment are translated to the t_k moment. Then on this basis, the two circular equations of t_k sum and t_{k+1} time can be obtained, and the equations can be obtained:

$$\begin{cases} r_k^2 = (x_{k+1}^S - Dx_{k,k+1}^S - x_k^M)^2 + (y_{k+1}^S - Dy_{k,k+1}^S - y_k^M)^2 \\ r_{k+1}^2 = (x_{k+1}^S - x_{k+1}^M)^2 + (y_{k+1}^S - y_{k+1}^M)^2 \end{cases} \tag{6}$$

Therefore, the measurement equation of the system can be obtained as:

$$\begin{aligned} z_{k+1} = h(x_k) + v_k &= \begin{bmatrix} r_k \\ r_{k+1} \end{bmatrix} + v_k \\ &= \begin{bmatrix} \sqrt{(x_{k+1}^S - Dx_{k,k+1}^S - x_k^M)^2 + (y_{k+1}^S - Dy_{k,k+1}^S - y_k^M)^2} \\ \sqrt{(x_{k+1}^S - x_{k+1}^M)^2 + (y_{k+1}^S - y_{k+1}^M)^2} \end{bmatrix} \end{aligned} \tag{7}$$

where v_k is the measurement noise, and the noise covariance matrix is expressed as:

$$R_k = E\left[v_k \, v_k^T \right] = \begin{bmatrix} \sigma_{r_k}^2 & 0 \\ 0 & \sigma_{r_{k+1}}^2 \end{bmatrix} \tag{8}$$

3 Multi-UUV Cooperative Navigation and Positioning Algorithm under Communication Delay

In the process of UUV collaborative navigation and positioning, because the master and slave UUVs exchange information through underwater acoustic communication, the current underwater acoustic communication technology cannot meet the requirements of high speed, real-time and reliability. In complex water environment, long-distance underwater acoustic communication will inevitably introduce the problem of communication delay, which will lead to the reduction of UUV positioning accuracy. Therefore, this section aims at this problem and improves the state-compensated EKF collaborative navigation and positioning algorithm to achieve multi-UUV cooperative navigation and positioning under time-delay conditions.

3.1 Multi-UUV Cooperative Navigation and Positioning Communication Delay Model

Different piloting methods correspond to different delay models.

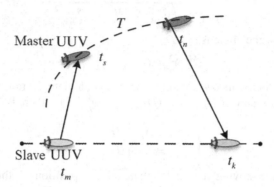

Fig. 2. Convergence curve of the algorithm

For the master-slave structure of single piloting in this paper, Fig. 2 below shows a schematic diagram of the underwater acoustic information transmission process between the master UUV and the slave UUV. The clocks between the master t_m and slave UUV t_s are synchronized. The slave UUV sends a request to the master UUV at time T. Underwater acoustic transmission, receiving information from UUV t_k and completing communication at any time.

The communication information sent by the t_k master UUV at time is received by the slave UUV at time t_s. Unlike the filter cycle without delay under ideal communication conditions, there is a time delay of filter cycles in this N process. The measurement information sent by the master UUV at any time is received by the slave UUV after being delayed by N filter t_s cycles, and the equivalent measurement is denoted as $Z^*(k)$. The status t_k update of the time system is completed by dead reckoning from the UUV itself, and the measurement update needs to use the position of the master UUV and the distance between the master and slave UUV. Due to the delay in non-ideal communication, the measurement update is not accurate in this case, and the state value used is the value measured before N filtering cycles.

3.2 Improved EKF Positioning Algorithm Based on State Compensation

When there is a delay, this section improves the EKF algorithm under ideal communication conditions based on state compensation. The basic principle of state compensation is to take the $t_s - N$ time system measurement matrix and noise variance matrix as t_s the corresponding quantity measurement at the time, and substitute these quantity measurements into the filter equation to start iterative calculation, and then at the t_k time, when the measurement information actually arrives from the UUV. When the state compensation amount is introduced to compensate $\delta \hat{X}_{k,k-1}$ the system state estimation error [2], thereby reducing the impact caused by delay.

Aiming at the characteristics of the above-mentioned time delay, the specific steps of the improved state compensation-EKF or DEKF are given below. First, the EKF filtering principle is used to linearize the multi-UUV co-location system:

$$X_{k+1} = A_k X_k + W_k \tag{9}$$

$$Z_{k+1} = H_k X_k + V_k \tag{10}$$

where k is the discrete time, X_k is the state vector, A_k is the state transition matrix, W_k is the t_k time noise sequence, Z_{k+1} is the t_{k+1} m-dimensional measurement at time, H_k is the $m \times n$ dimensional measurement matrix, and V_k is the m-dimensional measurement noise sequence. Linearize (9), (10) to obtain the Jacobi matrix:

$$A_k = \frac{\partial f}{\partial x} \delta t + I = \begin{bmatrix} 1 & 0 & -\delta t \cdot V(k) \cdot \sin(\varphi(k)) \\ 0 & 1 & \delta t \cdot V(k) \cdot \cos(\varphi(k)) \\ 0 & 0 & 1 \end{bmatrix} \tag{11}$$

$$H_k = \frac{\partial h}{\partial x}(\hat{x}_{k,k-1}) = \begin{bmatrix} \frac{(x^S_{k+1} - Dx^S_{k,k+1} - x^M_k)}{r_k} & \frac{(y^S_{k+1} - Dx^S_{k,k+1} - y^M_k)}{r_k} & 0 \\ \frac{(x^S_{k+1} - x^M_{k+1})}{r_{k+1}} & \frac{(y^S_{k+1} - y^M_{k+1})}{r_{k+1}} & 0 \end{bmatrix} \tag{12}$$

where δt is the dead reckoning period from the UUV itself. Under ideal conditions, the one-step prediction of the moment system state can be obtained through KF recursion: t_k

$$\begin{aligned}
\hat{X}'_{k,k-1} &= A_{k-1}\hat{X}_{k-1} \\
&= A_{k-1}[\hat{X}'_{k-1,k-2} + K'_{k-1}(Z_{k-1} - H_{k-1}\hat{X}'_{k-1,k-2})] \\
&= A_{k-1}[(I - K'_{k-1}H_{k-1})\hat{X}'_{k-1,k-2} + K'_{k-1}Z_{k-1}] \\
&= A_{k-1}[(I - K'_{k-1}H_{k-1})A_{k-2}\hat{X}_{k-2} + K'_{k-1}Z_{k-1}] \\
&= \cdots = [\prod_{i=1}^{N-1} A_{k-i}(I - K'_{k-i}H_{k-i})]A_s[\hat{X}_{s,s-1} + K'_s(Z_s - H_s\hat{X}_{s,s-1})] \\
&\quad + \sum_{j=3}^{N}[(\prod_{i=1}^{j-2}A_{k-i}(I - K'_{k-i}H_{k-i}))A_{k-j+1}K'_{k-j+1}Z_{k-j+1}] + A_{k-1}K'_{k-1}Z_{k-1}
\end{aligned} \tag{13}$$

The t_{k-1} system measurement at the time t_s in the ideal communication situation. If the delay is considered, the master UUV t_s sends measurement information at time, and after N filtering cycles, it arrives at the slave UUV at t_k time, and the t_s time system measurement is not known, then the above formula is:

$$\begin{aligned}
&\hat{X}'_{k,k-1} \\
&= [\prod_{i=1}^{N-1} A_{k-i}(I - K_{k-i}H_{k-i})]A_s\hat{X}_{s,s-1}
\end{aligned}$$

$$+ \sum_{j=3}^{N} [(\prod_{i=1}^{j-2} A_{k-i}(I - K_{k-i}H_{k-i}))A_{k-j+1}K_{k-j+1}Z_{k-j+1}] + A_{k-1}K_{k-1}Z_{k-1} \quad (14)$$

$$[\prod_{i=1}^{N-1} A_{k-i}(I - K'_{k-i}H_{k-i})]A_s[\hat{X}_{s,s-1} + K'_s(Z_s - H_s\hat{X}_{s,s-1})]$$

$$- [\prod_{i=1}^{N-1} A_{k-i}(I - K_{k-i}H_{k-i})]A_s\hat{X}_{s,s-1}$$

$$+ \sum_{j=3}^{N} [(\prod_{i=1}^{j-2} A_{k-i}(I - K'_{k-i}H_{k-i}))A_{k-j+1}K'_{k-j+1}Z_{k-j+1}]$$

$$- \sum_{j=3}^{N} [(\prod_{i=1}^{j-2} A_{k-i}(I - K_{k-i}H_{k-i}))A_{k-j+1}K_{k-j+1}Z_{k-j+1}]$$

$$+ A_{k-1}K'_{k-1}Z_{k-1} - A_{k-1}K_{k-1}Z_{k-1} \quad (15)$$

In the above formula (14), K' and K are both filter gains, the former is the gain under ideal communication conditions, and the latter is the gain under delay. If $t_s - N$ H and R are estimated at the moment t_s, and then they are substituted into the filter equation as the relative value of the moment for iterateve calculation, the above formula can be simplified as

$$\delta\hat{X}_{k,k-1} = [\prod_{i=1}^{N-1} A_{k-i}(I - K_{k-i}H_{k-i})]A_sK_s(Z_s - H_s\hat{X}_{s,s-1})] \quad (16)$$

It can be seen from the above formula that $\delta\hat{X}_{k,k-1}$ includes the system state matrix A_{k-i} of N filtering cycles from time t_s to time t_k. In addition, it is recorded $\delta\hat{X}_s = K_s\left(Z_s - H_s\hat{X}_{s,s-1}\right)$ as the t_s time system state estimation error, which $\delta\hat{X}_s$ is also $\delta\hat{X}_{s,s-1}$ a part of the composition, then the compensation formula is:

$$X_{k,k-1} = X'_{k,k-1} + \delta\hat{X}_{k,k-1} \quad (17)$$

Therefore, the state compensation of the system at the moment t_k can effectively compensate the filtering error H_s introduced by the moment t_s and the filter error R_s, so as to compensate the error of the UUV cooperative navigation and positioning system considering the communication delay.

Considering the delay, the following uses EKF to give the specific algorithm flow of the improved state compensation-EKF, and the one-step prediction of the system state can be expressed as

$$\hat{X}_{k+1,k} = A_k\hat{X}_k \quad (18)$$

The one-step prediction covariance matrix is

$$P_{k+1,k} = A_kP_kA_k^T + G_kQ_kG_k^T \quad (19)$$

The state estimation error compensation amount is

$$M_k = A_k(I - K_kH_k)M_{k-1}$$
$$\delta\hat{X}_{k+1,k} = M_kA_sK_s(Z_s - H_s\hat{X}_{s,s-1}) \tag{20}$$

The filter gain matrix is

$$K_{k+1} = P_{k+1,k}H_k^T(H_kP_{k+1,k}H_k^T + R_k)^{-1} \tag{21}$$

The state of the system is estimated to be

$$\hat{X}_{k+1,k} = X'_{k+1,k} + \delta\hat{X}_{k+1,k}$$
$$\hat{X}_{k+1} = \hat{X}_{k+1,k} + K_{k+1}(Z_s - H_k\hat{X}_{s,s-1}) \tag{22}$$

The estimated mean square error covariance matrix can be expressed as

$$P_{k+1} = (I - K_{k+1}H_{k+1})P_{k+1,k} \tag{23}$$

The above is the delay problem of the system considering the non-ideal communication situation. Based on the core idea of state compensation improvement, it is mainly to substitute the predicted main UUV measurement information into the filtering calculation process from the UUV at the time t_s, and then from the UUV in the accurate measurement When the information arrives t_k, the system state estimation error is compensated.

4 Simulation Analysis

In order to verify the effectiveness of the DEKF method based on state compensation in the case of communication delay, the following algorithm simulation test is carried out. The simulation time is set to 1 h, the sampling interval T is 1 s, and the random error of underwater acoustic communication delay δt follows a normal distribution, with a mean of 0s and a variance of $(0.1 \text{ s})^2$.

The corresponding measurement noise settings during the simulation are as follows: the variance of the speed measurement error is $\delta_{u,v}^2 = (0.2 \text{ m/s})^2$, the variance of the heading angle measurement error is $\delta_{u,\varphi}^2 = (0.2°)^2$, and the variance of the relative distance measurement noise between the master and slave UUV is $\delta_{u,l}^2 = (2 \text{ m})^2$, all of which are zero-mean Gaussian white noise. The main UUV sails along the direction parallel to the y-axis at a speed of 2 kn. The slave UUV first makes a circular motion and then sails in a straight line. The initial heading angle is 90° and the speed is 3 kn. Both start from their respective starting points at the same time.

The simulation results are as follows.

Figure 3 shows the comparison of EKF and DEKF state estimation values. The green dotted line represents the estimated value of DEKF algorithm, the purple dotted line represents the estimated value of EKF algorithm, and the black dot represents the real value. It can be seen from the comparison that the state estimation value of DEKF is closer to the real state because of the state compensation, while the state estimation

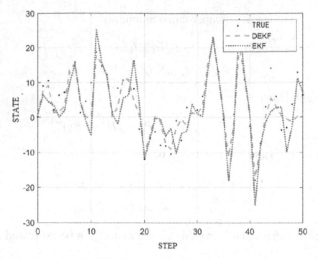

Fig. 3. State Estimates vs

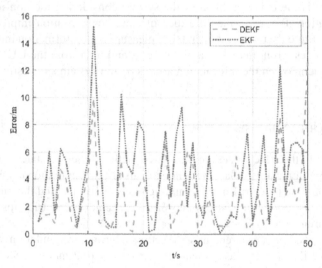

Fig. 4. Positioning error comparison

value of the EKF algorithm considering the communication delay is quite different from the real value.

Figure 4 shows the comparison of positioning errors between the two. The purple dotted line indicates the position error of the estimated path relative to the real path after the UUV corrects its own positioning error through the EKF algorithm under communication delay; the green dotted line indicates the corresponding path of the improved DEKF algorithm The position error of the estimate relative to the true path. It can be seen from the comparison that the EKF is affected by the communication delay, and there is a large deviation from the UUV path, the average positioning error is 8.3 m, the maximum positioning error exceeds 14 m, and the positioning effect is poor; compared with this, the DEKF path and the real The path offset is small, and the positioning error is basically within 5 m, which is better than EKF. At the same time, it can be seen from the variation range of the two positioning error curves that the error fluctuation range of the EKF estimated value is larger than that of the DEKF, and the DEKF has better stability.

5 Conclusion

In this paper, the master-slave structure of single pilot is determined as the research object, and the establishment of UUV motion model and measurement model is completed. During the modeling process, the mechanism of cooperative navigation and positioning is analyzed. Considering the delay constraints of underwater acoustic communication, the multi-UUV cooperative navigation and positioning communication delay model is analyzed, and the EKF cooperative positioning algorithm is improved based on state compensation, that is, the DEKF algorithm. The simulation analysis shows that the positioning accuracy of the DEKF algorithm considering the communication delay is improved compared with the EKF algorithm.

References

1. Wang, X., Chen, H., Chen, F.: An improved particle swarm optimization algorithm for unmanned aerial vehicle route planning. J. Phys. Conf. Ser. **2245**(1), 012013 (2022)
2. Wang, Z., Li, G., Ren, J.: Dynamic path planning for unmanned surface vehicle in complex offshore areas based on hybrid algorithm. Comput. Commun. **166**(17), 49–56 (2021)
3. He, C., et al.: Accelerating large-scale multiobjective optimization via problem reformulation. IEEE Trans. Evol. Comput. **23**(6), 949–961 (2019)
4. Li, P., Duan, H.B.: Path planning of unmanned aerial vehicle based on improved gravitational search algorithm. Sci. China Technol. Sci. **55**(10), 2712–2719 (2012)
5. Cao, H., Zhang, H., Liu, Z., et al.: UAV path planning based on improved particle swarm algorithm. In: 7th International Symposium on Mechatronics and Industrial Informatics (ISMII) (2021)
6. Xia, G., Han, Z., Zhao, B.: Local path planning for USV based on improved quantum particle swarm optimization. In: 2019 Chinese Automation Congress (CAC). IEEE (2019)
7. Hu, Z., Wang, Z., Yin, Y.: Research on 3D global path planning technology for UUV based on fusion algorithm. J. Phys. Conf. Ser. **1871**(1), 012128 (2021)
8. Lin, J., Pan, L.: Multiobjective trajectory optimization with a cutting and padding encoding strategy for single-UAV-assisted mobile edge computing system. Swarm Evol. Comput. **75**, 101163 (2022)
9. Hui, F., Liu, M., Xu, H.: Multi-target path planning for unmanned surface vessel based on adaptive hybrid particle swarm optimization. J. Huazhong Univ. Sci. Technol. (Nat. Sci. Ed.) **46**(6), 59–64 (2018)
10. Zhang, Y.G., Xie, W.J., Zhao, X.L.: Route planning of unmanned combat aerial vehicles based on improved particle swarm optimization algorithm. Mod. Comput. (2015)
11. Huang, P.Q., Wang, Y., Wang, K., Liu, Z.Z.: A bilevel optimization approach for joint offloading decision and resource allocation in cooperative mobile edge computing. IEEE Trans. Cybern. **50**(10), 4228–4241 (2019)

12. Zhen, X.U., Zhang, E., Chen, Q.: Rotary unmanned aerial vehicles path planning in rough terrain based on multi-objective particle swarm optimization. **31**(1), 12 (2020)
13. Zhou, Y., Wang, R.: An improved flower pollination algorithm for optimal unmanned undersea vehicle path planning problem. Int. J. Pattern Recognit. Artif. Intell. **30**(4), 1659010 (2016)
14. Bian, Q., Zhao, K., Wang, X., Xie, R.: System identification method for small unmanned helicopter based on improved particle swarm optimization. J. Bionic Eng. **13**(3), 504–514 (2016). https://doi.org/10.1016/S1672-6529(16)60323-2
15. Lin, J., He, C., Cheng, R.: Adaptive dropout for high-dimensional expensive multiobjective optimization. Complex Intell. Syst. **8**(1), 271–285 (2021). https://doi.org/10.1007/s40747-021-00362-5
16. Cheng, Z., Wang, E., Tang, Y., Wang, Y.: Real-time path planning strategy for UAV based on improved particle swarm optimization. J. Comput. **9**(1), 209–214 (2014)
17. Xin, J., Zhong, J., Li, S., et al.: Greedy mechanism based particle swarm optimization for path planning problem of an unmanned surface vehicle. Sensors **19**(21), 4620 (2019)

The Research of the Consistency Control Under the Condition of Time-Lag of Isomerism AUV Group Communication

Fan Ye[1], Heng Fang[2], and Hao Zhou[2(✉)]

[1] Chang Jiang Communication Administration, Wuhan 430014, China
[2] Wuhan University of Technology, Wuhan 430070, China
zhzmq@whut.edu.cn

Abstract. Concerning the issue of the formation control of the unmanned underwater vehicle (UUV). This thesis has done research on the algorithms of the Consistency Control under the Condition of Time-lag of Isomerism AUV Group Communication. Firstly, this thesis has done research on formation consistency control devices based on the idea of leading and following, under the condition of fixed topology. Under the connected condition of communication topology, the sufficient and necessary condition of consistency has been demonstrated. Based on Riccati equation, the methodology of the gain matrix of control protocols design has been illustrated in this thesis. On these bases, research on the design of the formation consistency control devices under the condition of communication delay of AUV has been carried out. The sufficient and adequate condition for the realization of AUV formation consistency has been discovered through the method of the consistency analysis and linear matrix inequality. Consequently, a stimulation experiments has been designed for the purpose of verification of the effectiveness and stability of the control algorithm mentioned in this thesis.

Keywords: Time-lag of communication · unmanned underwater vehicle · formation control · isomerism · leading and following

1 Introduction

Underwater vehicles (UUV) are mainly divided into Manned Vehicle (MV) and Unmanned Underwater Vehicle (UUV) [1]. UUV is usually divided into Remote Operation Vehicle (ROV) and Autonomous Underwater Vehicle (AUV) according to the existence of communication control cables between the vehicle and the mother ship [2]. The ROV is powered by a cable from the mother ship and operated remotely. The AUV has its own power, mainly relying on its own power for autonomous navigation [3]. The object to be studied in this paper is AUV, which can carry a variety of sensors to detect the target sea area and carry out missions such as seabed detection, mine detection, submarine detection and Marine hydro-logic environment exploration. AUV is an unmanned intelligent equipment platform for autonomous navigation and intelligent operation [4]. Because of its small size and flexible characteristics, it will play an increasingly important role in the future ocean exploration activities.

© The Author(s), under exclusive license to Springer Nature Singapore Pte Ltd. 2023
L. Pan et al. (Eds.): BIC-TA 2022, CCIS 1801, pp. 507–519, 2023.
https://doi.org/10.1007/978-981-99-1549-1_40

With the increasing demand for Marine tasks, a single AUV will be constrained by its own limitations, such as limited energy, limited task load capacity, low detection efficiency, poor redundancy, etc. Therefore, a single AUV can perform a limited range of tasks [5]. In order to make up for the deficiency of single AUV, multiple AUVs are usually selected to form a group and coordinate and stably execute tasks, which can improve the overall performance, reduce the task execution time, increase the redundancy of task data and the robustness of the system, avoid the failure of the entire task due to the failure of an AUV, and improve the success rate of the task.

AUV groups are mainly divided into isomeric and isomeric groups. Each AUV in isomorphic AUV group has the same type, and the form of the group is simple, which is convenient for research. Currently, most collaborative task research on AUV group is aimed at the same configuration group. However, in the actual task scenario, due to the different AUV manufacturers, their underlying technical structure, equipment use management, task load function and other differences, so the actual AUV groups are mostly heterogeneous groups. It is of important theoretical value and practical significance to conduct research on this type of groups. In the case of harsh hydro-logical environment and complicated exploration tasks in the exploration sea area, the collaborative exploration of heterogeneous AUV groups is highly intelligent and multi-functional and can complete the tasks that single AUV and homogeneous AUV groups cannot or are difficult to complete. Heterogeneous AUV groups are an inevitable trend in the AUV field.

Aiming at the formation control problem of heterogeneous unmanned underwater vehicles (UAVS) [6], the consistency control algorithm of second-order heterogeneous AUV groups and the consistency algorithm with time delay are studied in this paper. The main contents are as follows:

1) The system state of heterogeneous AUV is transformed into isomorphic form, and the consistency problem of heterogeneous AUV groups is solved by the consistency idea of isomorphic system.
2) The heterogeneous AUV group distributed formation control algorithm is implemented by combining the consistency algorithm with the leader-follower idea.
3) Considering the communication delay condition, the consistency control of heterogeneous AUV groups is realized.

2 AUV Modeling

The Cartesian coordinate system is used to study the motion of under actuated AUV. The coordinate system mainly includes fixed coordinate system and moving coordinate system, as shown in Fig. 1 [7]. The corresponding variables and symbolic meanings are shown in Table 1. The origin of the coordinates of the fixed coordinate system is located at a certain point on the sea level relative to the earth, where the axis points due north and the axis points due east, so the fixed coordinate system is also called the geodetic coordinate system or the northeast coordinate system [8]. The origin of the moving coordinate system coincides with the center of gravity of the AUV, where the axis is in the mid-vertical profile and points to the AUV bow. $x\ y$. The axis is perpendicular to the mid-longitudinal profile and points to the starboard side of the AUV.

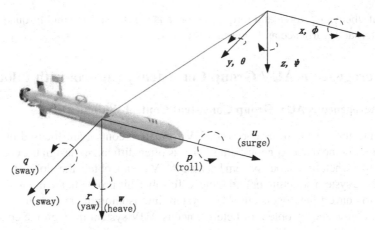

Fig. 1. Under-driven AUV model in inertial coordinate system and fixed coordinate system

Table 1. Model symbols

degree of freedom	position and euler angle	linear velocity and angular velocity	forces and moments
surge	x	u	X
pendulum	y	v	Y
heave	z	w	Z
roll	ϕ	p	K
pitch	θ	q	M
bow wave	ψ	r	N

The dynamic equation of AUV mainly considers the relationship between velocity and acceleration, that is, analyzes the relationship between the applied force and the motion state of AUV. The AUV studied in this paper is a rigid body with uniform mass distribution, and its shape structure can be approximated as left-right symmetry and up-down symmetry. The model of AUV is usually divided into two parts: kinematic model and dynamic model [9–12], the 6-DOF motion model of AUV can be expressed as follows,

$$\begin{cases} \dot{\eta} = J(\eta)v \\ M\dot{v} + C(v)v + D(v)v + g(\eta) = \tau + \omega \end{cases} \quad (1)$$

where $\eta \in \mathbb{R}^6$ is the space position and attitude of AUV in fixed Coordinate System. $v \in \mathbb{R}^6$ is the linear and angular velocity of AUV in moving coordinate system. $J(\eta)$ is the rotation transformation matrix from moving coordinate system to fixed coordinate system. M is the inertia matrix of the system (including the added mass). $C(v)$ is the Coriolis force matrix (including the added mass). $D(v)$ is the damping matrix, $g(\eta)$ is the gravity/buoyancy force and torque vector, τ is the thrust force and torque vector. ω is the external disturbance and internal disturbance.

3 Heterogeneous AUV Group Consistency Control with Pilot Craft

3.1 Heterogeneous AUV Group Consistent Control Without Delay

At present, most of the researches on AUV formation control are focused on the isomorphism. For the heterogeneous situation, it is often difficult to design the consistency protocol of the heterogeneous system because AUV cannot obtain the global information of the whole system structure or parameters. To solve this problem, the main idea of this paper is to convert heterogeneous AUV system into isomorphic system first, and then solve the consistency problem of heterogeneous AUV system through the consistency idea of isomorphic system.

Consider the equation:

$$B_e K_e^a = A_e \tag{2}$$

where

$$B_e = \begin{bmatrix} B_1 & -B_2 & & & \\ & B_2 & -B_3 & & \\ & & \ddots & \ddots & \\ & & & B_{N-1} & -B_N \end{bmatrix}, K_e^a = \begin{bmatrix} K_1^a \\ K_2^a \\ \vdots \\ K_N^a \end{bmatrix}, A_e = \begin{bmatrix} A_2 - A_1 \\ A_3 - A_2 \\ \vdots \\ A_N - A_{N-1} \end{bmatrix} \tag{3}$$

According to the AUV motion model, $A_i, B_i, i = 1, 2, \cdots, N$ have the same structural form:

$$A_i = \begin{bmatrix} \mathbf{0}_{6\times6} & \hat{J}_i \\ \mathbf{0}_{6\times6} & \hat{M}_i^A \end{bmatrix}$$

$$B_i = \begin{bmatrix} \mathbf{0}_{6\times6} \\ \hat{M}_i^B \end{bmatrix} \tag{4}$$

where $\hat{J}_i, \hat{M}_i^A, \hat{M}_i^B$ are full rank row matrices, then there must be $K_i^a, i = 1, 2, \cdots, N$ meeting the formula (3), hence,

$$A_1 + B_1 K_1^a = A_2 + B_2 K_2^a = \cdots = A_N + B_N K_N^a = A_f \tag{5}$$

And there must exist $K_i^b, i = 1, 2, \ldots, N, B_1 K_1^b = B_2 K_2^b = \ldots = B_N K_N^b = B_f$. Heterogeneous AUV formation consistency control protocol is designed, namely:

$$u_0(t) = K_0^a x_0(t)$$

$$u_i(t) = K_i^a x_i(t) + K_i^b a_{i0}(x_0(t) - x_i(t)) + K_i^b \sum_{j=1}^{N} a_{ij}(x_j(t) - x_i(t)) \tag{6}$$

where $i = 1, 2, \cdots, N$, K_i^a and K_i^b represents the gain matrix matched by dimension, $x_0(t)$ is the state of the pilot boat, $a_{i0} = 1$ demonstrates the communication which exists between the AUV and the pilot boat. $a_{i0} = 0$ demonstrates the communication which does not exist between the AUV and the pilot boat, a_{ij} is an element in the system adjacency matrix.

It is assumed that the following conditions are met during group movement:

Hypothesis 1: The communication topology between AUV groups includes spanning trees.

Hypothesis 2: All communication between follower boats is two-way.

By expressing the corresponding to the action topology, using L_α to express a corresponding Laplacian matrix with the action topology G_σ, since the pilot boat does not need to get information from the follower boat, hence, L_σ can be expressed as:

$$L_\sigma = \begin{bmatrix} 0 & \mathbf{0} \\ l_\sigma^{fl} & L_\sigma^{ff} \end{bmatrix} \tag{7}$$

where $l_\sigma^{fl} \in \mathbb{R}^N$ denotes the role of the pilot boat on the follower boat, $L_\sigma^{ff} \in \mathbb{R}^{N \times N}$ denotes the interaction between the follower boat.

Supposing

$$x(t) = \left[x_0^\mathrm{T}(t), x_1^\mathrm{T}(t), \cdots, x_N^\mathrm{T}(t) \right]^\mathrm{T} \tag{8}$$

under the action of protocol, the dynamic characteristics of AUV group system can be describe as,

$$\dot{x}(t) = (I_{N+1} \otimes A_f - L_\sigma \otimes B_f)x(t) \tag{9}$$

where I_{N+1} is the identity matrix of $(N+1) \times (N+1)$.

L_σ^{ff} is symmetric, then, there is an orthogonal matrix $\tilde{U}_\sigma \in \mathbb{R}^{N \times N}$, such that:

$$\tilde{U}_\sigma^T L_\sigma^{ff} \tilde{U}_\sigma = \Lambda_\sigma^{ff} = diag\left\{ \lambda_\sigma^1, \lambda_\sigma^2, \ldots, \lambda_\sigma^N \right\} \tag{10}$$

where $\lambda_\sigma^1 \leq \lambda_\sigma^2 \leq \cdots \leq \lambda_\sigma^N$ is the eigenvalue of L_σ^{ff}.

We can know that the active topology G_σ includes spanning trees, then we can know the structure of L_σ, $\lambda_\sigma^1 > 0$.

Supposing

$$U_\sigma = \begin{bmatrix} 1 & \mathbf{0} \\ 1_N & \tilde{U}_\sigma \end{bmatrix} \tag{11}$$

it yields,

$$U_\sigma^{-1} = \begin{bmatrix} 1 & \mathbf{0} \\ -\tilde{U}_\sigma^T 1_N & \tilde{U}_\sigma^T \end{bmatrix} \tag{12}$$

because

$$l_\sigma^{fl}+L_\sigma^{ff}1_N=0 \tag{13}$$

$$\left(U_\sigma^{-1}\otimes I_d\right)x(t)=\left[x_0^T(t),\,\zeta^T(t)\right]^T \tag{14}$$

System can be transformed into

$$\dot{x}_0(t)=A_f x_0(t) \tag{15}$$

$$\dot{\zeta}(t)=(I_N\otimes A_f-\Lambda_\sigma^{ff}\otimes B_f)\zeta(t) \tag{16}$$

3.2 Heterogeneous AUV Group Consistent Control with Time Delay

When there is a delay in information exchange between AUVs, the analysis method presented in the previous section cannot be used to ensure the consistency of the AUV group system. In this section, the consistency control protocol with time delay is designed according to the control protocol in the previous section, and the consistency stability analysis and linear matrix inequality criteria are given.

Under the time delay condition, consider the following heterogeneous AUV formation consistent control protocol with time delay, i.e.,

$$u_i(t)=K_i^a x_i(t)+K_i^b a_{i0}(x_0(t-\tau)-x_i(t-\tau))+K_i^b\sum_{j=1}^N a_{ij}(x_j(t-\tau)-x_i(t-\tau)) \tag{17}$$

where $i=1,2,\cdots,N$, K_i^a and K_i^b represents the gain matrix matched by dimension, $x_0(t)$ is the state of the pilot boat, $a_{i0}=1$ shows the communication between the AUV and the pilot boat, a_{ij} is the element in the system adjacency matrix, $0\le\tau\le\overline{\tau},\overline{\tau}$ is the known normal number.

Supposing

$$x(t)=\left[x_0^T(t),x_1^T(t),\cdots,x_N^T(t)\right]^T \tag{18}$$

the dynamic characteristics of AUV group system can be described as:

$$\dot{x}(t)=(I_{N+1}\otimes A_f)x(t)-(L_\sigma\otimes B_f)x(t-\tau) \tag{19}$$

According to the conversion method,

$$\left(U_\sigma^{-1}\otimes I_d\right)x(t)=\left[x_0^T(t),\,\xi^T(t)\right]^T \tag{20}$$

System can be converted to

$$\dot{x}_0(t)=A_f x_0(t) \tag{21}$$

$$\dot{\xi}(t)=(I_N\otimes A_f)\xi(t)-(\Lambda_\sigma^{ff}\otimes B_f)\xi(t-\tau) \tag{22}$$

4 Simulation Experiment

4.1 Time Delay Free Simulation

This section conducts simulation analysis for the consistency control algorithm proposed in 3. Set 1 virtual pilot boat and 4 follower boats to constitute heterogeneous AUV group, in which the virtual pilot boat is portable AUV, and the 4 follower boats are heterogeneous AUV in Sect. 2.3, No. 1 is portable, No. 2 is light, No. 3 is heavy, and No. 4 is giant. The AUV system parameters of each type are as follows:

1) Portable:

$$A_1 = \begin{bmatrix} \mathbf{0}_{6\times6} & diag\{1, 1, 1, 1, 1, 1\} \\ \mathbf{0}_{6\times6} & diag\{0.1668, 0.1991, 0.5766, 1.1637, 0.5294, 0.7492\} \end{bmatrix} \quad (23)$$

$$B_1 = \begin{bmatrix} \mathbf{0}_{6\times6} \\ diag\{-0.01236, -0.0087, -0.0087, -3.8790, -0.0771, -0.0772\} \end{bmatrix} \quad (24)$$

2) Light:

$$A_2 = \begin{bmatrix} \mathbf{0}_{6\times6} & diag\{1, 1, 1, 1, 1, 1\} \\ \mathbf{0}_{6\times6} & diag\{0.5253, 0.3690, 0.9395, 0.3457, 3.9322, 1.3020\} \end{bmatrix} \quad (25)$$

$$B_2 = \begin{bmatrix} \mathbf{0}_{6\times6} \\ diag\{-0.01162, -0.0068, -0.0070, -0.1070, -0.0150, -0.0043\} \end{bmatrix} \quad (26)$$

3) Heavy:

$$A_3 = \begin{bmatrix} \mathbf{0}_{6\times6} & diag\{1, 1, 1, 1, 1, 1\} \\ \mathbf{0}_{6\times6} & diag\{0.0882, 0.1532, 0.1364, 0.2740, 0.4463, 0.5041\} \end{bmatrix} \quad (27)$$

$$B_3 = \begin{bmatrix} \mathbf{0}_{6\times6} \\ diag\{-0.0007, -0.0004, -0.0004, -0.0271, -0.0005, -0.0002\} \end{bmatrix} \quad (28)$$

4) Huge:

$$A_4 = \begin{bmatrix} \mathbf{0}_{6\times6} & diag\{1, 1, 1, 1, 1, 1\} \\ \mathbf{0}_{6\times6} & diag\{0.1337, 0.1902, 0.1693, 0.0304, 0.4335, 0.4964\} \end{bmatrix} \quad (29)$$

$$B_4 = \begin{bmatrix} \mathbf{0}_{6\times6} \\ diag\{-0.0002, -0.0001, -0.0001, -0.0004, -0.0001, -0.0001\} \end{bmatrix} \quad (30)$$

It is assumed that the initial position of the pilot boat is randomly distributed in between [(0, 0), (10, 10)], and the initial position of each follower boat is randomly distributed in the interval of [(10, 10), (30, 30)], and the positions do not overlap. Assuming that the initial velocity is 5 m/s and the initial values of other state variables are set to 0. The communication topology is defined as an undirected connected graph. Figure 2 shows the topology.

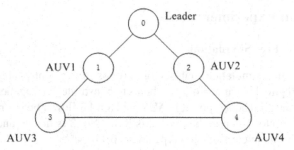

Fig. 2. Communication topology

The Laplacian matrix of the system is as follows:

$$L_\sigma = \begin{bmatrix} 0 & 0 & 0 & 0 & 0 \\ 1 & 2 & 0 & -1 & 0 \\ 1 & 0 & 2 & 0 & -1 \\ 0 & -1 & 0 & 2 & -1 \\ 0 & 0 & -1 & -1 & 2 \end{bmatrix} \tag{31}$$

The design formation is expected to be an equilateral triangle with the pilot boat as the center and four follower boats evenly distributed around it. The expected distance between each boat is 20 m. Simulation results can be obtained as follows, shown in Fig. 3 and Fig. 4.

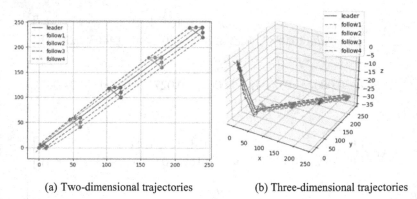

(a) Two-dimensional trajectories   (b) Three-dimensional trajectories

Fig. 3. Two-dimensional and three-dimensional graphs of motion trajectories of triangular formation

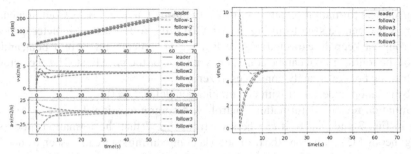

(a) x direction acceleration, velocity, position information (b) combined velocity information

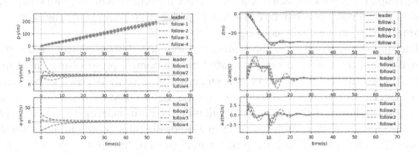

(c) Acceleration, velocity and position information in the y direction (d) Acceleration, velocity
and position information in the z direction

Fig. 4. Motion state information of triangular formation

As can be seen from Fig. 3, AUV groups maintain a stable triangular formation after
a period of time. In heterogeneous AUV group formation, the follower boat can sail at
the expected relative distance and Angle with the pilot boat, and as shown in Fig. 4.
The position of x, y and z, and speed of each follower boat in the formation converge
are inclined to reach the value of the pilot boat in approximately 30 s, and the value of
acceleration gradually converges to 0.

4.2 Simulation with Time Delay

AUV groups with time delay proposed in Sect. 3 is simulated and analyzed. The system
topology and initial conditions are the same as those in Sect. 3.3. The design formation
is expected to be an equilateral triangle with the pilot boat as the center and 4 followers
evenly distributed around it, 20 m apart from each other. When the heterogeneous AUV
group can sail according to the set triangular formation, it is said that the heterogeneous
AUV group reaches consistency.

For comparison, the formation control simulation experiments of heterogeneous
AUV groups with different time delays are designed below.

1) Simulation experiment with communication delay $\tau = 1.0$

Simulation results of formation control with communication delay, $\tau = 1.0$, are
shown in Fig. 5 and Fig. 6. As can be seen from Fig. 5, formation process of heterogeneous

AUV group and formation keeping sailing state can be seen from Fig. 6. Communication delay $\tau = 1.0$ can be seen from Fig. 6. After 20 s, the follower boat in heterogeneous AUV group can keep the expected relative distance movement with the pilot boat, and the speed of the follower boat can be consistent with that of the pilot boat. The acceleration of the trailing boat eventually converges to zero.

2) Simulation experiment with communication delay $\tau = 0.5$

Simulation results of formation control with communication delay, $\tau = 0.5$, are shown in Fig. 7 and Fig. 8. It can be seen from Fig. 7 that the formation process of heterogeneous AUV group and the formation keeping sailing state can be seen from Fig. 8 that the communication delay, $\tau = 0.5$. After 15 s, the follower boat in heterogeneous AUV group can keep the expected relative distance movement with the pilot boat, and the speed of the follower boat can be consistent with that of the pilot boat. The acceleration of the trailing boat eventually converges to zero.

The simulation results show the control performance of heterogeneous AUV groups under different time delay conditions based on the control protocol presented in this paper. The simulation results show that when the time delay decreases, the transient performance of the system will increase, and the steady-state error will also decrease. Under the condition of time delay, the following boats in heterogeneous AUV group can make use of the navigation state information of the pilot boats and the information between the lead boats to track the position and speed of the pilot boats more quickly. Meanwhile, they can also keep the position and speed consistent with other following boats, and finally form the desired formation and maintenance of the motion and their formation.

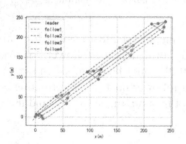
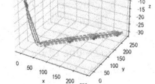

(a) Two-dimensional trajectories   (b) three-dimensional trajectories

Fig. 5. Path diagram of formation with delay 1.0 s

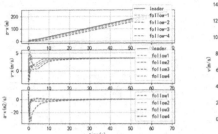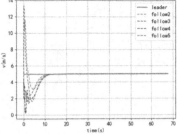

(a) acceleration, velocity, position information in the X direction (b) combined velocity information

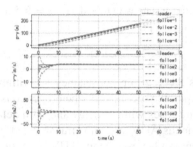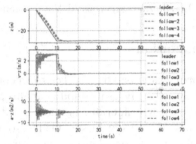

(c) Acceleration, velocity and position information in the y direction (d) Acceleration, velocity and position information in the z direction

Fig. 6. Heterogeneous AUV group consistency state with time delay ($\tau = 1.0$)

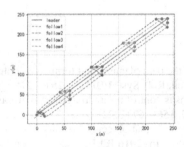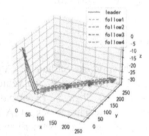

(a) Two-dimensional trajectories   (b) three-dimensional trajectories

Fig. 7. Path diagram of formation with time delay of 0.5 s

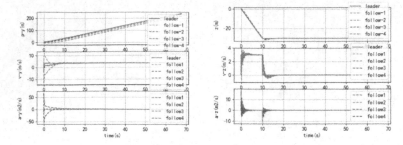

(c) Acceleration, velocity and position information in the y direction (d) Acceleration, velocity and position information in the z direction

Fig. 8. Heterogeneous AUV group consistency state with time delay ($\tau = 0.5$)

5 Conclusion

This paper mainly studies the heterogeneous AUV group consistency algorithm with time delay and analyzes its stability. In this paper, the second order heterogeneous AUV group consistency algorithm is studied, and the cases without delay and with delay are considered respectively. Firstly, the heterogeneous AUV system is transformed into an isomorphic system, and the consistency problem of heterogeneous AUV formation system is solved by using the consistency idea of isomorphic system. Then the consistency algorithm is analyzed based on the principle of linear matrix inequality and Lyapunov stability. Finally, simulation experiments are carried out respectively under the condition of time delay and no time delay, and the effectiveness of the proposed consistency control algorithm is verified.

References

1. Hadi, B., Khosravi, A., Sarhadi, P.: A review of the path planning and formation control for multiple autonomous underwater vehicles. J. Intell. Rob. Syst. **101**(4), 1–26 (2021)
2. Xie, M., Song, Y., Shen, S.: Event-based consensus control for multi-agent systems against joint sensor and actuator attacks. ISA Trans. **127**, 156–167 (2022)
3. Putranti, V., Ismail, Z.H., Namerikawa, T.: Robust-formation control of multi-autonomous underwater vehicles based on consensus algorithm. In: 2016 IEEE Conference on Control Applications (CCA), pp. 1250–1255, IEEE (2016)
4. Xie, X., Sheng, T., He, L.: Robust attitude consensus control for multiple spacecraft systems with unknown disturbances via variable structure control and adaptive sliding mode control. Adv. Space Res. **69**(3), 1588–1601 (2022)
5. Suleimanov, B.A., Rzayeva, S.J., Akberova, A.F., Akhmedova, U.T.: Self-foamed biosystem for deep reservoir conformance control. Pet. Sci. Technol. **40**(20), 2450–2467 (2022)
6. Lin, J., Pan, L.: Multiobjective trajectory optimization with a cutting and padding encoding strategy for single-UAV-assisted mobile edge computing system. Swarm Evol. Comput. **75**, 101163 (2022)
7. Wu, X., Jiang, D., Yun, J., Liu, X., et al.: Attitude stabilization control of autonomous underwater vehicle based on decoupling algorithm and PSO-ADRC. Front. Bioeng. Biotechnol. **10**, 843020 (2022)

8. Fossen, T.I.: Handbook of Marine Craft Hydrodynamics and Motion Control. Wiley, West Sussex (2011)
9. Ju, Y., Tian, X., Wei, G.: Fault tolerant consensus control of multi-agent systems under dynamic event-triggered mechanisms. ISA Trans. **127**, 178–187 (2022)
10. Prestero, T.T.J.: Verification of a six-degree of freedom simulation model for the REMUS autonomous underwater vehicle. Massachusetts Institute of Technology (2001)
11. Nekoo, S.R.: Tutorial and review on the state-dependent Riccati equation. J. Appl. Nonlinear Dyn. **8**(2), 109–166 (2019)
12. Zhang, X.M., Min, W., She, J.H., et al.: Delay-dependent stabilization of linear systems with time-varying state and input delays. Automatica **41**(8), 1405–1412 (2005)

Non-uniform Deformation Behavior of Magneto-Sensitive Elastomer Containing Uniform Sphere Particles with V-Shaped Arrangement

Tu Yu[1], Zhenkai Liang[2], Hanyang Zhang[2], Haiyu Zhang[3(✉)], Wei Gao[4], and Wentao Han[5]

[1] China Ship Development and Design Center, Wuhan 430064, China
[2] Hubei Key Laboratory of Theory and Application of Advanced Materials Mechanics, School of Science, Wuhan University of Technology, Wuhan 430070, China
[3] Green and Smart River-Sea-Going Ship, Cruise and Yacht Research Center, Wuhan University of Technology, Wuhan 430070, China
zhanghaiyu@whut.edu.cn
[4] School of Science, Lanzhou University of Technology, Lanzhou 730050, China
[5] College of Civil Engineering and Mechanics, Lanzhou University, Lanzhou 730000, China

Abstract. Magneto-sensitive elastomers can deform under the action of magnetic field, and this deformation can achieve specific deformation function through microstructure design. Since it is reported that the soft magnetic control robot made of magneto-sensitive elastomers can complete eighteen deformation modes, there are more and more research reports in this field in recent years. In order to explore the deformation rules of the Magneto-sensitive elastomers with specific arrangement of particles, spherical particles are used as the reinforcement phase in this research, and magneto-sensitive elastomers containing uniform sphere particles with kinds of specific arrangement are manufactured. The bending deformation behavior of this type of magneto-sensitive elastomer is studied through ANYS. The influence of microstructures on the magneto induced deformation behavior of magneto sensitive elastomers is mainly revealed.

Keywords: Magneto-sensitive elastomers · Bending deformation · Specific arrangement · Sphere particles

1 Introduction

Magneto-sensitive elastomers is a kind of flexible composite material composed of magnetizable particles added to the soft material matrix. Due to magnetization of particles under magnetic field, magnetic force will occur among particles. Through the stress transmission of flexible large deformation matrix, it can be significantly deformed under the action of magnetic field [1–3]. A large number of studies have found that this deformation can achieve specific deformation functions through microstructure design, such as one kind of the soft magnetic control robot made of magnetoelastic system can complete 18 deformation modes [4].

Microstructural design of magneto-sensitive elastomers is widely concerned in soft magnetic control robot design. Inspired by the response of natural petals to light stimulation, Wei Gao et al. studied the magnetic properties of the bionic magnetic sensitive intelligent gripper based on the magneto-sensitive elastomers with soft magnetic particles [5]. Driven by the external magnetic field, the gripper can grasp objects 4 times its own weight, and can work in complex environments such as acid, salt and alkali. For magneto-sensitive elastomers with programmable magnetic domain distribution, Miao Yu's team of Chongqing University has studied the influence of magnetic moment arrangement in soft matrix on the magnetic induced motion characteristics of bionic structures (foot toad, bat and soft claw) [6]. The results show that bionic structures with different magnetic moment distribution can realize various motion modes, such as walking, swimming, grasping, etc. Hu et al. studied the micro magnetic domain distribution inside the soft robot and the influence of the external magnetic field on its magneto-elastic properties for the magnetic anisotropy millimeter soft robot, and found that the soft robot can achieve the switching of deformation modes under the control of different magnetic fields, and can achieve complex actions such as crawling, jumping obstacles, and carrying in air, water, narrow pipes and other environments, The mechanism of human magnetic action of the soft machine is explained by theoretical model [4]. The micro-specific magnetic domain distribution can greatly improve the magnetic deformation characteristics and magneto-elastic properties of magnetic sensitive intelligent soft materials and expand their functional application fields. Xuanhe Zhao team of MIT studied the influence of specific magnetic domain distribution on the magnetic deformation of magneto-sensitive elastomers films [7]. The complex magnetic domain distribution can enable simple film structures to achieve complex three-dimensional deformation and functional applications. In addition, the team studied the omnidirectional steering and magnetic navigation of the soft robot under the cooperation of force, magnetic and affected part through the submillimeter scale self-lubricating flexible magnetic sensing continuum robot, which opens up a new way for remote operation and medical minimally invasive [8].

According to magneto-elastic behavior of magneto-sensitive elastomers, its magnetic deformation characteristics are continuously improved through the preparation process. In addition, corresponding models are established for the deformation characteristics of force magnetic coupling of different magnetism and structures, and the macroscopic force magnetic coupling deformation behavior is explained from the internal coupling mechanism of materials. At present, the theoretical research on the force magnetic coupling behavior is mainly based on the continuum medium model and the magnetic dipoles model [9–13].

However, theoretical model can only reveal the magnetoelastic interaction between two particles, or particles with uniform distribution. These methods cannot reveal the magnetoelastic behavior of magneto-sensitive elastomer with complex particle distribution, especially non-uniform particle distribution. This requires further systematic research by means of finite element analysis. In the finite element analysis of magnetoelastic behavior of magneto-sensitive elastomers, the magnetic dipole model is usually used to analyze the interaction between discrete magnets, and some simulations also use Maxwell stress to analyze.

In this paper, the performance of a simple robot made of spherical particles with V-shaped specific arrangement is studied through experiments, and its internal mechanism is briefly analyzed by means of finite element analysis. The influence of matrix modulus and the angle between chains in the V-shaped particle arrangement on the degeneration behavior of the robot has been focused on. The influence rules of some microstructures have been revealed.

2 Microstructure Selection

In order to select the appropriate bending deformation configuration, a variety of magnetic sensitive elastomers with specific non-uniform and regular particles arrangement have been discussed, containing triangular arrangement, "V" arrangement, "X" arrangement, "Y" arrangement, "H" arrangement. Among them, "V" arrangement is a widely used microstructure in the soft robot, such as reported by Hu [4]. By using this design idea for Hu, short chains of particles are used to instead of ellipsoids in 2D's form. The microstructures are shown in Fig. 1.

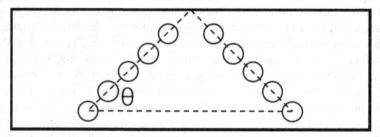

Fig. 1. Magento-sensitive elastomers containing uniform sphere particles with "V" arrangement

3 Preparation

PVC is used to make silica gel curing mold and punching mold respectively. The punching die can adjust its diameter and the inclination angle of the short chain according to the diameter of the filled particles, as shown in Fig. 2. After the matrix sample is solidified and molded, the particles are filled in the reserved channel of the matrix, and both ends are sealed with silica gel.

Finally, six experiment samples of magneto-sensitive elastomers were prepared by using silica gel with two different moduli (1.4 MPa and 2.1 MPa) as the matrix and three angular distributions (15°, 30° and 45°). The order of "A–F" from the dimension is shown in Fig. 4. Each specimen is 100 mm long, 20 mm wide and 20 mm thick. And the micro-parameters of these samples are shown in Table 1 and (Figs. 3 and 5).

Fig. 2. Curing mold and punching mold for silica matrix

Fig. 3. Bending deformation of magneto-sensitive elastomer under the action of pure gravity

Table 1. Micro-parameters of experiment samples of magneto-sensitive elastomers.

Sample No	A	B	C	D	E	F
Particle diameter	1mm	1mm	1mm	1mm	1mm	1mm
Gap between particles	1mm	1mm	1mm	1mm	1mm	1mm
Initial relative permeability of particles	1580	1580	1580	1580	1580	1580
Initial modulus of matrix	1.4MPa	2.1MPa	1.4MPa	2.1MPa	1.4MPa	2.1MPa
Intersection angle θ	15°	15°	30°	30°	45°	45°

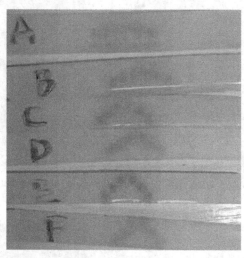

Fig. 4. Magneto-sensitive elastomers samples containing different microstructures

4 Experimenting

4.1 Instrument and Method

Parallel magnetic field device is used to generate a parallel magnetic field, which is made by parallel opposite poles electromagnet. This device is provided by solid mechanics of electromagnetic group of Lanzhou University. The device can generate a 200 mm parallel magnetic field area, meeting the requirements of magnetic field generation for magneto-sensitive elastomer experimenting.

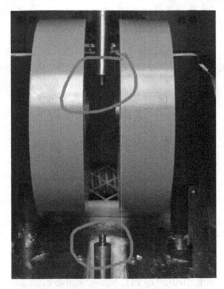

Fig. 5. Parallel magnetic field device

Fig. 6. Camps for magneto-sensitive elastomers samples.

Because the behavior of the flexible robot is to be imitated, there is no strong constraint on the two ends of the magneto sensitive elastomer, but only two clamps are used to control the magneto sensitive elastomer sample not to leave the test area and reflect its deformation characteristics. The clamps for magneto-sensitive elastomers samples are shown in Fig. 6.

The experiment effect is shown in Fig. 7. The end with screw controls the magnetic sensitive elastomer not to deviate from the area, and the section without screw can slip within a certain constraint range. As the deformation is relatively obvious, the study did not use high-precision infrared imager but used an external high-definition camera to record this behavior.

Fig. 7. Constraint for deformation of magneto-sensitive elastomers samples and effect.

4.2 Results

The bending deformation of the magnetic sensitive elastomer can be reflected by recording the displacement of the midpoint of the experiment simples of magneto-sensitive elastomer.

Due to the asymmetry of the particle distribution in the body of the manually prepared magnetic sensitive elastomer, different samples will twist when loaded to a certain magnetic field strength. The bending data of the samples before torsion are recorded in the experiment.

The experimental data of pure bending of six samples under the action of magnetic field are shown in Fig. 8. It can be seen from all the figures that with the increase of the magnetic field, position of the midpoint of the magnetic sensitive elastomer changes, indicating that the magnetic sensitive elastomer undergoes obvious bending deformation. The displacement increments of the magnetic sensitive elastomers (sample A, C, E) prepared on the soft substrate is smaller when the magnetic field just increased, and the displacement becomes smaller with the increase of the magnetic field; However, the displacement rules of magneto-sensitive elastomers (samples B, D, F) prepared from harder and softer substrates are different. This is because the modulus of silica gel will increase with the increase of deformation. When the deformation reaches the critical value of torsion, the shaking of the sample will make the displacement increment fluctuate, which is more obvious in the sample with a harder matrix. However, it is difficult to further reveal this phenomenon by using the finite element analysis method.

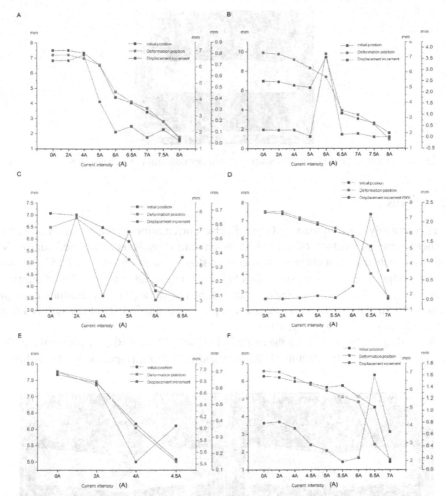

Fig. 8. Deformation of six simples under the strength of electrically controlled magnetic field: (A) Simple A; (B) Simple B; (C) Simple C; (D) Simple D; (E) Simple E; (F) Simple F.

5 Simulation

5.1 Verification

To verify some laws obtained in the experiments, ANSYS finite element analysis software is used to systematically simulate the law. The simulation work completed by ANSYS R15 is used here.

In order to verify the correctness of the model setting, two short chains are completely placed horizontally to form a chain, as shown in Fig. 9, and the deformation of the magnetic sensitive elastomer sample in this case is obtained. The results show that the deformation is basically consistent with the laws reported in many articles. This shows the rationality and accuracy of numerical modeling.

528   T. Yu et al.

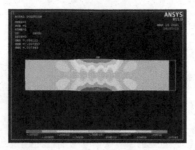

Fig. 9. Deformation of magneto-sensitive elastomer containing particles with single chain

5.2 Results

In order to analyze the influence of particle chain orientation on the deformation law of magneto-sensitive elastomers, finite element simulation analysis was carried out for "V" type particle distribution magneto-sensitive elastomers with dip angles of 15°, 30°, 40° and 50°, respectively. In addition to the chain inclination, and the modulus of matrix is set by 1.4 MPa, the other micro parameters of the magnetic sensitive elastomer sample can refer to the data of the experimental samples. The magnetic field strength is 1A/m. The results are as shown in Fig. 10.

According to the data, when the angle is 15°, 30°, 40° and 50°, the maximum deformation is 0.093 mm, 0.139 mm, 0.091 mm and 0.055 mm respectively. According to

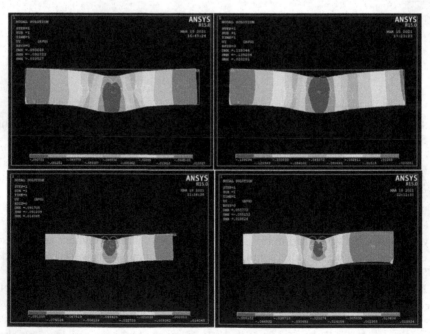

Fig. 10. Deformation of magneto-sensitive elastomer containing particles with different angles of the chain: (A) 15°; (B) 30°; (C) 40°; (D) 50°

the data, when the angle is 30 degrees, the displacement is maximum. This shows that when the included angle is 30 degrees, the magnetic control bending performance of the magnetic sensitive elastomer reaches the best state.

6 Conclusion

In this paper, the magnetically controlled denaturation behavior of magnetically sensitive elastomers with "V" particle arrangement is analyzed. Two conclusions have been obtained as follow:

1. The modulus of the matrix is one of the important factors that affect the magnetic deformation of the magnetic sensitive elastomer. The magnitude of the matrix modulus will directly affect the magnetic field critical value of the pure bending of the magnetic sensitive elastomer. When the magnetic field is close to the critical value that makes the magneto sensitive elastomer bend, the deformation enhancement of the magneto sensitive elastomer will have an obvious wave behavior.
2. The inclination angle of the particle chain is another microstructural factor that affects the deformation behavior of magneto-sensitive elastomers with "V" particle arrangement. It is not that the larger or smaller the dip angle is, the better, but there is an optimal value, which is very close to the angle where the gravitational interaction between particles is zero. This rule needs further verification.

References

1. Ivaneyko, D., Toshchevikov, V., et al.: Mechanical properties of magneto-sensitive elastomers: unification of the continuum-mechanics and microscopic theoretical approaches. Soft Matter **10**, 2213–2225 (2014)
2. Ivaneyko, D., Toshchevikov, V., et al.: Motion behaviour of magneto-sensitive elastomers controlled by an external magnetic field for sensor applications. Macromol. Symp. **338**(1), 96–107 (2014)
3. Ivaneyko, D., Toshchevikov, V., Saphiannikova, M., Heinrich, G.: Mechanical properties of magneto-sensitive elastomers in a homogeneous magnetic field: theory and experiment. J. Magn. Magn. Mater. **431**, 262–265 (2017)
4. Hu, W., Lum, G.Z., Mastrangeli, M., Sitti, M.: Small-scale soft-bodied robot with multimodal locomotion. Nature **554**(7690), 81–85 (2018)
5. Gao, W., Wang, L.L., et al.: Magnetic Driving Flowerlike Soft platform: biomimetic fabrication and external regulation. ACS Appl. Mater. Interfaces. **8**(22), 14182–14189 (2016)
6. Qi, S., Guo, H., et al.: 3D printed shape-programmable magneto-active soft matter for biomimetic applications. Compos. Sci. Technol. **188**, 107973 (2020)
7. Kim, Y., Yuk, H., et al.: Printing ferromagnetic domains for untethered fast-transforming soft materials. Nature **558**(7709), 274–279 (2018)
8. Kim, Y., Parada, G. A., et al.: Ferromagnetic soft continuum robots. Science Robotics, **4**(33), eaax7329 (2019)
9. Brigadnov, I., Dorfmann, A.: Mathematical modeling of magneto-sensitive elastomers. Int. J. Solids Struct. **40**(18), 4659–4674 (2003)

10. Borcea, L., Bruno, O.: On the magneto-elastic properties of elastomer-ferromagnet composites. J. Mech. Phys. Solids **49**(12), 2877–2919 (2001)
11. Davis, L.C.: Model of magnetorheological elastomers. J. Appl. Phys. **85**(6), 3348–3351 (1999)
12. Chen, L., Gong, X.L., Li, W.H.: Microstructures and viscoelastic properties of anisotropic magnetorheological elastomers. Smart Mater. Struct. **16**(6), 2645 (2007)
13. Ivaneyko, D., Toshchevikov, V.P., et al.: Magneto-sensitive elastomers in a homogeneous magnetic field: a regular rectangular lattice model. Macromol. Theory Simul. **20**(6), 411–424 (2011)

Study on the Influence of Exit Width Change on Heterogeneous Passengers Evacuation Based on the Social Force Model

Guangzhao Yang[1], Wei Cai[1], Min Hu[1(✉)] (iD), Cheng Li[2], and Donghua Pan[3]

[1] Green and Smart River-Sea-Going Ship, Cruise Ship and Yacht Research Center, Wuhan University of Technology, Wuhan, China
hu_min@whut.edu.cn
[2] School of Naval Architecture, Ocean and Energy Power Engineering, Wuhan University of Technology, Wuhan, China
[3] Jiangsu Automation Research Institute, Lianyungang, China

Abstract. Since the restaurant on cruise ships are filled with large numbers of passengers at mealtimes, the crowd in high-density may take the risk when they evacuate from the restaurant in emergent cases. The size of exits usually has a certain impact on the efficiency of passenger evacuation. It is necessary to study the correlation between the sizes of exits and evacuation efficiency, which could help to reasonably design the sizes of exits to ensure the safety of passengers. The AnyLogic software is used to establish the spatial model of the restaurant, and the sizes of restaurant exits are set as the independent variable. Then the evacuation performance of heterogeneous passengers under different exit sizes is simulated. The simulation results show that the evacuation time increases as the width of exits decreases, and the evacuation time of passengers is highly sensitive to the width of exits when the width is less than 3m.

Keywords: Cruise Ship Evacuation · Exit Width · Heterogeneous Passengers · Simulation

1 Introduction

There are many large public indoor spaces on a cruise ship such as large restaurants and theaters. Since these public spaces are usually large and enclosed, it is difficult for passengers to evacuate within a short time when emergency occurs. The key problem to be solved is how to ensure a safe and quick evacuation from the accident site.

Many scholars have studied on the subject of evacuation from large spaces through simulation, experiment and investigation. Wu et al. [1] aimed at the evacuation of large sports stadiums, they established a crowd route selection model based on cellular automata which showed that people's route selection was affected by the number and position of the alarms in the stadium. And the evacuation performance was also greatly influenced by the building's exit width and the pedestrians' walking speed and

L. Pan et al. (Eds.): BIC-TA 2022, CCIS 1801, pp. 531–539, 2023.
https://doi.org/10.1007/978-981-99-1549-1_42

respond capacity [2]. Cotfas et al. [3] built an Agent-based evacuation model with Net-Logo and proposed a self-adaptive exit-finding method which can help people in any position automatically find the nearest exit. Kurdi et al. [4] tried to solve the congestion problem in evacuating from multiple-exit facilities and came up with an evacuation balance algorithm, which optimizes the evacuation process and reduces the evacuation time by evenly distributing pedestrians among all exits.

Delcea et al. [5] considered the connection between room exit size and the evacuation time and discussed the influence of exit symmetry on evacuation efficiency. Son et al. [6] found that the pedestrians' evacuation time significantly reduced when the door opened outwards after analyzing their door-opening patterns when leaving the room. Kinateder et al. [7] conducted some crowd evacuation experiments with virtual reality equipment, the experiments' result showed that pedestrians were more inclined to follow the mainstream crowd's direction, and the more people moved to an exit, the exit became the more attractive to people in evacuation. Wang et al. [8] studied the bottleneck phenomenon in the evacuation process and found that the pedestrians' evacuation speed significantly increased with the increase of the exit width after analyzing the relationship between the spacing between pedestrians and the width of the room exit. Xiao et al. [9] built an emergency evacuation model considering multiple exits based on cellular automata, which considered the distance between pedestrian position and the exit and the pedestrian density near the exit. They found that the pedestrian evacuation time increased with the increase of the crowd density, and the evacuation efficiency was influenced by the exits' position. Song et al. [10]'s experiment described the pedestrians with the three-circle model and simulated pedestrians' steering behavior with active rotational torque (ART). Its result showed that his model provided precise prediction of the evacuation time of pedestrians at the exit. They also put forward a distribution strategy based on nonlinear exit capacities which can calculate more accurate queue time required by the crowd at the exits and balance the crowd numbers flowing to each exit, so as to improve the evacuation efficiency [11]. Li et al. [12] presented an improved pedestrian evacuation model for large venues with multiple exits, which took into account the influence of people's emotional contagion. They found that the leaders of small groups took different movement patterns from the others, they also found that the evacuation time was largely influenced by the width, numbers and location of the doors in the venue.

This paper focuses on large indoor spaces on cruise ships, studies the relationship between exit width and passenger evacuation efficiency by changing the exit width in the space to be evacuated. The rest of this paper is as follows. Section 2 is the description of the research scene. The evacuation process is simulated and the results are analyzed in Sect. 3. The last section presents the conclusions.

2 Scene

2.1 The Evacuation Space

This paper takes the restaurant on deck 3 of Vista cruise ship as the research object. The restaurant is 40 m long and 36 m wide and is typical as a large indoor space. The two exits are on the left and right sides of the restaurant, as shown in Fig. 1. There are lots of tables in the restaurant for the use of the passengers, and space between the tables constitute channels through where passengers move inside the restaurant and towards the exits.

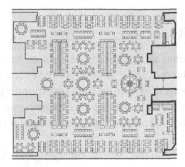

Fig. 1. The layout of the restaurant

2.2 Passenger Parameters

Table 1. The percentage and speed of different types of passengers

Passenger		Percentage (%)	Speed(m/s)	
Type	Characteristics		Minimum	Maximum
1	Women under 30	7	0.93	1.55
2	Women from 30 to 50	7	0.71	1.19
3	Women over 50	16	0.56	0.94
4	Women over 50, movement confined (1)	10	0.43	0.71
5	Women over 50, movement confined (2)	10	0.37	0.61
6	Men under 30	7	1.11	1.85
7	Men from 30 to 50	7	0.97	1.62
8	Men over 50	16	0.84	1.4
9	Men over 50, movement confined (1)	10	0.64	1.06
10	Men over 50, movement confined (2)	10	0.55	0.91

The ages of passengers on cruise ships are within a certain range, the evacuation process is greatly influenced by the passengers' age ranges because people's walking speeds vary in different ages and genders. In order to accurately simulate the passengers' evacuation process, it is necessary to set the corresponding speed for different types of passengers. Table 1 shows the passengers' composition types and expected speeds in use of evacuation analysis from Guidelines for Evacuation Analysis for New and Existing Passenger Ships [13] which was issued by the International Maritime Organization (IMO). As shown in Table 1, the passengers are divided into 10 types according to their ages and genders, and each type has its certain percentage of population and walking speed. In this model, the characteristics of heterogeneous passengers evacuation are analyzed by setting different parameters for different types of passengers.

2.3 Evacuation Procedure

The furniture in the restaurant affects the movement of passengers, and the widths of the exits on both sides of the restaurant also affect the evacuation of passengers. In this paper, the exit width is changed by setting the corresponding step length, so as to find the pattern between the exit width and the passengers evacuation time. The detailed procedures are shown in Fig. 2. The main procedures are described below.

(1) The step length is input into the evacuation scene and the exit width is determined according to the change of the step length.
(2) Passengers are loaded into the model.
(3) When loading the passengers, the passengers' attributes are initialized according to their parameters, including the population proportion, the speed and the starting position of each type of passengers.
(4) After all passengers are loaded into the space to evacuate, they enter the way-finding mode and start moving to the nearest exit.
(5) The evacuation process ends when the last passenger goes out of the exit.

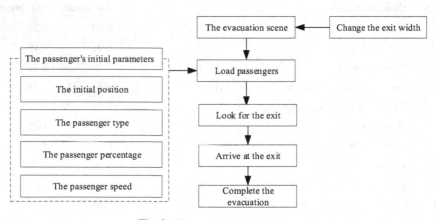

Fig. 2. The evacuation procedure

3 Evacuation Simulation and Result Analysis

3.1 Establishment of the Evacuation Model

Since the pedestrian evacuation process is dynamic, it is common for researchers to study it by establishing evacuation models which help simulate the pedestrians' behaviors.

The Social Force Model is based on Newtonian mechanics and assumes that pedestrians are driven by social forces. The model assumes that pedestrians are affected by three forces, namely the driving force, interactions between pedestrians and interactions with boundaries, and it regards the pedestrian as a particle following the laws of mechanical motion, using the force vector to describe the pedestrian's intrinsic motivation and the force been applied. The AnyLogic software based on social force model has been used by many scholars to conduct research on crowd evacuation [14, 15]. In this paper, AnyLogic is used to establish an evacuation scenario, and specific functions are written to set the relevant operation parameters.

Figure 3 shows the spatial model of the restaurant built with AnyLogic. Some fixed boundaries are formed through the simulation of the furniture's outline and the walls with AnyLogic's Wall module. The passengers in the model are not supposed to enter these boundaries as a simulation of the passengers in real world to bypass the furniture during evacuation. The exit widths on both sides are set as variables with the AnyLogic's parameter change function. In Fig. 3, the green modules on both sides represent the exits of the restaurant, and the red modules represent the walls controlling the width of the exit. Taking the left exit as an example, the width-controlling modules on both sides of the exit are ExitWall_LeftUp and Exitwall_LeftDown. Before each iteration operation, we move both ends of ExitWall_LeftUp and ExitWall_LeftDown 0.05 m towards the exit, making the width of the exit 0.1 m narrower than the previous operation. The initial widths of the exits on both sides are 5 m, and the iterative operations stop when the widths become less than 1m, then the calculation results will be output.

There are 750 passengers to be evacuated. In the initial stage of the evacuation, all passengers sit at the tables as a simulation of taking meals. When the evacuation begins, all passengers will move to the nearest exit. When all passengers evacuate from the restaurant, an evacuation simulation is complete. Then the process based on Any-Logic's parameter change function will change the widths of the exits automatically, starting another simulation again. And the simulations stop only when the iteration stop conditions are met, thus completing the whole simulation iteration process.

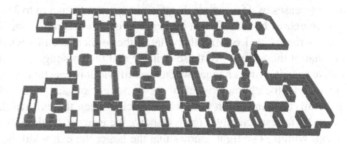

Fig. 3. The layout of the restaurant

3.2 The Result Analysis

Figure 4 is a snapshot of the passenger evacuation simulation, in which heterogeneous passengers are marked with different colors. As shown in the figure, all passengers are moving to the exits on both sides for evacuation.

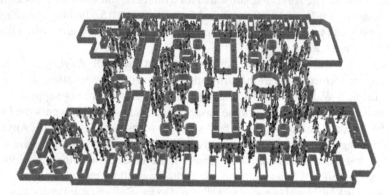

Fig. 4. The evacuation process of the passengers

Figure 5 shows the iteration diagram of evacuation simulation when the exit width changes. The figure shows that the passengers evacuation time gradually increases with the decrease of the exit width. And in some parts of the curve, the evacuation time slightly decreases with the decrease of the exit width. But they are reasonable minor fluctuations because the simulation targets on high-density crowd and is based on the social force model which reflects the interaction between people.

In addition, the figure shows that the slope of the curve is smaller when the abscissa is less than 2, and that phenomenon means that when the exit width is between 3 m and 5 m, its changes have lesser impact on the evacuation time. By contrast, the slope of the curve increases sharply when the exit width becomes less than 3 m. To be specific, the evacuation time increases by 156 s when the exit width drops from 5 m to 3 m, while the evacuation time increases by 244 s when the width drops from 3 m to 2 m. The latter's evacuation time increases by 1.6 times than that of the former when its width decrease is just a half of that of the former. All these indicate that the passenger evacuation time is highly sensitive to exit width changes when the exit width is less than 3 m.

In order to analyze the characteristics of the passenger's evacuation with different exit widths in detail, the evacuation results are selected when the exit widths are 5 m, 4 m, 3 m, 2 m and 1 m respectively. Figure 6 shows the evacuation time of passengers with different exit widths. The figure shows that the larger the exit width is, the higher the slope of the curve is, which means that the passengers can reach the exits in a shorter time through broader exits. In addition, the evacuation time is linear with the population size except at the end of the curve. It indicates that the passengers leave the restaurant at a rather constant speed. In the later stage of evacuation, as fewer passengers remain in the restaurant, fewer people leave per unit time, the curvature of the curve becomes smaller.

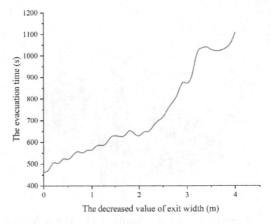

Fig. 5. The iteration diagram of the model simulation

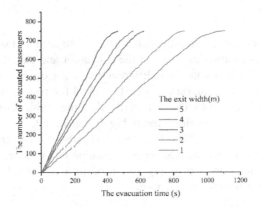

Fig. 6. The evacuation time with different exit widths

Figure 7 shows the distribution of evacuation time for heterogeneous passengers, which reflects passenger types and the relationship between exit widths and average evacuation time. The figure illustrates that the average evacuation times vary in different passenger types, among which the type 1, 2, 6, and 7 passengers take less time to evacuate than others. Those four types take shorter time to evacuate because they move at faster speeds which are listed in Table 1. The rest six types take relatively longer time to evacuate. Type 3 and type 7 passengers move faster than type 4 and type 8 passengers but take more time than the latter. The reason why type 3 and type 7 passengers take more time is that they suffer more congestions and obstacles for they are the most in population numbers according to Table 1. From the comparison of the results of different exit widths, it is found that the same type of passengers' evacuation times differs lesser when the width is 3 m, 4 m and 5 m, and that the evacuation times increase largely when the width is 1 m and 2 m. As shown in Fig. 7, each type of passengers' average evacuation time with the exit width of 2 m is remarkably longer than its time with the exit width of 3 m.

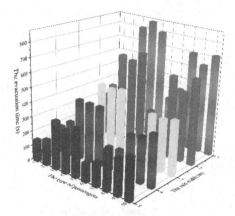

Fig. 7. The average evacuation time distribution for heterogeneous passengers

4 Conclusions

This paper takes the cruise restaurant which is a typical large indoor space as the research object, studies the evacuation capacity of the restaurant and calculates the evacuation time of passengers who evacuate from different width of exits by setting the widths of exits on both sides of the restaurant as variables. And the results show that passenger evacuation is less influenced by the exit width when the width is over 3 m. They also show that the passenger evacuation time is highly sensitive to the width of exit when the exit width is less than 3 m. Therefore, from the view of a constructor, although the larger the exits are, the easier the passenger evacuation is, oversize exits may affect the overall layout and other aspects, resulting a waste of space. It is also harmful to passenger security when the exit widths are too narrow.

References

1. Wu, Y., Kang, J., Wang, C.: A crowd route choice evacuation model in large indoor building spaces. Front. Archit. Res. **7**(2), 135–150 (2018). https://doi.org/10.1016/j.foar.2018.03.003
2. Wu, Y., Kang, J., Mu, J.: Assessment and simulation of evacuation in large railway stations. Build. Simul. **14**(5), 1553–1566 (2021). https://doi.org/10.1007/s12273-020-0754-7
3. Cotfas, L., Lancu, L., Ioanăş, C., Ponsiglione, C.: Large event halls evacuation using an agent-based modeling approach. IEEE Access **10**, 49359–49384 (2022). https://doi.org/10.1109/ACCESS.2022.3172285
4. Kurdi, H., Almulifi, A., Al-Megren, S., Youcef-Toumi, K.: A balanced evacuation algorithm for facilities with multiple exits. Eur. J. Oper. Res. **289**(1), 285–296 (2021). https://doi.org/10.1016/j.ejor.2020.07.012
5. Delcea, C., Cotfas, L., Bradea, I., Boloş, M., Ferruzzi, G.: Investigating the exits' symmetry impact on the evacuation process of classrooms and lecture halls: an agent-based modeling approach. Symmetry. **12**(4), 627 (2020)
6. Son, J.Y., Bae, Y.H., Kim, Y.C., Oh, R.S., Hong, W.H., Choi, J.H.: Consideration of the door opening process in pedestrian flow: experiments on door opening direction, door handle type, and limited visibility. Sustainability. **12**(20), 8453 (2020)

7. Kinateder, M., Warren, W.H.: Exit choice during evacuation is influenced by both the size and proportion of the egressing crowd. Phys. A **569**, 125746 (2021). https://doi.org/10.1016/j.physa.2021.125746

8. Wang, J.Y., et al.: Experimental study of architectural adjustments on pedestrian flow features at bottlenecks. J. Stat. Mech. **2019**, 083402 (2019). https://doi.org/10.1088/1742-5468/ab3190

9. Xiao, M., Chen, Y., Yan, M., Ye, L., Liu, B.: Exits choice based on cellular automaton model for pedestrians' evacuation. In: 2015 IEEE Advanced Information Technology, Electronic and Automation Control Conference, pp. 970–973.IEEE Press, Chongqing (2015)

10. He, C., et al.: Accelerating large-scale multiobjective optimization via problem reformulation. IEEE Trans. Evol. Comput. **23**(6), 949–961 (2019)

11. Song, X., Xie, H.N., Sun, J.H., Han, D.L., Cui, Y., Chen, B.: Simulation of pedestrian rotation dynamics near crowded exits. IEEE. Trans. Intell. Transp. **20**(8), 3142–3155 (2019). https://doi.org/10.1109/TITS.2018.2873118

12. Song, X., Sun, J., Xie, H., Li, Q., Wang, Z., Han, D.L.: Characteristic time based social force model improvement and exit assignment strategy for pedestrian evacuation. Phys. A **505**, 530–548 (2018). https://doi.org/10.1016/j.physa.2018.03.085

13. Li, F., Zhang, Y.F., Ma, Y.Y., Zhang, H.L.: Modelling multi-exit large-venue pedestrian evacuationwith dual-strategy adaptive particle swarm optimization. IEEE Access. **8**, 114554–114569 (2020). https://doi.org/10.1109/ACCESS.2020.3003082

14. International Maritime Organization. Guidelines for Evacuation Analysis for New and Existing Passenger Ships (2016)

15. Wang, Y., Kyriakidis, M., Dang, V.: Incorporating human factors in emergency evacuation – an overview of behavioral factors and models. Int. J. Disast. Risk. RE. **60**, 102254 (2021). https://doi.org/10.1016/j.ijdrr.2021.102254

A Review of Longitudinal Vibration and Vibration Reduction Technology of Propulsion Shafting

Wei Chen[1], Kangwei Zhu[2,3], and Haiyu Zhang[2(✉)]

[1] China Ship Development and Design Center, Wuhan, Hubei, China
[2] Green and Smart River-Sea-Going Ship, Cruise and Yacht Research Center, Wuhan University of Technology, Wuhan 430070, China
zhanghaiyu@whut.edu.cn
[3] Hubei Key Laboratory of Theory and Application of Advanced Materials Mechanics, School of Science, Wuhan University of Technology, Wuhan 430070, China

Abstract. The performance impact and damage of ship propulsion shafting mainly come from vibration. The main causes of the longitudinal vibration of the propulsion shafting are the propeller excitation force and the ship power mechanical excitation. Longitudinal vibration not only harms the ship's stealth capabilities but also shortens the life of the propulsion shafting components. Due to its significant advantages in low-frequency vibration control, active control has become an important vibration control strategy. This paper dis-cusses the basic principle and modeling method of active control of longitudinal vibration of shafting, summarizes the active control strategy of longitudinal vibration of shafting suitable for engineering application, and finally suggests things on the future development direction of active control technology.

Keywords: Propulsion shafting · Longitudinal vibration · Active vibration control

1 Introduction

In propeller driven ships, ship shafting is an important power transmission structure, which transfers propeller power to the hull in reverse, which is the driving mode of most ships in waterway transportation today. It includes the transmission shaft, bearings and components between couplings from the main engine output flange to the propeller, connecting the main engine and the propeller. When the ship is sailing in the water, it is inevitable to form an uneven wake field at the stern. When the propeller runs in the wake field, it will produce pulsating thrust, which will cause the longitudinal vibration of the ship's propulsion shafting, and it will be transmitted to the hull through the inference bearing and base, causing the structural vibration of the hull. Longitudinal vibration of ship shafting may result in fatigue damage to crankshaft shafting, wear of accelerating transmission gear, or vibration of other hull structures, reducing the ship's stealth performance [1]. Nowadays, structural acoustic performance has become an indispensable

L. Pan et al. (Eds.): BIC-TA 2022, CCIS 1801, pp. 540–553, 2023.
https://doi.org/10.1007/978-981-99-1549-1_43

design index for many industrial products, such as ships, aircraft, automobiles, etc., and its internal and external radiated noise has attracted much attention. Its acoustic performance has become the top priority in commercial competition and even military confrontation. The noise on the ship not only affects the riding comfort and the reliability of electronic equipment, but also interferes with the stealth performance of the submarine, which seriously affects the operational capability of the submarine. And, as ship sizes increase and requirements for stealth performance increase, longitudinal vibration of ship shafting becomes increasingly essential [2] (Fig. 1).

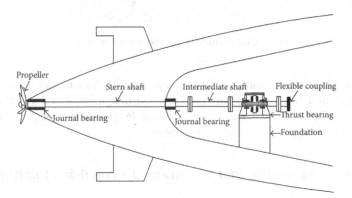

Fig. 1. Schematic diagram of submarine propulsion shafting system.

The longitudinal vibration of propulsion shafting can be controlled in three ways: (1) Improve the propeller design to reduce longitudinal pulsation force; (2) Reduce the transfer of propeller excitation force to the hull via the propulsion shaft, and (3) Improve the shell's stern structure to reduce acoustic radiation noise [3]. Methods 1 and 3 effectively solve the fundamental problem of shafting longitudinal vibration generation and acoustic radiation formation. However, additional mechanical problems that must be considered for upgrading the propeller and hull structure are more complex, making it impossible to tackle shafting vibration concerns by improving the propeller and hull structure. As a result, if the vibration source has been effectively managed or cannot be lowered further, it is critical to limit the influence of propeller excitation force on the hull via shafting (Fig. 2).

The premise of constructing the shafting longitudinal vibration control system is to ensure the basic operations of the propulsion shafting while not losing too much transmission efficiency. However, the existing propulsion shafting is intended primarily to perform the transmission function, so once the shafting structure is designed, it is difficult to modify, leading to numerous research on how to place some vibration control equipment on the shafting [21–25]. Although the passive control device is simple and reliable, it has limited ability to suppress broadband disturbances, and it is difficult to adjust it except for replacing the passive vibration control device. Active control has the advantages of strong adaptability and good control effect. At present, it has become an important way and research direction of axial vibration control [5].

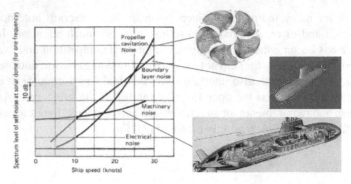

Fig. 2. Composition of noises of a submarine [1].

Therefore, this paper summarizes the existing shafting longitudinal vibration control schemes and modeling methods, outlines the shafting longitudinal vibration control strategies, and gives some suggestions on the future development of shafting longitudinal vibration control based on existing achievements.

2   Analysis of Longitudinal Vibration of Propulsion Shafting

The exciting force of shafting longitudinal vibration is mainly composed of the alternating thrust of the propeller. In the diesel engine, there is also the gas pressure in the cylinder and the inertia force of reciprocating motion [6, 7]. In addition, the torsional vibration of the shafting may also stimulate the longitudinal vibration, especially when the torsional vibration frequency is the same or like the longitudinal natural frequency, the coupling of this vibration is mainly completed by the crankshaft and propeller. In theory, the cyclotron vibration of shafting may also stimulate the longitudinal vibration, but the coupling between them is generally weak [8, 9].

As early as 1941, British scholar Poole [10] began to study the longitudinal vibration caused by the diesel engine crankshaft, In different fields, the research on longitudinal vibration of shafting caused by engines is also instructive in the field of ships, Murawski [11] considered the influence of the additional stress of the crankshaft on the longitudinal vibration in the vibration characteristic analysis, and also considered the crankshaft vibration caused by the reaction force of the propulsion bearing. Shu [12] analyzed the two main excitations that cause longitudinal vibration by studying the longitudinal vibration characteristics of the engine crankshaft of the high-speed train and the theoretical calculation method and tested the longitudinal vibration of the front end of the crankshaft using the self-designed three-dimensional vibration testing device, which verified the correctness of the theoretical model. The model is also applicable to the longitudinal vibration of ship shafting. However, the three-dimensional vibration model test designed is relatively simple, and the sensor design for measuring crankshaft longitudinal vibration needs to be more refined. However, this test also verifies the influence of coupling vibration in shafting longitudinal vibration. The theoretical research on longitudinal vibration caused by a diesel engine has become mature in the 1980s, and a large number of tests have been carried out to verify it [13]. This provides very important guidance for the design and installation of many ships shafting.

As one of the three major vibration sources, propeller has attracted scholars' attention as early as last century. In recent years, the degradation of submarine stealth performance caused by structural acoustic radiation caused by longitudinal vibration of shafting has been gradually a concern by scholars, and the research on unsteady spiral excitation force has also gradually increased [14, 15]. Kumai pointed out that 90% of the propeller bearing force indirectly transferred to the hull through the propulsion shafting was obtained by studying the transmission path of the propeller excitation force, and the excitation force of the longitudinal vibration of the propulsion shafting was mainly derived from the propeller excitation force [16]. Many scholars hope to optimize the propeller structure by changing the frequency of propeller excitation force to avoid the natural frequency of shafting. Rigby [17] proposed to increase the blade of the propeller to increase the frequency of the excitation force or change the position of the thrust bearing to reduce the natural frequency of the shafting, to avoid the normal working speed of the ship. However, this method can only avoid resonance or approach resonance frequency, which is difficult to balance under different working conditions or complex conditions. Carlton analyzed the propeller excitation force in the frequency domain and found that the propeller excitation force is mainly low-frequency excitation, while the high-frequency excitation counteracts the reaction force of the flow field, which has little impact on the propulsion shafting [1]. This discovery has played an important guiding role in the research on vibration reduction of axial vibration of shafting. Guide the design of shafting shock absorber to the direction of high frequency and wide frequency. on the other hand, based on the research on the longitudinal, transverse and torsional vibration forms of the propulsion shafting, some scholars found that a point on the shafting would not only deform in the direction of the external excitation force, but also in other directions, which also led many scholars to start discussing the coupling vibration of the shafting [18]. Davor Sverko thinks that the torsional longitudinal coupling vibration of the propeller will produce a large axial force, which will cause the longitudinal vibration of the shell [19]. This view has also been verified in many subsequent tests, and the longitudinal vibration caused by coupling vibration cannot be ignored. Dort and others made further theoretical research on torsional longitudinal coupled vibration [20]. Parsons analyzed the torsional shafting coupling vibration caused by the propeller and pointed out that propeller shafting includes inertial coupling and velocity coupling [18]. These studies have continuously improved the theoretical analysis of coupled vibration. More complex boundary conditions and influencing factors are considered in the coupled vibration analysis. Their work is of great significance to the longitudinal vibration analysis of ship shafting and the design of shafting and its shock absorber. However, there are still many limitations in the current theoretical analysis, such as the need to properly simplify the complex coupled system, which is difficult to ensure accuracy; As well as time-varying system is too complex, theoretical analysis is difficult to solve and so on.

3   Active Control Technology for Longitudinal Vibration of Propulsion Shafting

The active control of longitudinal vibration of propulsion shafting refers to the controller making the actuator generate a controlled force and apply it to the controlled object to counteract the vibration caused by external interference. Therefore, the key to the active control of shafting is the design of the actuator and controller. Based on theoretical analysis, the differential equation of longitudinal vibration of propulsion shafting is:

$$M\ddot{x}(t) + C\dot{x}(t) + Kx(t) = F(t) + F_a(t) \tag{1}$$

where M, C, and K represent the mass matrix, damping matrix, and stiffness matrix of the propulsion shafting respectively. $x(t)$ is the longitudinal displacement of the shafting. $F(t)$ is the longitudinal excitation force of the shafting, $F_a(t)$ is the longitudinal control force of the shafting. The difference between various vibration control methods lies in changing the different dynamic characteristic parameters of the propulsion shafting. The longitudinal vibration damping of the propulsion shafting can be increased by changing the damping matrix C; Changing the longitudinal excitation force $F(t)$ or applying force feedback can isolate the transmission of longitudinal excitation force to the hull; Changing the mass matrix M and stiffness matrix K can adjust the longitudinal vibration characteristics of the propulsion shafting to achieve transmission control.

3.1   Design of Active Control Actuator

To actively control the longitudinal vibration of the propulsion shafting, scholars try to arrange actuators at the thrust bearing of the propulsion shafting. In the 1980s, Lewis et al. of the United States proposed a vibration reduction method by paralleling an auxiliary magnetic thrust bearing with the original thrust bearing. Through closed-loop feedback control, the magnetic thrust bearing was adjusted in real-time, to achieve active control of the longitudinal vibration of the thrust shafting [21]-[23]. However, the test shows that although the magnetic thrust bearing can reduce the transmission of the axial pulse force on the shafting to the hull, its reaction makes the longitudinal vibration of the shafting larger. Baz et al. designed a pneumatic servo control system in 1990, using the air pressure balance to push the slider to provide control forces for the shafting. The experimental results of active control on the simulated shafting show that the control effect in the low-frequency band is remarkable [24]. However, the transmission equivalent of ship shafting is generally large, and the sensitivity of air pressure is far less than Lewis's magnetic bearing (Fig. 3).

Goodwin first proposed to use the resonant converter to control the axial vibration of the shaft system and studied the vibration characteristics of the shaft system and the influence of the resonant converter on the shaft system vibration through theoretical analysis [25]. The resonant converter is installed between the thrust bearing and the bearing base. The resonant condition of the shaft system is changed through its structure and materials, so as to avoid the resonant frequency and achieve the purpose of vibration reduction. Goodwin's research is very meaningful. Although the model is relatively simple, the design of resonant converter provides a reference for subsequent research.

By analyzing the approximate propeller shaft thrust bearing pedestal resonant converter shell model, Dylejko studied the problems of shell vibration and acoustic radiation caused by shell impedance and shafting vibration and obtained the resonant converter parameters of different shafting structures [26–28] (Fig. 4).

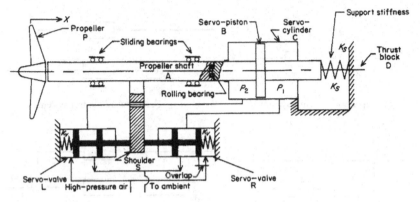

Fig. 3. Schematic diagram of the pneumatic active control system [24].

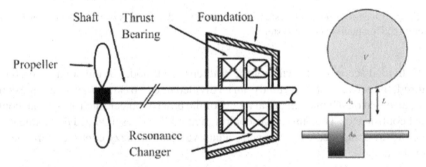

Fig. 4. Installation position and schematic diagram of resonant converter [25].

In addition to applying the controller in the bearing, Pan et al. analyzed the axial and bending vibration of the shafting by using the finite element method and established a simple experimental platform [29]. They further verified the influence of the bearing oil film stiffness on the shafting vibration by using the experimental method. They believed that the axial vibration of the shafting could be suppressed by actively controlling the oil film stiffness of the thrust bearing. However, the oil film stiffness changes with the speed of the shaft system, and the direct control of the bearing oil film stiffness becomes complex due to its time-varying characteristics, which makes it difficult to establish the model and algorithm of the controller.

3.2 Design of Active Control Controller

If the actuator is the body of axial vibration, the vibration control controller is the brain of the whole control system. The basic structure of the controller includes open loop control, feedforward control, feedback control and compound control (Fig. 5).

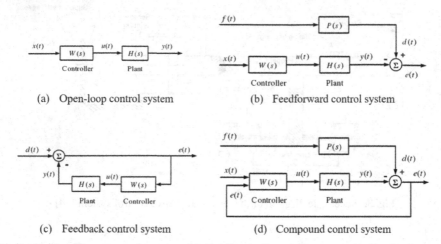

(a) Open-loop control system   (b) Feedforward control system

(c) Feedback control system   (d) Compound control system

Fig. 5. (a) Open-loop control system; (b) Feedforward control system; (c) Feedback control system; (d) Compound control system.

If divided according to whether the mathematical model of the controlled object is required, the active control methods can be divided into two categories: one is the control methods based on the mathematical model of the passive object, such as optimal control, robust control, etc.; The other is the control method that does not need the mathematical model of the controlled object, such as adaptive control, fuzzy control, neural network control, etc.

Modeling Method
Both the characteristic analysis of shafting vibration and the construction of some control algorithms need to model the entire shafting system. Among them, sensors or actuators are indispensable components in the process of active control of structural systems. Therefore, it is necessary to establish a corresponding dynamic model to analyze the impact of its layout on the vibration controllability. The control algorithm is a prerequisite to determine the characteristics of the controlled object, that is, the control algorithm needs to put forward requirements for the control ability of the controlled object when modeling. Generally, the control algorithm is proposed according to the vibration characteristics of the controlled object. Therefore, the control algorithm is an essential part in establishing a reliable model.

The system modeling methods include analytical method, numerical method, and numerical analytical combination method. The physical concept of analytical method is clear, simple, and practical, which is convenient to explain the essence of vibration problems and has high accuracy in the case of simple regular structure problems. Numerical methods include the finite element method, boundary method, etc. Due to the rapid development of computer technology, the application of numerical methods has become very extensive. When dealing with large and complex structural systems, it can assist analytical methods to obtain more reliable dynamic models.

In the modeling method of a complex continuous whole, the dynamic substructure method can deal with the continuity of the system well [30]. Nowadays, the most widely used method for longitudinal vibration analysis of shafting is the frequency domain dynamic substructure method [31]. The complex system is divided into independent substructures and then integrated according to the interface conditions after independent modeling. In this way, the influence of actuators on sensor placement can be divided, and the model size can be reduced without losing information.

For different research objects, modeling methods will be different. For the shafting vibration analysis coupled with the shell, attention will be focused on the dynamic modeling and vibration characteristics of the shell itself. When the shafting is considered as the key research object, the coupling effect between several different vibrations will be considered, Modeling methods will also vary [32–34].

Adaptive Control System.
The basic structure of the control system for shafting vibration includes switching [35], sliding mode [36, 37], predictive [38, 39], and internal model [40, 41]. As one of the most successful control methods in the field of active noise and vibration control (ANVC), adaptive control does not need an accurate model to update the controller directly or to conduct online identification and then use the model estimated online to design the controller, which is very consistent with the shafting vibration problem that is difficult to find an accurate model.

An adaptive controller includes a controller with adjustable parameters and a parameter-adjusting mechanism (see Fig. 6) [42]. The controller parameters are corrected by the error between the output response signal y(n) and the expected response signal d(n) so that the output response signal y(n) approximates the expected response signal d(n). The most attractive feature of adaptive control is the ability of self-adjustment and automatic tracking system changes realized by the combination of the two. The least mean square error criterion and the least square criterion are two adaptive error optimization criteria, The former has the minimum mean square value of the error signal when constructing the cost function, and the computational complexity is much smaller than the latter's recursive operation using the least squares [43, 44]. The result is easy to converge and has been implemented in hardware, so it is more widely used in engineering practice, Adaptive control methods include Filter-x LMS [45], Iterative feedback tuning (IFT) [46, 47], Model-free frequency domain iterative feedback method [48], etc.

90% of the excitation force of shafting vibration comes from the excitation force of the propeller, while the propeller will generate narrowband periodic and broadband random pulse force when moving in the non-uniform wake field [15]. In order to omit the interference caused by the relatively small broadband random pulse force in the active control of shafting vibration, a filter is often added to the controller. This also makes the Filter-x LMS method the most successful control algorithm (Fig. 7).

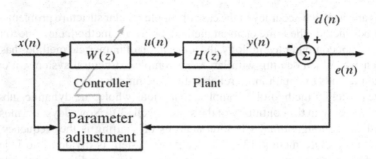

Fig. 6. Adaptive control system.

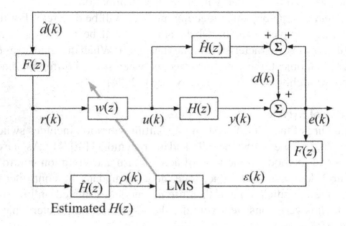

Estimated $H(z)$

Fig. 7. Block Diagram of LMS Control System with Embedded Tracking Filter [49].

Morgan first proposed the Filter-x LMS algorithm in 1980 [49], and then Burgess improved the active control theory based on the Filter-x LMS algorithm on this basis, which was used to control pipeline noise, with an obvious vibration suppression effect, but this system also faces the problem of large transmission delay [50]. Douglas et al. proposed a multi-channel Filter-x LMS fast algorithm and applied it to the active noise control system, under the condition that the optimization gradient algorithm greatly reduces the calculation cost, with an obvious vibration suppression effect [51]. Gong derived the differential equations of mean square and mean square transient convergence behavior of the multi-channel NANC system. The stability boundary of the multi-channel Filter-x LMS algorithm in mean and mean square sense is also derived [52]. Zhang proposed a multi-channel Filter-x LMS algorithm with variable step size based on a sampling function, which accelerated the convergence speed at the initial stage of iteration and improved the convergence accuracy of the algorithm. It is well applied in active vibration control of gear transmission systems [53]. A new architecture based on multi-parallel branch folding (MPBF) technology was proposed by Shi. It makes up for the shortage of excessive calculation load of the FFMC F-x LMS algorithm in multi-channel active noise control [54]. Fuller et al. have done a series of research work on adaptive feedforward control based on the Filter-x LMS algorithm, which has been applied to the active control of vibration and sound radiation with different results and achieved the beneficial effect of vibration and noise reduction in experiments [55–58].

Zhang et al. proposed an adaptive control method with an embedded tracking filter on the basis of Filter-x LMS algorithm for the vibration system with periodic vibration. The simulation and experimental results for the base show that this method can effectively suppress the vibration response of the system [59]. Zhang's research group has subsequently conducted a series of in-depth studies on Filter-x LMS algorithm with anti-saturation link embedded and has achieved good vibration and noise reduction effect in experiments. Among them, Wang Junfang proposed an adaptive feedforward control method that combines tracking filtering with LMS control algorithm with saturation suppression to solve the problem of controller output saturation when the vibration response is too large [60–63].

Control Strategy for Time-Varying Systems

It is mentioned in the vibration analysis in Sect. 2 that the oil film stiffness of the thrust bearing will change with the speed, and the corresponding dynamic characteristics of the shafting will also change, so the shaft system is time-varying. This characteristic will make the frequency signal transmitted by the control system different at different speeds, which will make the adaptive controller relying on the deterministic control model invalid. The adaptive filtering algorithm mentioned in the appeal can well handle the vibration caused by harmonic excitation in the time-invariant system. But for the control of time-varying systems, time-varying controllers are used in principle, and the parameters or structure of the controller generally need to be adjusted online [64]. In some time-varying systems, the change rule of time-varying parameters is easy to obtain, so it is not difficult to construct the controller pertinently. However, the change rule of the system model cannot be obtained in advance, so the controller can only be constructed with the help of prior knowledge such as experimental data.

In the current research situation, there are roughly two control strategies for time-varying systems:

1. The model-free adaptive harmonic suppression method is studied, and the controller is constructed using prior knowledge.
2. Study the adaptive harmonic suppression method based on model online estimation, such a model needs to identify the transmission of control channels simultaneously.

4 Conclusion

In this paper, the vibration mechanism, analysis modeling, control strategy and control algorithm of longitudinal vibration of propulsion shafting are summarized, and the widely used control algorithm and control strategy are summarized in detail. Based on the above research results, the following research suggestions are proposed:

1. The dynamic modeling of the propeller shaft hull coupling system needs more in-depth research, and the mechanical characteristics of the propeller and the vibration

and acoustic radiation characteristics of the hull also need more attention. This is not only helpful for the shafting design but also can further improve the optimization of the vibration control algorithm.

2. More shafting longitudinal vibration tests are still needed to verify many excellent actuator designs. The test data can not only further optimize the actuators, but also provide valuable prior knowledge for the active controller design of time-varying systems.

3. At present, the actuators used in most active control schemes need to be subsequently installed on the shafting system, which will affect the strength and impact resistance of the shafting. The integrated shafting control system should be the future development direction.

References

1. Carlton, J.: Marine Propellers and Propulsion. Butterworth-Heinemann, Oxford (2018)
2. Tamura, Y., Kawada, T., Sasazawa, Y.: Effect of ship noise on sleep. J. Sound Vib. **205**(4), 417–425 (1997)
3. Lin, T.R., Pan, J., O'Shea, P.J., Mechefske, C.K.: A study of vibration and vibration control of ship structures. Mar. Struct. **22**(4), 730–743 (2009)
4. Yuanchao, Z., Wei, X., Zhengmin, L., Jiangyang, H.: Review of the vibration isolation technology of submarine thrust bearing. In: 2019 25th International Conference on Automation and Computing (ICAC), pp. 1–6. IEEE (2019)
5. Soong, T.T., Masri, S.F., Housner, G.W.: An overview of active structural control under seismic loads. Earthq. Spectra **7**(3), 483–505 (1991)
6. Vizentin, G., Vukelić, G., Srok, M.: Common failures of ship propulsion shafts. Pomorstvo. **31**(2), 85–90 (2017)
7. Zhang, G., Zhao, Y., Li, T., Zhu, X.: Propeller excitation of longitudinal vibration characteristics of marine propulsion shafting system. Shock Vib. (2014)
8. Huang, Q., Zhang, C., Jin, Y., Yuan, C., Yan, X.: Vibration analysis of marine propulsion shafting by the coupled finite element method. J. Vibroeng. **17**(7), 3392–3403 (2015)
9. Zhang, Y., Xu, W., Li, Z., He, J., Yin, L.: Dynamic characteristics analysis of marine propulsion shafting using multi-DOF vibration coupling model. Shock Vib. (2019)
10. Poole, R.: The axial vibration of diesel engine crankshafts. Proc. Inst. Mech. Eng. **146**(1), 167–182 (1941)
11. Murawski, L.: Axial vibrations of a propulsion system taking into account the couplings and the boundary conditions. J. Mar. Sci. Technol. **9**(4), 171–181 (2004)
12. Shu, G.Q., Liang, X.Y., Lu, X.C.: Axial vibration of high-speed automotive engine crankshaft. Int. J. Veh. Des. **45**(4), 542–554 (2007)
13. Visser, N.J.: The axial stiffness of marine diesel engine crankshafts. Int. Shipbuild. Prog. **15**(168), 302–316 (1968)
14. van Wijngaarden, E.: Recent developments in predicting propeller-induced hull pressure pulses. In: Proceedings of the 1st International Ship Noise and Vibration Conference, pp. 1–8 (2005)
15. Sontvedt, T.: Propeller induced excitation forces. Eur. Shipbuild. 20(3) (1971)
16. Kumai, T., Tamaki, I., Kishi, J., Yumoto, H., Sakurada, Y.: On a method of measurement of propeller bearing force exciting hull vibrations. J. Soc. Naval Architects Japan **1970**(128), a85–a90 (1970)

17. Rigby, C.: Longitudinal vibration of marine propeller shafting. Trans. Inst. Mar. Eng. **60**, 67–78 (1948)
18. Parsons, M.G.: Mode coupling in torsional and longitudinal shafting vibrations. Mar. Technol. SNAME News **20**(03), 257–271 (1983)
19. Sverko, D.: Torsional-axial coupling in the line shafting vibrations in merchant ocean going ships (Doctoral dissertation, Concordia University) (1997)
20. Van Dort, D., Visser, N.J.: Crankshaft coupled free torsional-axial vibrations of a ship's propulsion system1. Int. Shipbuild. Prog. **10**(109), 333–350 (1963). https://doi.org/10.3233/ isp-1963-1010902
21. Lewis, D.W., Allaire, P.E., Thomas, P.W.: Active magnetic control of oscillatory axial shaft vibrations in ship shaft transmission systems part 1: system natural frequencies and laboratory scale model. Tribol. Trans. **32**(2), 170–178 (1989)
22. Lewis, D.W., Humphris, R.R., Thomas, P.W.: Active magnetic control of oscillatory axial shaft vibrations in ship shaft transmission systems part 2: control analysis and response of experimental system. Tribol. Trans. **32**(2), 179–188 (1989)
23. Baz, A., Gilheany, J., Steimel, P.: Active vibration control of propeller shafts. J. Sound Vib. **136**(3), 361–372 (1990)
24. Goodwin, A.J.H.: The design of a resonance changer to overcome excessive axial vibration of propeller shafting. Trans. Inst. Mar. Eng **72**, 37–63 (1960)
25. Dylejko, P., Kessissoglou, N.: Minimization of the vibration transmission through the propeller-shafting system in a submarine. J. Acoust. Soc. Am. **116**(4), 2569 (2004)
26. Dylejko, P.G.: Optimum resonance changer for submerged vessel signature reduction (Doctoral dissertation, UNSW Sydney) (2007)
27. Dylejko, P.G., Kessissoglou, N.J., Tso, Y., Norwood, C.J.: Optimization of a resonance changer to minimise the vibration transmission in marine vessels. J. Sound Vib. **300**(1–2), 101–116 (2007)
28. Pan, J., Farag, N., Lin, T., Juniper, R.: Propeller induced structural vibration through the thrust bearing. In: Proceedings of the Annual Conference of the Australian Acoustical Society, pp. 13–15 (2002)
29. Craig Jr, R.R.: Substructure methods in vibration (1995)
30. Jen, C.W., Johnson, D.A., Dubois, F.: Numerical modal analysis of structures based on a revised substructure synthesis approach. J. Sound Vib. **180**(2), 185–203 (1995)
31. Fahy, F.J., Gardonio, P.: Sound and Structural Vibration: Radiation, Transmission, and Response. Elsevier, Amsterdam (2007)
32. Parsons, M.G., Vorus, W.S., Richard, E.M.: Added mass and damping of vibrating propellers. University of Michigan (1980)
33. Jakeman, R.W.: Influence of stern tube bearings on lateral vibration amplitudes in marine propeller shafting. Tribol. Int. **22**(2), 125–136 (1989)
34. Sam, Y.M., Osman, J.H., Ghani, M.R.A.: A class of proportional-integral sliding mode control with application to active suspension system. Syst. Control Lett. **51**(3–4), 217–223 (2004)
35. Lan, K.J., Yen, J.Y., Kramar, J.A.: Sliding mode control for active vibration isolation of a long-range scanning tunneling microscope. Rev. Sci. Instrum. **75**(11), 4367–4373 (2004)
36. Hu, Q.: Sliding mode maneuvering control and active vibration damping of three-axis stabilized flexible spacecraft with actuator dynamics. Nonlinear Dyn. **52**(3), 227–248 (2008)
37. Yang, Z., Hicks, D.L.: Active noise attenuation using adaptive model predictive control. In: 2005 International Symposium on Intelligent Signal Processing and Communication Systems, pp. 241–244. IEEE (2005)
38. Wills, A.G., Bates, D., Fleming, A.J., Ninness, B., Moheimani, S.R.: Model predictive control applied to constraint handling in active noise and vibration control. IEEE Trans. Control Syst. Technol. **16**(1), 3–12 (2007)

39. Kuo, S.M., Morgan, D.R.: Active Noise Control Systems, vol. 4. Wiley, New York (1996)
40. Kinney, C.E., De Callafon, R.A.: An adaptive internal model-based controller for periodic disturbance rejection. IFAC Proc. **39**(1), 273–278 (2006)
41. Milic, L. (ed.): Multirate Filtering for Digital Signal Processing: MATLAB Applications. IGI Global, Hershey (2009)
42. Widrow, B., Walach, E.: Adaptive signal processing for adaptive control. IFAC Proc. **16**(9), 7–12 (1983)
43. Aström, K.J., Goodwin, G.C., Kumar, P.R. (eds.): Adaptive Control, Filtering, and Signal Processing, vol. 74. Springer Science & Business, Cham (2012)
44. Vér, I.L., Beranek, L.L. (eds.): Noise and Vibration Control Engineering: Principles and Applications. John Wiley & Sons, Hoboken (2005)
45. Pontana, F., et al.: Chest computed tomography using iterative reconstruction vs filtered back projection (Part 1): evaluation of image noise reduction in 32 patients. Eur. Radiol. **21**(3), 627–635 (2011)
46. Meurers, T., Veres, S.M., Elliot, S.J.: Frequency selective feedback for active noise control. IEEE Control Syst. Mag. **22**(4), 32–41 (2002)
47. Meurers, T., Veres, S.M., Tan, A.C.H.: Model-free frequency domain iterative active sound and vibration control. Control. Eng. Pract. **11**(9), 1049–1059 (2003)
48. Morgan, D.: An analysis of multiple correlation cancellation loops with a filter in the auxiliary path. IEEE Trans. Acoust. Speech Signal Process. **28**(4), 454–467 (1980)
49. Burgess, J.C.: Active adaptive sound control in a duct: a computer simulation. J. Acoust. Soc. Am. **70**(3), 715–726 (1981)
50. Douglas, S.C.: Fast implementations of the filtered-X LMS and LMS algorithms for multi-channel active noise control. IEEE Trans. Speech Audio Proc. **7**(4), 454–465 (1999). https://doi.org/10.1109/89.771315
51. Gong, C., Wu, M., Guo, J., et al.: Statistical analysis of multichannel F-x LMS algorithm for narrowband active noise control. Signal Proc. 108646 (2022)
52. Zhang, F., Sun, W., Liu, C., et al.: Application of multichannel active vibration control in a multistage gear transmission system. Shock Vib. **2022** (2022)
53. Shi, D., Gan, W.S., He, J., et al.: Practical implementation of multichannel filtered-x least mean square algorithm based on the multiple-parallel-branch with folding architecture for large-scale active noise control. IEEE Trans. Very Large-Scale Integr. (VLSI) Syst. **28**(4) 940–953 (2019)
54. Fuller, C.R., Rogers, C.A., Robertshaw, H.H.: Control of sound radiation with active/adaptive structures. J. Sound Vib. **157**(1), 19–39 (1992)
55. Vipperman, J.S., Burdisso, R.A., Fuller, C.R.: Active control of broadband structural vibration using the LMS adaptive algorithm. J. Sound Vib. **166**(2), 283–299 (1993)
56. Guigou, C., Fuller, C.R., Wagstaff, P.R.: Active isolation of vibration with adaptive structures. J. Acoust. Soc. Am. **96**(1), 294–299 (1994)
57. Cabell, R.H., Fuller, C.R.: A principal component algorithm for feedforward active noise and vibration control. J. Sound Vib. **227**(1), 159–181 (1999)
58. Zhang, Z., Huang, X., Chen, Y., Hua, H.: Underwater sound radiation control by active vibration isolation: an experiment. Proc. Inst. Mech. Eng. Part M: J. Eng. Marit. Environ. **223**(4), 503–515 (2009)
59. Zhang, Z., Hu, F., Wang, J.: On saturation suppression in adaptive vibration control. J. Sound Vib. **329**(9), 1209–1214 (2010)
60. Zhang, Z.Y., Hu, F., Hua, H.X.: Simulation and experiment on active vibration isolation with an adaptive method. Proc. Inst. Mech. Eng. Part M: J. Eng. Marit. Environ. **224**(3), 225–238 (2010)
61. Zhang, Z., Chen, Y., Li, H., Hua, H.: Simulation and experimental study on vibration and sound radiation control with piezoelectric actuators. Shock. Vib. **18**(1–2), 343–354 (2011)

62. Zhang, Z., Hu, F., Li, Z., Hua, H.: Modeling and control of the vibration of two beams coupled with fluid and active links. Shock. Vib. **19**(4), 653–668 (2012)
63. Åström, K.J., Wittenmark, B.: Adaptive Control. Courier Corporation (2013)
64. Bohn, C., Cortabarria, A., Härtel, V., Kowalczyk, K.: Active control of engine-induced vibrations in automotive vehicles using disturbance observer gain scheduling. Control. Eng. Pract. **12**(8), 1029–1039 (2004)

Design and Simulation of Heading Controller for Unmanned Boat Based on Fuzzy Neural PID

Yunpeng Su[1(✉)], Yong Chen[1], Xinlong Pan[2(✉)], Haipeng Wang[2], Ziwei Zhao[3],
and Hao Liu[1,4]

[1] China Ship Development and Design Center, Wuhan 430064, China
13073690139@163.com
[2] Naval Aviation University, Yantai 264001, China
airadar@126.com
[3] Wuhan University of Technology, Wuhan 430070, China
[4] Shanghai Jiao Tong University, Shanghai 200240, China

Abstract. The combination of intelligent algorithms and heading control of unmanned vehicles can improve its motion control performance, but most of them are currently in the experimental simulation stage, which cannot meet the anti-interference ability, stability and accuracy requirements of the intelligent control tasks of unmanned vehicles. In this paper, the nomoto model of the unmanned boat and the mathematical model of the steering gear are established, and a control algorithm combining fuzzy neural network and PID control is proposed, and the weight value in the fuzzy neural network model is adjusted by using the specified sample to learn in the BP neural network and the central value and width of the membership function, so as to optimize the controller. In the simulation, the heading control performance of conventional PID, fuzzy PID and fuzzy neural PID is compared at a given desired heading angle. The simulation results show that the fuzzy neural PID algorithm has the best heading control performance for the unmanned vehicle.

Keywords: PID · Fuzzy neural · Heading controller · USV

1 Introduction

An unmanned surface vehicle (USV) is an intelligent water platform that can use modern intelligent control algorithms to carry out autonomous planning and navigation based on navigation information, and complete tasks designated by shore personnel, such as information collection, security inspections, and operations in high-risk areas. Due to the fast speed and small mass of unmanned boats, they are easily disturbed by wind, waves and other factors, making them have the characteristics of large inertia and large lag. Therefore, there are higher requirements for heading control during sailing [1]. The course controller controls the steering gear according to the yaw signal to maintain the course of the unmanned boat. Its performance has an important impact on track tracking, autonomous navigation and safety. Therefore, the design of the course keeping controller becomes the key point of the motion control algorithm of the UAV. Classical PID is widely used in the early ship heading control, but the parameters of the classic PID autopilot do not change with the external environment, and cannot meet the increasingly complex military and civilian mission requirements in terms of anti-interference ability, stability and accuracy. With the extensive research on intelligent control algorithms by researchers from various countries in modern times, synovium control, neural network control, Backstepping control and other intelligent algorithms combined with the motion algorithm of the unmanned vehicle to improve its motion control performance has become a research hotspot. Various intelligent algorithms have their own advantages and disadvantages, and most of them are currently in the experimental simulation stage, and their practical application still needs to be further verified. Therefore, it is of great practical significance to conduct in-depth research on the course control of surface unmanned vehicles.

Most of the existing unmanned vehicle path planning methods are derived from the path planning of unmanned robots, and the global path planning capability is an important embodiment of its intelligence level [2]. Global path planning is to use the existing map information, combined with the task mission, and adopt the appropriate search algorithm to find the appropriate path in the global sense.

Rao Sen proposed a global path planning for unmanned craft by using hierarchical thought and genetic algorithm [3]. Fan Yunsheng et al. proposed a global path planning method combining electronic chart rasterization with genetic algorithm optimization, and adopted a comprehensive evaluation fitness function to improve the efficiency of path optimization [4]. Liu Jian designed the grid dynamic refinement method to build the obstacle model, and improved the artificial potential field method to improve the traditional potential field easy stagnation shortcomings [5]. Chen Chao, Tang Jian et al. designed A global path planning method combining viewable and A* algorithm to improve the inadaptability of the traditional viewable method to the environment [6]. Zhuang Jiayuan et al. designed a global path planning algorithm for UVs with distance optimization, which was combined with electronic chart to improve the accuracy of path planning [7]. Shu Zongyu adopted multi-objective particle swarm optimization algorithm to optimize the initial path searched by Dijsktra algorithm and introduced simulated annealing operator into the algorithm to obtain the global shortest path [8]. Liu Kun proposed an artificial potential field-ant colony path planning method to reduce the length and time of the planned path, aiming at the blindness of the artificial potential field method which is easy to fall into local optimal and ant colony algorithm search [9].

Path planning of unmanned craft is different from that of unmanned robot and unmanned vehicle [10]. First of all, the navigation environment of the unmanned boat is generally relatively wide, not restricted by the fixed traffic network, the number of obstacles is small, the volume of obstacles is large; In addition, the motion control of unmanned craft is not as real-time and simple as that of robot. The steering process also needs to consider the rotation radius and hull inertia and other factors. A* algorithm and Dijkstra algorithm need to build complex raster map, and need to search the raster part traversal, the efficiency is low. Ant colony algorithm (ACO) and genetic algorithm (GA) have many shortcomings such as difficulty in adjusting parameters [11]. Particle swarm optimization (PSO) is widely used in parameter optimization problems because of its advantages such as low adjustment parameters, high search efficiency, easy implementation and easy combination with other algorithms [12]. Therefore, particle swarm optimization (PSO) is used in this paper to study the global path planning of unmanned craft, and the particle swarm optimization algorithm is improved to meet the requirements of path planning for search accuracy and search speed.

M.Caccia et al. Conducted a lot of research on the "Charlie" double-hull unmanned ship, and used the self-oscillation model identification method to identify the parameters, and applied this to the heading of the combination of PID and Kalman filter. In the controller, its feasibility is verified by conducting sea experiments. Sarda et al. uses the nonlinear automatic driving technology based on the local control network to solve the control problem of the linear USV controller in the nonlinear, and Combining nonlinear local control network and model prediction algorithm, a nonlinear heading controller is built. Soohong Park et al. Combining conventional PID and intelligent fuzzy algorithm to control the course of the established USV linear control response model, the algorithm is better than conventional PID through simulation experiments control effect. L.Wang et al. took catamaran unmanned boat as the research object, combined with linear Gaussian control and fuzzy control theory, redesigned its heading control system. Witkowska et al. On the basis of considering the dynamic and nonlinear static characteristics of the steering gear, the course keeping control system was optimized based on the adaptive backstepping design method, and the parameters of the system were adjusted by using GA, and finally A full- scale simulation model was built to analyze and verify the control algorithm in time domain. Kurowski et al. [13] applied the extended Kalman filter method to a modular nonlinear heading control system in the form of a cascaded structure, estimated the parameters of the USV model through the subspace identification method, and finally carried out experimental verification [14–16].

Key technologies of surface unmanned vehicles, most of them require high manpower and material costs, and there are few practical applications of heading control algorithms for fuzzy neural PID. Therefore, this paper studies the path planning and heading control algorithms of unmanned boats, and applies the algorithms in the control system of unmanned boats to realize autonomous navigation in specific waters, and provide water quality monitoring, security inspections, and surface salvage for unmanned boats and other work to provide support.

2   Mathematical Motion Model of Unmanned Boat

2.1   Standard Particle Swarm Optimization

In order to describe and define these motions conveniently, the inertial coordinate system fixed on the earth and the appendage coordinate system fixed on the UAV are selected. The position is represented by x_0, y_0 and z_0 in the inertial coordinate system, and the yaw angle, heel angle and pitch angle around the corresponding axis are represented by Ψ, φ and θ. In the appendage coordinate system, u, v and w are used to represent the speed in the three directions of forward, drift and heave, and r, p and q are used to represent the angular speed in the three directions of yaw, heel and pitch.

For most unmanned boats such as water quality sampling, path planning and tracking, automatic obstacle avoidance, and water surface environment monitoring, the heave velocity, heel angular velocity and pitch angular velocity of the ship are very small, which can be ignored in actual analysis, so this paper only studies the forward, lateral drift and yaw motion of the unmanned vehicle on the plane, which is simplified to three degrees of freedom. The schematic diagram of the plane motion is shown in Fig. 1.

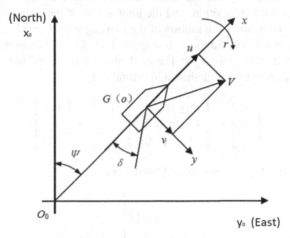

Fig. 1. Three degrees of freedom planar motion diagram of USV

According to the established dual coordinate system, analyze the force and moment of the unmanned boat when it moves in the plane, and establish the motion equation of the unmanned boat with three degrees of freedom, as shown in the formula (1) shows:

$$\begin{cases} m(\dot{u} - vr - x_G r^2) = \sum X \\ m(\dot{v} + ur + x_G \dot{r}) = \sum Y \\ I_{ZG}\dot{r} + mx_G(ur + \dot{v}) = \sum N \end{cases} \tag{1}$$

where m represents the mass of the UAV, x represents the abscissa of the center of mass of the UAV, I represents the moment of inertia around the z axis, u represents forward speed, r represents yaw rate, v represents lateral drift speed, X represents the force component in the X direction on the UAV, Y represents the force component of the UAV in the Y direction, N represents the moment that the UAV is subjected to around the Z axis.

In this paper, it is assumed that the origin and the center of mass of the coordinate system of the attached body of the unmanned vehicle coincide, and the formula (1) is linearized on the basis of linearization. Considering that the three movements are independent of each other and will directly or indirectly affect the forward movement when controlling the speed, the equation of motion is written in matrix form, as shown in formula (2):

$$\begin{bmatrix} m - Y_{\dot{v}} & (mx_G - Y_{\dot{r}}) \\ mx_G - N_{\dot{v}} & (I_{ZG} - N'_{\dot{r}}) \end{bmatrix} \begin{bmatrix} \dot{v} \\ \dot{r} \end{bmatrix} = \begin{bmatrix} Y_r & (Y_r - mu_0) \\ N_v & (N_r - mx_G u_0) \end{bmatrix} \begin{bmatrix} v \\ r \end{bmatrix} + \begin{bmatrix} Y_\delta \\ N_\delta \end{bmatrix} \tag{2}$$

where Y_v, Y_r, $Y_v\;'$, $Y_r\;'$, Y_δ, N_v, N_r, $N_v\;'$, $N_r\;'$ and N_δ are hydrodynamic derivatives.

The above hydrodynamic derivatives can be calculated through the known parameters of the unmanned boat, such as the length, width, displacement, square coefficient, rudder area, boat speed and draft, and then query the linear hydrodynamic regression formula obtained by Clarke fitting.

Through the analysis of the motion of the unmanned boat, the propeller at the tail is used to provide forward motion, and the joint action of the propeller and the rudder biased at the tail determines the motion of the two degrees of freedom, the lateral drift motion and the yaw motion. Suppose the expected heading of the unmanned boat is Ψr, and the deviation is $\Delta\Psi = \Psi r - \Psi$ The mathematical model of the state space of the unmanned vehicle is obtained as shown in formula (3):

$$\begin{bmatrix} \dot{v} \\ \dot{r} \\ \dot{\psi} \end{bmatrix} = \begin{bmatrix} m_{11} & m_{12} & 0 \\ m_{21} & m_{22} & 0 \\ 0 & 1 & 0 \end{bmatrix} \begin{bmatrix} v \\ r \\ \Delta\psi \end{bmatrix} + \begin{bmatrix} n_{11} \\ n_{21} \\ 0 \end{bmatrix} \delta \tag{3}$$

Gets $\Delta\Psi$ as the output is shown in formula (4):

$$\psi_\Delta = \begin{bmatrix} 0 & 0 & 1 \end{bmatrix} \begin{bmatrix} v \\ r \\ \Delta\psi \end{bmatrix} \tag{4}$$

The formula (3) and formula (4) into the transfer function formula of the heading $\Delta\Psi$ of the unmanned boat:

$$G(s)_{\delta\psi} = C[sI - A]^{-1}B = \frac{K(T_\varepsilon s + 1)}{s(T_2 s + 1)(T_1 s + 1)} \tag{5}$$

Considering that the transfer function of the third-order model is not easy to control, and the large inertia and large time-delay characteristics of the unmanned ship can be simplified to the second-order Nomoto transfer function model, as shown in formula (6):

$$\begin{cases} G(s) = \dfrac{\psi}{\delta} = \dfrac{K'}{s(T_0 s + 1)} \\ K' = \dfrac{n_{11} m_{21} - n_{21} m_{11}}{m_{11} m_{22} - m_{12} m_{21}} \\ T_0 = T_1 + T_2 - T_3 = \dfrac{m_{11} + m_{12}}{m_{11} m_{22} - m_{12} m_{21}} - \dfrac{n_{21}}{n_{11} m_{21} - n_{21} m_{11}} \end{cases} \tag{6}$$

In the formula, K' represents the gyration index, and $T0$ represents the steering index.

2.2 Steering Gear Mathematical Model

The unmanned boat in this paper uses steering gear to control the steering of the ship, and then change the course. Generally, the steering gear system is regarded as a first-order inertial process, which is expressed by formula (7):

$$T_E\dot{\delta} = \delta_E - \delta \tag{7}$$

In the formula: TE represents the time constant of the steering gear, generally 1–3 s, and 3 s in this paper;

δ represents the actual rudder angle, limited by mechanical saturation, generally $|\delta| \leq$ 35o, and also limited by the actual speed, generally $|\,'\delta| \leq 3o/s$;

δE indicates the target rudder angle, which is determined by the control part.

The above formula can be expressed in the form of transfer function as:

$$\frac{\delta}{\delta_E} = \frac{1}{T_E s + 1} \tag{8}$$

3 Design of Heading Controller Based on Fuzzy Neural PID

3.1 Heading Controller Structure

According to the algorithm idea of combining fuzzy control and neural network, the unmanned ship heading control system based on fuzzy neural PID is built. The heading of the human-boat can be well controlled during driving, and the block diagram of the heading control system of fuzzy neural PID is shown in Fig. 2.

The particle swarm optimization algorithm maps the solution of the problem to be solved to the position of each particle, guides the particle movement through the information transmission inside the particle swarm, and finally obtains the optimal solution of the problem to be solved. Particle swarm optimization can be used to describe the D-dimension problem to be solved as:

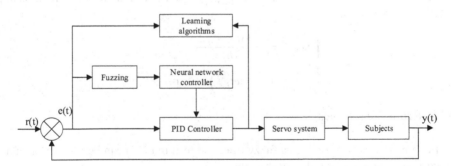

Fig. 2. Course control system based on Fuzzy Neural PID

The specific algorithm steps of building the fuzzy module in the neural network are as follows:

1) Determine the input layer

Take the deviation value e of the heading angle and the deviation change rate ec as input variables. These two variables are used as the input of the next layer, so the number of nodes in this layer is N1 = 2. The input-output relationship of this layer is shown in formula (9):

$$y_i^{(1)} = net_i^{(1)} = x_i, i = 1, 2 \tag{9}$$

2) Determine Linguistic Variables

Set the domain of discourse of the input variables to [-6, 6], and divide it into 7 parts, corresponding to NB, NM, NS, ZO, PS, PM, PB respectively. Each linguistic variable corresponds to a membership function μij. According to the scope of the variables e and ec of the simulation system and the set discourse domain, the quantization factors in the fuzzy system are determined as ke = 0.33 and kec = 0.004.

The deviation value e of the heading angle is selected as a Gaussian membership function, and the mathematical form is shown in the following formula:

$$\mu_1^j = e^{-\frac{(x_1 - c_{1j})^2}{\sigma_{1j}^2}}, j = 1, 2, \cdots, 7 \tag{10}$$

The deviation change rate ec of the heading angle is selected as a Gaussian membership function, and the mathematical form is shown in the following formula:

$$\mu_2^j = e^{-\frac{(x_2 - c_{2j})^2}{\sigma_{2j}^2}}, j = 1, 2, \cdots, 7 \tag{11}$$

In the formula, cij and σij represent the central value and width in the Gaussian membership degree, and the membership function is determined by these two parameters. The number of nodes in this layer is $N_2 = \sum_{i=1}^{2} m_i = 14$. The input-output relationship of this layer is as follows:

$$\begin{cases} net_j^{(2)} = -\frac{(x_i - a_{ij})^2}{(b_{ij})^2} \\ y_{ij}^{(2)} = f_j^{(2)}\left(net_j^{(2)}\right) = \exp(net_j^{(2)}) \end{cases} \tag{12}$$

3) Determine fuzzy rules

Each neuron corresponds to a fuzzy rule, and formula (13) can be used to calculate the corresponding applicability of the rule.

$$a_j = \mu_1^{i_1} \mu_2^{i_2}; i_1, i_2 \in \{1, 2, \cdots, 7\} \tag{13}$$

The number of nodes in the third layer is N3 = 49.

4) Normalized calculation

Nodes in the fourth layer are also 49, the purpose is to achieve normalization. The calculation formula is shown in formula (14):

$$\bar{a}_j = \frac{a_j}{\sum_{i=1}^{m} a_i}, j = 1, 2, \cdots, 49 \tag{14}$$

5) Determine the output layer

The purpose of the fifth layer is to achieve defuzzification, and the calculation formula is shown in formula (15):

$$y_j = \sum_{j=1}^{49} w_j \bar{a}_j, i = 1, 2, 3; j = 1, 2, \cdots, 49\} \tag{15}$$

Three output quantities in the output layer, which respectively represent the variation of proportional, integral and differential coefficients. According to the range of the three output quantities of the controller and the initial values of the proportional, integral and differential coefficients in the PID controller, the proportional factors in the fuzzy system are taken as kkp = 0.3, kki = 0.0025, kkd = 10.

3.2 Fuzzy Neural Network Training

Build the fuzzy neural network model as above to learn the weight wij of the output layer, the central value c_{ij} and width σ_{ij} of the input variable Gaussian membership function.
The given learning sample is:

$e = [6, 6, 6, 6, 6, 6, 6, 4, 4, 4, 4, 4, 4, 4, 2, 2, 2, 2, 2, 2, 2, 0, 0, 0, 0, 0, 0, 0,$
$-2, -2, -2, -2, -2, -2, -2, -4, -4, -4, -4, -4, -4, -4, -6, -6, -6, -6, -6, -6, -6];$
$ec = [6, 4, 2, 0, -2, -4, -6, 6, 4, 2, 0, -2, -4, -6, 6, 4, 2, 0, -2, -4,$
$-6, 6, 4, 2, 0, -2, -4, -6, 6, 4, 2, 0, -2, -4, -6, 6, 4, 2, 0, -2, -4, -6, 6, 4, 2, 0, -2, 4, 6];$
$k_{kp} = [6, 6, 4, 4, -4, 0, 0, -6, -4, -4, -4, -2, 0, 2, -4, -4, -2,$
$-2, 0, 2, 2, -4, -4, -2, 0, 2, 4, 4, -2, -2, 0, 2, 4, 4, 4, -2, 0, 2, 2, 4, 6, 6, 0, 0, 0, 2, 4, 4, 6, 6];$
$k_{ki} = [6, 6, 4, 4, 2, 0, 0, 6, 6, 4, 2, 2, 0, 0, 6, 4, 2, 2, 0, -2, -4, 4, 4,$
$2, 0, -2, -4, -4, 2, 2, 0, -2, -2, -4, -6, 0, 0, -2, -2, -4, -6, -6, 0, 0, -2, -4, -4, -6, -6];$
$k_{kd} = [6, 2, 2, 4, 4, 4, 6, 6, 2, 2, 2, 2, 2, 6, 0, 0, 0, 0, 0, 0, 0, 0, -2,$
$-2, -2, -2, -2, 0, 0, -2, -2, -4, -4, -2, 0, 0, -2, -4, -4, -6, -2, 2, 2, -4, -6, -6, -6, -2, 2];$

$$\tag{16}$$

The BP neural network training method is adopted, and the mean square error is used as the error performance index of the learning process. The transfer function type of the hidden layer neurons is selected as 'tansig', the transfer function type of the output layer neurons is 'purelin', and the backpropagation. The training function is 'trainingdm'. The maximum number of iterations is 5000, the learning rate is 0.04, the expected error is 0.005, and the momentum factor is 0.9.

After 5000 times of training, the training results of ΔKp, ΔKi and ΔKd are shown in Fig. 3, Fig. 4 and Fig. 5.

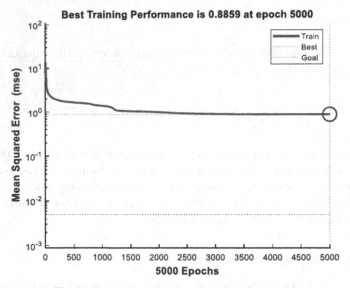

Fig. 3. Training results of neural network for ΔKP

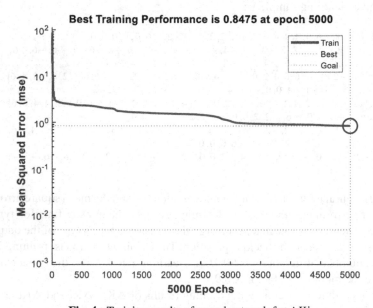

Fig. 4. Training results of neural network for ΔKi

4   Simulation and Analysis of Heading Controller for Unmanned Boat

Simulink simulation model of the heading control system of the unmanned boat is built according to the transfer function of the steering gear and the unmanned boat calculated

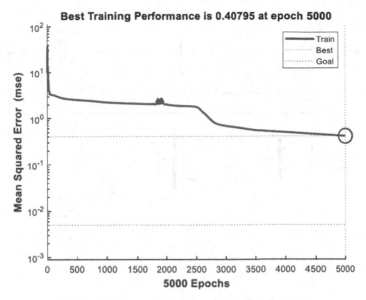

Fig. 5. Training results of neural network for ΔKd

above. According to the advantages of the fuzzy neural PID heading controller, three heading controllers including conventional PID, fuzzy PID and fuzzy neural PID are built at the same time according to the motion model of the unmanned boat and the mathematical model of the steering gear obtained above. These three heading controller simulation systems are shown in Fig. 6, Fig. 7, and Fig. 8 respectively.

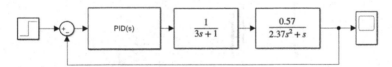

Fig. 6. Conventional PID course controller

Firstly, the PID tuning of the control system is carried out. According to the transfer function of the control object, this paper chooses the commonly used critical proportionality method to initialize the three parameters of the initial PID of the unmanned vehicle course angle control system. Let the control system adjust the size of the factor only under the action of the proportional factor until the response curve becomes constant amplitude oscillation, and obtain the initial value of the PID parameter according to the measured critical period and critical proportionality. After debugging, kp = 0.942, ki = 0.003, kd = 2.613. The input of the control system is the setting value of the heading angle, the simulation takes the heading angle as 15°, and the output is the output heading angle of the controlled unmanned boat. The simulation results are shown in Fig. 9 (Table 1).

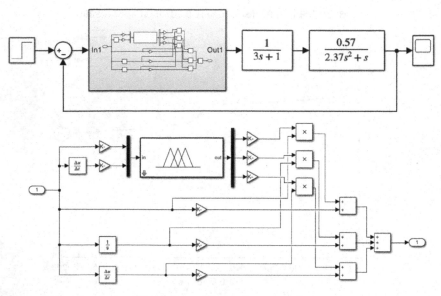

Fig. 7. Fuzzy PID heading controller

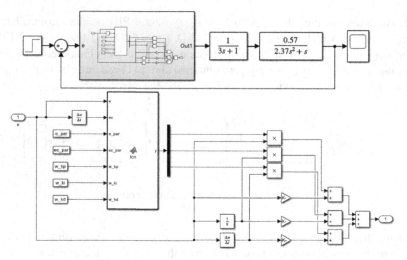

Fig. 8. Fuzzy neural PID heading controller

From the analysis of the effect of heading control and simulation results, it can be seen that the control performance of conventional PID, fuzzy PID and fuzzy neural PID on the heading angle is getting better and better, the overshoot is greatly reduced, and the rise time and adjustment time are also improved different degrees of improvement.

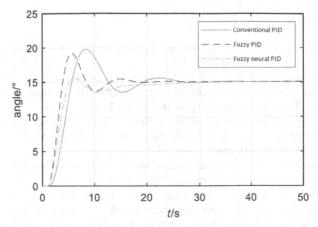

Fig. 9. Comparison of course control effect of three controllers

Table 1. Simulation results analysis of three kinds of course controllers

	Overshoot	Rise time(s)	Adjustment time ($\pm 5\%$)(s)
Conventional PID	31.9%	2.91	16.67
Fuzzy PID	28.2%	1.98	10.74
Fuzzy neural PID	4.1%	2.88	13.33

5 Conclusion

In this paper, the transfer function of the control system is firstly deduced according to the relevant parameters of the unmanned ship, and the fuzzy neural control framework is built by combining fuzzy and neural control. By learning the expected input and output samples, the BP neural network is used to control the The weights of the fuzzy neural PID heading controller, the central value and width of the input variable membership function are adjusted to realize the optimization of the controller module. Finally, in order to verify the control effect of the fuzzy neural PID heading controller, conventional PID, fuzzy PID and fuzzy neural PID heading angle control systems were respectively built in simulink and simulated and compared. The results show that the performance indicators of the fuzzy neural PID heading controller have been improved and promoted in the overshoot and adjustment time, and the superiority of the controller is reflected in the unmanned boat with large inertia and large lag.

Acknowledgement. This work is supported by the National Nature Science Foundation of China (62076249), the Key Research and Development Plan of Shandong Province (2020CXGC010701, 2020LYS11), and the Natural Science Foundation of Shandong Province (ZR2020MF154).

References

1. Zhao, D., Liu, X., Zhou, H., et al.: Design of fuzzy neural network heading controller for unmanned surface vehicles. J. Central China Normal Univ. (Nat. Sci.) (2018)
2. Khoud, K.B., Bouallègue, S., Ayadi, M.: Design and co-simulation of a fuzzy gain-scheduled PID controller based on particle swarm optimization algorithms for a quad tilt wing unmanned aerial vehicle. Trans. Inst. Meas. Control **40**(14), 3933–3952 (2018)
3. Zhang, D., Cai, Z., Lin, Q., et al.: Design of longitudinal stability controller for unmanned gyroplane based on fuzzy sliding mode theory. In: Proceedings of 2014 IEEE Chinese Guidance, Navigation and Control Conference. IEEE (2015)
4. Xia, Y., Yang, T.: CSIC: design of surface unmanned ship heading controller based on predictive function control. Ship Electron. Eng. (2019)
5. Ardeshiri, R.R., Nabiyev, N., Band, S.S., Mosavi, A.: Design and simulation of adaptive PID controller based on fuzzy Q-learning algorithm for a BLDC motor. Preprints (2020)
6. Si, H.X., Yang, S.L., Pan, S.: Optimization study of the unmanned craft's propulsion and intelligent control system based on simulated analysis. In: 2017 2nd International Conference on Machinery, Electronics and Control Simulation (MECS 2017), pp. 427–433. Atlantis Press (2016)
7. Li, M., He, Y., Ma, Y., Yao, J.: Design and implementation of a new jet-boat based unmanned surface vehicle. In: International Conference on Automatic Control and Artificial Intelligence (ACAI 2012), pp. 768–771. IET (2012)
8. Zeng, B., Song, Y., Liu, C.: Design and implementation of an unmanned boat visual target tracking system. In: 2020 Chinese Control and Decision Conference (CCDC), pp. 5225–5230. IEEE (2020)
9. Ngu, N.V., Hong-hua, W.: Design and simulation of fuzzy controller for air conditioners based on MATLAB. Sci. Technol. J. Agric. Rural Develop. (2010)
10. Lin, J., He, C., Cheng, R.: Adaptive dropout for high-dimensional expensive multiobjective optimization. Complex Intell. Syst. **8**(1), 271–285 (2021). https://doi.org/10.1007/s40747-021-00362-5
11. Du, J., Liu, G., Jia, T., Yan, H.: Design of formation controller based on BP neural network PID control. In: Wu, M., Niu, Y., Gu, M., Cheng, J. (eds.) ICAUS 2021. LNEE, vol. 861, pp. 2290–2298. Springer, Singapore (2022). https://doi.org/10.1007/978-981-16-9492-9_226
12. Li, H., Huang, F., Chen, Z.: Virtual-reality-based online simulator design with a virtual simulation system for the docking of unmanned underwater vehicle. Ocean Eng. **266**, 112780 (2022)
13. Wang, B., Wang, S., Peng, Y., Pi, Y., Luo, Y.: Design and high-order precision numerical implementation of fractional-order PI controller for PMSM speed system based on FPGA. Fractal Fractional **6**(4), 218 (2022)
14. Zhang, Y., Liu, L.: A design for preparation of textured piezoelectric ceramics based on the phase-field simulation. J. Adv. Dielectr. **12**(5), 2250015 (2022)
15. Wu, Y., Liu, H., Lu, P., Zhang, L., Yuan, F.: Design and implementation of virtual fitting system based on gesture recognition and clothing transfer algorithm. Sci. Rep. **12**(1), 18356 (2022)
16. Ramesh, T., Praveen, A.S., Pillai, P.B., Salunkhe, S.: Numerical simulation of heat sinks with different configurations for high power LED thermal management. Int. J. Simul. Multi. Design Optim. **13**, 18 (2022)

Flexible Formation Control of Multiple Unmanned Vehicles Based on Artificial Potential Field Method

Wei Wu[1](✉), Xuemei Qin[2], Jianhua Qin[3], Xiangyu Yu[1], and Qiongxiao Liu[1]

[1] China Ship Development and Design Center, Wuhan 430064, China
601600752@qq.com
[2] Wuhan Geomatics Institute, Wuhan 430022, China
[3] Wuhan Geotechnical Engineering and Surveying Co. Ltd., 430022 Wuhan, China

Abstract. Aiming at the problems of high complexity, strong coupling, and difficulty in carrying out experiments in the unmanned boat formation control algorithm at the present stage, this paper designs a multi-unmanned boat flexible formation controller based on the artificial potential field method, which will have nonlinear dynamic characteristics motion control problem of flexible formation of unmanned boats is transformed into the problem of tracking dynamic targets by unmanned boats. The controller adopts the idea of hierarchical cascading control, the planning layer introduces a virtual leader, and designs the swarm motion control mode under the condition of maintaining connectivity; the control layer uses a PID controller to help the unmanned vehicle track the dynamic continuous target generated by the planning layer, so as to realize the flexible formation motion control of multiple unmanned vehicles. Finally, the simulation verification of the flexible formation motion control algorithm is carried out for the static and dynamic virtual piloting situations.

Keywords: Artificial potential field · Flexible formation · Coordinated control

1 Introduction

The formation control problem of unmanned boats belongs to the field of multi-agent cooperative control. Its main goal is to control the system composed of multiple unmanned boats to maintain a predetermined geometric shape in the process of moving to a specific direction or target, so as to adapt environmental constraints. The formation control of unmanned boats can be divided into rigid formation control and flexible formation control. In recent years, the main research content of formation control includes the formation of unmanned boats to form a fixed formation and how to keep the formation stable during the movement. This part of the problem can be classified as rigid formation control problem, multi-UV formation switching problem, and collision avoidance and obstacle avoidance problem between UVs. These problems form the flexible formation control of the swarm movement of unmanned vehicles. The focus of solving the problem is how to switch between formations and how to change the formation motion

L. Pan et al. (Eds.): BIC-TA 2022, CCIS 1801, pp. 567–577, 2023.
https://doi.org/10.1007/978-981-99-1549-1_45

planning or formation structure to avoid obstacles. In addition, in a dynamic unknown environment, the self-adaptive problem of how the multi-UV system can automatically maintain or adjust the formation to adapt to the environment is also considered. In terms of formation control methods, it is mainly divided into centralized and distributed formation control. In terms of control strategies, there are leader-follower methods suitable for rigid formations, behavioral methods suitable for rigid and flexible formations, and artificial potential field methods mainly used for flexible formations.

Based on the leader-follower formation control strategy, a virtual formation reference point (FRP) is introduced to design the unmanned vehicle formation controller, which proves the Lyapunov stability and verifies the controller's robustness to input saturation through computer simulation. Arrichiello carried out research on the cooperative control of two unmanned boat formations under communication constraints, using a hierarchical control architecture to generate tasks, and using the NSB (null-space-based behavior) method to design behavior-based controllers for unmanned boat formations control, and carry out experimental testing. Compared with other behavior-based methods, NSB's mathematical representation is clearer and can ensure that high-priority actions are fully executed. The method has been tested in various formation and obstacle avoidance in unmanned systems such as ground robots and underwater robots, has high research value. Leonard introduced the virtual navigator, designed the artificial potential field function, designed the potential field force generated by the gradient of the position potential function and the velocity potential function, and realized the swarm motion control for the linear double integrator particle model, and carried out theoretical proof and simulation verification.

At this stage, the complexity and coupling of the UAV formation control algorithm are high, the actual experiment is difficult to carry out, and the cost of the platform used for the experiment is relatively high. Considering that the research on the flexible formation test for collision avoidance is not sufficient, this paper designs an easy-to-implement formation control algorithm based on the low-cost platform considering the flexible formation of unmanned boats for collision avoidance, and carries out actual unmanned boat formation navigation tests, which can make multiple unmanned boat formation control experiment is relatively efficient, easy and cost-effective.

2 Multi-UAV Formation Control Strategy

Inspired by the phenomena of birds flocking in pairs, fish swarms spontaneously gathering and swimming, and ant colonies cooperating in transportation, swarm motion control has attracted extensive attention from researchers in various fields. As early as 1987, Reynolds formulated a computational model of swarm movement, namely the Boids model, which defined three heuristic criteria of cohesion, separation and consistency. In order to meet the requirements of formation convergence, cohesion guarantees multi-agent distance between them should not be too far; the separation requires that the distance between the multi-agents should not be too close to avoid collisions; the consistency requires that the multi-agents finally achieve the same movement speed and direction, forming an overall cluster behavior. Based on the above model, scholars have done extensive and in-depth research on the swarm motion control theory, and one of the typical control strategies is the artificial potential field method.

The concept of artificial potential field was proposed by Khatib. The so-called artificial potential field method refers to the assumption that there is a "potential field" in the workspace of the control object, and the multi-agents in the potential field will generate mutual attraction or repulsion forces. Finally, the agent moves under the combined force of the potential field force. By constructing the artificial potential field function and minimizing the energy of the individual potential field, the purpose of multi-agent flexible formation control can be achieved. The advantage of the artificial potential field method is that the algorithm designed by it can solve the problem of collision avoidance very well and has strong real-time performance and strong online computing ability. Compared with the $l - \varphi$ rigid formation control coordination mode based on the leader-follower formation control strategy and the distance and angle between the unmanned vehicles are fixed, the artificial potential field method can make the multi-unmanned vehicle flexible formation movement more flexible and closer to in the real biological world. The main disadvantage of the artificial potential field method is that there may be a "zero potential energy point", all artificial potential fields cancel each other out, and the resultant force is zero, making it impossible for the controlled object to move forward. There are many ways to solve the defect of local minimum value. One way is to introduce a higher-level planner to make the controlled object maintain the ability of global planning. Another method can add a random disturbance when the controlled object falls into a local minimum, so as to jump out of the zero potential energy point.

3 Hierarchical Controller Design for Multiple Unmanned Boats

This section introduces the design of the multi unmanned vehicle flexible formation controller based on the artificial potential field method. The main realization idea is to divide the multi-unmanned boat is regarded as a particle, and a linear double-integral particle model is adopted. Based on the artificial potential field method and the condition of maintaining connectivity, a virtual navigator is introduced to design the swarm motion control law, and the swarm motion based on the linear double-integral particle model is realized as the planning goal of the flexible formation control of multiple unmanned vehicles. The control layer uses the PID controller to realize the unmanned vehicle with actual dynamic characteristics to track the dynamic continuous target generated by the planning layer, so as to realize the flexible formation motion control of multiple unmanned vehicles. The specific design methods of the planning layer and the control layer are introduced respectively below.

Considering a swarm motion control problem involving N agents, the dynamic equation of the linear double-integral particle model can be expressed as Eq. (1).

$$\begin{cases} \dot{r}_i = v_i \\ \dot{v}_i = u_i \end{cases} \tag{1}$$

where $i = 1, 2, \cdots, N$, $\dot{r}_i$ Represents the change in the position vector of the agent, v_i represents the velocity vector of the agent, and u_i represents the control input.

Figure 1 is a schematic diagram of the coordinates considering the swarm movement of N agents. In the relative coordinate system with the virtual leader as the reference point, $r_i = (x_i, y_i)^T$ represents the position vector of the agent, the position vector of the agent relative to the leader is $\hat{r}_i = r_i - r_L$, and the relative velocity vector is expressed as $\dot{\hat{r}}_i = v_i - v_L$.

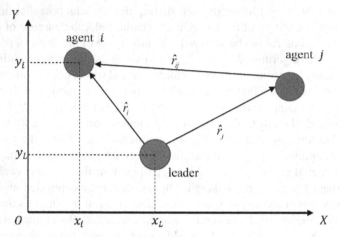

Fig. 1. Schematic diagram of multi-agent cluster motion coordinates

Then the kinematic equation of the agent i relative to the virtual leader can be expressed as formula (2)

$$\frac{d}{dt}\begin{pmatrix} \hat{r}_i \\ \dot{\hat{r}}_i \end{pmatrix} = \begin{pmatrix} \dot{\hat{r}} \\ u_i - \dot{v}_L \end{pmatrix} \tag{2}$$

Define the positional potential function v_{ij} between agents and the positional potential field force f_{ij} generated by the potential function as formula (3), (4).

$$V_{ij} = \begin{cases} a_1\left(\ln(\hat{r}_{ij}) + \dfrac{d_0}{\hat{r}_{ij}}\right), 0 < \hat{r}_{ij} < d_1 \\ a_1\left(\ln(d_1) + \dfrac{d_0}{d_1}\right), \hat{r}_{ij} \geq d_1 \end{cases} \tag{3}$$

$$f_{ij} = \begin{cases} \nabla_{\hat{r}_{ij}} V_{ij}, 0 < \hat{r}_{ij} < d_1 \\ 0, \hat{r}_{ij} \geq d_1 \end{cases} \tag{4}$$

where a_1, d_1, d_0 is a constant greater than 0, and the gradient of the positional potential function represents the positional potential field force between the agents. When $\hat{r}_{ij} < d_0$, $f_{ij} < 0$, a repulsive force is generated between the agent i and j by the positional potential function. When $d_0 < \hat{r}_{ij} < d_1, f_{ij} > 0$, there is an attraction between agents i and j by the position potential function, when $\hat{r}_{ij} \geq d_1, f_{ij} > 0$, there is no interaction between agents i and j.

Define the position potential function V_h between the agent and the virtual leader and the position potential field force f_h generated by the potential function, such as formula

(5), (6).

$$V_h = \begin{cases} a_2\left(\ln(\hat{r}_i) + \dfrac{h_0}{\hat{r}_{ij}}\right), 0 < \hat{r}_i < h_1 \\ a_2\left(\ln(h_1) + \dfrac{h_0}{h_1}\right), \hat{r}_i \geq h_1 \end{cases} \tag{5}$$

$$f_h = \begin{cases} \nabla_{\hat{r}_i} V_h, 0 < \hat{r}_i < h_1 \\ 0, \hat{r}_i \geq h_1 \end{cases} \tag{6}$$

where a_2, h_1, h_0 is a constant greater than 0.

In addition, the velocity potential field f_v between the agent i and the virtual leader is defined as formula (7). When the velocity and direction of the agent i are consistent with the virtual leader, f_v is 0.

$$f_v = -\alpha \dot{\hat{r}}_i, \alpha > 0 \tag{7}$$

Assuming the adjacency graph composed of agent clusters is connected, the design control is shown in Eq. (8).

$$u_i = -\sum_{\substack{j \in N_i \\ j \neq 1}} \nabla_{\hat{r}_{ij}} V_{ij} \hat{r}_{ij} - \nabla_{\hat{r}_i} V_h(\hat{r}_i) - \alpha \dot{\hat{r}}_i \tag{8}$$

where N_i represents the adjacency set of the agent i, V_{ij} is the position potential function between nodes, and V_h is the position potential function between nodes and the virtual leader.

Lemma 1: When a cluster has N An agent whose dynamics is described as formula (1) follows a virtual leader according to the control law (8). Suppose the adjacency graph of the cluster is connected and time-invariant, then all connected agents in the cluster system can achieve collision avoidance, and the speed of all agents will asymptotically converge to the speed of the virtual leader, and the total energy inside the cluster system will reach the minimum, the shape of the multi-agent will tend to be stable.

prove: k

Considering the Lyapunov function of the cluster system corresponds to the total kinetic energy and potential energy inside the cluster system, define the Lyapunov function of the following formula (9).

$$\Phi = \frac{1}{2} \sum_{i=1}^{N} \left(\dot{\hat{r}}_r^T \cdot \dot{\hat{r}} + i \sum_{\substack{j \in N_i \\ j \neq 1}} V_{ij}(\hat{r}_{ij}) + 2V_h(\hat{r}_i)\right) \tag{9}$$

when $\Phi \leq c, c > 0$, it can be seen from the continuity of the function that the set $\Omega = \left\{(\hat{r}_i, \hat{r}_{ij}) | \Phi \leq c\right\}$ is a closed set. Since $\Phi \leq c$ shows, $\hat{r}_i \cdot \hat{r}_{ij} \leq c$, then there is

$\left\| \dot{\hat{r}}_i \right\| \leq \sqrt{c}$. By the same token, $V_{ij} \leq c$. When the nodes i and j are adjacent nodes, $\Omega = \left\{ (\dot{\hat{r}}_i, \hat{r}_{ij}) | \Phi \leq c \right\}$ is known. From the connectivity of the graph, it can be known that the longest path between any nodes is N, so there is $\left\| r_{ij} \right\| \leq V_{ij}^{-1}(cN)$. The available set $\Omega = \left\{ (\dot{\hat{r}}_i, \hat{r}_{ij}) | \Phi \leq c \right\}$ is a compact set. Calculate the dericative of Φ with respect to time and combine the vertical formula (8) to get the formula (10):

$$\Phi = \sum_{i=1}^{N} \dot{r}_r^T \left(ui + \sum_{\substack{j \in N_i \\ j \neq 1}} \nabla_{\hat{r}_i} V_{ij}(\hat{r}_{ij}) + \nabla_{\hat{r}_i} V_h(\hat{r}_i) \right)$$

$$= -\alpha \sum_{i=1}^{N} \dot{\hat{r}}_r^T \cdot \dot{\hat{r}}_i \leq 0$$

(10)

It can be seen that Φ is semi-negative definite. From the above analysis, the system initially starts from the compact set Ω, and because of $\Phi \leq 0$, all the signals of the system still belong to the compact set Ω. It can be seen from formula (10) that if and only if $\dot{\hat{r}}_i = 0$, $\Phi = 0$, and from the LaSalle invariant set principle, the system trajectory will converge to the largest invariant set in the region $E = \{\hat{r}_i | \Phi = 0\}$, and the system will eventually converge to the equilibrium point at the origin asymptotically, it is noted that the compact set Ω can be arbitrarily large, so the semi-global asymptotic stability of the system can be guaranteed. At the same time, if there is a collision between interconnected nodes, it will cause $V_{ij} \to \infty$, and the system trajectory will leave Ω, resulting in a contradiction. Therefore, the designed control law meets the collision avoidance requirements between adjacent nodes.

The above swarm motion control problem is studied for the linear double integrator particle model. However, the actual unmanned vehicle dynamics equation has nonlinear characteristics and cannot be dealt with directly according to the linear particle model. In order to realize the flexible formation control of multiple unmanned boats, based on the idea of layered control, a layered controller for flexible formation of multiple unmanned boats is proposed. The hierarchical control idea decouples the flexible formation control problem of multiple unmanned vehicles into a multi-agent cluster control planning layer based on the artificial potential field method and a motion control layer of the dynamic target point tracking algorithm of the unmanned vehicle. The schematic diagram of the control idea is as follows: As shown in Fig. 2.

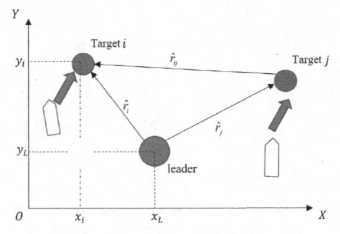

Fig. 2. Schematic diagram of control ideas

This section introduces the control law design of N particle models under the condition of maintaining connectivity under the guidance of a virtual navigator and proves the stability, ensures the collision avoidance between nodes, and realizes the cluster motion of N particle points in the planning layer. By obtaining the real-time position of the particle model in the planning layer, it is used as the tracking target of the unmanned vehicle in the control layer.

4 Simulation Verification and Analysis

In order to verify that the layered controller based on the artificial potential field method can realize the flexible formation control of multiple unmanned boats, this section conducts simulation experiments based on the Matlab platform. In the experimental verification, the unmanned vehicle dynamics model is introduced, and the multi-unmanned vehicle flexible formation control simulation is carried out by using the layered control method. The flexible formation includes two forms: linear flexible formation and curved flexible formation (Figs. 3, 4, and 5).

Add N unmanned boats in the simulation experiment. See the second chapter for the model and parameter description. Set the initial speed of the unmanned boat to 0 m/s, the initial heading to 0°, the initial position and the corresponding planning target initial position is the same. The virtual navigator starts from the origin and moves in a straight line along the y-axis at a fixed speed of 1 m/s. Under the condition that the virtual navigator moves along a straight line, the simulation results of the flexible formation control of multiple unmanned boats are as follows:

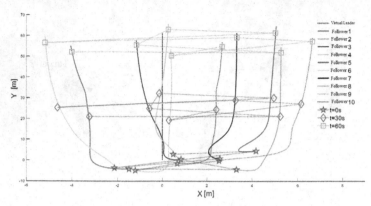

Fig. 3. Trajectory curve of multi-unmanned boat flexible formation in straight-line piloting mode

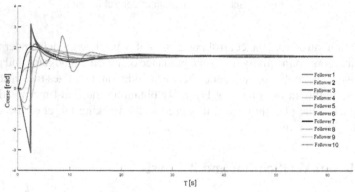

Fig. 4. The course change curve of the follower in the straight-line pilot mode of flexible formation of multiple unmanned boats.

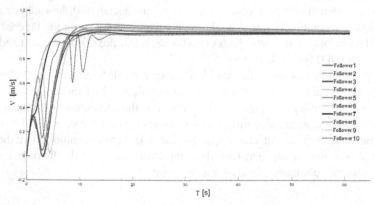

Fig. 5. Velocity change curve of multi-unmanned boat flexible formation in straight-line piloting mode

Further, for the case of curved flexible formation. Under the condition that other simulation conditions remain unchanged, the virtual navigator moves according to the kinematic trajectory of formula (11), and guides multiple unmanned vehicles to form a flexible formation according to the circular trajectory. The simulation results are as follows.

$$\begin{cases} u = 0.5\sin(0.01\pi t) \\ v = 0.5\cos(0.01\pi t) \end{cases} \tag{11}$$

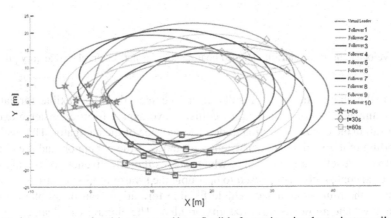

Fig. 6. Trajectory curve of multi-unmanned boat flexible formation circular trajectory pilot mode.

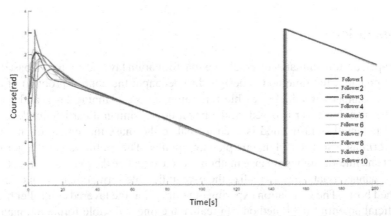

Fig. 7. The course change curve of multi-unmanned boat flexible formation circular trajectory pilot mode.

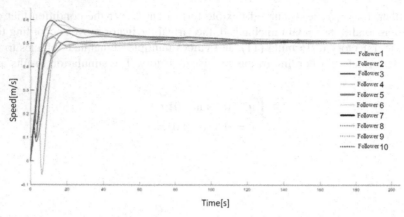

Fig. 8. Velocity curves of multi-unmanned boat flexible formation circular trajectory pilot mode

From the above simulation results, it can be seen that the flexible formation of multiple unmanned boats can achieve collision avoidance, form a cluster movement, and can sail in the desired direction under the guidance of the navigator. The final speed and heading of the vehicle are consistent with those of the navigator, and the output of the controller reaches a steady state, which proves the effectiveness of the algorithm. This layered control concept is used to decouple the swarm motion control consistency problem in the multi-unmanned vehicle flexible formation control problem from the unmanned vehicle motion control stability problem. The hierarchy is clear, concise, and easy to expand the algorithm and Realization has a certain degree of innovation and has guiding significance for engineering practice (Figs. 6, 7, and 8).

5 Conclusion

In this paper, a multi-unmanned vehicle flexible formation layered controller based on the artificial potential field method is designed to decouple the control problem of a large number of unmanned vehicle flexible formations into a planning layer and a control layer. The planning layer is based on the linear double-integral particle model, and the artificial potential field method is used to realize the mass movement of the particle model, forming a series of dynamic planning points; The motion control algorithm of the point and line-of-sight guidance method is used to realize the dynamic target tracking of the unmanned boat, so as to realize the overall flexible formation control of multiple unmanned boats. The simulation experiment verifies that the layered controller based on the artificial potential field method can realize the linear flexible formation navigation and the curved flexible formation navigation of multiple unmanned vehicles.

References

1. Zhang, J., Yan, J., Zhang, P.: Fixed-wing UAV formation control design with collision avoidance based on an improved artificial potential field. IEEE Access **6**, 78342–78351 (2018)

2. Hang, Y., Cam, L., Roy, U.: Formation control for multiple unmanned aerial vehicles in constrained space using modified artificial potential field. Math. Model. Eng. Prob. **4**(2), 100–105 (2017)
3. Zhai, H., Ji, Z., Gao, J.: Formation control of multiple robot fishes based on artificial potential field and leader-follower framework. In: Control and Decision Conference. IEEE (2013)
4. Lin, J., Pan, L.: Multiobjective trajectory optimization with a cutting and padding encoding strategy for single-UAV-assisted mobile edge computing system. Swarm Evol. Comput. **75**, 101163 (2022)
5. Zha, M., Wang, Z., Feng, J., et al.: Unmanned vehicle route planning based on improved artificial potential field method. J. Phys: Conf. Ser. **1453**(1), 012059 (2020)
6. Morais, C., Nascimento, T., Brito, A., et al.: A 3D anti-collision system based on artificial potential field method for a mobile robot. In: 9th International Conference on Agents and Artificial Intelligence (2017)
7. Hu, J., Wang, M., Zhao, C., Pan, Q., Du, C.: Formation control and collision avoidance for multi-UAV systems based on Voronoi partition. Sci. China Technol. Sci. **63**(1), 65–72 (2019). https://doi.org/10.1007/s11431-018-9449-9
8. Zhang, M., Liu, Z., Li, H., et al.: Leader-follower formation control of unmanned aerial vehicles based on active disturbances rejection control. In: 2019 4th International Conference (2019)
9. Xin, L., Zhu, D., Qian, Y.: A survey on formation control algorithms for multi-AUV system. Unmanned Syst. **2**(04), 351–359 (2014)
10. Wen, N., Zhao, L., Zhang, R.B., et al.: Online paths planning method for unmanned surface vehicles based on rapidly exploring random tree and a cooperative potential field. Int. J. Adv. Rob. Syst. **19**(2), 267–283 (2022)
11. Liao, Y., Jia, Z., Zhang, W., et al.: Layered berthing method and experiment of unmanned surface vehicle based on multiple constraints analysis. Appl. Ocean Res. **86**, 47–60 (2019)
12. Wang, P., Song, C., Dong, R., Zhang, P., Yu, S., Zhang, H.: Research on obstacle avoidance gait planning of quadruped crawling robot based on slope terrain recognition. Ind. Robot: Int. J. Robot. Res. Appl. **49**(5), 1008–1021 (2022)
13. De Silva, D.: Formation Control for Unmanned Aerial Vehicles. Technical University of Lisbon, Portugal, ISR & Instituto Superior Tecnico (2012)
14. Sabiha, A.D., Said, E., Kamel, M.A., et al.: Trajectory generation and tracking control of an autonomous vehicle based on artificial potential field and optimized backstepping controller. In: 2020 12th International Conference on Electrical Engineering (ICEENG). IEEE (2020)
15. Liu, X.: Two-dimensional path planning for unmanned aerial vehicles based on artificial potential field method. Ship Sci. Technol. **39**, 73–75 (2017)

Local Path Planning Combined with the Motion State of Dynamic Obstacles

Yuxin Hu[1], Jun Li[1], and Lin Jiang[2(✉)]

[1] Key Laboratory of Metallurgical Equipment and Control Technology, Wuhan University of Science and Technology, Wuhan 430081, China
[2] Institute of Robotics and Intelligent Systems, Wuhan University of Science and Technology, Wuhan 430081, China
jianglin76@wust.edu.cn

Abstract. Aiming at the problem that the current local path planning algorithm treats moving obstacles as transient static ones when avoiding obstacles, lacking initiative and security, a local path planning algorithm combining dynamic obstacle motion state is proposed. Firstly, the position information of all obstacles is obtained by a single scan LIDAR. Then, segmenting and linearly fitting all laser point clouds, removing the known obstacles in the map; and marking possible emerging dynamic obstacles with circles of appropriate size; then detecting the dynamic obstacles through the change of the coordinates of the center of the circle, obtaining the center coordinates of the dynamic obstacles continuously; thus, the equation of motion is solved by using the least squares. Finally, the cost map of dynamic obstacles is expanded in the direction of velocity, by combining DWA dynamic window method to realize avoiding dynamic obstacles safely. In this paper, through the simulation experiments and real environment experiments of robot meeting scenes from multiple angles, the results show that expanding the cost map of the motion direction of the dynamic obstacles provides more sufficient response space for the robot to avoid the dynamic obstacles. Compared with the original DWA, the algorithm in this paper can avoid dynamic obstacles more safely, which proves the security and feasibility of this algorithm.

Keywords: Obstacle Avoidance · Dynamic Obstacle Detection · Motion State Solution · Cost Map · DWA Algorithm

1 Introduction

People often divide path planning methods into global path planning [1, 2] and local path planning [3]. Local path planner controls the robot to avoid new static or dynamic obstacles in the environment. At present, the most widely used local path planning algorithms are: artificial potential field method, DWA (Dynamic Window Method) and TEB (Timed Elastic Band) algorithm. Artificial potential field method is to establish the potential field model of the environment artificially, assuming that the target point is the gravitational field model and the obstacles are the repulsive force field model, by adding their vectors, the calculated resultant force direction is the direction of the local

L. Pan et al. (Eds.): BIC-TA 2022, CCIS 1801, pp. 578–590, 2023.
https://doi.org/10.1007/978-981-99-1549-1_46

path. Fei Xu [4] proposed an improved artificial potential field method. Zhenzhong Yu [5] et al. proposed three improvements to the traditional artificial potential field method. MONTIELO [6] et al. proposed a path planning algorithm called bacterial potential field method for calculating the best path in static and dynamic environments. The dynamic window method is that the robot samples in the velocity space, obtains several paths in a sector area, scores the paths according to different constraints, and finally selects the path with the highest score as the local path. Hongbin Wang [7] et al. improved dynamic window method, which ensured that the final dynamic path was globally optimal. Yu Zhang [8] et al. also improved the dynamic window method, solved the problem that the tire vertical load was too small due to the robot's excessive acceleration. LIMADAD [9] et al. proposed a safe navigation method for unmanned vehicles. TEB path planning, it is a path formed by connecting local starting points and target points in the global path. At the same time, this path is endowed with deformable characteristics. The condition of deformation is the repulsive force acted by all obstacles, and the movement time constraint is defined between the starting point and the target point. KELLERM [10] et al. proposed a vehicle collision avoidance optimal trajectory planning problem based on time elastic band (TEB) framework. The algorithm was optimized for the target with multiple partial conflicts. Kailin Zheng [11] et al. improved the local path planner of TEB and used it in the motion planning of Ackerman robot. When avoiding moving obstacles, all the above improved algorithms regard them as instantaneous static obstacles for obstacle avoidance, which lacks initiative and security, and has no good adaptability to dynamic obstacles.

2   Calculate Laser Point Information on Obstacles

At present, the working principle of most laser radars is realized by TOF (Time of Flight) [12], its main components are laser transmitter, laser receiver and timer. When the transmitter emits a laser pulse, the timer starts counting. When the receiver receives the reflected pulse, the timer stops counting to get a time lag, and the distance between the obstacle and the radar is calculated by the flight time of light, so access to the laser radar Angle and distance information.

In order to further realize path planning, obstacle avoidance, navigation and other functions, the robot should first estimate the current pose information of the robot through matching the sensor data with the environment in real time. This article uses the AMCL (Adaptive Monte Carlo Localization) [13] algorithm for robot localization. After initialization of particles, state prediction, weight updating, weight normalization and resampling, the robot's own position estimation is obtained. Then, the absolute position information of the obstacles can be calculated according to the information returned by the lidar.

Laser point cloud is shown in Fig. 1, there is a simulation environment, which includes mobile robot with various sensors such as laser radar and a movable obstacle. The laser radar on the robot continuously releases laser data. In the environment display interface, and the laser hits the obstacle and the environment to form a laser point cloud. According to calculation, the position information of several laser points hitting the obstacle can be obtained. Considering that the larger the distance from the lidar, the sparser the laser points, only the laser data within a certain distance range are taken for calculation in the process of processing.

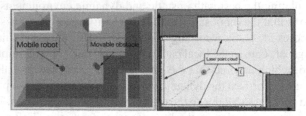

Fig. 1. Laser point cloud image

3 Judgment of Dynamic Obstacles and Solution of Motion State

3.1 Segmentation and Straight-Line Fitting of Laser Data

The coordinates of each laser point can be obtained from calculation above, and all laser points within the threshold range are retained to form laser point cloud $\left(X_{obs}^{i}, Y_{obs}^{i}\right) (i = 1, 2, \ldots, m)$, where m is the total number of laser points meeting the requirements, and all laser data are traversed. The square D_{k-1}^{k} of the distance between adjacent laser coordinates can be obtained from Eq. (1):

$$D_{k-1}^{k} = \left(\left(X_{obs}^{k} - X_{obs}^{k-1}\right)^{2} + \left(Y_{obs}^{k} - Y_{obs}^{k-1}\right)^{2}\right) \tag{1}$$

D_{k-1}^{k} represents the square of the distance between the k laser point and the $k-1$ laser point, at the same time set a distance threshold value $T_{k-1}^{k} = \left((r + p*\rho_{k})^{2}\right)$, in which r is a threshold constant, p is a proportional constant, and (ρ_{k}, θ_{k}) is the extreme distance of the k laser data, if $D_{k-1}^{k} > T_{k-1}^{k}$, the k laser data is judged as a breakpoint for the first time, iterate through all the data and separate all the laser points according to the breakpoint, set the minimum number of data points in each group to be P_{min}, laser groups with laser points less than P_{min} are removed.

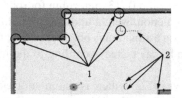

Fig. 2. Laser data grouping basis

As shown in Fig. 2, the grouping at this time only distinguishes the laser data of category "2" in the figure, and it is impossible to directly fit each group of laser data straight line, because the vertex of category "1" has not been judged. Let the k laser data be the breakpoint of the first set of laser coordinates in the above step, then the first set of data is (X_{obs}^i, Y_{obs}^i) $(i = 1, 2, \ldots, k - 1)$, and the least square [14] is used to achieve straight line fitting. Let the fitted straight line be $Y = a * X + b$, with parameters a and b, and solve it according to (2):

$$\begin{cases} a = ((k - 1)*C - B * D)/((k - 1)*A - B * B) \\ b = (A * D - B * C)/((k - 1)*A - B * B) \\ A = sum\left(X_{obs}^i * X_{obs}^i\right) \\ B = sum\left(X_{obs}^i\right) \\ C = sum\left(X_{obs}^i * Y_{obs}^i\right) \\ D = sum\left(Y_{obs}^i\right) \end{cases} \tag{2}$$

The two ends of the fitted straight line are removed, and the distance between $k - 3$ laser points in the middle of this set of data and the straight line is calculated. The distance between the nth point and the straight line is set as D_n, which can be obtained from Eq. (3):

$$D_n = \left| a*X_{obs}^n - Y_{obs}^n + b \right| / \sqrt{(a^2 + 1)} \tag{3}$$

Set the distance threshold $D_{max}^n = (\beta + p*\rho_n)$, where β is the threshold constant of the second segmentation, p is the scale constant, ρ_n is the distance extremum value of the nth laser data (ρ_n, θ_n), if $D_n > D_{max}^n$, the nth laser data is judged to be the "class 1" segmentation point [15] (Fig. 3).

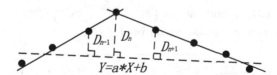

Fig. 3. "Class 1" laser data grouping basis

Loop the above two steps until all laser data are grouped, a straight line can be fitted to each group of laser data, and the straight-line equation of each group of data can be calculated according to the least square method in above step. In order to avoid the problem of over-segmentation in the above process, adjacent straight lines are taken for comparison, and L_m and L_{m+1} are set as the two adjacent straight lines. The equation of straight line L_m is: $Y = a_m*X + b_m$, and the last point on the straight line is (X_m^{last}, Y_m^{last}), and the equation of the straight line L_{m+1} is: $Y = a_{m+1}*X + b_{m+1}$, $(X_{m+1}^{first}, Y_{m+1}^{first})$ is the first point on the straight line, if $|a_m - a_{m+1}| < a_{max}$, and $|b_m - b_{m+1}| < b_{max}$ and $\left(\left(X_m^{last} - X_{m+1}^{first}\right)^2 + \left(Y_m^{last} - Y_{m+1}^{first}\right)^2\right) < T_{Lm}^{Lm+1}$, in which a_{max} is the slope threshold, b_{max} is the intercept threshold, $T_{L_m}^{L_{m+1}}$ is the distance threshold, $T_{L_m}^{L_{m+1}} = \left(\left(r + p*\rho_{m+1}^{first} \right)^2 \right)$, r is the threshold constant, p is the proportional constant, and ρ_{m+1}^{first} is the distance extreme value of the first laser data $(\rho_{m+1}^{first}, \theta_{m+1}^{first})$ in straight line L_{m+1}, then it is judged that L_m and L_{m+1} are the same, on the contrary, it is not processed. The line obtained after fitting is shown in Fig. 4.

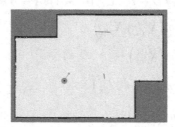

Fig. 4. Straight line obtained after laser point cloud fitting

3.2 Dynamic Obstacles Detection and Motion Solution

To detect dynamic obstacles, it should remove existing obstacles and new static obstacles in the environment. The straight lines obtained by fitting are compared with the established two-dimensional raster map. If the straight lines are on the raster map, they will not be processed below. In order to extract the center point of the possible dynamic obstacles, we assume that the obstacles are circular, and the real-time position of the obstacles can be obtained by calculating its center. Take the remaining lines, and let each straight line be one side of an equilateral triangle inside a circle. If there is a straight line $L_j : Y = a_j * X + b_j$, its two endpoints are $\left(X_j^{first}, Y_j^{first} \right)$ and $\left(X_j^{last}, Y_j^{last} \right)$, respectively. The radius R_1 of a circle can be obtained from Eq. (4):

$$R_1 = \frac{\sqrt{3}}{3} \sqrt{\left(X_j^{last} - X_j^{first} \right)^2 + \left(Y_j^{last} - Y_j^{first} \right)^2} \tag{4}$$

To ensure that the obstacles are fully contained, an additional boundary B_r is set, then the radius R of the final circle is:

$$R = R_1 + B_r = \frac{\sqrt{3}}{3} \sqrt{\left(X_j^{last} - X_j^{first} \right)^2 + \left(Y_j^{last} - Y_j^{first} \right)^2} + B_r \tag{5}$$

To filter out some excessive non-dynamic obstacles in the environment, the maximum allowable radius of the circle is set as R_{max}. If $R < R_{max}$, the center coordinate $\left(X_{center}^j, Y_{center}^j \right)$ of the circle is obtained, as shown in Eq. (6):

$$\left(X_{center}^j, Y_{center}^j \right) = \frac{\left(X_j^{last} + X_j^{first}, Y_j^{last} + Y_j^{first} \right)}{2} - \frac{R*\left(Y_j^{first} - Y_j^{last}, X_j^{last} - X_j^{first} \right)}{2\sqrt{\left(X_j^{last} - X_j^{first} \right)^2 + \left(Y_j^{last} - Y_j^{first} \right)^2}} \tag{6}$$

At this point, all unknown objects in the original environment which may be dynamic obstacles are enclosed in circles, as shown in Fig. 5.

Fig. 5. Marked possible dynamic obstacles unknown in the environment

The coordinates of the center of the circle of the marked obstacles and the corresponding time are obtained. Set the possible dynamic obstacles $\left(X_{center}^{j}, Y_{center}^{j}\right)$ at time T_1, the center coordinates are set as $\left(X_{center}^{j-T_1}, Y_{center}^{j-T_1}\right)$, and the coordinates of the center of the circle at time T_2 are set as $\left(X_{center}^{j-T_2}, Y_{center}^{j-T_2}\right)$. The velocity threshold is set as V_{max}, and the instantaneous velocity V_i is calculated.

$$V_i = \frac{\sqrt{\left(X_{center}^{j-T_2} - X_{center}^{j-T_1}\right)^2 + \left(Y_{center}^{j-T_2} - Y_{center}^{j-T_1}\right)^2}}{(T_2 - T_1)} \tag{7}$$

If $V_i < V_{max}$, it is judged as a static obstacle and is not processed. If $V_i > V_{max}$, it is judged as a dynamic obstacle, the latest first $G(G > 0)$ coordinate points are constantly updated, and the least square is used to achieve the linear fitting $Y_{center} = a_c * X_{center} + b_c$, and this equation is used as the motion state equation of the dynamic obstacles (See Fig. 6).

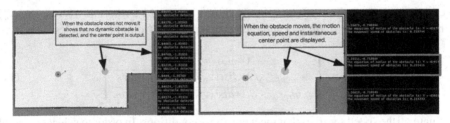

Fig. 6. Dynamic obstacles detection and motion solution

4 DWA Local Path Planner Combined Dynamic Obstacles Motion Information

4.1 Implementation of DWA Algorithm

The robot used in the following test is a differential moving robot, which can only advance and rotate, and the information of the code plate can be obtained for calculation within a short distance [16]. Assume that the motion of the robot before two adjacent points is a straight line, and the speed of the robot is $(v_{robot}, \omega_{robot})$. Within the time interval δt, the motion model of the robot can be expressed in Eq. (8).

$$
\begin{cases}
X_{robot} = X_{robot} + v_{robot} * \Delta t * cos(\sigma_{robot}) \\
Y_{robot} = Y_{robot} + v_{robot} * \Delta t * sin(\sigma_{robot}) \\
\sigma_{robot} = \sigma_{robot} + \omega_{robot} * \Delta t
\end{cases}
\tag{8}
$$

After the motion model of the robot is known, the trajectory space can be calculated according to its velocity. If sampling is conducted only according to the velocity $(v_{robot}, \omega_{robot})$ of the robot, there will be infinite sets of data, so the sampling space needs to be limited according to the characteristics of the robot itself and environmental restrictions.

The speed limit of the mobile robot is shown in Eq. (9), where v_{robot}^{min} and ω_{robot}^{min} are the minimum linear velocity and angular velocity of the robot, v_{robot}^{max} and ω_{robot}^{max} are the maximum linear velocity and angular velocity of the robot.

$$
V_m = \left\{
\begin{array}{c}
(v_{robot}, \omega_{robot}) | v_{robot} \in \left[v_{robot}^{min}, v_{robot}^{max} \right] \\
\wedge \, \omega_{robot} \in \left[\omega_{robot}^{min}, \omega_{robot}^{max} \right]
\end{array}
\right\}
\tag{9}
$$

The actual speed that the acceleration limit of the mobile robot can achieve is shown in Eq. (10), where v_c and ω_c are the current linear and angular velocity of the robot, $\dot{v}_b$ and $\dot{\omega}_b$ are the maximum deceleration velocity, and $\dot{v}_a$ and $\dot{\omega}_a$ are the maximum acceleration.

$$
V_d = \left\{
\begin{array}{c}
(v_{robot}, \omega_{robot}) \\
| v_{robot} \in [v_c - \dot{v}_b * \Delta t, \, v_c + \dot{v}_a * \Delta t] \\
\wedge \, \omega_{robot} \in [\omega_c - \dot{\omega}_b * \Delta t, \, \omega_c + \dot{\omega}_a * \Delta t]
\end{array}
\right\}
\tag{10}
$$

The braking distance limit of the robot is shown in Eq. (11), where $dist(v_{robot}, \omega_{robot})$ is the robot speed $(v_{robot}, \omega_{robot})$ and the nearest distance from the obstacles.

$$
V_a = \left\{
\begin{array}{c}
(v_{robot}, \omega_{robot}) \\
| v_{robot} \leq \sqrt{2 * dist(v_{robot}, \omega_{robot}) * \dot{v}_b} \\
\wedge \, \omega_{robot} \leq \sqrt{2 * dist(v_{robot}, \omega_{robot}) * \dot{\omega}_b}
\end{array}
\right\}
\tag{11}
$$

Evaluate all the speed samples that meet the above conditions. In order to select the speed combination with the highest score, the evaluation function is designed as Formula (12), where $head(v_{robot}, \omega_{robot})$ is the azimuth evaluation function, $dist(v_{robot}, \omega_{robot})$ is the distance evaluation function, $vel(v_{robot}, \omega_{robot})$ is the speed evaluation function, φ is the smoothing coefficient, and μ, β, and δ are the coefficient terms of the three evaluation functions.

$$
G(v_{robot}, \omega_{robot}) = \varphi \left(
\begin{array}{c}
\mu * head(v_{robot}, \omega_{robot}) \\
+ \beta * dist(v_{robot}, \omega_{robot}) \\
+ \delta * vel(v_{robot}, \omega_{robot})
\end{array}
\right)
\tag{12}
$$

4.2 Expand the Cost Map of Dynamic Obstacles Velocity Direction

In the process of navigation, the robot will expand the environment map and all obstacles that appear by a safe distance [17], that is the Cost map, however, there is no difference in the way that it deals with dynamic obstacles. In order to make DWA local path planner better realize the security of robot obstacle avoidance in the environment of dynamic obstacles, after obtaining the motion state of the obstacles, the expansion area of the speed direction of the dynamic obstacles is increased, and then the DWA dynamic window method is used for navigation [18].

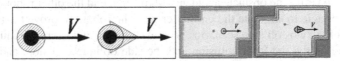

Fig. 7. Cost map of dynamic obstacles

Its expansion rules as shown in Fig. 7, with a larger area of the isosceles triangle overlap the original expansion, after the moving straight line equation of the obstacles being obtained, and the moving direction of the obstacles are taken as the upper vertex of the isosceles triangle area, the size of the triangle can be adjusted according to the moving speed of the obstacles, calculate the straight line equation of the three sides of the triangle, then, some points on the three sides and inside of the triangle are taken as expansion positions to simulate the artificial expansion area. The overall flow chart of the algorithm is shown in Fig. 8.

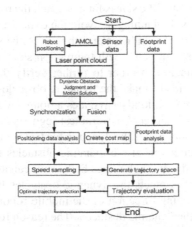

Fig. 8. Overall flow chart of the algorithm

5 Test Results and Simulation

5.1 Simulation Test and Results

To verify the feasibility of the algorithm proposed in this paper, a simulation experiment is carried out first. In the ROS (Robot Operating System) robot operating system under Ubuntu16.04 system, Gazebo and Rviz are used for co-simulation. In the simulation test, a circular chassis is used to build a wheeled differential robot, on which odometry information can be obtained. A lidar sensor is placed above the center of the chassis, and environmental information can be scanned by lidar [19]. In the simulation test, one robot is used to verify the obstacle avoidance algorithm, and the other robot only acts as a dynamic obstacle to simulate the encounter with the robot. To prove the universality of this algorithm, two collision scenarios, frontal collision, and lateral oblique collision, are simulated firstly, the front collision and the side oblique collision, and compared with the original DWA algorithm without any processing, and verified one by one.

The case of positive collision. According to the algorithm in this paper, after judging the motion state of the dynamic obstacles, the cost map of its velocity direction is expanded. When the obstacles are near the robot, the robot will have more sufficient judgment space and turn in advance to avoid the dynamic obstacles. If nothing is done to deal with the dynamic obstacles, the robot will treat the obstacles as static, so that when they are very close to each other, the robot will react, resulting in collision.

The oblique collision situation. When the robot meets the dynamic obstacles in front of the side, the original DWA will plan the avoidance path in its motion direction when approaching the obstacles, but the obstacles are still moving, and the robot will not react enough to cause the collision. After expanding the cost map of the speed direction of the dynamic obstacles, combined with DWA algorithm, the distance between the robot and the dynamic obstacles in front of the side is far away, and the robot has the reaction space to carry out the quadratic path planning to avoid the dynamic obstacles. It can be seen from the above-mentioned frontal collision and oblique collision simulation experiments that the local path planning combined with the motion state of dynamic obstacles can easily avoid dynamic obstacles, in order to further verify the algorithm's versatility, build a complex test scenarios to make the dynamic obstacles move back and forth in an interval, the mobile robot randomly navigates among several poses to simulate the normal encounter scene, and intercepts several test segments of the encounter scene.

In the first scene, the robot meets the dynamic obstacles in an oblique direction, the cost map of the moving direction of the dynamic obstacles is expanded. The mobile robot is far away from the dynamic obstacles body, so it can avoid easily. In the second encounter scenario, the robot meets the dynamic obstacles head-on. As the dynamic obstacles approach, the moving direction of the mobile robot is adjusted slightly, and finally it smoothly avoids the dynamic obstacles. The reason for no collision is that after expanding the dynamic obstacles cost map, the mobile robot detects that there is an obstacle in front of it at a distance from the obstacle, so there is more room to avoid it. In the third random encounter scenario. The mobile robot is in front of the dynamic obstacle, and they keep moving in the same direction, the mobile robot keeps a certain safe distance from the dynamic obstacle, the distance is the boundary of the cost map, which enlarge cost map to make the robot's movement more secure.

5.2   Real Environment Test and Results

After the simulation experiment proves the stability and feasibility of the algorithm in this paper, which is tested in the actual environment. The robot used in the real experiment is built by our laboratory, and it is a wheeled differential robot. It is mainly composed of three parts: mechanical part, sensor part and controller part, the mechanical part is mainly composed of glass fiberboard and other fasteners; the sensor part mainly includes SICKlms111 lidar, photoelectric encoder and Kinect2 camera; the controller part mainly includes Intel-NUC master control and drive control [19]. The laboratory scene is shown in Fig. 9. The mapping algorithm is used to construct a three-color grid map of the indoor environment, in which the grid map can be obtained by the robot automatically traversing the environment [20]. In the environment, one robot acts as a dynamic obstacle, and the other robot is used to verify the path planning algorithm in this paper.

Fig. 9. Overall environment of the laboratory.

Put the mobile robot in the environment and use AMCL algorithm to realize its own positioning, Fig. 10 (left) shows the original laser in the real experiment, it can be seen in the figure that laser data with the grid map has been successfully matched, the position information of each laser data can also be calculated. By judging and segmenting it, several groups of laser data are obtained. By the least-square method solve the linear equation of each group of data. At the same time, in order to avoid the problem of over-segmentation, two adjacent straight lines are judged and merged, and finally fitted. Compare the fitted straight line with the built two-dimensional grid map, take the straight line that is not on the grid map to solve the center of the newly added obstacle, and mark it with a circle, as shown in Fig. 10 (right).

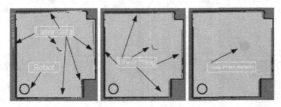

Fig. 10. Real environment laser data, fitted straight line and newly added obstacle marking diagram.

The center position of the newly added obstacle can be continuously obtained to determine whether it is a dynamic obstacle in real time. If it is a static obstacle, only its position coordinates will be displayed; if it is a dynamic obstacle, its motion equation, velocity, and position will be displayed in real time, as shown in Fig. 11.

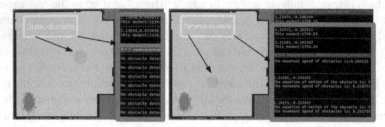

Fig. 11. Dynamic obstacle detection and motion solution in real environment.

If the dynamic obstacle is detected, the cost map can be expanded according to its moving direction. Figure 12 shows the original cost map of the dynamic obstacle in the actual scene and the cost map after expanding the speed direction of the dynamic obstacle.

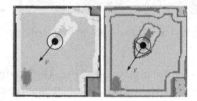

Fig. 12. Before and after expansion of dynamic obstacle cost map in real scene.

In the real test, the collision scenarios under the two scenarios were also simulated, including frontal collision and oblique collision on the side, and compared with the original DWA algorithm without any processing, verified one by one.

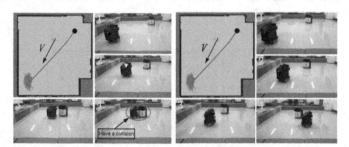

Fig. 13. Head-on collision simulation between the original DWA algorithm (left) and the proposed algorithm (right) in real environment.

The head-on collision test is shown in Fig. 13. The mobile robot is placed in a corner of the environment, and the moving obstacle is placed directly opposite the robot. The keyboard is used to control the obstacle to move in a linear. At the same time, the target point is clicked on the Rviz display interface of the mobile robot to simulate the frontal collision scene. As can be seen from the test process diagram, only obstacle avoidance through DWA local path planner leads to frontal collision between the robot and the obstacle, while the robot can avoid dynamic obstacles after the cost map is expanded by this algorithm.

Fig. 14. Oblique collision simulation between the original DWA algorithm (left) and the proposed algorithm (right) in real environment.

The oblique collision test is shown in Fig. 14. The mobile robot is placed in a corner of the environment, and the moving obstacle is placed in the left front of the robot. The keyboard is used to control the obstacle for linear movement. At the same time, the target point is clicked on the Rviz display interface of the mobile robot to simulate the oblique collision scene. From the test process diagram, the dynamic obstacles collide with the side of the robot only through the DWA local path planner. However, after the algorithm in this paper expands the cost map of the moving direction of the obstacles, the robot conducts the steering process in advance and avoids the dynamic obstacles.

6 Conclusion

On the basis of robot real-time positioning using AMCL algorithm, this paper uses a single two-dimensional laser radar to continuously obtain the position information of point clouds on obstacles, then divides them into groups according to the characteristics of point clouds, uses the least square to fit each group of laser data in a straight line, finally compares with the environment map, to remove the known obstacles, at the same time, the newly added obstacles are marked with a circle, and the dynamic obstacles are judged by the real-time change of its circle center. The experimental results show the feasibility of this method. Based on judging the dynamic obstacles, continuously record its center coordinate to solve the equation of motion and the velocity of the dynamic obstacles, and the cost map of the speed direction of the dynamic obstacles is innovatively expanded. Then combine with DWA algorithm for local dynamic obstacle avoidance. Simulation experiments and real-world experiments verify that after expanding the cost map of the speed direction of the dynamic obstacles. The mobile robot has more space to avoid the dynamic obstacles, which verifies the effectiveness of the proposed algorithm.

References

1. Zhu, D., Yan, M.: Survey on technology of mobile robot path planning. Control Decis. **25**(7), 961–967 (2010)
2. Duan, S., Wang, Q., Han, X.: Improved a-star algorithm for safety insured optimal path with smoothed corner turns. Chin. J. Mech. Eng. **56**(18), 205–215 (2020)
3. Liu, Z., Li, Y., Zheng, L.: Local path planning for autonomous vehicles based on sparse representation of point cloud in unstructured environments. Chin. J. Mech. Eng. **56**(02), 163–173 (2020)
4. Xu, F.: Research on robot obstacle avoidance and path planning based on improved artificial potential field. Comput. Sci. **43**(12), 293–296 (2016)
5. Yu, Z., Yan, J., Zhao, J.: Mobile robot path planning based on improved artificial potential field method. J. Harbin Inst. Technol. **43**(01), 50–55 (2011)
6. Montiel, O., Orozco-Rosas, U., Sepúlveda, R.: Path planning for mobile robots using bacterial potential field for avoiding static and dynamic obstacles. Expert Syst. Appl. **42**(12), 5177–5191 (2015)
7. Wang, H., Yin, P., Zheng, W.: Mobile robot path planning based on improved A* algorithm and dynamic window method. Robot **42**(03), 346–353 (2020)
8. Zhang, Y., Song, J., Zhang, Q.: Local path planning of outdoor cleaning robot based on an improved DWA. Robot **42**(05), 617–625 (2020)
9. Lima, D.A.D., Pereira, G.A.S.: Navigation of an autonomous car using vector fields and the dynamic window approach. J. Control Autom. Electr. Syst. **24**(1–2), 106–116 (2013)
10. Keller, M., Hoffmann, F., Bertram, T.: Planning of optimal collision avoidance trajectories with timed elastic bands. In: 19th IFAC World Congress 2014, pp. 4581–4586. Elsevier Ltd., Cape Town (2014)
11. Zheng, K., Han, B., Wang, X.: Ackerman robot motion planning system based on improved teb algorithm. Sci. Technol. Eng. **20**(10), 3997–4003 (2020)
12. Jiang, Y., Liu, R., Zhu, J.: A high-performance CMOS FDMA for pulsed TOF imaging ladar system. Opto-Electron. Eng. **46**(07), 67–74 (2019)
13. Guan, R.P., Ristic, B., Wang, L.: Kid sampling with gmapping proposal for Monte Carlo localization of mobile robots. Inf. Fusion **49**, 79–88 (2019)
14. Gong, X., Li, Z.: A robust weighted total least squares method. Acta Geodaetica Cartogr. Sinica **43**(09), 888–894, 901 (2014)
15. Choi, Y.H., Lee, T.K., Oh, S.Y.: A line feature based SLAM with low grade range sensors using geometric constraints and active exploration for mobile robot. Auton. Robot. **24**(1), 13–27 (2008)
16. Ballesteros, J., Urdiales, C., Velasco, A.: A biomimetical dynamic window approach to navigation for collaborative control. IEEE Trans. Hum.-Mach. Syst. (6), 1–11 (2017)
17. Sun, J., Chen, Z., Wang, P.: Multi-region coverage method based on cost map and minimal tree for mobile robot. Robot (004), 435–442 (2015)
18. Zhao, Q., Chen, Y., Luo, B.: A local path planning algorithm based on pedestrian prediction information. J. Wuhan Univ. (Inf. Sci. Edn.) **45**(5), 667–675 (2020)
19. Jiang, L., Li, J., Ma, X.: Voronoi path planning based on improved skeleton extraction. Chin. J. Mech. Eng. **56**(13), 138–148 (2020)
20. Jiang, L., Zhang, Y., Zhu, J.: Research on efficient algorithm of robot along the wall combined with historical motion state. Acta Autom. Sinica **46**(06), 1166–1177 (2020)

Simulation of Airflow Characteristics of a Seabird Following a Ship Based on Steady State

Chao Lu[1], Guangwu Liu[2(✉)], Wenguo Zhang[1], and Jian Wang[1]

[1] China Ship Development and Design Center, Wuhan 430064, China
[2] Green and Smart River-Sea-Going Ship, Cruise and Yacht Research Center, Wuhan University of Technology, Wuhan 430063, China
gliu@whut.edu.cn

Abstract. The seabirds have always been observed following the ships while sailing in a static condition. Despite the biological reason, there should be some physical profits for these following activities. This paper is trying to start an initial study of a seabird following a ship by Computational Fluid Dynamics (CFD) simulations on steady states. We have chosen a standard frigate simplified 2 (SFS2) standard computation model as simulation ship, while a classic seagull wing to simulate a seabird. The paper has inspected the accuracy of CFD with typical wind tunnel tests' and CFD simulation examples' results such as 7.62 m/sec in wind over deck (WOD) 0° and 10°. While the tests of wind tunnel had executed in the scale of 1:120, so the inspection geometric model is generated in the same scale. The results of inspection CFD are fitting well with the typical wind tunnel tests' and CFD simulation examples' results. Meanwhile a full scale SFS2's geometric model and a seabird wing's geometric model have been generated. A series of airflow simulations have been carried out then in steady states. Initial study of these simulations shows that in 0° WOD, the drag of the seabird's wing was reduced while following after the ship, compared with which situation that the ship was not followed by the seabird's wing. This paper has just started few typical studies of this interesting behaves of the seabirds, the following studies and simulations are being considered.

Keywords: Simulation · Airflow · Seabird · Steady States

1 Introduction

This paper focuses on the connection between airflow characteristics and seabirds biological phenomenon. While sailing on the ocean the ships are always been found followed by seabirds, even single on in groups. Despite the profit of predations, the following airflow should be a potential physical reason. The seabirds following behaves as shown in Fig. 1.

The phenomena of ship following by seabirds are always described by photographers and writers since the ocean sailing in the very beginning. To prove the hypothesis, a

L. Pan et al. (Eds.): BIC-TA 2022, CCIS 1801, pp. 591–604, 2023.
https://doi.org/10.1007/978-981-99-1549-1_47

Fig. 1. Seabirds group following the ship in Liaoning province [1]

small group from China Ship Development and Design Center (CSDDC) and Green& Smart River-Sea-Going Ship, Cruise and Yacht Research Center of Wuhan University of Technology (WUT) have started an initial work with Computational Fluid Dynamics (CFD) simulations and bionics references.

The initial work should include two steps. First, Inspection. The paper has inspected the accuracy of CFD with typical wind tunnel tests' and CFD simulation examples' results such as 7.62 m/sec in wind over deck (WOD) 0° and 10°.While the tests of wind tunnel had executed in the scale of 1:120, so the inspection geometric model is generated in the same scale. The results of inspection CFD are fitting well with the typical wind tunnel tests' and CFD simulation examples' results. Second, full scale simulation with seabird's wing model. A full scale SFS2's geometric model and a seabird wing's geometric model have been generated. A series of airflow simulations have been carried out then in steady states. Initial study of these simulations shows that in 0° WOD, the drag of the seabird's wing was reduced while following after the ship, compared with which situation that the ship was not followed by the seabird's wing.

Initial conclusions have been made at the end of the paper presents that the bird's following the ship could reduce the drag during flying.

2 Ship and Seabird Geometries

2.1 Ship Geometry

To investigate the effect of typical ship to the birds, the frigate class was considered as shown in Fig. 2. Instead of complex shape of the above ship structure, the so-called simplified frigate shape (SFS2) [2] was used in the wind tunnel test and the computational fluid dynamics (CFD) model. This geometry is a commonly used geometry for CFD benchmarking. The geometry was made into 1:120 scale model for numerical simulation matching the scale of the wind tunnel test model. Next, to simulate the airflow of frigate ship model with the commercial CFD code for current study.

Fig. 2. Frigate Ship Geometry [2]

While in full scale simulation step, a full-scale ship geometry has been adopted for the following condition. The full scale SFS2 geometry as show in Fig. 3 [4].

Based on the specification, this paper has generated a ship geometry as shown in Fig. 4 which could be scaled in 1:120 and full scale for two steps simulations.

594   C. Lu et al.

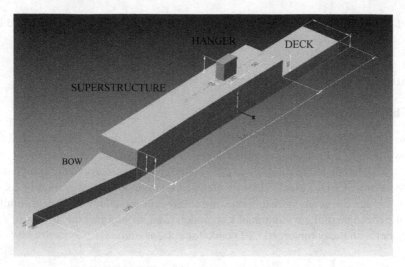

Fig. 3. Full-Scale Geometry of SFS2 (in Feet)

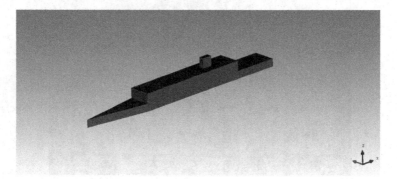

Fig. 4. SFS2 Geometry Generated in This Paper

2.2 Seabird Wing Geometry

To represent typical phenomena, this paper has chosen the seagull as the seabird, and simplified to a wing symmetry model. The wing type is a bionic seagull airfoil in full scale [3] with its section shown as in Fig. 5.

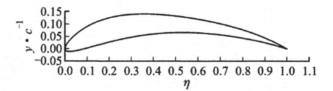

Fig. 5. Seagull Airfoil Typical Wing Section [3]

In this paper, a full-scale seabird wing model with 1 m span and 3° sweepback like a flying seagull spread is drawn. The wing section and full-scale seabird wing model are shown in Fig. 6.

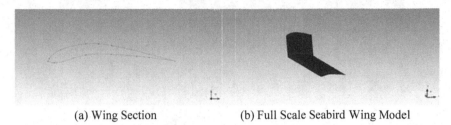

(a) Wing Section   (b) Full Scale Seabird Wing Model

Fig. 6. The Wing section (a) and Full-Scale Seabird Wing Model (b)

As metric system is a standard for international technologies, the geometries and model units are transferred in metric system as well as the calculation later [5].

3   Turbulence Type and Calculation States

3.1   Turbulence Type

The seabird following phenomenon is always happening in relatively long steady period. In this condition, the simulation should be in steady states. For ships airflow, the viscosity flow field could be set as a steady ship with steady turbulence flow. In Eq. (1), it shows, in steady states, forces such as gravity and trust or drugs.

$$\frac{\partial\left(u_i u_j\right)}{\partial x_j} = -\frac{1}{\rho}\frac{\partial p}{\partial x_i} + \frac{\partial}{\partial x_i}\left[\upsilon\left(\frac{\partial u_i}{\partial x_j} + \frac{\partial u_j}{\partial x_i}\right)\right] + g_i \tag{1}$$

In the equation, u_j represents a velocity component in coordinate x_j. $g_1 = 0$, $g_2 = 0$, $g_3 = 0$, g is gravity, p is flow pressure, ρ is flow density, v is velocity coefficient [6].

In engineering simulations, there are few turbulence types for common use. As classic turbulence type, a realizable $k - \varepsilon$ two-equation turbulence model were used with a pressure-velocity coupling scheme [8].

3.2   Calculation States

During inspection simulation, it is needed to compare the results with the example of SFS2 which had been tested in wind tunnel with PIV measurement and CFD calculation in specific position of specific states [2] Such as 7.62 m/sec, wind over deck (WOD) with 0°, 5°, 10°, 15°. The coordination of the calculation has been described in Fig. 7.

596   C. Lu et al.

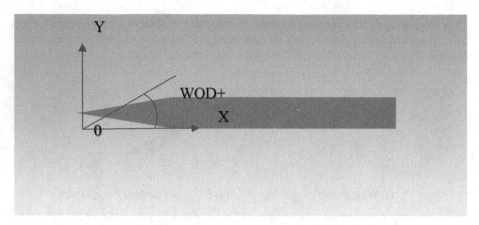

Fig. 7. The Coordination of the Calculation

This paper chose two typical states for both inspection simulations and full-scale simulations, both in steady states. The calculation states of this paper are shown in Table 1.

The wing fwd is set for calculation without the effect of the ship [6], and the wing aft is set after the ship. Both wings are 10 m high from bottom and 20 m before or after the ship.

Table 1. Summary of CFD States

CFD states	Wind X speed (m/sec)	WOD (°)	Scale
Inspection 1	7.62	0	1/120
Inspection 2	7.62	10	1/120
Full with 2 wings	7.62	0	1/1
Full with wing	7.62	10	1/1

The boundary conditions are kept same as the CFD examples shown in Table 2.

4 Inspection Simulations

The size and mesh structure of calculation domain of inspection simulations is shown in Fig. 8.

The example measurement locations have 3 plants above the helicopter deck as WP1, WP2 and WP3. WP1 was chosen for comparing, and the u speed would be measured at line $x/H = 0.95$. The WP1 location is shown in Fig. 9.

Table 2. Summary of Boundaries States

Regions	Boundary condition	Values	Ref
Inlet	Velocity inlet	x-direction:7.62	
Outlet	Pressure outlet	Ref. P = 0	
Ship	Wall	No slip	
Left/right	Wall	No slip	
Upper/bottom	Wall	No slip	
Wing fwd	Wall	No slip	With U-flow
Wing aft	Wall	No slip	With U-flow

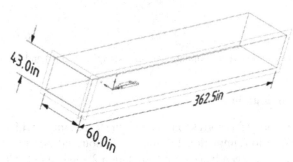

Fig. 8. The Size and Mesh Structure of Calculation Domain of Inspection Simulations

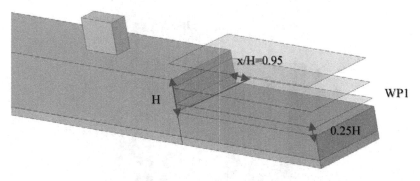

Fig. 9. WP1 Plant Location

4.1 Inspection Simulation 1

The inspection simulation 1 is corresponded with the test example with WOD = 0°, the elements of the inspection simulation 1 is about 660 thousand. After solver, the velocity vectors are compared between Inspection Simulation 1, example CFD and example's PIV [2] as shown in Fig. 10.

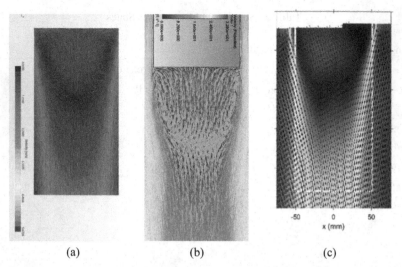

(a)   (b)   (c)

Fig. 10. Comparison between Simulation 1 (a), example CFD (b) and example PIV (c)

4.2 Inspection Simulation 2

The inspection simulation 2 is corresponded with the test example with WOD $= 10°$, the elements of the inspection simulation 2 is about 670 thousand. After solver, the velocity vectors are compared between Inspection Simulation 2 and example CFD [2] as shown in Fig. 11.

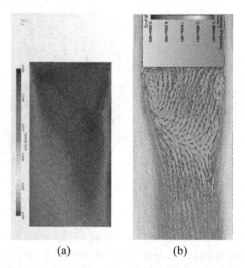

(a)   (b)

Fig. 11. Comparison between Inspection Simulation 1 (a) and Example CFD (b)

The u velocity magnitudes are collected online $x/H = 0.95$, and compared with wind tunnel (WT) test results, as shown in Fig. 12.

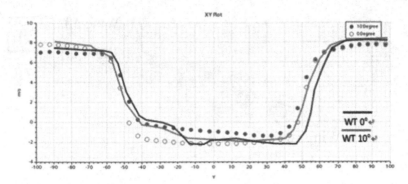

Fig. 12. The u Velocity Magnitudes Comparing with Wind Tunnel Tests (WT)

As show in Fig. 10 and Fig. 11, the air flow distribution form between Inspection Simulations, example CFD and wind tunnel PIVs are similar with each other. While the curves of Inspection Simulations comparing with wind tunnel tests also have similar shapes.

With the inspections above, the simulation code and programs should have the feasibilities to carry out further calculations with the seabird wings.

5 Full-Scale Simulations

As the wind speed and ship condition are in low Re scope, the scale has few effects for airflow trend [7], especially for the steady states. So, calculations in full-scale with consideration of relatively small dimensions of seabird wings are carried out in this paper.

5.1 Full-Scale Simulation 1

The full-scale simulation 1 has two wings before (wing fwd) and after (wing aft) the ship, both 10 m high from bottom. The wing fwd is 20 m before the bow to avoid the flow's effect on the ship.

The geometries of full-scale simulation 1 are shown in Fig. 13.

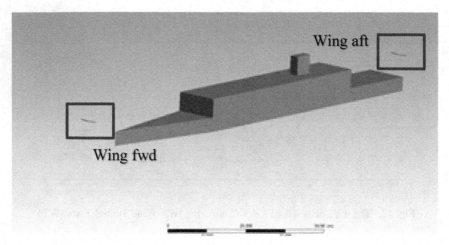

Fig. 13. The Geometries of Full-Scale simulation 1

The elements of full-scale simulation 1 is about 3.3 million. After solver, the surface pressures of wing fwd and wing aft are compared as shown in Fig. 14.

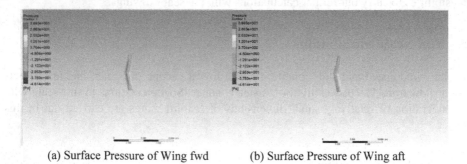

(a) Surface Pressure of Wing fwd   (b) Surface Pressure of Wing aft

Fig. 14. Surface Pressures Wing fwd and Wing aft

Then a total pressure of XZ section plane has been generated to compare the deference flow condition of the wing fwd and wing aft as shown in Fig. 15.

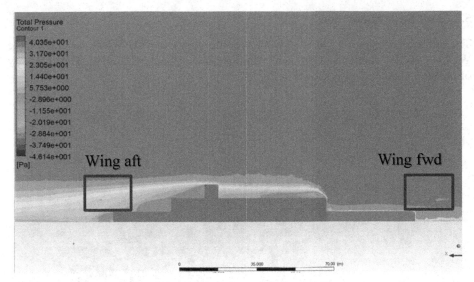

Fig. 15. Total Pressure of XZ Section Plane

A u drag force has been measured for both wing fwd and wing aft. Since the steady state has stabled aft 400 steps, the data from step 400 to step 500 was figured out in this paper, and average value has been calculated.

The solver monitors of u drag force is shown in Fig. 16.

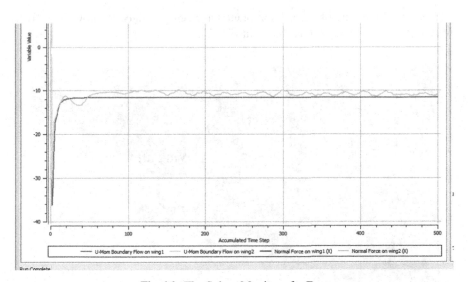

Fig. 16. The Solver Monitor of u Drag

5.2 Full-Scale Simulation 2

The full-scale simulation 2 has one wing after the ship (wing aft). When the ship has the WOD = 10° the elements of full-scale simulation 1 is about 3.2 million. The geometry of the ship and wing aft is shown in Fig. 17.

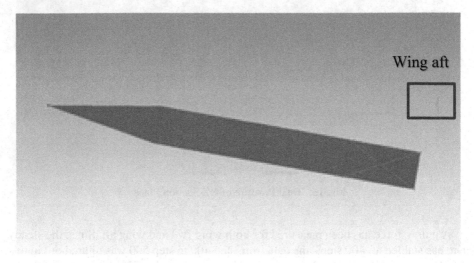

Fig. 17. The Geometry of the Ship and Wing aft at WOD = 10°

Then a total pressure of XZ section plane has been generated to show the different flow condition of the wing aft as shown in Fig. 18.

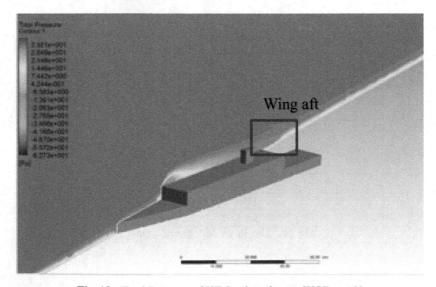

Fig. 18. Total Pressure of XZ Section plane at WOD = 10°

A u drag force has been measured for wing aft. Since the steady state has stabled after 400 steps, the data from step 400 to step 500 was figure out, and average value has been calculated. The solver monitors of u drag force of wing aft is shown in Fig. 19.

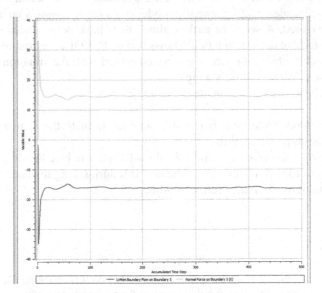

Fig. 19. The Solver Monitor of u Drag at WOD = 10°

The u drag average values are summed in Table 3.

Table 3. The u Drag Average Values

Conditions	WOD (°)	Average Values(N)
Wing fwd	0	11.63
Wing aft	0	10.92
Wing aft	10	15.97

As compared above in full-scale simulation 1, the u drag is reduced about 6.1% when the seabird flies following the ship. But when the ship is turning 10°, the situation would be changed, and the u drag would be increased.

6 Conclusions

This paper is the first attempt to study the phenomenon of seabirds following the ship. The initial work should include two steps. Firstly, the accuracy of CFD with typical wind tunnel tests and CFD simulation example results of velocity 7.62 m/sec and wind over deck (WOD) 0° and 10° was inspected. Because the test of wind tunnel had executed in the scale of 1:120, the inspection geometric model is generated in the same scale. The results of inspection CFD are fitting well with the typical wind tunnel tests and CFD simulation examples results. Secondly, full-scale simulation with seabird's wing model was done. A full-scale SFS2's geometric model and a seabird wing's geometric model have been generated. A series of airflow simulations have been carried out in steady states. Initial study of these simulations shows that in 0° WOD, the drag of the seabird's wing was reduced while following the ship, compared with the situation that the ship was not followed by the seabird's wing.

Such conclusions should be made in this initial step:

1. A CFD method could be a reasonable way to simulate the airflow relationship between the ship and seabirds.
2. The seabird's following the ship in WOD $= 0°$ could reduce the drag of the flying which means to save the energy of the seabirds during long distance fly. But other WOD conditions need to be further studied.

References

1. Li, X., Gu, X.: Active view of Chinese ocean: seabirds. Forest Humankind **02**, 36–51 (2020)
2. Choi, J., Miklosovic, D.S.: LES simulations using the moving mesh method with comparison to experimental results for a periodic ship airwake. In: AIAA Aviation 2020 Forum, pp. 1–23 (2020)
3. Wu, L., Wang, L., Liu, X., Ma, L., Xi, G.: Numerical simulation on the static and dynamic aerodynamic characteristics of bionic seagull airfoil. J. Xi'an Jiaotong Univ. **12**, 88–97 (2020)
4. Reddy, K.R., Toffoletto, R., Jones, K.R.W.: Numerical simulation of ship airwake. Comput. Fluids **29**, 451–465 (2000). https://doi.org/10.1016/S0045-7930
5. Mora, R.B.: Experimental investigation of the flow on a simple frigate shape (SFS). Sci. World J. **2014**, 1–8 (2014). https://doi.org/10.1155/2014/818132
6. Yuan, W., Wall, A., Lee, R.: Combined numerical and experimental simulations of unsteady ship airwakes. Comput. Fluids **30**(172), 29–53 (2018)
7. Kang, H., Snyder, M.R., Miklosovic, D.S., Friedman, C.: Comparisons of in situ ship air wakes with wind tunnel measurements and computational fluid dynamics simulations. J. Am. Helicopter Soc. **2**(61), 1–16 (2016)
8. Rahimpour, M., Oshkai, P.: Experimental investigation of airflow over the helicopter platform of a polar icebreaker. Ocean Eng. **15**(121), 98–111 (2016)

Numerical Simulation of Bionic Undulating Fin Surface Drag Reduction Structure

Hao Lu[1(✉)] and Mingyang Xu[2]

[1] Wuhan Digital Engineering Institute, Wuhan 430074, China
739937794@qq.com
[2] Green & Smart River-Sea-Going Ship, Cruise and Yacht Research Center, Wuhan University
of Technology, Wuhan 430070, China

Abstract. The fin surface configuration with low energy consumption and low resistance is one of the key factors to improve the performance of bionic undulation on fin propulsion robotic fish. In this paper, we designed six kinds of vertical flow resistance reduction structure based on quarter-arc and semi-arc configurations, with their design inspiration comes from the electric elfish and turbot. After dividing the mesh, we simulated the flow on the surface of the structure. The results show that the vertical flow direction structure can improve the thickness of the boundary layer and the flow pattern in the near-wall area. When the protruding structure increase to one third of the drag-reducing size range, the drag will increase, not reduce compared with smooth surface. Three kinds of grooves designed in this paper all achieve drag reduction rates of about 3.5% when groove slope angles are 15° and 45°.

Keywords: Undulating fins · Structural design · Bionic · Simulation analysis · Boundary layer · Drag reduction

1 Introduction

With the increasing prominent value of biomimetic engineering and the improvement of energy awareness in recent years, people gradually focus on the unique movement mechanism and highly adaptable surface structure of natural organisms. The application of biological characteristics in engineering practice will effectively improve the level of existing technology. At present, domestic and foreign scholars [1–4] have proposed a variety of new designs of underwater robotic fish that adopt MPF (the median and/or paired fin) propulsion mode [5]. Among them, the surface friction resistance of fluctuating fins is one of the important factors affecting the propulsion efficiency of fluctuating fins. If the fin surface resistance is effectively reduced, we will strongly promote the development of new underwater robotic fish.

The research on drag reduction of non-smooth surface has become a hot topic in recent years. Tao Min [6], Xu Zhong [7] et al. selected dung beetles to study bionic pitted non-smooth surface, demonstrated the influence of the shape, arrangement and depth spacing of pits on the optimal drag reduction through orthogonal test method.

L. Pan et al. (Eds.): BIC-TA 2022, CCIS 1801, pp. 605–617, 2023.
https://doi.org/10.1007/978-981-99-1549-1_48

Zhang Chengchun [8], Ren Luquan [9], Wang Ke [10] et al. studied the influence of pits and groove-shaped non-smooth surfaces on the drag reduction rate of the rotating body surface. However, because of many interference factors of manufacturing precision and drag reduction rate, non-smooth surface structure has not been applied in the production and manufacturing of underwater vehicles and robots on a large scale at present, and further exploration is needed to develop and apply it.

In this paper, two circular drag reduction configurations were designed from the perspective of bionic engineering by imitating the characteristics of the fin surface of the black ghost fish and Turbot, and six fin drag reduction schemes were given by combining the drag reduction mechanism of the non-smooth surface structure. We calculated the mechanical properties and drag reduction effects of the six schemes at low speed through the three-dimensional modeling and mesh division.

2   Collection and Calculation of Biometric Characteristics of Undulating Fins

There are a large number of fish species that adopt the median and/or paired fin (MPF) propulsion mode in nature. In this paper, two common and representative fish species are selected, namely, the Streak-feathered electric eel and the Turbot (Flounder), and the surface structure of their undulating fins is compared and observed. In the observation experiment, after biological anatomy treatment, external microscopic observation was used to extract the structural features of the fin surface, and the basis for the establishment of the model was obtained.

2.1   Physical Characteristics of MPF Model Fish

The streak-feathered electric eel, also known as "Black Ghost Fish", is native to South America, nocturnal and aggressive. The whole body is black knife shape, pelvic fin and anal fin connected, wavy to the tail. As shown in Fig. 1(a), the undulating pelvic fin of the eel has a convex non-smooth surface structure. Its surface is formed by connecting two small fins to the fin surface. The thickness and length of the fin structure are uniformly distributed, with small spacing and high density.

Turbot, which likes to lie under water, has a slightly diamond-shaped body, with two eyes on the left side of the head and wide dorsal and anal fins without stiff spines. As shown in Fig. 1(b), the undulating dorsal and anal fins of turbot are non-smooth surface structures with conical distribution of fin spines, which are formed by connecting two broad fins to the fin surface. The fin spines in the middle of the fish body are long, evenly distributed in thickness, with large spacing and low density.

2.2   Determination of Shape and Size of Fin Drag Reduction Element

According to biomimetic engineering technology, two relatively regular basic structure shapes, namely asymmetric quarter-arc configuration and symmetrical half-arc configuration, were extracted from the above two types of fish with typical characteristics by

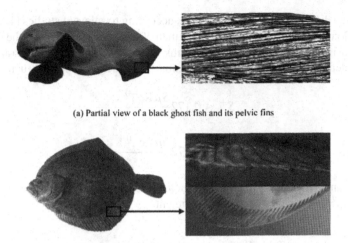

(a) Partial view of a black ghost fish and its pelvic fins

(b) Partial view of a turbot and its pelvic fins

Fig. 1. Biological shape characteristics of bionic objects.

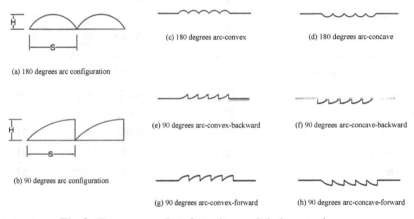

(a) 180 degrees arc configuration

(b) 90 degrees arc configuration

(c) 180 degrees arc-convex

(d) 180 degrees arc-concave

(e) 90 degrees arc-convex-backward

(f) 90 degrees arc-concave-backward

(g) 90 degrees arc-convex-forward

(h) 90 degrees arc-concave-forward

Fig. 2. Two structural configurations and six layout schemes.

fitting their undulating fins, and six different layout schemes were designed based on the current relatively mature groove drag reduction theory, as shown in Fig. 2.

Studies have shown that [11] there are two ways of groove placement: parallel flow-direction placement and vertical flow-direction placement, of which parallel flow-direction placement has better drag reduction performance. However, considering that vertical flow-direction placement is more conducive to the placement of driving components on the fluctuating fins of the robotic fish and the flexible swing of the undulating fins, vertical flow-direction placement is chosen in this paper for drag reduction research.

As for the determination of groove size, according to relevant studies [12, 13], when the dimensionless width s of groove is less than 30 viscosity lengths and the dimensionless depth h is less than 25 viscosity lengths, the groove surface has a drag reduction effect. We define the dimensionless dimension of the drag reduction structure as:

$$S^+ \leq 0.172 \frac{S\mu \mathrm{Re}^{-0.1}}{\gamma} \tag{1}$$

$$h^+ \leq 0.172 \frac{h\mu \mathrm{Re}^{-0.1}}{\gamma} \tag{2}$$

where μ stands for incoming flow velocity m/s; γ stands for kinematic viscosity coefficient m²/s; Re is the Reynolds coefficient.

Table 1. The parameters of the cell structure.

Structure of cell	Apex Angle α	Angle of attack β	H (mm)	S (mm)	curvature K
180° arc configuration	30°	15°	0.02	0.15	6.67
180° arc configuration	45°	22.5°	0.031	0.15	9.43
180° arc configuration	60°	30°	0.043	0.15	11.55
180° arc configuration	75°	37.5°	0.058	0.15	12.88
180° arc configuration	90°	45°	0.075	0.15	13.33
90° arc configuration	30°	15°	0.04	0.15	3.33
90° arc configuration	45°	22.5°	0.062	0.15	4.71
90° arc configuration	60°	30°	0.086	0.15	5.77
90° arc configuration	75°	37.5°	0.115	0.15	6.44
90° arc configuration	90°	45°	0.15	0.15	6.67

Through the investigation and comparison of the same type of underwater robotic fish designed by research teams in various universities, the results show that the speed setting of robotic fish using MPF propulsion mode is generally low, generally 0.5 m/s–4m/s. Therefore, the inlet speed set in this simulation is $\mu = 2$ m/s, and the cross-section diameter d = 2 mm. The kinematic viscosity of water is 1.0048×10^{-6} m²/s. The structural size range of cells that can achieve drag reduction effect is $S^+ \leq 0.201$ mm and $h^+ \leq 0.167$ mm according to formula (1) and (2). Based on the shape of the bionic structure above, by setting different size parameters, combined with different layout schemes, to explore the drag reduction effect within the range of size with the heigh, curvature, angle of attack change law, and then determine the size range of the optimal drag reduction scheme. The specific structural parameters are shown in Table 1.

3 The Model of Computation

3.1 The Establishment of Computing Domain

According to relevant studies [14], the length of turbulent transition section should be 25–40 times the width of the microstructure, and the height of the flow field should be greater than 10 times the height of the cell structure. Therefore, this paper sets the cross-section width of the flow field a = 2 mm, the height of the flow field b = 2 mm, and the length of the flow field c = 10.5 mm, and then the cross-section diameter of the flow field d = 2 mm. The flow field domain is shown in Fig. 3.

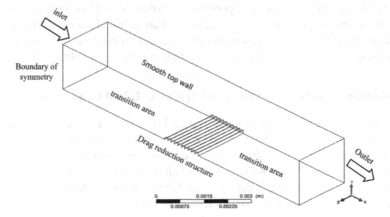

Fig. 3. The model of computing flow field domain.

3.2 Mesh Partitioning and Algorithm Selection

When using hydrodynamics method to analyze drag reduction performance of element structure, the meshing result of surface flow field of cell structure affects the correctness and accuracy of numerical simulation, and the accuracy and solving time are the

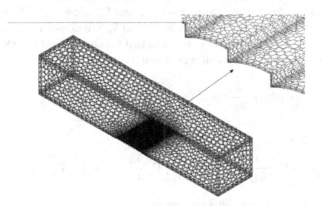

Fig. 4. Flow field grid division diagram.

key problems in the process of computational fluid dynamics numerical simulation. Therefore, a reasonable meshing processing is required.

We chose a meshing method depending on the characteristics of cell structure. As shown in Fig. 4, the surface of the two cell structures designed in this paper and the 36 groups of models used in the test are irregular or sharp corners, and different types of meshless elements are needed to solve various geometric shapes and flow patterns, if the transition between different types of meshing elements is not handled well, problems tend to occur in the solution, such as the non-conformal interface or tetrahedron that the transition area usually relies on. This meshing method may cause the grid quality of the transition area to decline, and generate too many cells, leading to the problem of lengthy solution time. In this paper, Octree algorithm and Ansys Mosaic technology are adopted to connect different types of mesh with general polyhedral elements. Mosaic grid has fewer elements and better quality, which is very suitable for models with complex shapes, and has the characteristics of high partition efficiency and short solving time.

3.3 Selection of Turbulence Model and Setting of Initial Conditions

Considering that the inlet flow velocity is small, the size of the element structure is small, which belongs to the low Reynolds number problem, and the main parameter to be considered is the near wall shear force, the shear stress transfer (SST) k-ω model is selected in FLUENT as the turbulence model of this simulation. SST k-ω is a deformation of the standard k-ω model. The equations and coefficients are derived from analytical solutions. The equations of the standard k-ω model are:

$$\frac{\partial}{\partial t}(\rho k) + \frac{\partial}{\partial x_i}(\rho k \mu_i) = \frac{\partial}{\partial x_j}\left(\Gamma_k \frac{\partial k}{\partial x_j}\right) + G_k - Y_k + S_k \tag{3}$$

and

$$\frac{\partial}{\partial t}(\rho \omega) + \frac{\partial}{\partial x_i}(\rho \omega \mu_i) = \frac{\partial}{\partial x_j}\left(\Gamma_\omega \frac{\partial \omega}{\partial x_j}\right) + G_\omega - Y_\omega + S_\omega \tag{4}$$

where G_k is the turbulent kinetic energy generated by the laminar velocity gradient. G omega is produced by the omega equation; Γ_k and Γ_ω show the diffusion of k and ω; Y_k and Y_ω are turbulence due to diffusion; S_k and S_ω are user-defined values.

The SST k-ω model is similar to the standard k-ω model, but the SST has a hybrid function designed for the near-wall region and its equation is:

$$\frac{\partial}{\partial t}(\rho k) + \frac{\partial}{\partial x_i}(\rho k \mu_i) = \frac{\partial}{\partial x_j}\left(\Gamma_k \frac{\partial k}{\partial x_j}\right) + G_k - Y_k + S_k \tag{5}$$

and

$$\frac{\partial}{\partial t}(\rho \omega) + \frac{\partial}{\partial x_i}(\rho \omega \mu_i) = \frac{\partial}{\partial x_j}\left(\Gamma_\omega \frac{\partial \omega}{\partial x_j}\right) + G_\omega - Y_\omega + D_\omega + S_\omega \tag{6}$$

where D_ω stands for orthogonal divergent term.

The difference between the two models is that in the standard k-ω model, α_∞ is constant. In the SST k-ω model, the α_∞ equation is as follows:

$$\alpha_\infty = F_1\alpha_{\infty,1} + (1 - F_1)\alpha_{\infty,2} \tag{7}$$

The SST k-ω model incorporates the cross-diffusion derived from the ω equation and takes the turbulent shear stress wave propagation into account in the turbulent viscosity, which makes the SST k-ω model more accurate and reliable than the standard k-ω model in a wide range of flow fields.

In view of the low flow velocity water area in natural environment, the fluid was set as incompressible, and the separation solver (the pressure-based solver) was selected, heat transfer was ignored, and the gravity acceleration was set at 9.81 m/s². The left boundary of the flow field is set as the velocity inlet boundary condition, the right boundary is set as the outflow boundary condition, the front and back of the flow field are set as the symmetric boundary condition, and the upper and lower boundary are set as the wall boundary condition. Due to the adoption of Ansys Mosaic technology in this paper, there are different types of grid elements after grid division, and the direction of fluid movement is different from the direction of grid division. In order to obtain more accurate calculation results, the second order upwind scheme is selected as the discrete scheme in this paper.

4 Analysis of Simulation Results

4.1 Validation of Turbulence Model Accuracy

In order to verify the accuracy and credibility of the selected SST k-ω model in simulation, according to the common verification method [15], it is necessary to compare and analyze the friction coefficient of smooth flat surface under different Reynolds coefficients with the results calculated by Pelant's empirical formula. If the error of both is within the specified range, The SST k-ω turbulence model selected in this paper is suitable for this simulation experiment. Prandtl empirical formula is:

$$C_f = 0.074 \, \mathrm{Re}^{-0.2} \tag{8}$$

The comparison results and rate of deviation are shown in Table 2. The maximum deviation between the simulated value of SST k-ω model and the theoretical value calculated by the empirical formula is 4.26%, and the deviation is kept at 3%–4% at low flow rate, which meets the basic accuracy requirements of simulation.

For ensure accuracy, the SST k-ω model requires the range of Y plus after calculation convergence to be $1 \le Y$ plus ≤ 5. The Y plus curve obtained by simulation verification in this paper is shown in Fig. 5, and the results show that it meets the requirements of the model.

By verifying the friction resistance coefficient and Y plus convergence curve of smooth flat surface, it shows that the SST k-ω model can better reflect the characteristics of near-wall turbulent flow field, and the relevant parameters set are in good agreement with the experimental requirements.

612   H. Lu and M. Xu

Table2. Comparison of friction coefficient of smooth surface with empirical formula.

Reynolds number	SST k-ω model	Prandtl's empirical formula	rate of deviation (%)	velocity v (m/s)
2000	0.016791	0.016182	3.76%	1.0048
4000	0.014666	0.014087	4.11%	2.0096
6000	0.012491	0.01299	3.84%	3.0144
8000	0.011751	0.012263	4.18%	4.0192
10000	0.011252	0.011728	4.06%	5.024
12000	0.010826	0.011308	4.26%	6.0288

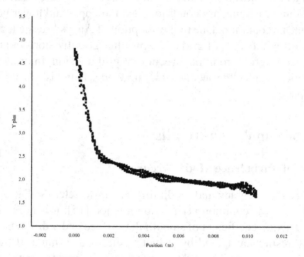

Fig. 5. Schematic diagram of Y plus.

4.2 Validation of Turbulence Model Accuracy

According to the drag reduction mechanism of turbulent boundary layer [16], the turbulent burst process will occur in the near-wall region of the fluid, and there is a significant velocity gradient along the flow direction. As shown in Fig. 6, turbulent flow occurs when the fluid passes through the smooth surface on the upper side, while stable flow state occurs when the fluid passes through the drag reduction structure on the lower side.

It can be seen from the analysis that the vertical flow arrangement of drag reduction structure in the near wall area not only increases the thickness of the boundary layer, causing the transition layer and logarithmic layer to move upward, but also makes the low-speed fluid flow inside the structure, which plays a certain lubrication role, avoids the direct contact between high-speed fluid and the wall in the flow field, and reduces the viscous resistance.

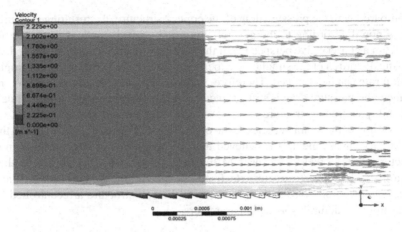

Fig. 6. The image of Velocity cloud and velocity vector.

4.3 Drag Reduction Mechanism and Analysis Method

According to the analysis method of drag reduction performance of non-smooth surface structures [15–17], the influence of pressure drag is usually ignored for drag reduction structures arranged in the parallel flow direction. After replacing friction force with shear force, the formula is used to calculate drag reduction rate. The formula of drag reduction rate is as follows:

$$Drag\ reduction\ rate = \frac{f_1 - f_2}{f_1} \times 100\% \tag{9}$$

where f_1 is the shear force of smooth flat surface, f_2 is the shear force of drag reduction structure.

In the vertical flow layout of drag reduction structure, the determination of drag reduction effect needs to take the combined force of pressure drag and viscous resistance as a reference basis, the size of the pressure drag depends on the wake width, is extremely sensitive to the shape of the drag reduction structure.

For the vertical flow arrangement of the convex structure, the water flow will have a roundabout movement after passing the convex structure, and there will be reverse pressure on the back surface of the convex structure. Therefore, there is a significant pressure drag between the facing surface and the back surface of each cell structure, and the water flow through multiple convex structures will produce a significant drag increasing effect. However, different convex structures due to the different surface curvature and angle of attack. There are some differences in the increase of the overall pressure drag of drag reduction structures.

For the vertical flow concave structure, the water flow into the concave structure at the same time will have a roundabout movement, resulting in the concave structure back flow surface reverse pressure. In addition, since the gravity factor is considered in the setting of this paper, there is partial contact between the water flow and the next oncoming wall after passing through the concave structure, which also generates a certain pressure drag.

Compared with the pressure drag of the convex structure, the resulting pressure drag can be ignored according to the general simplified analysis method. However, in this paper, from a conservative perspective, if the pressure drag is completely ignored, it may lead to the distortion of the simulation experiment, and the resulting artificially high drag reduction rate can be obtained. Therefore, the criterion of the resultant force of viscous resistance and pressure drag is still adopted as the analysis method for the vertical flow direction layout trench structure.

According to the characteristics of the operation of MPF mode robotic fish in the low flow velocity environment, it is believed that the conditions of large changes in speed parameters are less likely to occur in MPF mode robotic fish. Therefore, only the drag reduction effect of the cell structure at a fixed low speed value is examined, and the angle of attack, height, curvature and smoothness of the cell structure are controlled by the changes in the structure's top angle parameters.

4.4   Analysis of Drag Reduction Rate of Convex Structure

As can be seen from Fig. 7, the viscous drag reduction rate of the vertically flowing convex structure increases significantly with the increase of height, but the total drag reduction rate is inversely proportional to the height. The two convex layout schemes based on the quarter-arc configuration show better viscous drag reduction rate, but also show obvious drag increase effect. When the three kinds of convex structure arrangement reach the height of the drag reduction size range of 1/3, the total drag increasing effect begins to appear.

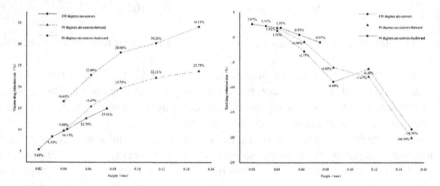

Fig. 7. Viscous drag reduction and total drag reduction of convex structures.

The results further verify that, although the viscous drag reduction rate is proportional to the height of the structure when the vertical flow direction is arranged, the convex structure has poor drag reduction performance due to the influence of pressure drag, but the small size of the convex structure still has good drag reduction performance. Although the large-size convex structure can reduce the viscous resistance more effectively, the pressure drags increases sharply due to the convex structure, which makes the total drag increase effect. Therefore, not all the cell structures in the range of drag reduction size have drag reduction effect.

4.5 Analysis of Drag Reduction Rate of Concave Structure

As can be seen from Fig. 8, the viscous drag reduction rate of vertical flow concave structure increases briefly with the increase of depth and then gradually tends to be stable. The total drag reduction rate is inversely proportional to the increase of depth at the beginning. The lowest point of total drag reduction occurs within the range of slope Angle 30°–40° and concave depth 1/3 to 1/2 of the drag reduction size. When the slope angle is 15° and 45°, the maximum total drag reduction rate of 3.5% is obtained, and then the total drag reduction rate tends to be stable with the increase of depth, showing a more obvious drag reduction effect.

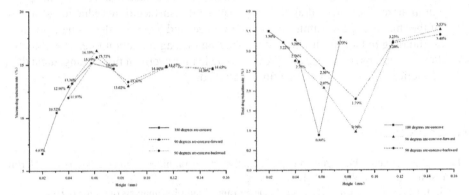

Fig. 8. Viscous drag reduction and total drag reduction of concave structures.

The results further verify that the viscous drag reduction rate tends to be stable with the increase of concave depth when the vertical flow direction is arranged, but the total drag reduction rate will be low within a certain size range, which provides a certain reference for the subsequent design of drag reduction structure size.

5 Conclusion

In this paper, through simulation experiments, the vertical flow layout drag reduction structure of quarter-arc and half-arc configuration was studied. We analyzed the influence of vertical flow layout drag reduction structure on wall resistance with the structure's top angle, height, curvature and other parameters as variables. Some layout schemes have achieved good drag reduction effect, and relevant conclusions are as follows:

(1) The vertical flow structure arrangement within a certain size range can also increase the thickness of the boundary layer, increase the boundary delamination, suppress the turbulent burst process, improve the flow pattern in the near wall area, and reduce the viscous resistance.
(2) When the depth of the vertical flow groove structure exceeds the drag reduction size range by 1/3, the influence on the viscous drag reduction rate tends to be stable, and the maximum viscous drag reduction rate is 15%. The three groove structure

layout schemes have good drag reduction effect when the slope angle is 15° and 45°, and the maximum total drag reduction rate is 3.54%.

(3) With the increase of the height of the structure, the vertical flow layout of the convex structure can obtain a larger viscous drag reduction rate. Among them, the convex structure designed in this paper can achieve 2.6% drag reduction rate when the angle of attack is 15°. However, when the height of the three convex structures reaches one third of the size range of drag reduction, the increased pressure drag cancels out the viscous drag reduction effect, resulting in the total drag-increasing effect of the drag reduction structures.

(4) In the simulation experiment, the layout of the drag reduction structure is small, and the total drag reduction rate is low compared with the actual situation. This paper summarizes the partial drag reduction law of the surface drag reduction structure of the fluctuating fin within a certain size range, and gives the shape characteristics of the drag reduction structure at the maximum drag reduction rate. The potential laws of viscous resistance and pressure drag in more vertical flow arrangement drag reduction structures should be further studied.

References

1. Gao, S.: Structural design and experimental study of bionic ray. Ph.D. Dissertation, Harbin Institute of Technology (2014)
2. Sun, T., Zhao, Q.J., Liang, D.T., et al.: Design and simulation analysis of bionic ray underwater robot. Manuf. Mach. **56**(09), 45–49 (2018)
3. Pang, S., Qin, F., Shang, W., et al.: Optimized design and investigation about propulsion of bionic Tandem undulating fins I: effect of phase difference. Ocean Eng. **239**, 109842 (2021)
4. Zeng, X., Xia, M., Luo, Z., et al.: Design and control of an underwater robot based on hybrid propulsion of quadrotor and bionic undulating fin. J. Marine Sci. Eng. **10**(9), 1327 (2022)
5. Wang, M.M., Yang, X.B., Liang, J.H.: A survey on bionic autonomous underwater vehicles propelled by median and/or paired fin mode. Robot **35**(03), 352–362 (2013)
6. Xu, Z., Yang, W.L., Wang, X.Z., et al.: Optimization design and anti-friction performance analysis of pit-shaped non-smooth surface. J. Harbin Eng. Univ. **34**(12), 1605–1610 (2013)
7. Tao, M.: Drag reduction and genetic optimization of pitted bionic non-smooth surfaces. Ph.D. dissertation, Jilin University (2007)
8. Zhang, C.C.: Study on drag reduction of bionic non-smooth surface by flow field control. Ph.D. dissertation, Jilin University (2007)
9. Ren, L.Q., Zhang, C.C., Tian, L.M.: Experimental study on drag reduction of rotary body using bionic non-smooth. J. Jilin Univ. **04**, 431–436 (2005)
10. Wang, K.: Study on drag reduction characteristics of underwater fringe grooves. Ph.D. dissertation, Northwestern Polytechnical University (2006)
11. Yang, H.X.: Numerical study on drag reduction of turbulent boundary layer in trench surface. Ph.D. dissertation, Harbin Institute of Technology (2008)
12. Walsh, M.: Turbulent boundary layer drag reduction using riblets. In: 20th Aerospace Sciences Meeting, p. 169 (1982)
13. Walsh, M., Lindemann, A.: Optimization and application of riblets for turbulent drag reduction. In: 22nd Aerospace Sciences Meeting, p. 347 (1984)
14. Yang, Z., Chang, Y.Y., Li, D.L., et al.: Functional surface design and drag reduction characteristics of underwater microstructures. Eng. Ship **42**(01), 141–147 (2020)

15. Zhao, L.S., Bai, X.Q., Fu, Y.F., et al.: Simulation analysis and experimental verification of drag reduction effect of Marine Trapezoidal microgroove. Eng. Ship **37**(10), 15–20 (2015)
16. Wang, J.J., Lan, SL., Miao, F.Y.: Study on drag reduction characteristics of turbulent boundary layer in trench surface. Ship Build. China (04), 4–8 (2001)
17. Dou, R.H.: Numerical study on drag reduction characteristics of bionic non-smooth surface. Ph.D. dissertation, Qingdao University of Science and Technology (2020)
18. Song, J.J.: Research on turbulent drag reduction and flow control on non-smooth surface. Ph.D. dissertation, University of Chinese Academy of Sciences (Institute of Engineering Thermophysics) (2012)
19. Feng, Y.: Numerical simulation of turbulence field on biomimetic non-smooth Surface and analysis of drag reduction mechanism. Ph.D. dissertation, Jilin University (2005)
20. Xie, H.: Research on bionic wall drag reduction mechanism and surface microstructure design method. Ph.D. dissertation, China Ship Research Institute (2017)
21. Zhu, H.Y., Hu, H.T., Yin, B.C., et al.: Drag reduction technology of high-speed train based on bump non-smooth surface. Railw. Locomot. Car **41**(01), 1–8 (2021)
22. Yang, S.B., Han, X.Y., Qiu, J.: Kinematics modeling and simulation of pectoral fin model of ray. J. Natl. Univ. Defense Technol. **31**(01), 104–108 (2009)
23. He, J.H., Zhang, Y.H.: Theoretical analysis and experimental test of the undulation propulsion velocity of biomimetic robotic fish fins. Hydrodyn. Res. Progr. **A30**(03), 330–337 (2015)
24. Zhang, Z. Q.: Drag reduction performance of bionic robotic fish on non-smooth surface. Ph.D. dissertation, Shandong University (2017)
25. Zhou, H., Hu, T.J., Xie, H.B., et al.: Computational and experimental study on dynamic behavior of underwater robots propelled by bionic undulating fins. Sci. China Technol. Sci. **53**(11), 2966–2971 (2010)

Molecular Computing
and Nanotechnology

Integrated Design and Absorbing Performance Analysis for Periodic Wave Absorbing and Bearing Structures

Weian Du[1], Runxin Wu[2], Haiyu Zhang[2(✉)], Chengyang Liu[1], and Yuxuan Du[3]

[1] China Ship Development and Design Center, Wuhan 430064, China
[2] Green & Smart River-Sea-Going Ship, Cruise and Yacht Research Center, Wuhan University of Technology, Wuhan 430070, China
zhanghaiyu@whut.edu.cn
[3] School of Science, Hubei University of Technology, Wuhan 430068, China

Abstract. The main development route in wave absorption technology is to develop new absorbing materials to achieve comprehensive requirements such as thin thickness, light weight, broadband coverage and high loss capability. However, as far as the current development trend of materials and structures concerned, it is nearly unrealistic to realize the integration of wave absorbing and load bearing or other compatible functions only through designs of multi-layer composites. Therefore, based on the concept of multi-scale collaborative design of materials and structures, this paper proposes a new type of honeycomb reinforced structure that can bear loads and absorb microwaves. The wave absorber designed in this structure is epoxy resin filled with carbon nanomaterials and carbonyl iron. According to the simulations, this design has broadband coverage and high loss capability, which can achieve full-band coverage of lower than − 0 dB within the 1–20 GHz frequency range, with its loss peak exceeding −30 dB, and the average level is about −20 dB; At the same time, this design may improve mechanical properties of the honeycomb structure to a certain extent due to its sheet structures inside, which is expected to achieve the integration of load bearing and microwave absorbing.

Keywords: Multi-scale collaborative design · Carbon nanomaterials and carbonyl iron composites · Integration of structures and functions · Superstructure design · Reinforced honeycomb

1 Introduction

After years of development, microwave absorbing technology has been widely used in various fields of military or domestic economy. The so-called stealth technology is to reduce the scattering cross section (RCS) of structures in incident directions, which could weaken the power of the scattering echo and greatly shorten its detection distance of radar detectors to achieve relative stealth. The simplest and most effective way to reduce RCS is to change the shape of equipment's surfaces and make it flatter and more integrated, so that microwaves can be reflected to a specific direction together and the RCS could be reduced from the incident direction. However, with the development and application of multi-static radars, radar detectors are distributed at several different locations and share a same information system at the same time [1]. Even if shaping stealth technology can reflect radar waves to other directions, the risk is that the reflected waves may enter into the visions of multi-static radars [2]. Therefore, the shaping stealth's effect will be tremendously slashed, so the main development route of the existing stealth technologies is to try to develop new materials to absorb incident waves and transfer its energy to thermal energy.

From the current developing trend of materials and structures in absorbing technologies, the advantage of absorbing materials lies in their loss capability [3–6], while wave absorbing structures have great advantages in impedance matching, broadband coverage, adaptability to incident angles, excellent mechanical performance or other comprehensive properties [7–10].

Therefore, based on the concept of multi-scale collaborative design of materials and structures, this paper proposes a honeycomb structure reinforced with a central cone and 12 sheet structures to achieve a design of bearing/absorbing integration (as shown in Fig. 1). Epoxy resin filled with carbon nanomaterials and carbonyl iron will be selected as the basic material in this design. According to the corresponding simulation verification, this design has broadband coverage and high loss performance, and can also achieve full-band coverage of lower than -10 dB within 1–20 GHz frequency range. The peak could exceed -30 dB, and the average level within 1–20 GHz range is about -20 dB; At the same time, this design may improve mechanical properties of the whole structure to a certain extent due to its reinforced structure inside, which is expected to better realize the integration of load bearing and wave absorbing.

2 Structural Design Principles

The core structure designed in this paper is a central cone, as shown in Fig. 1(b). According to the existing research [11], smaller angle can increase the number of contacts between the wall and the cone, which can make more energy enter the absorber and improve absorption performance. Based on this, this paper combines an absorbing cone with a honeycomb structure.

At the same time, in order to increase the strength and stability of the honeycomb structure, several sheets are added between the cone and the honeycomb wall to form a waveguide-like transmission channel. On the one hand, it can increase reflections and scatterings of incident waves, and on the other hand, it can form a transmission channel for those that cannot enter the cone or honeycomb wall so that they can be guided to the capture structure at the bottom, as shown in Fig. 1(c). After entering the capture structure, electromagnetic waves will contact the capture structure and the wall for many times until they are totally depleted.

The numerical simulation model and geometric parameters of structural design details are shown in Fig. 2.

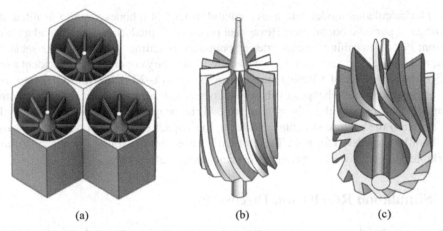

Fig. 1. (a) Overview (b) Central cone (c) Capture Structure

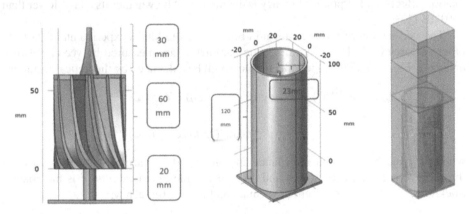

Fig. 2. Numerical simulation model and geometric parameters

3 Simulations

This paper uses COMSOL Multiphysics to conduct numerical simulations on the designed periodic structure. As there are many structural parameters and optimization indexes involved, only one set of geometric parameters are taken for simulation analyses, but not necessarily the optimal parameters. Optimization work needs to be further explored.

The geometric parameters are shown in Fig. 2. The central cone is divided into upper, middle and lower parts; The bottom radius of the central cone is 16 mm, the top radius is 5.5 mm, the slope angle of the cone is 10°, and the height is 60 mm; The lower part is 20 mm high; 12 sheet structures are formed by rotating 12 vertical lamellar structures at a certain angle.

The calculation model here uses a cuboid instead of a honeycomb to facilitate the settings of periodic boundaries. Hexagonal prism is adopted as periodic boundary condition; Perfect matching layer, scattering boundary condition and one port are set at the top, and ideal electric conductor is set as bottom boundary condition; The incident angle is 30° with its power of 1 W, and the frequency is set to 1–20 GHz with 1 GHz step size.

Adjustments and changes of these parameters will affect contact times of electromagnetic waves with the whole structure and the proportion of those which may be guided into the capture structure. However, the impact of TE/TM polarization waves will not be considered for now. Therefore, it is necessary to explore the impact of these variables on the wave absorption performance in the future work.

4 Simulation Results and Discussion

Firstly, the model settings and material parameters will be simulated and verified with 2mm flat structure. The verification model is shown in Fig. 4. It is generally considered that the effective absorption frequency is distinguished by whether its S11 is lower than −10 dB [12].

The verification results are shown in Fig. 4. The red line is experimental data from other research [13]. The electromagnetic parameters are measured by vector network analyzer (Fig. 3), and then the reflection loss will be calculated by theoretical formulas.

$$RL(f) = 20 \log 10 |(Z_{in} - Z_0) / (Z_{in} + Z_0) \qquad (1)$$

$$Z_{in} = Z_0 (\mu_r / \varepsilon_r)^{1/2} \tanh[i(2\pi f d / c) \cdot (\mu_r \varepsilon_r)^{1/2}] \qquad (2)$$

where Z_{in} and Z_0 are input impedance and air impedance respectively, ε_r and μ_r are the relative complex permittivity and relative complex permeability, d is the coating thickness, c is the speed of light in vacuum, and f is the frequency.

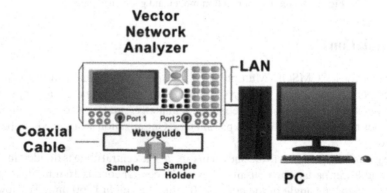

Fig. 3. Measurement of electromagnetic parameters [10]

The blue line (Fig. 4) comes from simulations in this paper. It can be seen that the two groups of data have an excellent coincidence, and both reach the peak at 5.5 GHz, respectively about −20 dB and −21 dB. However, it quickly returns to lower loss level in the other frequencies. Because of its narrow absorption bandwidth, it is difficult to achieve broadband absorption.

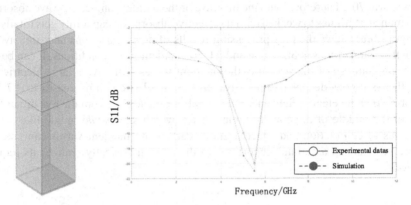

Fig. 4. Simulation verification and results (Color figure online)

From the above simulation verification, the settings of the calculation model and material parameters are correct and effective, so they can be used for simulations of the following designed structure.

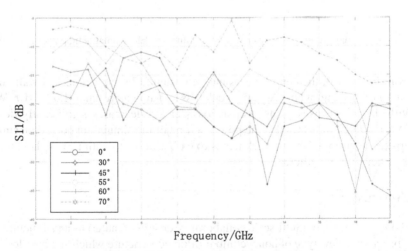

Fig. 5. S11 varying with incident angles

Figure 5 shows the S11 parameter distribution of the model varies with incident angles at 1–20 GHz. It can be seen that the loss performance of 0° and 30° angles are almost the best among them, achieving full-frequency coverage with its peak down to −36dB, and the average level is about −20dB. Moreover, with the angle increasing, the loss performance gradually becomes weaker, which is consistent with the characteristics of pyramid cone's wave absorption performance [11].

The interesting thing is that the effective frequency bandwidth can basically cover all frequencies even at 60° incident angle, but it gradually drops out of the effective bandwidth at 70°. Therefore, with the increase of the incident angle, the wave absorption performance continues to weaken until its effective absorption capability gets totally lost.

Figure 6(a) shows the post-processing results of normal sections and time average power-flow streamlines at 30° angle and 2 GHz incident wave conditions. It can be seen that at the entrance of the structure, the incident waves still have a large electric field strength. As it goes deeper into the structure, the field strength decreases to 3.7 V/m, indicating that the electric field energy is largely lost; Judging from the red streamlines, those in the middle of the cone become denser, which shows waveguide-like cavity's properties of contraction and concentration. Part of the incident waves can enter the capture structure in a certain proportion, so this design basically realizes its required functions.

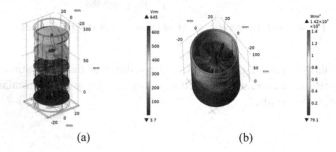

(a)   (b)

Fig. 6. (a) Normal section and power-flow streamline (b) Electromagnetic power loss density

From Fig. 6. (b), the energy is basically consumed by the cylindrical wall and the upper part of the central cone, and the peak loss density can reach to 1.42×10^5 W/m^3, but with the microwaves going deeper, loss density decreases significantly down to 79.1 W/m^3, which indicates that the power setting of the simulation port is too small, or perhaps the next optimization work needs to explore how to optimize the height of the whole structure more efficiently.

5   Conclusion

Based on the concept of multi-scale collaborative design of materials and structures, this paper proposes a new type of honeycomb reinforced structure which can bear loads and absorb microwaves. From the simulation results, the wave absorption performance of this structure can achieve broadband coverage and high loss capability. It can achieve full-band coverage of effective bandwidth at 1–20 GHz, with a peak value of −36 dB and an average level about −20 dB, which can meet the requirements of broadband coverage and high loss property. The parameters' optimization, structural bearing capability and multi-function compatibility based on this design need to be further explored through the future work.

References

1. Liang, H., Liu, J., Zhang, Y., Luo, L., Wu, H.: Ultra-thin broccoli-like SCFs@TiO2 one-dimensional electromagnetic wave absorbing material. Compos. Part B: Eng. **178**, 107507 (2019)
2. Zhang, Z., et al.: Radar-stealth and load-bearing corrugated sandwich structures with superior environmental adaptability. Compos. Sci. Technol. **227**, 109594 (2022)
3. Huang, L., et al.: Bionic composite metamaterials for harvesting of microwave and integration of multifunctionality. Compos. Sci. Technol. **204**, 108640 (2021)
4. Zhang, Z., et al.: Novel multifunctional lattice composite structures with superior load-bearing capacities and radar absorption characteristics. Compos. Sci. Technol. **216**, 109064 (2021)
5. Wang, P., Zhang, Y., Chen, H., Zhou, Y., Jin, F., Fan, H.: Broadband radar absorption and mechanical behaviors of bendable over-expanded honeycomb panels. Compos. Sci. Technol. **162**, 33–48 (2018)
6. Song, P., et al.: Obviously improved electromagnetic interference shielding performances for epoxy composites via constructing honeycomb structural reduced graphene oxide. Compos. Sci. Technol. **181**, 107698 (2019)
7. Huang, Y., et al.: Optimization of flexible multilayered metastructure fabricated by dielectric-magnetic nano lossy composites with broadband microwave absorption. Compos. Sci. Technol. **191**, 108066 (2020)
8. Fu, Z., Pang, A., Luo, H., Zhou, K., Yang, H.: Research progress of ceramic matrix composites for high temperature stealth technology based on multi-scale collaborative design. J. Mater. Res. Technol. **18**, 2770–2783 (2022)
9. Amanipour, V., Olfat, A.: CFAR detection for multistatic radar. Sign. Process. **91**(1), 28–37 (2011)
10. He, C., et al.: Accelerating large-scale multiobjective optimization via problem reformulation. IEEE Trans. Evol. Comput. **23**(6), 949–961 (2019)
11. Huang, Y., Song, W.L., Wang, C., et al.: Multi-scale design of electromagnetic composite metamaterials for broadband microwave absorption. Compos. Sci. Technol. **162**, 206–214 (2018)
12. Pometcu, L., Benzerga, R., Sharaiha, A., Pouliguen, P.: Combination of artificial materials with conventional pyramidal absorbers for microwave absorption improvement. Mater. Res. Bull. **96**, 86–93 (2017)
13. Fu, X., et al.: Electromagnetic wave absorption performance of Ti2O3 and vacancy enhancement effective bandwidth. J. Mater. Sci. Technol. **76**, 166–173 (2021)

Diffusion Optimization Algorithm Based on Membrane Calculation

Yanbing Zhang, Hao Wang$^{(\boxtimes)}$, and Liang Luo

Wuhan University of Technology, Wuhan 430063, China
hao_wang@whut.edu.cn

Abstract. In order to improve the efficiency and quality of solving complex optimization problems, a global optimization algorithm, called the Membrane Computation-based Diffusion Optimization Algorithm (MC-DOA), is proposed. Inspired by the dual membrane cell structure of chloroplasts and mito-chondria as well as the particle diffusion phenomenon, the algorithm uses a hybrid membrane structure system and four different particle search strategies to find the global optimum. Four types of benchmark test functions are selected to test the performance of MC-DOA, namely: single-peak function test, multi-peak function test and different dimensional optimization function test. The test results of MC-DOA are compared with various optimization algorithms (including some well-known intelligent optimization algorithms and some novel intelligent optimization algorithms), and the results show that MC-DOA has good solving ability and solving efficiency.

Keywords: Optimization algorithms · Membrane calculations · Particle diffusion strategies · Test Functions

1 Introduction

For optimization problems with large-scale optimization variables and complex design objectives, both traditional classical optimization algorithms and specific intelligent optimization algorithms have to fail to obtain excellent solution performance because of their own limitations. Therefore, it is of great importance to find an intelligent optimization algorithm suitable for complex optimization problems to improve the efficiency and quality of solutions. Emerging intelligent optimization algorithms include particle swarm algorithm, gray wolf optimization algorithm, whale optimization algorithm, artificial bee swarm algorithm, etc. Hou Xinpei [1] further analyzed the decomposition solution selection strategy and non-dominated sorting solution selection strategy on the basis of the original particle swarm algorithm and introduced multi-criteria variation to speed up the convergence of the algorithm. Chao Chen et al. [2] proposed an adaptive wolf swarm algorithm called three-level leadership and particle evolution equation in order to improve the convergence speed of the wolf swarm algorithm. Li Andong et al. [3] proposed an improved whale optimization algorithm using hybrid strategies to address the problems that the standard whale optimization algorithm is prone to fall into

L. Pan et al. (Eds.): BIC-TA 2022, CCIS 1801, pp. 628–646, 2023.
https://doi.org/10.1007/978-981-99-1549-1_50

local optimal solutions and slow convergence speed. Wang Jichao et al. [4] introduced the idea of "global optimum" in particle swarm algorithm into the improvement process of artificial bee swarm algorithm to form a particle swarm algorithm with faster convergence speed and higher optimization accuracy. Chao Wang et al. [5] proposed a new subcarrier and power adaptive allocation algorithm based on the fish swarm algorithm for the adaptive resource allocation problem of multi-user orthogonal frequency division multiple access systems.

Intelligent optimization algorithms incorporating a membrane computing framework have attracted attention in recent years. Membrane computing, abstracted from biology as a new branch of natural computing, provides a very suitable framework for distributed computing, aiming to create new computational ideas and computational models from the structure and function of biological cells and the collaboration of cell populations. Dongning Chen et al. [6] put six different particle swarm algorithms into each of the six basic membranes and proposed inter-membrane communication and particle update mechanisms. Nan Song et al. [7] proposed a particle swarm algorithm in a membrane framework for solving the unsupervised multi-target radar radiation source signal feature selection problem. Zhang Gexiang et al. [8] combined a differential evolutionary algorithm with an organization-based membrane computation for dealing with constrained optimization problems. Peijun Xie et al. [9] proposed a membrane computing particle swarm algorithm to improve the extreme learning machine (MCPSO-ELM) model for water and fertility prediction. Han Tao et al. [10] proposed a Fast SLAM algorithm based on membrane-computing particle swarm optimization for the problems of particle degradation and missing particle diversity in Fast SLAM algorithm. Liang Huang et al. [11] discussed several multi-objective optimization problems such as the loss of non-dominated solutions, the emergence of erroneous non-dominated solutions, and the necessity of online decision mechanisms. Hong-Yuan Gao et al. [12] proposed a novel multi-objective discrete combinatorial optimization algorithm based on quantum group theory and membrane computation, namely, the membrane-inspired quantum bee optimization algorithm [13]. Wang et al. [14] proposed a diffusion algorithm for continuous global optimization based on P-systems. Tingxi Liu [15] proposed an image segmentation algorithm combining membrane computation and Gaussian mixture model for the problems of color image segmentation. Fu Jie [16] proposed an adaptive membrane calculation optimization algorithm for adjusting the PID controller parameters, in view of the fact that the PID controller designed by the traditional method is difficult to meet the performance requirements of full working condition operation in the automotive anti-lock control system. Inspired by this, a new intelligent optimization algorithm MC-DOA (membrane computation based diffusion optimization algorithm) is used to improve the efficiency and quality of solving complex optimization problems.

2 Diffusion Optimization Algorithm Based on Membrane Calculation

Membrane computing, as a parallel distributed computing framework, is flexible and general. In this paper, a hybrid membrane structure system is abstracted from the dual membrane biological structure of chloroplast and mitochondria; and based on the particle diffusion phenomenon, four stochastic particle search models are constructed for specifying the evolutionary rules as computational objects in membrane structures.

2.1 Membrane Computing Framework

The hybrid membrane structure system proposed in this paper is a system in which two double membrane structures similar to chloroplasts and mitochondria are placed side-by-side, as shown in Fig. 1. Thus, the hybrid membrane structure system can be expressed as:

$$\Pi = (O, \mu, \omega_1, \omega_2, ..., \omega_5, (R_1, \rho_1), ..., (R_5, \rho_5)) \tag{1}$$

where O is the character set used to represent the computational objects, $O = \{R\}$; μ denotes the membrane structure of the hybrid membrane system Π, which can be obtained from Eq. (2); the degree of the membrane system proposed in this paper is 5, indicating that the hybrid membrane structure system consists of 5 layers of membranes, which are labeled as $\sigma_1, \sigma_2, \sigma_3, \sigma_4, \sigma_5$; $\omega_i (i = 1, 2, ..., 5)$, denotes the sets of computational objects within different membranes. $(R_i, \rho_i)(i = 1, 2, ..., 5)$, denotes the set of computational rules within the membrane σ_i and the corresponding set of rule partial order. The computational rules include evolutionary rules and transit rules.

$$\mu = [_1[_2[_4]_4]_2[_3[_5]_5]_3]_1 \tag{2}$$

In the search process, the solution of the optimization problem is considered as a computational object and the computational object representing the solution vector of the optimization problem is composed in a real number encoding. ω_{ij} denotes the jth solution in membrane σ_i and ω_i in membrane σ_i can be obtained from Eq. (3).

$$\omega_i = \{\omega_{i,1}, \omega_{i,2}, ..., \omega_{i,m}\} \tag{3}$$

where $\omega_{i,m}$ denotes that m computational objects are randomly generated in the membrane σ_i. In the hybrid membrane structure system proposed in this paper, the number of computational objects in the whole hybrid membrane structure system is $4m$ since no computational objects are generated in the σ_1 of the cell membrane. During the transit process, the computational objects in different regions will be exchanged and passed. A well-adapted solution will simulate the transport route of oxygen and gradually transport it inward. On the contrary, a poorly adapted solution will simulate the transport route of carbon dioxide and gradually transport it outward. Meanwhile the number of computational objects in each region is kept as m.

2.2 Particle Search Strategy

In MC-DOA, the primary computational object evolution rules are mainly used to generate new primary computational objects based on the existing primary computational

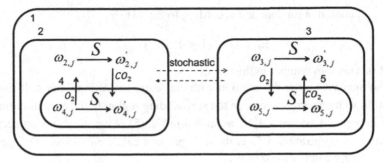

Fig. 1. Hybrid membrane structure system

objects. Moreover, the rules for generating new primary computational objects are different in different membranes. In this section, the four random search operators used in MC-DOA are described in detail.

Stochastic Search Operator Based on Concentration Versus Time Model

In the skin membrane σ_1, the stochastic search operator used in the main computational object evolution rule is abstracted from a special physical diffusion model. This diffusion model is mainly used to describe the law of concentration change with time when two solutions with different concentrations are suddenly in contact, as shown in Fig. 2.

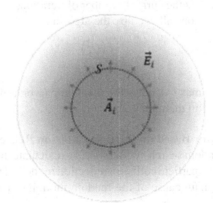

Fig. 2. Contact model for two different concentration solutions

Consider the special case shown in Fig. 2, where the region A is a higher concentration of a certain solution, and the surrounding region E is a lower concentration C_e of the same solution, and the two regions are separated from each other in the initial state. When the separation between the two regions is lifted, region A and region E will suddenly come into contact, and the substances in the solution will diffuse from A to E.

In the skin membrane σ_1, the main calculation object O_i^1 is abstracted as the concentration $\vec{A}_i$ of the solution in region A; the concentration $\vec{E}_i$ of the solution in the surrounding region E of region A is determined randomly by another main calculation object O^1_{rand} chosen at random and according to Eq. (4).

$$\vec{E}_i = [1 - (1 + T)R] \cdot O^1_{rand} \tag{4}$$

where R is a random number in the range.

At the moment t, after the solution with concentration $\vec{A}^t_i$ in region A diffuses with the solution with concentration $\vec{E}^t_i$ in the surrounding region E, the concentration of the solution in region A becomes $\vec{A}^{t+1}_i$, as shown in Eq. (5). After the diffusion occurs, the new solution concentration $\vec{A}^{t+1}_i$ is the new primary calculation object O^{1*}_i generated from the primary calculation object O^1_i.

$$\vec{A}^{t+1}_i = \vec{E}^t_i + \left(\vec{A}^t_i - \vec{E}^t_i\right)e^{-\alpha T} \tag{5}$$

In the above equation, t denotes the starting moment of diffusion, which is equal to the current number of iterations of the algorithm; T denotes the influence of the computational process on the random search, as shown in Eq. (6); $\vec{E}_i(t)$ is the value of the solution concentration in the region E around region A, as shown in Eq. (4); and the coefficient α is related to the fitness $f\left(O^1_i\right)$ of the main computational object O^1_i. The smaller the fitness corresponds the smaller value α, as shown in Eq. (7).

$$T = \frac{t}{N} \tag{6}$$

In the above equation, t is the current number of iterations of the algorithm; N is the maximum number of iterations allowed by the algorithm.

$$\alpha = \frac{f\left(O^1_i\right)}{max f} \tag{7}$$

In the above equation, $max f$ denotes the maximum (worst) fitness value of the main calculation object in the skin membrane σ_1.

Random Search Operator Based on Particle Collision Reflection Model
In the skin film σ_2, the random search operator used to calculate the main object evolution rule is abstracted from the "particle collision reflection model". The collisional reflection between particles is the main cause of the random motion of particles in the diffusion system.

The main object of calculation in the dermatome σ_2, O^2_i, represents the location region $\vec{P}_i$ of the particle i in the diffusion system, and the degree of adaptation $f\left(O^2_i\right)$ characterizes the concentration of the region where the particle i is located. Then, the particle collision probability C_i in the particle i region can be calculated according to Eq. (8).

$$C_i = \frac{f\left(O^2_i\right) - min f}{max f - min f} \tag{8}$$

In the above equation, $min f$ and $max f$ then denote the minimum and maximum fitness values of the main computational objects in the membrane σ_2, respectively. From

Eq. (8), it can be seen that the worse (large) adaptation corresponds to the region with high collision probability, while the better (low) adaptation corresponds to the region with low collision probability.

At the moment t, the particle i in the diffusion system is in the position region $\vec{P}_i^t$ and it is assumed that the particle i will move towards another particle $j(j \neq i)$ that is randomly determined. Since the probability of particle collision in the region $\vec{P}_j^t$ where particle j is located is C_j, particle i may be reflected by collision in the region $\vec{P}_j^t$, or it may continue to move in a straight line without collision. Figure 3 depicts the motion of a particle in a two-dimensional diffusion system without collisional reflection and with collisional reflection. When the particle i moves towards the region $\vec{P}_j^t$ without collision, the particle i moves in the new region $\vec{P}_i^{t+1}$ under the combined effect of the diffusion trend $\vec{D}_1$ towards the region $\vec{P}_j^t$ and the overall diffusion trend $\vec{D}_2$ of the system, as shown in Fig. 3(a).When the particle i moves toward the region $\vec{P}_j^t$ to collide, the particle i will be reflected by the region $\vec{P}_j^t$ to move to the new region $\vec{P}_i^{t+1}$, as shown in Fig. 3(b).The new location region $\vec{P}_i^{t+1}$ of a particle i is the new object O_i^{2*} in the skin membrane σ_2 obtained by the evolution of the main computational object O_i^2.

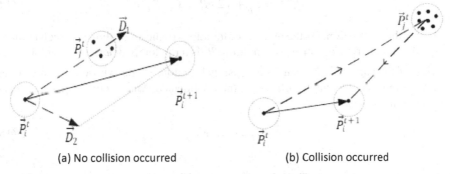

(a) No collision occurred   (b) Collision occurred

Fig. 3. Random collisional reflection of particles in a two-dimensional diffusion system

If no collision occurs, particle i will move in a straight line from region $\vec{P}_i^t$ to region $\vec{P}_i^{t+1}$ as follows.

$$\vec{P}_i^{t+1} = \vec{P}_i^t + \vec{D}_1 + \vec{D}_2 \tag{9}$$

In the above equation, $\vec{D}_1$ represents the tendency of the motion of particle i toward particle j, as shown in Eq. (10); $\vec{D}_2$ represents the influence of the overall diffusion trend of the diffusion system on the motion of particle i, as shown in Eq. (11).

$$\vec{D}_1 = R \cdot c \cdot \left(\vec{P}_j^t - \vec{P}_i^t \right) \tag{10}$$

$$\vec{D}_2 = 2R \cdot c \cdot \left(\vec{P}_{best}^t - (2R + 1) \cdot \vec{P}_{mean}^t \right) \tag{11}$$

In the above equation, R is a random number in the range $[0, 1]$; c is a directional coefficient whose value is calculated according to Eq. (12). $\vec{P}^t_{best}$ denotes the current lowest concentration (best fit) location region; $\vec{P}^t_{mean}$ is the geometric center of all particle location regions (the main calculation object) in the skin membrane σ_2, as shown in Eq. (13).

$$c = \begin{cases} -1, & \text{if } f\left(O_i^2\right) < f\left(O_j^2\right) \\ 0, & \text{if } f\left(O_i^2\right) = f\left(O_j^2\right) \\ 1, & \text{if } f\left(O_i^2\right) > f\left(O_j^2\right) \end{cases} \tag{12}$$

$$\vec{P}^t_{mean} = \frac{1}{n} \sum_{k=1}^{n} \vec{P}^t_k \tag{13}$$

In the above equation, n denotes the number of main calculation objects in the skin membrane σ_2.

If a collision occurs, the particle i will be reflected from the position region $\vec{P}^t_i$ to the new position region $\vec{P}^{t+1}_i$ by the collision, as shown in Eq. (14).

$$\vec{P}^{t+1}_i = \vec{P}^t_j + \left(2 \cdot \vec{R} - 1\right) \cdot S_{ij} \tag{14}$$

In the above equation, t is the number of iterations of the algorithm, and each iteration means that the particle performs one motion. $\vec{P}^t_j$ is a randomly determined region of the position of the particle j. $\vec{R}$ is an d-dimensional random vector whose elements are all random numbers in the range $[0, 1]$. S_{ij} is the Euclidean distance between particles i and j as shown in Eq. (15).

$$S_{ij} = \|\vec{v}_i, \vec{v}_j\|_2 \tag{15}$$

Random Search Operator Based on Brownian Motion

To increase the stochasticity of the MC-DOA global search and avoid prematurely falling into local optima, the main computational object evolution rule in the membrane uses a stochastic search operator designed inspired by Brownian motion. Brownian motion is the irregular motion of a particle suspended in a liquid or gas in a never-ending process. The Brownian motion process is a continuous-time stochastic process that belongs to the Lévy process.

In the membrane σ_3, the stochastic search operator used for the main computational object evolution rule is inspired by Brownian motion, so the main computational object O_i^3 represents the spatial position $\vec{P}_i$ of the particle i that makes Brownian motion in the diffusion system. Each iteration of the MC-DOA calculation indicates that the particles in the membrane σ_3 make one random motion, and the step size of each random motion approximately obeys the Lévy distribution.

At the moment t, the spatial position of the particle i in the diffusion system is P^t_i, and after a random motion, the particle i moves to the new position P^{t+1}_i, as shown in

Eq. (16).The new position P_i^{t+1} of particle i is the new primary computational object O_i^{3*} generated from the primary computational object O_i^3.

$$\vec{P}_i^{t+1} = \vec{P}_i^t + (UB - LB) \oplus Lévy(\alpha) \tag{16}$$

In the above equation, t is the number of iterations of the algorithm, and each iteration means that the particle performs a random motion; $LB = \left(x_l^1, \cdots x_l^d\right)$ and $UB = \left(x_u^1, \cdots x_u^d\right)$ denote the lower and upper limits of the variables in the d-dimensional optimization problem, respectively; α is a constant coefficient, which is recommended to be 1.5 according to the experimental test experience.

Random Search Operator Based on FICK's Diffusion Law

The same stochastic search model is used in skin films σ_4 and σ_5, which is designed and constructed based on Fick's law of diffusion. According to Fick's law, particles will move from the region of high concentration to the region of low concentration, so by tracking the diffusion trend of particles to adjust the location of observation points, it is finally possible to find an optimal observation location to observe the lowest particle concentration in the system.

The computational objects in the dermal films σ_4 and σ_5 represent the observation points in the diffusion system, while their corresponding adaptations represent the concentration gradients. For any observation points i and j, define the component of the diffusion flux from observation point i to observation point j in dimension k as shown in Eq. (17).

$$J_{ij}^k = -D \cdot \nabla C_j \cdot \frac{x_i^k \quad x_j^k}{R_{ij}} \tag{17}$$

In the above equation, D denotes the diffusion constant coefficient; ∇C_j denotes the concentration gradient after normalization, as shown in Eqs. (18) and (19). R_{ij} then denotes the Euclidean distance between observation points i and j, $R_{ij} = \|\vec{v}_i, \vec{v}_j\|_2$. ε is then a small amount added to avoid the denominator being zero. x_i^k, x_j^k denotes the position of observation point i on dimension k.

$$f'\left(O_j^4\right) = \frac{f\left(O_j^2\right) - \max f}{\min f - \max f} \tag{18}$$

$$\nabla C_j = \frac{f'(\vec{v}_j)}{\sum_{k=1}^{N} f'(\vec{v}_k)} \tag{19}$$

In the above equation, $f\left(O_j^4\right)$ is the adaptation of the calculated object O_{jj}^4, while $f'\left(O_j^4\right)$ indicates the calculated concentration gradient in the corresponding area; $\max f$ indicates the maximum adaptation of the calculated object in membrane σ_4 or σ_5. $\min f$ denotes the minimum fitness of the computed object in the membrane σ_4 or σ_5.

To increase the randomness of the search operator, a random weight is introduced into the calculation of the component J_i^k of the diffusion flux J_i in dimension k at a particular observation point i, as shown in Eq. (20).

$$J_i^k = \sum_{j=1, j \neq i}^{N} r_j J_{ij}^k \tag{20}$$

In the above equation, r_j is a random weight satisfying a uniform distribution in the range [0, 1].

Therefore, based on the diffusion trend at observation point i, a new observation point location $\vec{v}_i^t$ obtained from observation point $\vec{v}_i^{t+1}$ can be calculated, as shown in Eq. (21).

$$x_i^k(t+1) = x_i^k(t) + J_i^k(t) \tag{21}$$

In the above equation, t denotes the number of iterations.

In addition, in order to make this operator have a strong exploratory power at the beginning of the search process and a strong exploitation power at the end of the search process, the diffusion constant D is defined as an exponentially decreasing function shown in Eq. (22), and its function image is shown in Fig. 4.

$$D = 2Ne^{-\alpha \cdot \frac{t}{Maxiteration}} \tag{22}$$

In the above equation, N denotes the number of computed objects in membrane σ_4 or σ_5; α is a constant, which is recommended to be 17.5 based on testing experience; t denotes the current number of iterations; *Max iteration* denotes the maximum number of iterations allowed by the algorithm.

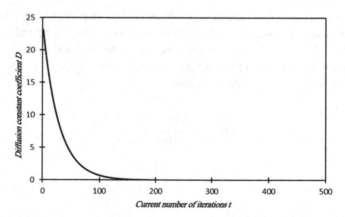

Fig. 4. Variation curve of diffusion constant coefficient with the number of iterations

3 MC-DOA Algorithm Test Experiment

3.1 Single-Peak Function Test Experiment

Test Function and Algorithm Parameter Setting
In the single-peak function test experiment, seven classical optimization functions were selected as benchmark functions to test the computational performance of MC-DOA, as shown in Table 1. The two-dimensional shapes of the single-peaked benchmark test functions are shown in Fig. 5. Each type of single-peaked benchmark test function simulates the solution space characteristics of various types of practical optimization problems by providing different shapes in the search space.

Table 1. Single-peak benchmark test function

Function	Expression	Range of values	F_{min}
Sphere	$F_1 = \sum_{i=1}^{n} x_i^2$	$[-100, 100]^D$	0
Schwefel 2.22	$F_2 = \sum_{i=1}^{n} |x_i| + \prod_{i=1}^{n} |x_i|$	$[-10, 10]^D$	0
Schwefel 1.2	$F_3 = \sum_{i=1}^{n} \left(\sum_{j=1}^{i} x_j \right)^2$	$[-100, 100]^D$	0
Schwefel 2.21	$F_4 = \max_{i} \{|x_i|, 1 \le i \le n\}$	$[-100, 100]^D$	0
Rosenbrock	$F_5 = \sum_{i=1}^{n} \left[100\left(x_{i+1} - x_i^2\right)^2 + (x_i - 1)^2 \right]$	$[-30, 30]^D$	0
Step	$F_6 = \sum_{i=1}^{n} \left[(x_i + 0.5)^2 \right]$	$[-100, 100]^D$	0
Quartic	$F_7 = \sum_{i=1}^{n} i x_i^4 + random[0, 1)$	$[-1.28, 1.28]^D$	0

In the single-peak function test experiment, the dimensionality (number of variables) D of all benchmark test functions is 30, and the MC-DOA test results are compared and analyzed with the computational results of four different intelligent optimization algorithms, PSO, GA, GWO and ALO. PSO and GA are two well-known intelligent optimization algorithms that have been widely applied to various engineering practical problems. The test results of PSO and GA are calculated based on the PSO and GA toolboxes in MATLAB 2019a. GWO and ALO are two novel intelligent optimization algorithms whose source codes used in the test experiments are obtained from http://www.alimirjalili.com/ALO.html and http://www.alimirjalili.com/GWO.html.

To ensure the objectivity of comparing the test results of different algorithms, the number of objective function calls is guaranteed to be equal for all algorithms in the test experiment, which is 30,000 times. For PSO, GA, GWO and ALO, the population size was set to 30 and the maximum number of iterations was set to 1000. Considering that MC-DOA contains 5 membranes and there is a relatively independent subpopulation in each membrane, the total population size of MC-DOA is set to $5 \times 12 = 60$, i.e., the subpopulation size in each membrane is 12, and the maximum number of iterations of the algorithm is set to 500 times in order to avoid the degradation of algorithm performance due to too few computational objects in each membrane. Each algorithm is run independently for 30 times.

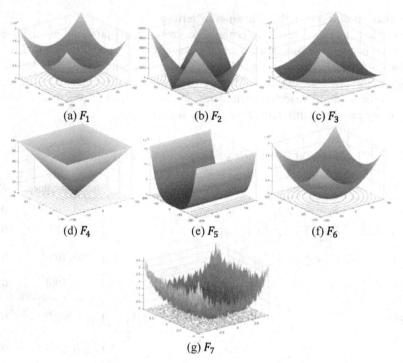

(a) F_1 (b) F_2 (c) F_3

(d) F_4 (e) F_5 (f) F_6

(g) F_7

Fig. 5. Two-dimensional single-peak benchmark test function image

Experimental Results and Analysis

The results of the single-peak function test experiments are shown in Table 2. From the test results, it can be seen that MC-DOA shows a strong competitiveness in dealing with single-peak optimization problems. For the seven single-peak benchmark test functions selected in the test trials, MC-DOA is able to provide higher solution accuracy for the benchmark functions F_1, F_2, F_3, F_4, F_5 and F_7 compared with the other four algorithms (PSO, GA, ALO, and GWO); for the benchmark function F_6, PSO has the highest solution accuracy, and the solution accuracy of MC-DOA is only better than that of GA and GWO.

The convergence curves of MC-DOA, PSO, GA, ALO and GWO for the seven single-peaked benchmark functions are shown in Fig. 6, from which it can be seen that the convergence performance of MC-DOA and GWO is significantly better than other algorithms. For the benchmark functions F_1, F_2, F_3 and F_4, MC-DOA and GWO can well avoid the premature convergence problem and show the convergence characteristic that convergence tends to accelerate with the increase of iterations, and the solution accuracy of MC-DOA is much higher than the other four algorithms. The test results show that for the benchmark test function F_5, MC-DOA is able to produce a result that is very close to the optimal solution, while the other four algorithms are unable to find an approximate solution close to the global optimum. Noise pollution is introduced in the benchmark test function F_7, and the test results show that MC-DOA can effectively overcome the influence of noise pollution on the search efficiency of the algorithm and exhibit better convergence characteristics than the other four algorithms. In the single-peak test, the solution accuracy of MC-DOA on the benchmark function F_6 is lower than that of PSO and ALO; it can be seen from the convergence curves that compared with

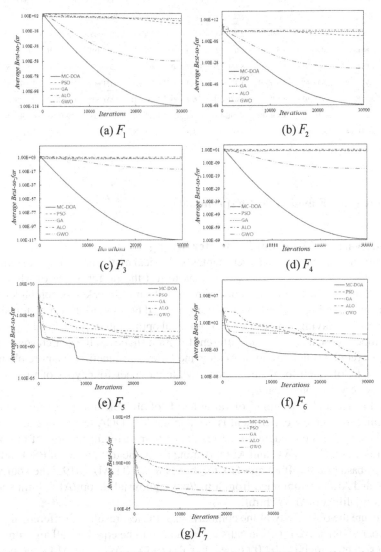

Fig. 6. Convergence curves of MC-DOA, PSO, GA, ALO, and GWO for single-peak benchmark

PSO and ALO, MC-DOA shows the phenomenon of premature convergence of the first pair, which is the main factor affecting the solution accuracy.

The single-peak test function is well suited for testing the global exploration capability of various optimization algorithms, and MC-DOA demonstrates superior global exploration capability in single-peak test trials.

Table 2. Comparison of test results of single-peak benchmark test function (D = 30)

Functions	MC-DOA		PSO		GA		ALO		GWO	
	Ave.	Std.	Ave.	Std.	Ave.	Std.	Ave.	Std.	Ave.	Std.
F_1	**6.70E-118**	**1.44E-117**	7.83E-09	2.23E-08	2.84E-02	3.98E-02	1.31E-05	9.23E-06	4.24E-59	6.98E-59
F_2	**3.89E-68**	**4.06E-68**	4.58E-04	1.07E-03	2.79E+00	1.59E+00	5.78E+01	5.42E+01	1.66E-34	3.94E-34
F_3	**1.05E-117**	**1.06E-117**	1.47E+01	8.16E+00	4.02E+00	1.89E+00	1.35E+03	6.13E+02	5.64E-15	1.80E-14
F_4	**4.77E-69**	**5.83E-69**	6.49E-01	1.77E-01	1.96E+00	4.21E-01	1.15E+01	2.47E+00	1.31E-14	1.35E-14
F_5	**2.91E-03**	**3.93E-03**	4.69E+01	3.17E+01	9.64E+00	2.08E+01	2.88E+02	4.16E+02	2.69E+01	7.69E-01
F_6	5.22E-05	3.21E-05	**1.13E-08**	**4.28E-08**	6.96E-02	1.52E-01	1.02E-05	8.89E-06	6.63E-01	2.93E-01
F_7	**2.41E-04**	**1.49E-04**	8.20E-02	3.01E-02	1.31E+00	3.13E-01	9.19E-02	2.36E-02	7.52E-04	4.16E-04

3.2 Multi-peak Function Test Experiment

Test Function and Algorithm Parameter Setting

In the multi-peak function test experiment, six classical mathematical optimization functions were selected as benchmark functions to test the performance of MC-DOA, as shown in Table 3. The two-dimensional shapes of the multi-peaked benchmark test functions are shown in Fig. 7. Each type of multi-peaked benchmark test function simulates the difficulty of solving various types of practical optimization problems by providing a large number of local optima and different function shapes in the search space. Among them, the overall gradient trend of Ackley and Griewank functions is more obvious, but there are a lot of small local fluctuations, while the other four types of functions have very many very obvious local optima.

The dimensionality (number of variables) D of all benchmark test functions in the multi-peak function test experiment is 30, and the MC-DOA test results are compared and analyzed with the computational results of four different intelligent optimization algorithms, PSO, GA, GWO and ALO. Among them, the test results of PSO and GA are calculated based on the PSO and GA toolboxes in MATLAB 2019a; the source codes of GWO and ALO are obtained from http://www.alimirjalili.com/ALO.html and http://www.alimirjalili.com/GWO.html.

To ensure the objectivity of the comparison of the test results of different algorithms, the number of objective function calls is guaranteed to be equal for all algorithms in the test experiment, which is 30,000 times. For PSO, GA, GWO and ALO, the population size is set to 30 and the maximum number of iterations is set to 1000. Considering that MC-DOA contains 5 membranes and there is a relatively independent subpopulation in each membrane, in order to avoid the degradation of algorithm performance due to too few computational objects in each membrane, the total population size of MC-DOA is set to $5 \times 12 = 60$, i.e., the subpopulation size in each membrane is 12. Each algorithm is run independently for 30 times.

Table 3. Multi-peak benchmark test functions

Function	Expression	Range of values	F_{min}
Schwefel 2.26	$F_8 = \sum_{i=1}^{n} - x_i \sin(\sqrt{|x_i|})$	$[-500, 500]^D$	$-419D$
Rastrigin	$F_9 = \sum_{i=1}^{n} \left[x_i^2 - 10 \cos(2\pi x_i) + 10 \right]$	$[-5.12, 5.12]^D$	0
Ackley	$F_{10} = -20 \exp\left(-0.2 \sqrt{\frac{1}{n} \sum_{i=1}^{n} x_i^2} \right)$ $- \exp\left(\frac{1}{n} \sum_{i=1}^{n} \cos(2\pi x_i) \right) + 20 + e$	$[-32, 32]^D$	0
Griewank	$F_{11} = \frac{1}{4000} \sum_{i=1}^{n} x_i^2 - \prod_{i=1}^{n} \cos\left(\frac{x_i}{\sqrt{i}} \right) + 1$	$[-600, 600]^D$	0
Penalized 1	$F_{12} = \frac{\pi}{n} \left\{ 10 \sin(\pi y_1) + \sum_{i=1}^{n-1} (y_i - 1)^2 \right.$ $\left[1 + 10 \sin^2 (\pi y_{i+1}) \right] + (y_n - 1)^2 \left. \right\} + \sum_{i=1}^{n} u(x_i, 10, 100, 4)$	$[-50, 50]^D$	0
Penalized 2	$F_{13} = 0.1 \left\{ \sin^2(3\pi x_1) + \sum_{i=1}^{n} (x_i - 1)^2 \left[1 + \sin^2(3\pi x_i + 1) \right] \right.$ $+ (x_n - 1)^2 \left[1 + \sin^2(2\pi x_n) \right] \left. \right\} + \sum_{i=1}^{n} u(x_i, 10, 100, 4)$	$[-50, 50]^D$	

$$\text{Remarks: } y_i = 1 + (x_i + 1)/4; \ u(x_i, a, k, m) = \begin{cases} k(x_i - a)^m, x_i > a \\ 0, -a < x_i < a \\ k(-x_i - a)^m, x_i < -a \end{cases}$$

Experimental Results and Analysis

The results of the multi-peak benchmark function tests are shown in Table 4. From the test results, it can be seen that MC-DOA shows significantly better solution accuracy and performance than the other four algorithms for all six multi-peak benchmark test functions. Among them, for the test functions F_9, F_{10} and F_{11}, MC-DOA is able to generate the theoretical optimal solution with 64-bit floating-point precision, while among the other four algorithms, only the GWO algorithm is able to generate the theoretical optimal solution for the test function F_{11}.

The convergence curves of MC-DOA, PSO, GA, ALO and GWO for the six multi-peak benchmark functions are shown in Fig. 8. From the figure, we can see that MC-DOA has relatively excellent convergence performance and can effectively avoid premature convergence while taking into account the search efficiency. For the test functions F_8, F_{12} and F_{13}, MC-DOA shows better search efficiency and faster convergence speed compared with the other four algorithms; for the test functions F_9, F_{10} and F_{11}, MC-DOA shows very good solution performance and can find the theoretical optimal solution before the other algorithms tend to converge.

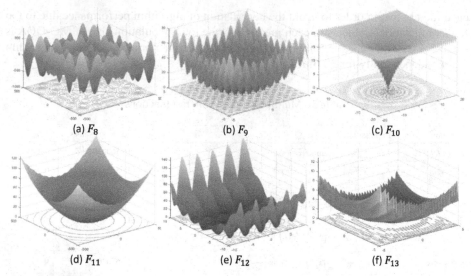

(a) F_8 (b) F_9 (c) F_{10}

(d) F_{11} (e) F_{12} (f) F_{13}

Fig. 7. Two-dimensional multi-peak benchmark test function image

Since the multi-peak test function has multiple local optima and the number of these local optima increases exponentially with the dimensionality of the function, the multi-peak function is very useful to test the local exploitation capability of various optimization algorithms. The results of the multi-peak testing experiments demonstrate that MC-DOA has very good exploitation capability.

3.3   Benchmark Function Test Tests with Different Dimensions

Test Function and Algorithm Parameter Setting
To test the effect of the size (dimensionality) of the optimization problem variables on the performance of MC-DOA, this algorithm comparison test experiment selected sphere (F_1), Rosenbrock (F_5), Rastrigin (F_9), Ackley (F_{10}), and Griewank (F_{11}) from the single-peaked and multi-peaked functions in Table 1 and Table 3 five expandable functions as benchmark test functions, and each benchmark test function contains five different dimensional variants of 5, 10, 30, 50, and 100 dimensions.

In this experiment, the test results of MC-DOA were compared with three intelligent optimization algorithms, HS, IBA and ABC, respectively. Among them, the computational results of HS, IBA, and ABC were obtained from the literature [17]. To ensure the objectivity of the test result comparison, MC-DOA is terminated with the maximum number of optimization function calls as the most terminating condition, and the maximum number of objective function calls is consistent with that of HS, IBA and ABC in the literature, which is 50,000 times, where the total population size is set to $5 \times 20 =$

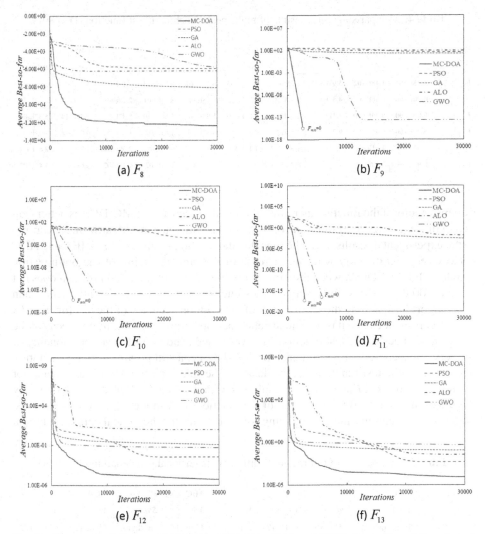

Fig. 8. Convergence curves of MC-DOA, PSO, GA, ALO, and GWO for multi-peak benchmark test functions

100 and the number of algorithm iterations is set to 500. The test results of MC-DOA are calculated from 30 independent runs obtained.

Experimental Results and Analysis

The results of the benchmark function test comparison tests in different dimensions (5, 10, 30, 50 and 100 dimensions) are shown in Tables 5. From the test results, it can be seen that MC-DOA is able to show better solution accuracy than HS, IBA and ABC on all four test functions except the test function F_5. As mentioned above, the test function F_5 is a single-peaked function that is very difficult to solve, and the solution accuracy of MC-DOA is worse than that of ABS only in the low dimension (5 dimensions); however, as the

Table 4. Comparison of test results of multi-peak benchmark test function (D = 30)

Functions	MC-DOA		PSO		GA		ALO		GWO	
	Ave.	Std.	Ave.	Std.	Ave.	Std.	Ave.	Std.	Ave.	Std.
F_8	1.25E+04	1.48E+02	5.97E+03	1.35E+03	8.13E+03	6.39E+02	6.22E+03	2.12E+03	5.79E+03	2.12E+03
F_9	0.00E+00	0.00E+00	4.57E+01	7.90E+00	1.79E+01	6.95E+00	8.40E+01	2.22E+01	1.14E-14	2.22E+01
F_{10}	0.00E+00	0.00E+00	3.12E-02	1.67E-01	1.86E+00	5.98E-01	1.86E+00	5.37E-01	1.52E-14	5.37E-01
F_{11}	0.00E+00	0.00E+00	9.77E-03	1.16E-02	2.58E-03	6.52E-03	1.54E-02	1.91E-02	0.00E+00	1.91E-02
F_{12}	5.63E-06	4.31E-06	3.46E-03	1.86E-02	1.19E-01	1.43E-01	9.22E+00	2.69E+00	4.81E-02	2.69E+00
F_{13}	8.06E-05	5.60E-05	4.76E-03	8.83E-03	1.11E-01	2.04E-01	4.22E-02	4.41E-02	4.66E-01	4.41E-02

dimensionality of the function increases, the solution accuracy of MC-DOA is better than that of the other three algorithms in 10, 30, 50 and 100 dimensions. Meanwhile, according to the experimental results, it can be seen that the solution accuracy of HS, IBA and ABC decays very significantly with the increase of the dimensionality of the optimization problem. For MC-DOA, when the dimensionality of the optimization problem does not exceed 100 dimensions, the decay of the solution accuracy is significantly better than other algorithms; among them, the decay of the solution accuracy of the test function F_5 is relatively obvious, but it is still much better than other three algorithms, the decay of the solution accuracy of the test function F_1 is very weak, and even if the dimensionality of the function increases to 100 dimensions, MC-DOA can still provide very high solution accuracy; for the test function F_1, the decay of the solution accuracy is very weak. For the test functions F_9, F_{10} and F_{11}, the solution accuracy of MC-DOA does not show any degradation when the dimensionality of the functions does not exceed 100 dimensions, and it always maintains the state of being able to produce the theoretical optimal solution.

Table 5. Comparison of test results of different dimensional benchmarking functions

Functions	D	HS		IBA		ABC		MC-DOA	
		Ave.	Std.	Ave.	Std.	Ave.	Std.	Ave.	Std.
F_1	5	3.20E-10	2.89E-10	3.91E-17	1.24E-17	4.30E-17	1.07E-17	3.95E-122	4.86E-122
	10	6.45E-08	3.07E-08	4.95E-17	2.30E-17	7.36E-17	4.43E-17	2.42E-120	2.67E-120
	30	7.21E+00	3.62E+00	2.92E-16	6.77E-17	4.69E-16	1.07E-16	2.33E-119	2.28E-119
	50	5.46E+02	9.27E+01	5.39E-16	1.07E-16	1.19E-15	4.68E-16	2.57E-119	4.60E-119
	100	1.90E+04	1.78E+03	1.45E-15	1.63E-16	1.99E-06	2.26E-06	6.84E-119	7.00E-119
F_5	5	5.94E+00	6.71E+00	4.55E-01	1.54E+00	2.33E-01	2.24E-01	8.75E-01	1.17E+00
	10	6.52E+00	8.16E+00	1.10E+01	2.55E+01	4.62E-01	5.44E-01	5.62E-04	1.54E-03
	30	3.82E+02	5.29E+02	7.57E+01	1.16E+02	9.98E-01	1.52E+00	9.36E-04	1.50E-03
	50	2.47E+04	1.02E+04	6.30E+02	1.20E+02	4.33E+00	5.48E+00	1.43E-03	1.82E-03
	100	1.45E+07	2.16E+06	6.42E+02	8.20E+02	1.12E+02	6.92E+01	1.11E-02	1.25E-02

(*continued*)

Table 5. (*continued*)

Functions	D	HS		IBA		ABC		MC-DOA	
		Ave.	Std.	Ave.	Std.	Ave.	Std.	Ave.	Std.
F_9	5	6.07E-08	5.52E-08	4.58E+00	2.31E+00	4.34E-17	1.10E-17	0.00E+00	0.00E+00
	10	1.05E-05	5.23E-06	2.20E+01	7.46E+00	5.77E-17	2.98E-17	0.00E+00	0.00E+00
	30	7.40E-01	7.00E-01	1.28E+02	2.49E+01	4.80E-05	2.43E-04	0.00E+00	0.00E+00
	50	3.76E+01	4.87E+00	2.72E+02	3.27E+01	4.72E-01	4.92E-01	0.00E+00	0.00E+00
	100	3.15E+02	2.33E+01	6.49E+02	4.52E+01	1.46E+01	4.18E+00	0.00E+00	0.00E+00
F_{10}	5	2.68E-05	1.24E-05	6.35E-10	9.77E-11	9.64E-17	5.24E-17	0.00E+00	0.00E+00
	10	2.76E-04	7.58E-05	6.71E-02	3.61E-01	3.51E-16	6.13E-17	0.00E+00	0.00E+00
	30	9.43E-01	5.63E-01	1.75E+00	9.32E-01	3.86E-15	3.16E-15	0.00E+00	0.00E+00
	50	5.28E+00	4.03E-01	8.43E+00	7.70E+00	4.38E-08	4.65E-08	0.00E+00	0.00E+00
	100	1.32E+01	4.90E-01	1.89E+01	8.50E-01	1.32E-02	1.30E-02	0.00E+00	0.00E+00
F_{11}	5	2.60E-02	1.38E-02	3.14E+00	1.41E+00	4.04E-17	1.12E-17	0.00E+00	0.00E+00
	10	0.00E+00	3.02E-02	1.04E+00	1.13E+00	6.96E-17	4.06E-17	0.00E+00	0.00E+00
	30	1.09E+00	3.92E-02	6.68E+00	6.43E+00	5.82E-06	3.13E-05	0.00E+00	0.00E+00
	50	3.76E+01	4.87E+00	2.72E+02	3.27E+01	4.72E-01	4.92E-01	0.00E+00	0.00E+00
	100	3.15E+02	2.33E+01	6.49E+02	4.52E+01	1.46E+01	4.18E+00	0.00E+00	0.00E+00

4 Conclusion

In this paper, a novel intelligent optimization algorithm (MC-DOA) is proposed by introducing a membrane computing framework into the intelligent optimization algorithm. Inspired by the CO2 and O2 cycles between chloroplasts and mitochondria in plant cells, a hybrid membrane structure is proposed and used as the computational framework for the optimization solution in MC-DOA. In addition, in order to be able to fully exploit the effect of the proposed computational framework, four stochastic search rules are abstracted from the particle diffusion phenomenon and mathematically modeled in turn for use in MC-DOA. In order to verify and test the computational performance of the proposed algorithm, a very detailed benchmark function performance test experiment is carried out in this paper. First, the MC-DOA is tested with conventional single-peak benchmark functions and multi-peak benchmark functions, and the computational results are compared with four representative intelligent optimization algorithms, namely PSO, GA, GWO and ALO. The results of the comparison test trials show that MC-DOA exhibits significantly better computational performance than the comparison algorithms for the vast majority of single-peak and multi-peak benchmark test functions. The test results also show that MC-DOA outperforms other comparative algorithms in large-scale optimization problems, and the decay of MC-DOA computational accuracy is smaller than other comparative algorithms as the size of the optimization problem increases. In the future, MC-DOA will be further developed and improved to solve more constrained optimization problems and multi-objective optimization problems.

Acknowledgements. This work was supported by Innovation Research Foundation of Ship General Performance of (30222221).

References

1. Hou, X.: Research on multi-objective particle swarm optimization algorithm incorporating multiple strategies and its application. Yanshan University (2018)
2. Chen, C., Zhang, L.: Fast adaptive wolfpack optimization algorithm with three-level leadership style. Comput. Eng. Appl. **55**(15), 59–68 (2019)
3. Li, A., Liu, S.: Hybrid strategy to improve whale optimization algorithm. Comput. Appl. Res. **39**(05), 1415–1421 (2022)
4. Wang, J., Li, Q., Cui, J.: An improved artificial bee colony algorithm–particle bee colony algorithm. J. Eng. Sci. **40**(07), 871–881 (2018)
5. Wang, Z., Li, Y., Chen, B.: Fish swarm algorithm based OFDMA adaptive resource allocation. J. Phys. **62**(12), 509–515 (2013)
6. Chen, D., Wang, Y., Yao, C.: Membrane computing multi-particle swarm algorithm. J. Mech. Eng. **55**(12), 222–232 (2019)
7. Song, N., Chen, T., Zhao, K.: Research on multi-target feature selection method of radar radiation source signal based on membrane particle swarm algorithm. J. Yunnan Univ. Nationalities (Nat. Sci. Ed.) **29**(05), 501–507 (2020)
8. Zhang, G., Cheng, J., Gheorghe, M., Meng, Q.: A hybrid approach based on differential evolution and tissue membrane systems for solving constrained manufacturing parameter optimization problems. Appl. Soft Comput. J. **13**(3), 1528–1542 (2013)
9. Xie, P., Zhang, Y.: Research on water and fertilizer prediction model with improved limit learning machine by membrane computing particle swarm algorithm. Chin. J. Agric. Chem. **42**(04), 142–149 (2021)
10. Han, T., Huang, Y., Zhou, N.: Improvement of fast SLAM algorithm based on particle swarm optimization for membrane computing. J. Xinjiang Univ. (Nat. Sci. Ed.) **37**(02), 156–162 (2020). (in English)
11. Huang, L., Su, H., Abraham, A.: Dynamic multi-objective optimization based on membrane computing for control of time-varying unstable plants. Inf. Sci. 181(11) (2010)
12. Gao, H., Li, C.: Membrane quantum swarm optimization for multi-objective spectrum allocation. J. Phys. **63**(12), 460–469 (2014)
13. Lin, J., He, C., Cheng, R.: Adaptive dropout for high-dimensional expensive multiobjective optimization. Complex Intell. Syst. **8**(1), 271–285 (2021). https://doi.org/10.1007/s40747-021-00362-5
14. Wang, H., Chen, S., Luo, L.: A diffusion algorithm based on P systems for continuous global optimization. J. Comput. Sci. **44**, 101112 (2020)
15. Liu, T.: Application of improved Gaussian mixture model based on membrane computing framework for image segmentation. Mod. Comput. **2020**(05), 61–65 (2020)
16. Fu, J., Zhao, J., Yu, L.: Adaptive membrane calculation algorithm and its application in ABS system. Control Eng. **26**(01), 155–161 (2019)
17. Karaboga, D., Akay, B.: Artificial bee colony (ABC), harmony search and bees algorithms on numerical optimization. In: Innovative Production Machines and Systems Virtual Conference (2009)

Tuning Geometric Conformations of Curved DNA Structures by Controlling Positions of Nicks

Chun Xie[1], Yingxin Hu[2], Kuiting Chen[1], Zhekun Chen[1], and Linqiang Pan[1(✉)]

[1] Key Laboratory of Image Information Processing and Intelligent Control of Education Ministry of China, School of Artificial Intelligence and Automation, Huazhong University of Science and Technology, Wuhan 430074, China
lqpan@mail.hust.edu.cn
[2] College of Information Science and Technology, Shijiazhuang Tiedao University, Shijiazhuang 050043, China

Abstract. DNA origami is one of the powerful techniques that utilize DNA as building blocks to synthesize nanostructures. Traditionally, through introducing different numbers of insertions and deletions of base pairs in DNA helices, the in-plane bending angle of curved DNA structures could be roughly tuned. Here, we explored a strategy that used the position patterns of nicks in staple strands to tune the geometric conformation of curved DNA origami structures, including in-plane bending, out-of-plane bending, and twisting angles. When the structure adopted different patterns of nicks positions, great difference appeared in the geometric properties. Further, by combining subunits of different nicks position patterns, the bending and twisting of the combined structure was effectively tuned. The strategy increases the design accuracy of curved DNA origami structures and expands the toolbox for designing DNA structures.

Keywords: DNA origami · chirality · molecular dynamics simulation

1 Introduction

Owing to the high fidelity and programmability of base-pairing, DNA has been utilized as building blocks to produce a wide range of nanostructures with pre-designed shapes [1]. DNA origami technique, folding a long DNA strand into a designed shape using hundreds of short oligo strands, has become a popular design strategy [2–4]. Various curved and twisted structures have been manufactured using DNA origami [5,6]. The curvature and twist rate have been shown to be important structural features for the functionality of DNA structures. For example, curved structures have been used to control the radius of liposomes [7,8], arrange plasmonic materials [9], and construct protein mimics [10,11]. Expanding the toolbox of controlling the conformation of DNA structures would enhance the functionality of DNA structures.

L. Pan et al. (Eds.): BIC-TA 2022, CCIS 1801, pp. 647–654, 2023.
https://doi.org/10.1007/978-981-99-1549-1_51

Traditional design parameters that affected the conformation of DNA structures included base-pair deletions or insertions, nicks in phosphate bonds. By applying different numbers and location patterns of base pair deletions or insertions in different helices, DNA structures of different twisting angle or curvature were manufactured [6,12,13]. Nicks of phosphate bonds in DNA strands, which abundantly existed in staple strands of origami structures, would greatly influence the bending and twisting stiffness of double strands [14,15]. The axial stiffness of rod-shape origami structures was controlled through varying the numbers of nicks [15]. By employing gaps at nicking positions of staple strands, the twisting of 6-helix origami structures was finely tuned [12]. In these researches, the impact of the two key design parameters, deletions/insertions of base pairs and nicks on the geometric properties of DNA structures were investigated separately. However, the associative impact of the two parameters on the conformation of DNA structures was not fully explored.

In this work, we proposed a strategy that tuned the geometric conformation of a curved DNA origami structure by controlling the nicks positions on different helices with base pair insertions or deletions. A 6-helix bundle curved structure was built by applying deletions and insertions. Three kinds of distribution patterns of nicks were designed, *i.e.*, nicks being deployed on helices without insertions or deletions, on helices with insertions, or on helices with deletions. A coarse-grained model named oxDNA [16,17], which has proven its reliability in analyzing DNA structures [18–20], was used to simulate and calculate the geometric conformations of the structures with different patterns. The three structures with different patterns of nicks showed similar in-plane bending angles, but great difference in out-of-plane chirality and twisting angles. Deploying nicks helices on helices with insertions caused the largest left-handed out-of-plane chirality and left-handed twisting angle. Deploying nicks helices on helices with insertions caused the largest left-handed out-of-plane chirality and left-handed twisting angle, while deploying nicks on helices without insertions or deletions caused the medium, and deploying nicks on helices with deletions caused the least. Further we combined subunits of different patterns in structure design, and the out-of-plane chirality and twisting angle was tuned with finer resolution. The strategy expands the designing toolbox for DNA structures.

2 Tuning DNA Structures by Varying Patterns of Nicks

2.1 Impact of Nicks on Conformations of DNA Structures

Nicks in DNA double strands (Fig. 1a) would cause changes in geometric and mechanical properties, like smaller bending and twisting stiffness of the double strands [15]. To test the feasibility of utilizing nicks to control the conformation of DNA structures, we designed two DNA structures by aligning two double strands parallelly and inserting two base pairs between staple crossovers. One structure contained nicks in staple strands, the other had no nicks in staple strands (Fig. 1b). We simulated the two structures using the finite element model for DNA origami, Cando [22–24]. In the structure with nicks, twisting and a global coiling appeared, while in the structure without nicks, only twisting appeared. The result demonstrated that nicks would affect the conformation of DNA structures.

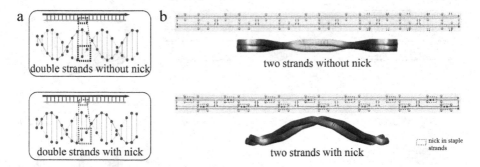

Fig. 1. The global conformation of DNA structures was affected by nicks in strands. (a) Schematic of double strands without or with nicks. Red dashed box indicated the existence of a nick. Black dashed box indicated that no nick existed. (b) Cando simulation of two strands without or with nicks in staples strands. Great difference in global confor-mations appeared between the two structures. Please note that short captions are centered, while long ones are justified by the macro package automatically.

2.2   Quantification of Bending and Twisting in Curved DNA Structures

To quantify the conformations of DNA structures, a 6-helix bundle curved origami structures with a specific nicks pattern was designed as a prototype (Fig. 2a). In each 21-base-pair-length block of the yellow regions of structure, two insertions were applied to the helices 0 and 1, and two deletions were applied to the helices 3 and 4. The flanking blue regions where no insertions or deletions functioned as geometrical direction indicators of the terminal parts in yellow curved regions. We used the in-plane bending angle θ_1, out-of-plane bending angle θ_2, and twisting angle θ_3 to represent the geometric conformation of the structure (Fig. 2b) [13]. Here, Positive (resp., negative) value of θ_2 indicated left-handed (resp., right-handed) out-of-plane bending, and positive (resp., negative) value of θ_3 indicated right-handed (resp., left-handed) twisting. The structure was simulated using a coarse grained molecular dynamic model, oxDNA. A snapshot among the simulation trajectories of the structure was shown in Fig. 2c left panel. The helices in blue regions were fitted into straight lines to calculate the angles. Trajectories of the three angles were shown in Fig. 2c middle panel, and the 500 snapshots were utilized to calculate the distributions of the angles (Fig. 2c right panel). The in-plane bending angle θ_1, out-of-plane bending angle θ_2, and twisting angle θ_3 of the prototype structure had values 108.0 ± 8.3, 30.0 ± 10.5, $-24.4 \pm 12.6°$.

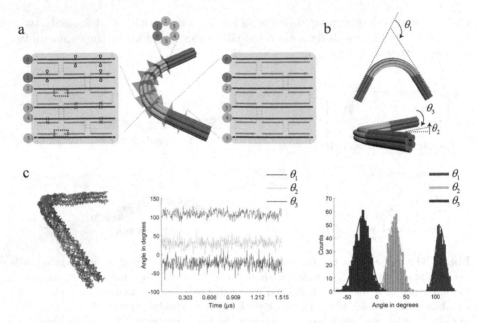

Fig. 2. Calculation of bending and twisting angles of the 6-helix bundle origami structure. (a) Design of a 6-helix bundle structure. The curved region (yellow) was composed of six 21-bp long blocks with insertions and deletions. Nicks in staples were deployed on the helices without insertions or deletions Straight regions (blue) with no deletions and insertions functioned as indicator of geometrical directions of the yellow region. (b) Schematic of the in-plane bending angle θ_1, out-of-plane bending angle θ_2, and twisting angle θ_3. (c) A snapshot of the simulation results. Trajectories of the bending and twisting angles (middle panel). Distributions of the three angles summarized from the trajectories (right panel).

2.3   Tuning the Conformation of DNA Structures by Varying Positions of Nicks

To tune the conformation by controlling the nicks patterns in DNA structures, we applied three nicks patterns to a 6-helix bundle DNA origami structure. In the three patterns, nicks were deployed on helices without insertions or deletions (pattern 1), helices with insertions (pattern 2), helices with deletions (pattern 3), respectively (Fig. 3a). Typical simulation snapshots of the three patterns were shown in Fig. 3b. Distributions of the in-plane bending angle θ_1, out-of-plane bending angle θ_2, and twisting angle θ_3 were shown in Fig. 3c. Structures with the three different patterns had similar in-plane bending angle θ_1 around 100°. The structure with nicks deployed on helices with insertions (pattern 2) had the largest left-handed out-of-plane chirality, $48.2 \pm 11.4°$, and left-handed twisting angle, $-52.2 \pm 17.9°$. The structure with nicks on helices with deletions (pattern 1) presented medium left-handed out-of-plane chirality, $-30.0 \pm 10.5°$, and left-handed twisting angle, $-24.4 \pm 12.6°$. The structure with nicks on helices with deletions (pattern 3) had the smallest left-handed out-of-plane chirality, $-1.0 \pm 12.6°$, and left-handed twisting angle, $-1.5 \pm 11.3°$. The out-of-plane chirality and twisting could be controlled by applying different nicks patterns in DNA structures.

To further increase the tuning resolution of the strategy, we combined design pattern 2 and pattern 3 in the same structure. The yellow curved region in the 6-helix origami structure was composed of 6 21-bp long blocks. Different numbers of blocks with pattern 2 and pattern 3 were designed in the structure. Structure with n pattern 2 blocks and m $(n + m = 6)$ pattern 3 blocks were names as *nimd*. By varying n from 6 to 0, 7 different structures were generated (Fig. 4a). Typical snapshots were shown in Fig. 4b. The values of in-plane bending angle θ_1, out-of-plane bending angle θ_2, and twisting angle θ_3 of the seven structures were shown in Fig. 4c. The twisting angle θ_1 of the structures was around 100°. The out-of-plane bending angle θ_2, and twisting angle θ_3 showed a roughly monotonous trend as n (the number of pattern 2 blocks) decreased.

3 Simulation Process of DNA Structures

3.1 Simulation Process of 2-Helix Structure Using Cando

The 2-helix structure used in this work were designed on the honeycomb lattice using caDNAno software [21] and simulated using Cando [22–24]. The caDNAno files were uploaded to an online website http://cando-dna-origami.org/. The analysis was conducted using the default geometric and mechanical parameters of DNA double strands. The axial rise per base-pair, helix diameter, and number of base-pairs per turn were set to 0.34 nm, 2.25 nm, and 10.5, respectively. The axial stretching stiffness, bending stiffness, and torsional stiffness of a double strand without nicks were set to 1100 pN, 230 pN nm^2, 460 pN nm^2. When a nick existed in the double strand, the bending and torsional stiffness were reduced by a factor of 100 whereas stretching stiffness was unchanged. The simulation results were directly downloaded from the website after simulation finished.

3.2 Simulation Process of 6-Helix Structures Using OxDNA

The 6-helix structures used in this work were designed on the honeycomb lattice using caDNAno software and simulated using oxDNA [16,17]. The caDNAno files were converted into oxDNA format files. The oxDNA format files were first relaxed through an adapted minimization algorithm. Then the molecular dynamics simulations of the relaxed file were conducted at 20° and 1 M Na$^+$ monovalent salt concentration. The simulation time step was set to 0.303 fs. Totally 1 trillion simulation steps were conducted for each structure. The trajectories of nucleotides were recorded every 1 million steps. The oxDNA trajectories and snap-shots were visualized using oxView [25,26]. The flanking blue helices were fit into straight lines and used to calculate the bending and twisting angles. Distributions of angles were calculated using the last 500 snapshots of each trajectory.

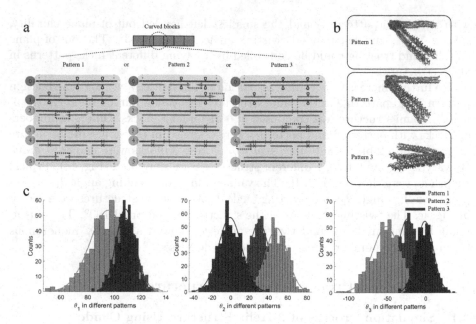

Fig. 3. Quantification of the bending and twisting angles of 6-helix structure with different nicks patterns. (a) Three nicks patterns were designed in the curved region. Pattern 1, nicks located on helices without insertions or deletions; pattern 2, nicks located on helices with insertions; pattern 3, nicks located on helices with deletions. (b) Snap-shots of structures with the three patterns. (c) Distributions of bending and twisting angles of structures with three different patterns.

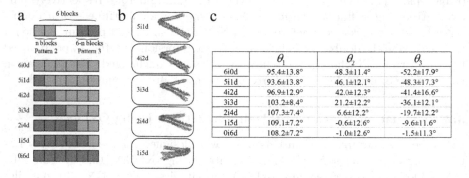

	θ_1	θ_2	θ_3
6i0d	95.4±13.8°	48.3±11.4°	-52.2±17.9°
5i1d	93.6±13.8°	46.1±12.1°	-48.3±17.3°
4i2d	96.9±12.9°	42.0±12.3°	-41.4±16.6°
3i3d	103.2±8.4°	21.2±12.2°	-36.1±12.1°
2i4d	107.3±7.4°	6.6±12.2°	-19.7±12.2°
1i5d	109.1±7.2°	-0.6±12.6°	-9.6±11.6°
0i6d	108.2±7.2°	-1.0±12.6°	-1.5±11.3°

Fig. 4. Tuning the conformation of 6-helix structures by combing subunits of different patterns. (a) By varying the number of blocks with pattern 2 and pattern 3, seven structures were generated. (b) Snapshots of the structures. (c) The bending and twisting angles of the seven structures. The in-plane bending angle θ_1 was around 100°. As n, the number of pattern 2 blocks, decreased, the out-of-plane bending angle θ_2, and twisting angle θ_3 showed a roughly monotonous trend.

4   Conclusion

In summary, we proposed a strategy that utilized different nicks patterns, where nicks of staple strands were deployed on different helices with base-pair insertions or deletions, to tune the conformation of curved DNA origami structures, such as out-of-plane chirality and twisting angle. Three kinds of nicks patterns were designed on a 6-helix curved DNA structure, i.e., nicks being deployed on helices without insertions or deletions, on helices with insertions, or on helices with deletions. Great difference in the out-of-plane bending and twisting angles of the structure appeared when the structure adopted different nicks patterns. Further, we combined subunit of different patterns in structure design, and realized a finer tuning of the out-of-plane bending and twisting angles by varying the ratio of different patterns. This strategy increases the design accuracy of DNA origami structures and expands the toolbox for designing DNA based nanostructures.

Moreover, the strategy in this work could be utilized to construct dynamic DNA nanodevices responsive to external stimuli. The generation of nicks in DNA strands could be manipulated by introducing external stimuli. By incorporating recognition sites of nicking enzymes [27] or photocleavable oligonucleotides [28] in DNA structures, local nicks in DNA helices could be generated after applying external stimuli, like nicking enzymes or light radiation, leading to the dynamic transformation of the global conformation.

Acknowledgment. The work was sponsored by the National Natural Science Foundation of China (62172171), Zhejiang Lab (NO. 2021RD0AB03), the Fundamental Research Funds for the Central Universities (HUST: 2019kfyXMBZ056), and the Science and Technology Project of Hebei Education Department (ZD2022098).

References

1. Pinheiro, A.V., Han, D., Shih, W.M., Yan, H.: Challenges and opportunities for structural DNA nanotechnology. Nat. Nanotechnol. **6**(12), 763–772 (2011)
2. Rothemund, P.W.K.: Folding DNA to create nanoscale shapes and patterns. Nature **440**(7082), 297–302 (2006)
3. Ke, Y., Ong, L.L., Shih, W.M., Yin, P.: Three-dimensional structures self-assembled from DNA bricks. Science **338**(6111), 1177–1183 (2012)
4. Douglas, S.M., Dietz, H., Liedl, T., Högberg, B., Graf, F., Shih, W.M.: Self-assembly of DNA into nanoscale three-dimensional shapes. Nature **459**(7245), 414–418 (2009)
5. Han, D., Pal, S., Nangreave, J., Deng, Z., Liu, Y., Yan, H.: DNA origami with complex curvatures in three-dimensional space. Science **332**(6027), 342–346 (2011)
6. Dietz, H., Douglas, S.M., Shih, W.M.: Folding DNA into twisted and curved nanoscale shapes. Science **325**(5941), 725–730 (2009)
7. Yang, Y., et al.: Self-assembly of size-controlled liposomes on DNA nanotemplates. Nat. Chem. **8**(5), 476–483 (2016)
8. Zhang, Z., Yang, Y., Pincet, F., Llaguno, M.C., Lin, C.: Placing and shaping liposomes with reconfigurable DNA nanocages. Nat. Chem. **9**(7), 653–659 (2017)
9. Urban, M.J., et al.: Plasmonic toroidal metamolecules assembled by DNA origami. J. Am. Chem. Soc. **138**(17), 5495–5498 (2016)

10. Franquelim, H.G., Khmelinskaia, A., Sobczak, J.-P., Dietz, H., Schwille, P.: Membrane sculpting by curved DNA origami scaffolds. Nat. Commun. **9**(1), 811 (2018)

11. Ketterer, P., et al.: DNA origami scaffold for studying intrinsically disordered proteins of the nuclear pore complex. Nat. Commun. **9**(1), 902 (2018)

12. Kim, Y.-J., Lee, C., Lee, J.G., Kim, D.-N.: Configurational design of mechanical perturbation for fine control of twisted DNA origami structures. ACS Nano **13**(6), 6348–6355 (2019)

13. Xie, C., Hu, Y., Chen, Z., Chen, K., Pan, L.: Tuning curved DNA origami structures through mechanical design and chemical adducts. Nanotechnology **33**(40), 405603 (2022)

14. Hays, J.B., Zimm, B.H.: Flexibility and stiffness in nicked DNA. J. Mol. Biol. **48**(2), 297–317 (1970)

15. Jung, W.-H., Chen, E., Veneziano, R., Gaitanaros, S., Chen, Y.: Stretching DNA origami: Effect of nicks and Holliday junctions on the axial stiffness. Nucleic Acids Res. **48**(21), 12407–12414 (2020)

16. Šulc, P., Romano, F., Ouldridge, T.E., Rovigatti, L., Doye, J.P.K., Louis, A.A.: Sequence-dependent thermodynamics of a coarse-grained DNA model. J. Chem. Phys. **137**(13), 135101 (2012)

17. Snodin, B.E.K., et al.: Introducing improved structural properties and salt dependence into a coarse-grained model of DNA. J. Chem. Phys. **142**(23), 234901 (2015)

18. Sharma, R., Schreck, J.S., Romano, F., Louis, A.A., Doye, J.P.K.: Characterizing the motion of jointed DNA nanostructures using a coarse-grained model. ACS Nano **11**(12), 12426–12435 (2017)

19. Snodin, B.E.K., Schreck, J.S., Romano, F., Louis, A.A., Doye, J.P.K.: Coarse-grained modelling of the structural properties of DNA origami. Nucleic Acids Res. **47**(3), 1585–1597 (2019)

20. Benson, E., Lolaico, M., Tarasov, Y., Gådin, A., Högberg, B.: Evolutionary refinement of DNA nanostructures using coarse-grained molecular dynamics simulations. ACS Nano **13**(11), 12591–12598 (2019)

21. Kim, D.N., Kilchherr, F., Dietz, H., Bathe, M.: Quantitative prediction of 3D solution shape and flexibility of nucleic acid nanostructures. Nucleic Acids Res. **40**(7), 2862–2868 (2012)

22. Castro, C.E., et al.: A primer to scaffolded DNA origami. Nat. Methods **8**(3), 221–229 (2011)

23. Pan, K., Kim, D.-N., Zhang, F., Adendorff, M.R., Yan, H., Bathe, M.: Lattice-free prediction of three-dimensional structure of programmed DNA assemblies. Nat. Commun. **5**(1), 5578 (2014)

24. Douglas, S.M., Marblestone, A.H., Teerapittayanon, S., Vazquez, A., Church, G.M., Shih, W.M.: Rapid prototyping of 3D DNA-origami shapes with Cadnano. Nucleic Acids Res. **37**(15), 5001–5006 (2009)

25. Poppleton, E., Bohlin, J., Matthies, M., Sharma, S., Zhang, F., Sulc, P.: Design, optimization and analysis of large DNA and RNA nanostructures through interactive visualization, editing and molecular simulation. Nucleic Acids Res. **48**(12), e72 (2020)

26. Bohlin, J., et al.: Design and simulation of DNA, RNA and hybrid protein-nucleic acid nanostructures with oxView. Nat. Protoc. **17**(8), 1762–1788 (2022)

27. Pan, L., Wang, Z., Li, Y., Xu, F., Zhang, Q., Zhang, C.: Nicking enzyme-controlled toehold regulation for DNA logic circuits. Nanoscale **9**(46), 18223–18228 (2017)

28. Jain, P.K., et al.: Development of light-activated CRISPR using guide RNAs with photocleavable protectors. Angew. Chem. Int. Ed. **55**(40), 12440–12444 (2016)

A Sensitive Nanothermometer Based on DNA Triplex Structure

Zhekun Chen[1], Yingxin Hu[2], Chun Xie[1], Kuiting Chen[1], and Linqiang Pan[1(✉)]

[1] Key Laboratory of Image Information Processing and Intelligent Control
of Education Ministry of China, School of Artificial Intelligence and Automation,
Huazhong University of Science and Technology, Wuhan 430074, China
lqpan@mail.hust.edu.cn
[2] College of Information Science and Technology, Shijiazhuang Tiedao University,
Shijiazhuang 050043, China

Abstract. DNA has been used to construct a variety of nanoscale thermometers due to its unique thermodynamic properties. It has proven to be one of the most promising, yet challenging aspects of DNA-based thermometers, to obtain significant signal changes when the temperature changes slightly. Here, we propose a strategy to construct nanoscale thermometers with sensitive temperature responses based on DNA triplex. The thermometers consist of a CT-rich DNA strand and stabilizing strands, whose conformations vary with temperature. By adjusting the sequence design or introducing mismatch bases in stabilizing strands, the temperature response interval of the thermometer can be reduced to $7\,°C$, which lead to higher sensitivity compared to the interval of 12–$15\,°C$ for conventional designs. The temperature responses of the thermometer are characterized by fluorescence experiments. The fluorescence kinetics experiments demonstrate the good repeatability. The design of the nanothermometer will be helpful for constructing advanced nanosystem with sensitive temperature responses as well as long service times.

Keywords: Nanothermometer · DNA nanotechnology · biosensor · triplex

1 Introduction

Temperature is an important parameter for many biological reactions [1,2]. With the development of nanoscale systems, it becomes imperative to develop tools capable of temperature sensing for nanosystem [3,4]. The nanothermometers with accurate temperature detection at nanoscale have been applied in various fields such as thermal-controlled drug delivery [5–7]. Over the past few years, a variety of nanothermometers were developed using different thermal-sensitive nanomaterials, such as nanoparticles [8], nanogels [9], polymers [10], nanomembrane [11], and carbon dots [12]. Due to the characteristics of the materials,

L. Pan et al. (Eds.): BIC-TA 2022, CCIS 1801, pp. 655–665, 2023.
https://doi.org/10.1007/978-981-99-1549-1_52

there are a number of drawbacks for these thermometers, such as poor bio-compatibility and complex synthesis process, that limits the practicality and applications [13–15].

Inspired by natural thermometers, DNA had been used to construct a lot of artificial thermal nanosystems [16–19]. Due to the reliable thermodynamic properties of DNA molecules, DNA-based nanothermometers exhibited predictable temperature responses [20,21]. DNA thermometers work by the melting transition of DNA molecules at different temperature, specifically, the conformation of DNA strands was formed at low temperature and denatured at high temperature [22–24]. By adjusting the sequences of the DNA strands, the melting temperature of conformations could be changed, as well as the thermal responses of DNA thermometers [25,26]. DNA hairpin-based thermometers have the predictable thermal responses determined by the stem sequence, and the response interval higher than $20\,°C$ [27]. By adding stabilizing strands, the thermal responses of the hairpin-based system could be adjusted, and the response interval could be reduced to $15\,°C$ [28,29]. In addition, a thermal strategy was proposed that worked by the temperature-dependent transitions of complex conformation of multiple strands. The closed triangle-like DNA thermometer, formed by three sequentially linked strands, display a steeper melting transition compared to the unimolecular stem-loop thermometer with a response interval of $12\,°C$ [28]. Another thermal system designed by Hahn et al. worked by cooperative strand displacement of multiple strands to fulfill the response interval of $13\,°C$ [30]. Recently, DNA triplex thermometers were developed with the response interval below $12\,°C$ [28,31]. The design of the triplex-based DNA thermometer could be further optimized for more sensitive temperature response by adopting different stabilizing strands.

In this study, we proposed a strategy to construct DNA thermometers with sensitive temperature responses. The thermometer consisted of one CT-rich DNA strand and the stabilizing strands. The CT-rich strand could bind to the stabilizing strand to form a stable triplex conformation. By changing the sequence pattern of the stabilizing strands or introducing mismatch bases, the thermal responses could be adjusted and the response interval could be reduced to $7\,°C$. The temperature responses of the thermometer are characterized by fluorescence experiments. The fluorescence kinetics experiments demonstrate the good repeatability. The design of the nanothermometer will be helpful for constructing advanced nanosystem with sensitive temperature responses as well as long service times.

2   Design of DNA-Based Thermometers

The DNA triplex thermometer consisted of a CT-rich long strand H and an AG-rich short stabilizing strand S, as shown in Fig. 1A. The strand H was modified with fluorophores (FAM) and quencher (BHQ1). Strand H consisted of two domains with symmetry sequence (gray) and one linking domain with random sequence (black). The stabilizing strand S (green) could bind to strand

H to form a stable triplex conformation in acidic buffer solution, that resulting the decreasing fluorescence signal at room temperature. As the temperature increased, the triplex conformation was gradually disrupted, and the quenched fluorescence signal was gradually restored. At high temperature, all DNA strands were in single-stranded conformation and the fluorescence signal reached the maximum. If the temperature decreased, this process would reverse.

2.1 DNA Thermometers with Stabilizing Strands of Different Design Pattern

To investigate the effect of stabilizing strands on the thermal responses, we tested several designs with different length and location for the stabilizing strands. The thermometers consisted of one dye-labeled hairpin strand H with two domains (27 bases, gray) and one stabilizing strand (27 bases, green) was shown in Fig. 2A. The stabilizing strand consisted of two domains, and their lengths were denoted by letters i (domain near the loop) and j (domain away from the loop), respectively. As shown in Fig. 2B, the length of the domain j away from the loop was set to be 14 nt, and the length i near the loop was set from 7 nt to 13 nt. As the length of domain i increased, the thermal response curves shifted to the right. As shown in Fig. 2C, the length of the domain i was set to be 5 nt, and the length j was set from 7 nt to 13 nt. As the length of domain j increased, the thermal response curves also shifted to the right. By comparing the fluorescence results in Fig. 2B and 2C, it could be found that the length change of the domain far from the loop had a greater effect on the thermal response. Figure 1D showed the thermal responses of the thermometers corresponding to the stabilizing strands with the same length but different locations. Design s11 represented the design with stabilizing strands located near the middle of the strand H, and s15 represented the design with stabilizing strands located at the end of strand H. As shown in Fig. 2D, for the stabilizing strands with the same length, the closer it located to the end of the strand H, the higher the thermal response temperature would be. The thermal response of the thermometer with stabilizing strand s15 had the highest response temperature, and the shortest response interval.

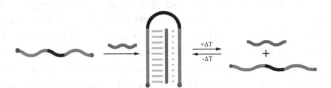

Fig. 1. The principle of the DNA triplex thermometer. (Color figure online)

To reduce the temperature response interval of the triplex thermometer, the stabilizing strand were separated into two short strands (see Fig. 3A). As shown in Fig. 3B, t1, t2 and t3 represented the design for one long stabilizing strand,

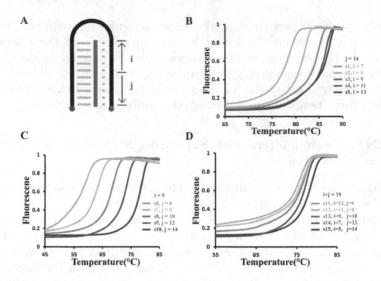

Fig. 2. DNA triplex thermometers with stabilizing strands of different design pattern. (A) Schematic of the quencher BHQ1 and fluorophore FAM modified DNA strands. Letters i and j represented the length of two domains on stabilizing strand. (B) The effect of the length of domain near the loop on the temperature responses. (C) The effect of the length of domain away from the loop on the temperature responses. (D) The effect of the location of stabilizing strands on the temperature responses. (Color figure online)

two stabilizing strands derived from long stabilizing strand, and the one short stabilizing strand, respectively. The length of stabilizing strands in designs t1 and t2 were the same, and the response curve for design t2 was shifted to the left. The temperature response ranges for the designs t2 and t3 were similar, but the response interval for design t2 is shorter than that for design t1. Figure 3C showed the thermal responses for the design with two stabilizing strands of different locations and the same total length. Designs t4 and t5 represented the design with two different length stabilizing strands, and design t6 represented the design with two same length stabilizing strands. The temperature response ranges for designs t4 and t5 were similar, but the response interval for design t5 was shorter than that for design t4. The response range for design t6 was lower than that for designs t4 and t5, indicating that two stabilizing strands with same length would lead to the leftward shift of the response curves. Figure 3D showed the thermal responses for the design with two stabilizing strands of the different total length. Designs t7, t8 and t9 represented the design with two stabilizing strands of different length 26bp, 24bp and 22bp, respectively. The response interval for designs t7 and t8 were similar, but the response range for design t8 was lower than that for t7. The response ranges for designs t8 and t9 were similar, but the response interval of t9 was wider than that for t8. This indicated that the weak decrease (two bases) in the total length of two stabilizing strands would lead to a leftward shift in the thermal response, while the large decrease (four bases) would lead to an increase in the response interval.

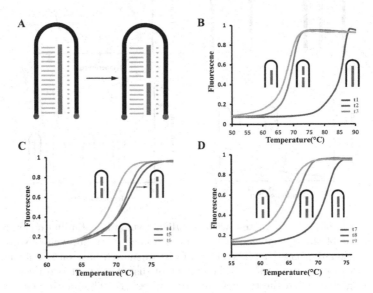

Fig. 3. DNA triplex thermometers with two stabilizing strands. (A) Schematic of the dye-labeled strand and two stabilizing strands. (B) The effect of the numbers of stabilizing strands on the temperature responses. (C) The effect of the different length of two stabilizing strands with the same total length on the temperature responses. (D) The effect of two stabilizing strands with the different total length on the temperature responses.

2.2 DNA Thermometers with Stabilizing Strands of Mismatch Bases

To increase the diversity of the temperature responses of the thermometers for different environments, the mismatch bases were introduced to the stabilizing strand design. Since the binding regions shorter than 6 bp would lead to the unstable hybridization of the stabilizing strand and strand H, it was impossible for the strategy using two stabilizing strands (see Fig. 3) to achieve abundant design with different thermal responses. The mismatch bases could be inserted at any site of the stabilizing strand, thus providing great flexibility of the mismatch base design. As shown in Fig. 4A, the yellow square on the green stabilizing strand represented the insertion of a mismatch base. The introduction of a mismatch base led to a decrease in the stability of the DNA triplex, thus affected the thermal responses of the thermometer. As shown in Fig. 4B, the response curves of the thermometer with one mismatch base were shifted to the left. The comparison of the response curves for m2 and m3 shows that the mismatch base near the loop could result in a shorter response interval than that for the

660   Z. Chen et al.

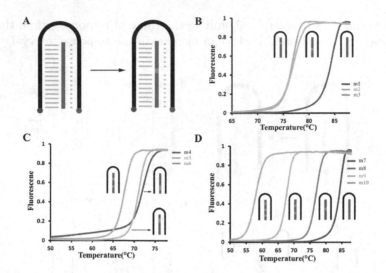

Fig. 4. DNA triplex thermometers with stabilizing strand containing mismatch bases. (A) Schematic of the dye-labeled strand and stabilizing strand. Yellow square represented the mismatch base. (B) The effect of one mismatch base on stabilizing strands on the temperature responses. (C) The effect of distribution of two mismatch bases on stabilizing strands on the temperature responses. (D) The melting curves of thermometers with different numbers of mismatch bases.

mismatch base far from the loop. If multiple mismatch bases were introduced, the distribution would affect the temperature responses. As shown in Fig. 4C, there were three different distribution patterns of the two mismatch bases. The adjacent mismatch bases near the loop could lead to a shorter response interval, which was consistent with the results for one mismatch base design in Fig. 4B. If two mismatch bases were far from each other, the thermal response curve would shift further to the left with the same response interval. If three separated mismatch bases were inserted, the thermal response curve would continue to shift to the left. As shown in Fig. 4D, the four response curves showed that the response curves shifted to the left by a range of 10 °C for each mismatch base, and the response intervals of these curves were all about 7 °C.

2.3   Fluorescence Kinetics of the Thermometers

To test the practicability and repeatable of the thermometer, the dye-labeled DNA strand H and stabilizing strand s8 were utilized for the kinetics and cycling tests. As shown in Fig. 2C, the fluorescence curve of the thermometer with H and s8 had the lowest fluorescence signal at 40 °C and highest at 80 °C, which were determined to be the operation temperature of the kinetics tests. The kinetics of the thermometer was characterized by measuring the rising fluorescence signal (from 40 °C to 80 °C) and the falling signal (from 80 °C to 40 °C), respectively. As shown in Fig. 5A and 5B, the rising process took 30 s to reached the 95%

of the maximum fluorescence, while the falling process took 15 s to reached the 95%. This proved the fast response time of the triplex-based DNA thermometer. A cyclic thermal program was set to be 2 min at 40 °C and then 2 min at 80 °C, which were tested for 100 times. As shown in Fig. 5C, there were no significant performance degradation after 100 cycles. This indicated that the system had reliable cycling behavior, and could be used to build repeatable thermal nanodevices with long service times.

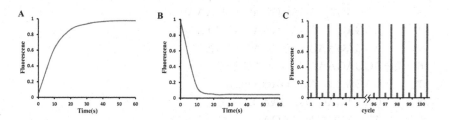

Fig. 5. Kinetics and repeated operations of the thermometers. (A-B) Fluorescence curves of kinetics at 40 °C (A) and 80 °C (B), respectively. (C) Repeated operations showing first and last five cycles of a 100-cycle experiment.

3   Implementation and Characterization of DNA-Based Thermometers

In this section, we describe the materials used, the preparation method and the characterization of the DNA nanothermometer.

The nanothermometers are composed of single strands of DNA. All DNA strands were purchased from Sangon Biotech Co., Ltd. (Shanghai, China). Unmodified DNA strands were purified by polyacrylamide gel electrophoresis, and modified strands with fluorophore or quencher were purified by high performance liquid chromatography. The sequences of all DNA strands were listed in Table 1. The DNA strands were dissolved in purified water as the stock solution and the concentrations were determined by measuring the absorbance at a wavelength of 260 nm using spectrophotometers (Thermo Fisher, NanoDrop One).

The DNA-based nanothermometers are assembled from DNA strands in particular buffer solution. All the DNA strands were prepared in 1 × TAE buffer (40 mM Tris-acetate; 1 mM EDTA; 10 mM magnesium chloride; pH = 5). The dye-labelled strand and the stabilizing strands were mixed together at a ratio of 1:1.5 in the buffer solution at room temperature. The final concentration of the DNA-based thermometers was 0.2 uM.

The temperature responses of the nanothermometer are verified by fluorescence experiments. All experiments were performed in 1 × TAE buffer using real-time fluorescence PCR (ABI, Step-one Plus). Reactions were prepared in 96-well plates at 20 uL/well volume to ensure rapid temperature change.

Table 1. DNA sequences

Strand	Sequence
H	TTCCTCTTTCTCCCTCTTCTTTCCCTTCTCTCTTCCCTTTC TTCTCCCTCTTTCTCCTT
s1	AGGAGAAAGAGGGAGAAGAAAGGGAAG
s2	AGGAGAAAGAGGGAGAAGAAAGGGA
s3	AGGAGAAAGAGGGAGAAGAAAGG
s4	AGGAGAAAGAGGGAGAAGAAA
s5	AGGAGAAAGAGGGAGAAGA
s6	AGGAGAAAGAGGGAGAAGA
s7	GAGAAAGAGGGAGAAGA
s8	GAAAGAGGGAGAAGA
s9	AAGAGGGAGAAGA
s10	GAGGGAGAAGA
s11	AGGAGAAAGAGGGAGAAGA
s12	GAGAAAGAGGGAGAAGAAA
s13	GAAAGAGGGAGAAGAAAGG
s14	AAGAGGGAGAAGAAAGGGA
s15	GAGGGAGAAGAAAGGGAAG
t1	AGGAGAAAGAGGGAGAAGAAAGGGA
t2	GAAAGAGGGAGAAGA
t3	AGGAGAAAGAGGG+GAAGAAAGGGAAG
t4	AGGAGAAAGAGGGAG+AGAAAGGGAAG
t5	AGGAGAAAGAGGG+GAAGAAAGGGAAG
t6	AGGAGAAAGAG+GAGAAGAAAGGGAAG
t7	AGGAGAAAGAG+GAGAAGAAAGGGAAG
t8	AGGAGAAAGAGGG+AGAAAGGGAAG
t9	AGGAGAAAGAG+AGAAAGGGAAG
m1	AGGAGAAAGAGGGAGAAGAAA
m2	AGGAGAAAGATGGAGAAGAAA
m3	AGGAGAAAGAGGGAGATGAAA
m4	AGGTTAAAGAGGGAGAAGAAA
m5	AGGAGAAAGTTGGAGAAGAAA
m6	AGGTGAAAGATGGAGAAGAAA
m7	AGGAGAAAGAGGGAGAAGAAA
m8	AGGAGAAAGAGGGAGATGAAA
m9	AGGTGAAAGATGGAGAAGAAA
m10	AGGATAAAGAGGGAGTTGAAA

4   Conclusion

In this study, the nanoscale thermometers with sensitive temperature responses were constructed. The thermometer consisted of a CT-rich DNA strand and the stabilizing strands, whose conformation varied with temperature. Compared to conventional designs, the thermometers based on DNA triplex structure have narrower temperature response interval. By adjusting the sequencing design of the stabilizing strands, the temperature interval of the thermal responses could be reduced to $7\,^{\circ}C$. The temperature responses of the thermometer were characterized by fluorescence experiments through labelling the CT-rich DNA strand with a fluorophore and a quencher. The fluorescence kinetics experiments demonstrate the good repeatability of the thermometer.

The DNA-based thermometer could be used to construct real-time nanoscale temperature monitor, such as the accurately nanosensor in intracellular enzyme reactions. Further, these thermometers could be helpful in building thermosensitive structures that might have applications in nanorobotics, drug delivery and advanced nano-assembly. For example, DNA thermometers could be incorporated into thermally targeted nanodevices to create temperature responsive nanosystems that perform specific functions by local temperature changes. The design of the nanothermometer will be helpful for constructing intelligent thermal-controlled nanosystem for advanced nanotechnology applications.

Acknowledgment. This work was sponsored by the National Natural Science Foundation of China (62172171), Zhejiang Lab (2021RD0AB03), the Fundamental Research Funds for the Central Universities (HUST: 2019kfyXMBZ056), and the Science and Technology Project of Hebei Education Department (ZD2022098).

References

1. Cramer, M.N., Gagnon, D., Laitano, O., Crandall, C.G.: Human temperature regulation under heat stress in health, disease, and injury. Physiol. Rev. **102**(4), 1907–1989 (2022)
2. Quint, M., Delker, C., Franklin, K.A., Wigge, P.A., Halliday, K.J., van Zanten, M.: Molecular and genetic control of plant thermomorphogenesis. Nat. Plants **2**, 15190 (2016)
3. Hossain, M.S., et al.: Temperature-responsive nano-biomaterials from genetically encoded farnesylated disordered proteins. ACS Appl. Bio Mater. **5**(5), 1846–1856 (2022)
4. Miotto, M., Armaos, A., Di Rienzo, L., Ruocco, G., Milanetti, E., Tartaglia, G.G.: Thermometer: a webserver to predict protein thermal stability. Bioinformatics **38**(7), 2060–2061 (2022)
5. Kang-Mieler, J.J., Mieler, W.F.: Thermo-responsive hydrogels for ocular drug delivery. Dev. Ophthalmol. **55**, 104–111 (2016)
6. Ahmed, K., Zaidi, S.F., Mati Ur, R., Rehman, R., Kondo, T.: Hyperthermia and protein homeostasis: cytoprotection and cell death. J. Therm. Biol **91**, 102615 (2020)

7. Song, K., Zhao, R., Wang, Z.L., Yang, Y.: Conjuncted pyro-piezoelectric effect for self-powered simultaneous temperature and pressure sensing. Adv. Mater. **31**(36), e1902831 (2019)
8. Guo, S., Yi, W., Liu, W.: Biological thermometer based on the temperature sensitivity of magnetic nanoparticle parashift. Nanotechnology **33**(9), 095501 (2021)
9. Uchiyama, S., et al.: A cell-targeted non-cytotoxic fluorescent nanogel thermometer created with an imidazolium-containing cationic radical initiator. Angew. Chem. Int. Ed. **57**(19), 5413–5417 (2018)
10. Qiao, J., et al.: Intracellular temperature sensing by a ratiometric fluorescent polymer thermometer. J. Mat. Chem. B **2**(43), 7544–7550 (2014)
11. Balcytis, A., Ryu, M., Juodkazis, S., Morikawa, J.: Micro-thermocouple on nano-membrane: Thermometer for nanoscale measurements. Sci. Rep. **8**(1), 6324 (2018)
12. Khan, W.U., Qin, L., Alam, A., Zhou, P., Peng, Y., Wang, Y.: Fluorescent carbon dots an effective nano-thermometer in vitro applications. ACS Appl. Bio Mater. **4**(7), 5786–5796 (2021)
13. Mrinalini, M., Prasanthkumar, S.: Recent advances on stimuli-responsive smart materials and their applications. ChemPlusChem **84**(8), 1103–1121 (2019)
14. Thapa, K.B., et al.: Single-metallic thermoresponsive coordination network as a dual-parametric luminescent thermometer. ACS Appl. Mater. Interfaces. **13**(30), 35905–35913 (2021)
15. Gong, P., et al.: In situ temperature-compensated DNA hybridization detection using a dual-channel optical fiber sensor. Anal. Chem. **93**(30), 10561–10567 (2021)
16. López-García, P., Forterre, P.: DNA topology in hyperthermophilic archaea: Reference states and their variation with growth phase, growth temperature, and temperature stresses. Mol. Microbiol. **23**(6), 1267–1279 (1997)
17. Lee, M.H., Lin, H.Y., Yang, C.N.: A DNA-based two-way thermometer to report high and low temperatures. Anal. Chim. Acta **1081**, 176–183 (2019)
18. Bu, C., Mu, L., Cao, X., Chen, M., She, G., Shi, W.: DNA nanostructure-based fluorescence thermometer with silver nanoclusters. Nanotechnology **29**(29), 295501 (2018)
19. Liu, X., et al.: Photothermal detection of microrna using a horseradish peroxidase-encapsulated DNA hydrogel with a portable thermometer. Front. Bioeng. Biotechnol. **9**, 799370 (2021)
20. Liu, T., Yu, T., Zhang, S., Wang, Y., Zhang, W.: Thermodynamic and kinetic properties of a single base pair in a-DNA and b-DNA. Phys. Rev. E **103**(4-1), 042409 (2021)
21. Chao, J., Liu, H., Su, S., Wang, L., Huang, W., Fan, C.: Structural DNA nanotechnology for intelligent drug delivery. Small **10**(22), 4626–4635 (2014)
22. Tashiro, R., Sugiyama, H.: The molecular-thermometer based on b-z transition of DNA. Nucleic Acids **48**(1), 89–90 (2004)
23. Bu, C., Mu, L., Cao, X., Chen, M., She, G., Shi, W.: Silver nanowire-based fluorescence thermometer for a single cell. ACS Appl. Mater. Interfaces. **10**(39), 33416–33422 (2018)
24. Drake, T.J., Tan, W.: Molecular beacon DNA probes and their bioanalytical applications. Appl. Spectrosc. **58**(9), 269A–280A (2004)
25. Miller, I.C., Gamboa Castro, M., Maenza, J., Weis, J.P., Kwong, G.A.: Remote control of mammalian cells with heat-triggered gene switches and photothermal pulse trains. ACS Synth. Biol. **7**(4), 1167–1173 (2018)
26. Yakovchuk, P., Protozanova, E., Frank-Kamenetskii, M.D.: Base-stacking and base-pairing contributions into thermal stability of the DNA double helix. Nucleic Acids Res. **34**(2), 564–574 (2006)

27. Xie, N., et al.: Scallop-inspired DNA nanomachine: a ratiometric nanothermometer for intracellular temperature sensing. Anal. Chem. **89**(22), 12115–12122 (2017)
28. Gareau, D., Desrosiers, A., Vallée-Bélisle, A.: Programmable quantitative DNA nanothermometers. Nano Lett. **16**(7), 3976–3981 (2016)
29. Ricci, F., Vallée-Bélisle, A., Porchetta, A., Plaxco, K.W.: Rational design of allosteric inhibitors and activators using the population-shift model: In vitro validation and application to an artificial biosensor. J. Am. Chem. Soc. **134**(37), 15177–15180 (2012)
30. Hahn, J., Shih, W.M.: Thermal cycling of DNA devices via associative strand displacement. Nucleic Acids Res. **47**(20), 10968–10975 (2019)
31. Yamayoshi, A., et al.: Selective and robust stabilization of triplex DNA structures using cationic comb-type copolymers. J. Phys. Chem. B **121**(16), 4015–4022 (2017)

Microwave Absorption Properties of Double-Layer Absorbing Material Based on Carbonyl Iron and Graphene Composites

Zhiwei Shao[1], Qin Zhao[2], Xiang Gao[2], and Fang Li[2(✉)]

[1] China Ship Development and Design Center, Wuhan, Hubei, China
[2] Hubei Key Laboratory of Theory and Application of Advanced Materials Mechanics, School of Science, Wuhan University of Technology, Wuhan 430070, China
liaaf@whut.edu.cn

Abstract. It is reported that developing microwave absorbers with impedance matching and high efficiency electromagnetic absorbing performance is still extremely challenging. Herein, we report the microwave absorption properties of single-layer and double-layer wave absorbent materials based on hydroxyl iron (CIP) and graphene nanosheets (GNSs) composites. The superior microwave absorption performance of the double-layer absorbers should be attributed to the greater impedance matching characteristic of the GNSs/CIP composite layer and the better dielectric loss ability of graphene layer as well as the increase of magnetic loss of hydroxyl iron (CIP). The results show that the double-layer absorbers, consisting of 35 vol% GNSs/CIP composite as matching layer and GNSs as absorption layer, with total thickness of 2 mm, exhibited a maximum reflection loss (RL) of −23.2 dB at 13.39 GHz, as well as an effective bandwidth below −10 dB of as wide as 5.54 GHz from 11.23 GHz to 16.67 GHz. These double layer structures based on carbonyl-iron (CIP) and graphene nanosheets (GNSs) composites could be helpful for designing new microwave absorbers with high absorption performance.

Keywords: Graphene · Hydroxyl iron · Microwave absorption property

1 Introduction

In recent years, with the rapid development of wireless communication in the industrial and commercial fields and the military field, electromagnetic pollution has inevitably entered every field of our lives, leading to the failure of various electronic devices and equipment, causing adverse and even fatal effects in space exploration, scientific calculation, military accuracy, surveillance and other aspects. And can cause a lot of human health problems. Therefore, the research of electromagnetic wave absorption materials has attracted wide attention in the world, and the development of high-performance electromagnetic wave absorption materials with broadband, thin, strong absorption and light weight has become the focus and hotspot of the research. An excellent absorbing material should have two characteristics at the same time, namely impedance matching characteristics and attenuation characteristics. Impedance matching requires that the impedance of the absorbing material matches that of the free space so that the electromagnetic wave can enter the material to the maximum extent. The attenuation characteristic is that through the properties of the material itself, such as dielectric loss and magnetic loss, the electromagnetic energy into the material is converted into heat energy and then dissipated. Absorbing materials are mainly divided into dielectric loss materials and magnetic loss materials. Dielectric loss materials have wide absorption band, poor impedance matching, and magnetic loss materials cannot guarantee the requirements of light mass. Therefore, in the design structure, the two will be combined to obtain good microwave absorption [1, 2].

Graphene is an excellent dielectric loss material with good conductivity and mechanical strength, high specific surface area and electron mobility, excellent thermal and oxygen stability and chemical stability, which show great scientific research value and application potential in the field of electromagnetic wave absorption [3–5]. In addition, the unique band structure of graphene allows electrons and holes to be separated, resulting in a new electron conduction phenomenon that facilitates microwave absorption. However, the loss mechanism of pure graphene is limited, and its high conductivity can also cause impedance mismatch of absorbing materials [6]. However, the introduction of magnetic materials into graphene can effectively improve the impedance matching and broaden the absorption band of the material. Magnetic metals such as Fe, Co, Ni and their oxides have strong magnetization, permeability and magnetic loss in the GHz frequency range. Excellent microwave absorption performance can be obtained by combining with graphene with strong dielectric loss in various ways [7, 8].

Feng et al. distributed CoNi nanocrystals uniformly on nitrogen-doped graphene sheets via a simple one-pot polyol route, which was then subjected to ultrasonic processing. The RL reached a minimum of -22 dB at 10 GHz for the composites with a thickness of only 2.0 mm, The effective absorption bandwidth of the composite is 14.4 GHz (3.6 GHz–18.0 GHz) [9]. Yang et al. first prepared $BaFe_{12}O_{19}/CoFe_2O_4$ composite nanoparticles by sol-gel method, and then dispersed them in ethylene diol with GO, and then prepared $RGO/BaFe_{12}O_{19}/CoFe_2O_4$ composite absorbent through the solubilizing agent. When the matching thickness is 3 mm, the maximum reflection loss is -32.4 dB, the effective absorption band is 4–7 GHz, and the absorption performance is better in the low frequency range [10]. Sun Yifeng et al. first prepared graphene oxide by Hummer's method, and then prepared lanthanum-doped Z-type barium ferrite $Ba_{2.7}La_{0.3}Co_2Fe_{24}O_{41}$ by sol-gel method. Then the ferrite and GO were combined to prepare $Ba_{2.7}La_{0.3}Co_2Fe_{24}O_{41}/$ reduced GO binary composite material. Finally, the conductive polymer polyaniline was used to wrap the binary composite material to prepare $Ba_{2.7}La_{0.3}Co_2Fe_{24}O_{41}/$ reduced GO/polyaniline terpolymer absorbing material. The average microwave absorption value of the composite reached -47.42 dB, and the absorption peak moved to the low frequency direction, and the absorption peak was -57.43 dB [11].

In the magnetic loss materials, compared with other metal powders, carbonyl iron powder has moderate conductivity, high saturation magnetization, high Curie temperature, and can be large-scale production, so it has been widely used in many fields.

A double-layer microwave absorption material based on carbonyl iron and graphene composites is proposed in this paper. The electromagnetic absorption and microwave absorption properties of single-layer and double-layer microwave absorbers are systematically studied. It has been shown that the thickness and concentration of graphene play a key role in the microwave absorption capacity and effective absorption bandwidth of the absorber.

2 Theoretical Model of Microwave Absorption

2.1 The Principle of Microwave Absorption

When an electromagnetic wave encounters an obstacle in the process of propagation, the surface of the electromagnetic wave will reflect and absorb the electromagnetic wave. Part of the electromagnetic wave can enter the inside of the absorbing body, and then through the interaction between the electromagnetic field and the internal molecular and electronic structure of the material, it will be converted into heat energy or offset in other forms, and finally cause energy attenuation.

In order to adjust and optimize the composition and structure of the absorbing material, it is necessary for the absorbing material to have impedance matching and attenuation characteristics. According to electromagnetic wave theory, when an electromagnetic wave passes through a medium where the impedance of two waves is equal, then the electromagnetic wave can pass from one medium to the other without reflection. Therefore, under the special condition that the wave impedance of the two media is equal, the electromagnetic energy incident into the absorbing material is the most, and the reflection coefficient R is the least. According to the transmission line theory, the free space impedance Z_0 (377 Ω), the material impedance Z_{in} and the microwave reflection coefficient RL have the following relationship [12, 13]:

$$Z_0 = \sqrt{\frac{\mu_0}{\varepsilon_0}} \tag{1}$$

$$Z_{in} = Z_0 \sqrt{\frac{\mu_r}{\varepsilon_r}} tanh\left[j\left(\frac{2\pi fd}{c}\right)\sqrt{\mu_r \varepsilon_r} \right] \tag{2}$$

$$\Gamma = \frac{Z_{in} - Z_0}{Z_{in} + Z_0} \tag{3}$$

where μ_0 is the permeability of free space; ε_0 is the dielectric constant of free space; μ_r is the complex permeability of the absorbing material. ε_r is the complex dielectric constant of the absorbing material. f is the test frequency; d is the thickness of the absorbing material; c is the speed of light. At present, the reflection loss R of the material is mainly used to evaluate the electromagnetic wave absorption performance of the material. When R = −10 dB, the electromagnetic wave energy attenuates by 90%, and the RL is reduced. Broadband at −10 dB is defined as effectively absorbed broadband.

2.2 The Theory Equivalent Medium

For heterogeneous materials, the theoretical method to determine the equivalent dielectric function is called effective medium theory. Currently, the commonly used effective medium theories mainly include Maxwell-Garnett theory and Bruggeman theory, which are similar in that they are quasi-static theoretical methods, and the prerequisite for their application is that the particle size of the inclusion phase is much smaller than the wavelength of the incident wave. This article mainly used - Garnett Maxwell theory, the permeability of μ_i randomly distributed evenly in the matrix spherical particles, particles of permeability with volume filling quantity respectively μ_m and f. Thus, effective permeability μ_{eff} values usually can be expressed as [14]:

$$\mu_{eff} = \mu_m + 3f\mu_m \frac{\mu_i - \mu_m}{\mu_i + 2\mu_m - f(\mu_i - \mu_m)} \tag{4}$$

2.3 The Model CST Calculation

Computer Simulation Technology (CST), a finite element analysis tool, was used for structural design and performance simulation. CST adopts the Finite Integration Technique (FIT) through the mesh division method of three-dimensional space, the microwave problem is equivalent to the calculation of N-port network S matrix, the structure is adaptively divided into finite element mesh, and the S-parameters are calculated from the reflection amount and transmission amount. Through Perfect Boundary Approximation (PBA), the speed and accuracy of CST software are further improved based on the FIT method.

3 Results and Discussion

3.1 The Verification of CST Calculation Method

The same electromagnetic parameters and model structure were used to study the reflection loss of graphene/natural rubber composites at different concentrations [15], and the simulation results were compared with the experimental results to explore the correctness of the design of simulation conditions. The results are shown in Fig. 1.

It can be seen that the simulation results using CST are basically consistent with the experimental results, and there is a slight difference at the peak value with the same variation trend. This is due to the error caused by the fitting of the imported electromagnetic parameters by the CST microwave studio, which does not affect the guiding role of the simulation results.

3.2 Characteristic Analysis of Single Layer Absorber

The real parts (ε') and imaginary parts (ε'') of the permittivity of GNSs and CIP and the real parts (μ') and imaginary parts (μ'') of the permeability are shown in Fig. 2.4 [17]. In the whole test frequency band, the imaginary part of the permeability of GNSs absorbent

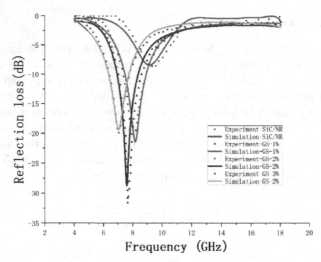

Fig. 1. Reflection loss of the graphene/natural rubber composites in experiment and simulation

is small, close to 0, but the imaginary part of the dielectric constant of GNSs is large, both above 10. On the contrary, the imaginary part of the permeability of CIP is large, higher than 1, while the imaginary part of the dielectric constant is small, close to 0. As we all know, the real part of dielectric constant and permeability represents the storage capacity of electromagnetic wave energy [16], while the corresponding imaginary part determines the loss capacity of electromagnetic wave energy. In other words, low permeability indicates that GNSs absorbers have poor magnetic loss performance to electromagnetic waves. Similarly, low imaginary part of dielectric constant indicates that CIP absorbers have poor dielectric loss performance to electromagnetic waves. In addition, the real part of the permittivity of GNSs and the real part of the permeability of CIP decrease gradually with the increase of frequency, showing an obvious frequency-dependent effect.

RL values of the single-layer absorbers based on GNSs over 2–18 GHz with various thicknesses are presented in Fig. 3a. It is found that the GNSs has a very poor microwave absorption performance, the maximum RL value only can reach −5.6 dB with a thickness of 2.0 mm, which should be ascribed to its poor impedance matching with free space. It can be seen from the Fig. 3a that the peak value of reflection loss moves to the low-frequency region with the increase of thickness, which is in line with the "1/4 matching model" [17], that is, the electromagnetic wave is consumed by physical interference at the interface of the absorbing coating. GNSs/CIP composite material was fabricated with GNSs as absorber and CIP as matrix. The thickness of single-layer absorber was 2 mm. As shown in Fig. 3b, the absorption peak shifted to the lower frequencies with the increase of the fillers of GNSs. The maximum RL value can reach −9.7 dB with a concentration of 35 vol% for GNSs/CIP composite material (Fig. 2).

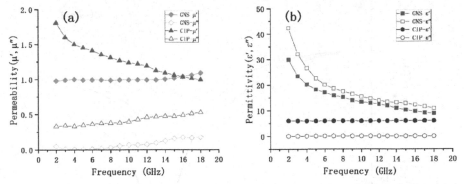

Fig. 2. Frequency-dependent electromagnetic properties of the GNSs and CIP, including (a) the complex permittivity and (b) the complex permeability.

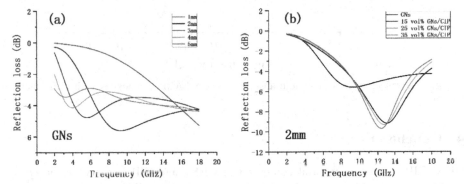

Fig. 3. (a)The reflection loss of single-layer absorbers of GNSs with a thickness arrangement of 1.0–5.0 mm; (b)The reflection loss of 2 mm thickness GNSs/CIP with different concentrations

3.3 Characteristic Analysis of Double Layer Absorber

As can be seen from Fig. 3, the absorption performance of single-layer absorbing structure is poor. For a double-layer absorber, a good matching layer can reduce the reflected energy as much as possible, and an excellent absorbing layer can attenuate the energy as much as possible [18]. Therefore, the impedance matching of the matching layer should be as close to the air as possible, and the absorbing layer should be composed of materials with high dielectric or magnetic loss.

As show in Fig. 4, reflection loss of the double-layer absorbent consisting of GNSs/CIP composites with different concentrations, the double-layer absorbers with a total thickness value of 2.0 mm, which are composed of the matching layer with 35 vol% GNSs/CIP composites and the absorption layer with GNSs. The thickness of impedance layer and matching layer is 1mm respectively, the RL value achieves a maximum value of −23.2 dB at 13.39 GHz, the effective absorption bandwidth below −10 dB can be as wide as 5.54 GHz, ranging from 11.23 GHz to 16.67 GHz. It is obviously that Obviously, the double-layer absorber with this structure exhibits a stronger microwave absorption performance.

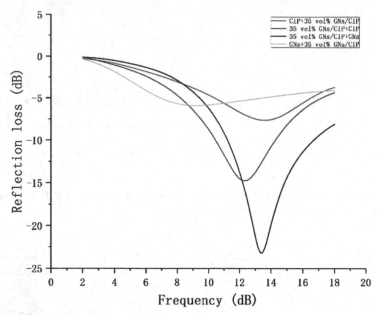

Fig. 4. The reflection loss of double absorbers of GNSs/CIP composites with a thickness of 2.0 mm

4   Conclusions

GNSs/CIP composite material composed of GNSs, and CIP was used to design the double-layer absorbing structure with different concentration. The microwave absorption performance of these double-layer absorbers consisting of impedance layer and matching layer of composites with different concentrations is studied in detail. Due to the high impedance matching of the matching layer made by 35 vol % GNSs/CIP composites and the high dielectric loss capacity of the absorbing layer made by GNSs, the double-layer absorbing layer has higher RL value, thinner thickness and wider absorption bandwidth than that of single-layer GNSs. When the thickness of the impedance layer and matching layer of the double-layer absorber is 1 mm, the maximum RL value at 16.9 GHz is −23.2 dB, and the effective bandwidth is less than −10 dB, the width is 5.54 GHz, and the range is 11.23 GHz to 16.67 GHz. These double - layer absorbers can meet the requirements of practical applications. The results of this study provide a promising method for the design of advanced microwave absorbers.

References

1. Fujiwara, O.: A globalized trend towards EMC technology. Trans. Inst. Electr. Eng. Japan, Part A **124-A**(1), 9–10 (2004)
2. Zhang, Y., Huang, Y., Zhang, T., et al.: Broadband and tunable high-performance microwave absorption of an ultralight and highly compressible graphene foam. Adv. Mater. **27**(12), 2049–2053 (2015)

3. Yang, H., Ye, T., Lin, Y., et al.: Preparation and microwave absorption property of graphene/BaFe12O19/CoFe2O4 nanocomposite. Appl. Surf. Sci. **357**, 1289–1293 (2015)
4. Sun, Y., Bao, L., Bao, G., et al.: Microwave absorption properties of La-doped Z-type barium ferrite/graphene/polyaniline composites. Chinese Rare Earths **40**(1), 52–58 (2019)
5. Nair, R.R., Blake, P., Grigorenko, A.N., et al.: Fine structure constant defines visual transparency of graphene. Science **320**(5881), 1308 (2008)
6. Shahzad, F., Yu, S., Kumar, P., et al.: Sulfur doped graphene/polystyrene nanocomposites for electromagnetic interference shielding. Compos. Struct. **133**, 1267–1275 (2015)
7. Vakil, A., Engheta, N.: Transformation optics using graphene. Science **332**(6035), 1291–1294 (2011)
8. Zang, Y., Xia, S., Li, L., et al.: Microwave absorption enhancement of rectangular activated carbon fibers screen composites. Compos. Part B-Eng. **77**, 371–378 (2015)
9. Lin, J., Lin, Z., Pan, Y., et al.: Polymer composites made of multi-walled carbon nanotubes and graphene nano-sheets: effects of sandwich structures on their electromagnetic interference shielding effectiveness. Compos. Part B-Eng. **89**, 424–431 (2016)
10. Chen, Y., Zhang, A., Ding, L., et al.: A three-dimensional absorber hybrid with polar oxygen functional groups of MWNTs/graphene with enhanced microwave absorbing properties. Compos. Part B- Eng. **108**, 386–392 (2017)
11. Liu, P., Huang, Y., Zhang, X.: Superparamagnetic NiFe2O4 particles on poly (3,4-ethylenedioxythiophene)-graphene: Synthesis, characterization and their excellent microwave absorption properties. Compos. Sci. Technol. **95**, 107–113 (2014)
12. Zhang, H., Xie, A., Wang, C., et al.: Novel rGO/a-Fe2O3 composite hydrogel: synthesis, characterization and high performance of electromagnetic wave absorption. J. Mater. Chem. **1**(30), 8547–8552 (2013)
13. Wang, Z., Yu, M., Pan, S.: Advances in theoretical research on electromagnetic parameter calculation of composite materials. Mater. Rep. **23**(14), 246–248 (2009)
14. Li, J.S., Hsu, T.C., Hwang, C.C., Lu, K.T., Yeh, T.F.: Preparation and characterization of microwave absorbing composite materials with GSs or FeCo/GS composites. Mater. Res. Bull. **107**(11), 218–224 (2018)
15. Liu, Y.: Study on Radar-infrared Stealth-compatible Performance of the Ferromagnetic Absorbent/Polyurethane Microwave Absorbing Coatings. Dalian University of Technology (2019)
16. Liu, J.R., Itoh, M., Horikawa, T., et al.: Complex permittivity, permeability and electromagnetic wave absorption of alpha-Fe/C (amorphous) and Fe2B/C (amorphous) nanocomposites. J. Phys. D-Appl. Phys. **37**(19), 2737–2741 (2004)
17. Liu, P., Yao, Z., Zhou, J., Yang, Z., Kong, L.B.: Small magnetic co-doped NiZn ferrite/graphene nanocomposites and their dual-region microwave absorption performance. J. Mater. Chem. C **4**, 9738–9749 (2016)
18. Wu, H., Wang, L., Guo, S., Shen, Z.: Double-layer structural design of dielectric ordered mesoporous carbon/paraffin composites for microwave absorption. Appl. Phys. A **108**(2), 439–446 (2012)

Reconfigurable Nanobook Structure Driven by Polymerase-Triggered DNA Strand Displacement

Kuiting Chen, Zhekun Chen, Chun Xie, and Linqiang Pan[(⊠)]

Key Laboratory of Image Information Processing and Intelligent Control
of Education Ministry of China, School of Artificial Intelligence and Automation,
Huazhong University of Science and Technology, Wuhan 430074, China
lqpan@mail.hust.edu.cn

Abstract. Due to its high degree of customization, DNA origami provides a versatile platform with which to engineer nanoscale structures and devices. Reconfigurable nanodevices driven by DNA strand displacement accomplish the task of transition between different conformations, endowing DNA origami with application values. Herein, we propose a strategy to regulate the conformation of DNA origami using the polymerase-triggered DNA strand displacement (PTSD) reaction. We design a book-shaped DNA origami structure consisting of four pages connected into a cuboid shape. The PTSD reactions initiated by different primer strands selectively remove the connecting strand, transforming the nano book into a two-page or a four-page conformation. We utilize three primer strands to remove thirty-five connecting strands and construct three conformations of the identical DNA origami, illustrating that the PTSD reaction is an effective tool for the reconfiguration of DNA origami. The statistical results of TEM images prove the effectiveness of the proposed method. Our work on the development of PTSD-driven reconfigurable nanostructure will offer a new way to create intelligent materials for advanced nanotechnology applications.

Keywords: DNA nanotechnology · DNA origami · Reconfigurable DNA nanostructure · Polymerase-triggered strand displacement

1 Introduction

The development of the DNA origami has directly inspired the idea of using DNA self-assemblies for constructing designer nanoscale devices. DNA origami has been successfully used to create static two-dimensional nanopatterns [1–3] or three-dimensional nanostructures [4–6], demonstrating its powerful spatial programmability. The emergence of dynamic nanodevices shows the broad application prospects of DNA origami. Application examples of dynamic DNA origami-based devices include nanorobotics [7], drug delivery tools [8], molecular-scale measuring devices [9], plasmonic systems [10], and information storage systems [11]. These dynamic DNA nanodevices generally contain two components,

L. Pan et al. (Eds.): BIC-TA 2022, CCIS 1801, pp. 674–683, 2023.
https://doi.org/10.1007/978-981-99-1549-1_54

i.e., the main structures and the reconfigurable units [12]. The conformation of the reconfigurable unit changes with the fuel molecules or external stimuli, leading to the particular movement of the nanostructure. Fuel molecules or external stimuli that used to drive dynamic DNA origami include DNA strands [13,14], metal ions [15], photos [16], thermal stimuli [17], pH [18], and endogenous small molecules [19]. The luxury of having a variety of molecular drives to choose from gives us the ability to design dynamic DNA nanodevices for a wide range of actions and applications.

Among these molecular drives, nucleic acid can reconfigure dynamic nanostructures through toehold-mediated strand displacement reactions (TMSD) [13,14]. Owing to the high programmability of TMSD, the dynamic nanodevices can achieve multi-step reconfigurations. For example, Ke *et al.* used TMSD to implement a three-step regulation that reconfigured a four-arm device into three conformations [9]. However, in multi-step transformation systems, it is required to design a large number of orthogonal DNA strand displacement reactions, which results in high costs of DNA sequence design and synthesis. Furthermore, the TMSD-driven reconfigurable structures must be annealed in a strict thermocycler when reconfiguring the pattern of DNA origamis [20]. These limitations hindered the application of DNA strand displacement in dynamic DNA nanodevices.

Our previous works demonstrated that polymerase-triggered strand displacement (PTSD), as a novel general method of regulating the conformation of DNA nanostructures, was a supplement to the traditional toehold-mediated strand displacement for transforming DNA nanodevices [21,22]. We successfully trimmed the rectangular DNA origami into six geometric patterns and proposed the concept of polymerase-driven DNA kirigami. This method allowed us to use only two primer strands to drive dozens-of-step PTSD reactions with a constant reaction temperature of 37°C. Herein, inspired by the kirigami strategy, we employed the PTSD reaction to drive the transformation of the reconfigurable DNA origami. Unlike DNA kirigami, which trimmed and changed the overall shape of the origami structure, this work reconfigured the conformation of DNA origami by removing a few staple strands without a sharp decrease in the DNA origami's molecular weight.

In this work, we constructed a reconfigurable book-like DNA origami structure. The pages of the nano book was flipped driven by PTSD reaction and the nano book could exhibit three conformations: one page with four layers, two pages each with two layers, and four pages each with one layer. The different conformations were clearly identified through transmission electron microscopy (TEM). The ratio of the three conformations of the nano book changed after adding different primer strands, proving that the PTSD reaction could drive the transformation of DNA nanodevices. We designed three primer strands to displace thirty-five staple strands, which required less designing and synthesis of DNA sequences than a TSMD-driven system with the same number of reaction steps. The PTSD reaction regulated the three conformations of the nano book, demonstrating that it could be used to drive multi-step dynamic nanostructures.

Our work on the development of PTSD-driven reconfigurable nanostructure will enrich the tools for regulating dynamic DNA nanodevices towards different applications.

2   Implementation and Characterization of DNA Nanobook

In this section, we describe the materials, the preparation and reconfiguration method of NB, the TEM characterization, and the statistical analysis methods.

The book-like origami (NB) was designed with the software CaDNAno. Staple strands and primer strands were synthesized by Sangon Biotech, Shanghai, China. All the strands were ordered via HAP purification. To generate NB, 8 nM of M13mp18 strands (Tilibit nanosystems, Garching, Germany) were mixed with 199 staples at a ratio of 1:12 in 1 × TAE buffer (Tris, 40 mM; acetic acid, 20 mM; EDTA, 2 mM; and magnesium chloride, 12.5 mM; pH 8.0). The mixtures were annealed from 80° to 24° within 36 h. The primers were then added to the mixtures in excess by 1.2-fold to the binding sites [21]. The sequences of Primer 1, Primer 2, and Primer 3 are 5'-TCTGCACATAGTAG-3', 5'-TACCTTGCTGTCTG-3', and 5'-TACTGAGATGAGTC-3', respectively. Then, the samples were annealed from 35° to 24° within 2 h. 500 μL of samples were purified through a 100 KDa ultracentrifugal filter (MWCO, Amicon, Millipore, Molsheim, France) to eliminate the excess staple strands and primer strands. The concentration of purified NB bound with primers was 15 nM. The DNA origami concentration was determined by the estimated extinction coefficient at 260 nm ($1.091 \times 10^8 \ M^{-1}cm^{-1}$).

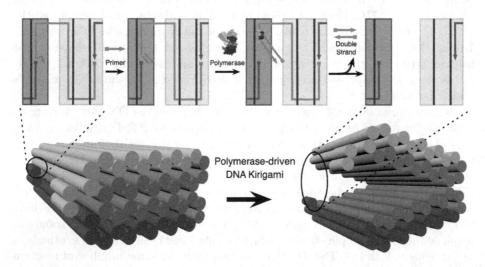

Fig. 1. Schematic of the PTSD-driven reconfiguration of the DNA nano book structure.

NB samples were mixed with 32 kU/L of Klenow polymerase (Thermo Fisher Scientific, Shanghai, China) in 1 × Klenow buffer. The usage of dNTP depends on different situations. Generally, the dNTP concentration is 5 times (including all four kinds of deoxyribonucleotides) that of the strands to be removed. The final concentration of the configured origami structure was 9 nM. Mixed samples were incubated for 6 h at 37 °C.

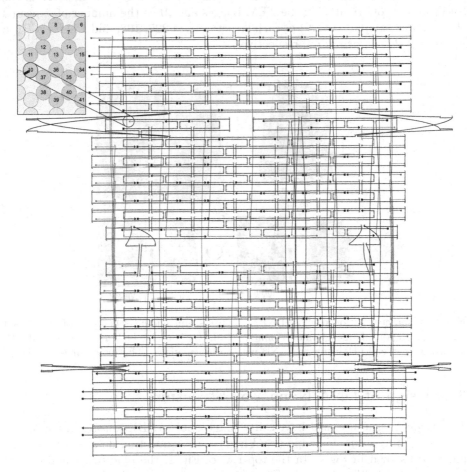

Fig. 2. Layout of NB. Blue line: scaffold (m13p18); black line: common staples; red line: connecting strand 1; orange line: connecting strand 2; green line: connecting strand 3. (Color figure online)

The mixed sample were then diluted to 2 nM with 1 × TAE buffer. Then, the diluted sample was absorbed on plasma cleaned (30 s oxygen plasma flash) carbon-coated copper grids (AZH300, Zhongjingkeyi Technology, Beijing) for TEM imaging. A 5 μL droplet of DNA origami solution was applied onto the carbon-coated side of the TEM grid, and the excess sample solution was blotted away with filter paper after an incubation of 2 min. The samples were stained using the uranyl acetate solution (Zhongjingkeyi Technology, Beijing). Excess stain solution was blotted away with filter paper after 100 s. After these procedures, the sample was left to dry under ambient conditions for at least 30 min before imaging. The TEM images were obtained using a FEI Tecnai G2 20 instrument operated at an acceleration voltage of 200 kV.

The yields of the structures were calculated as $p = M/N$, where N is the total number of origamis in the TEM images and M is the number of expected shapes. The sample size (N) of each structure was 225, 108, and 130 for a P1, a P2, and a P4 conformation, respectively.

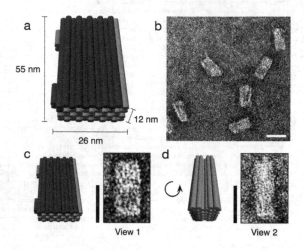

Fig. 3. Characterization of closed NB. a) Three-dimensional schematic of the closed NB DNA origami. Dimension: of 55 nm × 26 nm × 12 nm. b) TEM image of closed NB. c) and d) Typical TEM images of NB with different views. The scale bar is 50 nm.

3   Design of DNA Nanobook Structure

The origami structure, which we call a "nano book" (NB), consists of four pages connected into a cuboid shape (see Fig. 1). The connection details between the pages are shown in Fig. 1. In the top row of Fig. 1, the blue boxes and orange boxes represent different pages. The blue line in the boxes indicates the scaffold strand of DNA origami. The gray lines indicate the staple strand without toehold domain (common staples). The red staple indicates the connecting strand used to link pages. A 14-nucleotide-long toehold is extended at the 3' end of the connecting strand for recognizing the primer. A primer binds to the toehold of the connecting strand and triggers the PTSD reaction. As displayed in Fig. 1, the Klenow polymerase starts at the 3' end of the primer, using the red staple as a template for the polymerization reaction. After the PTSD reaction, the red connecting strand is removed, resulting in the separation of the two pages (see Fig. 1, bottom).

Figure 2 shows the complete design of NB. The NB structure was designed using the scaffolded DNA origami method based on the honeycomb lattice. We designed three groups of connecting strands, colored in red, orange and green. The red lines, orange lines and green lines show the positions of all the connecting strands. To facilitate the binding of the primer and polymerase, the sites where the connecting strands extend are elaborately designed to ensuring that the extension direction is towards the outside of the DNA origami (see Fig. 2, insert).

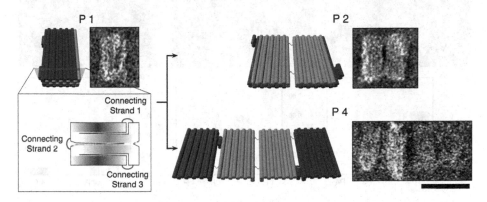

Fig. 4. Regulating the conformation of NB. Closed NB is shown as a one-page conformation (P1) in the TEM image. After the addition of Primer 2 and Klenow polymerase, NB is reconfigured into a two-page conformation (P2). When Klenow polymerase and Primer 1, 2, and 3 are present, NB is transformed into a four-page conformation (P4). Scale bar of the TEM images is 50 nm.

In order to avoid multimerization or aggregation of NB through blunt-end stacking interactions, the outward pointing ends of each extended common staples were passivated with 5-nt long single-stranded poly-T overhangs. After all the connecting strands are removed, each two pages are connected by two groups of hinges (single-stranded scaffold).

We first tested the conformation of the NB origami using transmission electron microscopy (TEM). The closed NB was designed with dimensions of 55 nm × 26 nm × 12 nm (see Fig. 3a). When the sample was absorbed onto copper grids, most of the origami structures landed toward the larger faces. As a result, in the TEM images, we observed a high percentage of particles with a size of 55 nm × 26 nm, before the PTSD reaction (see Fig. 3b and Fig. 3c). Figure 3c and Fig. 3d show the averaged negativestain TEM micrographs of NB. The NB particles in View 2 are thinner than those in View 1, with a size of 55 nm × 12 nm, demonstrating that the NB particles in View 2 are landing on the grid towards another face. In statistical analysis, the NB particles with both View 1 and View 2 were considered as a one-page structure (P1).

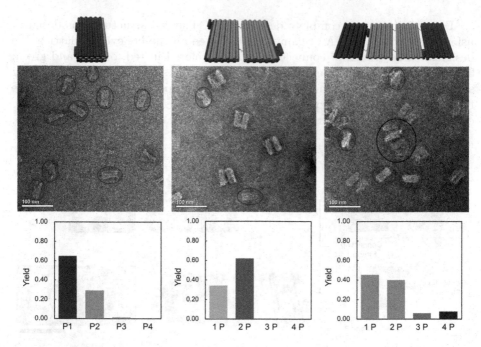

Fig. 5. Statistical analysis of the strategy to reconfigure NB using PTSD reaction. Top. Schematic of one-page, two-page, and three-page conformation. Middle. In the TEM images, we label one-page conformations with red circles, two-page conformations with yellow, three-page conformations with green, and four-page conformations with blue, respectively. Bottom. In the sample of P1, 65% of the NB is present in a one-page conformation. In the sample of P2, the percentage of two-page conformation is 62%. In the TEM images of P4, 14% of the NB displayed a four-page conformation or a three-page conformation. The sample size (N) of each structure was 225, 108, and 130 for a P1, a P2, and a P4 conformation, respectively. Scale bar of the TEM images is 100 nm.

4   Reconfiguration of NB Structure

To reconstruct the NB structure in a stepwise manner, the extended sequences of the three groups of connecting strands, *i.e.*, the toehold domain of the PTSD reaction, were designed orthogonally. The curves with different colors represent the connection strands (CS) containing different toehold (see Fig. 4, insert). The red CS1, orange CS2, and green CS3 are complementarily paired with Primer 1, Primer 2, and Primer 3, respectively. Of note, the schematic does not display the exact number of CSs. CS1 represents 11 connecting staples for linking the orange page to the gray page; CS2, which connects the two gray pages, consists of 14 connecting staples; and the green CS3 represents 10 connecting staples.

As shown in Fig. 4, before adding the primer strands and Klenow polymerase, the NB origami was closed, shown as a one-page conformation (P1) in the TEM image. After the addition of Primer 2 and Klenow polymerase, the PTSD reaction was initiated, taking CS2 as the substrate. CS2 was removed from NB, resulting in the separation of the two gray pages. NB was reconfigured into a two-page conformation (P2). When Klenow polymerase and Primer 1, 2, and 3 were present, all the CSs were removed after PTSD reaction, causing NB to be fully unfolded. The NB structure exhibited a four-page conformation (P4). The TEM results verified the three conformations of NB. It is worth noting that in the enlarged TEM images of the two-page conformation and the four-page conformation, we observed two joining sites between the pages, indicating the position of the hinge strands.

To quantitatively analyze the efficiency of the proposed method for regulating the NB structure, we calculated the percentage of expected conformations in the TEM images of different samples. In the typical TEM image of P1 sample, individual one-page particles were the expected conformation. In the typical TEM image of P2 sample, the expected two-page conformation was defined as an individual particle containing two closely parallel connected pages, whose hinges could be observed. The definition of the four-page conformation is similar to that of the two-page conformation. In these TEM images, we labeled one-page conformations with red circles, two-page conformations with yellow, and four-page conformations with blue, respectively (see Fig. 5, middle). In addition, some NB structures with incompletely unfolded three-page conformation were observed in the TEM, labeled with green circles.

The statistical results of different conformation samples showed that PTSD reaction could successfully reconfigure NB (see Fig. 5, bottom). In the TEM images of P1, 65% of the NB was present in a one-page conformation. While, in the images of P2, the percentage of one-page conformation decreased to 34%. The percentage of two-page shaped NB occupied 29% in P1 images, and the number increased to 62% in P2 images. These changes in percentages proved that the PTSD reaction initiated by Primer 2 played a role in transforming the closed NB into a two-page shaped NB. In the TEM images of P4, 14% of the NB displayed a four-page conformation or a three-page conformation. Compared to the statistical results of P1 and P2, although the percentage of expected conformation was not high, it was still evident that a part of closed NBs was transformed into a multi-page conformation after regulation of the PTSD reaction. From the statistical results, the yield of the two-page conformation was higher than that of the four-page conformation, using our regulation method. We guessed that to obtain the two-page shaped NB requires fewer steps of strand displacement reactions. Therefore, it was more accessible to reconfigure the closed NB into a two-page conformation in the same amount of time.

5   Conclusion

In conclusion, we introduced a strategy to regulate the conformation of DNA origami driven by polymerase-triggered DNA strand displacement (PTSD). We first design a book-shaped DNA origami structure consisting of four pages connected into a cuboid shape. The PTSD reaction initiated by different primer strands selectively removed the connecting strand, transforming nano book (NB) into a two-page or a four-page conformation. We utilized three primer strands to remove thirty-five connecting strands and constructed three conformations of the identical DNA origami, illustrating that the complex reconfiguration of DNA origami could be regulated with a small amounts of triggers using the PTSD reaction. The ratio of specific conformations in TEM results changed with the addition of corresponding primer strands, indicating the effectiveness of the proposed method.

In the PTSD reaction, a multi-step strand displacement network could be achieved by orthogonally designing the primer domains. Our study provides proof-of-concept for the PTSD-driven dynamic DNA origami device. In the future, PTSD-driven nanorobots may perform multiple tasks through multi-step DNA strand displacements initiated by orthogonal primer domains. Our work on the development of PTSD-driven reconfigurable nanostructure will offer a new way to create intelligent materials for advanced nanotechnology applications.

Acknowledgment. This work was supported by National Natural Science Foundation of China (62172171), Zhejiang Lab (2021RD0AB03), and Fundamental Research Funds for the Central Universities (HUST: 2019kfyXMBZ056).

References

1. Rothemund, P.W.K.: Folding DNA to create nanoscale shapes and patterns. Nature **440**(7082), 297–302 (2006)
2. Liu, W., Zhong, H., Wang, R., Seeman, N.C.: Crystalline two-dimensional DNA-origami arrays. Angew. Chem. Int. Ed. **50**(1), 264–267 (2011)
3. Woo, S., Rothemund, P.W.K.: Programmable molecular recognition based on the geometry of DNA nanostructures. Nat. Chem. **3**(8), 620–627 (2011)
4. Han, D., Pal, S., Nangreave, J., Deng, Z., Liu, Y., Yan, H.: DNA origami with complex curvatures in three-dimensional space. Science **332**(6027), 342–346 (2011)
5. Douglas, S.M., Dietz, H., Liedl, T., Högberg, B., Graf, F., Shih, W.M.: Self-assembly of DNA into nanoscale three-dimensional shapes. Nature **459**(7245), 414–418 (2009)
6. Ke, Y., et al.: Multilayer DNA origami packed on a square lattice. J. Am. Chem. Soc. **131**(43), 15903–15908 (2009)
7. Kopperger, E., List, J., Madhira, S., Rothfischer, F., Lamb, D.C., Simmel, F.C.: A self-assembled nanoscale robotic arm controlled by electric fields. Science **359**(6373), 296–301 (2018)
8. Li, S., et al.: A DNA nanorobot functions as a cancer therapeutic in response to a molecular trigger in vivo. Nat. Biotechnol. **36**(3), 258–264 (2018)
9. Ke, Y., Meyer, T., Shih, W.M., Bellot, G.: Regulation at a distance of biomolecular interactions using a DNA origami nanoactuator. Nat. Commun. **7**(1), 10935 (2016)
10. Kuzyk, A., Yang, Y., Duan, X., Stoll, S., Govorov, A.O., Sugiyama, H., Endo, M., Liu, N.: A light-driven three-dimensional plasmonic nanosystem that translates molecular motion into reversible chiroptical function. Nat. Commun. **7**(1), 10591 (2016)
11. Chen, K., Zhu, J., Bošković, F., Keyser, U.F.: Nanopore-based DNA hard drives for rewritable and secure data storage. Nano Lett. **20**(5), 3754–3760 (2020)

12. Song, J., Li, Z., Wang, P., Meyer, T., Mao, C., Ke, Y.: Reconfiguration of DNA molecular arrays driven by information relay. Science **357**(6349), eaan3377 (2017)
13. Gu, H., Chao, J., Xiao, S.-J., Seeman, N.C.: A proximity-based programmable DNA nanoscale assembly line. Nature **465**(7295), 202–205 (2010)
14. Kuzuya, A., Sakai, Y., Yamazaki, T., Xu, Y., Komiyama, M.: Nanomechanical DNA origami "single-molecule beacons" directly imaged by atomic force microscopy. Nature Commun. **2**(1), 449 (2011)
15. Gerling, T., Wagenbauer, K.F., Neuner, A.M., Dietz, H.: Dynamic DNA devices and assemblies formed by shape-complementary, non-base pairing 3D components. Science **347**(6229), 1446–1452 (2015)
16. Ryssy, J., et al.: Light-responsive dynamic DNA-origami-based plasmonic assemblies. Angew. Chem. Int. Ed. **60**(11), 5859–5863 (2021)
17. Chen, Z., Chen, K., Xie, C., Liao, K., Xu, F., Pan, L.: Cyclic transitions of DNA origami dimers driven by thermal cycling. Nanotechnology **34**(6), 065601 (2023)
18. Ijäs, H., Hakaste, I., Shen, B., Kostiainen, M.A., Linko, V.: Reconfigurable DNA origami nanocapsule for pH-controlled encapsulation and display of cargo. ACS Nano **13**(5), 5959–5967 (2019)
19. Douglas, S.M., Bachelet, I., Church, G.M.: A logic-gated nanorobot for targeted transport of molecular payloads. Science **335**(6070), 831–834 (2012)
20. Han, D., Pal, S., Liu, Y., Yan, H.: Folding and cutting DNA into reconfigurable topological nanostructures. Nat. Nanotechnol. **5**(10), 712–717 (2010)
21. Chen, K., Xu, F., Hu, Y., Yan, H., Pan, L.: DNA kirigami driven by polymerase-triggered strand displacement. Small **18**(24), 2201478 (2022)
22. Liao, K., Chen, K., Xie, C., Chen, Z., Pan, L.: Disassembly of DNA origami dimers controlled by programmable polymerase primers. Chem. Commun. **58**(92), 12879–12882 (2022)

An Analysis for Thermal Conductivity
of Graphene/Polymer Nanocomposites

Zirui Liang[1], Weigang Huang[2], Runzhe Rao[1], and Fang Li[1(✉)]

[1] Hubei Key Laboratory of Theory and Application of Advanced Materials Mechanics, School of Science, Wuhan University of Technology, Wuhan 430070, China
liaaf@whut.edu.cn
[2] China Ship Development and Design Center, Harbin, China

Abstract. Graphene/polymer nanocomposites have attracted much more attention due to its high thermal conductivity up to 5300 W/mK, as well as the lightweight, noncorrosive and elastic properties of the polymer matrix. At present, the existing experimental results have shown that the thermal conductivity of graphene/polymer nanocomposites can be effectively improved by increasing the aspect ratio of graphene, or increasing the transverse size and thickness of graphene at the same time under the condition of constant aspect ratio. In this study, the influence of the number of GNSs layers and size on the isotropic thermal conductivity of the composites are revealed based on Maxwell's effective medium theory by considering the effect of the number of GNSs layers on the interface thermal resistance and the thermal conductivity of GNSs. The expressions of equivalent thermal conductivity for the isotropic GNSs/polymer composites are obtained. The numerical results obtained in this paper are agree well with experimental data. And these results show that the thermal conductivity of the isotropy composites increases with the increase of GNSs layers, concentration and aspect ratio. However, the increment of thermal conductivity for the composites decreases gradually with the increase of GNSs layers.

Keywords: Graphene/polymer nanocomposites · Isotropy · Interface thermal resistance · thermal conductivity

1 Introduction

In recent years, with the miniaturization, integration and functionalization of electronic devices, as well as the emergence of new applications such as 3D chip stack structure, flexible electronic devices and light-emitting diodes, how to select appropriate thermal interface materials for effective heat dissipation to prevent overheating of equipment from affecting the reliability of electronic devices and shortening their service life has become a challenging problem [1, 2]. Traditional heat dissipation materials, such as composites prepared by filling thermal conductive fillers (such as silver and nickel) in polymer matrix, have a thermal conductivity of more than 1 W/mK as a thermal interface material, but because of the high load of fillers, the equipment is heavy, which

L. Pan et al. (Eds.): BIC-TA 2022, CCIS 1801, pp. 684–690, 2023.
https://doi.org/10.1007/978-981-99-1549-1_55

limits its application. As a special two-dimensional structural material, graphene has very good heat conduction performance. The thermal conductivity of pure and defect free single-layer graphene is up to 5300 W/mK, 13 times that of copper. In addition, its large specific surface area can effectively reduce the interfacial thermal resistance effect between graphene and polymer matrix. Therefore, graphene is a very ideal heat conduction material [2], which is compounded with polymer matrix, it can effectively improve the thermal conductivity of composites.

At present, domestic and foreign scholars have carried out a lot of research on improving the thermal conductivity of graphene/polymer nanocomposites. Gang Lian and other researchers used 0.92% graphene with a vertical parallel network structure as the filling phase to composite with the epoxy resin polymer matrix material, making the thermal conductivity of graphene/epoxy resin polymer nanocomposites reach 2.13 W/(mK), which is 1231% higher [3]. These phenomena prove that graphene can effectively improve the thermal conductivity of matrix materials, providing a new idea for thermal management. Besides, the Monte Carlo [4] and molecular dynamics methods of meso statistics are mainly used to analyze the equivalent thermal properties of composite materials, discuss the influence of the volume fraction and morphology of graphene in composite materials on the equivalent thermal conductivity considering the interface thermal resistance effect. Xi Shen discussed the effect of graphene layers number on the graphene thermal conductivity and interface thermal resistance based on the molecular dynamics simulation [5].

Generally, although the addition of graphene to the polymer matrix can effectively improve the thermal conductivity of thermal interface materials, the content of graphene is often too high. Therefore, it is necessary to reveal the action mechanism of graphene in thermal interface materials, and clarify the influence of such factors as the filling concentration, size of graphene on the thermal conductivity of composites, so as to provide a theoretical analysis basis for optimizing the thermal conductivity of graphene/polymer thermal interface materials. Therefore, based on the effective medium theory, this paper will focus on the analysis of the influence of different factors on the thermal conductivity of graphene/polymer nanocomposites, and discuss the thermal conductivity of graphene polymer matrix nanocomposites in different shapes under different layers and length width ratio.

2   Analytical Model

In this paper, the effective thermal conductivity for GNSs/polymer composites k_e will be obtained based on the effective medium theory considering the interface thermal resistance (Kapitza thermal resistance) R_I, that is determined by the number of layers of GNSs n [6],

$$R_I(n) = (7.83 - 1.24 \times (1 - \exp(-n/2.015)) \times 2) \times 10^{-9} \qquad (1)$$

Besides that, the effect of the number of layers of GNSs n on the graphene thermal conductivity k_g, which is also introduced and expressed as,

$$k_g(n) = 968.26 + 398.84 \times \left(1 - e^{-\frac{n}{11.75}}\right) + 481.9 \times \left(1 - e^{-\frac{n}{1.12}}\right) \qquad (2)$$

is also considered.

Based on the multiple scattering theory [7], the filled phase and its surrounding interface layer are taken as representative units of composite materials, effective thermal conductivity k_{ii}^c is expressed as

$$k_{ii}^c = k_s \frac{k_s + L_{ii}(k_g - k_s)(1 - v) + v(k_g - k_s)}{k_s + L_{ii}(k_g - k_s)(1 - v)}, \tag{3}$$

where $v = a_1{}^2 a_3 / (a_1 + \delta)^2 (a_3 + \delta)$, δ and k_s are the thickness and thermal conductivity of the interface layer between graphene and polymer matrix, respectively. a_1 and a_3 are the radii of ellipsoids for GNSs, respectively. L_{ii} is the geometric factor related to the shape of filled phase, which is given by the following formula:

$$L_{11} = L_{22} = \begin{cases} \dfrac{p^2}{2(p^2-1)} - \dfrac{p}{2(p^2-1)^{3/2}} cosh^{-1}p, p > 1, \\ \dfrac{p^2}{2(p^2-1)} + \dfrac{p}{2(1-p^2)^{3/2}} cos^{-1}p, p < 1, \end{cases} \tag{4}$$

$$L_{33} = 1 - 2L_{11} \tag{5}$$

In this equation, p is the length width ratio of ellipsoidal particles, $p = a_3/a_1$. Here, it is assumed that the interface is just a thin and low thermal conductivity area, and the interface thermal resistance is the limit case of the heat conduction between GNSs and polymer through the interface, as $\delta \rightarrow 0$ and $k_s \rightarrow 0$. Then the equivalent thermal conductivity (3) of composite material unit element can be rewritten as,

$$k_{ii}^c = \frac{k_g}{1 + \frac{\gamma L_{ii} k_g}{k_m}} \tag{6}$$

where,

$$\gamma = \begin{cases} (2 + 1/p)\alpha, p \geq 1 \\ (1 + 2p)\alpha, p \leq 1 \end{cases} \tag{7}$$

dimensionless parameter α is denoted as,

$$\alpha = \begin{cases} a_k/a_1, p \geq 1 \\ a_k/a_3, p \leq 1 \end{cases} \tag{8}$$

and the Kapitza radius α_k used to characterize thermal properties of zero thickness interface is expressed as, 6

$$a_k = R_I k_m \tag{9}$$

where R_I is the interface thermal resistance, k_m is the thermal conductivity of the matrix phase.

Due to the special two-dimensional structure of graphene, it is assumed that GNSs filled in the polymer matrix is a flat ellipsoid. Then, based on the effective medium theory, the effective thermal conductivity of composite material is obtained as $p < 1$,

$$k_e = k_m \frac{3 + f[2\beta_{11}(1 - L_{11}) + \beta_{33}(1 - L_{33})]}{3 - f[2\beta_{11}L_{11} + \beta_{33}L_{33}]} \tag{10}$$

in which,

$$L_{11} = L_{22} = \frac{p^2}{2(p^2 - 1)} + \frac{p}{2(1 - p^2)^{3/2}} cos^{-1}p, L_{33} = 1 - 2L_{11} \tag{11}$$

$$\beta_{ii} = \frac{k_{ii}^c - k_m}{k_m + L_{ii}(k_{ii}^c - k_m)} \tag{12}$$

$$k_{ii}^c = \frac{k_g}{1 + \frac{\gamma L_{ii} k_g}{k_m}} \tag{13}$$

$$\gamma = (1 + 2p)\alpha \tag{14}$$

$$\alpha = a_k / a_3 \tag{15}$$

3   Results and Discussion

In order to describe the influence of graphene concentration, size, thickness and interface thermal resistance on thermal conductivity of GNSs/polymer nanocomposites, The parameters used in the analysis are shown in Table 1 [5, 8].

Table 1. Material Properties of Graphene and Matrix Phase in Numerical Analysis.

n	k_m(w/mk)	k_g(w/mk)	$\alpha_k (10^{-9} m)$	$R_I (10^{-9} w^2/mk)$
1	0.2	1285.4	1.37	6.86
4	0.2	1551.7	1.14	5.69
5	0.2	1582.8	1.11	5.56
7	0.2	1628.2	1.09	5.43
10	0.2	1678.6	1.07	5.37

Firstly, in order to verify the accuracy of the analytical model in predicting the equivalent thermal conductivity for GNSs/polymer nanocomposites, the numerical results were compared with the experimental results, as shown in Fig. 1(a) and (b). The calculated results in this paper are in good agreement with the experimental results [12]. In Fig. 1(a), it is indicated that the thermal conductivity of composites shows a nonlinear increasing trend with the increase of GNSs concentration. The modified theoretical calculation results in this paper can also fully reflect this nonlinear increasing relationship, which verifies the accuracy of the modified theory. In addition, the relationship between thermal conductivity and concentration of GNSs at different layers is also compared with the experimental results [5], as shown in Fig. 1(b). The numerical results obtained in this paper are all good agreement with experimental results.

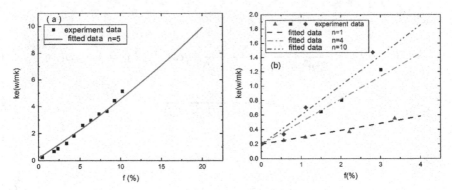

Fig. 1. Comparisons between numerical analysis and experimental results: (a) thermal conductivity varying with concentration as n = 5; (b) thermal conductivity changing with concentration of GNSs at different layers for GNSs.

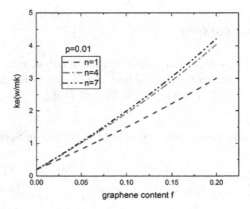

Fig. 2. Thermal conductivity of graphene/polymer nanocomposites varying with the volume fraction of graphene at different layers.

As shown in Fig. 2, the oblate spheroidal graphene inclusion particles with the length width ratio p = 0.01 are selected as the filler. The three different layers are single layer, four layers and seven layers respectively. The filler is randomly arranged. When the volume fraction of GNSs is zero, the thermal conductivity of the composites is epoxy resin, that is, 0.2 W/mk. With the increase of the volume fraction, the thermal conductivity of GNSs/polymer nanocomposites increases significantly, and presents a nonlinear feature. In addition, the larger the number of graphene layers, the higher the thermal conductivity at the same aspect ratio and volume fraction. This is because the larger the number of graphene layers, the greater the thermal conductivity of graphene, and the smaller the value of interface thermal resistance, both of which induces to the increasing of thermal conductivity for composite. It is worth noting that for the same number of layers, the difference for thermal conductivity between the first layer and the fourth layer is large, while the difference of thermal conductivity between the fourth layer and the seventh layer is less obvious than that between the first layer and the fourth layer. This is because that with the increase of the number of layers, the influence of the change of the number of layers on the thermal conductivity and interface thermal resistance gradually becomes less obvious when the number of layers is low.

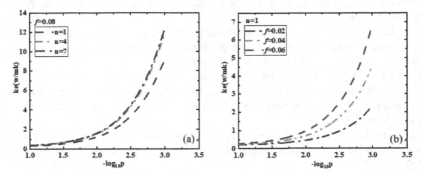

Fig. 3. The thermal conductivity of graphene/polymer nanocomposites varing with the aspect ratio of graphene: (a) different layers at 0.08 volume fraction; (b) different volume fractions for one layer of graphene.

The aspect ratio of graphene is an important factor affecting the performance of graphene [9, 10]. Figure 3(a) indicates the thermal conductivity along with aspect ratio p as the volume fraction of graphene is 0.08. It is found when the number of layers and the longitudinal size of graphene are all constant, the aspect ratio p is just decided by the transverse size of graphene. The smaller the transverse size is, the greater the aspect ratio is. It is obviously that the smaller the aspect ratio of the oblate ellipsoid for graphene, that is, the larger the transverse size, the greater the thermal conductivity of the composite. In addition, when the number of layers n is 1, the thermal conductivity changing with aspect ratio at different volume fraction is detected in Fig. 3(b). It is also shown that the thermal conductivity for graphene/polymer nanocomposites is still increasing with the increasing of volume fraction and aspect ratio of graphene.

4   Conclusions

When the number and shape of graphene are fixed, the effective thermal conductivity for graphene/polymer nanocomposites increases significantly with the increase of the filling concentration of graphene. However, when the filling concentration increases to a certain amount, the increase of its equivalent thermal conductivity becomes slow and finally reaches the saturation state. When the volume fraction and aspect ratio of graphene are constant, the thermal conductivity of the composite increases with the increase of the number of layers, and the rate of increase decreases with the increase of the number of layers. And when the number of layers is more than 7, the thermal conductivity gradually approaches saturation.

References

1. Nan, C.W., Birringer, R., Clarke, D.R., et al.: Effective thermal conductivity of particulate composites with interfacial thermal resistance. J. Appl. Phys. **81**(10), 6692–6699 (1997)
2. Balandin, A.A., et al.: Superior thermal conductivity of single-layer graphene. Nano Lett. **8**(3), 902 (2008)
3. Lian, G., Tuan, C.C., Li, L., Jiao, S., Wang, Q., Moon, K.S., et al.: Vertically aligned and interconnected graphene networks for high thermal conductivity of epoxy composites with ultralow loading. Chem. Mater. **28**, 6096–6104 (2016)
4. Meyer, H.A.: Symposium on Monte Carlo methods. J. R. Aeronaut. Soc. (1957)
5. Shen, X., Wang, Z., Wu, Y., Liu, X., He, Y.B., Kim, J.K.: Multilayer graphene enables higher efficiency in improving thermal conductivities of graphene/epoxy composites. Nano Lett. **16**(6), 3585–3593 (2016)
6. Su, Y., Li, J.J., Weng, G.J.: Theory of thermal conductivity of graphene-polymer nanocomposites with interfacial Kapitza resistance and graphene-graphene contact resistance. Carbon **137**, 222–233 (2018)
7. Rausch, J.B., Kayser, F.X.: Elastic constants and electrical resistivity of Fe3Si. J. Appl. Phys. **48**(2), 487–493 (1977)
8. Shahil, K.M., Balandin, A.A.: Balandin: graphene−multilayer graphene nanocomposites as highly efficient thermal interface materials. Nano Lett. **12**(2), 861–867 (2012)
9. Jiang, F., Zhao, W., Wu, Y., et al.: Anti-corrosion behaviors of epoxy composite coatings enhanced via graphene oxide with different aspect ratios. Prog. Org. Coat. **127**, 70–79 (2019)
10. Xing, Z., Sun, W., Wang, L., Yang, Z., Wang, S., Liu, G.: Size-controlled graphite nanoplatelets: thermal conductivity enhancers for epoxy resin. J. Mater. Sci. **54**(13), 10041–10054 (2019). https://doi.org/10.1007/s10853-019-03525-5

Simulation of Aflatoxin B1 Detection Model Based on Hybridization Chain Reaction

Rong Liu[1], Meng Cheng[2], Luhui Wang[2], Mengyang Hu[2], Sunfan Xi[2], and Yafei Dong[1,2(✉)]

[1] Department of Computer Science, Shaanxi Normal University, Xi'an 710119, China
dongyf@snnu.edu.cn
[2] Department of Life Science, Shaanxi Normal University, Xi'an 710119, China

Abstract. Aflatoxin B1 (AFB1) is one of the aflatoxin toxins, which is most common in peanut, walnut and other foods. It is the strongest carcinogen among known chemicals, and it is not easy to decompose under conventional heating conditions, posing a great threat to human health, so it has attracted people's attention. In this paper, a simple, highly selective, enzyme free and label free biosensor model was designed to detect AFB1. The addition of AFB1 in the system leads to the release of trigger sequence, and then two hairpins will be opened in turn and hybridized to form a continuous double chain structure rich in G-quadruplex. This structure will have high fluorescence after interacting with N-methylporphyrin dipropionate IX (NMM). In this paper, we use the software platform to analyze the thermodynamic properties of DNA sequences, simulate the reaction process and conditions of the model, and prove the feasibility theoretically. Finally, the feasibility of the model and sequence was verified by biological experiments.

Keywords: Hybridization Chain Reaction · G-quadruplex · Detection of AFB1

1 Introduction

Aflatoxin B1 (AFB1) is one of the secondary metabolites of Aspergillus flavus and Parasitic Aspergillus. Its acute toxicity is 10 times that of potassium cyanide and 68 times that of arsenic. Chronic toxins can induce canceration, and its carcinogenic capacity is 900 times higher than that of dimethylbenzidine [1]. Many foods are polluted by AFB1, which can cause liver cirrhosis, necrosis, canceration and other diseases in humans and animals, posing a major threat to human health [2]. Therefore, it is very important to develop an effective and highly sensitive aflatoxin B1 detection platform. So far, the common detection methods of AFB1 include High Performance Liquid Chromatography (HPLC), Thin Layer Chromatography (TLC) and Liquid Chromatography Mass Spectrometry (LC-MS) [3–6]. The above methods are most sensitive, but they also have the disadvantages of cumbersome sample pretreatment, expensive instruments, and high professional requirements for personnel, which limit the rapid detection of AFB1. The traditional immunoassays based on antigen antibody interactions have good sensitivity and specificity, but the protein probe molecules in this method are expensive and

L. Pan et al. (Eds.): BIC-TA 2022, CCIS 1801, pp. 691–702, 2023.
https://doi.org/10.1007/978-981-99-1549-1_56

vulnerable to environmental impact [7, 8]. Adapters are selected from random single stranded nucleic acid sequence library by Systematic Evolution of Ligands by Exponential Enrichment (SELEX), which are highly specific to target substances [9]. It has the same sensitivity as antigen antibody reaction, but it is easier to synthesize than protein and has better stability [10].

Hybridization chain reaction (HCR) is an isothermal, enzyme free signal amplification technology based on chain displacement reaction proposed by Dirks and Pierce in 2004 [11]. When target DNA is introduced, several self-stable hairpin structures will cascade to produce long strand DNA that can trigger the next level of hairpin, which can achieve signal amplification of target molecules. Due to the advantages of no auxiliary enzyme, good reproducibility and cost, this technology is widely used in the detection of proteins and small molecules [12, 13]. However, most HCR based detection methods require fluorescent groups and quenching groups, which increases the cost of detecting fluorescent background. To avoid this problem, G-quadruplex is introduced into our model. Under the action of cations, G-rich DNA sequences will accumulate together to form G-quadruplex [14]. Among many sensing models that use G-quadruplex as signal reporting, N-methyl mesomorpholine IX (NMM) is often used to assist G-quadruplex to generate strong fluorescence as signal output [15, 16]. NMM is an asymmetric anionic porphyrin, which has significant selectivity for G-quadruplex structure. Its own fluorescence is very weak, but after interacting with G-quadruplex, its fluorescence will increase 20 times (excitation wavelength and emission wavelength are concentrated at 399 nm and 608 nm respectively) [17, 18].

Before conducting biological experiments, computer technology can be used to predict the relevant nucleotide sequence structure and binding ability, and simulate the dynamic reaction process, so as to solve the shortcomings of time-consuming experiment, tedious process, waste of materials and so on [19–22]. Therefore, we combine computer simulation and biological experiments in our work to build an enzyme free and label free sensor model for AFB1 detection, which also improves experimental efficiency and efficiency. The schematic diagram is shown in Fig. 1. The model includes two hairpins and a double chain structure. Both hairpins (Hp1 and Hp2) contain G-rich sequences. Under the action of K+, the free G-rich sequences can form a G-quadruplex structure. One chain in the double chain structure is the aptamer sequence (Apt) of AFB1, and the other chain is the trigger chain sequence (S1). As the starting device of this model, S1 can open the hairpin and trigger the subsequent cascade reaction.

The sensing model is divided into three steps: (1) target recognition. When AFB1 exists, it will specifically bind to the aptamer sequence Apt in the double chain structure, and then release the trigger chain sequence S1. (2) Signal amplification. The S1 released in the previous step can be combined with the 5 'terminal free sequence of hairpin Hp1. Driven by thermodynamic stability, S1 combined with some Hp1 sequences to form a double chain, exposing the G rich sequence and 3 'end sequence in Hp1. However, the 3 'end sequence exposed by Hp1 was hybridized with the 3' end free sequence of Hp2 to open the hairpin Hp2, which also exposed the rich G sequence in Hp2. Subsequently, the exposed 5 'end of Hp2 will combine with the free sequence of Hp1 5' end, and then open the hairpin Hp1. In this process, Hp1 and Hp2 will be opened in turn and assembled into a continuous double chain structure. (3) Signal output. Through the above reaction steps, a large number of free G rich sequences are embedded in the resulting DNA continuous double strand structure. Under the action of K^+, the G-rich sequence exists as a G-quadruplex structure, which will interact with the fluorescent dye NMM to generate fluorescence signals.

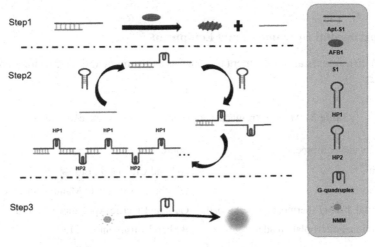

Fig. 1. Schematic diagram of aflatoxin B1 detection model based on hybrid chain reaction

2 Experimental Materials and Methods

2.1 Chemicals and Materials

The main materials and reagents involved in this experiment, as well as their configuration, preservation methods and sources are shown in Table 1.

Table 1. Main materials and reagents used in the experiment

Chemicals	preparation method	Storage method	Source
DNA	Ultrapure water dilution	$-20\,°C$ freezing	Bioengineering Co., Ltd
Tris-HCL Buffer	200mMKCl, 20nMMgCl2	Cryopreservation	Xi'an Jingbo Biotechnology Co., Ltd
AFB1	Dissolve with anhydrous methanol and dilute with Tris HCL buffer solution when using	Cryopreservation	Bioengineering Co., Ltd

(continued)

Table 1. (*continued*)

Chemicals	preparation method	Storage method	Source
NMM	Tris HCL buffer diluted to 50 μM	Low temperature and dark	Beijing Bellwether Technology Co., Ltd

2.2 Experimental Instruments and Equipment

The main instruments and equipment used in this experiment and their sources are shown in Table 2.

Table 2. Main instruments and equipment used in the experiment

instruments and equipment	Source
Pipette gun	Abbond China Co., Ltd
QL-901 Vortex mixer	Qilinbeier Instrument Manufacturing Co., Ltd
HH-2 digital display thermostatic water bath	Guohua Electric Appliance Co., Ltd
Multifunctional microplate reader	Bethen Instrument Co., Ltd

3 Fluorescence Detection of AFB1

Before the experiment, the Apt chain and S1 chain were mixed equally to form Apt-S1 double chain structure; Heat hairpin Hp1 and hairpin Hp2 at 90 °C for 5 min respectively, then rapidly cool them to room temperature and store them at 4 °C. First, mix an appropriate amount of AFB1, buffer solution and Apt-S1 solution, and react at 37 °C in a water bath for 30 min. Then, add Hp1 and Hp2 to the above solution, mix them uniformly with a vortex mixing oscillator, and react at 37 °C for 90 min. After the assembly of hybrid chain reaction, NMM was added to the centrifuge tube and reacted at 25 °C for 15 min. Finally, take 90 μL solution and put it into 96 microporous plates, and measure its fluorescence intensity with multifunctional microplate reader.

4 Results and Discussion

4.1 Sequence Design

The uniqueness of DNA is determined by the sequence of deoxynucleotides, double helix structure or super helix structure. Therefore, the design of nucleic acid sequence is very important. The nucleic acid sequence used in the model will be designed and analyzed with the help of computer software.

QGRS Mapper is a program based on recognized QGRS recognition and mapping algorithm, which is used to generate the composition and distribution information of

G-Rich sequence (QGRS) formed by predicted quadruplex in nucleotide sequence. In the QGRS Mapper, the assumed G-quadruplex is recognized by the following pattern: GxNy1GxNy2GxNy3Gx.

In the above mode, X represents the number of quadrants, and y1, y2, and y3 represent the gap length. Therefore, the sequence should be composed of four groups of equal length guanines separated by any nucleotide sequence, but with the following restrictions.

(1) The sequence must contain at least two tetrads ($x \geq 2$). Although the structure containing more G-tetrads is more stable, it is known that many nucleotide sequences and two G-tetrads can also form quadruplexes.
(2) The program provides users with search options for up to 45 base sequences.
(3) At most one gap is allowed to have a length of zero.

Table 3 shows some examples of effective QGRS.

Table 3. Example of QGRS

Sequence	QGRS parameters
GGACGGGGTTTGG	x = 2, y1 = 2, y2 = 0, y3 = 3
GGGTGGGTGGCAGAGCTGGGCTGGG	x = 3, y1 = 1, y2 = 10, y3 = 2
GGGGTGGGGTGGGGTGGGG	x = 4, y1 = 1, y2 = 1, y3 = 1

QGRS Mapper predicts that the designed nucleotide sequence can form G-quadruplex G-rich sequence, and the results are shown in Table 4. The prediction shows that the score of the ninth group is 42, which can form the most stable G-quadruplex.

Table 4. Prediction Results of G-quadruplex Formation by G-rich Sequences

Number	Position	Length	QGRS	G-Score
1	1	14	GGGAGGGAGGGAGG	21
2	1	14	GGGAGGGAGGGAGG	19
3	1	14	GGGAGGGAGGGAGG	19
4	1	14	GGGAGGGAGGGAGG	19
5	1	15	GGGAGGGAGGGAGGG	20
6	1	15	GGGAGGGAGGGAGGG	20
7	1	15	GGGAGGGAGGGAGGG	19
8	1	15	GGGAGGGAGGGAGGG	20
9	1	15	**GGGAGGGAGGGAGGG**	42
10	2	13	GGAGGGAGGGAGG	20

(continued)

Table 4. (*continued*)

Number	Position	Length	QGRS	G-Score
11	2	13	GGAGGGAGGGAGG	19
12	2	13	GGAGGGAGGGAGG	20
13	2	13	GGAGGGAGGGAGG	20
14	2	14	GGAGGGAGGGAGGG	19
15	2	14	GGAGGGAGGGAGGG	19
16	2	14	GGAGGGAGGGAGGG	19
17	2	14	GGAGGGAGGGAGGG	21

The secondary structure of nucleic acid is crucial for the uniqueness of the chain. Therefore, we introduced NUPACK software to analyze and design nucleic acid structures. When NUPACK generates the secondary structure of DNA, it will combine the thermodynamic model and dynamic programming algorithm to estimate the minimum free energy (MFE) of different nucleic acid structures and calculate the partition function to predict the thermal stability of nucleic acid chains. Specifically, the thermodynamic model decomposes the structures of DNA and RNA molecules into different rings according to the base pair diagram and calculates the enthalpy and entropy values according to the different types, sequences and lengths of ring structures. The free energy of any secondary structure of a sequence is equal to the sum of the free energies of various types of rings included. Then, according to the weight sum of the secondary structure set of this sequence, the pairing probability of any structure in its secondary structure set can be determined.

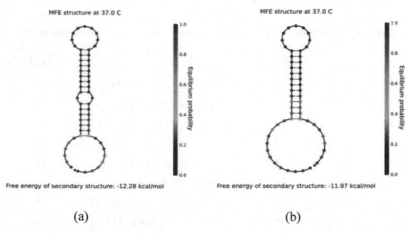

(a)   (b)

Fig. 2. MEF Structure Simulation Results of Hp1 (a) and Hp2 (b)

According to the above calculation method, NUPACK can estimate and simulate the minimum free energy and secondary structure of hairpins HP1 and HP2 designed in this model. As shown in Fig. 2, the secondary structure formed by the spontaneous folding of two sequences meets the expectation and has a low minimum free energy. The lower the minimum free energy, the more stable the structure is. The results show that the hairpin structure is feasible.

Download histogram data ⊘

Fig. 3. Equilibrium concentration of Apt-S1 forming double chain structure

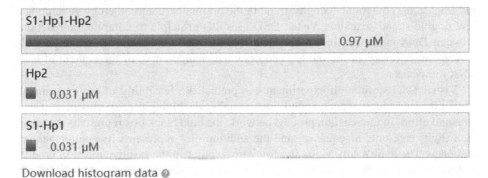

Download histogram data ⊘

Fig. 4. Equilibrium concentration of products in the presence of S1, Hp1 and Hp2

In the mixed biochemical reaction system, when the chain type in the system is greater than 1, the final equilibrium concentration of each product can be converted into solving a convex optimization problem, that is, the equilibrium concentration of each ordered complex can be calculated by using the calculated partition function and continuous concentration. First, simulate the equilibrium concentration of the double chain structure formed by Apt-S1. As shown in Fig. 3, the concentration of the double chain structure formed is much higher than that of the double chain structure formed by each single chain concentration, which is 1. Then, after the simulation S1 is released, the hairpin is triggered to generate the equilibrium concentration of each product. As shown in Fig. 4, a large number of S1-Hp1-Hp2 structures are generated, and the 5 'end of the formed structure Hp2 is free, which will open HP1. The simulation results are in line with expectations, indicating that the DNA sequence designed in the model is feasible in theory. Therefore, we preliminarily determined the sequence in the experiment as shown in Table 5, which can be verified again through biological experiments later.

Table 5. DNA sequence in aflatoxin B1 detection model based on hybrid chain reaction

Name	Sequence (5 '- 3')
Apt	GTTGGGCACGTGTTGTCTCTCTGTGTCTCGTGCCCTTCGCTAGGC
S1	GGAGGGCATGAGACACAGAGAGACAACACGT
Hp1	TCTCTCTGTGTCTCATGCCCTCCGGGAGGGAGGGAGGGCATCAGATACAAACAGGGTG
Hp2	GAGGGCATGACACACAGACAGGGTGGGAGGGAGGGCCACCCTGTTTGTATCTGATG

4.2   Visual DSD Emulation

Visual DSD is a software tool as well as a programming language, which is often used to design DNA computing models with extensive modeling and analysis capabilities. DNA strand replacement refers to the process in which a single strand of DNA reacts with some complementary double strands by combining bare DNA fragments to replace and release the constrained single strand in the original structure, thereby generating a new double strand structure. Visual DSD uses this principle to compile the reactions between DNA molecules into a chemical reaction network, demonstrate the possible movement trajectory of the model, and draw the change curve of the content of each DNA molecule.

Visual DSD simulation experiment can predict the feasibility of the experiment, predict the subsequent reaction conditions, simplify the biological experiment, and save time and effort. In the reaction process network, the bold black box represents the input, each digit represents a sequence, and the addition of * represents its complementary sequence. The hollow arrow and the solid arrow point to the positive reaction product and the reverse reaction product respectively. In theory, aflatoxin B1 will specifically bind to its aptamer Apt to release S1. Here, in order to facilitate computer input, AFB1 and the aptamer are represented by base complementary pairing. At the same time, in order to avoid complementation between some AFB1 sequences and hairpin sequences (in biological experiments, except for the aptamer sequence, there is no sequence that can be specifically combined with AFB1), the reaction network will be generated in two steps. The first step is the reaction of AFB1 and Apt-S1 to release S1. As shown in Fig. 5, (a) is a reaction network diagram. The bold black box is two inputs AFB1 and Apt-S1, which release S1 and generate sp4 after reaction. (b) shows the corresponding simulation results. The red line represents the trend of AFB1, the green line represents the trend of Apt-S1, and the blue line represents the trend of S1. It can be seen from the figure that, with the change of time, AFB1 and Apt-S1 decrease to 0, and S1 gradually increases. The simulation is in line with expectation.

The second step is that S1 triggers Hp1, and then turns on Hp2 to generate a reaction of continuous double chain structure. As shown in Fig. 6, (a) is a reaction network diagram, and S1, Hp1 and Hp2 are inputs. Ideally, S1 will trigger Hp1 to generate sp4; The exposed fragment of sp4 combines with the foothold of Hp2 and opens Hp2 to form sp5; Then the exposed fragment of sp5 combines with the foothold of Hp1 and opens Hp1 to form sp6; The exposed fragment of sp6 combines with the foothold of Hp2 and opens Hp2 to form sp9; Subsequently, the exposed fragment of sp9 combines with the foothold of Hp1 and opens Hp1 to form sp12. (b) shows the corresponding

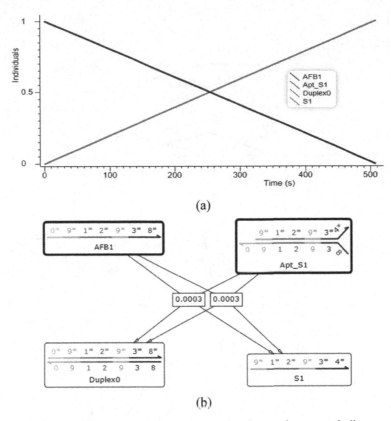

Fig. 5. First step reaction(a) Simulation results; (b) reaction network diagram

simulation results. When S1 (dark blue curve) drops rapidly, S1 and Hp1 combine to form a complex output0 (orange curve), which finally tends to zero. At the same time, output0 combines with Hp2 to form S1-Hp1-Hp2 complex: output1 (rose red curve), which disappears rapidly; Subsequently, the exposed segment of output1 can combine with Hp1 again to form a S1-Hp1-Hp2-Hp1 complex: output2 (sky blue), which finally disappears rapidly and produces longer hybrid double chains. Therefore, in the whole reaction process, Hp1 and Hp2 decreased gradually (red curve and green curve), and Hp1 and Hp2 hybridized into longer and more complex DNA double strand structure in turn.

The reaction process shows that the module we designed, and the reaction process are in line with expectations.

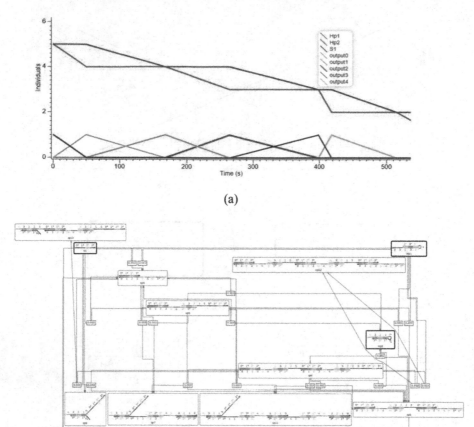

(a)

(b)

Fig. 6. Second step reaction (a) Simulation results; (b) reaction network diagram

4.3 Biological Experiment Verification

Finally, the feasibility of the proposed AFB1 detection method was verified by biological experiments. Figure 7 records the change of fluorescence intensity of different solutions. S1 Apt solution (curve 1), Hp1 solution (curve 2) and Hp2 solution (curve 3) show relatively low fluorescence values; Hp1 + Hp2 solution (curve 4) and S1 + Apt + Hp1 + Hp2 solution (curve 7) have higher fluorescence values; The fluorescence of S1 + Hp1 + Hp2 solution (curve 5) and AFB1 + S1 + Apt + Hp1 + Hp2 solution (curve 6) became significantly stronger. These results are consistent with the Visual DSD simulation results, which proves the feasibility of our AFB1 detection model.

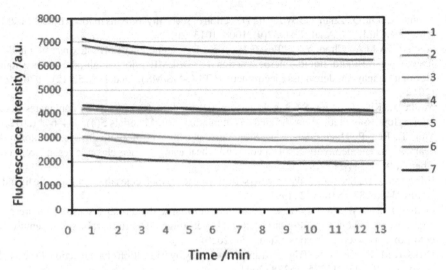

Fig. 7. Fluorescence intensity of different solutions interacting with NMM (wavelength = 608 nm). (1) S1 Apt solution; (2) Hp1 solution; (3) Hp2 solution; (4) Hp1+Hp2 solution; (5) S1+Hp1+Hp2 solution; (6) AFB1+S1+Apt+Hp1+Hp2 solution; (7) S1+Apt+Hp1+Hp2 solution.

5 Conclusion

In short, using the hybrid chain reaction technology, we proposed an enzyme free and label free fluorescence biosensor model for AFB1 detection, and verified the model through computer technology and biological experiments.

The method we developed has several advantages. First of all, this model does not need fluorescence groups and quenching groups, which makes the experiment more economical and simpler. Secondly, HCR is used for signal amplification instead of other protease and complex thermal cycle process, which makes the operation more convenient and controllable. Finally, by ingeniously designing the aptamer sequence of hairpin probe, this strategy has a good general adaptability for detecting other small molecules and proteins and provides a new idea for the future research of biology and computer science.

References

1. Dai, Y., Huang, K., Zhang, B., et al.: Aflatoxin B1-induced epigenetic alterations: an overview. Food Chem. Toxicol. **109**, 683–689 (2017)
2. Wang, C., Li, Y., Zhou, C., et al.: Fluorometric determination of aflatoxin B1 using a labeled aptamer and gold nanoparticles modified with a complementary sequence acting as a quencher. Mikrochimica Acta Int. J. Phys. Chem. Methods Anal. (11), 186 (2019)
3. Zhao, Z.Y., Han, Z., Yang, L.C., et al.: Simultaneous determination of five aflatoxins in ginkgo leaves by isotope dilution and HPLC-MS/MS. Acta Agric. Shanghai **30**(2), 54–59 (2014)
4. Gu, X., Lang, L., Wang, J., et al.: Determination of Aflatoxin B1 in Food by LC-MS/MS Method Based on Magnetic Bead-Aptamer. Chinese Journal of Applied Chemistry **37**(11), 1324 (2020)

5. Xiong, X., Liu, Q., Zhang, G.W., et al.: Detection of seven mycotoxins in foods by LC-MS/MS with QuEChERS. J. Anal. Test. **37**(9), 1008–1013 (2018)
6. Alsharif, A.M.A., Choo, Y.M., Tan, G.H., et al.: Determination of mycotoxins using hollow fiber dispersive liquid–liquid–microextraction (HF-DLLME) prior to high-performance liquid chromatography–tandem mass spectrometry (HPLC-MS/MS). Anal. Lett. **52**(12), 1976–1990 (2019)
7. Moon, J., Kim, G., Lee, S.: A gold nanoparticle and aflatoxin B1-BSA conjugates based lateral flow assay method for the analysis of aflatoxin B1. Materials **5**(4), 634–643 (2012)
8. Wei, T., Ren, P., Huang, L., et al.: Simultaneous detection of aflatoxin B1, ochratoxin A, zearalenone and deoxynivalenol in corn and wheat using surface plasmon resonance. Food Chem. **300**, 125176 (2019)
9. Ellington, A.D., Szostak, J.W.: In vitro selection of RNA molecules that bind specific ligands. Nature **346**(6287), 818–822 (1990)
10. Qian, M., Hu, W., Wang, L., et al.: A non-enzyme and non-label sensitive fluorescent aptasensor based on simulation-assisted and target-triggered hairpin probe self-assembly for ochratoxin a detection. Toxins **12**(6), 376 (2020)
11. Dirks, R.M., Pierce, N.A.: Triggered amplification by hybridization chain reaction. Proc. Natl. Acad. Sci. **101**(43), 15275–15278 (2004)
12. Chen, Y., Murayama, K., Asanuma, H.: Signal amplification circuit composed of serinol nucleic acid for RNA detection. Chem. Lett. **51**(3), 330–333 (2022)
13. Wu, J., Lv, J., Zheng, X., et al.: Hybridization chain reaction and its applications in biosensing. Talanta **234**, 122637 (2021)
14. Bochman, M.L., Paeschke, K., Zakian, V.A.: DNA secondary structures: stability and function of G-quadruplex structures. Nat. Rev. Genet. **13**(11), 770–780 (2012)
15. Xi, S., Wang, L., Cheng, M., et al.: Developing a DNA logic gate nanosensing platform for the detection of acetamiprid. RSC Adv. **12**(42), 27421–27430 (2022)
16. Zhang, X., Wang, J., Yang, H., et al.: A novel biosensor for detecting Vitamin C in milk powder based on Hg2+-mediated DNA structural changes. Curr. Anal. Chem. **18**(7), 845–851 (2002)
17. Yett, A., Lin, L.Y., Beseiso, D., et al.: N-methyl mesoporphyrin IX as a highly selective light-up probe for G-quadruplex DNA. J. Porphyr. Phthalocyanines **23**(11n12), 1195–1215 (2019)
18. Hu, M., Wang, L., Xi, S., et al.: A biosensor based on interchain reactions for the detection of acetamipirid and the construction of basics logic gates OR and AND. IEEE Trans. NanoBiosci. (2021)
19. Zadeh, J.N., Steenberg, C.D., Bois, J.S., et al.: NUPACK: analysis and design of nucleic acid systems. J. Comput. Chem. **32**(1), 170–173 (2011)
20. Fornace, M.E., Porubsky, N.J., Pierce, N.A.: A unified dynamic programming framework for the analysis of interacting nucleic acid strands: enhanced models, scalability, and speed. ACS Synth. Biol. **9**(10), 2665–2678 (2020)
21. Qiwang, W., Hong, S.: NUPACK prediction assisted of toehold induced strand displacement reaction and its application in SNPs genotyping by DNAzyme-catalyzed microfluidic chemiluminescence detection. Chem. J. Chin. Univ. Chin. **36**(12), 2386–2393 (2015)
22. Lakin, M.R., Youssef, S., Polo, F., et al.: Visual DSD: a design and analysis tool for DNA strand displacement systems. Bioinformatics **27**(22), 3211–3213 (2011)

Author Index

Printed in the United States
by Baker & Taylor Publisher Services